Ethnic Cleansing
in Twentieth–Century Europe

"Ethnic cleansing is a crime against humanity, regardless of who does it to whom. I support the work of the Institute for German American Relations as they continue to educate the public on the tragedy that displaced fifteen to seventeen million innocent German women and children; these most innocent souls who became victims of the worst period of ethnic cleansing in the history of the world: 'Ethnic Cleansing 1944-1950.'"

George W. Bush
President of the United States

Ethnic Cleansing in Twentieth-Century Europe

Editors
Steven Béla Várdy
Duquesne University
and
T. Hunt Tooley
Austin College

Associate Editor
Agnes Huszár Várdy
Robert Morris University

Foreword
Otto von Habsburg
Paneuropa-Union

SOCIAL SCIENCE MONOGRAPHS, BOULDER
DISTRIBUTED BY COLUMBIA UNIVERSITY PRESS, NEW YORK

2003

Copyright © 2003 by Steven Béla Várdy
ISBN 0-88033-995-0
Library of Congress Control Number 2003102445

Printed in the United States of America

TABLE OF CONTENTS

IV. SURVIVAL AND MEMORY: *VERTREIBUNG*

V. ETHNIC CLEANSING AND ITS BROADER IMPLICATIONS IN THE LAST THIRD OF THE TWENTIETH CENTURY

Foreword

Ethnic cleansing has been part of human history ever since classical times (witness the Babylonian Captivity of the Jews in the sixth century B.C.), but it did not become a commonly applied method of mass persecution until the twentieth century. Not until the century of Hitler and Stalin were tens of millions of people uprooted and forcibly ejected from their ancient homelands. This mass phenomenon affected many diverse nationalities and religious groups, among them the Armenians, Greeks, Turks, Hungarians, Jews, Germans, Croats, Serbians, Bosnians, and Crimean Tatars, as well as many others.

These groups all suffered because of the misdirected ideologies of Nazism, Communism, and national chauvinism. In terms of numbers, however, no one was more affected than the Germans, who had to witness the expulsion of fifteen to seventeen million of their co-nationals from the lands which they had settled, tilled, and cultivated for many centuries. These involved the Germans of former eastern Germany (now part of Poland and Russia), the Baltic countries, as well as the lands of the former Habsburg Empire. The latter included the Sudeten Germans and the Zipsers of short-lived Czechoslovakia, the Saxons of Transylvania, the Danubian Swabians of Hungary and former Yugoslavia, as well as various other scattered German ethnic islands throughout Eastern Europe.

I agree with President George W. Bush, who stated that "Ethnic cleansing is a crime against humanity, regardless of who does it to whom." I also agree with the United Nations Draft Declaration on Population Transfer, which states that "Every person has the right to remain in peace, security and dignity in one's home, or one's land and in one's country." It is this basic human right that has been grossly violated by those responsible for the expulsion of the Germans, Hungarians, Greeks, Armenians, and many others from their ancient homelands. And while much of this cannot be undone, the enormity of this crime should be recognized and acknowledged. The world should also learn about these violations of the basic human rights of many millions; if for no other reason, then to prevent any future expulsions on such a horrendous scale.

I applaud the editors of this book, and the authors of the enclosed studies, for taking the effort to make the world aware of the suffering of many tens of millions of forcibly displaced innocent people.

OTTO VON HABSBURG

Preface

This book is the result of the Conference on Ethnic Cleansing in Twentieth-Century Europe at Duquesne University (Thirty-Fourth Annual Duquesne University History Forum, held November 16-18, 2000). The Conference was cosponsored by the Institute for German-American Relations [IGAR] and Austin College (Sherman, Texas). Financially the conference was underwritten by grants from Duquesne University and Austin College, with additional funding by IGAR. The latter Institute is also responsible for the partial subsidy for the publication of this book in Columbia University Press's East European Monographs Series. The three-day conference attracted about sixty scholars and survivors of ethnic cleansing from seven different countries, who presented a total of forty-eight papers. Most of these papers have found their way into the present volume.

The conference also hosted two keynote speakers: Lieutenant-General Michael Hayden, Director of the National Security Agency (a graduate of Duquesne University), and Dr. Géza Jeszenszky, Hungary's Ambassador to the United States (as well as a published scholar-historian). The essays stemming from their conference addresses are printed at the beginning of this volume.

The organizers of this conference were Professor Steven Béla Várdy (Director, Duquesne University History Forum) and Professor T. Hunt Tooley (Austin College). We were supported by Professor Agnes Huszár Várdy (Robert Morris University) and Dr. Marianne Bouvier (Executive Director, IGAR). The first three of these also undertook the task of editing this volume, with Professor Tooley assuming the additional burden of formatting the manuscript into a camera-ready copy.

The papers collected in this volume represent more than the sum of what the contributors brought with them to the conference. It seemed to approach the scholarly ideal of a balance between discussion and presentation. The topic encompassed a vast array of case studies, behaviors, origins, and patterns, but it remained in the end highly focused. The presenters, discussants, and audience were engaged, perceptive, and lively. We were historians in the majority but profited from the presence of a wide range of specialists in other fields as well as many students and "laymen," many of them with personal or family histories involved in the events under discussion. As the conference progressed, we worked quite consciously to define and refine the topic under discussion, both in formal sessions and countless informal talks. This challenging and stimulating atmsophere led most of the presenters to work and rework their papers in the following months. The result is a measure of coherence and collective purpose in this volume which is gratifying to the editors.

It is a pleasure to thank publicly those persons and institutions who contributed both to the Conference on Ethnic Cleansing in Twentieth-Century Europe and to the book itself. We would like to thank Paula Jonse, formerly Humanities Division Administrative Assistant at Austin College, who was of great assistance at many stages of this project. The Information Technology office of Austin College provided much support in the production of this book. Karen Tooley helped in the organizational phase of the conference and also worked on the preparation of the manuscript. Special thanks is also due to Martha Kent and Alfred de Zayas, who not only contributed articles to the book but also helped in many other ways both to bring this effort to its conclusion and to improve the quality of its end result.

The two coeditors and the associate editor would like to thank their respective institutions for the support they have received. The hospitality and generosity of Duquesne University and in particular of the Duquesne Department of History played a crucial role in creating a comfortable and collegial venue for the meeting. We are grateful as well to the Institute for German-American Relations for cosponsoring this conference and the resulting volume. The coeditors would also like to thank Professor Stephen Fischer-Galati for publishing this work in Columbia University Press's internationally known East European Monographs Series.

The Editors
Pittsburgh, Pennsylvania, and Sherman, Texas
December 2002

Introduction

Ethnic Cleansing in History

STEVEN BÉLA VÁRDY and T. HUNT TOOLEY

As pointed out by José Ortega y Gasset (1883-1955) in his epoch-making work *The Revolt of the Masses* (1929)[1], during the nineteenth and twentieth centuries the Western World had witnessed the emergence of the common populace to a position of economic and political influence in human society. Being essentially of republican sympathies, and sympathizing with the exploited underclasses of Western Civilization, Ortega readily recognized the positive implications of this mass phenomenon for the people in general. At the same time, however, he feared that this ascendance of the uncouth, boorish, and unwashed masses might lead to civilization's relapse into a new form of barbarism. He was convinced of the "essential inequality of human beings," and consequently he believed in the unique role of the "intellectual elites" in the shaping of history.[2]

As a disciple of the *Geistesgeschichte* view of human evolution, Ortega was convinced of the primacy of spiritual and intellectual factors over economic and material forces in historical evolution. Given these convictions, he feared that the emergence of a mass society—dominated, as it was, by economic and material considerations—would result in the reemergence of barbarism on a mass scale.

That Ortega's fears were partially justified can hardly be doubted in light of the mass exterminations witnessed by several twentieth-century generations of human beings. As we all know, in the second quarter of the past century six million Jews and many thousands of non-Jewish people were exterminated at the orders of a lowly corporal turned into the unquestioned leader of Germany (Hitler). At the same time, forty to fifty million innocent human beings fell victim to the twisted mind of a Caucasian brigand turned into the "infallible leader" of the homeland of socialism (Stalin). Moreover, since the end of World War II, the world has also stood witness to mass killings, expulsions, and genocides in such widely scattered regions of the world as Cambodia in Southeast Asia, Rwanda in Central Africa, and in Bosnia in former Yugoslavia.

---

[1] *La rebellion de las masas* (1929; English translation, 1932).

[2] *Academic American Encyclopedia* (Princeton, 1980), 14: 449.

In looking at these terror actions against an ethnic group, religious denomination, or nationality—be these mass expulsions, partial exterminations, or genocides—we are often confused how to categorize them. Scholars and publicists are particularly confounded at the distinctions or alleged distinctions between "genocide" and "ethnic cleansing." The first of these terms came into common use in conjunction with the Jewish Holocaust of the World War II period, while the second term gained currency in the inter-ethnic struggles of Bosnia during the early 1990s.

This obfuscation and bewilderment became even more pronounced recently, particularly in consequence of the belated application of one or another of these terms to such earlier events as the so-called "Armenian Holocaust" of 1915,[3] and the various "ethnic cleansings" that took place in wake of the two world wars and the simultaneous changes in political borders. We know that most ethnic cleansings involve some physical abuse and some number of intended or unintended deaths. We also know that none of these so-called Holocausts were able to exterminate all members of a particular group. Consequently, in actual practice or application, the meaning of these two terms often tend to merge. At times it is really difficult to differentiate between the two, particularly in light of the fact that the application of violence in some ethnic cleansings often reaches the point of mass killings, thus turning that event into a potential genocide.

While we recognized the difficulty of distinguishing between these two phenomena, in the conference upon which this book is based we tried to limit our attention to the historical events that could clearly be classified as „ethnic cleansing." We were able to do this, because we equated "genocide" with the Jewish Holocaust,

---

[3] The ex post facto application of the term "Holocaust" or "genocide" to the forced transfer of many of Ottoman Turkey's Armenian population from Turkish Armenia in the north to Cilicia or Lesser Armenia in the south is a hotly debated issue. Many scholars view it as a population transfer that should be called "ethnic cleansing." Others, emphasizing the large percentage of the transferees who died, prefer to classify it as a "Holocaust." For the Armenian side of the story see Robert Melson, *Revolution and Genocide: On the Origins of the Armenian Genocide and Holocaust* (Chicago, London, 1992); Vahakn N. Dadrian, *The History of the Armenian Genocide: Ethnic Conflict from the Balkans to Anatolia to the Caucasus* (Providence, Oxford, 1995); and Vahakn N. Dadrian, "The Role of Turkish Physicians in the World War I Genocide of Ottoman Armenians," *Holocaust and Genocide Studies* 1 (Autumn 1986): 169-192. For the Turkish and American side of that story see Mim Kemal Öke, *The Armenian Question, 1914-1923* (Oxford, 1988) and Stanford J. Shaw and Ezel Kural Shaw, *History of the Ottoman Empire and Modern Turkey* (Cambridge, 1976-1977), 2: 314-317. See also Ronald Suny, "Rethinking the Unthinkable: Toward an Understanding of the Armenian Genocide," in Ronald Grigor Suny, *Looking Toward Ararat: Armenia in Modern History* (Bloomington, Ind., 1993), 94-115.

which was definitely a case of ethnic cleansing on a mass scale, but which was also more. We tried to start with a definition of "genocide" as "the planned, directed, and systematic extermination of a national or ethnic group." At the same time our working definition of "ethnic cleansing" was more focused on the process of removing people from a given territory: "the mass removal of a targeted population from a given territory, including forced population exchanges of peoples from their original homelands as well as other means." We did this in part because had we included "genocide" as a topic of our conference, in the sense of the National Socialist war against the Jews, much of the attention of the participants would have been taken up by the Holocaust. There is, of course, hardly a more significant twentieth-century historical topic than Hitler's efforts to exterminate the Jews. But precisely for that reason, during the past half a century, it has been in the focus and awareness of hundreds of scholars, who have produced thousands of volumes on this topic. Not so the topic of "ethnic cleansing," which has largely been ignored until the Bosnian crisis of the 1990s.[4] In any event, most of the presentations at the Conference on Ethnic Cleansing in Twentieth–Century Europe at Duquesne University, as will be seen, worked to examine and refine the definition of the phenomenon of ethnic cleansing. At the conference itself, the participants devoted much time and energy to discussions of definitions. The results will be seen in the expanded and refined conference papers which make up the chapters of this book.

Although the term "ethnic cleansing" has come into common usage only since the Bosnian conflict, the practice itself is almost as old as humanity itself. It reaches back to ancient times. An early example of such an ethnic cleansing was the "Babylonian Captivity" of the Jews in the sixth century B.C. (from 586 to 538 B.C.). After capturing Jerusalem in 586 B.C., King Nebuchadnezzar II (r. 605-561 B.C.) of Babylonia proceeded to

---

[4] On ethnic cleansing in general see Andrew Bell-Fialkoff, "A Brief History of Ethnic Cleansing," *Foreign Affairs* 72 (Summer 1993):110-120; Andrew Bell-Fialkoff, *Ethnic Cleansing* (New York, 1996); Dražen Petrović, "Ethnic Cleansing: An Attempt at Methodology," *European Journal of International Law* 5 (1994): 342-359; Robert M. Hayden, "Schindler's Fate: Genocide, Ethnic Cleansing, and Population Transfer," *Slavic Review* 55 (Winter 1996): 727-748; Jennifer Jackson Preece, "Ethnic Cleansing as an Instrument of Nation-State Creation: Changing State Practices and Evolving Legal Norms," *Human Rights Quarterly*, 20 (1998): 817-842; Norman M. Naimark, *Ethnic Cleansing in Twentieth Century Europe* [The Donald W. Treadgold Papers, no. 19] (Seattle, 2000); and Norman Naimark, *Fires of Hatred: Ethnic Cleansing in Twentieth–Century Europe* (Cambridge, Mass., 2001).

deport the Judeans to his own kingdom, and in this way he "cleansed" the future Holy Land of most of its native inhabitants.

Similar ethnic cleansings took place in the period of the so-called "barbarian invasions" of the fourth through the sixth centuries, when large, nation-like tribes—Germans, Slavs, and various Turkic peoples—moved back and forth between Western and Eastern Europe, and even Central Asia. They forcibly displaced one another, and in this way reshaped the ethnic map of the European continent. This so-called *Völkerwanderung* ("wandering of nations")—which, in some instances, stretched into the late Middle Ages—brought such peoples as the Huns, Avars, Bulgars, Magyars, Pechenegs, and Cumans into the very heart of Europe. Its aftereffects were felt as late as the thirteenth century, when the Mongols or Tatars invaded Europe, conquered the eastern half of the continent, and then settled down there to rule over the East Slavs for several centuries.

Although this process of forcible relocations has been practiced for millennia, ethnic cleansing as an official policy did not come into being until more recent times. In the Western world, large-scale forcible relocation of a specified "people' was introduced in the early nineteenth century United States, as the official policy of the United State government. Informally, scores of Indian tribes had "emigrated" from their lands as a result of European pressure, at times escaping from direct violence by European settlers, at times looking for food, at times being pushed by other native groups reacting to direct pressure from European settlers. The process of removal became standardized federal policy in 1830, when Congress passed the "Indian Removal Act." Some of the saddest manifestations of this policy, implemented during Andrew Jackson's presidency (1829-1837), was the decimation and expulsion of the affiliated Sac and the Fox tribes from the Upper Mississippi region (Black Hawk War of 1832), the forcible relocation of the Creek, Choctaw, Chickasaw, and Cherokee nations from the Old Southwest to Indian Territory (Trail of Tears, 1838-1839), and the similar expulsion of the Seminole Indians from Florida to future Oklahoma (Second Seminole War, 1835-1843). The process of forcible removal to reservations was repeated countless times from the 1820s through the 1880s.[5]

---

[5] See Francis Paul Prucha, *The Sword of the Republic: The United States Army on the Frontier, 1783-1846* (Lincoln, 1987); Black Hawk, *An Autobiography* (1833 ed.), ed. Donald Jackson (Urbana, Ill., 1990); Ronald N. Satz. *American Indian Policy in the Jacksonian Era* (Lincoln, 1975); and Grant Foreman, *Indian Removal: The Emigration of the Five Civilized Tribes of Indians* (Norman, Okl., 1986).

In Europe itself, the rise of "national" awareness at the end of the eighteenth century spread across Europe from West to East. By the middle of the century, some nationalist leaders and thinkers were already thinking in terms of an exclusivist doctrine calling for the "nation" to correspond with the "state," that is, to make political borders correspond with ethnic or linguistic borders. Since precise ethnic boundaries hardly existed anywhere in Europe, any planning for such new "nations" necessitated thinking about what to do with individuals from other ethnic groups who were left on the inside of someone else's national state.

Carrying out such exclusivist ethnic nationalism was approached in a number of ways in various settings, and a number of small states moved toward policies of ethnic exclusion in the nineteenth century. It was on the peripheries of the great European empires (including especially the Ottoman Empire) that a sharp-edged, ethnicity-oriented policy led to a variety of policies of forced assimilation, expropriation of property, violence, and in several cases, mass killing.[6]

Following the destruction or mutilation of such established European empires as Austria-Hungary, Ottoman Turkey, Russia, and Germany in wake of World War I, and the simultaneous creation of nearly a dozen allegedly national, but in fact mostly multinational small states, the policy of ethnic cleansing was introduced into modern Europe as a regular policy and in a certain sense "legitimized." The newly created, reestablished, or radically enlarged "successor states"—particularly Czecho-slovakia, Yugoslavia, and Romania in the center; Bulgaria, Greece, and Turkey in the south; and to a lesser degree Poland and Lithuania in the north—expelled hundreds of thousands of minority inhabitants from their newly acquired or restructured territories.[7] Many of these expulsions also involved military encounters among several of these nations and newly created states. The most violent of these encounters was the Greek-Turkish War of 1920-1923, which resulted in a forced population

---

[6] On the rise of this hard-shelled, ethnic nationalism in the nineteenth century, see the excellent anthology, John Hutchinson and Anthony Smith, eds., *Nationalism* (Oxford, 1994), esp. 160-195.

[7] See the classic work by C. A. Macartney, *Hungary and Her Successors: The Treaty of Trianon and its Consequences, 1919-1937* (Oxford, 1937) [new ed. 1968]; István I. Mócsy, *The Effects of World War I: The Uprooted: Hungarian Refugees and their Impact on Hungarian Domestic Politics, 1918-1921* (Boulder and New York, 1983). One of the most comprehensive of the relevant scholarly volumes, which includes studies by over thirty scholars, is B. K. Király, P. Pastor, and I. Sanders, eds., *War and Society in East Central Europe: Essays on World War I: A Case Study of Trianon* (New York, 1982).

exchange that compelled 1.3 million Greeks to leave Anatolia, and 350,000 Turks to evacuate Greek-controlled Thrace.[8]

The climax of this policy of ethnic cleansing came in the wake of World War II, when—based on the erroneous principles of collective guilt and collective punishment—perhaps as many as sixteen million Germans were compelled to leave their ancient homelands in East-Central and Southeastern Europe. At the Yalta and Potsdam Conferences, the leaders of the victorious great powers agreed to truncate Germany and transfer Eastern Germany's ethnic German population to the remaining portions of the country. They likewise agreed to expel the 3.5 million Sudeten Germans from the mountainous frontier regions of Bohemia and Moravia (the Czech state)—a region that these Germans had inhabited for over seven centuries. A similar policy of expulsion was also applied, although less stringently, to the smaller German ethnic communities of Hungary, Romania, and Yugoslavia. All in all, 16.5 million Germans may have fallen victim to this officially sponsored policy of ethnic cleansing.[9]

Although the Germans were the primary victims of this new policy, the Hungarians were also been subjected to it, especially in Eduard Beneš's reconstituted Czechoslovakia. In the course of 1945-1946, over 200,000 thousand of them were driven across the Danube, most of them in the middle of the winter and without proper clothing and provisions. This so-called "Košicky Program"—which became the Czechoslovak government's official policy vis-à-vis the Hungarians[10]—was a smaller version of the

---

[8] Cf. Arnold J. Toynbee, *The Western Question in Greece and Turkey: A Study in the Contact of Civilizations*, 2nd ed. (London, Bombay, 1923); Stephen Ladas, *The Exchange of Minorities: Bulgaria, Greece and Turkey* (New York, 1932); Dimitri Pentzopoulos, *The Balkan Exchange of Minorities and its Impact upon Greece* (Paris and the Hague, 1962).

[9] On the post-World War II expulsion of the Germans see especially Alfred M. de Zayas's classic work, *Nemesis at Potsdam. The Anglo-Americans and the Expulsion of the Germans*, rev. ed. (London, Boston, 1979); see also Gerhard Ziemer, *Deutscher Exodus: Vertreibung und Eingliederung von 15 Millionen Ostdeutschen* (Stuttgart, 1973); *Die Vertreibung der deutschen Bevölkerung aus der Tschechoslowakei*, 2 vols. (Munich, 1984). Heinz Nawratil, *Vertreibungs-Verbrechen an Deutschen* (Munich, 1987).

[10] Concerning Hungarian expulsions and the fate of Hungarian minorities in the surrounding "successor states" see Elemér Illyés, *National Minorities in Romania: Change in Transylvania* (Boulder and New York, 1982); John Cadzow, Andrew Ludányi, and Louis J. Elteto (eds.) *Transylvania: The Roots of Ethnic Conflict* (Kent, Ohio, 1983); Stephen Borsody (ed.), *The Hungarians: A Divided Nation* (New Haven, 1988); and Raphael Vágó, *The Grand Children of Trianon: Hungary and the Hungarian Minority in the Communist States* (Boulder and New York, 1989).

"ethnic cleansing" that had cleared Czechoslovakia of its German citizens. It is to the credit of Václav Havel, the President of the Czech Republic, that in his former capacity as the last President of Czechoslovakia he acknowledged the immorality of the policy.

The most recent manifestations of ethnic cleansing—at least as far as Central and Southeastern Europe are concerned—were cases which were experienced recently by the former Yugoslav provinces of Bosnia and Kosovo.[11] These were the actions that popularized the expression "ethnic cleansing" and gave it a definition as distinct from the term "genocide," a term which, as seen above, implies not only the displacement, but also the mass extermination of a targeted ethnic group.

The papers presented at the "Conference on Ethnic Cleansing" at Duquesne University (November 16-18, 2000) survey much of the process of forced population exchanges in twentieth-century Europe. It seems a strange twist of fate that this first-ever conference on ethnic cleansing should have taken place at an institution, which itself came into being in consequence of a kind of "ethnic cleansing." And that was Otto von Bismarck's anti-Catholic crusade known as the *Kulturkampf* (1872-1878), which drove the Religious Order of the Holy Ghost out of Germany, and brought them to a hill in the middle of Pittsburgh, Pennsylvania, where in 1878 they founded an institution of higher learning, known today as Duquesne University.

One of the primary functions of the Conference on Ethnic Cleansing in Twentieth-Century Europe was to bring to light the state of new research on the various episodes of ethnic cleansing in modern Europe. In doing this, the organizers and contributors hoped to explore the historical and legal aspects of ethnic cleansing and to look comparatively at the experiences of populations expelled and the process of ethnic cleansing itself. Many commonalities emerged in the process of putting the conference together, and from these we developed a number of themes which seemed to be uppermost in the minds of those participating. We stated these before the conference as follows:

1. **Definition.** What is ethnic cleansing? Should all cases of population transfer be conceptualized as the same phenomenon? Has "ethnic cleansing" been diluted in terms

---

[11] Norman Cigar, *Genocide in Bosnia: The Policy of "Ethnic Cleansing"* (College Station, 1995); and Christopher Bennet, "Ethnic Cleansing in Former Yugoslavia," in *The Ethnicity Reader: Nationalism, Multiculturalism and Migration,* ed. Monserrat Guibernau and John Rex (Cambridge, 1997), 122-135.

of meaning? Should we adopt another terminology in dealing with the varieties of forced transfers of populations?

2. **Origins**. Ethnic cleansing has existed in some form from antiquity, but it has never been practiced with more variety or intensity than in the twentieth century. Where do we look for the origins and roots of this outburst? Nationalism? State building? Popular movements? Economic conditions?

3. **Consequences**. Clearly, many twentieth-century politi-cal leaders have opted to engage in forced population transfers and related behaviors. Misery to those trans-ferred has been one result, and that should not in any way be minimized. We should also ask, however: "What have been the long-term results?"

4. **Processes**. Looking at many cases of ethnic cleansing, how do the processes, carried out over the century and in many different regions, compare? Can we see a con-tinuity? Or do the cases of ethnic cleansing tend to exhibit highly specific, or particular, characteristics centered around local conditions and history?

We tried to define our topic in such a way as to make it clear that the Holocaust would be a crucial and constant background consideration. Yet in a sense, our conference was an attempt to assess a particular historiography—the analysis of ethnic cleansing—at a relatively early stage of development, while the scholarship on the Holocaust is not only enormously larger, but also more mature in a historiographical sense.[12] As one can see in the following book, many conference participants drew on the historiography of the Holocaust for both analytical and comparative purposes, even as they focused on historical episodes that have been studied much less. Indeed, some of the episodes addressed in the following articles are hardly known at all in the English-speaking world, except among small groups of descendants of the "cleansed" people and a few scholars.

Although the contributors to this book therefore approach some of these cases for almost the first time in a scholarly way, we

---

[12] For an intensive introduction to the historiography of the Holocaust, see Michael R. Marrus, *The Holocaust in History* (New York, 1987).

should make it clear that the long list of topics is in no way meant to be absolutely comprehensive. Numerous cases of ethnic cleansing in twentieth-century Europe are not represented here. Their absence, it must be said, is not due to any desire of the organizers of the conference and the editors of the book to exclude or suppress one case or another of the abhorrent practice of ethnic cleansing. Our treatment is limited in this sense by the practical reason that we were restricted to those scholars who answered the various calls for papers in the usual scholarly sources.

In the end, participants included individuals from seven countries in two continents. Most were historians, but many were political scientists, literary scholars, sociologists, and legal scholars. Presenters ranged from those still engaged in graduate study to scholars who have published many books in the field of European history, law, and politics. Among the presenters were four survivors of ethnic cleansing whom we asked to write explicitly about their own experiences.

The articles investigate dozens of cases of ethnic cleansing in the twentieth century. A number of contributors deal with ethnic cleansing in the period of World War I, in particular those episodes arising from the clash between Greece and Turkey at the end of the war. The period of World War II gave full play to the deadly policies of Stalinist Russia in the East and Hitler's Third Reich, as well as many cases which clearly spun off of these policies of ferocious ethnic cleansing. Hence, contributors deal with ethnic cleansing of Poles by Ukrainians, Romanians by Russians, Germans by Titoist Yugoslavia, and others.

The vast case of the ethnic cleansing of Germans, the "Expulsion," forms an important part of the book. Fifteen contributors deal with the expulsions of Germans from East Central Europe from one standpoint or another. This extensive share seems in no way out of place, since the expulsion of sixteen million ethnic Germans from half a dozen European countries, at a loss of over two million lives, constitutes an episode which surely merits attention but which has been neglected by all but a handful of historians.[13] The contemporaneous ethnic cleansing of

---

[13] On the recent historical writing on the expulsions of Germans, see Alfred de Zayas, *A Terrible Revenge: The Ethnic Cleansing of the East European Germans, 1944-1950* (New York, 1993) (originally published in Germany as *Anmerkungen zur Vertreibung der Deutschen as dem Osten* [Stuttgart, 1986]). Also of interest is the special issue of *Deutsche Studien* devoted to "Flucht und Vertreibung der Ostdeutschen und ihre Integration"—see for an overview the introduction by Gerhard Doliesen, "Der Umgang der deutschen und polnischen Gesellschaft mit der Vertreibung," *Deutsche Studien* 126/127 (1995): 105-110. The late 1990s saw interest in the topic grow in several directions. See, for example, Philipp Ther,

Hungarian populations, mentioned above, will be a familiar topic to even fewer readers.

In dealing with the history of post-Cold War ethnic cleansing in the Balkans, contributors pay much attention to processes and patterns, but they also do much to explore concepts and definitions. Multiple approaches to the complexities of the Balkans and the violence which has marked the dissolution of Yugoslavia prove most useful in conceptualizing as well as recounting ethnic cleansing in the region that seems to have named it.

The studies of ethnic cleansing which follow represent an earnest attempt to make sense of a terrible aspect of the twentieth century, a century whose reputation for barbarity, when viewed in total, goes beyond even the pessimistic vision of Ortega y Gasset. Recording the erosion of indvidual autonomy and dignity, the frequent lapse of the rule of law, and the blatant disregard of the ideals of justice long considered to be at the heart of the European tradition does not tell the whole story. That story must also include the rise of a new set of barbarous practices and behaviors that ignored the pleas of individuals for a homeland, rejected the rights of individuals to their own property and the fruits of their labor, and in many cases denied the right of the targeted peoples to live.

Deutsche und polnische Vertriebene: Gesellschaft und Vertriebenenpolitik in der SBZ/DDR und in Polen 1945-1956 (Goettingen, 1998); Mathias Beer, "Im Spannungsfeld von Politik und Zeitgeschichte: das Grossforschungsprojekt: 'Dokumentation der Vertreibung der Deutschen aus Ost-Mitteleuropa,'" *Vierteljahrshefte für Zeitgeschichte* 46 (1998); Detlef Brandes, *Der Weg zur Vertreibung, 1938-1945: Pläne und Entscheidungen zum "Transfer" der Deutschen aus der Tschechoslowakei und aus Polen* (Munich, 2001); Philipp Ther and Ana Siljak, eds., *Redrawing Nations: Ethnic Cleansing In East-Central Europe, 1944-1948* (New York, 2001). Norman Naimark has approached the German expulsions comparatively in *Fires of Hatred.*

From "Eastern Switzerland" to Ethnic Cleasing: Is the Dream Still Relevant?

GÉZA JESZENSZKY
**Hungarian Ambassador
to the United States**

Ten years ago, in the bliss of *annus mirabilis*, the term "ethnic cleansing" was unknown. Today we devote a conference to that subject only to discover how many criminal events in history may fall under that hideous category. While the organizers of the conference were wise in pointing out the difference between forced migration, population exchange, and mass extermination aiming at genocide, we haven't yet adopted a definition of the term "ethnic cleansing." I would like to point out, however, that in my opinion the term is worse than euphemistic; it is misleading. It has nothing to do with cleanliness, purity; on the contrary, it is a codename for killing and/or expelling people who are undesirable because of their national and/or religious identity, or because of the language they speak. The term really means "ethnic killing," ethnocide, which is indeed different from genocide, but it is related to it in being totally unacceptable, something to be prosecuted by the international community.

On one of my first trips to Western Europe I came across a book entitled *Katyn: A Crime without Parallel*.[1] Since then I learned that, sadly, the brutal murder of tens of thousands of Polish POWs by the Soviet NKVD was not without parallel; it was surpassed only too often. Before the Balkan horrors of the 1990s I shared the belief of so many people that after the crimes of Hitler's Nazism and Stalin's Communism similar actions could not happen again. We were mistaken, and that is how the title of the present conference was born.

In the course of our proceedings we heard scholarly accounts of little-known forced re-settlements, mass murders, and war crimes: the fate of the Greeks of Asia Minor, that of the Crimean Tatars and other smaller peoples in the Soviet Union, and the largely untold suffering and eventual elimination of the Germans who used to live in Bohemia, in Transylvania, in the Banat and the Vojvodina. Another hardly known story, described in four papers, is the massive reduction in the size and proportion of the Hungarian populations in Slovakia, Romania, and Yugoslavia.

---

[1] Louis Fitzgibbon, *Katyn: A Crime Without Parallel* (Dublin, 1971, 1975).

My message is not simply to highlight another sad story of abuse. My aim is more positive: it is an effort to show that the horrors of Bosnia and Kosovo were not inevitable, that the coexistence and peaceful cooperation of peoples who live side-by-side, often intermingled, is possible, that there are promising models for this kind of arrangement both in the past and in the present. In my view the most viable way to such cooperation is the Swiss model of arranging the various ethnic groups of a country into separate *Kantons*.

The *Confederatio Helvetica* as Model

Louis Kossuth, the leader of the War for Hungarian Independence, writing in exile in 1862, proposed a Confederation of the Danubian Nations as the best guarantee against interference and domination by the nearby great powers and as a solution for handling the conflicting territorial claims of the many smaller peoples living in the Danube Basin. He ended his essay with a peroration: "Unity, agreement, fraternity among Hungarians, Slovaks and Romanians! Behold, my most ardent desire, my sincerest advice! Behold, a blithesome future for us all!"[2] More than fifty years later, at the end of World War I, Oszkár [Oscar] Jászi, a long-time advocate of the rights of the non-Hungarian nationalities of Hungary, soon an exile in Austria, and later a Professor at Oberlin College, Ohio, wrote a book which advocated the replacement of the dual Austro-Hungarian Monarchy by a federation of German Austria, Czech Bohemia, Hungary, Polish-Ukrainian Galicia and Croatia, making up "The United States of the Danube." He called for the adoption of the Swiss *Kanton* system, where the administrative units of a region would reflect the ethnic composition of the population.[3] In this way the historic Kingdom of Hungary was to become "a kind of Eastern Switzerland," where the various nationalities would enjoy territorial and/or cultural autonomy. Jászi was not a lonely dreamer: from the nineteenth–century Czech František Palacky to the Romanian

---

[2] Ferencz Kossuth, ed., *Kossuth Lajos* Iratai [The Papers of L. Kossuth], vol. 6 (Budapest, 1898): 9–12.

[3] Oszkár Jászi, *A Monarchia jövője: a dualizmus bukása és a Dunai Egyesült Államok* [The Future of the Monarchy: the Fall of Dualism and the United States of the Danube] (Budapest, 1918; new edition, 1988). Cf. István Borsody, "Oszkár Jászi's Vision of Peace," in *Hungary's Historical Legacies: Studies in Honor of Steven Béla Várdy*, ed. Dennis P. Hupchik and R. William Weisberger (Boulder, *Col.*, 2000), 116-129.

Aurel Popovici and the Austrian Otto Bauer and Karl Renner in the 1900s, many Central European political thinkers believed that the best way to assure the peaceful coexistence of the eleven national groups living in the Habsburg Monarchy lay in national autonomies, a kind of compromise between total separation (i.e., independence) and enforced unity.[4]

A plan similar to Jászi's was submitted to the British Foreign Office in October, 1918, by Leo Amery, an adviser to the British Prime Minister, Lloyd George. Its conclusion was: "Permanent stability and prosperity could best be secured by a new Danubian Confederation comprising German Austria, Bohemia, Hungary, Yugoslavia, Rumania and probably also Bulgaria.... In any case the various nationalities of Central Europe are so interlocked, and their racial frontiers are so unsuitable as the frontiers of really independent sovereign states, that the only satisfactory and permanent working policy for them lies in their incorporation in a non-national superstate."[5] Almost simultaneously the American team preparing peace, the Inquiry, when drawing up proposals for new states to emerge in Central Europe, proposed the transformation of Austria-Hungary into a federation.[6] Even when the allies decided to work for the break-up of the Monarchy, the final U.S. recommendation came to the conclusion that "the frontiers supposed [*sic*] are unsatisfactory as the international boundaries of sovereign states. It has been found impossible to discover such lines, which would be at the same time just and practical.... many of these difficulties would disappear if the boundaries were to be

---

[4] Ignác Romsics, "Plans and Projects for Integration in East Central Europe in the 19th and 20th Centuries: Toward a Typology," in *Geopolitics in the Danube Region: Hungarian Reconciliation Efforts, 1848–1998,* ed. Ignác Romsics and Béla K. Király (Budapest, 1999), 1–4. Cf. Éva Ring, ed., *Helyünk Európában* [Hungary's Place in Europe] (Budapest, 1986), esp. vol. 1: 577–579.

[5] "The Austro-Hungarian Problem," 20 October, 1918, Public Record Office, London, FO 371/3136/17223. See Géza Jeszenszky, "Peace and Security in Central Europe: Its British Programme during World War I," *Etudes historiques hongroises 1985* (Budapest, 1985), 457-482.

[6] Charles Seymour, "Austria-Hungary Federalized Within Existing Boundaries," May 25, 1918, Inquiry Document 509, RG 256, National Archives, Washington, D.C. (hereafter, NA). See Géza Jeszenszky, "A dunai államszövetség eszméje Nagy-Britanniában és az Egyesült Államokban az I. világháború alatt" [The Idea of a Danubian Federation in Great Britain and the United States during World War I], *Századok* (1988), 659. Cf. Magda Ádám, "Plan for a Rearrangement of Central Europe, 1918," in *Hungarians and Their Neighbors in Modern Times, 1867-1950,* ed. Ferenc Glatz (Boulder, Col., 1995), 77-83.

drawn with the purpose of separating not independent nations, but component portions of a federalized state."[7]

The idea of using the Swiss model for managing the ethnic mosaic of a large part of Central and Eastern Europe was revived during the Second World War. One of the many plans for a fair postwar settlement and a solution of the problem of Transylvania, traditionally a bone of contention between Hungary and Romania, was drawn up by a member of the Hungarian Parliament, Endre Bajcsy-Zsilinszky.[8] In his book published in 1944 in English on the future of Transylvania,[9] he argued for the cantonal model in an independent state, having separate as well as mixed Hungarian and Romanian territorial units, thus preserving ethnic peace in that historic province, where religious tolerance was enacted already in the sixteenth century. A recent study by a Hungarian writer recalls the past plans for Transylvania to be reconstituted along the Swiss model and argues for a system of autonomous regions.[10]

The Ethnic Mosaic in Central Europe

About a thousand years ago Central Europe embraced Christianity and four kingdoms emerged: Poland, Bohemia (the land of the Czechs), Hungary, and Croatia. Serbia and two Romanian principalities, Wallachia and Moldova, following the rite of the Eastern Orthodox Church, were established two or three centuries later. While in all these states one national group (and language) was dominant, they contained several ethnic minorities, and there was a constant influx of new settlers: mainly Germans and Turkic peoples (especially Cumanians), but also Romanians (then called Wallachs) and some Jews. In the first group Latin, in the second Old Church Slavic was the language of learning and administration. In everyday life each person was free to use whichever lan-

---

[7] Inquiry Doc. 514, RG 256, NA. Géza Jeszenszky, "Peace and Security in Central Europe," 662. Cf. Tibor Glant, *Through the Prism of the Habsburg Monarchy: Hungary in American Diplomacy and Public Opinion During World War I* (New York, 1998), 220-23.

[8] Endre Bajcsy-Zsilinszky (1886-1944), Hungarian author and politician. He was a courageous opponent of Nazi Germany, a leader of the Hungarian resistance, who was executed by the Hungarian hirelings of Hitler on Christmas Eve, 1944.

[9] Endre Bajcsy-Zsilinszky, *Transylvania: Past and Future* (Geneve, 1944).

[10] Béla Pomogáts, "A Chance Missed: Transylvania as 'Switzerland of the East,'" *Minorities Research* No. 2. (2000): 132-145.

guage he or she preferred, and many people spoke several languages. Loyalty to the state, however, was not based on language or ethnicity, but rather on allegiance to the sovereign King, who reigned *Dei gratia*, by the grace of God. The settlers, *hospes*, whether importing skills and trades to the towns; mining gold, silver, or salt; or working in the fields, were granted royal privileges, including the free use of their language and the running of their own local affairs. When two or more ethnic groups lived in the same town they usually inhabited separate quarters, and the mayor or judge as well as the councilors were selected on a rotational basis. Ethnic strife was rare, and the Kings were ready to protect the minorities, whose taxes were an important source of revenue. The various ethnic islands, primarily the German "Saxons" in Transylvania and the *Zipser* in North-Eastern Hungary, were able to preserve their linguistic identity until the twentieth century.

Wars and epidemics naturally took a heavy toll of the population, but systematic killing based on language or ethnic identity was rare. When the multinational Ottoman Empire, dominated by Islamic Turks, but often managed by Greeks and Albanians, conquered the Balkan Peninsula, the conflict undoubtedly had a religious connotation. The ensuing centuries of warfare had a most detrimental impact on the evolution of Southeastern Europe, including decimating the population. The armies of the Sultan (often Tatar, Slavic or Albanian auxiliary troops) killed, plundered and took people into slavery by the tens of thousands, but the mercenaries of the Habsburg kings were not much better in their treatment of the population of Hungary either. Contemporary travelers like Lady Mary Montague described the devastations in graphic terms. Following the expulsion of the Ottomans from Hungary in the early eighteenth century, big changes occurred in the ethnic composition of Central Europe. The devastated and depopulated southeastern part of the Great Hungarian Plain saw massive organized colonization mainly by Germans (called Swabians) arriving from the West, Slovaks from the North, and Rusyns from the Northeastern Carpathians, moving to the fertile and "free" land of today's Vojvodina and the Banat. Serbs and Romanians arrived, also in great numbers, escaping poverty, mismanagement, and—occasionally—religious intolerance under Ottoman rule. The Habsburg Emperor-Kings established a military borderland to protect their possessions, the local Catholic Croats being joined by Orthodox Serbs—thus inadvertently laying the foundation for the brutal conflicts of the twentieth century. Both the older and the new ethnic islands had substantial territorial as well as religious autonomy. Croatia had its own Diet, the *Sabor* in Zagreb; the military border from the Austrian Alps to the eastern Carpathians had its own military administration; and the Orthodox Serb Patriarch of Karlowitz as well as the Romanians of Transylvania enjoyed religious freedom and autonomy.

The Origins of Ethnic Conflicts

It was largely due to the Ottoman wars and subsequent colonization that by the late eighteenth century the Hungarians became a minority in the Kingdom of Hungary, but that became an issue only from the 1830s on, with the rise of nationalism, when language and its concomitant, ethnic identity, became the primary basis for group loyalty. "National awakening" was bound to lead to trouble wherever populations were ethnically mixed: in Hungary, in Poland-Lithuania, and all over the Balkans. The towns were often inhabited by one national group, while the villages nearby by another. The situation was complicated by the cleavage between landowning nobles, industrious burghers of towns, and the poor serfs, who started to move to the urban and industrializing areas. The most typical case was in eastern Galicia, with a Polish-speaking gentry and a Ukrainian rural population, where the towns had also sizeable German and Yiddish-speaking Jewish population. A similar, but even more complicated, pattern could be observed in Transylvania, where the nobility was Hungarian (absorbing also some Romanians), the urban element divided between Hungarians and Germans, while the peasants were both Hungarian (called Székely or Szekler in the southeastern corner of the principality), German (or more properly *Sächsisch* in the *Königsboden*), and Romanian, the latter moving down from mountain pastures to the destroyed Hungarian villages, or arriving from Moldavia and Wallachia over the Carpathians. Social tensions also grew with industrialization setting in.

Contrary to the commonly held belief that "ethnic cleansing," or in a more general way, present-day ethnic conflicts go back to centuries of animosities and hostilities, there were far fewer wars between the peoples of the Danubian Basin than in Western or Northern Europe. The first serious nationalist clashes between them occurred only in 1848, during "the springtime of nations," when the common desire for liberty, equality, and national freedom floundered on the territorial issue, on the conflicting claims to the same territory, e.g. Transylvania. It was particularly difficult to separate peoples along national lines when they lived mixed, overlapping each other. The call was for "Home Rule," national independence, or at least for very substantial territorial autonomy. The new government of Hungary, in accordance with the principles of contemporary liberalism, believed that equality before the law and the new constitutional system would satisfy the non-Hungarian citizens of the country, and rejected demands for federalizing the country along national lines. The reactionary advisers of the Habsburg King sent in an army to suppress Hungary, and by skillfully manipulating the Croatian, Serbian, and Romanian peasantry, led by loyal, *"Habsburgtreu"* priests and officers, they induced these "nationalities" to rebel against the government. The Hungarians were supported by the vast majority of the Slovak, German, and Rusyn nationalities and by all the Jews of the kingdom, as well as by a large number of Polish, Austrian and Italian volunteers. The outcome was a long war, with brutal atrocities committed against Hungarian civilians (not returned in kind), and it ended only with the Russian Czar sending a 200,000–strong army to help his Imperial colleague, crushing not only the Hungarians, but also the aspirations of the Slavs and Romanians for self-government and the hopes of the Poles for restoring their independence.

The Hungarians learned the lesson, and when in 1867 they made peace with the Habsburgs and the laws of 1848 were restored, they made an honest effort to placate the non-Hungarian citizens by passing the first Law on National Minorities in the world. Despite its liberal principles it did not go far enough. In particular, it did not accept the demands for territorial and ethnic autonomy, except in the case of the historic province of Croatia. This liberal piece of legislation was also increasingly disregarded, especially from the 1890s on. The autonomy of the Romanian and Serbian Churches, however, was recognized, including their right to have an independent educational system, so the ethnic composition of Hungary did not change dramatically. "Magyarization," a policy to turn non-Hungarians into Hungarians through assimilation, may have been desired by many Hungarians, but a change in one's language and identity was observable only in the towns and in the industrial zones, affecting only the urban German and Jewish population. Assimilation among them was voluntary and often enthusiastic, while the rural Slovak, Serb, and Romanian communities were hardly affected.[11]

---

[11] The "national question "—interethnic relations in Austria-Hungary, policies towards the national minorities, changes in the ethnic composition of the Austro-Hungarian Monarchy—has been treated in a large number of books and articles. C. A. Macartney, *The Habsburg Monarchy* (Oxford, 1968) offers probably the most comprehensive and balanced picture, while Hugh Seton-Watson, *Nations and States* (Boulder, Col., 1977) deals specifically with the issue. For a recent summary, with statistical figures and maps, see Stephen Borsody, ed., *The Hungarians: A Divided Nation* (New Haven, 1988).

With verbal intolerance growing on both sides, at the end of World War I, encouraged by the Entente Powers and the United States, the non-Hungarian population of Hungary opted for secession. The unity of the Romanians, as well as the unification of the Czechs and Slovaks, and also of the Serbs, Croats and Slovenes was thus accomplished. Sadly, the peacemakers were not content with realizing the principle of self-determination, espoused and advocated by President Wilson. The new borders did not reflect ethnic lines, even where it would have been quite easy to do so. One third of all Hungarians, more than three million, were incorporated in Czechoslovakia, Romania, and Yugoslavia against their will, a decision that denied the call for plebiscites. Czechoslovakia acquired three million Germans and one million Hungarians; Poland one million Germans and more than six million Ukrainians and Belorussians; Romania 4.5 million non-Romanians (Hungarians, Germans, Ukrainians, Russians, Bulgarians etc.). One third of the population of the new Central Europe did not belong to the majority nations; they were national minorities.

Radical Efforts to Change the Ethnic Map

The new or vastly enlarged states were aware of the precariousness of their new holdings. Rather than trying to win over their minorities with kindness, with autonomy (in many cases solemnly promised before the peace treaties were signed), they did not even bother to keep their international obligations, the Minority Treaties signed with the Great Powers in 1919, meant to protect minorities. The aim of the new masters was to reduce the number and proportion of the national minorities by expulsion, harassment leading to "voluntary" emigration, assimilation, or just by statistical maneuvering. Of course there was considerable difference in methods between the various countries, but even the supposedly exemplary democracy of Masaryk's Czechoslovakia tried to dilute the compact bloc of Hungarians in Southern Slovakia by creating colony-like settlements, sending in Slovak and Czech army veterans and other carpetbaggers.

In interwar Central and Eastern Europe, one cannot speak of "ethnic cleansing" in the horrible, post–1990 version of the term. But by looking at the figures, at the changes in the actual number of the minorities, even more in their proportion on the national level and in the various localities, I cannot but speak of "creeping ethnic cleansing," a gradual but steady tendency of changing the ethnic composition of the regions inhabited by national minorities. This was by no means a natural process, simply the outcome of industrialization and its concomitant, urbanization; it was induced and maintained artificially, using the resources of the state.

The Second World War brought sufferings and brutality unparalleled in modern history to most of the peoples and individuals living in Central and Eastern Europe. It is enough to mention the murder of six million Jews, the deportation and forced resettlement of millions of non-Russians by Stalin, aptly called the nation-killer by Robert Conquest, or the expulsion and elimination of most of the Germans from the *Südostraum*. The papers read at the present conference provide chilling pictures of inhuman behavior. Whereas some groups and individuals must bear a far heavier responsibility for those crimes than others, with the Nazis taking the lead, it must be a sad conclusion that there was no nation in Europe which did not produce individuals who deserved a place in a court for the investigation of war crimes. Brutality begets brutality, and the mass expulsions, forced population transfers and vengeance directed at the innocent, culminating in mass murders, did not come to an end until 1948, in the Soviet Union until the death of Stalin 1953. The imposition of Communism did stop open ethnic cleansings but inaugurated decades of massive and large-scale changes in the ethnic composition of the territories previously disputed, especially in that of the cities.

The pattern was the same everywhere. The peasants were forced into the collective farms; after that it was easy to send the unhappy and redundant people to work in the newly created heavy industries and mines. Ugly urban settlements mushroomed everywhere, and the new arrivals often spoke a language different from that of the old inhabitants or of the surrounding area. That is how millions of Russians moved into the Baltic States, annexed by the Soviet Union in 1940, and, starting much earlier, into the Caucasian region or to Central Asia. The postwar Communist governments, using the extensive powers of the state, launched new campaigns to send Serbs to Kosovo and the Vojvodina, Romanians to Transylvania, and Slovaks to the southern rim of Czechoslovakia, re-annexed from Hungary. At first glance that was the opposite of "ethnic cleansing"—it was rather "ethnic mixing"; but this rapid and centrally coordinated process created a reservoir of ill feeling and mutual distrust, especially since the newcomers, the colonists, were given the top jobs, and thereby making up the local police and the secret police. Usually they did not even try to learn the language of the minority, which often was the local majority.

It is not the task of the present essay to give an account of how Hungarians became victims of such an undeniable, deliberate changing of the ethnic composition of the land of their birth. I can only present the various types of those detestable policies. The unilateral, unauthorized expulsion of tens of thousands of Hungarians from Czechoslovakia was followed by an organized "population exchange," involving about 150,000 people, agreed to by Hungary as a lesser evil than outright expulsion, which became the lot of the three million Germans in Western Czechoslovakia. But at least those policies were carried out more or less in an organized way, and while quite a few people died during the process, outright murder was rare. In Transylvania, Romanian irregular troops carried out a number of "executions" of innocent Hungarian peasants in several villages in late 1944, but the number of victims was much smaller than the toll of deaths caused by the inhuman conditions of the POW-camps or the forced labor camps in the Danube estuary, or delta. Tens of thousands of Germans (Saxons from Transylvania and Swabians from the Banat) were sent to the Soviet Union for labor; many did not return. In Subcarpathia—a province ceded by Hungary (under duress) to Czechoslovakia only to be passed on to the Soviet Union in 1945—practically the whole male Hungarian population was rounded up and sent to forced labor inside the Soviet Union in late 1944, with extremely high rate of mortality. The clearest case of ethnic cleansing occurred in the Vojvodina, the province which was returned to Hungary in 1941. At the end of 1944 the Communist partisans of Tito tortured and murdered about 40,000 Hungarian civilians, including women, children, and priests. From the late 1940s on the brutal methods gave way to gradual ethnic change described above. By 1990 the results were quite astounding, as shown in the following tables and charts.

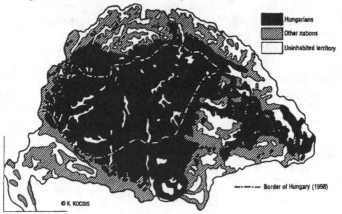

Hungarians
Other nations
Uninhabited territory

— · — · — Border of Hungary (1998)

© K. KOCSIS

Figure   Ethnic map of Hungary (late 15th century)

Figure 7. Ethnic map of the Carpathian Basin (around 1980)

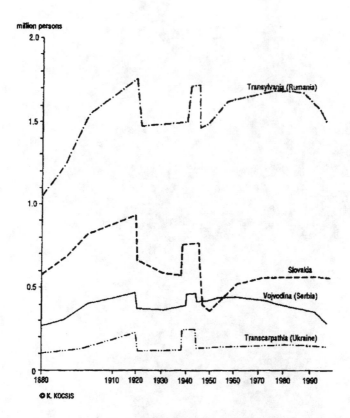

Change in the number of ethnic Hungarians in Transylvania, Slovakia, Vojvodina and Transcarpathia according to the census data (1880–1990)

Table. Change in the number and percentage of the Hungarian minorities in different regions of the Carpathian Basin (1880–1991)

Year	Slovakia		Transcarpathia (Ukraine)		Transylvania (Rumania)		Vojvodina (Yugoslavia)		Croatia		Transmura Region (Slovenia)		Burgenland (Austria)	
	number	percent	number	percent	number	percent	number	percent	number	percent	number	percent	number	percent
1880	574,862	23.1 M	105,343	25.7 M	1,045,098	26.1 M	265,287	22.6 M	49,560	1.9 M	13,221	17.7 M	11,162	4.2 M
1910	881,326	30.2 M	185,433	30.6 M	1,658,045	31.7 M	425,672	28.1 M	119,874	3.5 M	20,737	23.0 M	26,225	9.0 M
1930	585,434	17.6 N	116,584	15.9 N	1,480,712	25.8 M	376,176	23.2 M	66,040	1.7 M	15,050	–	10,442	3.5 M
1941	761,434	21.5 M	233,840	27.3 M	1,711,851	28.9 M	456,770	28.5 M	64,431	–	16,510	20.1 M		–
1950	354,532	10.3 N	139,700	17.3 N	1,481,903	25.7 M	418,180	25.8 N	51,399	1.4 N	10,246	10.8 N	5,251	1.9 U
1961	518,782	12.4 N	146,247	15.9 N	1,616,199	25.9 N	442,560	23.9 N	42,347	1.0 N	9,899	11.0 N	5,642	2.1 U
1970	552,006	12.2 N	151,122	14.5 N	1,625,702	24.2 M	423,866	21.7 N	35,488	0.8 N	9,064	10.0 N	5,673	2.1 U
1980	559,801	11.2 N	158,446	13.7 N	1,691,048	22.5 N	385,356	18.9 N	25,439	0.6 N	8,617	9.5 N	4,147	1.5 U
1991	567,296	10.8 N	155,711	12.5 N	1,604,266	20.8 N	339,491	16.9 N	22,355	0.5 N	7,636	8.5 N	6,763	2.8 U

Sources: Census data (Slovakia: 1880, 1910, 1930, 1941, 1950, 1961, 1970, 1980, 1991; Transcarpathia:1880, 1910, 1930, 1941, 1959, 1969, 1979, 1989; Transylvania : 1880, 1910, 1930, 1941, 1948, 1956, 1966, 1977, 1992; Vojvodina, Croatia, Transmura Region: 1880, 1910, 1931, 1941, 1948, 1961, 1971, 1981, 1991; Burgenland: 1880, 1910, 1934, 1951, 1961, 1971, 1981, 1991).

Remark: Hungarians include the Székelys (Secui) and Csángós (Ceangái).

Abbreviations: M– mother (native) tongue, N– ethnicity, E– ethnic origin, U– every-day language ("Umgangssprache")

cities and towns of the present-day Slovakia (1880 – 1991)

Year	Total population		Slovaks		Hungarians		Germans		Others	
	number	%	number	%	number	%	number	%	number	%
Párkány - Štúrovo										
1880	3,547	100	41	1.2	3,340	94.2	54	1.5	112	3.2
1900	4,424	100	10	0.2	4,397	99.4	12	0.3	5	0.1
1910	4,578	100	26	0.6	4,509	98.5	39	0.8	4	0.1
1919	4,989	100	257	5.1	4,703	94.3	17	0.3	12	0.2
1921	5,137	100	316	6.1	4,722	91.9	31	0.6	68	1.3
1930	6,145	100	1,431	23.3	4,046	65.8	123	2.0	545	8.9
1938	5,233	100	97	1.8	5,099	97.4	5	0.1	32	0.6
1941	5,868	100	69	1.2	5,634	96.0	41	0.7	124	2.1
1991	13,347	100	3,310	24.8	9,804	73.5	3	0.0	230	1.7
Léva - Levice										
1880	7,597	100	1,316	17.3	5,806	76.4	451	5.9	24	0.3
1900	9,786	100	1,242	12.7	8,286	84.7	198	2.0	60	0.6
1910	10,816	100	948	8.8	9,618	88.9	208	1.9	42	0.4
1921	11,556	100	3,382	29.3	7,462	64.6	215	1.9	497	4.3
1930	13,975	100	6,886	49.3	5,432	38.9	216	1.5	1,441	10.3
1938	13,608	100	2,052	15.1	11,246	82.6	216	1.6	94	0.7
1941	14,150	100	1,555	11.0	12,338	87.2	162	1.1	95	0.7
1980	26,502	100	22,100	83.4	4,010	15.1			392	1.5
1991	33,991	100	28,126	82.7	5,165	15.2	6	0.0	694	2.0
Losonc – Lučenec										
1880	6,471	100	1,551	24.0	4,449	68.8	404	6.2	67	1.0
1900	10,634	100	1,441	13.6	8,800	82.8	278	2.6	115	1.1
1910	14,396	100	2,055	14.3	11,646	80.9	471	3.3	224	1.6
1921	13,798	100	6,713	48.7	5,760	41.7	594	4.3	731	5.3
1930	17,186	100	9,953	57.9	4,411	25.7	907	5.3	1,915	11.1
1941	16,641	100	1,987	11.9	14,023	84.3	335	2.0	296	1.8
1970	21,308	100	17,570	82.5	3,514	16.5			224	1.0
1980	24,770	100	20,520	82.8	3,803	15.4			447	1.8
1991	28,861	100	23,272	80.6	4,830	16.7	13	0.0	746	2.6
Rimaszombat – Rimavská Sobota										
1880	7,339	100	1,473	20.1	5,484	74.7	185	2.5	197	2.7
1900	8,048	100	741	9.2	7,197	89.4	73	0.9	37	0.5
1910	9,166	100	880	9.6	8,014	87.4	92	1.0	180	1.9
1921	9,296	100	2,750	29.6	6,164	66.3	123	1.3	259	2.8
1930	11,221	100	4,734	42.2	4,736	42.2	130	1.2	1,621	14.4
1941	9,947	100	997	10.0	8,828	88.8	50	0.5	72	0.7
1970	16,238	100	9,220	56.8	6,770	41.7			248	1.5
1980	19,205	100	11,000	57.3	7,800	40.6			405	2.1
1991	24,771	100	14,256	57.6	9,854	39.8			661	2.7
Rozsnyó – Rožňava										
1880	5,226	100	482	9.2	4,374	83.7	285	5.4	85	1.6
1900	5,748	100	369	6.4	5,123	89.1	195	3.4	61	1.1
1910	7,119	100	570	8.0	6,234	87.6	177	2.5	138	1.9
1921	6,937	100	1,163	16.8	5,514	79.5	150	2.2	110	1.6
1930	7,413	100	2,930	39.5	3,472	46.8	191	2.6	820	11.1
1941	7,676	100	530	6.9	7,025	91.5	90	1.2	31	0.4
1961	9,557	100	6,500	68.0	3,040	31.8			17	0.2
1970	10,980	100	7,380	67.2	3,570	32.5			30	0.3
1991	18,647	100	12,271	65.8	5,826	31.2	10	0.0	540	2.9

Bratislava City in 1940). 1921. 1930, 1961, 1970, 1980, 1991: Czechoslovakian census data /ethnicity/.

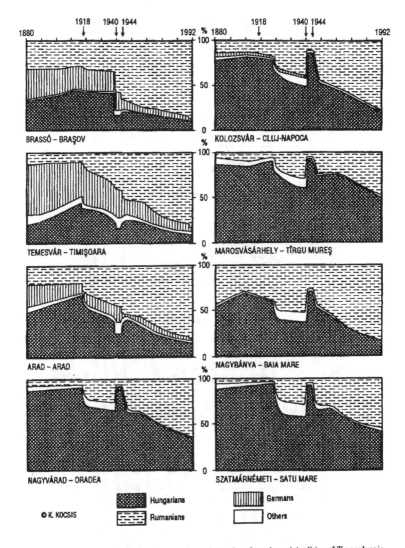

Figure   Change in the ethnic structure of population in selected municipalities of Transylvania
(1880–1992)

Table 26. Ethnic structure of the population of the present territory of Vojvodina (1880–1991)

Year	Total population number	Serbs number	%	Hungarians number	%	Germans number	%	Croats number	%	Montenegrins number	%	Slovaks number	%	Rumanians number	%	Ruthenians, Ukrainians numb.	%	Others number	%
1880	1,172,729	416,116	35.5	265,287	22.6	285,920	24.4	72,486	6.2			43,318	3.7	69,668	5.9	9,299	0.8	10,635	0.9
1890	1,331,143	45,7873	34.4	324,430	24.4	321,563	24.2	80,404	6.0			49,834	3.7	73,492	5.5	11,022	0.8	12,525	1.0
1900	1,432,748	483,176	33.7	378,634	26.4	336,430	23.5	80,901	5.6			53,832	3.8	74,718	5.2	12,663	0.9	12,394	0.9
1910	1,512,983	510,754	33.8	425,672	28.1	324,017	21.4	91,016	6.0			56,690	3.7	75,318	5.0	13,497	0.9	16,019	1.1
1921	1,528,238	533,466	34.9	363,450	23.8	335,902	22.0	129,788	8.5			59,540	3.9	67,675	4.4	13,644	0.9	24,773	1.6
1931	1,624,158	613,910	37.8	376,176	23.2	328,631	20.2	132,517	8.2									172,924	10.6
1941	1,636,367	577,067	35.3	465,920	28.5	318,259	19.4	105,810	6.5									169,311	10.3
1948	1,640,757	827,633	50.4	428,554	26.1	28,869	1.8	132,980	8.1	30,531	1.9	69,622	4.2	57,899	3.5	22,077	1.3	42,592	2.7
1953	1,701,384	867,210	51.0	435,210	25.6			127,040	7.5	30,532	1.8	71,191	4.2	57,219	3.4	23,040	1.3	89,942	5.2
1961	1,854,965	1,017,713	54.9	442,560	23.9			145,341	7.8	34,782	1.9	73,830	4.0	57,259	3.1			83,480	4.4
1971	1,952,533	1,089,132	55.8	423,866	21.7	7,243	0.4	138,561	7.1	36,416	1.9	72,795	3.7	52,987	2.7	25,115	1.3	106,418	5.4
1981	2,034,772	1,107,375	54.4	385,356	18.9	3,808	0.2	119,157	5.9	43,304	2.1	69,549	3.4	47,289	2.3	24,306	1.2	234,628	11.6
1991	2,013,889	1,143,723	56.8	339,491	16.9	3,873	0.2	98,025	4.9	44,838	2.2	63,545	3.2	38,809	1.9	22,217	1.1	259,368	12.8
1996	2,213,000	1,422,000	64.3	285,000	12.9	3,000	0.1	62,000	2.8	46,000	2.1	60,000	2.3	34,000	1.5	21,000	0.9	280,000	13.1

Sources: 1880, 1890, 1900, 1910, 1941: Hungarian census data (mother/native tongue), 1921, 1931: Yugoslav census data (mother /native tongue), 1948, 1953, 1961, 1971, 1981, 1991: Yugoslav census data (ethnicity), 1941: combined Hungarian (in Bácska 1941) and Yugoslav (in Banat and Syrmia/ Szerémség/ Srem 1931) census data. 1996: estimation of K. Kocsis based on "Census of Refugees..... Belgrade, 1996.

Remarks: Data between 1880 and 1910 include the settlements of Tompa, Kelebia, Csikéria of the present-day Republic Hungary at that time belonging to the administrative area of Szabadka/Subotica City. The Croats include the Bunyevats, Shokats and Dalmatinian ethnic groups and the "Serbs of Roman Catholic religious affiliation" in 1890.

Table: Ethnic Structure of the population living in the regions of Southeastern Europe (1920, 1980)

Regions	Year	Hungarians	Germans	Romanians	Slovaks / Czechs	Russians, Ukrainians, Ruthenians	Serbs, Montenegrins	Croats	Muslims	Bulgarians	Macedonians	Albanians	Turks, Tartars, Gagauzes	Jews	Gypsies
SLOVAKIA 49,025 km	1921	22.0	4.9	-	68.4	3.0	-	-	-	-	-	-	-	2.5	0.3
	1980	11.2	0.1	-	87.8	0.3	-	-	-	-	-	-	-	-	-
HUNGARY 93,033 km	1920	89.6	6.9	0.3	1.9	0..	-	-	-	-	-	-	-	-	0.1
	1980	98.8	0.3	0.1	0.1	-	-	-	-	-	-	-	-	-	0.3
TRANS-CARPATHIA 12,800 km	1921	17.9	1.7	1.7	3.1	60.2	-	-	-	-	-	-	-	12.9	-
	1979	13.7	0.3	2.3	0.8	81.4	-	-	-	-	-	-	-	0.3	0.5
TRAN-SYLVANIA 103,093 km	1920	25.5	10.6	57.3	0.6	0.4	0.9	-	-	0.3	-	-	-	3.5	0.8
	1977	22.0	4.3	70.9	0.3	0.6	0.5	-	-	0.1	-	-	-	0.1	0.6
ROMANIA (PROPER) 134,407 km	1920	1.4	0.8	91.6	-	1.0	-	-	-	0.7	-	-	0.5	1.1	0.8
	1977	0.1	0.0	98.8	-	0.2	-	-	-	0.0	-	-	0.5	0.1	0.2
MOLDAVIA 33,700 km	1926-30	-	1.3	68.4	-	15.5	-	-	-	2.2	-	-	3.1	8.1	0.0
	1979	-	0.0	63.9	-	27.0	-	-	-	2.0	-	-	3.5	1.5	0.0
CROATIA 56,538 km	1921	2.4	2.9	-	1.2	-	17.4	68.1	0.5	-	-	-	-	-	-
	1981	0.6	0.0	-	0.5	-	11.8	75.1	-	-	-	-	-	-	-
BOSNIA-HERCEGOVINA 51,564 km	1921	-	0.9	-	-	0.7	43.5	21.5	30.9	-	-	-	-	-	-
	1981	-	0.0	-	-	0.1	32.4	18.4	39.5	-	-	-	-	-	-
VOIVODINA 21,506 km	1921	24.4	21.0	4.5	3.9	1.3	34.8	7.7	-	0.1	0.9	-	-	-	-
	1981	18.9	0.2	2.3	3.5	1.2	56.6	5.4	-	0.1	-	0.2	-	-	1.0
SERBIA (PROPER) 50,968 km	1921	0.1	0.5	5.3	0.1	0.4	87.0	0.3	2.5	1.7	0.7	0.7	-	-	-
	1981	0.1	-	0.6	0.1	-	86.8	0.6	2.7	0.5	0.5	1.3	-	-	1.0
KOSOVO 10,887 km	1921	-	-	-	-	-	20.5	-	6.3	-	-	65.8	6.3	-	2.2
	1981	-	-	-	-	-	13.2	-	3.7	-	-	77.4	0.8	-	-
MONTENEGRO 13,812 km	1921	-	-	-	-	-	75.8	5.8	12.3	-	-	5.5	-	-	-
	1981	-	-	-	-	-	71.9	1.2	13.4	-	-	6.5	-	-	-
MACEDONIA 25,713 km	1921	-	-	0.3	-	-	2.3	-	5.2	-	62.4	13.9	12.7	-	2.3
	1981	-	-	-	-	-	2.5	-	2.1	-	67.0	19.8	4.5	-	-
BULGARIA 110,912 km	1920	-	-	1.3	-	0.2	-	-	-	79.3	2.3	-	15.0	0.6	1.7
	1980	-	-	0.0	-	0.1	-	-	-	85.5	2.5	-	8.4	0.0	2.6
ALBANIA 28,748 km	1923	-	-	1.2	-	-	-	-	-	-	0.9	90.4	-	-	1.2
	1980	-	-	0.4	-	-	-	-	-	-	0.4	97.0	-	-	0.4
CARPATHO-BALKAN-AREA toto 796.706 km	1920	22.1	4.1	27.9	5.3	2.2	11.1	6.7	1.7	9.7	1.3	2.7	2.3	1.3	0.5
	1980	18.0	0.5	29.4	6.1	2.9	11.7	5.9	2.7	10.4	2.1	5.8	1.4	0.1	0.7

Looking at the charts and figures helps one understand why the national minorities living in Central and Eastern Europe are so keen on the right to self-government and language rights. What they fear is nothing less than gradual elimination, creeping and silent ethnic cleansing. A comparison of their numbers and proportion in the total population in 1920 and 1991, respectively, indicates that this fear is well founded.

Is the Dream Relevant After the Horrors?

In late 1991 the Yugoslav People's Army launched an attack on Croatia and following heavy fighting overran Eastern Slavonia and the Knin region, in Central Croatia. The civilian population, Croats and Hungarians, who did not flee or hide, were killed in large numbers. The Serbs living in Western Slavonia became targets of retribution and escaped to Serbia. In 1992 the "ethnic cleansing" of the Muslims and Croats of Bosnia commenced. Those who were not killed or expelled were subject to systematic torture and rape. The horrors continued until in 1995 NATO finally stopped what had become a war of mutual extermination. Thereafter, the Croatian army occupied the "Serb Krajina" in the middle of Croatia, inducing its inhabitants to flee. The latest chapter of mass expulsions and killings took place in Kosovo in 1999, and while it was stopped and reversed thanks to the bombing campaign of NATO, the conflict between the Serbs and the Kosovar Albanians, and the sporadic killings have not stopped as of this writing.

Winston Churchill proved more right than he ever imagined: independence brought terrible sufferings to the peoples of the one-time Austro-Hungarian Monarchy (Bosnia included!).[12] Is there an end to all that? Is there a way back to ethnic peace and justice? Can the nightmare of "ethnic cleansing" be replaced by ethnic survival and revival? Since the international community seems to be committed to this goal, a point testified to by the numerous recommendations and conventions of the U.N., the OSCE, and the Council of Europe, the answer must be in the affirmative. Since both NATO and the European Union made it clear that the observation of minority rights is one of the preconditions of admitting an applicant country into its ranks, one

---

[12] "...there is not one of the peoples or provinces that constituted the Empire of the Habsburgs to whom gaining their independence has not brought the tortures which ancient poets and theologians had reserved for the damned." Sir Winston Churchill, *The Gathering Storm* (London, 1948), 14.

should be optimistic. But whereas there is a general commitment not to allow new ethnic cleansings, there is no similar commitment for preventing the slow, creeping version of ethnic change and all the concomitant suffering and conflicts.

In my opinion the best guarantee for the survival of the remaining national minorities lies in re-discovering and adopting the cantonal model. This is not what the international community is striving for. There is much talk about the need for a multiethnic society in the former Yugoslavia and beyond that. Yet the history of the last centuries and the events of the last ten years show us that the way to achieve and preserve such a model lies not in mixing peoples artificially in a reckless way, neither in returning all the expelled, "cleansed," victims to the scene of the crime, expecting them to get along with their tormentors in good spirit. It lies in following the Swiss model of autonomous units bound together by geography, common traditions, and economic interests. It has been found working not only in Switzerland, but also in South Tyrol and in Catalonia, and hopefully it will work in Corsica. That model should be introduced in Serbia, particularly in the Vojvodina and the Sanjak. That is what the millions of Hungarians in Slovakia and Romania are striving for. An Eastern Switzerland in the Danubian Basin was a dream once. It should be turned into a reality.

Conclusion

I would like to submit four preliminary conclusions from the present study of the subject "ethnic cleansing."

1. Historians should strive to present these facts to the political leaders, to the general public, and particularly to their students. A healthy indignation about the horrors witnessed over Bosnia and Kosovo should direct attention to the earlier versions of those policies.

2. It is worth noting that those crimes do not pay in any way. Even apart from signs that the indicted war criminals are sooner or later brought to court, the practice of getting rid of the "undesired" national groups has proved to be detrimental, even to those who thought they would benefit from it. The expulsion and elimination (or in the case of Ceausescus's Romania occasionally the outright selling) of the Germans, Jews, Hungarians, Greeks, Muslims—who were all diligent, hard-working, often better-than-average educated people—was a very serious loss to the countries concerned. Those majorities that remained found themselves poorer in many ways.

3. It is essential to come to terms with the past. The International War Crimes Tribunal in The Hague has its important tasks, but the crimes must be faced also by the people in whose name they were committed. *Responsibility, remorse, and reparation* — these three Rs are the precondition for the fourth, *reconciliation*. I am proud to say that my country, Hungary, has carried out these steps over its painful twentieth century and is willing to continue on that path.

4. I recall the very first sentence of my Latin textbook, *Historia est magistra vitae*. In a sophisticated translation, the sentence says that history is a philosophy which teaches through examples. This use of history is even more ambitious than Henry Kissinger's aim of "shedding light on the likely consequences of comparable situations,"[13] and I do hope that the terrible practice of "ethnic cleansing" offers us more than one lesson. It goes beyond the command "thou shalt not"; the study of ethnic cleansing should also show that ethnic diversity is not bound to lead to violent clashes. Perhaps that study can also help us realize that the best way for creating or re-creating a multiethnic society might be found in the cantonal model.

---

[13] Henry Kissinger, *Diplomacy* (New York, 1994), 27.

Ethnic Cleansing

Lt. Gen. MICHAEL V. HAYDEN, USAF
Director, National Security Agency

Many thanks for that kind introduction. It's great to be back in Pittsburgh and back at Duquesne. I was here last June at an event and had the opportunity to comment that one could take the boy out of Pittsburgh but that it was a lot harder to take the Pittsburgh out of the boy. Apropos my Balkan focus for this evening, I still remember watching the Browns and Steelers in a 1995 play-off game from the basement of the Marine house in Tirana, Albania.

There was also the time in 1994 when I walked into the office of Branco Krga, the Chief of Intelligence of the Serbian army, and heard him proudly announce that his grandfather had lived here in Pittsburgh. I really appreciate this opportunity to share a few thoughts this evening and especially thank Dr. Várdy for the invitation. I believe the records will show that I was in the first class Dr. Vardy taught at Duquesne, the History of Western Civilization in the old theater.

I have three purposes this evening.

1) To pass on some personal experience since I left Duquesne; I am not a victim of ethnic cleansing, unless one counts my family's Irish immigrant experience of the mid-nineteenth century, but I was able to see ethnic cleansing and its effects, and perhaps something of its causes, in the Balkans of the 1980s and 1990s;

2) To commend you for what you are doing here, that is, putting events into a larger context; and finally

3) To both commend you and exhort you to "follow the facts" no matter where they might lead.

Before I begin discussing these, let me first of all remind you that what I say here tonight is as a scholar and an observer, not as a spokesman for the United States government. And what I will talk about tonight is a product NOT of my current work as the Director of the National Security Agency, but rather of my experience in previous positions as the air attaché in Bulgaria and the Chief of Intelligence for U.S. Forces in Europe. All that said, these remarks were passed through the Department of Defense for security and policy review.

In July of 1994, I was in an UNPROFOR helicopter with Michael Rose, the British commander of UN forces in Bosnia. I was escorting General Jack Galvin, former NATO commander, and Rose was intent on our getting a full picture of the war. We took off from Zetra Stadium in Sarajevo; you may recall Zetra as the venue for the opening ceremony for the 1984 Winter Olympics. The land between it and the nearby indoor rink—site of the speed skating events—had already been turned into the largest cemetery in Bosnia. As we flew over central Bosnia (utterly spectacular scenery), Rose pointed out destroyed homes within otherwise intact villages—the destruction not the by-product of combat, but rather of the purposeful criminal attack of neighbor against neighbor. In other cases we could see clusters of villages with one—almost always the smallest and the one at highest altitude—methodically destroyed by its former neighbors. A destroyed mosque or church provided the context for who was attacking whom in each instance.

Later in the day we arrived at Mostar, that city on the Neretva that may exhibit the best late twentieth-century example of cultural warfare. Mostar had been a multi-ethnic city. The modern quarter on the west bank of the river was largely Croat. Muslims were centered in the old quarter on the east bank. Serbs generally occupied the hills to the east. As we circled for a landing, Rose pointed out the ruins of an Orthodox church in the hills and said that the priceless mosaics within it had been ground to dust by Muslim attackers. In the heart of the city itself, one could walk through the old Turkish quarter facing the river and see Muslim defensive positions in the old houses where centuries-old stone walls had been reduced by Croatian *small arms fire*. And then one could look into the green waters of the Neretva and see the remains of the magnificent Ottoman bridge from which the town took its name, destroyed by Croatian mortar fire because it was a symbol of Islam, the Ottomans, and their former Muslim neighbors in the city.

Later I walked the streets of Sarajevo, into the former municipal library (Archduke Franz Ferdinand's last stop on earth), and kicked the charred rubble created by Serb artillery fire. From near the library, one could easily look up into the hills and see the Serbian side of the confrontation line. I remember asking myself at the time, "What manner of man would pull the lanyard on his tube and send a shell into such a dense urban area?"

The answer came slowly, over time and over many visits to the war zone. And the answer came with disturbing clarity. Who could do such things? *Just about anyone.* The disturbing thing about Bosnia for me was not that these people were so evil, but that the veneer of civility that I had lived under all of my life was so thin. I don't mean to establish moral equivalency here, claiming that all are equally guilty or that everyone has done such hideous deeds. I just want to point out that I have met many of the people responsible for *these* deeds and—well—they're pretty average.

In the Serb leadership, Radovan Karadžić was a failed poet and the sports psychologist for the Sarajevo soccer club. Nikola Koljević, who committed suicide after the war, was a Shakespearian scholar. Ante Roso, commander of Croatian forces in Bosnia, was an NCO in the French Foreign Legion. The most urbane of Bosnian leaders, Foreign Minister Harris Siladžić, had been an Islamic scholar, and Eyup Ganić, the Bosnian vice-president, had taught at a university in Michigan.

During the war I had dinner in Belgrade with an officer named Dimitreyevich, head of the military security services. He had known Ratko Mladić before the war. Given the prominent role Mladić was playing, I asked him what Mladić had been like before the dissolution of Yugoslavia. He looked me in the eye and responded, "Nothing special." Mladić, by the way, the head of the Bosnian Serb Army, was married to a Macedonian and had enrolled his nationality in the last census as "Yugoslav."

Two nights after my dinner with Dimitreyevich, I had dinner with Milo Panić. By then he had retired, but before he left service Panić had overseen what everyone agrees was one of the most brutal offensives of the entire war, the Serb attack across the Sava River on Vukovar in Croatia. I think most of you have seen the pictures. Serb artillery pounded the town's defenders mercilessly, destroying most of the city, and—when resistance collapsed—Serb irregulars finished the job. As the story goes, the last Croat killed as the Serb irregulars savaged the town was working a bulldozer uncovering Serbian graves in a churchyard.

All that said, Panić was a delightful dinner companion. His round shape reminded me of the Pillsbury doughboy and his observations on the war were balanced and thoughtful. I looked across the table and attempted to impute guilt to this aging, over-weight pensioner. By and large I was successful. But much later I thought of Panić trying to take a well and fiercely defended city with his artillery-rich and infantry-poor force (which pretty much describes the Yugoslavian army), and opting to play to his strength (bombardment and indirect fire) rather than his weakness (closing with an enemy in fortified positions). I thought of him as I walked the streets of Dresden on the fiftieth anniversary of the Allied fire-bombing raid. And I thought of him when a casualty-averse NATO force kept its strike aircraft in Kosovo above 15,000 feet to avoid casualties to its pilots.

So that's what really scared me about Bosnia. These weren't universally evil people. In fact, they didn't seem to be any more or less evil than others I had met in my career. But there sure was evil afoot in the Balkans. I could see it everywhere I looked. So what caused it? I refused to believe it was inevitable either because of the nature of these people or of their history.

I think that's where you come in. And that's why I earlier commended your work here this week. Americans are an ahistoric people. We come to new situations asking, "what's the problem?" when a more appropriate inquiry might be, "what's the story?" And the way we now get news—as it is provided to the public and as it is provided to our policymakers—strengthens this tendency toward the here and now at the expense of a longer view.

James Fallows put this quite well when he pointed out that: "In the real world, events have a history. Part of the press' job is to explain that history, although that goes against TV's natural emphasis on the now. In real life, events have proportions.... Part of the press' job is to keep things in proportion. TV's natural tendency is to see the world in shards. It shows us one event with the air of utmost drama, then forgets about it and shows us the next."[1] This need for deeper historical and cultural context applies even to strictly military analysis. I recall one instance during the fierce fighting in Mostar that I mentioned earlier. A young intelligence briefer at European Command headquarters was explaining the fighting when a senior officer asked, "Who was on the offensive here?"

"The Croats," the young officer accurately replied.

"How far will they go?" continued the senior.

"They will fight to the Neretva."

"Why? Is that a defensible line?"

"Very defensible. The banks of the river drop off some twenty meters. But that's not why they'll stop there."

"Then why will they stop?"

"Because in the great schism of Christendom in the eleventh century, the dividing line between Orthodoxy and Catholicism was the Neretva. To be Croat is to be Catholic. They'll go to the river and stop."

And so they did.

And if historical context can help military analysis, how much more important must it be for questions of policy and broad strategy. For example, how would historical analysis have helped

---

[1] James Fallows, *Breaking the News: How the Media Undermines American Democracy* (New York, 1996), 53.

American understanding of the unconscionable Serb violence against Croats in the Krajina region following the Western recognition of Croatia? Given the World War II experience of Croatia's Serbian minority, could we have anticipated such violence as these Serbs' constitutional ties to Serbia were severed without their consent by a Croatia governed by a holocaust revisionist historian who had already resurrected the place names, flag, shield and currency of Ustasha Croatia? And if history could have taught us to anticipate this violence, could we have done anything to prevent it?

With a deeper appreciation of history, we might have understood that calls for sovereignty over disputed land could be based on historical claims, legal claims, constitutional claims, or even the demands of economic or security logic rather than our preferred formula—the popular will of the current residents of the land. There may be no reason to change our formulation—but perhaps we could better anticipate the course of action we were incentivizing for those who subscribed to one of the other formulas.

Others remember history! I recall one sunny afternoon at an UNPROFOR outpost overlooking what we called the Posavina corridor, that narrow stretch of Serb-controlled territory that connected the Serb lands in the Drina Valley with those further to the west near Banja Luka. A young man and his elderly father passed by the outpost with their sheep. Michael Rose stopped them and asked a few pertinent questions of the younger man: was he in the army; how long a time was he at home before he returned to his unit; did he take his weapon home or did he leave it for his replacement? Rose, more to be polite than anything else, asked the old man if he too was a soldier. "Da," he replied, raising his shirt, to reveal an ugly scar on his side and back, and announcing "Moskva." He then stepped backward, assumed the position of attention, extended his right arm and hand, and proclaimed, "Heil Hitler"-- obviously a veteran of the Bosnian Muslim brigade that served in the German Wehrmacht's Waffen SS.

This call for more historical context is not, by the way, a cry to accept all versions of the past as equally true. Ratko Mladić, commander of the Bosnian Serb forces, remembers history in a peculiar way. I recall sitting across the table from him at Lukavitsa barracks in a Serb-controlled suburb of Sarajevo as he recounted his version of the Serbian past—Serbia the suffering, Serbia the savior of Western civilization, Serbia the ally of the United States in two world wars. Mladić concluded by telling our party what the war in Bosnia was all about: this was a clash of civilizations (and we Christians needed to stick together); and this was also about

European big power politics (and we Americans needed to keep our eyes on the *Germans*).

Before you condemn this view as uniquely myopic, let me tell you about a similar meeting in Zagreb a few weeks later where Franjo Tudjman, President of Croatia, asked the U.S. delegation the same rhetorical question: What was this war all about? His answer: this was a clash of civilizations (and we Christians needed to stick together); and this was also about European big power politics (and we Americans needed to keep our eyes on the *English*).

In 1994, I was on a flight from Sarajevo with a senior American diplomat who had started to read Noah Malcolm's recently published history of Bosnia. You may recall that one of Malcolm's tenets was the historical thread and hence legitimacy of an independent Bosnian state. The diplomat looked at me and asked what I thought the book meant. Without, I believe, intending to discount Malcolm's impressive scholarship, he answered his own question: "History can be used to justify almost anything."

And that brings me to my third major point. Follow the facts. On that same trip to Sarajevo our C-21 military transport got stuck in a major snowstorm on the taxiway when we returned to Frankfurt. I can still recall, as we were digging the jet out, asking a diplomat friend of mine to pay careful attention to the facts on the ground in Bosnia. There were so many pressures at work: donor fatigue when it came to humanitarian assistance; general distaste for the conduct of the war, especially Serbian conduct; the impact the festering crisis was having on overall US-European relations; the general health of NATO; the burden that continued fighting put on improving US-Russian relations. These are, by the way, in my personal calculus, sufficiently weighty reasons for using U.S. military power. But I concluded with my diplomat friend by emphasizing that if a particular event on the ground seemed to be driving the United States toward intervention, I'd like to get a phone call. In other words, I wanted to help him make sure he had the facts right.

Visitors to our headquarters in Germany were often surprised when we briefed them on the Balkans and offered observations that seemed to contradict the view of the war that they brought with them. For example, the observation that from the Serb attack on Gorazhde in April, 1994, to the Serb attack on Srebrenica in July, 1995—a period of some fifteen months—the Bosnian government was on the strategic offensive in Bosnia, dictating the timing, location and intensity of all the fighting. Or that it was the Bosnian army that consistently violated and finally ended the ceasefire brokered by President Jimmy Carter in December 1994.

Or our judgement that a strategic equilibrium existed in the Bosnian war, that neither the equipment-rich but manpower-poor Serbs, nor the infantry-heavy but artillery-light Muslims could move the confrontation lines—a judgement that had some serious implications for the image of the Serbs as an unstoppable military juggernaut and some equally serious implications for the then-raging debate on whether or not to arm the Muslim forces with heavy weapons. Or that, as nearly as we could tell, in both relative and absolute numbers, more Bosnian Serbs had been displaced by the war than either Bosnian Croats or Bosnian Muslims. Or that, despite the undeniable suffering caused by the war, we had no earthly idea of the source of the Bosnian government claim (reported globally) of 250,000 dead or missing (a number that is now halved even by official Bosnian government estimates).

Or that, as great as the suffering truly was, the siege lines around Sarajevo were as porous as many of the siege lines in our own Civil War; that a thriving black market and foreign assistance combined to create bizarre economic circumstances —like lemons being available in the market of a "besieged" European city or gas going for less per liter (for those who had currency) than it did in Stuttgart; like all the factions shelling the Sarajevo airport at one time or another in order to interrupt relief flights and thus drive up the black market cost of foodstuffs; or the pathetic images of Sarajevo citizens lining up—often under the threat of sniper fire—to secure water even though the Soros Foundation had developed a way to pump water from the Miljacka into makeshift purification units and then into the city's water lines. This latter effort was rejected by the Bosnian government allegedly because water purity could not be guaranteed. Aide workers told me that a further reason could have been the alternative uses that Bosnian government officials put the fuel supplied to them by the UN to deliver water throughout the city. Or that, in our view, the overall result of outside humanitarian assistance had been to prolong the conflict with aide donors largely acting as the quartermasters of the warring factions and shielding the leadership from the natural consequences of continued fighting at the expense of normal economic activity.

My runway conversation with my friend had been motivated by events that had transpired a few weeks before, in late November 1994, when a crisis over Bihać brought us to the brink of intervention.[2]

---

[2] Although the following account includes much of my personal experience, it draws heavily on research done by Liam Hayden, "Myths and Reality: Bihać and the War in Bosnia," Yale University (unpublished), April 1997.

Bihać was one of the most complex battlefields in what was already an incredibly complex war. There were two Muslim armies inside this UN-proclaimed "safe area." The Bosnian V Corps was loyal to the government in Sarajevo. The second Muslim force belonged to Fikrit Abdić, a local businessman, opportunist, and the top vote-getter in Bosnia's presidential election even though he deferred to Aliya Izetbegović for the latter position. The Sarajevo government wanted to defang Abdić, who had become all too friendly with neighboring Serbs and Croats (thereby creating one of the few examples of interethnic cooperation still alive in Bosnia). In a brilliant campaign filled with daring and deception, the Bosnian V Corps commander, A. F. Dudaković, defeated Abdić's forces.

Then, on 24 October, Dudaković turned to the Serbs, attacked southward, and by the beginning of November, had captured some 250 square kilometers of territory.

The "story"of Bihać was prominent for a time, but coverage rarely noted that the violence was, at least initially, intra-Muslim. U.S. media accounts of the "enclave" also rarely discussed the fact that Bihać served not only as safe area, but as a military stronghold, a base for successful Bosnian government military operation.[3]

When it came under attack, press coverage highlighted the town's civilian character. Characteristic was *The New York Times* observation that "nationalist Serbs have ignored Security Council resolutions declaring it a 'safe area.'"[4] CNN observed that "Bihać, … declared a safe haven by the United Nations, . . . is anything but safe."[5]

According to one UNPROFOR participant, it seemed that as the V Corps offensive met with its initial success, "the press was [sic] more cheerleaders of the success of the V corps than reporters of fact."[6]

Television seemed to have an especially hard time relaying the complexity of the situation, especially as the Serbs regrouped and their counterattack approached Bihać. Coverage described the im-

---

[3] Lt. Col. Marc Rebhun, UNPROFOR Deptury G-2 (Head of Intelligence) March 1994-Feb. 1995; interview by Liam Hayden, February-March, 1997.

[4] Chuck Sudetić, "Serb Pounding of Bihac Wounds 4 UN Peacekeepers," *New York Times*, 13 December 1994, p. 8.

[5] "Fighting in Bihać Continues to Escalate," *CNN News* 3:02p.m. ET (17 November 1994) Transcript #710-1.

[6] Lt. Col. Rebhun, interview by Liam Hayden, February, March 1997.

pending violation of the "safe haven" without identifying Serb movements as a counteroffensive. "CNN coverage was full of references to the threat to the 'safe area' or 'safe haven' of Bihać," a marked reversal from its earlier description of Bihać as an "enclave."[7]

For more than a week in late November, "CNN reported that the Serbs were entering the safe haven of Bihać. A 22 November account reported that there was hand-to-hand fighting on the outskirts of Bihać. On the twenty–third, CNN reported: 'the Serbs show no sign of stopping their assault...' The account lamented that 'once again the Bosnian Muslims say it is too little, too late from the policy that's more concerned with UN peacekeepers than it is with civilians.' Another twenty–third broadcast reported, 'defenses around that area, that city, are collapsing.'"[8]

News coverage on the twenty–fourth exemplified the confusion. Christiane Amanpour reported, "Parts of the safe haven have now fallen to the Serbs, particularly the last line of defense on the south side of the city."[9] Later the same day, one headline read, "Bihać taken by Serbs."[10] By 10:00 p.m. that night, Christiane Amanpour led with "Bihać teetering on the brink."[11] On the twenty–fifth coverage stated that "with reports tonight that Serb tanks have actually entered Bihać itself, the fate of the town appears to be beyond UN control and in the hands of the Serbs."[12]

On the twenty–sixth, the Serbs were again described as "advancing" on the town, which was described as "on the brink"; Amanpour reported, "the mayor of Bihać...is appealing for UN intervention to prevent what he calls a 'certain massacre' by Serb

---

[7] Jack Shymanski, "Fighting Breaks Out in Bihać, Bosnia," *CNN News* 3:19p.m. ET (27 October 1994): Transcript #704-3. October 29 and November 4 coverage confirmed this trend; cf. Transcript #916-2 and Transcript #710-1. Quoted in Hayden, 31.

[8] Jim Clancy, *CNN NEWS* (22 November 1994): Transcript #721-1; Jill Dougherty, "US Officials Consider Humanitarian Air to Bosnia," *CNN NEWS* (23 November 1994): Transcript #941-3; *CNN NEWS* (23 November 1994): Transcript #722-3. Quoted in Hayden, 32.

[9] *CNN NEWS* 12:03p.m. (24 November 1994): Transcript #819-1.

[10] *CNN NEWS* 3:02p.m. ET (24 November 1994): Transcript #723-1.

[11] "Bosnia Cease-file Talks Belie Fierce Ground Combat," *CNN News* 10:02p.m. ET (24 November 1994) Transcript #1053-3.

[12] "Serb Tanks Cruise Bihać in Flagrant U.N. Violation," *CNN News* 6:03 p.m. ET (25 November 1994): Transcript #943-1.

forces advancing on his town"[13] On the twenty–seventh CNN reported "the Serbs are poised, if they should want to, to go into
Bihać at will."[14] By the twenty–eighth, viewers were told "Serb
forces are now engaged in hand-to-hand combat with the defenders."[15]

On the twenty–ninth, "the UN safe haven of Bihać is in jeop-
ardy of being overrun at any time. The mayor of the city, says,
most recently, 'Bosnian Serbs are within 500 yards of the town
center--perhaps closer.'"[16]

Then without warning, the hysterical tone and accompanying apocalyptic predictions disappeared. Bihać was off the scope.
What happened?

First of all, press coverage of the conflict around Bihać lacked
a fundamental understanding of geography. The terrain would
not support the kind of attack the Serbs were described as ready to
mount. As a Marine infantryman put it: "The press was playing it
up that the BSA was going into Bihać [but] they couldn't do it, it
wasn't physically possible and the V Corps knew it....[It was] impossible to attack over the Grabež [plateau] with artillery and
tanks."[17]

Although press accounts continually reported that the Serbs
were on the "verge," those who knew the terrain knew that they
would never cross that threshold.

Secondly, most of the American intelligence community (at
least those at European Command headquarters in Stuttgart and
those serving with UNPROFOR in Zagreb and Sarajevo) believed
that the V Corps was strong enough to resist any attempt to take
the town. We predicted that the Serbs would return to the old confrontation lines--lines based on terrain well suited to defense.

Third, the "human catastrophe" that many had predicted
simply did not materialize. There were stories parallel to those on
the fighting that the civilians in the Bihać pocket were on the

---

[13] "Bosnians Blame United Nations for Serbian Victories," *CNN News* 6:07 p.m. ET
(26 November 1994): Transcript #944-2.

[14] *CNN News* 11.57 a.m. (27 November 1994): Transcript #885-1.

[15] Christiane Amanpour, *CNN News* 6:06 p.m. (28 November 1994): Transcript
#946-2.

[16] "Douglas Hurd Says NATO Efforts Aimed at Bosnian Peace," *CNN News* 8:25
a.m. (29 November 1994): Transcript #935-8.

[17] Col. Kent Koebke, United States Marine Corps, interview by Liam Hayden,
March 1997.

verge of starvation and that UNPROFOR had to intervene to permit UNHCR relief convoys to reach the enclave. I visited UNHCR headquarters in Sarajevo shortly after the first relief convoy had arrived at the enclave—in March 1995, more than four months after the fighting, and found that the convoy had brought in seed for the spring planting, not food!

The tone of reporting was aggravated by the fact that reporters were located, not in Bihać, but in Sarajevo! Throughout the war reporters demonstrated a willingness to accept Bosnian government reports without independent confirmation. This set in motion a bizarre circle of logic. After the fighting, when asked about Bihać specifically, a U.S. official explained that "we had no first hand view of events... we relied on the press-- *the CNN people there.*"[18]

Of course, the CNN people were not *there*; they were in Sarajevo.

I think it is unarguable that reporting on the war—press reporting and official reporting—was affected by an overall moral tone that itself was created by the conduct of the conflict, especially the conduct of the Bosnian Serbs. And I would even argue that the press played an important role as moral conscience. *Wall Street Journal* Pentagon correspondent Tom Ricks believes that U.S. coverage "kept policymakers from being able to follow their inclination to sweep the whole damn problem under the rug."[19] But facts—even detailed facts—do matter. Slogans and labels cannot be allowed to pass for thought. The world is almost hideously complex and while this complexity cannot be allowed to freeze us into inaction, our actions need nonetheless to be fact-based.

When I was first assigned to EUCOM in 1993 I began an almost frantic search for English-language scholarship pertinent to the ongoing war in Yugoslavia. One of the works that I came across was a massive 1930s travelogue of the Balkans written by Dame Rebecca West. The work, *Black Lamb and Grey Falcon*, has been criticized as being unbalanced in its historiography, too favorable to a Serbian view of the past. But in its early pages it contains a wondrous warning that I kept close to my heart (and my desk):

> English persons, therefore, of humanitarian and reformist disposition constantly went out to the Balkan peninsula to see

[18] Unidentified State Department official knowledgeable about the Balkans, interview by Liam Hayden, 18 March 1997.

[19] Interview by Liam Hayden, February 1997.

who was in fact ill-treating whom and being by the very nature of their perfectionist faith unable to accept the horrid hypothesis that everybody was ill-treating everybody else, all came back with a pet Balkan people established in their hearts as suffering and innocent, eternally the massacree and never the massacrer.[20]

Tony Cordesman reminds us of an "American tendency to divide the world into good and evil, and to sanctify useful states while demonizing [others]," warning that "[w]e live in a gray world and we cannot succeed by trying to make it black and white."[21]

It is a momentous thing to use American military power. The lives of Americans, allies, adversaries, and frequently innocents are put at risk. And when we engage our military force we almost inevitably embrace a moral responsibility for the final outcome of whatever conflict or issue we have touched.

I began by saying that I wanted to relate some personal experiences. But frankly that was just a come-on to get you to listen to my real message: *your* role of putting events into a larger context, and *your* responsibility to "follow the facts" no matter where they might lead.

---

[20] Rebecca West, *Black Lamb and Grey Falcon*. (New York: Penguin Books, 1994), 4.

[21] Anthony Cordesman, "The Quadrennial Defense Review and the American Threat to the United States," Center for Strategic and International Studies, 14 January 1997 (CSIS : http://www.csis.org/stratassessment/reports/qdr.html.).

I.

THE RISE OF TWENTIETH–CENTURY ETHNIC CLEANSING: ORIGINS AND PRECONDITIONS

Redrawing the Ethnic Map in North America: The Experience of France, Britain and Canada, 1536-1946[1]

N. F. DREISZIGER

Defenseless minorities, vanquished peoples, society's peripheral elements, can be mistreated in a variety of ways. Their maltreatment may range from mild harassment to genocide. Almost at the apex of these forms of abuse is ethnic cleansing, which might be defined as the alteration of the ethnic map of a region through the application of force and/or threats of violence.

Ethnic cleansing is quite familiar to the peoples of Eurasia and Africa, but the Old World does not have a monopoly of this phenomenon. The driving out of peoples from their ancestral lands and the forcible removal of residents from their homes, have also happened in the New World. While the most obvious examples of such events had taken place in Latin America and the lands that now comprise the United States, the half-continent north of the present US-Canada boundary has not been free of ethnic cleansing in one form or another.

This statement might come as a surprise to those who know little about the history of Canada, a country that is often regarded as one of the most peaceful nations of the world. But, surprising as it may sound to a present-day audience, Canadian history is not free of ethnic cleansing or of its various, perhaps less pernicious mutations. The circumstances under which ethnic cleansing took place in Canada--or, before 1867, in British North America, and before 1760, in New France--varied from time to time, and the victims had ranged from aboriginal populations to racial and ethnic minorities. This paper will examine the experience of New France, British North America, and Canada in this shadowy human practice. It will also attempt to compare briefly what happened in this part of the continent to what took place elsewhere, particularly south of the present-day US-Canadian border. The paper will also

[1] The author would like to thank the Social Sciences and Humanities Research Council of Canada and the Arts Research Program of the Department of National Defence of Canada for supporting his researches on related subjects during the past two decades. He is also indebted to Jane Errington for encouragement and valuable advice concerning this particular project.

speculate on the question of why in this part of the world ethnic cleansing in particular--and the mistreatment of peripheral groups in general--tended to be less frequent and less severe than in many other parts of the world, especially in the nineteenth and twentieth centuries.

The contempt that Europeans had for the basic rights of North America's aboriginal peoples manifested itself very early in the history of White-Indian relations in the lands that later became Canada. At the end of his second voyage to the St. Lawrence River in 1535-36, French explorer Jacques Cartier returned to France with a handful of Indian captives, including Donnacona, the chief of the Iroquois village of Stadacona. Thus it happened that Europeans began their occupation of North America by taking Indians from their land, and ended up taking the land from the Indians.[2]

It is not the purpose of this paper to examine in detail the Europeans' treatment of North America's native populations. From the early part of the seventeenth century on, when the first permanent European settlements began appearing on mainland North America,[3] there began an experiment in European expansion, which can be seen as the largest-scale ethnic cleansing that had ever happened in modern history. As a result of this process, within less than three centuries, an entire continent was virtually cleansed of its original population and became settled by new-comers. Researchers have estimated that the coming of Europeans resulted in a reduction of the continent's original population "at least by two thirds, ... *perhaps by as much as 95 percent.*" [4] Though historians often refer to genocidal wars that happened in this pe-riod in several North American localities, the greatest damage was done by the diseases the newcomers had brought with them. One

---

[2] The practice of stealing Indians from their lands had started with Columbus as early as 1496 when he took thirty slaves with him back to Europe. The Portuguese followed his example and made the capture of New World aborigines a veritable commercial enterprise, though never on the scale of the African slave trade. Samuel Eliot Morison, *The European Discovery of America: The Northern Voyages, a.d. 500-1600* (New York, 1971), 215f, 420f.

[3] Basque and English fishermen had established fishing stations in Newfoundland some decades before, as had the Vikings of Greenland, as early as the first decades of the eleventh century. The history of these settlements, and their relations with the local aboriginal populations, are very poorly documented. Ibid., 32-336 *in passim*, especially chapter seven.

[4] Kirkpatrick Sale, *The Conquest of Paradise: Christopher Columbus and the Columbian Legacy* (New York, 1990), 304f, the emphasis is Sale's.

researcher enumerated 93 medical conditions, ranging from smallpox to chronic alcoholism, which played roles in the elimination of some Indian tribes and the decimation of others.[5]

Interestingly enough, New France's record in the Great Lakes-St. Lawrence region is not nearly as bad as that of the English colonies on the Atlantic seacoast. This generalization is valid despite Cartier's dastardly deed of stealing Indians from Iroquois lands in the 1530s, and Samuel Champlain's military expeditions against the Mohawks soon after the establishment of the first permanent French settlement in the St. Lawrence Valley early in the seventeenth century. The fact was that the French, being few and being dependent above all on the fur trade, were more likely to view the Indians as trading partners and military allies than did their European neighbors to the south. Indeed, with the exception of some of their deadly wars with the Iroquois Confederacy, which was bent for some time on driving the French from the continent, the French were much more circumspect--and even respectful--in their relations with aboriginal tribes than were most of the English colonists.[6]

The French-Iroquois wars remind us that ethnic cleansing was, on occasion, attempted not only by Europeans but also by some of the native nations of North America. The Iroquois certainly tried it, first on their Indian neighbors and then on the French settlers of the St. Lawrence Valley. In the process, whole nations were wiped from the map of eastern North America, the most obvious being the Hurons of the Georgian Bay region. In a few lightning-fast raids, the Hurons were driven from their traditional villages in the late 1640s. Those among them that survived the onslaught, were scattered--only to be exposed to the ravages of famine and accompanying disease. An argument that has been made in defense of the Iroquois is that it was their dire economic situation that forced them to lash out against their rivals in the fur trade, and that before they became dependent on white man's

---

[5] Russell Thornton, cited ibid., 305. In the lands that later became Canada one tribe that suffered heavy losses from smallpox during the early seventeenth century was the Huron. The disease was unintentionally carried to Huronia by French missionaries. For a massive history of the Hurons see Bruce G. Trigger, *The Children of Aataentsic: A History of the Huron People to 1660*, 2 vols. (Montreal, Kingston, 1976).

[6] This fact is pointed out by James Axtell, *The Invasion Within: The Contest of Cultures in Colonial North America* (New York, 1985), 4f. For an insightful survey of French-Indian relations see Bruce G. Trigger, *Natives and Newcomers: Canada's "Heroic Age" Reconsidered.* (Kingston, Montreal, 1985).

goods, they had not engaged in total warfare against their neighbors.

Having eliminated the Hurons as a commercial and strategic factor in the Great Lakes region, the Iroquois turned against the French. Their aim seems to have been the driving out of the French from the St. Lawrence Valley. French fur traders and settlers were ambushed. The lucky ones were slaughtered, the unlucky were carried back to Iroquois lands where most of them were tortured to death. The news of the dangers of life in New France spread to France and immigration to the colony stopped. It was only the arrival, in the mid-1660s, of regular troops—France's Carignan-Salières Regiment—that saved the colony from certain decline and possible disappearance.

Once New France had the military might to fend off Iroquois attacks, it went on the offensive. French troops, however, were no match for the Iroquois who could melt into the wilderness as soon as columns of marching French regulars approached. In fact, the first incursion by the French into the lands of the Mohawks was a disaster. Undertaken in the winter of 1665-66, it brought only privations and death to European soldiers unused to the harsh North American climate and wilderness. More successful was the expedition of the autumn of 1666 when, though encountering no Mohawk warriors, the French regulars managed to find, sack and burn several of their villages. Now it became the Frenchman's turn to begin the task of ethnic cleansing, which the soldiers undertook "with a certain gusto and expertise."[7] As winter approached, however, the French decided to return to their bases in the St. Lawrence Valley, leaving incomplete their task of cleansing the Mohawk River basin of the Iroquois. Within a little more than a year, the Carignan-Salières Regiment was recalled to France. Its most important task had been achieved: the Iroquois Confederacy was forced to sue for peace. In the years that followed New France would create a militia, augment its population, and further expand its fur-trading empire. Though conflict with the Iroquois

---

[7] Jack Verney, *The Good Regiment: The Carignan-Salières Regiment in Canada, 1665-1668* (Kingston, 1991), 82. It is possible that some members of the regiment had acquired—or bolstered—their expertise in sacking villages in Hungary, in the war against the Turks of 1664, where they might have served with the Habsburg's Imperial Army under commanders that included the Hungarian general Miklós Zrínyi (1620-1664). Verney has not found documentation as to how many of the regiment's soldiers had fought the Turks in these campaigns before their dispatch to Canada, ibid., 189, note 18. On the war against the Turks see Georg Wagner, *Das Türkenjahr 1664* (Eisenstadt, 1964).

would resume in the 1680s, never again would the colony have to face the prospect of annihilation by Indians.[8]

Later, the French became more adept at North American warfare when they adopted the methods that had been used against them by their Iroquois adversaries. In the wars of the eighteenth century the French, often accompanied by their Indian allies, would use these guerilla tactics against the English settlements of New England and New York. The aim of the French in applying such often deadly tactics was neither genocide nor ethnic cleansing. Their savage raids against frontier outposts were usually designed to impress Indian friend and foe with the ruthlessness and boldness of the French, and to convince them of France's determination to carry warfare into enemy territory. A secondary aim was to force the English colonial militia to stay home, to guard their farms and loved ones, thereby to deny the governors of the English colonies the chance to mount major military expeditions against French targets. The exception to this pattern of behavior was the attempt by the French on the eve of the Seven Years' War to cleanse the Ohio Valley of English fur traders and pioneers.

By the middle of the eighteenth century the Anglo-Americans had started to penetrate the lands west of the Allegheny Mountains. The fact that these territories served as traditional homes for many Indian nations was no deterrent to English fur traders and would-be settlers, or to speculators trying to lay claims to large tracts of land. From their posts in the Great Lakes Basin, the French watched these developments with increasing concern. At first they tried to convince the region's Indians that they should oppose the advance of the English. When persuasion failed, the French resorted to cajolery and bribes; still, the tribes of the Ohio refused to resist vigorously the Anglo-American influx.

In 1752 the French court decided to enforce France's claim to the region. It appointed Ange de Menneville, the marquis de Duquesne, as the new governor of New France and instructed him to drive the English from the Ohio Valley. Duquesne and his men undertook their task with ruthless determination. In short order and at heavy cost to themselves, they established a strong French presence south of Lake Erie. In 1754 they began the construction of Fort Duquesne, at the junction of the Allegheny and Monongahela Rivers, deep in what only a year before had been English-held territory. They also threatened the local Indians with annihilation should they decide to support the English. Attempts by Anglo-

---

[8] Ibid., chapter 9, "Conclusion," especially 125-28; W.J. Eccles, *The Canadian Frontier, 1534-1760* (New York, 1969), 64-67.

American militiamen to resist these developments led only to their disgraceful eviction, along with their commander, the young George Washington.[9] It took a major war between England and France to alter this situation and to banish France forever from the interior of the continent.

While in the northern half of the continent, at least up to the middle of the eighteenth century, a more-or-less symbiotic relationship between the French and native peoples predominated, in the south almost the opposite tendencies came to prevail. In the British North America of the seventeenth and early eighteenth centuries, conflict between Indian and white populations became endemic. A large part of the explanation lies in the prevailing economic conditions in the English colonies. There, agriculture was an important factor that resulted in the relentless advance of the agricultural frontier into the interior of the continent. This expansion brought conflict with the Indians. What developed in the English colonies, first on the Virginia frontier, but by the end of the seventeenth century also in New England, was a genocidal war that for the native tribes of eastern North America was, in the words of one historian, "a savage struggle for the survival not just of individuals but of entire cultures."[10]

While the competition for land might explain the development of this war, its brutality was prompted by other circumstances. Part of the explanation probably lies in the emergence among the English colonists of the image of the Indian as a savage creature with no redeeming human qualities. This image was no doubt promoted by those who had a stake in driving the Indian from his ancestral lands. Interestingly, it was not always subscribed to by the white men who bore the brunt of the conflict on the edge of the frontier.[11] Another factor making for ruthlessness in this conflict was the fact that many of colonial America's leaders were hardened veterans of England's wars. As one astute observer of early American society had remarked, these people had "lived by the sword" before coming to North America and they often

---

[9] Eccles, *The Canadian Frontier*, 157-65.

[10] John Ferling, *A Wilderness of Miseries: War and Warriors in Early America* (Westport, 1980), quoted in Bruce White, "The American Army and the Indian," in *Ethnic Armies: Polyethnic Armed Forces From the Time of the Habsburgs To the Age of the Superpowers*, N.F. Dreisziger, ed. (Waterloo, Ont., 1990), 69.

[11] Bruce White cites the example of Robert Rogers, the famous Indian fighter, who in his accounts of the Indian wars (published in the mid-1760s) expressed admiration for many of the traits and traditions of his adversaries. White, 70.

knew no other ways of dealing with other cultures than by the force of arms.[12]

The English colonies' governors did not limit inhuman treatment only to their Indian neighbors. One of them at least, was ready to apply ruthless methods in his dealings with white, but non-English, settlers within his own jurisdiction. This happened in a British colony that later became a part of Canada. That colony was Nova Scotia, and it became the place where the first ethnic cleansing of a white, Christian population in North America would take place.

The deportation of much of Nova Scotia's French-speaking population during the French and Indian Wars, as the Seven Years' War was known in North America, had complex origins rooted in developments that took place in the early eighteenth century.[13]

The French began settling Acadia, what the English would later call Nova Scotia, in the early part of the seventeenth century. By 1700 they had a number of stable, largely self-sustaining settlements there. In the following years Acadia changed hands several times until 1713 when the French ceded the region to the British as part of a general peace settlement known as the Treaty of Utrecht. For many years, Acadia's new rulers made no efforts to promote English settlement in these lands and were satisfied with keeping them out of the hands of the French. Change began to come to Nova Scotia in the middle of the eighteenth century, especially with the establishment of Halifax as Britain's new naval center in the North Atlantic. These changes were accompanied by the arrival in the colony of thousands of English and German settlers. Though no longer a majority in their traditional lands, the Acadian population continued to be an important economic, political and even strategic factor in the colony's life.

From the very beginning of their rule in Nova Scotia, the colony's governors had been concerned with the strategic implications of the existence of substantial French-speaking and Catholic populations within this outpost of British power in a region still controlled mainly by France. What made the concerns of the colony's governors even more legitimate was the fact most of the Acadian-inhabited lands were close to the ill-defined demarcation line between French and English power. For these reasons, Nova

---

[12] White, "The American Army," 70f.

[13] On this subject see N. E. S. Griffiths, *The Contexts of Acadian History, 1686-1784* (Montreal & Kingston, 1992).

Scotia's first governors insisted on an oath of unconditional loyalty to Britain from the Acadians. They, however, offered only a promise of neutrality in any conflict between France and Britain. After years of wrangling, the English authorities seem to have accepted this offer of limited loyalty.

Tension arose in the region in the early 1750s when the French decided to build Fort Beauséjour near what they considered to be Nova Scotia's border. The English retaliated by building Fort Lawrence just a few miles to the east. When hostilities broke out signaling the start of the French and Indian Wars in North America, the English, in an attempt to strengthen their position against the French, besieged Fort Beauséjour. The capture of the fort's defenders—and among them, a few Acadians—resulted in the decision by Colonel Charles Lawrence, Nova Scotia's lieutenant governor, to deport the local Acadian population. The settlers were herded onto British ships and transported to other British colonies or England.

The main motive for the deportation of the Acadians of the Bay of Fundy region was probably strategic: to improve the position of the British command in Nova Scotia for the expected war with France. The continuation of the deportations during the first three years of the Seven Years War, especially in 1758 when it was the Acadian population of Ile Saint-Jean that bore the brunt of British war policy, was conceivably similarly motivated. Why the deportations continued after the British had ousted all French forces from North America in 1760, is not easy to explain from the strategic perspective. The local British commanders might have feared the return of the French, or possibly, the abandonment by the British government of the conquered territories during the peace negotiations.[14]

Though no attempt was made by the British to deport the population of their newly conquered colony in the St. Lawrence Valley, certain leaders in London did envisage a future Canada that was predominantly English-speaking and Protestant. These expectations were embodied in the Royal Proclamation of 1763

---

[14] In 1745, an armed force outfitted and manned mainly by New Englanders, captured the French fortress of Louisbourg, located on Île Royale (present-day Cape Breton Island, NS). Much to the displeasure of the New Englanders, Britain returned the fortress to France in the Treaty of Aix-la-Chapelle in 1748. The fate of Louisbourg must have weighed heavily on the minds of those who were responsible for the post-1760 Acadian deportations. It should be added that, during the years of the Acadian deportations—as is often the case during times ethnic cleansing—thousands of people fled their homes to avoid deportation. They sought refuge in territory still held by the French.

that established English law in the new colony of Quebec, and even promised an English-style assembly. These hopes were dashed when no mass migration of English settlers to Quebec materialized, and the colony's post-war military governors preferred to rule without an assembly. In any case, trouble south of the border soon brought a change in British policies.

After the French danger to North America's English colonies disappeared, the stage was set for the historical processes that culminated in the American Revolution. In a quirk of fate, what had been New France before now became British North America; and what had been British North America became the new United States—with some exceptions, such as Nova Scotia, which remained British throughout both wars of the second half of the eighteenth century.

These new strategic realities brought about a fundamental change in Britain's policy toward the new BNA's French-speaking subjects and its native populations. The British, now relegated very much to the periphery of the North American continent, needed the support of both the French and the Indians. The new policy toward the people of Quebec was embodied in the Quebec Act of 1774 that, among other things, restored the French civil code and granted the Catholic Church influence not enjoyed elsewhere in the British Empire. For at least two generations there would be no new attempts at changing the ethnic map of Quebec, while the decades that followed the outbreak of rebellion in the Thirteen Colonies would see England cultivate the friendship of Indians, especially in the Great Lakes Basin that served as a common frontier for BNA and the young American Republic.

While the British cunningly cultivated the friendship of most Indian tribes, one tribe of North American natives underwent extinction in England's easternmost North American colony: Newfoundland. But the gradual decline and eventual disappearance of the Beothuks cannot be ascribed to any genocidal designs or policies. It was more the result of what might be called "over-hunting" by local whites combined with the ravages of white man's diseases.[15] While one Indian tribe was heading for extinction in Newfoundland, in the British North America that almost a century

---

[15] The Beothuks, according to Samuel E. Morison, "...were hunted like wild beasts and treated with the utmost cruelty by both French and English settlers...." Morison, *European Discovery of America*, 216. Morison might have added the Irish settlers and even the Micmac Indians. See also L.F.S. Upton, "The Extermination of the Beothuks of Newfoundland," *Canadian Historical Review*, 58 (1977): 133-153, as well as Frederick W. Rowe, *Extinction: The Beothuks of Newfoundland* (Toronto, 1977). The last of the Beothuks died in 1829 of tuberculosis.

later became Canada there were no incidents of ethnic cleansing in the last decades of the eighteenth century. In fact the BNA of the times became the beneficiary of one of the most blatant examples of ethnic cleansing in North American history. That event was the departure of the Loyalists from the triumphant young Republic during and immediately after the War of Independence. The Loyalists were not deported. Nevertheless, conditions became intolerable for them in their former homeland where they were harassed, dispossessed, and even tarred and feathered on occasion. Incidentally, the arrival of mainly English-speaking loyalists in Canada had changed the ethnic map of British North America in a more fundamental way than any British policy of ethnic cleansing or forcible assimilation might have done, had such policies been implemented.

The ethnic map of Canada continued to change during the decades that followed the coming of the Loyalists. This process was precipitated predominantly through immigration. The Loyalists were followed by the "late Loyalists," newcomers from the American Republic who came to British North America mainly for economic reasons. There was also immigration from the British Isles and even from continental Europe, especially in the new century. In the meantime, immigration was reshaping the ethnic map of the United States, along with a policy of relocating the native's tribes from their lands east of the Mississippi to the Far West. It took nearly two decades of cajoling, threatening, and applying force—by federal or state authorities or by local ruffians anxious to hasten the advance of European settlement—to achieve a near-total cleansing of this vast region of its original inhabitants. The Indians who had taken what became known as the "trail of tears" were often decimated by hunger and disease, en route to or in their new homelands.

In sharp contrast to the United States, the nineteenth century brought few if any examples of overt ethnic cleansing in the history of BNA. During the War of 1812 between Britain and the United States, several hundred American settlers left Upper Canada and returned to the Republic. A future historian who will examine the circumstances of this small population displacement might come to the conclusion that this was the result of covert ethnic cleansing, resembling the coming of the Loyalists to BNA a generation earlier, only in reverse. What research there is on this war to date, however, does not warrant this conclusion.[16]

---

[16] The most important work in this respect is George Sheppard, *Plunder, Profit, and Paroles: A Social History of the War of 1812 in Upper Canada* (Montreal & Kingston 1994). Another work, dealing mainly with the ideological consequences of the war,

The next crisis in Canadian history, the Rebellions in Upper Canada and Lower Canada in 1837 and 1838, also brought harsh measures against segments of the population. Though the rebellions, especially in Upper Canada, were by European standards minor affairs and were easily suppressed, they were followed by severe repression. In Lower Canada the post-rebellion persecutions were designed to cow the population into submission. In Upper Canada, an ulterior motive was probably the desire of the colony's Tory establishment to get rid of advocates of reform by scaring them into leaving the province for the United States; and thousands did just that.[17]

Repression and even a form of covert ethnic cleansing reemerged in Canada in the 1870s and 1880s. In 1869-70, and again in 1885, Canada experienced uprisings by its half-breed—*métis* (French-Indian)—population. Known as the First and Second Riel Rebellions respectively, after their leader Louis Riel, these disturbances were part of the troubles the young Dominion of Canada experienced in its expansion into the Great Plains of North America. The first disturbances broke out soon after the announcement of the transfer of this vast region from the authority of the London-based Hudson Bay Company to the newly created Canadian federation. They were triggered by the refusal of the Red River Valley's *métis* population to acknowledge the terms of this transfer which had been arrived at without consultation with the local population. Louis Riel formed a provisional government. The government in Ottawa dispatched troops and began negotiations with Riel and his advisors. Before the troops arrived, the crisis was resolved through a political settlement. No bloodshed followed; nevertheless, relations remained strained between the region's white, English-speaking and Protestant settlers on the one hand, and the mixed-blood, Catholic, largely French-speaking *métis*, on the other. Riel fled to the United States. In the years that followed,

---

is Jane Errington, *The Lion, the Eagle, and Upper Canada: A Developing Colonial Ideology* (Montreal & Kingston, 1987).

[17] As a contemporary Toronto-area resident put it, "a great many are daily leaving the Province..."; some were induced to do so by "the dread of punishment," while others were "disgusted" with the government's policies. These words are from the diaries of Isaac Wilson, 30 April 1838, Public Archives of Ontario, excerpted in *The Rebellion of 1837 in Upper Canada: A Collection of Documents*, Colin Read and Ronald J. Stagg, eds. (Toronto, 1985), document E 93 (pp. 417-18). The suppression of the more serious rebellion in Lower Canada is told in Allan Greer, *The Patriots and the People: The Rebellion of 1937 in Rural Lower Canada* (Toronto, 1993), chapter 11, especially p. 353.

most of his former supporters were gradually driven from their traditional lands.[18] They moved to areas beyond the furthest reaches of the Canadian agricultural frontier.

As so often happened in the process of European settlement of the North American continent, in little more than a decade the *métis* found themselves again on the edge of the expanding agricultural frontier. They wanted recognition of their titles to the new lands they came to occupy, but the Canadian government was slow to respond to their demands. The result was another rebellion, which this time was joined by some of the Plains Cree Indians inhabiting the region. Once again, the rebels were led by Riel who had returned from exile. As in 1869-70, troops were dispatched from Central Canada to crush the rebellion. Unlike in 1870 when the troops arrived after a political settlement had been reached, on this occasion the militia had to quell the uprising by force. In addition to crushing *métis* resistance, the North-West Rebellion had another casualty. In defeating the rebels, an incipient movement by the Plains Cree for autonomy, was also extinguished. Though the Cree had given only limited support to the rebels, the local authorities used the rebellion to force them into submission. One long-term legacy of the rebellion was that not only the *métis* of the North-West, but the Plains Cree as well, became what social historians have described as "broken people."[19] While these episodes of the mistreatment of non-white minorities in Canada in the late nineteenth century might be ascribed to Canada's growing pains, little can be said to justify such incidents in the Canada of the twentieth century.

There can be no doubt that Canadian society of the late nineteenth and early twentieth centuries was blatantly racist. The only reason that it accepted large numbers of non-European immigrants was because they were desperately needed for work that was so backbreaking and dangerous that no Europeans could be found to undertake it. Other non-Europeans had to be admitted because they came from the Indian sub-continent, an integral part of the British Empire of the times, and held British passports. Some Canadians, including many of their leaders, were determined to limit Asiatic immigration. Canada "must be white" was

---

[18] Gerald Friesen, *The Canadian Prairies: A History* (Toronto and London, 1987), 197-201.

[19] Ibid., 235-36.

their rallying cry, and to a large extent, they were successful in keeping non-white immigration to a minimum.[20]

A few of Canada's politicians wanted to go even further. They had plans for those non-European immigrants who were in the country already. George Black, a prominent member of the Conservative Party after World War I, proposed not only that Asiatics should be excluded from the country, but argued also that those who were already here should be deported and their property confiscated.[21]

While no mass deportation of undesirable immigrants was ever undertaken during the interwar period, Canada's governments continued their exclusionist immigration policy throughout the 1920s amd especially the depression–ridden 1930s. Not only were non-whites barred from entering the country, but other minorities such as Jews—in particular those from Eastern Europe and other non-preferred lands—were also on the proscribed list. The restrictions on Jewish immigration, it might be added, continued throughout the Nazi era when Jewish refugees from Nazi-controlled lands tried in vain to seek refuge in Canada.[22] Only in the post-war era were Jews admitted, along with thousands of the other "displaced persons" from Europe.

Wars usually pose a great threat to individual liberties. This was no different in Canada. While World War I saw no attempts at the cleansing of any part of the country of "undesirable" populations, it did give rise to the internment of a great many recent immigrants who had the misfortune of having been born in the

---

[20] One proponent of a "White Canada" was W. L. Mackenzie King, a high-ranking bureaucrat in Ottawa in the early 1900s. He helped to negotiate a 'gentlemen's agreement' with Japan through which the Japanese government undertook to restrict emigration to Canada from Japan. He also negotiated with Great Britain to curb the coming of East Indians to the Dominion. On King's views on Orientals see my *The Wartime Origins of Ethnic Tolerance in Canada* (Toronto: Robert F. Harney Professorship & Program in Ethnic, Immigration and Pluralism Studies, U. of Toronto, Lectures and Papers in Ethnicity no. 29, 1999), 2f. Mackenzie King later became Canada's longest-serving Prime Minister.

[21] George Black, member for the Yukon, speaking in the House of Commons in 1922. Canada, Parliament, *Debates of the House of Commons*, vol. 2 (1922): 1521. Black later became Speaker of the House.

[22] Irwing Abella and Harold Troper, *None is too Many: Canada and the Jews of Europe, 1933-1948* (Toronto, 1983). Ironically, some Jews were shipped to Canada during the war as POWs from Britain, where they had been accepted as refugees before the war's outbreak and where they were later interned as "enemy aliens." Many of these wartime internees were eventually released into Canadian civilian life.

territories of the Central Powers.[23] During the Second World War, the internment of enemy aliens was re-instituted, but on a more limited scale. In fact, in late 1941 it was decided to exempt immigrants from the minor Axis countries (Finland, Hungary and Rumania), not only from internment, but also from other wartime restrictions.[24] While some European groups were more leniently handled during this war than had been the case in World War I, the same could not be said of the Canadian government's treatment of Canada's Japanese population.

Anti-Asiatic sentiments in general and anti-Japanese feelings in particular, had deep roots in Canadian society, especially in the province of British Columbia, the home to most of Canada's Asian immigrants. They pre-dated World War I, and had caused ugly outbursts of mob violence not unlike the anti-Jewish pogroms in contemporary Russia. The Great War did not acerbate these feelings, in part because Japan was Britain's ally and Japanese warships guarded British Columbia's coastline against Germany's Pacific fleet.

Anti-Japanese passions failed to subside during the interwar years and intensified with the approach of the Second World War. The outbreak of actual conflict in the Pacific in December 1941, served only to heighten these feelings. Fearing an invasion of mainland North America by the Empire of Japan, British Columbia's public clamored for the removal of the Japanese from coastal areas. The government in Ottawa tried to appease these sentiments. At first it embarked on the internment of some Japanese enemy aliens. When this move failed to calm public concerns, Ottawa offered to remove all Japanese-Canadian adult males. When that proposal also disappointed popular expectations, the government agreed to the removal of the entire Japanese-Canadian population of coastal British Columbia.[25] The fact that the US gov-

---

[23] While some "enemy aliens" were interned, others were "paroled" and had to put up with the restrictions of a parole regime for he war's duration. Ironically, many of those affected were Ukrainians, Rumanians, Czechs, etc. from the lands of Austria-Hungary, whose sympathies were certainly not with the Central Powers. For an introduction to this subject see John H. Thompson, *Ethnic Minorities during Two World Wars* (Ottawa, 1991).

[24] On this subject see my article: "7 December 1941: A Turning Point in Canadian Wartime Policy Toward Enemy Ethnic Groups?" *Journal of Canadian Studies* 32, no. 1 (Spring 1997): 93-111.

[25] A succinct introduction to the subject is W. Peter Ward, *The Japanese in Canada* (Ottawa, 1982), especially 13-14. For a more detailed treatment see Patricia E. Roy, J.L. Granatstein, Masako Iino, and Hiroko Takamura, *Mutual Hostages: Canadians and Japanese during the Second World War* (Toronto, 1990), chapter 4.

ernment had just issued an order for the "evacuation" of the Japa-nese-American population of America's Pacific Coast,[26] was probably not the cause of Ottawa's decision, but it made it easier for the Canadian government to act in a similar manner.

By the autumn of 1942 all Japanese Canadians had been re-moved from western British Columbia to camps in the interior of the province and elsewhere in Canada. To add insult to injury, or injury to insult as some might see it, their possessions—even sav-ings—were confiscated. Furthermore, in March of 1945, the "evacuees" were given a choice. After war's end they could either resettle somewhere in Canada other than British Columbia, or they could opt for accepting the government's offer of "assisted passage" to Japan. Over 10,000 of them chose this second option. Some felt pressured to do so while others were by then disillu-sioned with Canada—for them going to Japan appeared the lesser of two evils.

The conclusion of hostilities in September 1945, could not be followed by a start of the repatriation program: conditions were chaotic and suitable ships could not be found. In the meantime, many of the Japanese Canadians who had earlier agreed to go to Japan, changed their minds. Still, the government in Ottawa per-sisted in its plans for the deportation of those who had accepted the repatriation scheme and had not changed their minds *before* the war's end. Though a court challenge to the deportation order failed, a public outcry by liberal elements of Canadian society forced the government of Mackenzie King to reconsider its policy. Furthermore, by this time it had become evident that the cleansing of British Columbia of Japanese Canadians might be accomplished without large-scale deportations: in the course of 1946, 4,700 for-mer Japanese-Canadian residents of the province had re-settled in other parts of Canada. Given this turn of events, early in 1947 the government in Ottawa agreed to restrict "repatriation" to those members of the Japanese-Canadian community who were still interested in going. In the end, not 10,000 but fewer than 4,000 Japanese Canadians left Canada to start life anew in war-ravaged Japan.[27]

---

[26] An excellent book on the "evacuation" of Japanese Americans is Peter Irons, *Justice at War: The Story of the Japanese American Internment Cases* (New York, 1983).

[27] Roy et al., *Mutual Hostages*, chapter 6, especially 157-91. Ward points out that the children accompanying the returnees probably left Canada against their own wishes (Ward, *The Japanese,*. 15f).

The "evacuation" of Canada's Japanese-Canadian population from the West Coast in 1942 and their "internal relocation" and "voluntary repatriation" during the war's aftermath, were the last of the incidents in Canadian history that resembled the ethnic cleansings that have taken place in many other parts of the world--both before and after World War II. That war, and in particular, the suffering inflicted during it on many peoples by the Nazis and the Japanese military, discredited racism. In most democratic countries of the world, including Canada, official ethnic and racial tolerance was not long in coming.[28] In the end, though after many delays, Japanese Canadians were offered a government apology along with compensation for their wartime losses and sufferings.

This study has touched on the subject of how and to what extent the ethnic map of the lands that eventually became parts of Canada were transformed by ethnic cleansings. In describing these processes, it has treated ethnic cleansing by (inadvertent) germ warfare, the tomahawk, the flintlock, tarring and feathering, threats of imprisonment, administrative fiat, and so on. This essay has also hinted that not all the transformations in the ethnic map of BNA and Canada had been affected by ethnic cleansing or other forms of violent mistreatment of peripheral groups. Even in colonial times, as that astute observer of early America, James Axtell, has remarked, "the contest [for the continent] was fought largely... with weapons other than flintlock and tomahawk."[29]

Immigration policy has been a very important means of controlling the evolution of the ethnic makeup of BNA, Canada, and of course, also the United States. The immigration of appropriate peoples--in most cases White, Protestant, Anglo-Saxon, or at least Nordic--could be used to swamp aboriginal populations or even, as some WASPs had hoped, the Catholic French. It has been mentioned above that, after the British conquest, the assimilation of the St. Lawrence Valley's French population by immigration from the Thirteen Colonies was envisaged by some, only to be abandoned when troubles arose for the British in these colonies.

The idea of assimilating Canada's French population through large-scale immigration, this time mainly from the British Isles and northern Europe, re-surfaced in Canada after the rebellions of the late 1830s and manifested itself above-all in the creation of the United Province of the Canadas in 1841. These aspirations were again disappointed and were abandoned in 1867 when Quebec

---

[28] On this subject see my work, *The Wartime Origins*, cited above.

[29] Axtell, *Contest of Cultures*, 4.

received a large degree of autonomy as a partly self-governing province of the new federal Dominion of Canada. The advocates of a predominantly WASP Canada, through the employment of appropriate immigration and settlement policies, were more successful in determining the ethnic makeup of other parts of their country, especially the newly developing Canadian West. They were, in fact, able to restrict the demographic and political-social influence of both the aboriginal peoples and the French-speaking Catholics of this region. Likewise, carefully controlled immigration policies and procedures made sure that Canada continued to be a predominantly "white man's country" until the coming of official racial tolerance to the nation in the 1960s and the 1970s. Reinforcing "appropriate" immigration policies to achieve these ends, were such instruments as residential schools for aboriginal children, which were used for decades to assure the integration—and the assimilation—of native children into mainstream Canadian society.[30]

In part because the Canadian state had these policies at its disposal to mold the ethnic makeup of the nation, it rarely felt compelled to resort to more drastic measures. But there were even more important reasons for the relative rarity of ethnic cleansings in BNA and Canada. These had to do with strategic realities on the continent. In the first century of the existence of the young American Republic as a neighbor, imbued as it was with the idea of "manifest destiny," British North America lived in fear of invasion from the south and could ill-afford to alienate minorities, whether Indians or French Canadians, whose support would be essential in times of conflict. Similar considerations existed, perhaps with less intensity, during the wars of the twentieth century when the impulse to treat enemy aliens harshly was counterbalanced by the awareness that their support for the war effort was an asset to the nation. Another factor making for the more humane treatment of ethnic and religious groups in Canada was the existence of people with strong humanitarian convictions. These people usually succeeded in restraining those who, in their fear of people of different cultures and in their panic at times of national and international

---

[30] The residential schools were operated by the Churches. No doubt, books could be written about the Churches' role in the assimilation of native populations, or the preservation of French culture in Quebec.

crisis, favored harsh measures against "potentially dangerous" minorities.[31]

The most important reason, however, for the relative rarity of ethnic cleansings and other brutal handling of minorities in BNA and Canada in the nineteenth and twentieth centuries was no doubt the absence of conditions of total war on Canadian soil. The War of 1812 was a limited war, made so by the military weakness of the belligerents. Only during its last phases were there episodes of ruthless conflict and these were localized in their impact. The internal strife experienced in Upper and Lower Canada in the late-1830s and in the Canadian West in the mid-1880s, were limited civil conflicts not at all comparable to the American Civil War or revolutionary struggles elsewhere in the world in the nineteenth century. And, for Canada, the two world wars of the twentieth century--which had brought untold suffering to minorities throughout the Old World--were wars that were fought, almost exclusively, on foreign soil.

While Canada's past is not free of crimes against minorities, quite justifiably, Canada today is known more for having offered refuge to a great many victims of ethnic cleansings throughout the ages than for having created such victims. As we all know, the refugees of persecutions started to come to the country in the eighteenth century and continued to come throughout the nineteenth and the twentieth. Among them were Indian nations--the Iroquois in the eighteenth century and the Dakota or Sioux in the nineteenth--fleeing the American Army, English- German- and Gaelic-speaking Loyalists fleeing mob violence in the newly-born American Republic, Armenians and Jews who were made homeless by the holocausts of their peoples, and in the second half of the twentieth century, victims of persecutions in communist-controlled lands--the writer of these lines among them.

---

[31] I have explored this subject in somewhat greater detail in my article, "December 7, 1941," (especially p. 105) and in a paper I gave at a Canadian military history conference in Ottawa in May, 2000.

World War I and the Emergence of Ethnic Cleansing in Europe

T. HUNT TOOLEY

Ethnic cleansing and other processes for manipulating, eradicating, or moving large populations are age-old.[1] At least by the end of the second millennium B.C., Mesopotamian regimes had adopted complex and gruesome techniques for "cleansing" recalcitrant or inconvenient groups of people, and similar techniques have reemerged at many points in time and space since then. Yet the twentieth century produced substantial developments in the history of population manipulation. Obviously, the ability of the twentieth-century state to displace or eradicate much larger numbers of people—by use of both new technologies and new modes of organization—presents a truly significant change. Moreover, at the moment of the state's enhanced abilities in this area, a new doctrine of nationalism based on an almost biological concept which we might call the "ethnic state" helped bring about new ways of identifying the target segment of a given state's population. Taken together, technological and conceptual changes seem to have created a new variation of the ancient process of population manipulation. If we consider the changes both in terms of nature and of scope, we might even call the twentieth-century ethnic cleansing a whole new phenomenon.[2]

In many ways, the emergence of a specific form of population manipulation—that is, mass removal or killing of groups based on ethnic status—in Europe in the early twentieth century seems anomalous. Restrictions upon the cruelty of governments, espe-

---

[1] I will use the term "ethnic cleansing" to mean any attempt to remove a given group defined in part or in whole by ethnicity from a specific area. Transfer or removal might, in theory, relatively calm process, but given the attachment of most individuals to their property, home, and homeland, violence—or at the very least the open threat of it—will be an integral part of the process. At worst, of course, murder and other direct violence might accompany the process of ethnic cleansing. Some terms with a more neutral patina—like "forced migration," or "forced population transfer"–do not express the targeting mechanism of "ethnic cleansing."

[2] Andrew Bell-Fialkoff, *Ethnic Cleansing* (New York, 1996), 1-27; Norman Naimark, *Fires of Hatred: Ethnic Cleansing in Twentieth-Century Europe* (Cambridge, Mass., 2001), 1-16. Another view of the origins and nature of twentieth-century ethnic cleansing is Philipp Ther, "A Century of Forced Migration: The Origins and Consequences of 'Ethnic Cleansing,'" in *Redrawing Nations: Ethnic Cleansing in East-Central Europe, 1944-1948*, ed. Philipp Ther and Ana Siljak (Lanham, Md., 2001), 43-72.

cially in wartime, had been greatly tightened by the time of the nineteenth century, and thoughtful people in the European world envisioned even more amelioration of the tradtionally harsh ways of government. Indeed, to some extent a tendency toward the control of cruelty in war also seemed to "civilize" politics, international affairs, and ethnic relations. Further, the extraordinary practical individualism which marked modern Europe worked against forced removal of anyone, since the high value placed on property as an individual right made it seem the more atrocious to drive individuals away from property in which or on which they lived, or with which they had "mingled their labor," as John Locke put it. One does well not to exaggerate, but even if it was not a golden age of ethnic relations, liberal Europe—the Europe of the mid-nineteenth century—nonetheless thought it had seen the last of many of the past's cruel norms, including the forced removal of "undesirable" people.

Even in wartime, the killing of noncombatants and the destruction of their property were increasingly deplored and where possible punished. Rape, pillage, and murder occurred in all Napoleonic-era armies, for example, but it is significant that commanders like Wellington expended much time and effort in enforcing more civilized behavior among their troops.[3] Patterns of wartime behavior became still more subject to codification and control from the period after the 1860s, when European states created elaborate "rules" of warfare, many of which were aimed at protecting non-combatants and limiting war's destructiveness. Enlightenment teachings against cruelty in wars and punishments had helped create a Europe in which egregious cruelty was much less frequent than in previous centuries, and in which, as modern scholarship has shown, both the level of violence and the total number of deaths from war—military and civilian—had declined.[4]

---

[3] See Gordon Corrigan, *Wellington: A Military Life* (Hambledon and London, 2001). A case of rape during the Napoleon's Russian campaign and a harsh punishment for the crime is described in Heinrich von Brandt, *In the Legions of Napoleon*, trans. and ed. Jonathan North (London, 1999). 101; on the subject, see as well p. 192.

[4] For an overview and analysis of the quantitative literature on levels of violence, see Johan M. G. van der Dennen, "On War: Concepts, Definitions, Data: A Review of the Macroquantitative Research Literature up to 1980," prepared for the course "Political Violence," Interuniversity Center for Postgraduate Studies, Dubrovnik, Yugoslavia, January 1980 (Internal Publication of the Polemological Institute, Groningen, and published online by the Centrum voor Recht & ICT, at http://rint.rechten.rug.nl/rth/dennen/unesco1.htm (accessed 29 July 2002); see also A. J. Jongman and J.M.G. van der Dennen, "The Great 'War Figures' Hoax: An Investigation in Polemomythology," *Bulletin of Peace Proposals* 19 (1988): 197-203.

Yet in the period of World War I, strictures against violence directed at noncombatants eroded rapidly, and violence against civilians, both enemy and friendly, rose sharply.[5] Indeed, European governments justified violence against civilians at home increasingly in terms of the war effort abroad, much as radical Jacobins called on Parisian revolutionaries to move against "the enemy within" in September 1792. One measure of this twentieth–century "preemptive" violence against domestic enemies was the driving of whole communities and even populations—if identified as enemies—from their homes and preferably out of the country. This practice grew to such an extent just before and during World War I as to make the massive population transfers of the Lausanne Treaty in 1923 appear to be a kind of practical norm, acceptable not only to the governments directly involved, but suitable also to the rest of the European powers. A sea-change had occurred.[6]

Where do we look when searching for the causes and conditions which led to such a change, a change which seemed to reverse a trend of the civilizing tendencies reaching back to the seventeenth century? I propose to comment on the genesis of modern population politics by discussing six "theses." Underpinning these theses, and to some extent tying them together, is a series of historical observations, in short form as follows. Since the rise of the modern state in the sixteenth and seventeenth centuries, the state's monopoly of violence has been associated with the internal security of the state's population. States have likewise tended to identify as war even the domestic use of state violence.[7] State violence, or the threat of it, is crucial in manipulating specific "domestic" population groups. Since population groups within a country must be identified and defined, it has been the habit of nation-states

---

[5] I am positing a rise in violence both from the standpoint of expanding the pools of potential targets (see, for example, William R. Hawkins, "Imposing Peace: Total vs. Limited Wars, and the Need to Put Boots on the Ground," *Parameters*, Summer 2000, 72-73) and from the standpoint of the enormous rise of civilian casualties from the nineteenth century to the twentieth. See also Dennen, "On War"; and R. J. Rummel, *Death By Government* (New Brunswick, N.J., 1994), especially chapters 1 and 2. See also the wide–ranging discussion by Mark Levene, "Why is the Twentieth Century the Century of Genocide?" *Journal of World History* 11 (2000): 305–336.

[6] This change is examined in Irving Louis Horowitz, *Taking Lives: Genocide and State Power*, 5th ed. (New Brunswick, N.J., 2002).

[7] This is clear from Machiavelli's *The Prince* and from Hobbes's careful depiction of the limitation of war to the status of being state-controlled in *Leviathan*. See the excellent recent discussion of this issue, including the significant aspect of the rise of police "forces," in Martin van Creveld, *The Rise of the State* (Cambridge, 1999), 155-188.

since early in their history to define the targeted population as "enemies" in some way. Hence, the treatment of noncombatants in wartime is closely related to the state's treatment of targeted internal "enemies."[8]

Thesis One

Some of the most atrocious European cases of ethnic cleansing, forced migration, and even mass murder since World War I have been associated with new borders or with minorities living in a borderland. In the fifty years before World War I, Central and Western European regimes dealt with many issues related to ethnic minorities. Many of these problems concerned new minorities resulting from wake-of-war boundary changes, precisely the situation which in the twentieth century tended to lead to violence and often ethnic cleansing. Yet *until the eve of the First World War, the European international system and Europe's domestic regimes tended to work out the related problems of ethnic minorities and changing borders within a non-violent framework, except in Southeastern Europe.*

We can look, for example, at the arrangements made in the wake of the first two wars of German unification (1864 and 1866). Denmark lost all of Schleswig-Holstein in the 1864 war, and the northern duchy of Schleswig was ceded to Prussia without regard to the strong Danish majority of the population in the north of the region. About a hundred thousand ethnic Danes were thus brought under the Prussian state and later the German Empire. The Treaty of Prague, which ended the Austro-Prussian War in 1866, allowed for this Prussian annexation of territory, but on the condition that a plebiscite would take place in the northern districts—called South Jutland by the Danes and North Schleswig by the Germans—before the Prussians took over, and that a Danish referendum victory would mean that this northern territory would remain Danish.

The Prussians quickly carried out annexation without a plebiscite. The Danish government protested, and Prussia went through the motions of arranging for a plebiscite after the fact, but the negotiations eventually foundered. Later, in 1878, when Austria-Hungary needed German support in connection with the occupation of Bosnia-Herzegovina, the Germans took advantage of the

---

[8] For critiques of the state's monopoly of violence and the manipulation of populations, see Ludwig von Mises, *Nation, State and Economy*, trans. Leland Yeager (New York, 1983 [orig. published in German, 1919]); Horowitz, *Taking Lives: Genocide and State Power*; Creveld, *The Rise of the State*.

situation to extract from Austria a clause in the Treaty of Vienna which released the Germans from the obligation to hold a referendum in North Schleswig. Just after Prussian takeover, some 60,000 Danes had emigrated (opted) to go either to Denmark or elsewhere, many of them young men avoiding the Prussian military draft. When the Treaty of Prague provided for a plebiscite, many of these "optants" moved back to Germany, though without German or Prussian citizenship. For these "optants," 1878 meant being stuck in Germany, though without German citizenship.[9] Still, though many observers viewed this episode as a shabby one, and though this ethnic injustice was compounded from the 1880s onward by Prussian policies of Germanization, no violence marred any of these events. No Danes lost their property; indeed, because of German development, property values of even the Danish farmers tended to rise.[10]

In the famous case of Alsace-Lorraine, newly united Germany, fresh from the victory of the Franco-Prussian War in 1871, incorporated a region whose inhabitants spoke a German dialect. Yet if the Alsatians were theoretically ethnic confreres, their dialect and their commitment to Germandom were seen as problematic at best throughout the period from 1871 to 1918. Governed from Berlin, the region displayed much low-level dissatisfaction. Still, even in the worst cases of friction—and the worst were various clashes between civilians and army officers stemming from harsh treatment of Alsatian army recruits—the German government not only apologized for the behavior of the army officers, but even outlawed the use of a particular ethnic epithet in the army.[11]

We find even less violence when we turn to other multiethnic regions of Central and Western Europe. Though not the result of any recent border changes, the eastern borderlands of Germany exhibited many tensions which had the potential of turning violent. This region would, as well, come to be closely related to some of the major outbreaks of ethnic cleansing in the mid-twentieth century. Here, in a borderland characterized by a large Polish

---

[9] Sarah Wambaugh, *Plebiscites Since the Great War*, 2 vols. (Washington, D.C., 1933), 1: 49-52.

[10] Niels Hansen, "L'évolution de la politique nationale en Slesvig de 1806 à 1918," in *Manuel historique de la question du Slesvig 1906-1938* (Copenhagen and Paris, 1939), 4-5; the so-called Optant Treaty is printed in the same work, 62-63.

[11] Jena M. Gaines, "The Politics of National Identity in Alsace," *Canadian Review of Studies in Nationalism* 21 (1994): 99–109. See also the comments of the German Chancellor, Theobald von Bethmann-Hollweg, in *Stenographische Berichte über die Verhandlungen des Reichstages*, 1913-1914, 291: 6155-58.

population, ethnic tension between Germans and ethnic Poles grew from the 1880s onward. From that time debates in the German *Reichstag* and the Prussian Parliament and German press repeatedly tried to clarify the status of Polish-speaking "Germans." From the eighties, the Prussian administration discriminated in both open and secret ways against Poles in the civil service, in school matters, and in many other areas. Some violence pitted ethnic Poles against the police in the prewar years, and some of the strikes of the Polish unions looked very much like ethnic conflict, since especially in the industrial section of Upper Silesia, the work force was predominantly of Polish ethnicity. Even the German Socialist Party, the SPD, withdrew its affiliation from the Polish Socialist Party, claiming that the true allegiance of the Polish socialists was to Poland rather than socialism. Further, a Polish nationalist party gained enough votes to send deputies to the German *Reichstag* in 1903. In some ways, ethnic conflict underlay much of public life in the east, and many Prussian subjects of Polish extraction were no doubt reevaluating their own identity as "Polish Germans" long before the First World War.[12]

These eastern borderlands of Germany were also the scene of vigorous activity by both state and private interests to settle more German farmers in the border areas, efforts usually called the Settlement Movement. The process involved buying out the Polish owners, or buying out penurious German nobles and selling at low prices to German "settlers" from elsewhere. No force or violence was involved, but the overtones of this Germanizing movement are obvious, and the movement itself was associated with "integral nationalist" and Social Darwinist writers from that period who were quite vocal in their anti-Semitism and racialist ideology.[13]

Similar patterns may be discerned in other Central and Western European areas: tension, especially as various minorities "awakened" to their identity in the nineteenth century, but no large-scale violence. This generalization would hold for the Catalonian movement and similar ethnic movements in the West as well as throughout the Habsburg Empire, whose congeries of nationalities jostled with each other but mostly avoided violence,

---

[12] Richard Blanke, *Prussian Poland in the German Empire, 1871-1900* (Boulder, Col., 1981); Laurence Schofer, *The Formation of a Modern Labor Force: Upper Silesia, 1865–1914* (Berkeley, 1975); T. Hunt Tooley, *National Identity and Weimar Germany: Upper Silesia and the Eastern Border, 1918-1922* (Lincoln, Neb., 1997), 1–23.

[13] See Richard Wonsor Tims, *Germanizing the Poles: The H-K-T Society of the Eastern Marches, 1994-1914* (New York, 1941).

overseen by a monarchy whose "live and let live" policies aimed for anything but violent confrontation.[14]

In non-Habsburg Southeastern Europe, the disintegration of the Ottoman Empire in the course of the nineteenth century led to the emergence of new states with new borders. It was at the same time home to many, often intermingling, ethnic groups and to increasing tensions and fears that minority populations might sympathize with the "enemy." The creation of the new states was usually violent, and the wars which accompanied these events were brutal. The accelerated changes resulting from the Russian defeat of the Turks in the 1870s war necessitated decisions about all border and minority-related issues, as groups of people found themselves in new states, often as a part of a minority. At the same time, the Ottoman government responded to secession and the threat of it with terror and violence against specific groups, measures which made the Ottoman regime famous throughout Europe for a number of "massacres." As papers in this volume demonstrate, Balkan states likewise moved in many cases against their Muslim populations (often called "Turks") as well, driving out and slaughtering large numbers. As a result of the Balkan Wars of 1912/13, the competing neighboring states in the Balkans did even more to clear up their borders in the ethnic sense.[15]

When faced with ethnic tensions resulting from changed borders before World War I, Central and Western European states tended to normalize relations without violence, even if the minorities in question still had grievances. In Southeastern Europe, border changes stemming from the Russo–Turkish conflict of the 1870s generated massacres, displacement, and other elements we would identify as behaviors related to ethnic cleansing. Overall, outbreaks of violence were rare until just before 1914.

Thesis Two

In spite of the "humaneness" of most European governments in dealings with undesired populations, the level of violent solutions to unwanted populations began to rise slowly after 1870. Most of the cases that consti-

---

[14] István Deák, *Beyond Nationalism: A Social and Political History of the Habsburg Officer Corps, 18148-1918* (New York, 1992).

[15] André Gerolymatos, *The Balkan wars: Conquest, Revolution, and Retribution from the Ottoman Era to the Twentieth Century and Beyond* (New York, 2002); Richard C. Hall, *The Balkan Wars, 1912-1913: Prelude to the First World War* (London, 2000); in the present volume, see below, Dennis P. Hupchick, "'Bulgarian 'Turks': A Muslim Minority in a Christian Nation-State."

tuted this slow acceleration of violence occurred on "peripheries" of Europe. All of these cases were related to nationalism and empire.

Some striking exceptions to the policies and atmosphere of "humaneness" emerged late in the nineteenth century, especially in peripheries of the European continent itself, and in the various European overseas empires. This behavior could be aimed at religious minorities, but increasingly by the end of the century, violent population politics seemed to be focused on ethnic minorities and took place in the western and southern zones of the Russian Empire, in the European components of the Ottoman Empire, and in neighboring states in Southeastern Europe.

The mid-1880s seems to have formed a watershed in the reversal of the civilizing processes of the Enlightenment when it came to governmental solutions to minority problems. Throughout Europe, one might argue, during this crucial decade a new and hardcore variety of ethnic nationalism moved to the center stage of policy-making. In the Russian Empire, repressive nationalistic policies by Alexander II were already in the works by the time of his assassination in 1881 (especially in Poland and Ukraine). The eighties signaled the beginning of the Russification program on most of the peripheries of the Russian Empire. These policies were promoted by the famous official of church and state, K. P. Pobedonostsev, and aimed at both ethnic and religious minorities. Plans for the Jews played a very significant part in the Russification program. As Pobedonostsev declared: "One-third will die out, one-third will become assimilated with the Orthodox population, and one-third will emigrate."[16]

In the European segments of the Ottoman Empire, too, the last decades of the nineteenth century seemed to foreshadow the violent ethnic politics of the twentieth. The most famous victims of Ottoman state policy in this period were the Bulgarians and Armenians, who from the eighties on were experiencing violent pressure from the Ottoman regime under the harsh Sultan Abdul Hamid II. This pressure (including periodic "massacres") certainly began to look something like ethnic cleansing. In the recently independent states in the Balkans, the two Balkan wars of 1912-13 produced the first clear-cut ethnic cleansing in twentieth-century Europe as the competing nationalities in the areas broken off from the Ottoman Empire fought each other and the Ottomans to satisfy their national pride. In this process, belligerents tried to remove persons of one ethnicity from a given region in order to justify

---

[16] David MacKenzie and Michael W. Curran, *A History of Russia, the Soviet Union, and Beyond,* 4th ed. (Belmont, Cal., 1993), 400-402.

control over it on ethnic grounds. Hence, the European zone that gave the twentieth century its last ethnic cleansings in the 1990s, also gave it its first cases.[17]

Another periphery of both Europe and empire at this period pointed toward brutality in dealing with unwanted populations. The "new imperialist" colonies formed a different kind of imperial periphery, but a periphery nonetheless. Certainly on this periphery, Europeans practiced widespread violent manipulation of populations. It is worth contemplating that the "New Imperialism" of European powers (the new-style political control of overseas empires in the last third of the century) coincided precisely with the chronological framework of the new, hard-shelled, one might say Social Darwinistic, nationalism. Indeed, connections between imperial conquest and domestic militarism, cruelty, or authoritarianism constituted one of the central tenets of the anti-imperialist movement in Europe and the United States.[18] American historian Carlton J. H. Hayes suggested some very direct connections between overweening nationalism and imperial cruelty in his classic history of Europe during this period, *An Age of Materialism*.[19] Certainly, if we are looking for cases which combine the inflicting of violence and death on given populations for the purpose of "cleansing" a given area, we find a great many such cases "on the ground" in overseas empires of the European Great Powers.[20]

---

[17] See citations in note 15, above.

[18] See, for example, Booker T. Washington, "Cruelty in the Congo Country," *The Outlook* 78 (Oct. 8, 1904); Henry Van Dyke *The American Birthright and the Philippine Pottage: A Sermon Preached on Thanksgiving Day, 1898* (New York, n.d. [1898]); Anti-Imperialist League, "Platform of the American Anti-Imperialist League," given in Carl Schurz, *The Policy of Imperialism*, Liberty Tract No. 4 (Chicago, 1899). On the Anti-Imperialist League, see Jim Zwick, "The Anti-Imperialist League: A Brief Organizational History" [http://www.boondocksnet.com/ail/ailhist.html]; and Jim Zwick, ed., *Anti-Imperialism in the United States, 1898-1935* [http://www.boondocksnet.com/ail98-35.html (Feb. 24, 2001)]. On the concentration camps, see Brian McAllister Linn, *The U.S. Army and Counterinsurgency in the Philippine War, 1899-1902* (Chapel Hill, 1989), 25-37, 154-169.

[19] Carlton J. H. Hayes, *A Generation of Materialism, 1871-1900* (New York, 1941).

[20] My approach to this thesis was greatly stimulated by an April 2000 discussion on population politics and empire on the email list "Forced Migration History List" (Listowner, Nick Baron) FORCED-MIGRATION-HISTORY@JISCMAIL.AC.UK, an email discussion list associated with the project "Population Displacement, State-Building and Social Identity in the Lands of the Former Russian Empire, 1918-1930" at the University of Manchester (http://www.art.man.ac.uk/ HISTORY/ ahrbproj/details.htm). I am grateful to Jeff Handmaker, Jonathan Bone, and Peter Holquist for their comments and analysis in this discussion.

Since its beginning European imperialism had found it necessary to move various groups to various places: American Indians
to Spanish farms and mines, or Africans to sites of intensive agriculture, etc. In the same way, European colonial empires had likewise killed off or driven away populations so as to use their land,
though in many regions of the world, disease proved to be a much
more efficient "ethnic cleanser."

One important tool for carrying out such tasks, and one that
would become central to nearly all subsequent cases of ethnic
cleansing and genocide, was the concentration camp. The late-
nineteenth-century burst of large-scale imperialist conquest and
war made it essential to control populations, identify enemies,
separate certain subgroups from each other, and hold groups together for disposal. At some point, the technology of barbed wire,
a cheap and easily used product invented in the United States in
the 1870s to fence in cattle, offered itself as a cheap and practical
way of fencing in human beings.

This new use of barbed wire seems to have been applied first
by the Spanish Army while fighting the Cuban insurrection. The
Spanish forces were apparently using barbed wire to help section
off the island of Cuba by the mid-1890s. In 1896/97 the newly appointed governor, General Valeriano Weyler, adopted something
like the reverse of what the term "concentration camp" would later
denote. Spanish soldiers rounded up some 300,000 Cuban civilians
in the attempt to divide the peaceful Cubans from the insurgents.
His idea was to feed and protect the "good" civilians in what were
called "reconcentration camps," though his plans for providing
food and shelter seem to have been very inadequate. In the event,
over thirty percent of the those "reconcentrated"—upwards of a
hundred thousand people—starved or died of disease.[21]

During the Boer War (1899-1902), the British Army used
barbed wire enclosures to create "concentration camps" more
closely approximating the modern meaning of the term: the enclosures were used for imprisoning Boer (South Africans of Dutch
descent) women and children so that the Boer guerilla fighters, the
kommandos, would surrender. General Roberts (Frederick Lord
Roberts) originally initiated the system in 1900, partly as a response to the growing number of refugees resulting from the
British attempt to starve the Boers out by burning their homes
(some 30,000 of them), partly with the idea of using the incarceration of Boer women and children to force the Boers to stop fighting. The British built some forty concentration camps, containing

---

[21] Sebastian Balfour, *The End of the Spanish Empire, 1898–1923* (Oxford, 1997), 13ff.

about 116,000 prisoners, most of them women and children. Starvation and diseases stemming from unhygienic conditions killed enormous numbers of these "inmates." Over about a year and half, just under 30,000 Afrikaners died, 22,000 of them children under sixteen. Of a total Boer population of about 200,000, fifteen percent died, and the disproportionate mortality level among young people presents a genuine demographic catastrophe. Contemporary arguments that the local mortality rates were normally high do not account for a death rate this high. Perhaps less known, well over a hundred thousand black Africans, at this juncture forming an "inconvenient" population which found itself wandering and homeless because the British "scorched earth" policy, were likewise rounded up into concentration camps. Over 14,000 died of disease and poor conditions, about 12,000 of those children.[22]

At about the same time, the United States launched into fullfledged imperialism in the Philippines (as a result of the Spanish-American War, 1898), taking the vast area from Spain and imposing its rule on the local peoples. In the ensuing resistance, American commanders (some of them veterans of American "Indian wars" aimed at the expulsion of specific groups of Indians) carried out a "dirty" war against Filipino forces, in the end losing about 5,000 troops against an enormous death toll of over 200,000 Filipino deaths. In the course of this war of conquest and "pacification," General Jacob Smith, a "veteran" of the Wounded Knee Massacre in 1890, sent to a subordinate the order: "I want no prisoners, I wish you to kill and burn, the more you kill and burn the better it will please me. I want all persons killed who are capable of bearing arms (ten years of age and above) in actual hostilities against the United States." He emphasized what might be called the demographic function of his mission: "The interior of Samar must be made a howling wilderness."[23] Smith was later court-martialled, but American efforts throughout the war of conquest and the subsequent "pacification" were consistent with Smith's behavior. General Leonard Wood ordered an attack on six hundred Moros huddling in a crater in the mountains, fugitives from new taxes being collected by the United States. After the "battle," he reported that all were killed: men, women, and children. Concentration camps made their appearance here too, again constructed from the practi-

[22] S. B. Spies, "Women in the War," in *The South African War: The Anglo-Boer War 1899-1902*, ed. Peter Warwick (New York, 1980), 161-185; Peter Warwick, "Black People and the War" in *The South African War*, 186-209.

[23] Steven M. Gillon and Cathy D. Matson, *The American Experiment* (Boston, 2002), 864-865.

cal technology of barbed wire, at about the same time that the British were employing them in South Africa.[24]

Slightly later, in the "protectorate" of German Southwest Africa, later to become Northern Namibia, the German government was directly involved in "cleansing" areas of the Herero people, whose revolt against German colonizers had left several hundred German settlers dead. To deal with the problem, Berlin sent General Adolf Lebrecht von Trotha, a soldier known for his inflexible and draconian policies in German East Africa and in China during the Boxer Rebellion, a war in which the Kaiser had enjoined his officers to act like "Huns." Sent to chastise the Hereros, Trotha commented upon his arrival in 1904: "I know the tribes of Africa.... They only respond to force. It was and is my policy to use force with terrorism and even brutality. I shall annihilate the tribes in revolt with streams of blood and streams of gold. Only after a complete uprooting will something emerge." In October 1904 he issued an order along the lines of Smith's: "Every Herero found within German borders, with or without guns, with or without livestock, will be shot. I will not give shelter to any Herero women or children. They must return to their people, or they will be shot." No male prisoners were to be taken. Women and children were to be harried into the wasteland. His army followed these orders and in the process killed outright or by starvation between 50,000 and 70,000 Hereros.[25]

Since classical times, commentators on imperial systems have insisted that the behavior of empires on its extremities tends to affect the behavior of the imperial government toward its home base. Hence, if we are looking for links between ethnic cleansing and large-scale violence, or even mass killing, we should look to the peripheries of the European (and other Western) imperialist states of the late nineteenth century. To connect such ideas with conceptions of violence discussed below, it is worth noting that

---

[24] See Helen C. Wilson, *Reconcentration in the Philippines* (Boston [Anti-Imperialist League pamphlet], 1906); reproduced on the web by Jim Zwick at http://www.boondocksnet.com/ai/ailtexts/wilson060121.html as a component of Jim Zwick, ed., "Anti-Imperialism in the United States," 1898-1935. http://www.boondocksnet.com/ai/ (Aug. 30, 2002).

[25] Jon Bridgman, *The Revolt of the Hereros* (Berkeley, 1981); Günther Spraul, "Der Völkermord an den Herero: Untersuchungen zu einer neuen Kontinuitätsthese," *Geschichte in Wissenschaft und Unterricht* 12 (1988): 713-739. See also Jon Bridgman and Leslie J. Worley, "Genocide of the Hereros," in *Genocide in the Twentieth Century: Critical Essays and Eyewitness Accounts*, ed. Samuel Totten, William S. Parsons, and Israel W. Charny (New York, 1995), 3-48. For the official view, see *Die Kämpfe der deutschen Truppen in Südwestafrika*, 2 vols. (Berlin, 1906-1907).

almost all commanding generals in the First World War—Joffre, Falkenhayn, Pershing, and many others—had spent part of their careers planning and carrying out colonial wars.[26]

Thesis Three

Even the new rules of war allowed combatants to commit violence against enemy civilians in the enemy's country under some circumstances (spying, guerilla or non–uniformed combat, and some other cases). If a state could kill or forcibly move civilian enemies in another country, the existence of "enemies" at home meant that the state should be able to kill or forcibly remove these enemies as well. In countries where the definition of citizenship became increasingly connected with belonging to a given ethnic group, members of ethnic minorities could become classified as enemies and therefore targets of ethnic cleansing.[27] The classic example of this process of "defining" domestic ethnic enemies is of course the Holocaust,[28] in the beginning stages of which various offices of the SS went to extraordinary lengths to define exactly who was a Jew. Yet this process had gone on less rigorously in many places in East Central Europe. It was a process closeley associated with the origins and practice of ethnic cleansing. At the same time, legal norms of warfare sanctioned violence against civilians in warfare only if the enemy civilians jeopardized their non–combatant status. In cases of ethnic cleansing, twentieth–century states have had to assume that all members of a given minority were enemies by virtue of their ethnicity. Moving against these enemies violently then became a matter of course. *Twentieth-century ethnic cleansing hinged on the related processes of defining ethnic minorities as legally different from the majority and of declaring members of the defined minority as internal enemies of the nation. Therefore, the standards of European behavior toward external enemies of the nation, especially the civilian or noncombatant enemy*

---

[26] See the essays in Karen Barkey and Mark von Hagen, eds., *After Empire: Multi-Ethnic Societies and Nation-Building* (Boulder, Co., 1997), 102.

[27] Especially instructive here is Bell-Fialkoff's chapter, "Cleansing as a Metonym of Collective Identity," in *Ethnic Cleansing, 57-115.*

[28] Omer Bartov discusses this issue as it relates to the Holocaust in "Defining Enemies, Making Victims: Germans, Jews, and the Holocaust," *American Historical Review* 103 (June 1998): 771–786; Raul Hilberg deals with the question extensively in *Destruction of the European Jews*, 3 vols., rev. ed. (New York, 1985).

populations, becomes a central issue in the history of the emergence of ethnic cleansing.

After the Thirty Years War (1618-1648), and partly as a result of its horrific brutality, European states began to work out a number of agreements—some written, some tacit—aimed at ameliorating the disastrous effects of war on noncombatants. The eighteenth-century Enlightenment spurred on such developments, so that while armies became larger throughout the eighteenth and nineteenth centuries, and while weapons became more powerful, many of the dreadful consequences of war for noncombatants were lessened.

In the second half of the nineteenth century, European powers entered into a number agreements which would, it was hoped, regulate war even more strictly, adding more protections for civilians, regulating conditions for prisoners of war, and the like. In the 1860s European powers ratified the Geneva Convention, addressing many of these issues, and the Hague Declarations signed by most powers between 1899 and 1907 further "civilized" the prosecution of war, at least in theory. Certainly, even general anecdotal evidence from nineteenth-century wars shows that for whatever reasons, European powers tried to play according to a fairly strict set of rules with regard to civilian populations, at least European populations.

Yet on the sharp edges of empire, civilians could indeed be redefined as fair objects of violence even if they had not engaged in the explicit guerilla behavior proscribed by international agreement. In war on the edges of the great overseas empires, it was relatively easy for Europeans to see "the brown people" as fair game, since Social Darwinist modes of thinking tended to dehumanize them in any event. In the setting of empire in non–European territories, we can also trace a process in which European powers interpreted the relatively new international rules of warfare and protection of individuals in such a way as to redefine civilian populations of the enemies as in a sense outside the rules of civilized warfare, as refractory and hidden "enemies." We see this redefinition most directly in connection with irregular warfare.

The new international regulations were quite strict about defining combatants as those in uniform.[29] This definition aimed at

---

[29] The Hague Convention annex entitled "Regulations Respecting the Laws and Customs of War on Land" included among its first regulations that combatants must "have a fixed distinctive emblem recognizable at a distance" and would "carry arms openly"; Hague Convention with Respect to the Laws and Customs of War on Land (Hague II), signed 29 July 1899, entry into force 4 Sept. 1900.

protecting civilian males not in military formation and therefore, like women and children, *hors de combat*. Still, by the same definitional logic, persons in civilian clothing who committed acts of war (ambush, sabotage, spying) were risking immediate execution, since this behavior was inimical to the agreed-upon system of European warfare. The 1899 Hague Convention regulations on land warfare were excplicit: irregular fighters could not be clearly identified on the battlefield as enemy troops and hence could be on either side, forcing uniformed soldiers to be more on guard and more trigger-happy when facing civilians. The irregular partisans or guerillas "hid" in the civilian population, making the population that "harbored" them seem guilty of harboring secret enemies. There was no way to ferret out irregular fighters or even identify them. And since these fighters had set aside the rules of "civilized warfare" by non-sanctioned fighting, the armies faced with such uncivilized behavior outside the law had no choice but to fight back with uncivilized behavior outside the law. Hence, if a region produced insurgents or guerillas, then the civilian population harboring and aiding those insurgents was fair game for reprisal, for removal, or for other means of control.

This whole line of reasoning is hardly new to the twentieth century. Irregular tactics emerged occasionally in eighteenth–century warfare, and the French Army had faced well-coordinated Spanish guerillas during the Peninsular War after 1808.[30] The French sometimes responded in the ways depicted in Goya's famous series of paintings about the war, in particular the shocking image of "The Execution of the Rebels 3rd May, 1808." At the end of the century, virtually at the moment of heightened hopes of "civilizing" warfare's cruelties toward civilian populations, European armies on imperial frontiers—as we have seen above—came to conclusions much to the disadvantage of civilians in wartime. Indeed, they emerged from the nineteenth century armed with definitions which multiplied the ways in which a population could put itself outside the rules of warfare and therefore, in a sense, "deserve" the consequences. The new definitions also multiplied the methods which could be used to control, police, and punish such large numbers of refractory enemy civilians.[31]

---

[30] For a primary account of dealing with guerillas, see the excellent memoir by Heinrich von Brandt, *In the Legions of Napoleon: The Memoirs of a Polish Officer in Spain and Russia 1808-1813*, cited above.

[31] A good review of the American approach to the question is John M. Gates, "Indians and Insurrectos: The U.S. Army's Experience with Insurgency," *Parameters* 13 (1983), 59-68.

Armies put these definitions to use in enemy country during World War I. The Russian armies which surprised the Germans by bursting into East Prussia a few weeks into the war, far ahead of any schedule the Germans thought possible, carried on a grim kind of war, targeting civilian housing in particular, burning and looting on a very wide scale, and killing civilians, over 6,000 in the end. Probably most of those killed were regarded as "spies" or seemed to be giving other ununiformed help to their country's army. At the same time, and more famously, the German armies which invaded Belgium and France in August 1914 seemed so casual about making the calculation that the remedy for civilian resistance was terror and reprisal that historians have not yet decided whether the policy of terror or *Schrecklichkeit* trickled down from above or rose up from below. Here the water is muddied somewhat because of the well-known propaganda compendium called the Bryce Report, a commentary of German atrocities in Belgium issued in 1915. Much of the Bryce Report seems to have been fabricated, but some historians are now reexamining German violence against civilians in Belgium. One way or the other, over 6,000 Belgian civilians were killed by the Germans in roughly the same weeks that about the same number of Germans was being killed by the Russians. The property damage in Belgium, in the end, was less than that in East Prussia. The Germans burned almost 16,000 houses and buildings, almost all of this during the first month of the war, while the Russians destroyed some 42,000 buildings in East Prussia and damaged many more. Still, the Germans carried out the most senseless property crime of the war, when they systematically burned parts of the Belgian city of Louvain (Leuven), including one of the great libraries of Europe, that of the University of Louvain.[32]

In both Belgium and East Prussia, refugees fled from the approaching armies and from their depredations. From Belgium over a million people fled to Great Britain, France, or the Netherlands. The majority returned after German guarantees of safety, but some 300,000 Belgian refugees remained abroad. In East Prussia, two separate invasions by the Russians early in the war sent some 870,000 (out of a population of two million) Germans fleeing from the invading army. These were all housed in

---

[32] E. H. Kossmann, *The Low Countries 1780-1940* (New York, 1978), 522-523; Robert B. Asprey, *The German High Command at War* (New York, 1991), 52. Significant studies of the issue of German atrocities are in John N. Horne and Alan Kramer, *Atrocities, 1914: A History of Denial* (New Haven, 2001); and Trevor Wilson, "Lord Bryce's Investigation into Alleged German Atrocities in Belgium, 1914-1915," *Journal of Contemporary History*, 14 , no 3 (July 1979): 370.

Germany, but property destruction was so high that for the next years most had no place to live if they did return. Before the Russians were pushed out of German territory, they sent some 11,000 German civilians (predominantly women, children, and elderly men) to Siberia.[33]

The final connection between putting enemy populations outside the law and the emergence of ethnic cleansing within one's own borders involves a very short step. If a segment of one's own population, or a particular region, has likely connections (perhaps ethnic but not necessarily) with a regularly organized enemy army, then redefining this group as an enemy population is not difficult. If partisan fighting is involved in any way—or merely the threat or possibility of it—then this enemy population "within" can be set outside the bounds of law. All measures taken against them seem justified, even "terror," since upon these measures the winning or losing of the war could hang.[34]

Thesis Four

It has become a cliché of modern urban culture that violence seems to beget violence: and we might well transfer this unfortunate insight to thinking about mass brutality in the twentieth century. One place to look for the emergence of brutal population politics in twentieth-century Europe is on the battlefields of World War I. Another is in the ideological dictatorships that emerged during or as a result of the war. *World War I, its bloody battlefields, and its total war privations brought brutal violence closer to the core of normative or expected behavior. At the level of the state, the new "total war" regimes, in particular the Bolshevik regime, made mass violence and terror a routine instrument of policy.*

If warfare had become more controlled in many ways in the nineteenth century, European armies were nonetheless already drifting into a "modern" disregard for limitations of violence in the

---

[33] Stephen Pope and Elizabeth-Anne Wheal, *The Dictionary of the First World War* (New York, 1995), 65. The East Prussian figure comes from an article by a German official who had been involved in the administrative provision of food and shelter for the East Prussian refugees: *Landesrat* Meyer [sic], "Die Flüchtlingsfürsorge," in *Der Wiederaufbau Ostpreussens: eine kulturelle, verwaltungstechnische und baukünstlerische Leistung*, ed. Erich Göttgen (Königsberg, 1928), 120. The introduction to the same book by Adolf von Batocki (pp. 11-15) also contains some useful material on the invasion. Batocki was the wartime and postwar *Oberpräsident* (governor) of East Prussia.

[34] Horne and Kramer discuss the *"franc tireur,"* or irregular, issue intensively for the German case in *German Atrocities 1914*, 94-174.

early stages of the First World War. The fictional quality of the famous Bryce Report and other wartime propaganda aside, we have seen that the invading soldiers of both Germany and Russia were quite ready to shoot civilians in 1914. Even at the outset, the possibility exists that the very substantial changes in policies regarding "enemy" civilians we have discussed above do seem to have affected the behavior of European armies fighting other Europeans.

But things did get worse. The process of warfare—both Western Front warfare and the many other kinds of World War I combat—led in many directions, but one of these was in the direction of seeing violence as an end in its own right. This led to putting into practice the ideas of prewar thinkers who extolled violence as a kind of purifier of society, such as the French "social thinker" George Sorel. Once the war started, many soldiers were repelled by the violence of modern war, but some were attracted by it. Responding directly to his experience in the trenches, German Western Front fighter Ernst Jünger expressed something like a Sorelian praise of violence in *War as Inner Experience*. The positive acceptance of violence as a new and creative force is likewise demonstrated by the veterans who returned to Italy, Germany, and Russia to continue lives of violence as *Freikorps* members and SA men, as *squadristi* toughs, soldiers in Red or White armies, or as soldiers of fortune in whichever factions of whichever civil war they happened to fall. Living with violence on the war fronts likewise had a more general cultural impact. In an important recent assessment, Vejas Gabriel Liulevicius has shown that the Eastern Front was a transformative experience for many German soldiers and officers. The vast and to Germans "strange" region lent itself, Liulevicius argues, to the reconceptualizing of violence as acceptable. Indeed, he sees the Eastern Front of World War I as an important episode in the history of the brutal Hitler orders for special ruthlessness before the invasion of Poland in 1939 and Russia in 1941.[35] It is also pertinent to point out here that Europe's most famous ethnic cleanser, Adolf Hitler, held his first job as soldier for four and a half years on the Western Front.

A growing literature is beginning to trace the "mobilization" of home fronts in much greater detail, and indeed the "totalization" of the civilian populace. These studies look at ways in which the val-

---

[35] On postwar violence, see Robert G. L. Waite, *Vanguard of Nazism: The Free Corps Movement in Postwar Germany 1918-1923* (New York, 1969). For the eastern front, see Vejas Gabriel Liulevicius, *War Land on the Eastern Front: Culture, National Identity and German Occupation in World War I* (Cambridge, Mass., 2000).

ues of the war really were brought home to the belligerent populations, parallel perhaps to the processes designed to whip up hatred of the enemy in George Orwell's fictional social critique,
1984.[36] Indeed, an important aspect of this research is the emphasis
on "mobilization" as a process initiated by the state, and with the
tools of the state, such as the army, and including war itself. It is
important to note that in spite of such rising levels of hatred and
brutality fostered among populations at war, one can also trace a
persistent revulsion against war among participants on the battle
fronts and a opponents to the encompassing patterns of war at
home. We may indeed generalize from the evidence that the war
produced more violent proclivities (after the pattern of modern
inner city violence), but we must also be careful to identify the
sources of the mobilization process itself.

Another wellspring of modern ethnic cleansing is Marxism–Leninism. Recent scholarly studies, including those in this
volume dealing with the Soviet Union's history of ethnic cleansing,
demonstrate that the conquest of the Russian Empire by Marxist-
Leninists contributed quite directly to the emergence of ethnic
cleansing and other population manipulations.

The massiveness of Stalin's ethnic cleansing is evident. Robert
Conquest rightly subtitled his biography of Stalin, "Breaker of Nations."[37] Lenin had originally promoted Stalin within the party
partly because the Georgian packaged himself as an expert on nationalities, and indeed Stalin was listed in Lenin's first Bolshevik
government as Commissar for Nationalities. Stalin's first work as
Nationalities Commissar was to word a Decree on Nationality (15
November 1917) which outlined the "Free development of national
minorities and ethnic groups inhabiting Russian territory." Stalin's
knowledge of nationalities later combined with his paranoia to
create ethnic cleansing on a massive scale on many peripheries of
the former Russian Empire. In scope of repression and killing, Stalin was the twentieth century's "greatest" ethnic cleanser.[38]

---

[36] See the essays in John Horne, ed., *State, Society and Mobilization in Europe during the First World War* (Cambridge, 1997).

[37] Robert Conquest, *Stalin: Breaker of Nations* (New York, 1991), 70-71.

[38] Both Conquest and A. M. Nekrich studied this question as early as the 1970s. See Nekrich, *The Punished Peoples: The Deportation and Fate of Soviet Minorities at the End of the Second World War* (New York, 1978). A number of Soviet scholars have pro-
duced studies of Stalinist ethnic cleansing since 1991. For a summary of this work and some statistical materials, see J. Otto Pohl, *Ethnic Cleansing in the USSR, 1937-1949* (Westport, Conn., 1999). See, more recently, Terry Martin, *The Affirmative Action Empire: Nations and Nationalism in the Soviet Union, 1923-1939* (New York,

82   The Emergence of Ethnic Cleansing

Yet the contribution of Marxism-Leninist Communism to the emergence of ethnic cleansing in Europe goes beyond the crimes of Josef Stalin. It is important to remember that in terms of the brutalization process associated with the First World War, Bolshevik brutality and terror was a carefully planned expedient. In even more direct ways than the violence of the Western Front, intentional Bolshevik terror and violence against targeted individuals and groups, even in the first months of the Bolshevik regime, helped create an atmosphere of extraordinary brutality that marked the history of the Soviet Union and radiated outward to influence even more of the twentieth-century world. The groundbreaking collection entitled *The Black Book of Communism*, first published in France in 1997, demonstrates that violence was not simply the result of Stalin's paranoia or even yet some *ad hoc* and desperate strategy of Lenin. Extraordinary and exemplary violence was the considered policy of the Bolshevik leadership from the beginning.[39] Recent work by political scientist A. James Gregor on the relationship between Marxism to Fascism suggests that we might cast the net even more broadly. In his view, even the Italian Fascist praise of and proclivities toward political violence is traceable in some degree to its parentage on the Marxist left.[40] Recent studies have also pointed to the links between the concept of "collective guilt" and Marxist-Leninist regimes, from Lenin to Ceaușescu.[41]

In sum, at the level of "mobilizing" societies and individuals, we can think of the war as kind of inoculation against the horrors of violence. At the same time, the rise of Bolshevik terror and a whole new complexion of brutality against targeted domestic groups, including routine spectacular cruelty in war to troops and

---

2001); and Ronald Grigor Suny and Terry Martin, eds., *A State of Nations: Empire and Nation Making in the Age of Lenin and Stalin* (New York, 2001).

[39] Stéphane Courtois et al., *The Black Book of Communism: Crimes, Terror, Repression*, trans. Jonathan Murphy and Mark Kramer (Cambridge, Mass., 1999); see especially the long section by Nicolas Werth, "A State against Its People: Violence, Repression, and Terror in the Soviet Union," 33-268.

[40] A. James Gregor, *The Faces of Janus: Marxism and Fascism in the Twentieth Century* (New Haven, 2000).

[41] See below, Nicolae Harsányi, "The Deportation of the Germans from Romania to the Soviet Union: 1945-1949"; László Hámos, "Systematic Policies of Forced Assimilation Against Rumania's Hungarian Minority, 1965-1989" ; and Tamás Stark, "Ethnic Cleansing and Collective Punishment: Soviet Policy Towards Prisoners of War and Civilian Internees in the Carpathian Basin."

civilians alike, helped prepare the way for the ethnic cleansing that would become part and parcel of twentieth-century politics.

Thesis Five

In spite of some counterindications, it was, by and large, state policy which touched off violence leading to "ethnic cleansing" or forced migration in the period of World War I. The multiform emergence of ethnic cleansing, mass deportations, and even genocide during the First World War was associated almost exclusively with the *state* rather than with popular attitudes or with spontaneous crowd violence.

The period before the war certainly witnessed episodes of popular violence toward ethnic minorities and other groups deemed to be "enemy" populations. As mentioned above, in the Romanian peasant uprising of 1907, violence against Jews seems to have been mostly local and individual in nature, in part the settling of old scores. Russian pressure against Jews in the Pale of Settlement was marked by the frequent collaboration of local populations who joined in or imitated direct violence by the government. In the immediate wake of the war, to take another example, risings of large numbers of Poles in Germany's eastern borderlands were perhaps prepared and financed by the reviving Polish state, but the uprisings in Posen/Poznań (December 1918) and Upper Silesia (August 1920, May-July 1921) featured much enthusiasm and impetus "from below."[42]

But most cases that emerged *during* the war were associated with the state, encouraged by the state, and in most cases carried out by the forces of the state.[43] We will do well, therefore, to link the history of ethnic cleansing in the twentieth century to the theoretical work on nationalism by John Breuilly and John Hutchinson, both of whom have emphasized the role of the state in the creation of political nationalism and politicized national groupings since the seventeenth century.[44] Certainly it has been the states, and indeed the most "total" states of Europe, which have engaged in violent population politics on the largest scale.

---

[42] On the Polish uprisings, see Tooley, *National Identity and Weimar Germany*, 63-78, 183-188, 241-255.

[43] Naimark comments on this issue in *The Fires of Hatred*, 8-11.

[44] John Hutchinson, *The Dynamics of Cultural Nationalism* (London, 1987), 12-36; John Breuilly, *Nationalism and the State* (Manchester, 1982), 335-350.

A full survey of the manipulations of unwanted populations *during* World War I in Europe would exceed the limits of this essay, but a rapid overview is necessary.

Let us begin with the zone in which population politics had already been linked on a regular basis with violence and disregard for lives and property, namely, the zone of Anatolia and the Balkans northward through the western regions of the Russian Empire. There is no doubt that in this region the state carried out population removal as a direct result of the war. Both the Armenians in the Ottoman Empire and the Jews in the Russian Empire had already felt the force of the state directed against them in various pogroms, confiscations, and massacres over the years. World War I brought physical transfer on a much larger scale and outright mass killing as well.

In the Ottoman Empire, the Armenians—who stood out as a prosperous, commerce-based, and Christian minority within agrarian, Muslim Turkey—were increasingly singled out for harsh treatment in the decades before the war, periodically harassed and killed by government forces. Once the war started, Armenians found themselves on the very edges of the battle front across which the Russian armies faced the Turkish armies from the Black Sea to the Caspian. Hence, the existing ideological and ethnic goals of the Young Turk leadership dovetailed with potential dangers of sabotage. Some small pockets of anti-Ottoman activity did in fact break out. The result was a genocidal massacre directed by the "Special Organization," a government agency which managed the removal of a vast number of Armenians from their homes and the rapid and brutal killing of well over a million of them. Enver Pasha, one of the triumvirate which ruled Turkey during the war, declared openly in May 1916: "The Ottoman Empire should be cleaned up of the Armenians and the Lebanese. We have destroyed the former by the sword, we shall destroy the latter through starvation."[45]

In Russia, with the outbreak of the war, authorities sensed and feared betrayal by dissident groups—especially non-Russian groups—all over the empire. Significant sectors of the Russian intelligentsia, apparently with broad popular support, regarded the Jews as natural allies to the Germans and reported on wide-

---

[45] Vahakn N. Dadrian, "The Role of the Special Organisation in the Armenian Genocide during the First World War," in *Minorities in Wartime: National and Racial Groupings in Europe, North America and Australia during the Two World Wars,* ed. Panikos Panayi (Oxford, 1993), 51. See also Dadrian's *The History of the Armenian Genocide: Ethnic Conflict from the Balkans to Anatolia to the Caucasus* (Providence, R.I., 1995).

spread acts of sabotage and betrayal behind the front. Much indiscriminate killing of Jews took place at the beginning of the war. By 1915, as the Russian armies fell back in disarray after a series of spectacular defeats, Russian civilian and military authorities began expelling Jews in the rear of the battle fronts. Almost a million had been shipped off in boxcars within a few months.[46]

The second wave of violence and deportation broke over the Jews after the Bolshevik Revolution, during the Civil War. Especially intensive in Ukraine, this tendency to kill Jews by most of the factions involved in the Russian Civil War led to killings amounting to 100,000 Jews. The driving of many more Jews away from homes and shelter led to the death by exposure, starvation, and disease of 100,000 more. Although Russian Jews did not suffer the kind of systematic and genocidal program meted out to the Armenians during World War I, as Mark Levene has pointed out, the progression from over a hundred–thousand deaths in the period 1919/1920 to the 500,000 estimated to have died in Ukraine during Operation Barbarossa is an increase, indeed, in a gruesome process, but not one of magnitude.[47]

Pogroms and ethnic cleansing against the Jews only begin the story of the Russian Empire at war, however. German settlements in the various regions of the Empire present a parallel case, suffering large-scale forced removal at the same time as the deportations of the Jews. It has to stand as one of the great ironies of the history of ethnic cleansing that two groups perhaps most associated in the popular mind with ethnic cleansing in the twentieth century (the Germans as perpetrators and the Jews as victims) were some of the first victims of twentieth-century-style ethnic cleansing–at the same time, at the hands of the same regime, for the same reason.

Most of the German settlements in Russia owed their origin to the invitation of Catherine the Great, in the 1760s, and many Germans settled crown lands in the southern territories newly won from the Ottoman Empire. The Volga Germans, for example, started with about 30,000 immigrants in the 1760s and accounted for 1.7 million souls by 1914. German Mennonites, who landed, along with other German groups, in Ukraine, were happy not only to practice their religion, but to send out colonies, and hence, by 1909 they had produced settlements which extended from the

---

[46] See Peter Gattrell, *A Whole Empire Walking: Refugees in Russia During World War I* (Bloomington and Indianapolis, 1999); and Mark Levene, "Frontiers of Genocide: Jews in the Eastern War Zones, 1914-1920 and 1941," in *Minorities in Wartime*, 95.

[47] Levene, "Frontiers of Genocide," 87.

Dnieper to Tomsk in Siberia and included settlements in Turke-
stan.

These Germans in Russia noticed hostility from Russian Sla-
vophiles before World War I, and Russification impacted many of
the groups, but unlike the Jews in the Pale of Settlement (the western section of the empire, to which most Jewish settlement was
limited), they remained in control of their lives and property until
World War I. The war brought popular distrust and animosity, but
as with the Jews, it was the advance of the armies of the Central
Powers which turned animosity to a state program of population
manipulation. The state immediately passed a range of anti-
German measures, including the outlawing of German books and
the use of German in public ceremonies. Almost at the outset of
the war, German settlements with any proximity to the front (in
Poland, for example) were deported to south Russia. The so-called
Volhynian Germans, in northern Ukraine, were deported in 1915
to the Volga region, and in late 1916 the Duma passed the Laws of
Liquidation, meaning total deportation. While these goals were
overtaken by revolutionary events, the laws served as an effective
threat, and they would be carried out by Stalin in 1941.

The Bolshevik regime seemed at first far less antagonistic to
the Germans than the Tsarist regime, even setting up a Volga
German Workers' Commune in 1918. Meanwhile, though not targeted for ethnic reasons, the Volga Germans were among the
hardest hit groups in the famine of the Civil War years, losing
some 166,000 to starvation. Their numbers were likewise reduced
in the forced collectivization and created famine of 1928 through
1933. Eventually, a second invasion by Germany, that of June 1941,
would bring on the deportation of the Volga Germans by Stalin.[48]

The Mennonites were at first more or less willing to cooperate
with any regime, since as an autonomous, pacifist, and devoutly
religious community they had traditionally tried (and managed) to
be on good terms with the government regardless of its complexion. After the Bolshevik Revolution, the Mennonites in the core
settlements in Ukraine were prepared to work with the Bolsheviks
if it came to that, but in 1918, the anarchist army of Nestor Makhno
descended on the region and carried out the most horrific depredations of murder, rape, torture, and plunder. Whether or not the
atheist Makhno had a specific hatred for the Mennonites, it was the
Mennonite towns and villages that bore the brunt of his raids. The
pacifist Mennonites in fact organized armed self-defense units, the
Selbstschutz, to ward off Makhno's terror. In fact, it was after the

---

[48]Levene, "Frontiers of Genocide," 83-117. See also footnote 38, above.

Selbstschutz units fired on some Bolshevik troops, mistaking them for Makhnovites, that the Bolshevik Regime turned against the Mennonites. In the longrun, one might well suppose that the autonomous nature of these farming communities would have resulted in their destruction by the Soviets. Almost immediately the Mennonites started arranging for emigration, and by 1930 about 21,000 had arrived in Canada. Over a hundred–thousand Mennonites remained behind to undergo forced collectivization, purges, exile, and eventually deportation. After Operation Barbarossa, they were evacuated to Germany, and then after World War II, they were repatriated to Russia by the Western Allies.[49]

As noted, the Soviet regime cleansed more ethnic groups than any regime in history. We should also point out that the deportations and other "cleansing" activities began not with Stalin in the late twenties, but with Lenin at the very beginning. They stretch from the deportation of the Don Cossacks in 1920 to the destruction of some six million Ukrainians by intentional famine in the early thirties, and then to the deportations of Poles, Ukrainians, Baltic peoples, Moldavians, Tatars, and many others during World War II. Needless to say, these events fit very closely into the categories of state-directed ethnic cleansing on the western and northern edges of the Ottoman Empire and the western and southern edges of the Russian Empire.

If we move from the Ottoman and Russian borderlands to the German-Polish border region—a zone where intensive ethnic cleansing would later be carried out by Germans, Poles, and Soviets—we find a much more ambiguous picture. As described above, in the areas to the east of the Eastern Front, Russian authorities were engaged in the removal of potential German allies—that is to say, both Germans and Jews—by early 1915 at the latest. Once the great struggles in East Prussia were over, the front stabilized, giving the German eastern command (called in the argot of the time, *Ober Ost*) an occupational zone, including much of the Baltic area as well as Poland. Especially when they formed the immediate rear areas of a battle front, occupational zones were bound to undergo the most authoritarian kind of control, since individuals were completely at the mercy of the occupiers. The *Ober Ost* was managed by Erich von Ludendorff, the brains behind Hindenburg, until mid-1916. Using the same team that would later introduce

---

[49] C. Henry Smith, *Smith's Story of the Mennonites*, 5th ed. (Newton, Kans., 1981), 356; see also Dietrich Neufeld, *A Russian Dance of Death*, ed. and trans. Al Reimer (Winnipeg, 1977). I am indebted to my Austin College colleague, Professor Todd Penner, for these citations and for bringing the story of the Russian Mennonites to my attention.

forced labor in occupied Belgium and France and create the draconian, proto-totalitarian Hindenburg Program of 1916, Ludendorff
was careful to appoint a "political section" of the staff, which engaged in longrange planning for the states of the east. This new
European East would be based on German-controlled dependent
states in a region marked by new German settlements. The solutions for clarifying the borders of these new states resorted, almost
casually, to population transfers as a part of larger and more complex plans for the future.[50]

From the same European region only a year or two later—the
war having just ended–one of the earliest uses of the term
"cleansing" in the context of unwanted ethnic groups appeared in
Posen/Poznań, where local Polish nationalists set up a "Central
Organization for Cleansing Poznań (city) of Jews and Germans"
(*Centralna Organizacja dla Oczyszczenia Poznania od Żydów i Niemców*). And indeed, the first postwar migrations of any sort in the
German-Polish border region resulted from the Poznań uprising of
December 1918, supported by the new government of Poland,
which led to an exodus of Germans and Jews. The "Posen Pattern"—meaning a violent overthrow of the local authorities sanctioned by the victorious Allies—echoes through much of the German official record about the border throughout the immediate
postwar period.[51]

At the same time, though plans for introducing the forcible
moving of populations were in the air, there could be exceptions
within the region. The Slavic minority in East Prussia, for example,
the Masurians, far from being the object of ethnic antagonism,
were the object of intensive government favor. Mazuria had suffered much destruction during the Russian invasion, and the German government went to heroic lengths to mollify the Masurians
and even win their hearts. The Russians had hardly been pushed
out of Masuria when the Reich government spearheaded a massive campaign—in the midst of a desperate World War—to rebuild
the war-torn province.[52] Other minorities fared less well, and much
dissatisfaction arose among Poles and Danes from the German

---

[50] See the important recent work by Liulevicius, *War Land on the Eastern Front,* especially chapter 3, "The Movement Policy."

[51] I am indebted for the information on this Central Committee to Tomasz Ka-
musella, who also pointed me to this citation: Joachim Rogall, *Die Deutschen im
Posener Land und in Mittelpolen* (Munich, 1993) 125-131. On the "Posen Pattern," see
Tooley, *National Identity and Weimar Germany,* 30-31.

[52] On the rebuilding of East Prussia during the war, see Göttgen, ed., *Der Wiederaufbau Ostpreussens.*

borderlands. Polish coal miners in Silesia were an important component of the growing incidence of strikes after 1916. Still, these groups were neither excluded from civic life nor separated from their property.[53]

If we move our survey to other parts of the European or Western world, we will indeed find evidence of increased tendencies toward violent treatment of populations. Most of them were similar in essence to the cases discussed above, but with much less violence involved. In the United States, where millions of German immigrants had settled since the eighteenth century, ethnic Germans experienced no general harassment, but much anti-German sentiment surfaced in spite of the large German share of the American ethnic make-up and in spite of the short time during which the United States was actually at war. Mob violence in a few cases led to lynch-type activities directed against Germans, and certainly the propaganda against the Germans was intense enough to make many families change their names, stop speaking German at home, and otherwise demonstrate their loyalty to the United States. Anti-German wartime measures against "enemy aliens" tended to be even less vitriolic. Much closer to the conflict, in France, where hatred of Germany had been in some measure part of the national creed since 1871, the French government did everything in its power to woo the Alsatian population, regardless of the fact that they were ethnically part of Germandom.[54]

The efforts of the Australian government against German Australians presents a case farther up the scale of violence. In Australia the government carried out a sustained attack on so-called enemy aliens, many of them fifth generation Australians with relatives fighting at Gallipoli and on the Western Front. One might argue that, unlike Russia's Germans in the rear of the Russian armies, these Germans were hardly in a position to sabotage and betray, but public anti-German sentiment was surprisingly strong, and as Gerhard Fischer has shown in his studies of the subject, economic influences were crucial. In 1916, in the wake of the arrival of large contingents of the Australian/New Zealand forces on the Western Front (shipped there after the Allies departed Gallipoli), anti-German incidents began to multiply in

---

[53] Tooley, "Fighting Without Arms: The Defense of German Interests in Schleswig, East and West Prussia, and Upper Silesia, 1918-1921" (Dissertation: University of Virginia, 1986), 5-33.

[54] On Germans in the United States: Frederick C. Luebke, *German–Americans and World War I* (Dekalb, Ill., 1974); on Alsace, Paul Smith, "The Kiss of France: The Republic and Alsatians during the First World War," in *Minorities in Wartime*, 27-49.

Australia. By 1917 the government was arresting Australian citizens of German origin and interning them in concentration camps. Ultimately, tens of thousands were interned, and most of the property of this prosperous merchant class was confiscated by the state. Eventually, the bulk of these Germans departed from Australia.[55]

Extending the comparison of ethnic politics and population manipulation to the wake of World War I, attitudes appear to have changed in the course of the war. In the peace settlements, the Allied peacemakers gave much thought to various ways to draw Europe's new borders, but in the end, ethnicity became the default consideration. Whatever else Woodrow Wilson's attachment to "self-determination" represented, at its core it was less a Renanian call for a "daily plebiscite" than it was a call for all ethnic groups to have their own exclusive state. In Wilson's thinking, Poles should want to be together in a country with all other Poles, Latvians should all want to be in a country with other Latvians, and so forth. The treaties allowed for setting some of the disputed postwar border by consulting the populations in referenda, but when there were plebiscite results indicating that not all "Poles" preferred the new Poland over Germany, for example, the Allies accepted with bad grace, or, alternatively, allowed the issue to be settled by short-term violence, as in the case of Upper Silesia, on the German-Polish border.[56]

Though the United States dropped out of the peacemaking process, the whole conception of "purifying" ethnic states by means of international agreements, a conception which has roots in some versions of the idea of self-determination, continued to captivate the Allies. Though Ottoman Turkey had been on the side of the Central Powers in the war and Greece with the Entente, postwar history effected a startling change in relation– ships. Turkish nationalists headed by Mustafa Kemal (later renamed *Atatürk*) fought off Entente invasions, including one by Greece in the west, then did away with the Ottoman Sultan and Empire, and then reformed and partly modernized the country. For numerous reasons the Allies now agreed to renegotiate the Treaty of Sèvres. The result was the new Treaty of Lausanne and a smaller, ethnically purer, Turkish Republic of Turkey, built more along the lines

---

[55] Gerhard Fischer, *Enemy Aliens: Internment and the Homefront Experience in Australia, 1914-1920* (St. Lucia, Queensland, 1989). See also Fischer, "Fighting the War at Home: The Campaign against Enemy Aliens in Australia during the First World War," in *Minorities in Wartime*, 263-286.

[56] Tooley, *National Identity and Weimar Germany*, 218-270.

of European-style ethnic unity. The negotiating powers enhanced this result by agreeing to exchange minority populations, Greeks in Turkey for Turks in Greece.

This was the result on its face. In reality, the 1.3 million Greeks in Ottoman Turkey had suffered deportation, concentration camps, and a high mortality rate during the war, for reasons directly related to the Armenian horrors. At the end of the war, when Greece invaded western Anatolia, the Turkish authorities killed more local Greeks. In the end, when the Lausanne Treaty of 1923 arranged for the exchange of 1.3 million Greeks and 356,000 Turks, the actual number of Greeks not already dead, kidnaped to the interior of Anatolia, or driven out of Turkey was about 290,000. The Turkish Army and police drove them from their homes with great ferocity, sending them under the worst conditions and with no provisions to ships which would take them away. The Turks and other Muslims in Greece were kicked out immediately and with great compulsion. Treaty provisions for compensation of individuals were never implemented.[57]

Many of the cases examined here point up the financial benefits which accrued to the state in connection with ethnic cleansing policies. In the case of the Armenians, the property of the prosperous victims was directed into state coffers or went as payment to various groups and individuals.[58] Indeed, when we remember that the fiscal pressures of the war on governments was one of the war's salient elements of political economy,[59] we will understand that the wealth and property of the thousands of individuals driven from their homes and murdered during the First World War played no insignificant role in government plans.

Looking at all of these cases, we might discern a continuum of intensities in population politics in the period of World War I. The most violent and relentless case was undoubtedly the genocide in Armenia. At the other end of the spectrum would be the belligerent countries where neither state programs nor popular pressures led to violent activity against some identified ethnic group. We will find ethnic cleansing at the more violent, the more intensive

---

[57] A brief view is given in Naimark, *Fires of Hatred*, 42-56. For more extensive treatment, see Dmitri Pentzopoulos, *The Balkan Exchange of Minorities and Its Impact Upon Greece* (Paris, 1962).

[58] See Lorna Touryan Miller and Donald Eugene Miller, *Survivors: An Oral History of the Armenian Genocide* (Berkeley, 1999).

[59] An introduction to the war economy among the Western Front powers is in Hunt Tooley, *The Western Front: Battleground and Home Front in the First World War* (London, 2003), chapter 4.

end of the spectrum, but we will see gradations of activities throughout the war period that derive from the same source and display many of the same traits.[60]

Thesis Six

Most cases of ethnic cleansing in Central Europe in the period of World War II and after stemmed from or were in some way related to the territorial aspects of the Paris Peace Settlement of 1919/1920.

The peace settlement in the wake of the First World War is still a controversial and at times daunting topic. It encompasses ideas about modern liberalism and democracy, issues of international balance and the size of states in an international system, collective guilt, and a host of other subjects. The historical craft has been generally critical of the peacemakers and the peace they produced, though historians of foreign relations have, from time to time, tried to rehabilitate both the Paris Peace and its architects.[61] Still, criticism of the settlement has tended to remain the dominant attitude. Most historians and scholars of international affairs would agree that the particular ways in which Europe was "reconstructed" led to many problems in the interwar period and beyond. This is especially true for historians of the border settlements and of the subsequent ethnic history of East Central Europe.

One might point out, on the positive side that where border-related problems are concerned, the Paris peacemakers were not completely oblivious to the minorities they were creating with their new borders. In fact, they set up a series of minority treaties which were to protect those ethnic minorities created by the new borders (even if their successors showed little enthusiasm for protecting minorities after the war). The territorial settlement has been defended in detail in various points for various reasons.[62]

Yet a review of the specific border changes and their consequences makes it difficult to deny that a majority of the ethnic conflicts which would lead toward genocide and ethnic

---

[60] For an early, and incisive, analysis of the interconnected problems of nation, state, and minority, see Mises, *Nation, State and Economy*, 39-55.

[61] Examples are Paul Birdsall, *Versailles Twenty Years After* (New York, 1941); and more recently, Clifford R. Lovin, *A School for Diplomats: The Paris Peace Conference of 1919* (Lanham, Md., 1997).

[62] See the important study by Richard Blanke, *Orphans of Versailles: The Germans in Western Poland, 1918-1939* (Lexington, Ky., 1993).

cleansing in the period of World War II either originated with the territorial aspects of the peace settlement of 1919/1920 or were at least spurred on by the treaties of the Paris suburbs.[63] Ethnic borders througout East Central Europe were confused and confusing, both to outsiders and to the inhabitants of this complicated region. Before the war, four great empires had presided over the collection of dozens of nationalities stretching from the Baltic to the Balkans. With the collapse of these empires, East Central Europe partly remade itself and partly was remade in the form of national states, some based on preexisting states, some based on areas with no recent history of autonomous existence. No border lines could be conceived which could have divided people into true "ethnic states." The ethnic groups in this whole zone of Europe were mixed and mingled at the edges and often not just at the edges. Hence, the creation of the new "successor states" after World War I necessitated "cleaning up" ethnically. The Western powers tried to enforce minority treaties in all these states, but with little in the way of leverage. All these states eventually discriminated in one way or another against the outgroups and minorities inside their borders, leaving a legacy of interethnic bitterness. Some of the greatest bitterness came from peoples who had fought for the Central Powers and lost: Germans and Hungarians in particular. Both saw the borders redrawn to include them in outgroups in the new national states run by other national groups. But many minorities in East Central Europe faced becoming targets of expropriation and marginalization during the interwar period. Hence, where ethnic nationalism in the old empires had produced only occasional bitterness, in the interwar period ethnic hatred and bitterness grew apace as a result of the peace treaties which drew and confirmed the new borders.[64]

---

[63] A number of recent studies and collections tend to support this assertion. See Carole Fink and Peter Baechler, eds., *L'Etablissement des Frontières en Europe après les Deux Guerres Mondiales* (Bern, 1996); Seamus Dunn and T. G. Fraser, eds., *Europe and Ethnicity: World War I and Contemporary Ethnic Conflict* (New York, 1996); Christian Raitz de Frentz, *A Lesson Forgotten: Minority Protection Under the League of Nations: The Case of the German Minority in Poland* (New York, 1999); Carole Fink, "The League of Nations and the Minorities Question," *World Affairs* 157 (Spring 1995): 197-205; Manfred F. Boemeke, Gerald Feldman, and Elisabeth Glaser, eds., *The Treaty of Versailles: A Reassessment After 75 Years* (Cambridge, 1998). From the standpoint of legal scholarship, see Jennifer Jackson Preece, "Ethnic Cleansing as an Instrument of Nation-State Creation: Changing State Practices and Evolving Legal Norms," *Human Rights Quarterly* 20 (1998) 817-842.

[64] One of the best short accounts of the territorial changes and their effects is still Leonhard von Muralt, *From Versailles to Potsdam*, trans. Heinrich Hauser (Hinsdale, Ill., 1948).

In a sense, following the collapse of the four empires no conceivable set of new boundaries could have satisfied all concerned. But the outlook of the principle Allied peacemakers contributed to many of the problems. The polestar of border–drawing at Paris was the phrase "self–determination." Whatever this expression meant in general parlance then or means now, Woodrow Wilson, his expert advisors of the "Inquiry," and many European peacemakers used the phrase to mean simply that the new borders of Europe should correspond to ethnic boundaries, and further that–given the opportunity—all right-thinking European individuals would choose to live in a state together with their ethnic relatives. When the big three discussed the determination of the Upper Silesian border, for example, both Clemenceau and Wilson insisted that a plebiscite was completely unnecessary, since "statistics" had revealed that the "great majority of Upper Silesia" was Polish. The minutes show Lloyd George replying that the Four should not toss aside the German claim that many Polish Upper Silesians preferred Germany: "Surely the clause just read did not mean that if the Poles preferred to remain under Germany, they would have to become Polish because they were of Polish race." Wilson replied: "We have no doubt about the ethnographic fact.... what I said in the Fourteen Points does not compel us to order a plebiscite in Upper Silesia."[65]

Many agendas were at work with Wilson and other peacemakers of 1919, but it is nonetheless the case that they tended to assume that ethnic belonging implied political allegiance, that "ethnography" determined loyalty. This assumption would be strikingly disproven in all three plebiscites on Germany's eastern borders (Allenstein and Marienwerder in West Prussia and Upper Silesia), since substantial portions of non-"German" populations voted to remain in Germany.[66]

---

[65] See the very useful comparison of the different versions of minutes taken during Council of Four meetings (in particular the discusson of 3 June 2002), done by Henry Schwab of Princeton, New Jersey, in a web-published essay, "The Paris Peace Conference 1919: Woodrow Wilson, Hankey and Mantoux," dated 2001 and accessed on the web on July 28, 2002, at http://www.schwab-writings.com/hi/wi/index.html. The published versions of the Council of Four discussions from June 1919 are in Arthur S. Link, ed., *The Papers of Woodrow Wilson* (Princeton, N.J., 1989); and Paul Mantoux, *The Deliberations of the Council of Four (March 24-June 28, 1919)*, trans. and ed. Arthur S. Link with the assistance of Manfred F. Boemeke (Princeton, N.J., 1992). I have used quotations from both versions, which may be consulted most conveniently in Schwab's collation.

[66] On the most disputed plebiscite area, Upper Silesia, see Tooley, *National Identity and Weimar Germany*, 218-252; and Tooley, "German Political Violence and the Border Plebiscite in Upper Silesia, 1919-1921," *Central European History* 21 (March 1988): 56-98.

And the attempt to create a Europe of ethnically determined borders drew all parties into the ensuing struggles. New states favored by the Allies, the vanquished states (or parts of them) which faced losing territories long connected them, the Allies themselves—all found themselves in struggles in which the well–being of local people had little or no weight at all. Again, the Upper Silesian case provides an example. When the Allies eventually arranged for the plebiscite in Upper Silesia, both sides sent plenipotentiaries to deal with the Interallied Plebiscite Commission. The German representative was Prince Hermann von Hatzfeldt, whose humane views were overwhelmed in what might be called the ethnification of the border. Once on the ground, Hatzfeldt suggested to Berlin that several of the easternmost regions just be given up to Poland, since the struggle was producing so much bitterness. The German Foreign Office quickly put an end to any thoughts of conceding territory. On the other hand, Hatzfeldt tried to persuade the Interallied Commission to set an early date for the plebiscite itself: "The longer the plebiscite is put off, the greater grows mutual embitterment; and in the end, even after the plebiscite—let it turn out how it will—the Germans and Poles will find themselves with the task of living beside and with each other." Unheeding, the Commission put the vote off until nearly a year later, in March 1921.[67] Governments on both sides of the question marginalized the interests of local regions and individuals. The Allies, in effect, saw their main interest in creating "ethnographic" boundaries, in Wilson's terminology. The Germans saw theirs as holding onto as much territory as possible, even if the local people did want to be a part of the new Poland.

These are aspects of a single case, but the peace settlement encompassed border changes throughout Central and East Central Europe. Germany, Austria, and Hungary (the vanquished) felt keenly the loss of national territory and breakup of longstanding associations. The countries who had sided with the victors did not lose territory, yet all these countries wanted more than they held at the end of the peacemaking. Hence, though the successor states wanted to maximize territory, they also aimed to create the hardshell ethnic states prescribed by late nineteenth-century nationalism. These two goals in fact contradicted each other. The larger the territory, the more likely it was that sizeable minorities

---

[67] Tooley, *National Identity and Weimar Germany*, 148; for some important patterns of the plebiscites, see Tooley, "The Internal Dynamics of Changing Frontiers: The Plebiscites on Germany's Borders, 1919-1921," in *L'Etablissement des Frontières en Europe après les Deux Guerres Mondiales*, ed. Carole Fink and Peter Baechler (Bern, 1996).

would be included. The internal policies of these states would invariably discriminate against minorities. So Hungarians in Transylvania suffered under Romanian rule; Croatians disliked their new Serbian ruling class; Slovaks, Hungarians, and Germans felt marginalized in Czechoslovakia; Lithuanians, Ukrainians, Germans, and others in Poland faced inequities and sometimes worse at the hands of the Polish state; this list could go on and on. Many Germans, Austrians, and Hungarians hoped to regain national irredenta at some future date, and perhaps in the process settle old scores. One might also add to this general picture that the national states of the new East Central Europe were careful to keep the fires of bitterness smoldering between the wars, since the leaders of these states in essence agreed that impermeable ethnic borders and centralized national cohesion formed the goal of all of modern statecraft.

Many peacemakers foresaw troubles with this confusion of motives and aims—Harold Nicolson and his classic account of work on the South Slav Committee at Paris come to mind[68]—but the confusion of aims and the desire to cut the losers down to size exacerbated what would have been a terribly confusing set of national borders in any case. The resulting imbalances helped engendered bitterness that would reemerge literally with a vengeance twenty or twenty-five years later. In the context of the present essay, therefore, it is important to distinguish the treaties of Paris as a very specific cause of the emergence of twentieth-century Europe's particular brand of ethnic cleansing.

This is not to say that the peace treaties were the only causative factors of ethnic cleansing, or even that ethnic cleansing would never have taken place under some other international settlement. But a mere perusal of the chapters of the present collection of studies demonstrates the extent to which bitterness, hatred, and ethnic cleansing resulted when the Allies engineered ethnic minorities into hostile national states by the Allies in 1919, and when such engineering provoked predictable countermeasures.

Conclusion

In the period of World War I, activities and processes which we might call ethnic cleansing remained limited to a specific zone of Europe. Since the French Revolution, all European states have at one time or another seen their ethnic minorities as roadblocks to national cohesion and national fulfillment. Yet it is important to recognize that until after World War I, ethnic cleansing in the nar-

---

[68] Harold Nicolson, *Peacemaking 1919* (New York, 1939).

row sense only took place within a zone of Europe whose western boundary we might describe as passing southward from Riga to Warsaw, thence southeastward, skirting the lands of the Austrian Empire, to Bucharest, then southwest across Serbia and directly south to the sea. In terms of the continuum of governmental behaviors we have reviewed, ethnic cleansing took place only eastward of this line until the period leading to World War II.

Still, if we reorganize our search by breaking down ethnic cleansing into some basic components, we find that the combination of national state and late nineteenth-century empire—with a strong dose of Social Darwinism intermingled in both—capable of engaging in many of the behavioral components of what we call ethnic cleansing. We also find that the war served as an accelerator of ethnic cleansing in several ways. On a practical level, the war destroyed old boundaries and raised up new "ethnic states" throughout East Central Europe; and the ethos of these states created situations in which ethnic cleansing presented itself as an obvious solution to problems. At the same time, it is important to keep in mind that states have been much more implicated in ethnic cleansing than undirected populaces—it has tended to be states or quasi-states that have begun, urged on, and directed ethnic cleansing. Since World War I represented an acceleration in the scope and power of government, and in its need for centralization and homogenization, it is not surprising that the intensity of governmental action against unwanted ethnic minorities accelerated as well. This rapid and—in a terrible way—innovative acceleration in the aggrandizing state's war against its domestic "enemies" really imparts the defining characteristic to the twentieth century's population politics. The quest for a homogenous nation, the pressures of the growing state, the need to transfer "enemy" wealth to the state, the Social Darwinist imperatives of "demographic" solutions to the struggle for survival: all these elements of later ethnic cleansing were ground–tested during the First World War, a conflict we might well describe as the midwife of modern population politics.

"Ethnic Cleansing," Emigration, and Identity: The Case of Habsburg Bosnia-Hercegovina

PETER MENTZEL

There is a significant linkage between "ethnic cleansing," emigration, and the formation of ethnic identity. This point seems so obvious that it is sometimes not even explicitly mentioned. The "Ethnic Cleanser" must have a clear idea of who it is who must be expelled (or murdered, robbed, or brutalized). But in the process of doing so, the perpetrator of these crimes will also strengthen his victim's sense of ethnic identity or ethnic distinctiveness. This phenomenon seems to have been a feature of the 1992-95 war in Bosnia-Hercegovina. In that case, many Bosnian Muslims became more conscious of their ethno-national and/or religious identities as a result of being the targets of violence.[1] The Bosnian Muslim national identity that had existed officially in Yugoslavia since 1968 was itself renamed "Bosniak" in 1993.[2]

Many scholars point to earlier periods of crisis or tension as important in the development of a distinctive Bosnian Muslim national identity. This paper will examine the interplay between emigration and ethnic identity using the case of Habsburg ruled Bosnia-Hercegovina (1878-1918). It will argue that the emigration of Bosnian Muslims during this period can provide us with interesting clues as to the development of a Bosnian Muslim ethnonational self-consciousness. In particular, the information available to us about Muslim emigration seems to confirm the hypothesis (advanced by Francine Friedman, among others) that a distinct Bosnian Muslim national identity developed in earnest after the formal annexation of Bosnia-Hercegovina by the Habsburgs (i.e., after 1908).

This paper also offers some implicit suggestions on the usages and limitations of the term "ethnic cleansing." While most students of the subject would probably not consider the Muslim emigration from Habsburg Bosnia-Hercegovina to be a *bona fide* ex-

---

[1] Francine Friedman, "The Muslim Slavs of Bosnia and Herzegovina (with Reference to the Sandžak of Novi Pazar): Islam as National Identity," *Nationalities Papers* 28 (2000): 177-178.

[2] Pedro Ramet, "Die Muslime Bosniens als Nation," *Die Muslime in der Sowjetunion und in Jugoslawien: Identität, Politik, Widerstand, ed.* Andreas Kappeler, Gerhard Simon, and Georg Brunner (Cologne, 1989), 107. Robert Donia, "The New Bosniak History," *Nationalities Papers* 28 (2000), 351.

ample of this phenomenon, it does share some important similarities with more famous cases of "ethnic cleansing," and indeed with the events in Bosnia during the 1992-1995 war. For example, as in other cases of "ethnic cleansing," the emigration of Bosnian Muslims between 1878 and 1918 occurred within the context of war and occupation by a foreign power and resulted in a significant change in the demographic makeup of the province. Likewise, as is common in most examples of the phenomenon, the Habsburg authorities pursued a policy of colonizing their new territory with immigrants from the rest of the Dual Monarchy.

The question of Muslim emigration can, it seems to me, be broken down into three discrete subjects: first, the numbers and characteristics (socio-economic, geographic, etc.) of the emigrants; second, their reasons for emigration; and, third, whether the story of Muslim emigration tells us anything about the complex issue of Bosnian Muslim identity. Finally, as part of an overall critique of the concept of "ethnic cleansing," this paper will make some observations regarding the usefulness (or lack thereof) of this term in the case of Habsburg Bosnia-Hercegovina.

The Crisis of the Seventies and Muslim Emigration

The emigration of Muslims from Bosnia-Hercegovina between 1878 and 1918 certainly did not occur in a vacuum. Indeed, it was part of a much broader phenomenon in which millions of Balkan Muslims emigrated, almost all of them to the shrinking territory of the Ottoman Empire. Although this is not the place to detail this story, its broad outlines are worth reviewing. In 1878 and 1879 and again between 1912 and 1914 much of the Balkan Muslim population was "ethnically cleansed" by the new states in the region. In parts of Bulgaria and Greece the Muslim population almost entirely disappeared as the result of fleeing from or being expelled by the new rulers.[3]

The Habsburg occupation of Bosnia-Hercegovina in 1878 had been preceded by three years of chaos in the province. A peasant uprising in the spring of 1875 had rapidly taken on the appearance of an anti-Ottoman rebellion. The rebellion, in turn, led to war between Montenegro and Serbia and the Ottoman Empire and, finally, in April 1877, to a war between Russia and the Ottomans. During the rebellion and subsequent wars, the Muslim population

---

[3] An exhaustive overview of this subject is Justin McCarthy, *Death and Exile* (Princeton, 1995). A more concise treatment by the same author is, "Muslims in Ottoman Europe: Population from 1800-1912," *Nationalities Papers* 28 (2000).

of Bosnia-Hercegovina was subjected to violence and persecution, and some Muslims must have fled the province at this time. The Russians' (hard-won) victory (March 1878) in the war led to a diplomatic crisis finally resolved by the Congress of Berlin in June 1878 and its resulting Treaty (signed 13 July 1878). Among its other provisions, Article 25 of the treaty left Bosnia-Hercegovina legally within the Ottoman Empire but provided for Austro-Hungarian occupation and administration of the province.[4] In effect, Bosnia-Hercegovina was made a Habsburg colony.

The Habsburg occupation of the province began in the summer of 1878. Seventy-two thousand men and officers were eventually needed in the invasion. Fighting was extremely fierce and continued even after the occupation of Sarajevo in August. The Imperial and Royal forces finally declared victory in October 1878. The guerrillas were mainly Muslim peasants and craftsmen, although some Serbs at times participated as well.[5] Although the Ottoman government officially called upon its officials and soldiers to take no part in the resistance, many Ottoman soldiers and officers in fact joined the anti-Habsburg forces.

Soon after the Imperial and Royal forces of Austria-Hungary had succeeded in subduing the resistance, the Habsburg authorities, the local Bosnian political forces, and the Ottoman government reached a series of agreements enshrined in the Novi Bazar Convention, also known as the April Convention of 1879. This document was supposed to regulate Austro-Hungarian rule in the province. In it, the Austro-Hungarian authorities "recognized the right of Turkish functionaries to retain their posts, the right of Bosnian Muslims to unimpeded communication with their spiritual leaders in the Ottoman Empire, and the right of Turkish currency to circulate in Bosnia. [The occupying authorities] also promised to honor all customs and traditions of the Bosnian Muslims."[6]

A detailed account of the history of Habsburg occupation is beyond the scope of this paper. Yet it is worth noting that most accounts of the period of occupation (1878-1918) generally de-

---

[4] However, "...in the minds of the ... governments which accepted it [the proposed occupation of the province], there was no expectation that the occupation would be otherwise than permanent." Bernadotte E. Schmitt, *The Annexation of Bosnia, 1908-1909* (Cambridge, 1937), 1.

[5] Robert Donia, *Islam under the Double Eagle* (Boulder, 1981), 31.

[6] Ibid, 10-11.

scribe it as relatively benign, especially as far as relations between the Austro-Hungarian authorities and the Muslims were concerned. Such characterizations are potentially at odds with the information available on the protests, petitions, and even armed rebellions against the Habsburg authorities. The emigration of a substantial part of the Muslim population should be understood within this context.

The Numbers and Social Make-up of the Emigrants

Emigration from Habsburg ruled Bosnia-Hercegovina was largely a Muslim phenomenon, although some Serbs also left. Serbs emigrated almost exclusively to the Kingdom of Serbia, while Bosnian Muslims went overwhelmingly to the Ottoman Empire.[7] Bosnian Muslim emigrants settled in or near most of the bigger towns and cities throughout the Empire. Important destinations included places as diverse as Istanbul, Karamursal, Inegol, Afyonkarahisar, Damascus, Bursa, and Ismit.[8] Many other Bosnian Muslims emigrated to Ottoman Balkan provinces, especially Monastir, Kosova, Edirne, and Selanik.[9]

While emigration seems to have gone on during the entire Habsburg period, different scholars have noted several peak periods, usually associated with crisis situations. According to one historian, "in all about 140,000 emigrated to Turkey between 1878 and 1918."[10] Other authors have arrived at much higher figures (as high as 500,000).[11] The general consensus seems to be that approximately 150,000 Muslims emigrated. The matter is complicated by the fact that the Habsburg authorities only began to compile figures after the anti-conscription riots of 1882. The first published figures on emigration were released only in 1906.

---

[7] Some emigrants went to other countries, including the USA. By 1911 there were communities of Bosnian Muslims in Chicago, Detroit, Gary (Indiana), and Los Angeles. Mustafa Imamović, *Bošnjaci u Emigraciji*, (Sarajevo, 1996), 68-69.

[8] Vojislav Bogičević, "Emigracije Muslimana Bosne i Hercegovine u Tursku u doba Austro-Ugarske Vladavine, 1878-1918," *Historijski Zbornik*, III (1950), 187.

[9] McCarthy, "Population," 36.

[10] Mark Pinson, "The Muslims of Bosnia-Herzegovina under Austro-Hungarian Rule, 1878-1918," in *The Muslims of Bosnia-Herzegovina: Their Historic Development from the Middle Ages to the Dissolution of Yugoslavia*, ed. Mark Pinson (Cambridge, 1994), 94.

[11] For a discussion on determining the number of emigrants, see Imamović, 61-62.

The timing of emigration waves, and the number of Muslims who left during each period, are debated. One scholar identified four periods of peak emigration: immediately after the occupation of 1878; after the announcement of general conscription in 1881; after the political upheavals of 1900-01; and after formal annexation in 1908.[12] Another pointed to three main periods of emigration: 1878-79, 1881-1883, and 1899-1901 (the last date marking a high point), after which emigration decreased.[13] A third scholar divided Muslim emigration into four periods, corresponding roughly to the four decades of Habsburg control. According to this view, the peak period of emigration occurred between approximately 1900 and 1910, after which the numbers of emigrants dropped sharply. The same scholar noted further that the numbers of emigrants can also be analyzed to reveal two main periods of accelerated emigration. The first occurred during the first decade of Habsburg rule (c. 1878-1888), the second during the annexation crisis (1908-1910).[14]

Thus, although the studies of Muslim emigration from Bosnia differ on certain specifics, they are all largely in agreement that emigration declined sharply after 1908/1909, i.e., after the annexation crisis. Interestingly, not only did emigration decline sharply after 1909, but there is evidence that some of those who emigrated earlier wanted to move back to Bosnia. In 1912 Bosnian Muslim emigrants filed a petition for a "mass return" to Bosnian-Hercegovina with the Austro-Hungarian authorities.[15]

The steady emigration of Muslims from Bosnia-Hercegovina changed the demographic makeup of the province. In the years immediately before the Habsburg occupation, Muslims made up almost 48 percent (722,000 out of a total population of approximately 1,510,000) of the population of Bosnia and Hercegovina (not including the Sandžak of Novi Bazar). After an initial drop immediately after the occupation, the number of Muslims in Bosnia-Hercegovina increased during the Habsburg period, although the population never regained its pre-occupation numbers. Furthermore, Muslims actually declined steadily as a percentage of

---

[12] Pinson, "The Muslims of Bosnia-Herzegovina," 94.

[13] Aydin Babuna, *Die nationale Entwicklung der bosnischen Muslime: Mit besonderer Berücksichtigung der österreichisch-ungarischen Periode,* (Frankfurt/Main, 1996), 48.

[14] Bogičević, 181.

[15] Pinson, "The Muslims of Bosnia-Herzegovina," 125.

the total population. Muslims made up 37 percent of the total population in 1885, 35 percent in 1895, and only 32 percent by 1910.[16]

Almost all social and economic classes were represented among the emigrants. In 1878 and 1879 the initial emigration of Muslims was limited largely to former Ottoman officials and their families and close supporters. Between 1891 and 1897 (not, it should be noted, a period of intense out–migration), 72 percent of the emigrants were somehow connected with agriculture (whether as farmers or landlords), 11 percent were craftsmen, workers and day laborers made up 7 percent, merchants 5 percent, free professions 2 percent and "other" 3 percent (figures are rounded).[17] Likewise, in the years between 1903 and 1906 Muslim free peasants accounted for the overwhelming majority of emigrants.[18] These figures reflect the agricultural nature of Bosnian society. In the 1895 census, 88 percent of the population was engaged in agriculture.[19]

Reasons for Emigration

Despite the relatively large numbers of emigrants and the cross-section of Bosnian society they represented, the reasons for these out-migrations are not clearly understood. Several scholars note economic factors as reasons for emigration. The Habsburg authorities themselves pointed to the change from an economy based largely on payments in kind (*Naturalwirtschaft*) to a money economy (*Geldwirtschaft*) as the most important reason for emigration.[20] The economic situation in Bosnia-Hercegovina was indeed profoundly disrupted by the occupation and subsequent annexation of the province. Craftsmen and merchants were especially affected due to the influx of manufactured goods from Austria-Hungary and the establishment of new channels of trade and

---

[16] The Muslim population was 493,000 in 1885, 549,000 in 1895 and 612,000 in 1910. McCarthy, "Population," 32, 38; Robert J. Donia and John V.A. Fine, *Bosnia & Hercegovina: A Tradition Betrayed* (New York, 1994), 86-87. See also Imamović, 62.

[17] Imamović, 60; Babuna, *Entwicklung*, 49.

[18] Babuna, *Entwicklung*, 51.

[19] Donia and Fine, 75-76.

[20] Ibid.

distribution.[21] A general price inflation due to the increased money in circulation was also economically disruptive to many Bosnians.[22]

One study noted six factors that contributed to Bosnian Muslim emigration. Among them were outside activity and agitation (presumably from the Ottoman Empire), fear of persecution, religious fanaticism, a feeling of disgust at the prospect of living under non-Muslim rule, opposition to conscription, and economic factors.[23] Other scholars have downplayed the importance of economic factors. They note that the emigrants included wealthy landowners whose quasi-feudal relationship to the peasants was not altered by the Austro-Hungarian administration.

While most accounts admit that economic factors played some role, almost all studies stress a cause which one scholar has described as the religious or "psychoreligious" reasons for emigration. Some Muslim religious leaders seem to have argued that the Quran forbade Muslims to live under Christian rule.[24] Many Bosnian Muslims might well have agreed with this position or simply felt awkward or uncomfortable living under the rule of a Christian power. Similarly, some Muslims emigrated as a self-conscious act of political protest. These emigrants settled in Istanbul, whence they issued manifestoes and proclamations decrying Habsburg rule of the province.[25] A final reason that might help to explain Muslim emigration was the fear of persecution, or even death, at the hands of the new rulers. The Habsburg invasion was quite violent and occurred (it should be remembered) immediately after a major Balkan war in which many thousands of Muslims in Bulgaria and elsewhere were killed, brutalized, or forced to flee.[26]

Indeed, despite the official pronouncements of goodwill by the Habsburg authorities, the military and civil occupation forces often violated those aspects of the Novi Bazar Convention that aimed at safeguarding the Muslims. Petitions sent by Muslims to

---

[21] Imamović, 60.

[22] Babuna, *Entwicklung*, 50.

[23] Bogičević, 184.

[24] McCarthy, "Population," 35.

[25] Donia, 30-31.

[26] McCarthy, "Population," 36.

the Imperial and Royal authorities (sometimes even to the Emperor-King Franz Josef himself) included a wide range of grievances. Among the most frequent were complaints that Muslim customs and property rights were not being respected.[27] In fact, the authorities consistently acted against the spirit, and frequently the letter, of the Novi Bazar Convention. As one scholar of Habsburg Bosnia wrote: "In spite of these attempts to maintain some semblance of Turkish sovereignty [i.e., the provisions of the Convention], Turkish money was subsequently excluded, the provinces were brought within the Austro-Hungarian customs union, passports for the inhabitants were issued from the Austro-Hungarian embassy in Constantinople, the administration was assimilated to that of the [Habsburg] Monarchy, the capitulations were abolished... and the Austro-Hungarian military service was introduced."[28] It was this last action that sparked the first violent protest against Austro-Hungarian rule since 1878.

The law introducing military conscription was introduced in November 1881 and was immediately opposed by all the ethno-religious groups in the province. Besides constituting a challenge to the continuation of Ottoman suzerainty, many Muslims also were unsure if they could serve in the military of a Christian state. Bosnians of all religions opposed conscription on the grounds that the law made no provisions to exempt from service sons whose families relied on them economically. The law was actively resisted throughout the province and actually led to an armed uprising centered in eastern Hercegovina in January 1882. With considerable difficulty the Austro-Hungarian armed forces finally succeeded in quelling the revolt in March, although the area remained unsettled until the mid 1890s.[29]

The second period of profound opposition to the occupation, between approximately 1899 and 1902, was sparked by the conversion of a young Muslim woman, Fata Omanović, to Catholicism. While earlier cases of conversion had angered and provoked the Bosnian Muslim community, the case of Fata Omanović developed within the context of increasing Muslim political activism and frustration and accordingly led to a serious crisis.

Fata lived in a mixed (Muslim and Catholic) village near Mostar, in Hercegovina. On the night of May 2-3, 1899, she fled from

---

[27] Donia, 74, 100-101, 120, 140.

[28] Schmitt, 2; Donia, 10.

[29] Donia, 33. Imamović, 52.

her family's home with the aid of some Catholic neighbors to escape marriage to a Muslim suitor she disliked. She subsequently
disappeared.[30] Despite the apparently good-faith efforts of the
Austro-Hungarian authorities to find Fata and punish those who
helped her flight, Muslim leaders were able to use this incident to
form strategic alliances between different Muslim groups and to
galvanize anti-Habsburg sentiments throughout the province.[31]
The Muslim opposition forces drew on pan-Islamic symbols, as
well as support from the emigré community in Istanbul, to help
unite the Muslim community in Bosnia-Hercegovina and formu-
late a set of demands aimed at cultural autonomy.[32]

In the face of the growing power of the Muslim opposition,
the Habsburg occupation authorities tried to split the unity of the
Muslims and also began to use increasingly oppressive measures
to silence the activists, including arrests and imprisonment. The
authorities also tried to break the links between the Bosnian Muslim opposition and their supporters in Istanbul by instituting a
law in October 1901 that allowed the government to bar reentry to
Bosnia-Hercegovina of any emigrant suspected of anti-Habsburg
activities. The use of this law effectively stranded several important opposition leaders in Istanbul and temporarily weakened the
activities of the Bosnian Muslim activists.[33]

Significantly, the periods between 1881 and 1882 and then
1899 and 1902 marked episodes of intensive emigration from Bosnia-Hercegovina to the Ottoman Empire.[34] During each of these
periods the Habsburg occupation authorities seemed eager to facilitate this out-migration of Muslims. Applications for emigration
were usually granted, especially after the beginning of Benjámin
Kállay's administration (1882–1903) following the 1882 revolt.[35]

---

[30] It was generally believed that she had fled to Dalmatia, in Austria, but other
reports put her in Montenegro, the Sandžak of Novi Bazar, or Macedonia. She
eventually surfaced in Slovenia, where she had married a Catholic and taken the
name Ema Prijatelj; Donia, 114. See also Aydin Babuna, "The Emergence of the
First Muslim Party in Bosnia-Hercegovina," *East European Quarterly*, XXX, 2, 1996.
145.

[31] Donia, 114-115.

[32] Ibid., 123; Babuna, "Emergence," 146.

[33] Donia, 164; Babuna, "Emergence," 146.

[34] Imamović, 52, 55.

[35] Ibid., 52. Bogičević, 184.

The authorities also assisted emigrants with transportation and occasionally even money.[36] Furthermore, the 1901 emigration law noted above could be, and indeed was, used to keep emigrants from returning to the province. The Austro-Hungarian administrators might have been motivated by the desire to keep order in the province, yet they might also have tried to facilitate emigration in order to make room for colonists from the rest of the Dual Monarchy.[37] Similarly, certain groups in Bosnia encouraged emigration with the intent of buying up inexpensively the former owners' property.[38] The Ottoman government for its part sought to help the Bosnian Muslim emigrants who moved into Imperial territory while at the same time tried to encourage them to remain in Bosnia.[39]

"Ethnic Cleansing" and Emigration from Bosnia-Hercegovina

One of the aims of this brief essay has been to test the usefulness of the term "ethnic cleansing." In particular, can the process of Muslim emigration from Habsburg Bosnia-Hercegovina be referred to by this name? As noted in the introduction to this paper, the conditions in which many thousands of Muslims abandoned their homes and set off to uncertain futures bore an uncanny resemblance to other, more widely cited, cases of "ethnic cleansing," both in the Balkans and elsewhere. Many, if not most, Bosnian Muslims felt insecure and oppressed by the Habsburg civil and military forces. The frequent petition drives calling attention to the on-going violations of the Novi Bazar Convention, not to mention the crises of 1881/82 and 1899-1902, are evidence of a feeling of animosity and uneasiness on the part of the Bosnian Muslims.

On the other hand, the Austro-Hungarian occupation authorities did not treat their Muslim subjects in Bosnia-Hercegovina with anything approaching the violence of the post-

---

[36] Imamović, 58.

[37] Babuna, *Entwicklung*, 52. The numbers of these immigrants were not insignificant. By late December 1880 there were 4,510 Austrian and 11,765 Hungarian subjects in Bosnia-Hercegovina. By 1895 those numbers had increased to 24,018 Austrians and 42,358 Hungarians. Bogičević, 178.

[38] Babuna, "Emergence," 132. Babuna does not comment upon the identity of these groups (he calls them "Kreise") except to mention that some of them were Muslims themselves. Babuna, *Entwicklung*, 51.

[39] Babuna, *Entwicklung*, 50. Imamović, 58.

1878 Balkan states, not to mention the appalling atrocities committed during World War II or as part of the "Wars of Yugoslavian Secession." Specifically, the occupation authorities do not seem to have driven large numbers of Muslims out of the province using force or the threat of force (with the possible exception of the violent occupation of the country in 1878).

At the same time, it seems as though for most of the period of occupation these same authorities did little to hinder emigration and might have on occasion actually encouraged it. The promulgation in 1901 of the "illegal emigrant" law certainly could work to reduce the overall numbers of Bosnian Muslims, no matter what its actual intent. As noted earlier, the percentage of Muslims in the overall Bosnian population indeed decreased during the Habsburg period.

What the history of these events might be pointing to, at least as far as defining "ethnic cleansing" is concerned, is that intent is a significant factor in deciding upon the appropriate use of the term. That is, if there were compelling evidence that the Habsburg forces intended that many or most of the Bosnian Muslims emigrate, it might seem less jarring to think of the Bosnian Muslim emigres as victims of "ethnic cleansing." As it is, the Austro-Hungarian administrators in general do not seem to have been motivated by this desire. Rather, as this essay will now go on to argue, the motivation for emigration stemmed largely from the changing self-identity of the Bosnian Muslims themselves.

Emigration and Identity

What, if anything, can this information about the emigration of the Muslims tell us about the formation of a Bosnian Muslim national identity? It is first of all worth noting that most students of the subject agree that when Habsburg rule began in Bosnia-Hercegovina in 1878 the Bosnian Muslims did not have a national self-consciousness but that by 1918 (when Habsburg rule ended) at least some Muslims were thinking of themselves in ways that very strongly resemble national identity.[40] When Habsburg rule in Bosnia began Bosnian Muslims called themselves variously "Turks," "Muslims," or "Bosniaks," more or less interchangeably according to circumstances. By 1918 the use of the term "Turk" for

---

[40] There is, of course, considerable debate as to when exactly Bosnian Muslims developed a national self-consciousness. Furthermore, as Robert Donia has recently pointed out, Bosnian Muslim history is "replete with evidence of a well developed *group* consciousness" (emphasis in original); Donia, "Bosniak History," 357.

self-ascription had given way almost entirely to "Muslim" or "Bosniak."[41]

If we correlate the peak periods of emigration with these changes in nationally descriptive terminology we find some very interesting patterns. While, as noted above, different scholars disagree about the details, most seem to point to emigration reaching its zenith either just before 1908 (i.e., shortly after the political upheavals of 1899-1902) or immediately after 1908 (i.e., in 1909, after the Ottoman government's formal acknowledgment of the annexation).

The economic hypothesis for emigration does not seem to be of much explanatory use here. The economic situation in Bosnia-Hercegovina remained turbulent during the entire period of Habsburg rule. The "psychoreligious" and political explanations for emigration seem to bear much more interesting relationships to the actual emigration data. Bosnia-Hercegovina was *de jure* part of the Ottoman Empire until 1908. The activities of many of the emigrants in Istanbul make it clear that at least some of them maintained an identity that was closely bound up with the Ottoman Empire. Likewise, the Bosnian Muslims who remained in the province were increasingly troubled by what they perceived as Habsburg insensitivity to their religious beliefs. This discomfort was epitomized by the crisis of 1899-1902, precipitated as it was by the conversion of a Muslim to Catholicism.

The Ottoman Empire was the obvious choice for emigrants motivated by religious feelings and a convenient locus of Bosnian Muslim loyalty. The Empire was at the time the biggest independent Muslim-ruled country in the immediate vicinity of Bosnia (or, indeed, in the world). Furthermore, since the reign of Abdülaziz (1861-1876) the Ottoman Sultans had more and more frequently called themselves "caliphs" and made use of pan-Islamic propaganda and symbols.[42] The Empire could therefore function as a source of both religious and political identity.

---

[41] Significantly, the Bosnian Muslim self-descriptive name rendered in English as "Turk" ("Turci") was different from the name applied to Turkish speakers or Anatolians, i.e., "Turkuši"); Dennison Rusinow, "The Ottoman Legacy in Yugoslavia's Disintegration and Civil War," in *Imperial Legacy: The Ottoman Imprint on the Balkans and the Middle East*, ed. L. Carl Brown (New York, 1996), 93. Ivo Banac, "Bosnian Muslims: From Religious Community to Socialist Nationhood and Postcommunist Statehood, 1918-1992," in *The Muslims of Bosnia-Herzegovina: Their Historic Development from the Middle Ages to the Dissolution of Yugoslavia*, ed. Mark Pinson (Cambridge, 1994), 133. See also Friedman, "Muslim Slavs," 168.

[42] Stanford J. Shaw and Ezel Kural Shaw, *History of the Ottoman Empire and Modern Turkey*, Vol.II (Cambridge, 1977), 157-158.

In October 1908 the Habsburgs annexed Bosnia-Hercegovina, at least partially in response to the Young Turk revolution of that same year. The new government in Istanbul responded to the annexation with diplomatic pressure and a boycott of Austro-Hungarian products and services. After months of negotiations, the crisis was finally settled in February 1909 on the basis of the payment to the Ottoman Empire of an indemnity of 2.5 million lira.[43]

The annexation crisis triggered what was apparently the last major wave of emigration of Bosnian Muslims to the Ottoman Empire. It also seems to have coincided with what many scholars describe as the development of a secular Bosnian Muslim national identity. As one scholar put it, the annexation proclamation broke the "thin but psychologically real" connection many Bosnian Muslims felt between Bosnia-Hercegovina and Istanbul.[44] Many Bosnian Muslims felt betrayed or abandoned by the Ottoman government and may have concluded that they needed to reach an accommodation with the Habsburg authorities.[45] This furthered the importance of Bosnian Muslim political parties (especially the Muslim National Organization, founded in 1906) which, although dominated by the Begs and Agas, represented, at least theoretically, the interests of all of Bosnia's Muslims. Indeed, by 1910 the Muslim elite had secured from the Habsburg authorities most of the cultural autonomy and economic privileges for which they had been struggling. It was also during the post-annexation period that many Bosnian Muslims, especially intellectuals, abandoned the term "Turk" to describe themselves and adopted "Muslim."[46]

The official annexation of Bosnia-Hercegovina by the Habsburgs (and the corresponding "abandonment" of the province by the Ottoman Empire) thus might have helped to change the ways in which Bosnian Muslims thought about themselves. The practicality of a continued identification with the Ottoman Empire was effectively ended and the strength and usefulness of the Sultan-Caliph's pan-Islamic program was also seriously called

---

[43] Donald Quataert, *Social Disintegration and Popular Resistance in the Ottoman Empire, 1881-1908* (New York, 1983), 121.

[44] Imamović, 56.

[45] Pinson, 126; Babuna, "First Muslim political party," 135.

[46] Francine Friedman, *The Bosnian Muslims: Denial of a Nation* (Boulder, 1996), 75.

into question. With ties to both the Ottoman Empire and the world's Islamic community thus seriously compromised, Bosnian Muslims turned inward and began to construct an identity more closely linked to Bosnia and their own unique heritage. The precipitous drop in the numbers of emigrants and the vigorous growth of Bosnian Muslim political activity in the province are therefore linked as evidence of the growth of a Bosnian Muslim national identity.

"Neither Serbs, nor Turks, neither water nor wine, but odious renegades": The Ethnic Cleansing of Slav Muslims and its Role in Serbian and Montenegrin Discourses since 1800[1]

CATHIE CARMICHAEL[2]

Physical destruction of the Islamic communities of the Balkans is a process that has taken place over the last two hundred years or so. During the period from 1821 to 1922 alone, Justin McCarthy estimates that the ethnic cleansing of Ottoman Muslims led to the death of over five million individuals and the expulsion of a similar number.[3] Hundreds of thousands of Muslims were also killed, primarily on the grounds of ethnicity, during the Second World War and the Yugoslavian Wars of Dissolution. Ideological marginalization of Islamic communities accompanied the decline and fall of the Ottoman Empire, but as an ideology or a series of related ideas, it draws upon far older prejudices going back to the Middle Ages against the "Turk" and the religion of Islam in pan-European discourses, which have used as a justification to persecute Muslims alongside Jews.[4] As the Ottoman Empire weakened and rival European powers encouraged the development of nationalist ideologies among the subject peoples, the Muslims in the Balkans sometimes became viewed as a kind of ethnic "fifth column,"[5] left over from a

---

[1] Jovan Cvijić, *La péninsule balkanique: Geographie humaine* (Paris, 1918), 353, quoting a saying about what he refers to as "Islamicized Serbs" attributed to Orthodox Serbs.

[2] I would like to thank the participants at the Duquesne Conference for their comments, particularly John Schindler, Ben Lieberman, Brian Williams, and the editors of this volume. I am very grateful for additional suggestions made by Djordje Stefanović, Glenda Sluga, Mehmet Ali Dikerdem, Florian Bieber, Marko Živković and Dejan Djokić.

[3] Justin McCarthy, Death *and Exile, The Ethnic Cleansing of Ottoman Muslims 1821-1922* (Princeton, NJ, 1996), 338.

[4] Michael Sells, *The Bridge Betrayed: Religion and Genocide in Bosnia, 2ⁿᵈ ed.* (Berkeley, 1998), 119.

[5] Milica Bakić Hayden refers to this negative view of Muslims as a "betrayal syndrome" in her article "Nesting Orientalisms: The case of the Former Yugoslavia," *Slavic Review* 54 (1995), 927.

previous era, who could never be integrated successfully into the planned future national states.

In this article, I explore hostile attitudes towards Muslims, particularly Slav Muslims, and their role in Serbian and Montenegrin discourses. I argue that much of the hostility towards Muslims at the level of popular culture has been distilled and then used in the repertoire of extreme nationalists since the early nineteenth century to create an artificially extreme distrust of the Muslims who continued to live in the region, particularly in Bosnia, Serbia, Montenegro and Kosovo. Ideas that reside on the fringes of societies during times of peace, which are generally of interest only to social misfits and degenerates, can take on tremendous significance during times of crisis, sometimes preparing the ground for war and acts of ethnic fury. Obviously as Noel Malcolm remarks, "between low level prejudices on the one hand and military conflict and mass murder on the other, there lies a very long road."[6] Extreme nationalist ideas are especially pervasive because they are based on very crude notions of difference, which seem to lodge themselves in the subconscious and then emerge during crises. The attacks on the United States in September 2001 by Islamic extremists also gave a further boost to anti-Islamic discourses in the Balkans, with newspaper editorials in Serbia reminding their readers about links between Osama bin Laden and the Bosnian Muslim militants.[7]

Early Serbian nationalism and the production of a national idea depended very largely on the production of a small number of individuals in the late eighteenth and nineteenth centuries such the Vuk Karadžić and the ruler of Montenegro, the *Vladika* (Bishop) Petar II Petrović Njegoš. Many of the early Serb intellectuals received their training and developed their ideas under the aegis of the Habsburg Monarchy, so their ideas can be very directly linked to the growth of Romantic nationalism in Central Europe with its emphasis on the national spirit and the authenticity of the common people. However, although the Romantic notion of the *srpski narod* (the Serb nation) should have included all South Slavs speaking the language which was referred to as Serbo-Croat from 1850s until 1990s by many

---

[6] Noel Malcolm, *Kosovo: A Short History* (London, 1998), xxviii.

[7] See, for example Božidar Dikić, "Bin Laden na Baščaršiji" [Bin Laden in the Sarajevo bazaar], an article published by the Belgrade daily *Politika* on 4 October 2001, which claimed high levels of support for Islamic fundamentalists in Bosnia; http://www.politika.co.yu/2001/1004/01_12.htm on 4 June 2002.

commentators and linguists,[8] nationalist ideas which developed in the nineteenth century tended to exclude Muslims from the nation, because by adopting Islam they were perceived to have become de facto Ottomans (and were sometimes referred to indiscriminately as "Turks").[9] Like many of his contemporaries, Karadžić had historical, linguistic and racial views as to what constituted a nation.[10] He stated that there were five million people who spoke the same language (the Serbian dialect favored by early Serbian nationalists and called Stokavian), but they were divided by religious confession. He added, "only the three million Orthodox consider themselves as Serbs...[Muslims] think that they are true Turks and call themselves that, despite the fact less than one in a hundred of them knows Turkish."[11] His view of Serbdom combined notions of primordial ethnicity with Herderian linguistic consciousness. To his mind all five million were Serbs whether they knew it or not and what had happened in the medieval Empire of Dušan was as important as the intervening centuries, although he specifically did want to take out Turkish loan words from the Serbian language.[12] He was, however, one of the first writers to use the word "cleanse" (*očistiti*), with all its Christian overtones of the redemptive powers of

---

[8] On the development of Serbo-Croat, see, Cathie Carmichael, "A People exists and that People has its language": Language and Nationalism in the Balkans," in *Language and Nationalism in Europe*, ed. Stephen Barbour and Cathie Carmichael (Oxford, 2000), 221-239.

[9] South Slav Muslims themselves were slow to adopt European nationalism, although there were some Islamic fascists in Bosnia during the Second World War; on this see Francine Friedman, *The Bosnian Muslims: Denial of a Nation* (Boulder, Col., 1996), 122-25. Many Slav Muslims migrated to Turkey after periods of significant political change such as the 1870s or 1910s. There was also never a significant Muslim South Slav diaspora agitating for a separate "Bošnjak" state, and even in 1992 the many Muslims in Bosnia saw their primary alliance with other Bosnians (Serbs and Croats) who accepted the legitimacy of the Izetbegović government. On Muslim identity, see Florian Bieber, "Muslim National Identity in the Balkans before the Establishment of Nation States," *Nationalities Papers* 28 (2000): 13-28.

[10] Aleksandar Pavković, "The Serb National Idea: a Revival 1986-92," *Slavonic and East European Review* 72 (1994), 444.

[11] Vuk Karadžić, "Serbi sve i svuda," in *Etničko Čišćenje: Povijesni dokumenti o jednoj srpskoj ideologiji* [Ethnic Cleansing: Historical documents concerning a certain Serbian Ideology], ed. Mirko Grmek, Marc Gjidara and Neven Simać (Zagreb, 1993), 29.

[12] Asim Peco, *Turčizmi u Vukovim Rječničima* [Turkisms in Vuk's Dictionaries], (Belgrade, 1987).

baptism, to describe the killing of Muslims in Belgrade in 1806.[13] The insurrection led by Karadjordje was accompanied by deliberately targeted acts of atrocities against Muslims in Serbia who were to be driven out in the wake of independence. The Serbs then began to (successfully) destroy all the architectural heritage of the Ottomans and now only the Barjakli Džamija mosque from this earlier period remains.[14]

Karadžić, who spent many years as a protégée of the Slovene linguist and Imperial librarian Jernej Kopitar in Vienna in the early nineteenth century, was one of the scholars responsible for codifying and thus elevating the scattered ballads of the guslars (players of a stringed instrument, the *gusle*) into a national literary canon.[15] One of the themes of these epic poems was the struggle against Ottoman domination and many had ancient themes and motives, often from the Middle Ages. Interest in South Slav epic poetry, which was exceptionally well-preserved at the turn of the nineteenth century, was found all over Europe at the time.[16] Walter Scott translated the ballad of *Hasanaginica* into English, and Jacob Grimm read and favorably reviewed Karadžić's work.[17] The appeal of epic poetry was not just cultural. To some extent Serb anti-Ottoman sentiment and activity suited the geopolitical interests of both Habsburgs and the Romanovs at that time. As a cultural artifact it has left a long-lasting mark on Serb and Montenegrin national consciousness.

---

[13] Tim Judah, *The Serbs: History, Myth and Destruction of Yugoslavia* (New Haven and London, 1997), 75.

[14] Andrei Simić, "Nationalism as Folk Ideology: The Case of the Former Yugoslavia," in *Neighbors at War: Anthropological Perspectives on Yugoslav Ethnicity, Culture and History*, ed. Joel M. Halpern and David A. Kideckel (State College, Penn., 2000), 112.

[15] Malcolm, 79-80.

[16] Ivo Žanić, in his study *Prevarena povijest. Guslarska estrada, kult Hajduka i rat u Hrvatskoj i Bosni i Hercegovini 1990-1995: Godine* [The Falsification of History: The Elevation of the Guslar, the Cult of the Hajduk and War in Croatia, Bosnia, and Hercegovina] (Zagreb, 1998), has argued that the symbol of the gusle has remained vital in nationalist imagery and was revived in the 1990s. He reproduces a photograph of Radovan Karadžić' (p. 390) proudly holding a gusle at the birthplace of his forebear Vuk in 1992.

[17] Celia Hawkesworth, "The Study of South Slav Oral Poetry: a Select Annotated Bibliography of Works in English (1800-1980)," in *The Uses of Tradition: A Comparative Enquiry into the Nature, Uses and Functions of Oral Poetry in the Balkans, the Baltic and Africa*, ed. Michael Branch and Celia Hawkesworth (London, 1994), 37-8 (School of Slavonic and East European Studies Occasional Paper, no. 6).

Another important text to encourage the idea of a historical betrayal by Slavs who had converted to Islam was the poem *Gorski Vijenac* (The Mountain Wreath) published in 1847 by Petar II Petrović Njegoš, who was the Prince-Bishop of Montenegro from 1830 until his early death in 1851. The main theme of the poem is the supposed dilemma faced by his predecessor *Vladika* Danilo (1696–1737) about what to do with Montenegrins who had become Muslim. The poem contains many references to smiting Slav Muslims, including the threat of Vojvoda Batrić that: "we will burn down Turkish homes so that no trace of the dwellings of our home-grown faithless devils could be known."[18] Vladeta Popović described Njegoš's poem as revealing the "essence and substance of a race that has had to go through many tribulations and fight against many difficulties,"[19] thereby giving it the status of a genuine historical account rather than the poetic vision of an educated man who had read Ossian as well as Ivanhoe and the Greek classics.[20] *Gorski Vijenac* was read by subsequent generations of Montenegrins and other South Slavs, achieving canonical status very rapidly. It has been called a "true breviary of interethnic hatred."[21] It is probably the chief textual link between the discourse about Islam and everyday life for the people themselves. For Milica Bakić Hayden, Njegos's "depictions of the converts as traitors whose weakness and opportunism deprived them of the religious and cultural identity bequeathed to them by their forefathers in Kosovo are reflected in popular--if tacit--perception of Muslims among Serbs and Montenegrins."[22] Although it is not entirely fair to isolate a single text from the context in which it was written (in this case, the context of a Montenegro, which was almost an island of non-Ottoman government in the Balkans), it is also fair to say that Montenegro's identity, particularly its quintessentially non-Islamic character was manipulated by its intellectuals and other nineteenth century writers of pan-Serbian sympathies.

In invoking the influence of *Gorski Vijenac* or any folkloric text,

---

[18] Petar Petrović Njegoš, *Gorski Vijenac* [The Mountain Wreath] (Sarajevo, 1990), 155 (l. 2604-2606).

[19] Vladeta Popović, "Introduction," in *The Mountain Wreath of P. P. Nyegosh, Prince Bishop of Montenegro, 1830-1851*, trans. James William Wyles (London, 1930), 11.

[20] Ibid., 17.

[21] Grmek et al., eds., *Etničko Čišćenje*, 25.

[22] Bakić Hayden, 927-8.

it would be as well to be careful to avoid essentialization about the nature of Serb or Montenegrin culture. Folklore and the analysis of other texts from popular culture are very often used in the literature on the Balkans to make rather sweeping assertions about national character. Ivo Rendić-Miocević characterizes the Dinaric Serbs as suffering from "projection," "narcissism," and "paranoia," which he dubs the "Prince Marko syndrome"[23] with its obvious idea of the continuity of folk traditions from the epic songs of the guslari to the present day. Branimir Anzulović is also undeterred about making pronouncements about the link between popular culture and violence. "In Balkan highland culture, violence is often taken for granted, without any sense of guilt or sorrow for the victims.... A high level of violence and the condoning of the most vicious cruelty as just punishment can be observed in the Serbian folk song 'Grujo's Wife's Treachery.'"[24] If we were to try to create other personality types on this basis, we could state that the Americans had a morbid fascination with drowning because they sing "Clementine" or that the English had a fixation with decapitation because they sing "Oranges and Lemons." Recent critics of the political use of folklore in the 1990s have pointed to ways in which folklore has been misused by the Croatian and Serbian governments.[25] A realization of the part that governments, nationalist ideologues, and the media have played should deter scholars from making unguarded comments about popular culture.

The many foreign writers who have visited the Balkans since the early nineteenth century also helped to circulate and replicate the singular myth of Ottoman oppression versus Serb or Montenegrin heroism and were often moved by the idea of the perpetual struggle against the Turks, although many expressed their horror at brutal spectacles such as the dozens of severed and desiccated Turks' heads surrounding the *Vladika* residence in Cetinje.[26] In his poem "Montenegro," Tennyson sought to

---

[23] Ivo Rendić-Miocević, *Zlo velike jetre: povijest i nepovijest Crnogoraca, Hrvata, Muslimana i Srba* [The Evil of the Enlarged Spleen: History and Non-History of the Montenegrins, Croats, Muslims and Serbs], (Split, 1996), 126-7.

[24] Branimir Anzulović, *Heavenly Serbia: From Myth to Genocide* (London, 1999), 49.

[25] Dunja Rihtman-Auguštin, "Ugledna etnologinja i antropologinja govori o instrumentalizaciji folklora i teroru mitologijom" [A prominent anthropologist speaks out about the instrumentalization of folklore and the terror of mythology], *Feral Tribune*, 23 studenoga, 1998, 22-23.

[26] Xavier Marmier, *Lettres sur L'Adriatique et le Montenegro*, vol. 2 (Paris, 1853), 120.

encapsulate many of the contemporary European images of a perpetual and valorous struggle between Orthodox and Muslim communities:

> They kept their faith, their freedom, on the height
> Chaste, frugal, savage, arm'ed by day and night
> Against the Turk....
> O smallest among peoples! rough rock–throne
> Of Freedom! warriors beating back the swarm
> Of Turkish Islam for five hundred years....[27]

For many of these foreign writers, the Ottomans, their culture, and the lands that they ruled or had ruled for centuries were constructed within a literary trope, which has been described by Edward Said as "Orientalism," which was set up as opposite and inferior to the supposed values of Europe or the Occident. As part of a general rejection of the past, South Slav writers began to use the same tropes to describe the Ottomans as other European writers (although it is also fair to state that for many foreign writers Southeastern Europe still had many "oriental" characteristics or more particularly Balkan features, which set them apart from Europe[28].

The emphasis so often found in this nationalism on collective suffering under the Turks has often been seen by commentators as paranoid in psychoanalytical terms. As two Slovene writers commented in 1989: "Everything, from the assault by the Turks on Europe and the bombardment of Belgrade to Communist takeover of the government in the heart of Serbia all amount to a single conspiracy forged by the papists, Sultan Murat, Franz Josef and the Albanians against the [Serb] nation."[29] Of particular importance is the role played by the death of the medieval King Lazar at the battle of Kosovo Polje in 1389, which became an act of faith for nationalists. Michael Sells has argued that during the nineteenth century, "Serbian nationalist writers transformed Lazar into an explicit Christ figure, surrounded by a group of disciples ... and betrayed by a Judas.... In this story the Ottoman Turks play the role of the Christ killers. In the nationalist myth, (the betrayer)

---

[27] Alfred Lord Tennyson, *Poetical Works* (London, 1926), 533-4.

[28] This concept of Balkanism is explored at some length by Maria Todorova, *Imagining the Balkans* (New York, Oxford, 1997).

[29] Ervin Hladnik-Milharčič and Ivo Standeker, "Tako v nebesih kot na zemlji" [As it is in heaven, so on earth], *Mladina*, 7 June, 1989, 8.

Vuk Branković, represents the Slavs who converted to Islam under the Ottomans and any Serb who would live with them or tolerate them."[30] Some contemporary nationalists have presented Kosovo as a symbol for an almost Manichaean battle with the forces of contemporary Islam. On 28 June 1989, the poet Matija Bećković announced that Kosovo Polje should be seen as "a Jerusalem in which the whole of Europe has its churches." Earlier that year he had stated: "Six hundred years after the battle of Kosovo it is necessary for us to declare: Kosovo is Serbian and that fact depends neither on Albanian natality or Serbian mortality. There is so much blood and holy relics there that it will be Serbian even when not one Serb remains there."[31] More recently the Macedonian government (and particularly hardline minister Ljube Boškovski) has attempted to create a link between the Albanian militant NLA and Islamic fundamentalists, while at the same time presenting itself as a kind of Christian rampart against terrorism.[32]

It would be unfair to state that all "memory" of Turkish persecution was paranoid in its character. As Vera Mutafchieva reminds us the "hegemonic" practices of the Ottomans in the Balkans were accompanied by many individual and collective cruelties, such as forced conversion to Islam, the Janissary system and discrimination in taxation[33]. In Serbia, the revolt against the Ottomans had a specific cause linked to the misrule of the Dahis and other Balkan populations had similar grievances at this time. Had Ottoman rule been entirely acceptable to the Balkan populations, it is unlikely that traditions of banditry would have ever developed along the extensive border areas or that uprisings would have occurred in the way that they did.[34] However, it is also possible to deconstruct many of the main tenets of anti-Turkish mythologies: Božidar Jezernik has examined the case of the representations of eyes gauged out of holy murals or even Ćele

---

[30] Sells, 31.

[31] Both quotations from Bećković are from Robert Thomas, *Serbia under Milošević: Politics in the 1990s* (London, 1999), 49.

[32] Paul Anderson, 'Macedonia's Shaky Peace', at http://news.bbc.co.uk/hi/english/world/europe/newsid_1869000/1869822.stm, 12th March 2002.

[33] Vera Mutafchieva, "The Notion of the 'Other' in Bulgaria: The Turks. A Historical Study," *Anthropological Journal on European Cultures* 4, no. 2 (1995): 53-74.

[34] The role of the border and banditry in South Slavonic history is discussed by Xavier Bougarel, in "La 'revanche des campagnes', entre réalité sociologique et mythe nationaliste," *Balkanologie*, 2, no. 1 (1998): 17-36.

kula and argued that many of the claims of nationalists were fraudulent.[35] One could argue that there was a great deal of synthesis and peaceful cohabitation between Serb and Ottoman culture over many centuries. Use of aromatics and spices in cuisine, rituals surrounding the drinking of coffee and the recreational use of tobacco as well as the melodies of folk tunes and the wearing of amulets and talismans have all been linked to the Turkish legacy. Jovan Cvijić even argued that Dinaric fatalism was linked to Turkish notions of *ksmet* (destiny).[36] However, since anti-Islam is an ideology, any deconstruction of myth has the mere status of a historical opinion. Perhaps it also misses the main point, namely that historical myths represent one truth about the past, abandoning attempts to tell the whole truth, in the way that Picasso during his blue period abandoned attempts to convey the whole truth about color. Ivan Čolović employs a phrase taken from the work of the ethnographer Veselin Cajkanović, *klicanje predaka* (the cheer of ancestral voices) to illustrate how the dead are summoned up to serve the political purposes of the living.[37] Ottoman oppression was certainly one fact about the Serb and Montenegrin experience of the past, but not the only one and it was open to conscious or unconscious manipulation by nationalists. As a metaphor for oppression, it was often invoked and remained active within the repertoire of nationalist writers. When the Republic of Serbia's power was deemed to be threatened by the granting of autonomous region status to Kosovo and Vojvodina after 1974, Dobrica Čosić stated: "The Republic of Serbia has been reduced to a pashalik of Belgrade, given up (inféodéc) to the begs of Pristina and the archdukes (*voïvodes*) of Novi Sad,"[38] invoking very emotive and highly inaccurate political terms from the past.

---

[35] Božidar Jezernik, *Dežela, kjer je vse narobe: Prispevki k etnologiji Balkana* [A Land where Everything is Topsy Turvy: Contributions to the Anthropology of the Balkans] (Ljubljana, 1998), 79-114, has argued that it was the Slav Christians themselves who scratched out the eyes of saints to make medicinal poultices. He has also cast doubt upon the validity of the tower of skulls (Ćele kula) built in Niš after 1809 to celebrate a Turkish victory over the Serbs, that found its way into the repertoire of Serbian nationalist mythology (see Ibid., 165-66).

[36] Cvijić, 351.

[37] Ivan Čolović, "Vreme i prostor u savremenoj politickoj mitologiji" [Time and place in contemporary political mythology], in *Kulture u tranziciji* [Culture in Transition], ed. Mirjana Prosić-Dvornić (Belgrade, 1994), 124.

[38] Dobritsa Tchossitch, *L'Effondrement de la Yougoslavie: Positions d'un Résistant* (Lausanne, 1994), 41, cited in Florian Bieber, *Serbischer Nationalismus vom Tod Titos*

Another "truth" about relations between Christians and Muslims is that they often had much closer and certainly more complex relations than nationalist myths might lead us to believe. The names of many well known Serbs and Montenegrins are etymologically partly Turkish (Karadžić, Asanović, etc). We cannot rule out other forms of interpersonal relations. As Mark Mazower remarks: in 1815, the rebel Miloš Obrenović "first conducted his Muslim blood-brother, Ashin Bey, to safety and then proclaimed the opening of a new 'war against the Turks.'" Messages were sent round the country that the inhabitants should kill anyone they encountered wearing green clothes--the sign of a Muslim."[39] Zorka Milich questioned a number of centegenarian women in Montenegro and uncovered a wide range of beliefs about and prejudices against Muslims. Some of them exhibited mistrust and dislike of "Turks," others only a mild awareness of difference. Jovana, aged 102, stated that "The Turks were evil. When they saw a good Serb, they did everything in their power to kill him." She remarked that the custom allowing a widow to marry her husbands brother was "disgusting" and the use of cosmetics by the women made them "stink." She also felt that the wearing of a veil was a "strange custom," prefacing this remark with "who knows why?"[40] Another informant, Ljubica aged 112, when asked her opinion of *poturice* (Christian converts to Islam), replied: "Most of the Turks round here are our people.... They should be ashamed of themselves. Their religion is not better than ours."[41] Misha Glenny has argued that the very closeness between religious communities in Bosnia is an important factor in explaining pattern of ethnic violence: "the Bosnian Serbs, Croats and Muslims have been adorned with many different cultural uniforms over the centuries, by which they identify one another as the enemy when the conflict breaks out. Despite this, underneath the dress they can see themselves reflected.... The only way that fighters can deal with this realisation is to exterminate the opposite community. How else does one explain the tradition of facial mutilation in this region?"[42] John B. Allcock has argued,

---

zum Sturz Milosevics, (Unpublished Doctoral Dissertation, University of Vienna, March 2001), 206.

[39] Mark Mazower, *The Balkans* (London, 2000), 80.

[40] Zorka Milich, *A Stranger's Supper: An Oral History of Centegenarian Women in Montenegro* (New York, London, 1995), 100-2.

[41] Milich, 40.

furthermore, that "traditional codes of morality require that individuals be ready to kill their neighbours, with whom they might be 'in blood.'"[43]

Visitors to Bosnia before the war were often struck by the ease with which the different religious groups socialized with each other. Stereotypes and jokes about *Bosanci* by other former Yugoslavs usually emphasized their laid-back attitude and the fact that they were happy to celebrate all the religious festivals in their republic, thus gaining far more days off work. All groups socialized together, but Andrei Simić has argued there was an "invisible psychological wall" between neighbors and a "superficial cordiality, more often than not masked a deep sense of alienation, suspicion, and fear."[44]

During times of peace interethnic tensions were difficult to discern. In Kosovo in 1999, a massacre of ethnic Albanians in the town of Suva Reka was apparently instigated by a local man, Zoran Petković, a "Serb who was friendly with the Albanians ...even as relations deteriorated between the two ethnic groups."[45] A local policeman Islam Yashlari, when asked of his opinion of Petković after the massacre, replied: "I don't know what happened to him. He was just a guy who didn't like to work too much, then when the war started he changed. He wanted to be somebody."[46] The survivors were able to recount their story because they were sheltered by a local Serb family. It is also probable that during the wars in Bosnia and Kosovo that young Serb men were forced on pain of death to kill their neighbors in a cynical act of spreading guilt and responsibility for ethnic crimes by nationalist extremists, which has really nothing whatsoever to do with either traditional morality or consciousness of ethnicity.[47]

---

[42] Misha Glenny, *The Fall of Yugoslavia: The Third Balkan War*, (Harmondsworth, 1992), 169.

[43] John B. Allcock, *Explaining Yugoslavia* (London, 2000), 390.

[44] Simić, 115.

[45] Maggie O'Kane, "One family's story of the terror inside Kosovo: And of the friendly bus driver who turned into a mass murderer," *Guardian Weekly*, June 27, 1999, 1.

[46] Ibid., 12.

[47] The relationship between "local rivalries" and the "terrorizing tactics of outside extremists" is discussed briefly by Susan Woodward in her, *Balkan Tragedy: Chaos and Dissolution after the Cold War*, (Washington, 1995), 242-3.

Serbian and/or Montegrin popular culture is therefore not *per se* anti-Islamic although elements of mistrust between religious communities may have primordial characteristics. *Gorski Vijenac* is of its era and belongs more to modern discourses about nationalism rather than to popular culture, despite it quasi-epic format. Nevertheless, by 1878 and the Congress of Berlin, which recognized the sovereignty of Serbia and Montenegro, a putative hatred of "Turks," which by inference could also include Slav Muslims, was seen by nationalist writers as a defining Serb trait. The lack of chronological coherence to this myth can be seen by the comment attributed to a Serb bishop Duchitch (Dučić) by the American John Reed in 1915: "In Serbia, we do not trust too much to God. We prayed to God for five centuries to free us from the Turks and finally we took guns and did it ourselves."[48] Another aspect of the rejection of the "Turkish yoke" and Islam is the invocation of the idea of a "Turkish taint" (the shame of having cohabited with an "Oriental" culture for many centuries and its legacy in popular culture and mentalities). Marko Živković illustrates this idea with the example of a televised session of the Serbian parliament in 1994, when a member of the opposition group DEPOS played a tape-recording of Iranian pop music alongside turbofolk.[49] When he had proved that the melodies were very similar, he quoted Vladimir Dedijer: "We Serbs sometimes behave as if we were made [i.e. begotten] by drunken Turks."[50] Živković describes this as "deep self-recrimination ... couched in the idiom of self-Orientalization."[51] It is as if the Turkishness that has been so forcefully repudiated can never really go away and continues to define the mentalities and culture of the Serbs, despite themselves. An example of the cynical use of self-Orientalization can be found in the conversation between Stojan Protić and Ante Trumbić of the Yugoslav Committee in 1917, reported by Ivan Mestrović in his memoirs. Protić is quoted as saying that "When our army crosses the Drina, it will give the Turks twenty four or even forty eight hours to return to the faith of their ancestors. Those who are unwilling will be struck down [*posjeci*] as we have done on other occasions in Serbia."

---

[48] John Reed, *War in Eastern Europe: Travels through the Balkans in 1915* (London, 1999), 26.

[49] Turbofolk is a kitsch genre of popular music in Serbia which utilizes traditional "oriental" tunes and traditional gender imagery.

[50] Marko Živković, "Too much character, too little *Kultur*: Serbian Jeremiads 1994-95," *Balkanologie* 2 (1998), 77.

[51] Ibid, 79.

Trumbić was silent, but Mestrović could see that his hands were trembling. Then he asked Protić if he was serious, he replied: "Very serious, Mr Trumbić. In Bosnia with the Turks one cannot use European methods, but must use ours [*po naski*]."[52]

During the twentieth century anti-Islamism continued to be an essential *leitmotiv* within Serbian and Montenegrin nationalist discourse. To some it remained the unsolved problem of Yugoslavian politics. In 1933, the President of the Council of Ministers, Milan Srskić stated: "I cannot stand to see minarets in Bosnia; they must disappear."[53] The notorious lecture given by Vasa Čubrilović in 1937 to the Srpski Kulturni Klub (Serbian Cultural Club) about the ethnic cleansing of Albanians from Kosovo[54] contains a distillation of his anti-Islamic prejudices and self-Orientalization. Drawing on the popular and highly influential ideas of the geographer Jovan Cvijić about the power-seeking and violent "Dinaric" personality,[55] Čubrilović thought that Montenegrins could be used to drive Albanians out of Kosovo[56] since they exhibited many of the necessary violent traits to do the job.[57] But there was another reason why the mountain people might be of some use in this case: according to Cvijić, "the Dinaric has an ardent desire to avenge Kosovo... and to resuscitate the Serbian Empire...even in circumstances where the

---

[52] Grmek et al., eds., 82-3. Given the behavior of the Croatian military in Bosnia in 1993 in particular, the comment by Marcus Tanner *Croatia: A Nation Forged in War*, (New Haven, 1997), 116, that this dialogue is "an ominous reminder of the difference between Croat and Serb political culture" is erroneous.

[53] Quoted in Norman Cigar, *Genocide in Bosnia: The Policy of Ethnic Cleansing* (College Station, Texas A&M University Press, 1995), 18.

[54] Vasa Čubrilović, "Iseljavanje Arnauta" [The Expulsion of Albanians], in *Izvori velikosrpske agresije* [The Origins of Serbian Aggression], ed. Bože Čović (Zagreb, 1991), 106-24.

[55] This concept of a distinct personality type for the population of the mountainous Dinaric region is discussed by Marko Živković in his article "Violent Highlanders and Peaceful Lowlanders: Uses and Abuses of Ethno-Geography in the Balkans from Versailles to Dayton," *Replika* (1997): 107-120. The Influence of Cvijić and ethno-psychology is also discussed by Bojan Baskar in "Made in Trieste: Geopolitical Fears of an Istrianist Discourse on the Mediterranean," *Narodna Umjetnost* 36 (1999): 121-134.

[56] He was also instrumental in drafting a Yugoslavian-Turkish agreement in 1938 for the relocation of Muslims/Turks, which was never enforced because of the outbreak of war.

[57] Čubrilović, 112-3. On this see also, Bojan Baskar, "Anthropologists facing the collapse of Yugoslavia," *Diogenes* 47 (1999), 60.

less courageous or a man of pure reason would have despaired. Betrayed by circumstances and events, abandoned by all, he has never renounced his national and social ideal."[58]

Kosovo and Bosnia were the sites of particularly vicious interethnic fighting from the period from 1941 until 1945. Elsewhere in Yugoslavia during the Second World War, Muslims were targeted on the grounds of their faith and ethnicity.[59] It is estimated that between 86,000 and 103,000 Slav Muslims were killed in Bosnia and Sandžak. Many perished at the hands of Serb nationalist Četnici. One extreme Četnik ideologue Stevan Moljević, who advocated "cleansing the land of all non-Serb elements," believed that the government-in-exile in London "should resolve the issue [of emigration] with Turkey,"[60] although with the Communist ascent to power and the defamation of the Četnik movement, the plan remained a theory, seemingly relegated to the past. However, Moljević's ideas didn't entirely disappear in the wake of relatively good interethnic relations after the Second World War. They were resurrected in 1991 by Vojislav Seselj, who had been jailed for his nationalist views in the mid-1980s, when he told the German newspaper *Der Spiegel*, that Muslims were Islamicized Serbs, whom he would drive out of Bosnia to Anatolia, if they opposed any attempt to take away their status as a nation.[61]

In the postwar era, the Communist regime in Yugoslavia attempted a fine balancing act between the nationalities within its borders with various levels of success. In the 1980s, after the death of Tito, Yugoslavia's Muslims were again remarginalized by certain Serb nationalists such as Miroljub Jevtić, who attempted to link the rise of Islamic "fundamentalism" in the wider world but particularly Iran to the Muslims in his own country.[62] According to Norman Cigar, Serbian scholars specializing in Oriental Studies

---

[58] Čvijić, 282.

[59] On the ethnic cleansing of Muslims during the Second World War, see Vladimir Dedijer, *Genocid na muslimana 1941-45: Zbornik documenta ili svjedocenja* [Genocide against the Muslims from 1941-45: A Collection of Documents and Testimonies] (Sarajevo, 1990).

[60] Philip J. Cohen, *Serbia"s Secret War: Propaganda and the Deceit of History* (College Station, 1996), 440.

[61] Seselj, cited in Grmek et al. eds., 203.

[62] Miroljub Jevtić, *Savremeni džihad kao rat* [The Current Jihad as War], (Belgrade, 1989), 42ff.

(including Jevtić), "contributed considerably to making hostility towards the Muslim community intellectually respectable among the broad strata of the Serbian population."[63] The Orthodox Church also played a significant role in this process of radicalization.[64] Elsewhere the denial of a Bosnian spirit of mutual respect between religious communities by Seselj,[65] who echoed with his characteristic lack of originality some of the negative sentiments about Islam and the cultural life of Bosnia's Muslims propagated by the writer Ivo Andrić,[66] was a flagrant manipulation of history and a genuine tradition of peaceful coexistence in that republic. It was also during this period that interethnic relations deteriorated considerably in Kosovo,[67] although the Albanians were not considered to be traitors to the Serbs in the way that Slavonic speaking Muslims were. The novelist Vuk Drasković, whose own political attitude towards the Muslims has been marked by inconsistency, also broke "the mould shaped by Tito"[68] with the publication of *Noz* (The Knife) in 1982,[69] which has notable anti-Islamic sentiments in its depiction of interethnic relations during the Second World War in Hercegovina.

The link between the propaganda of the 1980s and the fighting of the 1990s has been well documented. After the breakdown of Communist authority between 1987 and 1990, official media in Serbia, Kosovo, Montenegro and Croatia became filled with "hate-filled panic-mongering rhetoric,"[70] which clearly prepared the populations of certain regions and republics for interethnic strife. In Belgrade in 1989, a journalist informed Sabrina Ramet that, "they [i.e. Bosnian and Kosovo Muslims] have big families in order to swamp Serbia and Yugoslavia with Muslims and turn

---

[63] Cigar, *Genocide in Bosnia*, 27.

[64] Ibid, 30-2.

[65] Vojislav Šešelj, *Pravo na istinu* [Directly to the Truth], (Belgrade, 1988), 8.

[66] Bieber, "Serbischer Nationalismus," 21.

[67] On interethnic and gender relations, Wendy Bracewell, "Rape in Kosovo: Masculinity and Serbian Nationalism" in *Nations and Nationalism*, 6 (2000): 563-90.

[68] Judah, 79.

[69] Allcock, 398, comments that "Yugoslavs have spontaneously talked about people being 'put to the knife' as a synonym for ethnic extermination."

[70] Mark Thompson, A *Paper House: The Ending of Yugoslavia* (London, 1992), 130.

Yugoslavia into a Muslim republic. They want to see Khomeni in charge here."[71] Prior to the Bosnian elections of 1992, the SDS (Serb Democratic Party) told voters that "if Bosnia became independent they would once again be subjected to the laws of the Muslim landlords, agas, begs and pashas, and that independence represented a rollback of everything Serbs had died for since 1804, if not 1389... [Serbs] were told that for hundreds of years they had been Bosnia's single largest community and that in the last twenty-five years the Muslims had suddenly 'outbred' them."[72] In such rarified circumstances, it appears that certain numbers of people simply abandoned their experience and sense, becoming motivated by "instinct," however chimerical this might be in practice. In April 1992, a Serb soldier, Miloš, fighting in the siege of Sarajevo told journalist Ed Vulliamy: "Their [Muslim] women are bitches and whores. They breed like animals, more than ten per woman.... Down there they are fighting for a single land that will stretch from here to Tehran, where our women will wear shawls, where there is bigamy ...."[73]

The practice of genocide does not organically erupt from within a society. It is a planned affair, announced in advance, its practice intimately linked to a small number of individuals who see it as either a desirable or unavoidable part of their wider political concerns. Hate texts feed into a low level awareness of difference, which exists at the level of popular culture, but would never come to widespread violence if individuals with power did not provoke it. In a sense the relationship between elements within popular culture and extreme, virulent nationalism is rather similar to that between class hatred and Marxist revolutionary ideas. That is not to say, however, that these ideas that exist within popular culture are superficial or ephemeral. Four hundred Serbs asked to characterize Muslims in a questionnaire in 1997, described them as "primitive," "mendacious," "hostile to other nations," "dirty," "uncultivated," "squabbling," "stupid," "cowardly," and "lazy."[74] Adopting a more "scientific" tone, psychiatrist and Krajina Serb activist Jovan Rasković told *Intervju*

---

[71] Sabrina P. Ramet, "Islam," in *Balkan Babel: The Disintegration of Yugoslavia from the Death of Tito to Ethnic War*, ed. Sabrina P. Ramet (Boulder, Col., 1996), 185.

[72] Judah, 199.

[73] Ed Vulliamy, *Seasons in Hell: Understanding Bosnia's War* (London, 1994), 49.

[74] Dragan Popadić & Miklóš Biró, "Autostereotipi i Heterostereotipi Srba u Srbiji" [Autostereotypes and Heterostereotypes of Serbs in Serbia], *Nova Srpska Politicka Misao*, Nr. 1-2 (1999), 98-99, cited in Bieber, "Serbischer Nationalismus," 493.

magazine in September 1989 that "Muslims [are] fixated in the anal phase of their psychosocial development and [are] therefore characterized by general aggressiveness and an obsession with precision and cleanliness."[75] To state that there is a link between academic and popular culture can in no way diminish the responsibility of the individual. Ideologies are like ajar doors, which any individual more or less chooses to walk through.

Certainly anti-Islamism was not confined to Serbs and Montenegrins: it was also commonplace in the popular culture of the other republics of Yugoslavia and was not a problem that the Communists ever dealt with despite a commitment to brotherhood and unity. A poem which glorifies the achievements of the anti-Ottoman rebels in Serbia at the beginning of the nineteenth century attributed to Filip Visnjić has been "taught in Serbian schools ever since."[76] *Gorski Vijenac* was read as a school textbook in Communist Yugoslavia in every republic and available in Slovene and Macedonian translations as was *Smrt Smail-Age Cengijica* (The Death of Smail Age Cengijić) by the Croat poet Ivan Mazuranić, which explores the theme of Slavic Christian resistance to Turkish oppression. That is not to say that every person who read these texts was "contaminated" by their poetical content. Tim Judah discusses a conversation between Aleksa Djilas and his cousin: "How did the Muslims in your class react when they had to read *The Mountain Wreath* and learn parts of it by heart?" His cousin was dumbstruck: "It had never crossed his mind to ask his Muslim classmates such a question—even though some were his close friends. Clearly he did not connect them with the Muslims against whom Njegoš wrote."[77] Tolerance and multiculturalism were also a significant part of the culture of the former Yugoslavia, especially in cosmopolitan towns like Sarajevo and Belgrade. In Banja Luka, an American journalist recalls a conversation with a Serb, Spasoje Knezević, "In history, progress is only possible with the mixing of nationalities," Spasoje said as though speaking to a jury. "I'm not only mad and embarrassed about this stupid war, but disappointed. What is being done here is not in the favor of Serbs. We are losers too. Look at the number of Serbs who have died or been forced out of their homes. Look at

---

[75] Quoted in Ramet, 185.

[76] Misha Glenny, The *Balkans 1804-1999: Nationalism, War and the Great Powers* (London, 1999), 11.

[77] Judah, 78.

the destruction. We used to go to Italy for vacation. Now we can't afford gasoline to drive our cars across town. After this war is finished, it will be a shame for someone to be a Serb."[78]

Mistrust of Muslims, both Slav Muslims and Kosovars, existed alongside tolerance at an unofficial level, but it needed to be awakened by nationalist intellectuals after 1974 to have a significant political impact. Moreover, the ineffective response of the Communist authorities to the national question, only belatedly allowing Muslims to define themselves "in the ethnic sense" and allowing them forms of ethnic individuation such as the wearing of traditional dress and the building of new mosques was too little too late to prevent the drift into rival extreme nationalisms. After the death of Tito, many lost their inhibitions about openly nationalist politics. During the mobilization of Serbian nationalism in the late 1980s by Slobodan Milošević, his supporters would turn up to rallies with placards with provocative anti-Islamic slogans: "Oh Muslims, you black crows, Tito is not around to protect you," or "I'll be the first, who will be the second to drink some Turkish blood?"[79]

In general it is not really possible to have ethnic cleansing without ideologies of ethnic cleansing. The anti-Serbism of Croat nationalists has been as destructive as Serb anti-Islamism, especially during the Second World War. As an ideology it also appears to have weak roots in popular culture.[80] It is often stated that many of the perpetrators of ethnic crimes in Yugoslavia were very ordinary people before the war: it was the crisis made them into murderers. "Sakib Ahmić, a Muslim villager [from Bosnia] testified that he had watched the [Catholic] Kupreskić brothers 'grow up into decent people' until the fighting began in their village. They broke into his home and murdered his son Naser and daughter-in-law Zehrudina, as well as their children...."[81] We are all familiar with instances of the torture and abuse of people on the grounds of religious difference, particularly Muslims

---

[78] Peter Maass, *Love thy Neighbor: A Story of War.* (New York,1996), 106.

[79] Norman Cigar, "The Serbo-Croatian War," in *Genocide after Emotion: The Postemotional Balkan War*, ed. Stjepan Meštrović (London, 1996), 57.

[80] Cigar, "The Serbo–Croatian War," 67, characterizes Serb nationalism in 1990-91 as being led by a "small nucleus [of] hard-core [activists]." Similarly he notes, p. 59, that the neo-fascist HSP in Croatia gained only 5 percent of public support in the 1990 elections.

[81] Richard Norton-Taylor, "Croats jailed for Ethnic Slaughter," *Guardian Weekly*, January 20-26, 2000, 4.

during the recent wars in Bosnia and Kosovo. We also know that these abuses follow clear cultural patterns, which might allow us to begin to construct links between ideological manipulation and human behavior in extremis, a kind of "method in madness."[82] In the way that the Ustaša were reputed to ask their potential victims to make the sign of the cross to establish whether they were Orthodox or Catholic, violence against Muslims during the war in Bosnia often emphasized very obvious differences such as the traditional Islamic aversion to eating pork[83]. Peter Maass recorded the way in which Christian motives were perverted during the war in Bosnia: "a teenage girl explained to me how one of the Muslim men in her village had been nailed to the front door of the Mosque, so that he was like Christ on the cross, and he was still alive at the time."[84] Many other eyewitnesses recorded similar occurrences, which were committed by individuals whose main link with each other was shared hate texts--newspapers, literature as well as radio and television broadcasting. Muslims in Bosnia were raped[85] or made to urinate in the mosque and had crosses carved into their flesh.[86] It is well recorded that Serbian and Croatian nationalists targeted the material (i.e. symbolic) culture of the Muslims in Bosnia[87]. These examples suggest that clear patterns of anti-Muslim behavior were played out during the Yugoslavian Wars of Dissolution between 1991 and 1999. They also suggest that a mental dehumanization of Muslims took place

---

[82] The Hamlet metaphor is used by Vulliamy, 85.

[83] "We stumbled upon a gathering of some eighty opprobrious Chetniks camped down at Prijedor police station in September 1992 .... They had little enlightening to say, except to register that in the town of Kluj, 'The Turks' cannot eat pigs, but they run like pigs and squeal like pigs." Vulliamy, 54.

[84] Maass, 7.

[85] Violations of women have been interpreted as particularly humiliating to the traditional culture of Muslims. Similar patterns of behavior were also recorded in Bulgaria in 1878. In the village of Oklanli, Turkish women were raped over several days and then burnt alive. The atrocities were carried out by their Christian neighbors in communities that had lived in the same villages for centuries. See McCarthy, 72.

[86] Cornelia Sorabji, "A Very Modern War: Terror and Territory in Bosnia-Hercegovina," in *War: A Cruel Necessity? The Bases of Institutionalized Violence*, ed. Robert A. Hinde and Helen E. Watson (London, 1995), 83.

[87] On the history and demise of the old Turkish bridge at Mostar, see Božidar Jezernik, "Qudret Kemeri: a bridge between barbarity and civilisation," *Slavonic and East European Review* 73 (1995): 470-484.

(quite apart from the actual physical destruction). Robert Hayden discusses the case of a Serb soldier forcing a fez on the head of a "distraught" Muslim prisoner in Banja Luka in 1994. He explains the episode thus: "the visible mark of Islamic culture ensured that 'Muslim' was more than simply a label of difference, but rather indicated a culture not only apart, but in the Orientalist rhetorical structure dominant in Europe, including the Balkans, also inferior to that of Europe."[88] Primarily, this process of constructing Muslims as inferior and different was acquired culturally and as such has a history and genealogy. The nationalities of the former Yugoslavia were not destined to play out "ancient hatreds," nor was the multinational character of the state a primary cause of its collapse.[89] The story of Yugoslavia could have had many different endings.

---

[88] Robert Hayden, "Muslims as 'Others' in Serbian and Croatian Politics," in *Neighbors at War*, ed. Halpern and Kideckel, 123.

[89] Ethnic complexity is often singled out as an explanatory cause of Yugoslavia"s collapse. John D. Treadway states that "four years of warfare, accompanied by ethnic cleansing and flights of population, have 'simplified' the arrangement and composition of its population. The ethnic and religious crazy quilt that was the 'old' Yugoslavia is not quite as intricate or as confusing as in years past. But the current regional admixtures of peoples and confessions constitute a witch's brew – the stuff of which powder kegs were (and are) made." "Of Shatter Belts and Powder Kegs: A Brief Survey of Yugoslav History," in *Crises in the Balkans: Views from the Participants*, ed. Constantine P. Danopoulos and Kostas G. Messas (Boulder, Col., 1997), 40-1.

Bulgaria's "Turks": A Muslim Minority in a Christian Nation–State, 1878–1989

DENNIS P. HUPCHICK

While the term "ethnic cleansing" was coined only in the last decade of the twentieth century, the genocidal-like activity that it denotes—the ideologically motivated territorial eradication of one particular group of people by another[1]—has occurred repeatedly throughout history. Although instances abound in the historical record of foreign conquerors razing entire cities and massacring their inhabitants to intimidate enemies, the most common occurrences of such genocide were driven by religious beliefs. This was particularly true after the rise of monotheism, which engendered among its adherents a strong sense of group superiority—the "Chosen People"—relative to non-believers. The Jews' original acquisition of the "Promised Land" and the Crusaders' actions in the "Holy Land" were but two of the multitudinous examples of religious-based genocidal atrocities. Ethnicity played little or no role in historically documented cases of such activity until the emergence of nationalist concepts in post-Reformation Western Europe and their consolidation during the nineteenth century as the primary fountainhead of group and state identity.

Western Europe's position of global imperial dominance resulted in the export of its own nationalist-based sense of group identity and statehood to every corner of the world. As a result, most attempts to study territorial genocide do so from the current political, economic, social, and ethnic perspectives shaped by Western Europe. They also tend to downplay or ignore the religious or civilizational components of this phenomenon. The fact that the secular, scientific realities of Western Europe play a preponderant role in world affairs today (and in recent centuries), however, does not mean that the emotional, irrational cultural forces institutionalized by religious belief operating among groups of people no longer are important. The cultural motivations involved in territorial genocide are just as important as any strictly political, economic, or social factors—and perhaps more so.[2]

---

[1] For an in-depth study of the "ethnic cleansing" process on a global stage, see Andrew Bell-Fialkoff, *Ethnic Cleansing* (New York, 1996), which provides a useful working definition of the term on pp. 1-4.

[2] See Dennis P. Hupchick, *Culture and History in Eastern Europe* (New York, 1994), 5-6, for a more complete definition of "culture" used in this context.

Until the eighteenth-century spread among Western Europeans of secularism grounded in the Scientific Revolution, language and religion were the fundamental components of human group self-identity. With the exception of those who comprised the elite leadership elements in societies (such as rulers, aristocracies, and prelates), group self-identity linked to territory (*i.e.*, the state) did not extend much beyond the confines of local villages or regions. Language provided a more concrete basis for group identity but, so long as the majority of any given society essentially remained illiterate and existed in provincially limited mundane worlds, one hardly can speak of their possessing a strong ethnic self-awareness, in the modern sense, grounded in language alone. Rich oral folk traditions of tales, festive songs, and agrarian instructions passed down through the generations did preserve languages and often were expressions of local community pride, yet the linguistic identity that such folklore imparted remained somewhat vague and rudimentary. Other factors being relatively equal, if one person could understand another person's speech, then the other person was recognized as a fellow in some indefinite group sense; if not, the other person was a foreigner. While such a basic group awareness was necessary for eventually building ethnic consciousness, it played a minor role in the daily lives of most people before they were exposed to modern Western European ethnonational concepts.[3]

By far, religious affiliation provided the most concrete basis for conscious group self-identity prior to the appearance of modern Western European ethnonational perceptions. This was particularly the case for followers of monotheistic faiths, whose European adherents included Western Roman Catholic and Protestant Christians, Eastern Orthodox Christians, Jews, and Muslims. For monotheists in the pre-Western European-dominated world, correct religious belief was one of life's paramount considerations, since it alone guaranteed the ultimate reward of eternal paradise after death. Institutionalized churches provided the framework for correct belief, in which membership bestowed fundamental group identity. Unlike linguistic distinctions, which often were blurred by shadings in spoken dialects, religious differences were basic and absolute. If one person recognized that another held the same religious belief (no matter the languages involved), then that other person was a

---

[3] Studies of "ethnicity" abound. A useful edited collection of selections from the more important literature is John Hutchinson and Anthony D. Smith, eds., *Ethnicity* (Oxford, 1996).

fellow "true believer". If this was not the case, the other definitely was a foreigner and often quite possibly a threat to "true" belief. So definitive has been the religious sense of group self-identity that most civilized societies can be labeled culturally by their primary form of religious expression.[4]

As a consequence of their unique historical development, by the end of the eighteenth century the Western Europeans had evolved a secularist primary cultural expression grounded in enlightened principles and science. Out of this emerged a new sense of group identity—nationalism—merging long-standing political traditions with Liberal and Romantic precepts, and forged into practice by revolutions in North America and France.[5] Western European nationalism, an abstract supposition that a unique group of people (a self-identified "nation") possessed the right to sovereign, independent political existence within a territorially defined state and under a government considered innately its own and distinct from all others, expressed itself in two forms of group consciousness. One constituted identities originating in the traditional, increasingly secularized monarchical states that had been modified by the rationalistic principles of science and the Enlightenment. It was state- and civic-oriented, with the emphasis placed on the militant support of the state's collective citizenry and the sanctity of the state's territorial sovereignty. The other form sprang from the late eighteenth- and nineteenth-century Romantic Movement, which rejected the predominance in Western European realities of rationalism rooted in Renaissance, scientific, and Enlightenment traditions. Instead, it emphasized identities based on irrational and emotional factors, such as ethnicity and religion, essentially grounded in "Dark Age" medieval traditions. In the Romantic (often termed "ethnic") form

---

[4] See Hupchick, *Culture and History*, esp. 7-8, for the contention that religion usually is the *definitive* cultural expression of civilized society, as language is of the ethnic group. Support for this argument is provided throughout the four essays comprising the body of that work. Despite the fact that modern Western European civilization now consciously embraces a scientific philosophy as its primary mode of cultural expression, until the maturation and general acceptance of Enlightenment secular ideals at the end of the eighteenth century by Western Europeans, the term "Europe" itself was used only as a geographical label and not in the cultural sense as is common today. The term used to identify human "Europe" until that time was "Christendom," which was recognized as being divided between "Western" (Roman Catholic and Protestant) and "Eastern" (Orthodox) varieties (hence, two—Western and Eastern—"European" civilizations). See Norman Davies, *Europe: A History* (Oxford, 1996), 7-10, for a general overview of the "Europe"/"Christendom" relationship.

[5] See Hupchick, *Culture and History*, 122f, for an overview of that evolution.

of national consciousness, human group identity fundamentally was a function of the group's unique culture and history within a specific territory rather than of generic scientific, political, social, or economic development.

The former kind of group national consciousness, conceived of in terms of sovereign state citizenship ("civic" nationalism), emerged among similar or related populations living within existing regional monarchical states (for instance, France, Spain, and Britain). The latter form ("Romantic" or "ethnic" nationalism) enjoyed its greatest appeal among those inhabiting either territorially large, multiregional, and multiethnic imperial states (such as the Habsburg and Russian empires) or distinct regions with similar or related populations divided among numerous small, subregional states (the Italian and German states, for example). By the middle of the nineteenth century the two commonly operated in tandem, with "ethnic" identities being used to reinforce "civic" nationalist precepts and "Romantic"/"ethnic" nationalists claiming that groups possessing unique ethnicities and histories required unique "homeland" states of their own, whose purpose was to represent and defend that uniqueness politically, socially, economically, culturally, and territorially. The end result was the formation of the "nation-state" concept in Western Europe during the second half of the nineteenth century, which ultimately found justification in the early twentieth-century Wilsonian principle of "the national self-determination of peoples."[6]

Nation-state nationalism was a cultural concept developed out of unique evolutionary historical circumstances within Western European civilization alone. Most nineteenth- and twentieth-century Western Europeans sincerely believed that it truly represented, in Voltaire's words, the "best of all possible worlds." If so, it also was inherently divisive, potentially violent, and a generally unattainable ideal. Nation-state nationalism's intrinsic territorial imperative frequently spawned competitive animosities among neighboring states over possession of land with its

---

[6] Just as for "ethnicity," the literature treating the concept of "nationalism" and the "nation-state" is extensive and growing increasingly larger since the collapse of European communism. John Hutchinson and Anthony D. Smith, eds., *Nationalism* (Oxford, 1994) provides a collection of selections from the most pertinent studies of the subject. Geoff Eley and Ronald G. Suny, eds., *Becoming National: A Reader* (New York, 1996) also includes useful selected edited studies. Also see George W. White, *Nationalism and Territory: Constructing Group Identity in Southeastern Europe* (Lanham, 2000), particularly the first two chapters. Of interest regarding "ethnic" nationalism is the introductory overview in Loring M. Danforth, *The Macedonian Conflict: Ethnic Nationalism in a Transnational World* (Princeton, 1995), 13f.

inhabitants and material resources. Just as monotheism readily lent justification for "true" belief becoming "superior" belief, nation-state nationalism often easily slipped from defending a nation's "uniqueness" to exerting its "superiority," especially after Darwinian theories of "natural selection" and the "survival of the fittest" crept into Western European political thought during the second half of the nineteenth century. Such ultranationalist thinking fed a series of national wars in Europe during that period and led to two world wars and the cold war during the twentieth century. Moreover, human demography itself precluded any nation-state from including a single, unique group of people alone, rendering the ideal of "one people ('nation'), one state" little more than a fiction in most cases. At no time and in no place could nation-state borders neatly and definitively separate any two neighboring ethnically self-defined nations. Numbers of each inevitably have been left living on the "wrong side" of virtually every nation-state border, creating the internal problem of ethnic/national minorities in all supposedly unitary nation-states, and frequently resulting in those people's discriminatory reduction to inferior, second-class status within the states.[7]

Despite the divisiveness, violence, and discrimination engendered by nation-state nationalism, Western Europeans have persisted in perceiving it as the inevitable and best (if flawed) vehicle for expressing group self-identity. Other components of their own cultural makeup, such as liberal-democratic political principles, humanistic ideals, and the needs and rewards of industrial capitalism, tended to dampen for them the frequency (but not the intensity) of its negative human consequences. While this arguably may have been the case for Western European societies, the export of nationalism (particularly in its Romantic "ethnic" form) and the nation-state-ideal to non-Western European societies with different cultural traditions and historical developments was generally harmful for the recipients. The injection of Western-style ethnonationalism and the nation-state into the traditionally autocratic and theocratic Ottoman Balkans during the nineteenth century was a telling case in point.

If ever a region of Europe epitomized religious belief as the most fundamental expression of human group identity, it was the Ottoman Balkans. Since the Ottoman conquest of the Balkan Peninsula took place between the mid-fourteenth and early sixteenth centuries, the region—today composed of the modern

---

[7] See Hupchick, *Culture and History*, 131-32.

states of Albania, Bosnia-Hercegovina, Bulgaria, Greece, Macedonia, and Little-Yugoslavia (Serbia and Montenegro), and parts of Croatia, Romania, and Turkey—lay under direct theocratic Islamic control for a period of 250 to 500 years.[8] In traditional Islamic society, no separation existed between religious and secular matters (they intertwined and often were synonymous), and religious considerations predominated in all matters of state, from administrative organization to social classification. The Ottoman Empire's Muslim rulers adhered to the fundamental Islamic worldview that divided humankind into "believers" and "nonbelievers." In states controlled and governed by "believers" (Muslims), all law was religious, grounded in sacred Islamic precepts found in the Koran and the traditions *(sunnas)* and sayings *(hadiths)* of the Prophet Muhammad—on which is based the Holy Law or *Şeriat*.[9]

Similar to most Islamic imperial states before it, the Ottoman Empire was created through conquests of extensive territories largely inhabited by non-Muslim populations. Until the expansionary campaigns of Ottoman Sultan Selim I the Grim (1513-20) in Islamic West Asia and North Africa, the empire's Balkan possessions constituted about half of all Ottoman territories, whose non-Muslim subjects equaled or outnumbered the Muslims in the empire as a whole. Even after Selim's and his immediate successors' sixteenth-century conquests in the Islamic Middle East tipped the empire's demographic scales definitively in favor of Muslims, Balkan non-Muslims continued to represent over a fifth of the total subject population.[10]

---

[8] The two Romanian Principalities (Wallachia and Moldavia) were exceptions. Although technically under Ottoman suzerainty for over three centuries as autonomous vassal states, they experienced little direct Ottoman control over their domestic affairs, thus escaping the Ottomans' Islamic theocracy.

[9] Concerning the *şeriat, hadith* and *sunnah*, see E. I. J. Rosenthal, *Political Thought in Medieval Islam* (Cambridge, 1962), 21-22; and Kenneth Gragg, *The Call of the Minaret* (New York, 1964), 98. Useful introductions to the Ottomans' fundamental Islamic worldview can be found in: Halil Inalcik, *The Ottoman Empire: The Classical Age, 1300-1600*, trans. by Norman Itzkowitz and Colin Imber (London, 1973), chaps. 9-10; Norman Itzkowitz, *Ottoman Empire and Islamic Tradition*, 2nd ed. (Chicago, 1980); Justin McCarthy, *The Ottoman Turks: An Introductory History to 1923* (New York, 1996); Stanford J. Shaw and Ezel K. Shaw, *History of the Ottoman Empire and Modern Turkey*, vol. 1, *Empire of the Gazis, 1280-1808* (Cambridge, 1976); and Wayne S. Vucinich, *The Ottoman Empire: Its Record and Legacy*, reprint ed. (Huntington, 1979).

[10] See Dennis P. Hupchick, *The Bulgarians in the Seventeenth Century: Slavic Orthodox Society and Culture Under Ottoman Rule* (Jefferson, 1993), 13-14, for a general demographic overview.

The significant number of non-Muslim Ottoman Balkan subjects posed legal difficulties for the Islamic governing authorities.[11] In an Islamic state based on the *Şeriat*, that law was applicable only to Muslims and possessed no validity among non-Muslim subjects. The very process of conquest in the Christian Balkans innately carried with it increasing administrative problems for the Ottomans, as each new territorial acquisition brought with it increased numbers of subjects who lay outside of the law. Initially, the Ottomans dealt with the problem on an *ad hoc* basis by incorporating laws already in force in acquired territories into secular legal decrees issued by the sultans and by enlisting local Christian or Jewish religious leaders to settle mundane legal issues among their respective coreligionists. By the reign of Sultan Mehmed II the Conqueror (1451-81), however, the number of non-Muslim subjects had swelled to such an extent that he found it necessary to create some sort of organized structure for administratively integrating them into the theocratic Islamic Ottoman state. In 1454, one year after taking Constantinople, Mehmed instituted the *millet* system of administration for the empire's non-Muslims.

The *millet* system further institutionalized the strictly religious nature of group identity within the theocratic Ottoman Empire by dividing its subject population into *millets* ("nations") based solely on religious affiliation, with each subject to its own ecclesiastical and traditional laws and administered by its highest religious authorities. All non-Muslim subjects initially were distributed among three *millets,* representing the three most important non-Islamic faiths within the empire: The Orthodox Christians, headed by the patriarch of Constantinople and representing the largest and most important non-Muslim group; the Jews, who were of great commercial and cultural significance for the Ottomans, headed by an elected representative of the rabbinical council in

---

[11] The situation also possessed fiscal advantages. Islamic precepts included toleration of so-called "People of the Book" (*e.g.*, Christians and Jews, among a few others): Monotheists who possessed collections of written scriptures containing divine revelations. They were extended official "protection" *(zimma)* by the state authorities on condition that they acknowledged the domination of Islam and of its temporal representative, the sultan. That religious toleration was not extended unconditionally. The *zimmis* ("protected ones"), as the Ottomans' Christian and Jewish subjects were classified by this arrangement, were compelled by the Hanifid school of Islamic religious law to pay discriminatory taxes (the most important of which was the *cizye* poll tax), to which Muslim subjects were not liable, as well as all other general taxes. Given the significant number of Christians and Jews in the empire, the additional discriminatory taxes paid by them constituted one of the most lucrative sources of revenue for the Islamic state's treasury.

Istanbul; and the Armenian (Gregorian Monophysite) Christians, headed by an Armenian patriarch of Istanbul appointed by the sultan, who also represented the empire's Roman Catholic subjects. The Muslims, by virtue of their representation through the Islamic state, constituted a *de facto* fourth *millet*.[12]

Each non-Muslim *millet* was responsible for representing its membership before the Ottoman court and for its own internal administration. They all were granted the rights to tax, judge, and order the lives of their members insofar as those rights did not conflict with Islamic sacred law and the sensibilities of the Muslim ruling establishment. The religious hierarchies of the *millets* thus were endowed by the Ottoman central authorities with civil responsibilities beyond their ecclesiastical duties, and their head prelate was held accountable for the proper functioning of their internal affairs. In effect, each *millet,* personified by its religious administrators, became an integral part of the empire's domestic administration, functioning as a veritable department of the Ottoman central government. In return for ensuring the smooth administration of its non-Muslim subjects, the government granted each *millet* a considerable amount of autonomy in the spheres of religious devotion and cultural activity, judicial affairs not involving Muslims, and in local self-government.

Although the term *millet* involved the idea of "nation" in the Turkish language, it shared little in common with the Western European concept of nationality. In the first place, it identified people solely on the basis of their religion; ethnicity played no role. The Ottomans made no distinctions among Greek, Bulgarian, Serbian, Macedonian, or Romanian Orthodox believers; they all were lumped together in a single *millet,* even though it is certain that the Muslim authorities were aware that ethnic differences did exist among them. Given the theocratic nature of the Islamic state, ethnicity was considered relatively unimportant in the fundamental scheme of things by the Muslim Ottoman rulers. It made little difference to them that the head of the Orthodox *millet* was, and would remain, a Greek.

In the second place, by eliminating all consideration of ethnicity, *millet* identification lacked the territorial connotations

---

[12] Although originally only three in number, by the mid-nineteenth century the number of *millets* had proliferated to include such Christian denominations as Nestorians, Syrian Orthodox, Maronites, Catholics, and assorted Uniate (such as Greek and Armenian Catholics) and Protestant groups. Often, the Ottomans extended *millet* status to protect Christian converts from one denomination to another from persecution by their former coreligionists. See McCarthy, *The Ottoman Turks,* 130-31.

subjects in the late eighteenth century, they first were digested within a *millet* context. Centuries of *millet* existence within an imperial state ruled by "foreigners" determined that the Romantic "ethnic" form of Western nationalism, with its emphasis on religion, language, and historic "homeland," rather than the "civic" variety, won acceptance. Over the course of the first half of the nineteenth century, the Balkan Christian proponents of "ethnic" nationalism were quick to create the histories and vernacular-based literary languages, linked to religious affiliations, necessary to draw the required distinctions for unique ethnic group identities. Not surprisingly, given the theocratic nature of Islam and their historically dominant position in the Islamic state, most Balkan Muslims rejected Western-style nationalist contentions and held firmly to their religious identities, despite the fact that large numbers of them (perhaps as many as one half) were converts, or descendants of converts, to Islam and "ethnically" related to their non-Muslim neighbors.[15]

Aided by the anarchy, inflation, corruption, and crushing taxation in an Ottoman society destabilized by constant, direct contact with an increasingly dynamic and technologically superior Western Europe, the Ottomans' non-Muslim subjects came to embrace the "ethnic" nationalist and nation-state ideals of the Westerners and, between 1804 and 1913, succeeded in carving out ethnically-identified states of their own from the empire's Balkan possessions.[16] That process, however, demonstrated the

---

[15] Balkan Christian conversions to Islam are a contentious issue. For attempts at objective studies (*i.e.*, generally free of nationalist biases), see: Speros Vryonis, "Religious Changes and Patterns in the Balkans," in H. Birnbaum and S. Vryonis, eds., *Aspects of the Balkans* (The Hague, 1972), 151-76; B. G. Spiridonakis, *Essays on the Historical Geography of the Greek World in the Balkans during the Turkokratia* (Thessaloniki, 1977); Wayne S. Vucinich, "Islam in the Balkans," in A. J. Arberry, ed., *Religion in the Middle East: Three Religions in Concord and Conflict*, vol. 2 (Cambridge, 1969), 236-52; and Peter F. Sugar, *Southeastern Europe under Ottoman Rule, 1354-1804* (Seattle, 1977), 49f.

[16] Although the Ottoman Empire traditionally has been described as having slipped into "decline" from the early seventeenth century until its demise in 1922, Justin McCarthy makes a good case for it having been progressively "destabilized" during that period by outside Western European pressures, arguing that "decline" implies the preponderance of internal factors in causing the condition. He finds little evidence that, without the external Western forces, Ottoman society would have experienced the fate that it suffered (though what the alternative would have been is anybody's guess). See his *The Ottoman Turks*, chaps. 5-6. Charles and Barbara Jelavich provide a good English survey of the Balkan nation-building process in their *The Establishment of the Balkan National States, 1804-1920* (Seattle, 1977), as does Stavrianos, *Balkans Since 1453*. Also useful is Stevan K. Pavlowitch, *A History of the Balkans, 1804-1945* (London, 1999).

devastating human impact that Western European-style ethnic-based nation-state nationalism possessed among societies whose own historical development differed from that of Western Europe. The Balkan recipients of ethnonational political culture lacked the West's liberal-democratic, Renaissance-based humanist, and industrial-capitalist traditions that served as limited restraints on nationalism's inherent divisiveness, violence, and discrimination.

Beginning with the 1804 Serbian Revolution, intensifying with the 1821 Greek Revolution, and continuing through the 1875-76 Serbo-Turkish War and the 1877-78 Russo-Turkish War, and culminating in the 1912-13 Balkan Wars, the Balkan Christians' militant efforts to implement Western-style ethnonationalism were accompanied by massive atrocities perpetrated on the Muslim populations in the Balkans, collectively representing a welter of "ethnic cleansing" the magnitude of which far surpassed that which occurred in late-twentieth-century Bosnia and Kosovo. By the time that the Ottomans were expelled from the Balkans (except for a portion of southeastern Thrace) at the close of the Second Balkan War (1913), an estimated 1.7 million Balkan Muslims had been killed and close to 1 million permanently forced into Ottoman Thrace and Anatolia as refugees.[17]

Millet and ethnonational identities virtually were indistinguishable to the majority of Balkan Christians who executed those genocidal atrocities. They retained the traditional association of Muslims with (Ottoman) "Turks," whom the small number of nationalist agitators among them often succeeded in painting as the religious, more than the strictly national, enemy *par excellence.*[18] In whipping up among the Christian villagers' fear and hatred of the Muslims/"Turks" as the religious/national enemy, often linking those emotions to the villagers' general greed for more land, the leaders of the Balkan nationalist movements

---

[17] These figures are derived from Table 30, p. 339, in Justin McCarthy, *Death and Exile: The Ethnic Cleansing of Ottoman Muslims, 1821-1922* (Princeton, 1995). See chaps. 1, 3, and 5 of that work for the grisly details of the genocide.

[18] Until after independent or autonomous national states were founded, nationalists (in the Western European sense) were restricted to the relatively small middle class elements in Balkan Christian societies. Their majorities remained rural, generally illiterate, and provincial in mentality until state-run education systems were founded that could indoctrinate them with ethnonationalist precepts. In some cases, that indoctrination did not take hold until the late nineteenth and early twentieth centuries. For the role of education in nationalist indoctrination, see James F. Clarke, "Education and National Consciousness in the Balkans," in Dennis P. Hupchick, ed., *The Pen and the Sword: Studies in Bulgarian History by James F. Clarke* (New York, 1988), 24-57.

created the emotionally charged mass followings that permitted them to advance their ethnonational agendas leading to the creation and consolidation of ethnic-based nation-states. The Balkan Muslims suffered accordingly.

Once the nationalists succeeded in winning their nation-state goals, all of the new states found themselves strapped with minority populations representing either groups ethnically related to nationals in neighboring states or Muslims who had survived the holocaust of the Christians' national liberation struggles. Lacking strong Western-style liberal-democratic humanist traditions of governance, the leaderships in the Balkan states, representing the ethnonational majorities, commonly instituted domestic policies aimed at either eradicating their minority populations altogether through ethnonational assimilation or reducing their numbers to insignificance so that the "one people ('nation'), one state" ideal was attained and neighboring states' claims to portions of territories that they possessed were foiled.

The ways in which the dominant majorities in the Balkan states dealt with their minorities varied in detail from state to state but, in general, minorities were looked down on as second-class citizens, who only could enjoy full citizenship benefits by renouncing their native ethnonational cultures and accepting those of the ruling majorities. Within the context of ethnonationalism, this entailed either forsaking their languages, religions, manners of dress, and customs for those of the majorities or face being expelled from the state entirely. The model for the Balkan majorities' assimilatory/discriminatory actions against their minorities included: closing minority schools; making the majority's language mandatory in education, the civil service, and the courts; persecuting the minority's religion (if different from the majority's) or placing it under the jurisdiction of the majority's ecclesiastical primate (if the same); obstructing the expression of the minority's literary and artistic culture; relegating the minority to the lowest rungs of the state's social and economic ladders; and frequently exerting force on the minority to emigrate.[19] Modern Balkan national history presents a sad chronicle of such policies.

The Balkan minority most universally affected by the nation-state ethnonationalism of the Christian Balkan states was the Muslim/"Turk." The case of the Muslims/"Turks" of Bulgaria sheds light on the human problems engendered by the implementation of Western European-style ethnonationalism in Balkan society.

---

[19] An informative overview of minority treatment is Edward Cháaszár, *The International Problem of National Minorities* (Indiana, Penn., 1988).

Although Bulgaria declared official independence from Ottoman suzerainty in 1908, the country effectively enjoyed independent freedom of action after its acquisition of autonomous status at the Congress of Berlin in 1878. This success was the result of a botched national uprising in 1876, which was repressed so violently by Ottoman irregular forces that European outrage over stories of Christians being butchered permitted Russia to undertake a war against the Ottoman Empire expressly for creating a Bulgarian national state. Religious antagonisms ran at fever pitch throughout those events. The Bulgarian rebels achieved little beyond staging patriotic rallies in the few villages that they held and massacring their Muslim neighbors. The so-called "Bulgarian Horrors" inflicted on the rebels in putting down their uprising were perpetrated by irregular forces *(başıbozuks)* mostly recruited from Muslim villagers in the rebellious regions, many of whom lost relatives and friends at the Christian rebels' hands or had heard their own horror stories of such events. During the Russo-Turkish War of 1877-78, Russian troops conducted a policy of anti-Muslim genocide in Bulgarian-inhabited Balkan territories and Bulgarians, both military volunteers and civilians, frequently acted as their accomplices. By the time that the fighting stopped in March 1878, some 260,000 of those Muslims who had inhabited the lands that ultimately came to comprise modern Bulgaria (*i.e.*, Bulgaria Proper [lands north of the Balkan Mountains but including the Sofia region to their south] and Eastern Rumelia [lands south of the Balkan Mountains, including the Plovdiv and Sliven regions]) were killed or had died of maltreatment. Another 500,000 were driven out, becoming refugees in Ottoman Thracian and Anatolian territories. In all, some 17 percent of the fifteen million Muslims who lived in the "Bulgarian" lands prior to the rebellion and war were dead and 34 percent were expelled permanently.[20]

---

[20] For the Christian Bulgarian and Russian acts of anti-Muslim genocide during the events described, see McCarthy, *Death and Exile,* chap. 3. Close to 1 million Muslims initially fled their homes as refugees but approximately half of them later returned to claim their abandoned properties and businesses. The mortality and permanent refugee figures of "Bulgarian" Muslims are given on pp. 88-91. Bilâl Şimşir, *The Turks of Bulgaria* (London, 1988), 4-5, places the 1876 population of Bulgaria proper at 1.13 million Christian Bulgarians and 1.12 million Muslims (mostly Turkish speakers), and that of Eastern Rumelia at 483,000 Christian Bulgarians and 681,000 Muslims (both Turkish and Bulgarian speakers), giving a pre-rebellion and pre-war total of 1,613,000 Christian Bulgarians and 1,801,000 Muslims (mostly Turkish speakers).

The Berlin Treaty creating an autonomous Bulgarian Principality included provisions protecting the life and property of the new state's Muslim citizens, as did the state's highly liberal-democratic constitution written under Russian auspices. Despite those legal constraints, however, local Bulgarian authorities and Bulgarian villagers, animated by inflamed religious ethnonational emotions (combined with land greed), continued to attack and uproot Muslim villagers with the tacit approval of the state government. Such actions reached their culmination in the heightened national feelings surrounding the union of Eastern Rumelia with Bulgaria Proper in 1885, after which concerns over international recognition put a damper on overt anti-Muslim activities. Thereafter, the governing nationalists found ways within the law to subtly reduce the political and social positions of the Muslim minority and to pressure them into emigrating so that their lands and businesses could be expropriated for fellow Christian, Slavic-speaking Bulgarians and the state made more reflective of the ethnonational nation-state ideal.[21]

Land laws were passed placing vacant land into government hands (with monetary compensation paid to the absentee owners, thus providing an incentive for buying off Muslim owners), which then was sold cheaply to desirous peasants (who invariably were Christian Slavic Bulgarians). Courtroom decisions involving cases pitting Christian Bulgarians against Muslims nearly always were decided in favor of the former. Mountains of red tape helped render permanent the illegal expropriation of Muslim properties by Bulgarians during the turmoil of 1877 through 1885, while the Bulgarian-controlled government increasingly encroached on the operations of the Muslims' own institutions (especially education, local community councils, and religious courts) in the name of increasing centralized administrative efficiency. A reversion to more blatant anti-Muslim actions accompanied the Bulgarians' heightened nationalist fervor during the Balkan Wars of 1912-13, when military operations brought into the state further Thracian territory with a predominantly Turkish-speaking population. Some Muslims were forced to change their Islamic Turco-Arabic names for Christian Slavic ones. By the end of the Balkan Wars, the Muslims' presence in Bulgaria's total population had been reduced from the 26 percent that they held in 1878 to 14 percent

---

[21] The following studies provide a detailed overview of Bulgaria's Muslim minority policies: Richard J. Crampton, "The Turks in Bulgaria, 1878-1944," in Kemal H. Karpat, ed., *The Turks of Bulgaria: The History, Culture and Political Fate of a Minority* (Istanbul, 1990), 43-78; Ali Eminov, *Turkish and Other Muslim Minorities in Bulgaria* (New York, 1997); and Şimşir, *The Turks of Bulgaria.*

(with Turkish speakers comprising 10.8 percent of the total population) because of near constant emigration. Higher birth rates among Muslims relative to those among Christian Slavic Bulgarians kept the minority's share of the total population fairly constant at around 13 percent (and at approximately 10 percent for Turkish speakers) into the 1980s.[22]

Bulgaria's Muslim minority was comprised of various ethnic components. The largest was the Turkish speakers, who generally constituted between 75 and 89 percent of the Muslim minority during any given census year between 1887 and 1992.[23] They were concentrated in the eastern, northeastern, and southern regions of the state. Pomaks (or Bulgarian-speaking Muslims), who mainly inhabited the central and western Rhodope region in southern Bulgaria, represented the second largest component of the Muslim minority, ranging from 12.8 percent (in 1920, the first year in which Muslim ethnic identity was included in the census) to 14.7 percent in 1992.[24] The remaining Muslims were divided among Gypsies (found scattered throughout the state), Tatars (living mostly in Bulgarian Dobrudzha), and Albanians (found in a smattering of urban centers).

Until the 1920s, all of the Muslims generally persisted in living their mundane lives within a *millet* context. While abiding by Bulgaria's state laws, they also enjoyed a good deal of local community autonomy, administered by regional *muftis* (Islamic legal experts) through *Şeriat* courts and under the general authority of a chief *mufti* (all of whom ultimately were tied administratively to the Ottoman *Şeyülislam* in Istanbul). Community councils operated private Islamic schools teaching in the languages (Turkish, Bulgarian, or Romany) of their constituents and staffed by teachers trained in *medreses* (Islamic institutions of higher education). They also oversaw such traditional Islamic institutions as mosques *(camiis)*, foundations *(vakıfs)*, and charitable institutions *(imarets)*, while sponsoring religious publications, journals, and newspapers. After Mustafa Kemal Atatürk established the secularized national state of Turkey in 1923-24, the Turkish-speakers in Bulgaria progressively espoused a Turkish ethnonational identity and, in 1928, they adopted the new Latin-based Turkish alphabet created in Turkey

---

[22] The population figures are found in: Crampton, "The Turks in Bulgaria," 47; Eminov, *Turkish Minorities*, 71, 79, 81, tables; and Şimşir, *The Turks of Bulgaria*, 5-6.

[23] Eminov, *Turkish Minorities*, 72, table; Şimşir, *The Turks of Bulgaria*, 5, table.

[24] From figures quoted in Eminov, *Turkish Minorities*, 71, table and p. 100.

associated with the Western European concept of the nation. No matter where one lived within the empire, no matter how mixed the population, *millet* affiliation governed one's life. Neighbors could be Muslims, Orthodox Christians, Jews, or Armenian Christians; they all were members of their own self-contained administrative communities, complete unto themselves, with no claims on the others. For all Ottoman subjects, their homeland lay anywhere within the borders of the empire. This fact led to increasingly mixed ethnic populations throughout the Balkans.[13]

Until the close of the eighteenth century, when Western European ideas of ethnonationalism first began penetrating the Ottoman Balkans, *millet* affiliation—that is, religious belief—was the fundamental source of group identity among all of the empire's Balkan subjects. For some, such as the Macedonian Slavs and especially most Muslims, who lay scattered throughout the peninsula from Bosnia to Bulgaria, it continued to be so long after nationalism made its appearance. Any concrete sense of ethnic self-awareness developed in conjunction with *millet* identity. Members of all non-Muslim *millets* recognized the obvious separation between themselves and the empire's Muslim inhabitants, whom they generally labeled "Turks" because of the Muslims' association with the theocratic state's rulers. Moreover, since they were free to practice their faiths and to cultivate their religious cultures under the auspices of their respective *millets*, the empire's Balkan non-Muslim subjects naturally came to identify not only with their respective religions but also with the form of language used in their devotions. As early as the seventeenth century, the Slavic Orthodox Christians were aware that, although they were Orthodox, they did not worship in the same language as did the Greek hierarchy that controlled the *millet*, but group differences among themselves were not so clear cut.[14]

When Western European nationalist concepts began seeping into the Balkans among the empire's Christian non-Muslim

---

[13] For overviews of the *millet* system as a whole, see: Hupchick, *Culture and History*, 146-47; McCarthy, *The Ottoman Turks*, 127-32; L. S. Stavrianos, *The Balkans Since 1453* (New York, 1958), 89-90, 103-5, and Shaw and Shaw, 151-53, among others.

[14] The difference between the Greek and Slavic liturgical languages was readily recognizable, but the differences among the three forms of Church Slavic used by Balkan Orthodox Slavs—Bulgarian, Russian, and Serbian ("Resava")—were not. All three were present in some number among texts used by all Orthodox Slavs. Since all three generally were so far removed from the various vernaculars spoken by the faithful, however, the differences were of little consequence other than contributing to blurring the lines distinguishing the assorted Slavic ethnic groups. See Hupchick, *The Bulgarians in the Seventeenth Century*, 71, 112f, 196.

(becoming the first to do so outside of that state). Yet even then, the solidarity of the *millet* identity within the Muslim minority as a whole continued.

The Christian Bulgarian state authorities, though insidiously whittling away at the Muslims' autonomous religious (hence, ethnonational) institutions by imposing on them increasing varieties of state regulations that directly interfered in their operations, essentially recognized the Muslims' *millet* status. During the period of Agrarian Party rule in Bulgaria, under the generally internationalist (and antinationalist) leadership of Aleksandŭr Stamboliiski (1919-23), such recognition was formalized. The governmental legal system instituted by the Agrarians in 1919 overtly followed the traditional Ottoman *millet* approach regarding the Muslim minority, legitimizing their religious educational and administrative institutions, granting them cultural autonomy, and guaranteeing them freedom in their private family lives. No systematic anti-Muslim or anti-Turkish policies were pursued, but mounting regulations on Muslim religious cultural expression (*e.g.*, oversight of education and teacher certification by the central Ministry of Education, censorship of textbooks, the introduction of mandatory teaching in Bulgarian, and strict licensing of publications) and many Bulgarian authorities' persistence in expressing their anti-Muslim nationalist sentiments by overstepping their legal bounds kept continuous pressure on the Muslims to emigrate.[25]

The Bulgarians' consistent efforts to force Muslim emigration from the state were grounded in the Romantic "ethnic" nation-state nationalist principles espoused by Bulgaria's original revolutionary leadership, which were embraced by all succeeding Bulgarian administrations (with the Agrarian exception). Those principles demanded that the state represent an ethnonationally homogeneous population inhabiting a historically defined native "homeland." Such nationalist policy caused Bulgaria trouble with all of its neighboring states in territorial squabbles over Macedonia, Thrace, and Dobrudzha, and determined its ultimately unfortunate alignments with the losing sides in both

---

[25] For Stamboliiski and the Agrarians, see John D. Bell, *Peasants in Power: Alexander Stamboliiski and the Bulgarian National Union, 1899-1923* (Princeton, 1977). On the *millet* approach taken by the Bulgarian government, see Eminov, *Turkish Minorities*, 49-50. The intensity of the pressure exerted on the Muslims to emigrate ebbed and flowed, depending on the national priorities of successive governments, until the Communist takeover of Bulgaria in 1944. Between 1925 and 1940 an average of between 10,000 and 12,000 Muslims fled the state annually. See: Crampton, "The Turks in Bulgaria," 61-62; and Şimşir, *The Turks of Bulgaria*, chap. 3.

world wars. It also led to assimilation or liquidation efforts directed against all minority populations within the state (not only the Muslims but also Greeks, Romanians, Vlahs, and non-Muslim Gypsies). The official definition constructed to determine "Bulgarian" ethnicity declared that all descendants of medieval Slavicized Thracians, Slavs, and Turkic Proto-Bulgars who espoused Orthodox Christianity and once inhabited territories controlled by the First Medieval Bulgarian State were ethnic Bulgarians. All others were "foreigners" and not accepted as part of the Bulgarian ethnic nation.[26]

Although language and history played important roles as components of Bulgarian ethnonational identity, religion (*i.e.*, Orthodox Christianity) was a key ingredient. In fact, prior to the events of 1876-78 that brought a small group of nationalist revolutionaries to power in the newly created state, the Bulgarian national movement largely had operated within a *millet* context. The majority of Bulgarian nationalists originally sprang from a middle class that benefited from the Ottomans' granting them the commercial concessions within the empire formerly held by Greeks until the Greek Revolution in the 1820s. Little interested in harming their economic well-being by undertaking violent revolutionary actions against the Ottomans, they espoused an evolutionary process that they believed would lead to peaceful national independence through winning Ottoman recognition of a Bulgarian church organization free of Greek patriarchal control—essentially creating a separate Bulgarian Orthodox *millet*. The "Bulgarian Church Question," a movement that they initiated in 1860, expressed their nationalism primarily in religious terms, and it was only after they won the struggle against the Greek church in 1870 that the minority firebrand revolutionary nationalists emerged to force the religious success more rapidly into political victory.[27]

---

[26] Kemal H. Karpat, "Introduction: Bulgarian Way of Nation Building and the Turkish Minority," Karpat, *The Turks of Bulgaria*, 2-3; and Eminov, *Turkish Minorities*, chap. 1, provide overviews of the Bulgarian nationalist perspective. According to Karpat, "Introduction," 3, the Muslims generally suffered less than did the others until the post-1944 Communist era in Bulgaria both because of their numbers and because neighboring Turkey (both Ottoman and national) consistently served as a deterrent.

[27] The standard work on the Bulgarian national revival is Nikolai Genchev, *Bŭlgarsko vŭzrazhdane* [Bulgarian National Revival], 3rd ed. (Sofia: Fatherland Front, 1988). For the "Bulgarian Church Question," see: Zina Markova, *Bŭlgarskata ekzarhiya, 1870-1879* [The Bulgarian Exarchate, 1870-1879] (Sofia: Bulgarian Academy of Sciences, 1989); *idem, Bŭlgarskoto tsŭrkovno-natsionalno dvizhenie do Krimskata voina* [The Bulgarian Church-National Movement to the Crimean War]

Until a public education system grounded in an official national history stressing Bulgaria's medieval greatness and the so-called national calamity of the five-century "Turkish Yoke" indoctrinated the majority rural population in Bulgarian ethnonational principles (a process that did not approach completion until the mid twentieth century), the state's national leadership essentially played on fundamental Orthodox religious identity in carrying out their antiminority discriminatory policies. Ethnically Bulgarian Catholics, who inhabited northwestern regions of the state since the seventeenth century (and some of whose ancestors could be traced back to the twelfth), were heavily pressured to turn Orthodox (thus becoming full-fledged "Bulgarians") and persecuted if they resisted.[28] Bulgarian-speaking Pomaks, generally perceived as "Turks" by the Bulgarians because of their religion and despite their language, were treated in the same fashion as all other Muslims in the state.[29]

After the Communists came to power in Bulgaria in 1944, the *millet* approach traditionally taken by the government toward the Muslim minority briefly was continued. In efforts to win the support of the significant Muslim minority for their new regime by overtly recognizing their rights as a separate religious entity, the Bulgarian Communist leadership initially encouraged Muslim education, cultural activities (including publications, such as a bilingual newspaper, and a Turkish-language radio program), and a certain amount of political expression (representation in the National Assembly, participation in local government, and freedom to stage public meetings and marches). Although the Communist leadership prided itself on being "enlightened" in the political and social spheres, it soon proved unable to rise above

---

(Sofia, 1976); and Toncho Zhechev, *Bŭlgarskiyat velikden ili Strastite bŭlgarski* [The Bulgarian Easter or The Passionate Bulgarians] (Sofia, 1976).

[28] Karpat, *The Turks of Bulgaria*, p. 3. For the origins of the Bulgarian Catholics, see Hupchick, *The Bulgarians in the Seventeenth Century*, 73f.

[29] Most specialists consider the Pomaks descendents of Slav Bulgarians who converted to Islam during the period of Ottoman rule over the Bulgarian lands. The Pomaks themselves possess an ambiguous notion of their ethnic identity—maybe Slavs, maybe Turks, maybe Thracians—but remain unequivocal in their Muslim identity. In recent years some Pomaks have attempted to create an ethnic identity of their own that would separate them from both Bulgarians and Turks. See Eminov, *Turkish Minorities*, 108f, for details. Religion even played a role in discriminating against non-Bulgarian-speaking Orthodox minorities (*i.e.*, Greeks, Romanians, and Vlahs) since they obviously were not part of the "Bulgarian" Orthodox congregation, the identity of which possessed long *millet* traditions.

the old chauvinistic national minority policies that characterized the pre-World War II years.[30]

The Muslims' situation abruptly changed for the worse in 1946 when party leader Georgi Dimitrov announced a new policy aimed at eradicating all traces of the Ottoman legacy within the state in the name of Marxist-Leninist social reconstruction. Thereafter, the *millet*-like approach toward the Muslim minority rapidly gave way to an attempt to replace traditional group identities (both religious and ethnic) with class consciousness. In 1946 all private (*i.e.*, religious) schools were nationalized. A 1949 law on religion placed all churches, including the Islamic, under direct governmental regulation, thus effectively eliminating the Muslims' semiautonomous religious status, and, in that same year, an intensive state-wide collectivization program was initiated.[31]

The Muslims, whose religion was more than merely a faith but a way of life, resisted the policy of the Communist leadership. Already possessing long-standing cultural ties to Turkey, most overtly adopted a Turkish ethnic identity, whether they were Turks (*i.e.*, originally Turkish speakers) or not. The former early concessions granted them were reviled as mere token gifts hiding the government's underlying motive of facilitating their eventual assimilation into the dominant Slavic Bulgarian majority. When in 1950 the collectivization drive reached the predominantly Muslim-inhabited tobacco-growing regions, the locals correctly saw it as a further escalation of anti-Muslim actions on the part of the government, and as a threat to their Islamic way of life. The authorities, desirous of controlling the Muslims' productive lands, gave them the choice of either leaving the state or having their property confiscated outright. The situation rapidly deteriorated into the forced emigration to Turkey of Bulgarian Turks/Muslims by the government authorities in numbers so large (*ca.*, 155,000 between 1950-51) and in such poverty that their plight became an international scandal. The period of Muslim emigration forced by collectivization ended only when, because of its economic and political inability to absorb any further influx of emigrants, Turkey closed its border to refugees from Bulgaria.[32]

---

[30] See: Eminov, *Turkish Minorities*, 5.

[31] Ibid., 82f; Şimşir, *The Turks of Bulgaria*, 131f, 145f.

[32] See: Huey Louis Kostanick, *Turkish Resettlement of Bulgarian Turks, 1950-1953* (Berkeley: University of California Press, 1957), esp. chaps. 3 and 4; and Eminov, *Turkish Minorities, passim*.

Those Muslim Turks and Pomaks (who originally were not considered "Turks" by Turkey) who remained within Bulgaria—some 700,000 to 820,000—enjoyed a brief period during which the authorities attempted to placate them by permitting a rejuvenation of their cultural activities (in hopes that pro-government support would result). After 1956, however, when the Communist leadership embraced the tenets of "Communist nationalism"—theoretically nationalist in form but socialist in practice—the Muslims' position again deteriorated. During the 1960s the government initiated a campaign attacking Islamic religion, education, culture, and mores in a misguided effort to create a homogeneous socialist Bulgarian national identity within the state. Instead, the Turkish-speaking Muslims reacted by strengthening their identity (grounded in language and religion) and the Pomaks increasingly identified themselves as "Turks."[33]

During the 1970s the top Communist party leaders, including the head of state Todor Zhivkov, secretly decided to implement a policy of "unifying" the state's population into a single, homogeneous, socialist Bulgarian "nation" by assimilating or otherwise eradicating all non-Bulgarian minorities. They promulgated a new constitution in 1971 that declared Bulgaria a unitary nation-state comprising a culturally homogeneous "Bulgarian socialist" population of either "normal" or "non-normal" Bulgarian ethnic ancestry.[34] To support the new ethnic

---

[33] Muslims were pressured to cease attending mosque (which the Communist leadership viewed as "reactionary" and "unproductive"), to adopt modern, Westernized dress, and to renounce a number of their traditional religious customs. Instruction in the Bulgarian language alone, stressing the inculcation of Marxist-Leninist ideals and precepts, was made mandatory universally in the schools and all religious-oriented subjects were banned. See Eminov, *Turkish Minorities*, 84f. To counter the Pomaks' growing pro-Turkish assertions, a series of Bulgarian historical studies were published to emphasize and extol their "Bulgarian" ethnicity. See, for example, Hristo Hristov and Veselin Hadzhinikolov, eds., *Iz minaloto na bŭlgarite mohamedani v Rodopite* [On the Past of the Muslim Bulgarians in the Rhodopes] (Sofia: Bulgarian Academy of Sciences, 1958). Hristov was a leading light of the Academy's Historical Institute and a native of the Rhodope region.

[34] Eminov, *Turkish Minorities*, 7-8, 84-85. On pp. 92f Eminov discusses the reasons underlying the party leadership's decision to eradicate the Muslim minority, positing that the motives were: a) the population growth among the Muslims outstripped that of the Slavic Bulgarians and threatened the "Bulgarian" character of the state; b) a fear that Turkey's improving economy would attract increased numbers of Muslim emigrants from Bulgaria should Turkey demand a more lenient interstate emigration agreement, thus threatening Bulgaria with a potential loss of a million workers; c) to keep "Turk" numbers officially as low as possible in the upcoming 1985 census; and d) to suppress the spread of fundamentalist Turkish and Muslim propaganda among the Muslim minority, who were in close

policy, official history was rewritten to validate the claim of the population's ethnic unity. In a complete rejection of the previously accepted historical contention that the majority of Turkish-speakers in Bulgaria were descendents of Anatolian Turks settled by the Ottomans in the Bulgarian lands during the fourteenth through sixteenth centuries, the new official history declared that virtually all Muslims in Bulgaria (Turkish-speakers included) actually were descendents of Christian Slavic Bulgarians who had been converted forcibly to Islam during the Ottoman period. A series of historical studies expounding the new interpretation were published, collections of supporting sources (some authentic but many fabricated) were issued, and school textbooks were revised throughout the 1970s and 1980s.[35]

By 1973 the Pomaks and most Muslim Gypsies had succumbed to the government's pressures and publicly assumed

---

touch with developments in Turkey where Islamic fundamentalism and ultranationalism were on the rise. Some more cynical Bulgarian intellectuals caustically attribute the leadership's decisions to the effects of drunken stupor!

[35] The Bulgarian historiography of the 1970s and 1980s not only "wrote" ethnic Turks out of Bulgarian history but denied the Turkish ethnicity of most all of the Balkan Muslims in general. See ibid., 8-9. Avid party ideologist and historical hack Petür H. Petrov played a leading role in the process with his studies: *Sŭdbonosni vekove na bŭlgarskata narodnost: Kraya na XIV vek—1912 godina* [Fateful Centuries of the Bulgarian Nationality: The End of the 14th Century-1912] (Sofia: Science and Art, 1975), and *Obrazuvane na bŭlgarskata dŭrzhava* [Formation of the Bulgarian State] (Sofia: Science and Art, 1981); and his edited collections of sources: *Asimilatorska politika na turskite zavoevateli: Sbornik ot dokumenti za pomohamedanchvaniya i poturchvantya, XV-XIX v.* [Assimilation Policy of the Turkish Conquerors: Collection of Documents for Islamization and Turkification, 15th-19th Centuries] (Sofia: Science and Art, 1964), and *Po sledite na nasilieto: Dokumenti za pomohamedanchvaniya i poturchvaniya* [On the Traces of Violence: Documents for Islamization and Turkification] (Sofia: Science and Art, 1972 [2nd ed., in 2 vols., 1987-88]). The Rhodope Pomaks were singled out for specific treatment in illustrating the alleged coercive nature of conversions to Islam in all of the party-line "revisionist" studies. The validity of the sources used to support the forced Rhodope conversions argument was first tentatively and briefly questioned by a Western scholar in Dennis P. Hupchick, "Seventeenth-Century Bulgarian *Pomaks*: Forced or Voluntary Converts to Islam?," in Steven Béla and Ágnes H. Várdy, eds., *Society and Change Studies in Honor of Béla K. Király* (New York: East European Monographs, Columbia University Press, 1983), 305-14. More in-depth, critical Western and Turkish studies supporting and expanding on Hupchick's questioning of the sources later appeared in Karpat's study, *op cit.*, and in Michael Kiel, *Art and Society in Bulgaria in the Turkish Period* (Maastricht: Van Gorcum, 1985), esp. p. 5 and 5n. Following the fall of the Zhivkov government, some Bulgarian historians felt themselves free to question the Rhodope forced conversions' sources: See Antonina Zhelyazkova, "The Problem of the Authenticity of Some Domestic Sources on the Islamization of the Rhodopes, Deeply Rooted in Bulgarian Historiography," in *Etudes balkaniques*, no. 4 (1990): 105-11.

Bulgarian ethnic identities. The Turks, however, refused to do so voluntarily. Their continued resistance led the Communist leadership to assemble a special committee to study the problem and find ways to assimilate the intransigent Turks successfully. Its final recommendations included policies for raising the Turks' cultural level, disseminating socialist values among them, overcoming religious obstacles to assimilation, and gaining the Turks' trust and confidence in the government. Although the committee's recommendations were accepted, they never were implemented. Instead, Zhivkov and his hard-line crony Georgi Atanasov pushed for stronger, more direct measures against the Turks and the elimination of any resistance to them by force. It was not until late 1984 that the hard-liners managed to have their approach accepted by the general Communist leadership.[36]

After careful planning and arranging close coordination among the party leaders, the police, and the army, a two-phased concerted campaign to change all Muslim Turco-Arabic names officially to Christian Slavic Bulgarian ones was initiated in December 1984. In the first phase (December 1984 through January 1985), some 310,000 Rhodope and southern Bulgarian Turks were ordered before their local authorities and forced to sign official identification documents renouncing their given names and adopting the new ones, which frequently were bestowed on them by the local administrators. The Turks in the northeastern regions of the state were lulled into accepting the events in the south by government assurances that the southern Turks were actually "Islamized" Bulgarians being brought back to their rightful ethnic fold while they (the northeastern Turks) were the only "true" Turks. Once the southern Turks had been "Bulgarized" officially, the government turned to the northeast in February 1985 and carried out phase two of the name-changing campaign. In both cases, the Turks reacted with violent resistance, requiring strong police and army action to be suppressed. The number of Turks killed, wounded, incarcerated, or exiled at that time still is unknown.[37]

---

[36] Eminov, *Turkish Minorities*, p. 85.

[37] Ibid., 86f. Changing names was deemed easier than eradicating Turkish culture. The cynical religious ethnonational animosity of the authorities toward Muslims was aptly demonstrated in the fact that many of the official names given to the Turks were intentionally insulting to their Islamic faith and identities. For example, Turks whose given name was Abdullah were forced to accept the name Hristo. Through the entire process, the authorities publicly claimed that all of the name changes were voluntary.

To reinforce the permanence of the name changes, a policy of eradicating any expression of traditional Islamic culture was instituted. The public use of the Turkish language was prohibited. To prevent any communication of the Turks with Turkey or with other potentially supportive outside agencies (*e.g.*, Radio Free Europe), their radios and tape recorders were confiscated. All traditional Muslim or ethnic Turkish public festivals or celebrations (*e.g.*, weddings and funerals) were banned under pain of imprisonment. Even being overheard speaking Turkish in private residences often resulted in police searches, interrogations, and incarcerations. A number of Muslim buildings were expropriated by the government for other uses or demolished.[38]

Despite the repression, however, the Turks persisted in clinging to their religious and ethnic identity, which intensified rather than diminished. Their resistance to the authorities' policies continued, and whole areas of the Rhodopes and northeastern Bulgaria were cordoned off by the police and the military, and martial law was enforced there. Outsiders of any kind were forbidden. Finally, in May 1989 public demonstrations of Turks erupted in northeastern Bulgaria and quickly spread to those in the southeast and in the Rhodopes. The government's response was immediate and brutal. Turkish leaders were arrested but swiftly went on a hunger strike. As news of the Turkish resistance leaked to Bulgarian intellectuals, environmentalists, and other social leaders who were growing disenchanted with the existing Communist-run state regime, as well as to the outside world, the Communist leadership decided to end their "Turkish Problem" by initiating a mass deportation of the Turkish minority.[39]

Amid a highly organized, militantly Bulgarian nationalist campaign of public demonstrations and media coverage extolling the "Unity of the Bulgarian Nation (People)," the Turk leaders were expelled to Turkey and passports were issued to any Turk who wished to "vacation" in Turkey. Throughout the rest of May and into August 1989, over 350,000 Turks applied for the proffered passports (many of whom did so at gunpoint) and left Bulgaria. The exodus ended only when, as in 1991, Turkey was constrained to close its border with Bulgaria because it lacked the capability of absorbing any additional refugees.[40]

---

[38] Ibid., 90-91. The author of this paper personally witnessed a number of instances of anti-Turkish discrimination during the period after the name changing.

[39] Ibid., 96-97.

[40] This author was having dinner with his wife at a cafe in central Edirne the evening that the first Bulgarian Turk deportees arrived in Turkey. They rode

Although the Zhivkov government may have attained its aim to reduce the Turkish minority by the 1989 deportations, they also caused severe economic and social dislocations throughout the state. Students and members of party-controlled bureaucratic and professional agencies had to be conscripted to bring in the harvests from fields abandoned by their former Turkish owners. The situation only added to the rrising resentment of political opposition groups who discovered a more public voice after the Turks' exodus. The Turks, by their persistent, vocal, and active opposition to the authorities ironically served as an example for the Bulgarians, many of whom shared the traditional nationalist aversion to Islam and the Muslim minorities but also disapproved of the discriminatory minority policy as it had been enacted. That misguided policy was one of the main factors leading to the downfall of the Zhivkov government in November 1989, since Turkish human rights (a concept born of Western nationalist struggles) became the umbrella under which all Bulgarian anti-Communist opposition movements finally were able to unite. The question of the Turks in Bulgaria—whether they are *millet* or ethnic in identity—remains a burning issue in the post-Communist period and poses a continuing problem for Bulgarian coalition governments, which find themselves representing a Christian Slavic nation-state but dependent on the support of a crucial Muslim (non-Bulgarian, in *millet* terms) political party.[41]

---

around the town center in two pickup trucks, waving the Turkish flag, shouting their story, accompanied by musical instruments, and, every hundred meters or so, stopping to dance wildly and joyfully in the middle of the street to the enthusiastic cheers and applause of all onlookers. Not even the police, who grudgingly joined in the celebration, minded the massive traffic jam that resulted. Later, back in Sofia, the author watched in amazement the television coverage of the Nazi-like torchlight nationalist military rituals staged by the government in the city's central square, the daytime nationalist parades and speeches extolling the "Unity of the Bulgarian nation," and the daily news coverage of cars, carts, wagons, and other assorted vehicles backed up for miles from the border crossing checkpoint leading to Turkey, all of which were piled high with what appeared to be all of the passengers' worldly possessions that could be attached in some manner. The Bulgarian news interviewers asked them such absurd questions as: "Will this be your first *vacation* in Turkey?" If the author had been of a mind to do so, he could have sold his Renault 5 at least five times to Turks who left polite notes inquiring about that possibility on its windshield. As a postscript, approximately half of the 1989 émigrés returned to Bulgaria from their "vacations" in the decade following the collapse of communism in the state.

[41] For the political role that Muslims played in the early post-Communist years in Bulgaria, see John D. Bell, "The Democratization of Bulgarian Political Life," in *Balkanistica*, vol. 9 (1996), spec. ed., *Bulgaria Past & Present: Transitions & Turning Points*, ed. Dennis P. Hupchick and Donald L. Dyer, 3-16.

The Twentieth Century's First Genocide: International Law, Impunity, the Right to Reparations, and the Ethnic Cleansing Against the Armenians, 1915-16

ALFRED DE ZAYAS[1]

Genocide is not an invention of modern times, nor is mass expulsion or what has come to be known as "ethnic cleansing." Many extermination campaigns against different ethnic or religious groups have taken place in the course of history.

The gradual recognition of human rights and the humanistic impetus of eighteenth century Enlightenment could have conceivably led to greater tolerance and understanding for other human beings whose culture, language, ethnicity or religion may be different. It did not. The nineteenth century saw mass extermination of indigenous peoples, notably the displacement, spoliation and annihilation of hundreds of thousands, if not millions, of the native population of the North American continent, wrongly referred to as "Indians," pursuant to the United States policy of "manifest destiny." European colonialism in Africa and Australia also led to the decimation of the native population of many territories, which Europeans curiously referred to as "terra nullius," as if the land had been unpopulated or had belonged to no one prior to the arrival of the Europeans.

The twentieth century has seen many more genocides. The first was perpetrated against the Armenians, a people who have inhabited the Caucasus and the Anatolian peninsula for thousands of years. As an ethnically and religiously different community in a frequently hostile environment, it has suffered various forms of persecution over the ages.

Already in the late nineteenth century an atmosphere of emerging nationalism and an obsession with achieving ethnic or religious homogeneity led to a number of pogroms against the Armenians in 1894-96 under Sultan Abdul Hamit. As many as 30,000 were killed in those early massacres.

The First World War provided the opportunity for a deliberate attempt by the Ottoman authorities to eliminate the Armenian

---

[1] The views expressed in this article are those of the author in his personal capacity and do not necessarily reflect those of organizations with which he is associated.

minority. Arnold Toynbee lived through this period and studied the phenomenon. In 1969 he recalled: "the massacre of Armenian Ottoman subjects in the Ottoman Empire in 1896 ... was amateur and ineffective compared with the largely successful attempt to exterminate [them] that was made during the First World War in 1915."[2] The earlier massacres under Sultan Abdul Hamit had lacked planning and experience and thus were not carried out in genocidal scale.[3]

Before the term "genocide"[4] was coined by the Polish jurist Raphael Lemkin during the Second World War, the international community had spoken of massacres, mass killings or exterminations, which, if occurring in time of war would necessarily violate the provisions of The Hague Regulations on Land Warfare, and give rise to State responsibility under article 3 of the Hague Convention (IV) of 18 October 1907, including an obligation to grant compensation in cases of breach.[5]

Decades of persecution and smaller massacres of Christian Armenians in the Ottoman Empire culminated in the systematic deportation and extermination campaign conducted by the Young Turk *Ittihad* Government[6] in 1915-16. There are various estimates of the number of persons who directly or indirectly lost their lives in the wake of the genocidal campaign. At least one million Armenians, more probably one and a half million, were put to death, while Europe looked on.[7]

The massacres began on 24 April 1915, and already some five weeks later, on 28 May 1915, the Governments of France, Great

---

[2] Arnold Toynbee, *Experiences* (London, 1969), 241.

[3] Vahakn N. Dadrian, "The Historical and Legal Interconnections Between the Armenian Genocide and the Jewish Holocaust: From Impunity to Retributive Justice," *Yale Journal of International Law* 23 (1998), 503-559 at 508.

[4] The label genocide was first used officially in the indictment of the Nuremberg Trials as one of the crimes covered under the general concept war crimes, which constituted point 6(b) of the indictment.

[5] "A belligerent party which violates the provisions of the said Regulations shall, if the case demands, be liable to pay compensation. It shall be responsible for all acts committed by persons forming part of its armed forces."

[6] Union and Progress Party.

[7] Vahakn N. Dadrian, "Genocide as a Problem of National and International Law: The World War I Armenian Case and its Contemporary Legal Ramifications," *Yale Journal of Intenrational Law* 14 (1989), 221-334; Permanent Peoples' Tribunal, *A Crime of Silence: The Armenian Genocide* (Paris, 1985).

Britain and Russia issued a joint declaration denouncing the Ottoman Government's massacre of the Armenian population in Turkey[8] as constituting "crimes against humanity and civilization for which all the members of the Turkish Government would be held responsible together with its agents implicated in the massacres."[9] This statement was also quoted in the Armenian Memorandum presented by the Greek delegation to the 1919 "Commission of Fifteen Members" at the Paris Peace Conference on 14 March 1919. The Commission's report contained an Annex I with a table of war crimes including the massacres of Armenians by the Turks and the deportation of survivors to concentration camps in Syria and Mesopotamia, where hundreds of thousands perished. The American Ambassador at Istanbul, Henry Morgenthau, kept careful diaries of his meetings with high government officials, including the Minister of Interior and Chief of the *Ittihad* party, Pasha Talat. In a note dated 4 November 1915 he reported to Washington that the *Ittihad* had "frightened almost everyone into submission.... There is no opposition party in existence. The Press is carefully censored and must obey the wishes of the Union and Progress Party... They have annihilated or displaced at least two thirds of the Armenian population."[10]

Punishing crimes against humanity and genocide were concerns of the international community at least since the Paris Peace Conference of 1919. In the case of the Treaty of Sèvres between the victorious Allies and Turkey, which Sultan Mohammed V had signed on 10 August 1920, but which was never ratified by Turkey, the Allies announced in article 230 their intention to punish those responsible for the genocide against the Armenians.[11] How-

---

[8] Cherif Bassiouni, *Crimes Against Humanity* (Dordrecht, 1992), 168-9.

[9] Egon Schwelb, Crimes Against Humanity, 23 *British Yearbook of International Law* (1946), 178-226. at 181; Arthur Beylérian, *Les Grandes Puissances, L'Empire Ottomon et les Arméniens dans les archives françaises 1914-1918* (Paris, 1983), 29.

[10] Vahakn N. Dadrian, *The History of the Armenian Genocide: Ethnic Conflict From the Balkans to Anatolia to the Caucasus* (Providence, R.I., 1995), 323-24.

[11] "Le gouvernement ottoman s'engage à livrer aux Puissances alliées les personnes réclamées par celles-ci comme responsables des massacres qui, au cours de l'état de guerre, ont été commis sur tout territoire faisant, au 1er août 1914, partie de l'Empire ottoman. Les Puissances alliées se réservent le droit de désigner le tribunal qui sera chargé de juger les personnes ainsi accusées, et le gouvernement ottoman s'engage à reconnaître ce tribunal...." The Allies also reserved their right to bring the accused before a special tribunal of the League of Nations. See André Mandelstam, *La Societé des Nations et les puissances devant le Problème Arménien*, 2nd ed. (n.p., 1970), chapter 4.

ever, the emergence of the Soviet Union, the geopolitical development after the war, coupled with the rise of the Turkish Nationalists under General Mustafa Kemal (later given the name *Atatürk*) and the abolition of the Sultanate in November 1922, eventually led to the abandonment of the Treaty of Sèvres and renegotiation between the Allies and the new Turkish Government under Kemal, at the Conference of Lausanne from 2 November 1922 to 4 February 1923 and from 23 April to 24 July 1923. The resulting Treaty of Lausanne abandoned the idea of ensuring punishment for the crimes committed against the Armenians, and the concept of an independent Armenia, which would have granted self-determination to a people as deserving as other ethnicities who did establish their own States pursuant to the treaties of Versailles, St. Germain, and Trianon.[12]

Some war crimes trials against Turkish Unionists and top leaders of the Young Turk *Ittihad* Party responsible for the genocide against the Armenians did, however, take place before Turkish courts martial in Istanbul.[13] A parliamentary Committee conducted investigations, and an Inquiry Commission was established on 23 November 1918, which delivered to the prosecution separate dossiers concerning 130 suspects. The Key indictment focused on the Cabinet Ministers and the top leaders of the ruling *Ittihad* party, including former Justice Minister Ibrahim, the War Minister Enver, and Interior Minister (later Grand Vizier) Talat. The indictment sought to establish that "the massacre and destruction of the Armenians were the result of decisions by the Central Committee of the *Ittihad*."[14] The prosecution relied solely on the Ottoman Penal Code, in particular articles 45 and 170.

Most of the prosecution documents consisted of decoded telegrams sent to and from the Interior Minister, the Third and Fourth Army Commanders, the Deputy Commanders of the Fifth Army Corps and the Fifteenth Division from Ankara province. Documentary evidence was adduced to substantiate the charge that *Ittihad* party Chief Talat had given oral instructions to interpret the order for "deportation" as an order for "destruction." On the basis of the documents, the Court martial established pre-

---

[12] J.F. Willis, *Prologue to Nuremberg: The Politics and Diplomacy of Punishing War Criminals of the First World War* (Westport, Conn., 1982); Henry Morgenthau, *Secrets of the Bosporus* (London 1918), especially Chapter 24, "The Murder of a Nation," 198-214.

[13] Dadrian, "Genocide as a Problem of National and International Law," 291-317.

[14] Dadrian, *History of the Armenian Genocide*, 324.

meditation and intent. Two Turkish officers were convicted and executed, others were sentenced to long terms of imprisonment. Other members of the Young Turks were condemned to death in absentia. The trials were, however, discontinued after Kemal Ataturk came to power.

Meanwhile, Great Britain proposed the establishment of an International War Crimes Tribunal and proceeded to collect evidence for the trials, but the political developments after the First World War did not allow its establishment.[15]

The Armenians were not the only victims of deportation and massacres at the hand of Ottoman forces, which committed grave crimes against other ethnic groups, notably the Kurds and the Greeks. Not only did these crimes go unpunished, they even received a measure of approval from the international community. A part of the Lausanne settlement of 1923 provided for the official approval of the compulsory population "exchange" between Greece and Turkey[16] and the establishment of a bilateral Mixed Commission which was charged with the administration of the exchange. Some observers consider that this part of the settlement amounted to approval of ethnic cleansing, another form of genocide.

The Prosecution of Genocide Since the Nuremberg Trials

Many genocides have occurred in history. All have gone unpunished with the exception of the Nazi genocide against the Jews and more recently the genocide by the Hutus against the Tutsis in Rwanda. The military actions carried out by Serbian forces against Muslim Bosniacs and against ethnic Albanians of Kossovo are currently under examination by the International Criminal Tribunal for the Former Yugoslavia, before which former Yugoslav President Slobodan Milošević stands indicted on the crime of genocide.

The Nuremberg indictment established an important precedent in creating the notion of crimes against humanity, a concept

---

[15] The war crimes committed by the German armed forces during the First World War should also have been tried by international criminal tribunals, and the Allies demanded the handing over of nearly one thousand persons pursuant to articles 227 to 230 of the Treaty of Versailles with Germany. No international trials, however, were ever held, and only a handful of military persons were tried before the German Supreme Court at Leipzig in 1921. See Alfred de Zayas, "Der Nürnberger Prozess," in Alexander Demandt, *Macht und Recht* (Munich, 1990), 247-270.

[16] League of Nations, *Treaty Series* 32: 75.

that includes genocide. Moreover, in its statement of the offense, under Count III, War Crimes, the Nuremberg indictment charged the accused with conducting "deliberate and systematic genocide, viz., the extermination of racial and national groups, against the civilian populations of certain occupied territories in order to destroy particular races and classes of people and national, racial or religious groups, particularly Jews, Poles, Gypsies and others."[17]

Thus, the Nazi extermination of millions of Jews became the first genocide that was subject to international judicial scrutiny, and many of those who perpetrated it or were involved in aspects of its execution were held personally liable, tried and punished as common criminals. They were brought not only before international courts like the International Military Tribunal at Nuremberg or before the twelve American Tribunals at Nuremberg, but also before countless tribunals of the victorious Allies such as France and England, and even before the courts of States that did not exist at the time of the genocide, e.g. Israeli Tribunals, and before German and Austrian tribunals to our own day.

Genocide is not an ordinary crime subject to periods of statutory limitation for purposes of prosecution. Pursuant to the United Nations Convention on the Non-applicability of Statutory Limitations to War Crimes and Crimes Against Humanity, which was adopted by the General Assembly on 26 November 1968 and entered into force on 11 November 1970, no statute of limitations shall apply to genocide irrespective of the date of commission. The same principle of non-application of statutes of limitations to genocide was expressed by the General Assembly in its Resolution 2391 of 26 November 1968.

The provisions of the Convention apply to representatives of the State authority and private individuals who, as principals or accomplices, participate in or who directly incite others to the commission of any of those crimes, or who conspire to commit them, and to representatives of the State authority who tolerate their commission.

This Convention strengthens the provisions of the Convention on the Prevention and Punishment of the Crime of Genocide, which was adopted by the General Assembly on 9 December 1948, one day before the adoption of the Universal Declaration of Human Rights, and which entered into force on 12 January 1951. As of 14 December 2001, there were 133 States parties, including Armenia and Turkey.

---

[17] *Trials of the Major War Criminals Before the International Military Tribunals* (Nuremberg, 1947), I: 43 (Indictment Count Three A).

Article 2 of this Convention defines the crime of genocide as follows:

Genocide means any of the following acts committed with intent to destroy, in whole or in part, a national, ethnical, racial or religious group, as such:

(a) killing members of the group;

(b) causing serious bodily or mental harm to members of the group;

(c) deliberately inflicting on the group conditions of life calculated to bring about its physical destruction in whole or in part;

(d) imposing measures intended to prevent births within the group;

(e) forcibly transferring children of the group to another group.

As indicated before, Turkey is a State party to this Convention (accession 31 July 1950, in force 30 October 1950), but it has not acceded to the Convention on the non-applicability of statutes of limitation.

Among international lawyers there is consensus that genocide has always been a justiciable criminal offense and that the Convention on the Prevention and Punishment of the Crime of Genocide is declarative of international law in its core provisions and therefore has retroactive effect. Reference in this connection can be made to the Nuremberg prosecutions against the major war criminals of the Second World War, who were convicted of having committed crimes against humanity including genocide. The Nuremberg Trials took place prior to the adoption and entry into force of the Convention.

Also among historians there is consensus that in the years 1915 and 1916 the Armenian population of Turkey was subjected to measures which fall within the definition of genocide, and that approximately one and a half million Armenians perished as the result. Perhaps only half a million Armenians from Turkey survived, and it is thanks to the Armenians living in Russia and to the Armenian diaspora that the Armenian people, its language, music und culture have survived. The effects of genocide, however, cannot be erased.

Article VI of the Genocide Convention stipulates: "Persons charged with genocide ... shall be tried by a competent tribunal of the State in the territory of which the act was committed, or by such international penal tribunal as may have jurisdiction with respect to those Contracting Parties which shall have accepted its jurisdiction."

Article VIII provides: "Any Contracting Party may call upon the competent organs of the United Nations to take such action under the Charter of the United Nations as they consider appropriate for the prevention and suppression of acts of genocide...."

Article IX provides: "Disputes between the Contracting Parties relating to the interpretation, application or fulfilment of the present Convention, including those relating to the responsibility of a State for genocide... , shall be submitted to the International Court of Justice at the request of any of the parties to the dispute."

Inter-State Complaints

Forty-five years would elapse since the adoption of the Convention until a State invoked article IX of the Convention before the International Court of Justice. On 20 March 1993 the Government of Bosnia and Herzegovina filed a case with the Court and demanded interim measures of protection against the Federal Republic of Yugoslavia and reparations for the damage already caused. Such measures were granted under article 41 of the Statute of the Court. In several subsequent decisions, the ICJ rejected the contentions of the Federal Republic of Yugoslavia according to which the Convention did not apply to the conflict. In further counter-claims the Federal Republic of Yugoslavia argued that it was the Government of Bosnia and Herzegovina that had committed acts of genocide against the Serbs in Bosnia. Claims and counterclaims were deemed admissible by the Court. However, on 20 April 2001 Yugoslavia withdrew its counterclaims, and the Court accepted the withdrawal on 10 September 2001.

The Convention against Genocide would also allow any State party to the Genocide Convention, e.g. Armenia, to file a case against another State party, e.g. Turkey, in order to obtain an interpretation of the Convention and a determination that genocide had occurred. Such a determination would constitute a basis for a possible remedy, which could take the form of reparations for the destruction of Armenian churches and monasteries and the illegal confiscation of privately owned Armenian lands.[18]

---

[18] Kévork K. Baghdjian, *La Confiscation, par le gouvernement turc, des biens arméniens* (Montréal, 1987).

Impunity

Is impunity for genocide acceptable under international law? Since the last decades of the twentieth century, it is clear that international law abhors the concept of impunity. This is why the Statute of the International Criminal Court was adopted in Rome in July 1998. It took less than four years for the required sixty ratifications to be received, the last three being deposited on 11 April 2002. The Statute has thus entered into force, but the Court itself has not yet been established. Armenia has signed but not yet ratified it. Turkey has neither signed nor ratified it. Nevertheless, the international community and many non-governmental organizations and human rights activists are committed to ensure that the Court does make a difference for future generations. The concepts of State immunity or personal immunity for acts of State are not available as a defense against charges of war crimes or crimes against humanity. In its article 6, the statute of the ICC defines genocide as an offense *erga omnes*, an offense which is subject to prosecution, even if the victims are a State's own population. In this context the Pinochet precedent is an example and at the same time a warning to other heads of state. But for Pinochet's advanced age and state of health, he could have been tried in the United Kingdom, Spain, or in any number of other States that sought his extradition.

The United Nations Human Rights Committee has repeatedly insisted on the right of victims of human rights violations to receive reparations and on the obligation of States parties to punish perpetrators of human rights violations. In rejecting impunity and amnesty laws, the Committee stated in its "Views" in case No. 322/1988 Hugo Rodríguez v. Uruguay:

> The Committee reaffirms its position that amnesties for gross violations of human rights and legislation such as Law No. 15,848, *Ley de Caducidad de la Pretensión Punitiva del Estado*, are incompatible with the obligations of the State part under the Covenant. The Committee notes with deep concern that the adoption of this law effectively excludes in a number of cases the possibility of investigation into past human rights abuses and thereby prevents the State party from discharging its responsibility to provide effective remedies to the victims of those abuses. Moreover, the Committee is concerned that, in adopting this law, the State party has contributed to an atmosphere of

impunity which may undermine the democratic order and give rise to further grave human rights violations.[19]

The ad hoc International Criminal Tribunal for the Former Yugoslavia was established 1994 at The Hague pursuant to Security Council Resolution No. 827. Several convictions for crimes against humanity have been handed down, including for the crimes of rape and enslavement.[20] On 2 August 2001 Trial Chamber I of the Tribunal rendered the first judgement convicting an individual of having committed the crime of genocide. General Radislav Krstić was sentenced to forty-six years of imprisonment for his involvement in genocide, forced transfer and deportation committed between July and November 1995, in particular for his responsibility for the crimes committed by Serbian forces in the town of Srebrenica.[21] On 23 November 2001 the UN Tribunal indicted Slobodan Milošević for committing genocide against the Bosnian people. His trial for crimes against humanity committed during the Serbian crackdown on ethnic Albanians in Kosovo and during the war in Bosnia and Croatia opened in February 2002. He is the first head of state to stand trial for genocide.

The International Criminal Tribunal for Rwanda was established in 1995 at Arusha, Tanzania, pursuant to Security Council Resolution No. 955. On 2 September 1998 the first verdict interpreting the Genocide Convention was handed down by the Arusha Tribunal in the judgment against Jean-Paul Akayesu, who was held guilty on nine counts for his role in the 1994 Rwandan genocide.

Universal jurisdiction for certain crimes such as torture and genocide has also emerged. This is, *inter alia*, reflected in the Princeton Principles on Universal Jurisdiction, which were drafted and revised by a number of senior professors of international law, notably Cherif Bassiouni, Chair of the drafting committee of the Princeton Project on Universal Jurisdiction.

---

[19] *Annual Report of the Human Rights Committee to the General Assembly, 1994 Report,* U.N. Doc. A/49/40, vol. 2, Annex IX B, para. 12.4 at p. 10.

[20] Judgement of Trial Chamber II in the Kunarac, Kovać and Vuković case, 22 February 2001.

[21] *"Prosecutor v. Radislav Krstić," Chambre de 1ère instance, la Haye, 21 août 2001.*

Ubi jus, ibi remedium

For the surviving victims of genocide and mass expulsion, the foremost right is the right to truth, which entails recognition of their suffering, of their status as victims, of the right to their history. These remedies may appear immaterial, but they are indispensable to one's identity and to the will to continue living. Other, more concrete remedies are the right to return to one's homeland and the right to restitution of or compensation for wrongfully confiscated property. Such remedies were stipulated in articles 142 and 144 of the Treaty of Sèvres, which the Sultan signed but Kemal Atatürk never honored.

Although the United Nations has adopted countless resolutions and decisions concerning the right of refugees and expellees to return, this right has been exercised only in a few cases, notably by some 300,000 Crimean Tatars and their descendants, who had been deported by Stalin to Uzbekistan and who have been returning to the Crimea with the help and blessing of the Office of the High Commissioner for Refugees, the European Union, and the Organization for Security and Cooperation in Europe. This is a good precedent and not without relevance for Armenians.

Other examples of return are the voluntary repatriation of Hutus to Rwanda, under the auspices of UNHCR, and the voluntary repatriation of Bosnian Muslims pursuant to Annex 7 to the Dayton Accords of December 1995. The Human Rights Chamber established under Annex 6 of the Dayton Accords has examined thousands of cases and also ordered restitution and compensation to the victims of ethnic cleansing.[22]

Reparations

It should be remembered that in addition to individual criminal responsibility for genocide, the 1948 Convention also establishes State responsibility, that is, international liability vis-à-vis other States parties to the Convention in cases of breaches of its provisions. Parties to the Convention can bring a case before the International Court of Justice in order that the Court make a determination of State responsibility for acts of genocide.

On the basis of such a finding and relying on general principles of international law, a State can demand reparations from another State, and, for instance, mixed compensation commis-

---

[22] Leif Berg and Ekkehard Strauss, *The Human Rights Chamber of Bosnia and Herzegovina* (Sarajevo, 2000).

sions could be established to determine the appropriate level of compensation.

In this context let us remember that whereas the non-ratified Treaty of Sèvres had demanded reparations from Turkey in 1920, the revised peace settlement contained in the Treaty of Lausanne of 1923 dispensed with reparations. Turkey was thus the only State among the defeated Central Powers and their allies not required to pay reparations. To this extent Turkey would not be able to escape current claims for reparation by reference to any prior reparations actually paid.

In the context of the Turkish invasion of Northern Cyprus in July 1974 and the expulsion of over 175,000 Cypriots of Greek ethnicity to the south of the island, the Government of Cyprus filed several claims with the European Commission on Human Rights. In its first decision the Commission found in 1976 that "the transportation of Greek Cypriots to other places, in particular the excursions within the territory controlled by the Turkish army, and the deportation of Greek Cypriots to the demarcation line ... constitute an interference with their private life, guaranteed in article 8(1) which cannot be justified on any ground under paragraph 8(2)"[23] The Commission furthermore considered that the prevention of the physical return of Greek Cypriots who had been driven out from their homes in the north of Cyprus amounts to an infringement of their right to respect of their homes as guaranteed in article 8(1). The Commission further noted that the acts violating the Convention were directed exclusively against members of the Greek Cypriot community and concluded that Turkey had failed to secure the rights and freedoms set forth in the Convention without discrimination on the grounds of ethnic origin, race, and religion as required by article 14 of the Convention.[24]

Individuals can also demand remedies, and over the past thirty years Human Rights Treaties such as the European Convention on Human Rights and the International Covenant on Civil and Political Rights and its Optional Protocol have created important jurisprudence granting remedies to victims of human rights violations. For instance, based on Protocol I to the Euro-

---

[23] "Cyprus v. Turkey," Application Nos. 6780/74 and 6950/75, Report of 10 July 1976, application No. 8007/77, Report of 4 October 1983.

[24] In 1983 the European Commission again dealt with the issue and found that the displacement of persons, separation of families and discrimination violated the European Convention. See Christa Meindersma, "Population Transfers in Conflict Situations," *Netherlands International Law Review* 41 (1994): 31-83 at 71-72.

pean Convention on Human Rights, an individual like Mrs. Titina Loizidou could obtain a judgment in Strasbourg entitling her to compensation from Turkey for illegally confiscating her property during the invasion and expulsions of 1974. The Court held by eleven votes to six that the denial of access to the applicant's property and consequent loss of control thereof is imputable to Turkey.[25] In a further judgment of 28 July 1998, the Court ordered restitution. By fifteen votes to two, the Court rejected Turkey's claim that Mrs. Loizidou had no entitlement to an award under article 50 of the Convention. By fourteen votes to three it decided that Turkey was obliged to pay to Mrs. Loizidou within three months, 300,000 Cypriot pounds for pecuniary damage. By thirteen votes to four the Court held that Turkey had to pay to Mrs. Loizidou 137,084 pounds for costs and expenses.

It is important to note that Turkey formulated a preliminary objection *ratione temporis*, which the Court considered and rejected by a vote of eleven to six. Thus, a preliminary objection against a case presented by the Government of Armenia against Turkey or by an Armenian descendent of the genocide might have, by analogy, a chance of being considered, notwithstanding the principle that treaties cannot be applied retroactively. Moreover, there is an argument to be made concerning the continuing violation arising from the impossibility for Armenians today to return in safety and dignity to the homes and properties that were confiscated in connection with the genocide.

In matters of restitution and/or compensation for property confiscated arbitrarily and in violation of the principle of equality and non-discrimination, the United Nations Human Rights Committee has held in seven "Views" concerning many more applicants that the Czech Republic had violated the applicants' rights under article 26 of the International Covenant on Civil and Political Rights and that they were entitled to restitution and/or compensation.[26] It should be noted that whereas the International Covenant on Civil and Political Rights entered into force on 23 March 1976, most confiscations complained of took place in the years 1945-1950. Nevertheless, it was the arbitrariness of the restitution legislation itself and its discriminatory application that led the Committee to make a finding that the overarching princi-

---

[25] Case of Loizidou v. Turkey (Merits) 40/1993/435/514, Judgment of 18 December 1996.

[26] See *inter alia* Case No. 516/1992 *Simunek v. Czech Republic*, Case No. 586/1994, *Adam v. Czech Republic*, 857/1996 *Des Fours Walderode v. Czech Republic*.

ple of equality enshrined in article 26 of the Covenant had been violated.

The Human Rights Committee, however, would not be able to entertain a claim against Turkey, because Turkey has not yet become a party to the International Covenant on Civil and Political Rights or to the Optional Protocol thereto. Armenia is a State party to both instruments.

Before examining a hypothetical case concerning the Armenian genocide, the European Court of Human Rights in applying the European Convention would have to address the *ratione temporis* obstacle, since the events occurred long before the entry into force of the Convention. However, there are continuing effects, including the trauma that has affected generations and the profound sense of loss of one's homeland,[27] bearing in mind that the Armenian community of Anatolia had lived and flourished in this land for some two thousand years. Let us also not forget that because of the destruction of the Armenian homeland in Eastern Anatolia, the survivors had to emigrate and their children and grandchildren gradually and irreparably lost part of their culture and language--i.e. their identity. There is also the aspect of cultural genocide in that monasteries, churches and other monuments attesting to the cultural heritage of the Armenians of Anatolia have been destroyed or allowed to decay. Indeed, there is an international moral obligation to make amends for crimes such as genocide, and a political imperative for the international community to remove sources of tension, so as to build peace on the foundation of just reconciliation.

From the strictly legal standpoint, it is worth stressing that when a State becomes a party to a Convention, it intends in good faith to take all necessary legislative, judicial and administrative measures to give life to the treaty and implement its provisions. Thus, one may ask, what has Turkey done since it acceded to the Genocide Convention in 1950 in order to make amends for the genocide against the Armenians? The general principle *pacta sunt servanda* would require Turkey not only to refrain from committing genocide in the future but surely implies that the victims of earlier genocides should be granted adequate reparations.

From the political standpoint, let us remember that after the collapse of the Soviet Union, the new democratic governments of the successor republics returned countless confiscated churches

---

[27] Viz. Yael Danieli, *Differing Adaptational Styles in Families of Survivors of the Nazi Holocaust* (Washington, D.C., 1981); Abraham J. Peck, *The Children of Holocaust Survivors* (New York, 1983).

and monasteries to the representatives of the respective Churches. In the case of the 451 Armenian monasteries and 2538 Armenian churches in Anatolia, the Turkish government should be persuaded by the European Union and NATO to return as many as were not destroyed and still exist, transformed into mosques, museums, prisons, sport centres, etc.[28] to the Armenian Patriarch at Istanbul. The Turkish government should also be persuaded to finance the reconstruction of other Armenian historical buildings.

In any event, alone the submission of a case to the European Court of Human Rights would focus media attention on the problem.

State Succession

Old States disappear and new States emerge as a result of wars, dismemberment, federation, secession, decolonization etc. The recent past has delivered numerous examples, e.g. as a result of the dissolution of the Soviet Union.

When States get cornered, they sometimes try to escape by reinventing themselves. Most recently we have seen that the Federal Republic of Yugoslavia ceased to exist and that a new Federal Republic of Yugoslavia was admitted into the United Nations as a new State on 1 November 2000.

What consequences, if any, does this have with regard to the liability of the new Yugoslav State for the obligations incurred by the prior State? The International Court of Justice did not accept the argument of the new Yugoslav State that for purposes of the Genocide Convention, it was not identical with the former Yugoslav State under Milošević and thus not responsible for the actions, which had led to the filing of the case Bosnia and Herzegovina against the Federal Republic of Yugoslavia in 1993.

After the Second World War the Nazi German government disappeared, the territory of the Reich was occupied by four victorious powers, and it was not clear whether and how a German state would be allowed to reemerge. Eventually the Allies gave a third of Germany's territory to neighbouring States and allowed two German States to emerge, which, however, were not absolved of the international liabilities of the Reich. In the fifty-seven years following the end of the war, the Federal Republic of

---

[28] Dickran Kouymjian, "La Confiscation des biens et la destruction des monuments historiques comme manifestations du processus génocitaire," in *L'Actualité du Génocide, Actes du colloque*, ed. Comité de Défense de la Cause Arménienne (Paris, 1999), 219-230.

Germany has set the example for other countries by fully assuming the moral and legal responsibility for the crimes perpetrated by the Hitler government, not only for the genocide but also for the forced labour of millions of civilians from occupied countries and for the confiscation of private property.

At the end of the First World War, the Ottoman Empire collapsed and the Sultanate was abolished. But the new leaders of Turkey assumed sovereignty over formerly Ottoman lands and successfully obtained the abandonment of the Treaty of Sèvres and its replacement by the Treaty of Lausanne. No serious international lawyer would contend that the Turkey that signed the Treaty of Lausanne was a new State free of the legal obligations of the Ottoman empire.

It is accepted international law that a revolution or any kind of overthrowing of a government does not result in the emergence of a new State devoid of inherited responsibilities. The Vienna Convention on Succession of States in Respect of State Property, Archives and Debts of 8 April 1983,[29] like the Convention on the Prevention and Punishment of the Crime of Genocide, is only partly constitutive but rather essentially declarative of international law. New are provisions that reflect the emergence of many new States as a result of decolonization and the recognition of the right to self-determination.

But *tabula rasa* with regard to crimes against humanity and genocide would be wholly incompatible with the Vienna Convention and violate other general principles of international law. Moreover, the concept of automatic succession to human rights treaties has been espoused by an increasing number of international lawyers and the parallel principle of the continued applicability of human rights treaties to successor States is reflected in the General Comment No. 26 of the Human Rights Committee concerning "the continuity of obligations under the International Covenant on Civil and Political Rights."[30]

Article 36 of the Vienna Convention provides that a succession of States does not "as such affect the rights and obligations of creditors". Thus, the claims of the Armenians for their wrongfully confiscated properties did not disappear with the change from the Sultanate to the regime of Mustafa Kemal (Atatürk).

---

[29] UN Doc. A/CONF. 117/14.

[30] UN Doc. A/53/40, vol. I, Annex VII (1998).

Negationists

The systematic annihilation of Armenians in 1915-16 was recognized by Ottoman courts martial shortly after World War I, which sentenced some of the Turkish perpetrators to death.[31] Official Turkish policy since the 1920s has simply denied the genocidal intent of these mass murders.[32]

No serious historian would today question the fact that there was a genocide against the Armenians in 1915-16. Some may argue that the number of victims exceeded one million and a half, while others might postulate the figure of one million, and others still might minimize it further. The facts, however, are known and have been known since 1915. Every history school book should reproduce the text of articles 142, 144, and 230 of the Treaty of Sèvres, which clearly show what the Allies believed and what the Sultan acknowledged in 1920. It is disgraceful that some politicians prefer to ignore these facts and pretend that the issue is open to serious historical doubt.

This reminds us of the way many politicians treated the murder of 15,000 Polish prisoners of war by Stalin's NKVD in 1940. All capitals knew who was responsible since 1943, and yet they preferred to remain silent, or even to lend credence to Stalin's brazen accusation that it was the Germans who had committed this particular crime. The British Government, for instance, refused to send any representative to the unveiling of the monument at Gunnesbury Cemetery in London in September 1976, so as not to compromise relations with the Soviet-bloc countries. The official British statement was that "it has never been proved to Her Majesty's Government's satisfaction who was responsible." The London *Times*, however, observed: "Enough has been published to convince anyone who is not a dedicated defender of the Soviets that the massacre did take place in 1940, when Katyn was under Soviet and not German control."[33] Only when in 1990 Gorbatchev accepted Soviet responsibility and expressed regret to the

---

[31] Vahakn N. Dadrian, "Documentation of the Armenian Genocide in Turkish Sources," in *Genocide: A Critical Bibliographic Review*, ed. Israel Charny, vol. 2 (London, 1991).

[32] Vahakn N: Dadrian, "Ottoman Archives and the Denial of the Armenian Genocide," in *The Armenian Genocide*, ed. Richard Hovanissian (New York, 1992), 280-310.

[33] *The Times*, 17 September 1976; Alfred de Zayas, *The Wehrmacht War Crimes Bureau* (Lincoln, Neb., 1989), 276.

Polish Prime Minister Jaruzelski did the last political diehards in the West publicly admit what everyone already knew.

Current Initiatives and Recommendations

*An important resolution adopted by the French Parliament on 18 January 2001 was followed by Law No. 2001-70 of 29 January 2001 which recognizes the historical fact that the Armenian people were subjected to genocide by the Ottoman Turks in 1915.[34] This is an important beginning that opens the way to political and legal developments.[35] This resolution should be given wide dissemination and be properly reflected in school curricula.

*The decision of the European Court of Human Rights of 25 September 2001, declaring inadmissible *ratione personae* complaint No. 72657 by the Turkish lawyer Sedat Vural, claiming that the French law constituted defamation of the Turkish people and demanding reparation thereof, deserves wide dissemination.

*The reality of the genocide against the Armenians has been recognized by His Holiness Pope John Paul II (27 September 2001) and by many heads of state, including Konstantinos Stefanopoulos, President of Greece (10 July 1976), the Prime Minister of Canada Jean Chretien (24 April 1966), and the French President Francois Mitterrand (1984). Many national Parliaments have also adopted relevant resolutions, including the Canadian Senate (13 June 2002), Italian Chamber of Deputies (16 November 2000), the European Parliament (15 November 2000), the Lebanese Parliament (11 May 2000), the Swedish Parliament (28 May 1998), the Council of Europe Parliamentary Assembly (24 April 1998), the Belgian Senate (26 March 1998), the US House of Representatives (Resolution 3540, 11 June 1996), the Hellenic (Greek) Parliament (25 April 1996), the Canadian House of Commons (23 April 1996), the Russian Duma (14 April 1995), the Argentinean Senate (5 May 1993), the European Parliament (18 July 1987), the U.S. House of Representatives (Joint Resolution 247, 12 September 1984), the Cypriot House of Representatives (29 April 1982), the Uruguayan

---

[34] "L'Assemblée nationale et le Sénat ont adopté, Le Président de la République promulgue la loi dont la teneur suit: **Article unique**: La France reconnaît publiquement le génocide arménien de 1915. La présente loi sera exécutée comme loi de l'Etat."

[35] 'Un entretien avec Louis Joinet"—"Lutte contre l'impunité: Le temps des questions," *Droits fondamentaux*, no. 1, (July-December 2001), www.revue-df.org.

Senate and House of Representatives (20 April 1965), etc. Similarly, among the State and Provincial Governments that have recognized the genocide should be mentioned the Parliament of New South Wales (Australia), the Legislature of Ontario (Canada), the Consiglio Provinciale di Roma (Italy), the Alaska State Senate as well as the state legislatures of 26 states of the United States, including New York.

*A concerted effort is being made by a variety of nongovernmental organizations and human rights activists to obtain recognition of the genocide against the Armenians through the Parliaments of other States, particularly European States and the United States. Most recently, on 10 December 2001, human rights day, the Conseil d'Etat de Genève issued a proclamation recognizing the genocide against the Armenians in 1915.[36] Bearing in mind that Geneva is the seat of the European UN Office and, in particular, of the Office of the High Commissioner for Refugees and the High Commissioner for Human Rights, and headquarters of countless international organizations, this Geneva proclamation has the potential of placing the Armenian genocide on the agenda of many of them.

*The official recognition of the genocide against the Armenians by the French Parliament should be followed by the adoption of appropriate legislation in France e.g. to provide sanctions against negationists.

*The official recognition of the genocide against the Armenians by the Parliaments of many States should inspire Turkcy to recognize its historical responsibility and to enact legislation to provide a measure of reparation to the descendants of the survivors of the genocide.

*An international collegium of experts in international law could be established in order to provide solid legal argument to support future initiatives in international fora. Such a collegium could draw upon the expertise of the participants in the Tribunal

---

[36] "En reconnaissant le génocide arménien de 1915, Genève se situe, en respect des textes internationaux sur le génocide, dans la lignée de la reconnaissance de ce crime par de nombreux pays, par l'Organisation des Nations Unies et par le Parlement européen."

Permanent des Peuples, which on 13-16 April 1984 held its special session on the Genocide against the Armenians.[37]

*Armenians should join forces with those international lawyers and scholars demanding a convention to protect the right to one's homeland, partly on the basis of the UN study of ICJ Justice Awn Shawkat Al-Khasawneh and its 13-point draft declaration.[38]

*As mentioned above, a more recent victim of Turkish nationalism is the ethnic Greek people of northern Cyprus, subjected to a mass expulsion in 1974 and deprived of their lands. In today's world, such actions are condemned by the relevant organs of the United Nations, but unless the Security Council imposes sanctions, little happens by way of enforcement of the right of return of the Cypriots to their homes in Northern Cyprus, nor have they received any compensation for clearly discriminatory and illegal confiscations.

*The Government of Cyprus has filed and argued several cases before the European Commission and Court of Human Rights, which have repeatedly held that Turkey has violated several provisions of the Convention and is under an obligation to respect the right to return and to provide compensation. However, Turkey has consistently refused to implement the decisions of the European Commission and Court. The Committee of Ministers has followed up with resolutions urging Turkey to comply with the judgment without delay.[39]

*There exists an obvious commonality of interests between the Armenians and the Cypriots, and their human rights claims

---

[37] Tribunal Permanent des Peuples, *Le Crime de Silence: Le Génocide des Arméniens,* (Brussels, 1984).

[38] E/CN.4/Sub.2/1997/23.

[39] Résolution Intérimaire ResDH (2001)80 du 28 juillet 1998 dans l'affaire Loizidou, adoptée le 26 juin 2001. "Le Comité des Ministres ... déplorant très profondément le fait que, à ce jour, la Turquie ne se soit toujours pas conformée à ses obligations découland de cet arrêt ... se déclare résolu à assurer, par tous les moyens à la disposition de l'organisation, le respect des obligations de la Turquie en vertu de cet arrêt, en appelle aux autorités des Etats membres à prendre les mesures qu'elles estiment appropriées à cette fin." Thus, States parties to the European Convention on Human Rights are encouraged to apply appropriate political pressure on Turkey, which may entail legitimate economic sanctions, to ensure compliance with the Court judgment.

would strengthen each other if advanced vigorously in all European and UN fora. In particular the European Parliament should follow-up on its 1987 resolution providing that recognition by Turkey of the genocide against the Armenians is a *sine qua non* to Turkish membership in the European Union. Moreover, the European Parliament should be again seised of the question and some sort of reparation, at least symbolic, to the descendants of the victims should be offered.

*There exists also a commonality of interests between the descendants of the survivors of the Armenian genocide and the survivors of the expulsion of 15 million ethnic Germans 1944-1948 from East-Prussia, Pomerania, Silesia, East Brandenburg, Sudetenland, Hungary, Yugoslavia, etc., a form of "ethnic cleansing" carried out by the Soviet Union, Poland, Czechoslovakia, Yugoslavia, etc., in the course of which more than two million ethnic Germans perished.[40]

*The issue of the genocide of the Armenians is a matter that should be mainstreamed in the relevant United Nations mechanisms. I should mention the famous study on genocide prepared by Mr. Ben Whitaker in 1985 for the Sub-Commission on Prevention of Discrimination and Protection of Minorities,[41] referring in its paragraph 24 to the genocide against the Armenians. I have already mentioned the study of Justice Al Khasawneh on the right to one's homeland and should now mention the study of the French magistrate Louis Joinet for the U.N. Sub-Commission entitled "The question of impunity of authors of violations of human rights."[42] As to UN supervisory mechanisms, mention should be made of the work of the Special Rapporteur on Freedom of Religion and Belief, Prof. Abdelfattah Amor, who has incorporated the Armenian concerns in his reports, most recently his report of 25 October 2000 submitted to the UN General Assembly on his visit to Turkey[43] specifically referring to the plans of the Young Turks concerning the elimination of the Armenian

---

[40] Alfred de Zayas, *A Terrible Revenge: The Ethnic Cleansing of the East European Germans 1944-1950* (New York, 1994).

[41] Report E/CH.4/Sub.2/1985/6.

[42] E/CN.4/Sub.2/1997/20/Rev. 1.

[43] Report A/55/280/Add.2 available on the Internet: http://www.unhchr.ch.

community in the Ottoman Empire and the widespread confiscation of their properties.[44]

*More jurisprudence is needed on the subject. The European Court of Human Rights could, for instance, be seised with cases concerning aspects of the Armenian genocide including the negation of the genocide.

*Turkey should be urged to create a special fund to indemnify the descendants of victims of the genocide against the Armenians, as e.g. Germany has created several such funds and as countries such as Switzerland have created to return savings deposited in Swiss banks by Jewish persons who perished in the Holocaust.

*Although excellent scholarly publications have been produced on the subject of the genocide against the Armenians, there remain many issues to be researched and developed further. Professors of history, law, sociology should be encouraged to assign to their students dissertations based on aspects of this genocide.

*The implications of the Armenian genocide deserve public discussion and constitute a legitimate subject not only for scholarly but also for artistic expression. The 2002 film *Ararat* by the Oscar–nominated Canadian director/writer Atom Egoyan focuses on the historical event, its continued negation by Turkish officials, and the traumatic impact of the demographic catastrophe on subsequent generations of Armenians in the homeland and in the diaspora. The film merits a wide audience.

*The international organizations and mechanisms have not been sufficiently engaged. Certainly Armenian culture deserves international protection, and the UNESCO could give it greater attention, e.g. by financing the reconstruction of destroyed Armenian monasteries and churches in Anatolia and the recognition of the old Armenian capital Ani as a common heritage of mankind.

*Finally, the 24th of April should be recognized world-wide as the day of remembrance of the genocide against the Armenians, the first genocide of the twentieth century. A resolution to this

---

[44] Jean-Marc Bernard, UN Rapport Officiel de l'ONU fait etat pour la Toute première fois dans les Annales Onusiennes de la Confiscation des Biens Armeniens en 1915 en Turquie, in Journal France-Arménie, novembre 2000.

effect should be proposed to the UN General Assembly and accompanied by the requisite level of lobbying.

Conclusion

Although the Armenians can legitimately invoke many treaties and general provisions of international law to support their claims, political reality is such that progress often cannot be achieved by legal means alone. Even a judgment of the International Court of Justice is of limited value as long as it is not enforced.

Enforcement of international law, including international case law, depends on political will. Therefore, emphasis must be placed in creating the conditions that will reinforce a political will to do justice to the Armenians.

Surely the principles contained in the Convention Against Genocide and the rule of the non-applicability of statutes of limitations to genocide are binding not only on States parties, but on all countries, as a form of *jus cogens* or peremptory international law. Already the Advisory Opinion of the International Court of Justice in the 1951 case concerning Reservations to the Convention concluded that the principles underlying the Convention were part of customary international law and therefore binding all States.

The Law of Universal Conscience demands that the genocide against the Armenians be officially recognized and that the State responsible for this genocide make atonement and provide for the restitution of stolen properties to the descendants of the survivors of this great crime against all of humanity. In a sense, as long as the Turkish State has not officially asked for forgiveness for the genocide, as Gorbachev did with respect to Katyn, and as long as no reparation has been offered to the survivors of the victims, the genocide continues.

Indifference and ignorance can be dangerous. As Hitler is reported to have remarked on repeated occasions, as early as 1931: "Who, after all, remembers the annihilation of the Armenians?"[45]

---

[45]Kevork B. Bardakjian, *Hitler and the Armenian Genocide* (The Zoryan Institute Special Report Number 3) (Cambridge, Mass., 1985), 1; W. Baumgart, "Zur Ansprache Hitlers vor den Führern der Wehrmacht am 22. August 1939," *Vierteljahrshefte für Zeitgeschichte* 16 (1968): 127-128, 139; Yves Ternon, "La qualité de la preuve," in *L'Actualité du Génocide des Arménien,* 135-140; E. Calic, *Unmasked: Two Confidential Interviews with Hitler in 1931* (London, 1971), 11, 80, 81; Diane Orentlicher, "Genocide," in *Crimes of War: What the Public Should Know,* ed. Roy Gutman, David Rieff, and Kenneth Anderson (New York, 1999), 156.

Appendix—*Two Poems*

Literature can help survivors of ethnic cleansing to come to grips with the horror and to work for a renaissance of that which one loves most, looking toward the future, but without forgetting the past. This positive force—the love for one's homeland, of one's roots, and of one's family—is expressed in the following two poems. They are reproduced here with the kind permission of the authors.

The Uprooted

Nous les déracinés,
arrachés de nos terres
des siècles et des années,
refusons de nous taire.

Nous les déracinés,
implantés dans le Monde
partout où nous sommes nés,
nos racines sont profondes.

Nos racines arrachées
à nouveau s'enracinent
sans jamais se détacher
de nos vraies origines.

Des siècles et des années
nous qui savons souffrir,
bien que déracinées,
refusons de mourir.

—Shamiram Sevag

Arménie

C'est mon pays. Je l'aime
de telle sorte
qu'avec moi je l'emporte
où que j'aille. Il est tout petit,
en vérité,
comme une vielle mère épuisée par la vie,
comme un enfant qui vient de naître.
Et sur la carte,
il n'est pas plus gros qu'une larme ...
C'est mon pays. Il m'est si cher
que pour ne pas soudain le perdre,
j'ai fait de mon coeur son logis.

—Hovhannès Krikorian
(translated by Vahe Godel)

Ethnic Cleansing in the Greek-Turkish Conflicts from the Balkan Wars through the Treaty of Lausanne: Identifying and Defining Ethnic Cleansing

BEN LIEBERMAN

From the First Balkan War in 1912 through the Greek-Turkish War of 1919-1922, Greek and Turkish civilians suffered years of terror marked by theft, arson, deportation, and massacres. Waves of flight culminated in the population exchange of millions after the final Greek defeat in 1922. Even as large numbers of Greeks and Turks lost their homes, the conflict also brought a struggle to interpret and define civilians' experiences. Many, including foreign observers, found it difficult to comprehend what they saw. Greek and Turkish political and military leaders, for their part, competed to define what had happened as they sought to win international support. Thus, the preface to the first of a series of accounts of Greek atrocities published in 1922 by the Turkish General Staff of the Western Front explained the goal of making known Greek "cruelties...for the edification of the civilised world."[1]

The publicity, both Greek and Turkish, devoted to atrocities and accounts by foreign observers offer a guide to early-twentieth century perceptions of ethnic cleansing and raise several related questions for historians. How closely, first of all, did these sources demonstrate what historians now define as ethnic cleansing? However, since contemporaries did not in fact speak of ethnic cleansing, how did they identify the abuses and atrocities that they described? For early twentieth century observers of forced migration and mass flight, how was ethnic cleansing different from extermination?

The Balkan Wars

To explain the causes of the Greek-Turkish War of 1919-1922 and the subsequent exchange of Greek and Turkish populations set up by the Treaty of Lausanne in 1923, Greeks, Turks, and foreign observers looked back at least as far as the Balkan Wars if not earlier.

---

[1] Second Section of the General Staff of the Western Front, *Greek Atrocities in Asia Minor First Part* (Constantinople, 1922), 5.

The First Balkan War in 1912 brought defeat for Ottoman armies and the flight of large numbers of Muslim refugees from areas in Macedonia and Thrace lost to the Balkan allies. In a period in which Muslim and Turkish identity were beginning to merge in the Balkans, Muslims did not flee exclusively from the Greek army. Indeed the largest number may well have fled before the advancing Bulgarians or left their homes to escape the terror inflicted by irregular units (*Komitadjis*), or even their neighbors, and others left their homes in areas conquered by Serbia. Still, Muslims fleeing the Greek army shared in the common experience of refugees in 1912. Observers of the columns of refugees depicted a landscape filled with peasants leaving their homes with whatever possessions they could carry. Reporting on the aftermath of the Greek victory at Yenidge-Vardar, west of Salonika, the British journalist W. H. Crawfurd Price, met "a troop of barefoot Moslem peasants leading donkeys upon which were piled mattress and quilt and coffee-pot, all they had saved in the rush from Yenidge, when Greek guns set fire to the rude huts."[2]

Was the suffering of civilians, Muslim, Greek, Bulgarian, Serbian, Albanian, or Macedonian, in the Balkan Wars simply terrible misfortune or the result of deliberate policy? The most decisive judgment came from the International Commission established by the Carnegie Endowment for International Peace to investigate the "causes and conduct" of both the First Balkan War and the Second Balkan War, which broke out in 1913 between the first war's victors. The Commission observed that "the object of these armed conflicts, overt or covert...was the complete extermination of an alien population."

The Ottoman Greeks from the Balkan Wars to 1918

Yielding victory and territory for Greece, the Balkan Wars increased insecurity for the many Greeks who remained in Ottoman territory. Indeed, Greek sources charged that mistreatment of Greek civilians began amidst Turkish defeat. Thus, Archmandrite Alexander Papadopoulos, in a 1919 booklet devoted to publicizing persecution of Greeks before the First World War, recounted destruction of villages and murders of Greeks by "Turkish" military forces after key Turkish defeats in 1912. With Bulgaria's defeat at the hands of its erstwhile allies, Greece and Serbia, along with Romania, in the Balkan War of 1913, the "Turkish army" along

---

[2] W.H. Crawfurd Price, *The Balkan Cockpit: The Political and Military Story of the Balkan Wars in Macedonia* (London, 1915), 97.

with irregulars, then moved back toward Adrianople/Edirne and "reduced entire Greek settlements in Thrace to ruins."[3]

Peace came in 1913, but relations remained tense between Greece and Turkey through 1918, even though Greece did not formally enter the First World War until June 1917. The Turkish government began to deport its own Greek civilians from selected regions as early as the spring of 1914. During the war years, Turkish authorities forcibly moved Greeks from regions including Eastern Thrace, the Sea of Marmara, the Aegean coast of Asia Minor, and the Black Sea Coast. Some were expelled to Greek offshore islands, while others were sent into the interior of Asia Minor. Literature publicizing the Greek case described a Turkish campaign of persecution that included plunder, burning of homes, destruction of towns and villages, murders, and flight.[4] As many as 240,000, 450,000, or even more, according to some Greek sources, were forced to leave their homes.[5]

For individual Ottoman Greeks, the war brought experiences similar to what we now know as ethnic cleansing. Thus, an account of the expulsion of Greeks from the Isle of Marmara in June 1915 recounted how civilians learned of their fate. On June 5, 1915 the Governor informed Greek councilors that that Greeks were to be sent to Asia Minor, and government officials at midnight on June 6 "began knocking at the doors of the houses telling the occupants to light up and make their preparation for departure." The first steamer of deported Greeks left the very next day. Within days, the Greek community had been expelled. Perhaps this was not quite equivalent to modern ethnic cleansing in that the island's Greeks were not expelled from Turkey itself, but they saw little or no chance for return since virtually "all moveable and immovable property was left in the possession of the government."[6]

Deportations of Greeks clearly occurred in several regions of Turkey during the First World War, but how were these perceived in their entirety? Was there a general program of ethnic cleansing

---

[3] Alexander Papadopoulos, *Persecution of the Greeks in Turkey before the European War*, trans. Caroll Brown (New York, 1919), 25, 31.

[4] Papadopoulos, *Persecution of the Greeks in Turkey before the European War* and *Persecutions of the Greeks in Turkey since the Beginning of the European War*, trans. by Carroll N. Brown and Theodore P. Ion (New York, 1918).

[5] *Persecutions of the Greeks in Turkey since the Beginning of the European War*, 64-65; and *Lausanne Conference on Near Eastern Affairs 1922-1923: Records of Proceedings and Draft Terms of Peace* (London, 1923), p. 24.

[6] *The Tragedy of the Sea of Marmora*, introduction by N. G. Kyriakides, 5-6.

or a more limited campaign? Some of the foreign observers of Turkey described a massive if incomplete campaign to move Greeks from their homes. George Horton, the American Consul at Smyrna, an observer sympathetic to Greeks, distinguished between the treatment of those with Greek citizenship, who did not suffer "extreme persecution," and the Ottoman Greeks who "were massacred, robbed, driven out of their homes, ravished" or drafted and sent to forced labor without "food or clothing, until many of them died...."[7] Ambassador Henry Morgenthau depicted a Young Turk drive to persecute Greeks on the coast to encourage their flight before war. By spring 1914, "whole settlements of Greeks in Asia Minor were rounded up by the Turkish troops . . . and deported."[8] Arnold Toynbee saw "Turkish reprisals against the West Anatolian Greeks" as "general" in spring 1914, though the practice of driving "entire Greek communities" from their homes did not then extend to larger towns and cities. Persecution intensified in spring 1916 to include "deportation, first partial and then wholesale, of the Greek population along the Aegean and Marmora littoral."[9]

War brought an expanding campaign of persecution of Ottoman Greek civilians that perhaps ultimately extended to ethnic cleansing in some regions, but contemporary Greek publicity often described Turkish policy as a campaign against Hellenism. A 1914 memorandum by the Ecumenical Patriarch declared that "Hellenism . . . is face to face with a systematic persecution and the beginning of the application of a program of annihilation."[10] Publicity for the Greek cause in 1918 described a continuing "program put into operation by the Young Turks in the year 1913, with the object of annihilating Hellenism."[11]

Outlining an assault against Hellenism, did Greek sources perceive a campaign to dismantle the Greek cultural presence in Turkey or an effort to drive Greeks from the country? The boundary, if any, between a campaign against Greek culture in Turkey,

---

[7] George Horton, *Recollections Grave and Gay: The Story of a Mediterranean Consul* (Indianapolis, 1927), 224.

[8] Henry Morgenthau, *I was sent to Athens* (Garden City, New York, 1929) 16-17, 19-20.

[9] Arnold J. Toynbee, *The Western Question in Greece and Turkey* (London, 1922) 140-42.

[10] Papadopoulos, *Persecution of the Greeks in Turkey*, 141.

[11] *Persecutions of the Greeks in Turkey since the Beginning of the European War*, v.

and an effort to destroy Greeks themselves was not always clear. At times, Greek publicity depicted an effort to eradicate Greeks, or as one publication put it, "all this horrible treatment has one object, namely the annihilation of the Greeks in Turkey who must disappear as have the Armenians."[12] Still, Greek publicity, despite stressing massive suffering by civilians burned out of their homes, did not describe any such campaign as having been completed by the war's end. Significant Greek communities remained in Turkey in 1918, though many Ottoman Greeks had been forced from their homes or had fled.

1918-1923

The end of the Great War brought defeat for Turkey, the arrival of an allied military administration in Constantinople in December 1918, the landing of Greek forces in Turkey in 1919, and war between Greece and the new Turkish nationalist movement, which created a provisional government at Ankara in 1920. From the Greek landing at Smyrna/Izmir in 1919 to the final Nationalist victory in 1922, civilians suffered immense hardship during the new Greek-Turkish conflict. There is little dispute over the overall course of the ill-fated Greek campaign in Turkey, yet Greek and Turkish Nationalist publicity offered very different interpretations of the causes for and the extent of civilian flight between 1919 and 1922.[13] As the British Admiral Webb observed, it was "almost a physical impossibility for one side tell the truth about the other" after the Greek occupation of Smyrna.[14]

Persecution of Greeks

Accounts of Greek civilian suffering during the war of 1919-1922 isolated several key stages of civilian flight. The war itself made the position of Turkey's Greeks increasingly precarious in areas of the interior far from the Greek army, especially on the Black Sea coast. Greek publicity collected evidence from missionaries and journalists, among other sources, of deportations of Greeks from the Black Sea in 1921-1922. Receiving numerous Greek complaints,

---

[12] *Persecutions of the Greeks in Turkey since the Beginning of the European War*, 55.

[13] For an authoritative overview see Michael Llewellyn Smith, *Ionian Vision: Greece in Asia Minor 1919-1922* (London, 1973).

[14] *Documents on British Foreign Policy 1919-1939, First Series 4*, ed. by E. L. Woodward and Rohan Butler (London, 1952), 733.

the Allied High Commissioners in Constantinople asked the Turkish Nationalists to investigate the reports of deportations.[15]

There was no doubt that deportations occurred, but at what level and with what ultimate goal? Greek sources and sympathetic observers described persecution of Greeks during the war as nothing less than extermination. Thus the Locum Tenens of the Greek Patriarchate and his political advisor told the British High Commissioner in Constantinople, Sir. Horace Rumbold, in October 1921 that "they could not stand by and see the Greek race exterminated."[16] Major F. D. Yowell of Near East Relief concluded that "the policy of the Turkish toward the Greeks who were, and are still, being deported . . . seems to be one of extermination."[17]

As the struggle to define the events of the war intensified, Nationalist leaders rejected Greek charges of a general campaign to exterminate Greeks. Instead they depicted any punishment, including deportations, as a justified response to treason. Pressed by the Allied High Commissioners to investigate charges of deporting Greeks, the Turkish Nationalists represented by Minister of Foreign Affairs Yussuf [Youssouf] Kemal blamed Greeks themselves for any such actions near the Black Sea. Sir Horace Rumbold reported in September 1921 that Yussuf Kemal attributed Nationalist measures to the "actions and activities of the Greek population itself." In November Yussuf Kemal, according to Rumbold, rejected comparisons between "Greek massacres . . ." and "legal and impartial sentences on Turkish subjects in Samsoun region who plotted the dismemberment of Turkey...."[18] In February 1922 Rumbold reported that Yussuf Kemal charged the "Ottoman Greeks" with seeking to "engineer a rebellion" and added that "no Ottoman Greeks had been deported into the interior" until Greek warships bombarded Samsoun.[19]

Much like the Greek publicity it countered, the Turkish Nationalist response to the Allied High Commission depicted a

---

[15] Lysimachos Oeconomos, *The Martyrdom of Smyrna and Eastern Christendom* (London, 1922) 26, 32, 36; and *Documents on British Foreign Policy 1919-1939 First Series Volume XVII*, ed. by W.N. Medlicott, Douglas Dakin, and M.E,. Lambert (London, 1970), 379-81.

[16] *Documents on British Foreign Policy 1919-1939, First Series 4*, 430.

[17] Edward Hale Bierstadt with Helen Davidson Creighton, *The Great Betrayal: A Survey of the Near East Problem* (New York, 1924), 53.

[18] *Documents on British Foreign Policy 1919-1939 First Series Volume XVII*, 397, 498.

[19] *Documents on British Foreign Policy 1919-1939 First Series Volume XVII*, 632.

country of separate and hostile national communities. Greeks could only betray the government as "Ottoman Greeks", an identity that crossed ethnic and national lines, yet the Turkish Nationalist themselves threatened any fragile Ottoman authority. The very logic used to refute Greek charges showed Nationalist acceptance of the principle of expelling large portions of Greeks as an alien and intractably hostile nation within Turkey. Yussuf Kemal described removal of Greeks from the Black Sea coast to Rumbold as "measures of precaution" to prevent a possible Greek landing, but such measures "consisted in the deportation of all Greeks who were not natives of Pontus, in the transfer into the interior of all capable of bearing arms and consequently of actively assisting the Hellenic forces....."[20] The line between civilians and combatants blurred as all Turkey's Greeks who might be able to bear arms became subject to possible removal from their homes.

With the defeat of the Greek army in Turkey in 1922, the flight of Greek refugees attracted widespread international attention. As the Greek army retreated, Greek civilians clogged the roads leading into Smyrna, and the burning of most of the city in September then prompted further flight of panicked refugees. Greek and Turkish sources contested only responsibility for the fires at Smyrna. While Greeks and their foreign sympathizers cited reports blaming Turkish soldiers for setting the fire, Turkish sources blamed the Greeks themselves, or Armenians, with setting fires in Smyrna.[21]

With the fire and flight from Smyrna, much of the remaining Greek population began an exodus. The allies agreed that Eastern Thrace would return to Turkey, and soon hundreds of thousands of Greeks from Thrace packed up whatever possessions they could carry, many of which ended up in or on the sides of roads, and made their way by train, by car, or by foot out of Eastern Thrace before the arrival of the Turkish Nationalists.

As many of Turkey's Greeks left their homes (a significant though declining number remained in Constantinople/Istanbul), Greek and Turkish leaders and publicists disputed the cause for the exodus. The reason for flight was clear to Venizelos, leader of the Greek delegation at the Lausanne Peace Conference, who said that Greeks "fled for fear of reprisals."[22] Foreign reporters wit-

[20] *Documents on British Foreign Policy 1919-1939 First Series Volume XVII*, 397.

[21] *Greek Atrocities in Asia Minor Part 4* (Constantinople, 1922), 3-5; and *Lausanne Conference on Near Eastern Affairs*, 625-26.

[22] *Lausanne Conference on Near Eastern Affairs*, 627.

nessing the march of civilians out of Eastern Thrace offered similar explanations. Reporting on a twenty-mile long column of refugees crossing the Maritza River in October 1922, Ernest Hemingway wrote, "They don't know where they are going. They left their farms, villages and ripe, brown fields . . . when they heard the Turk was coming." W. T. Massey of the *Daily Telegraph* described refugees discarding their possessions "in order that they might not lose the race to the goal, to reach which before the Turk appeared on the skyline...."[23]

Turkish sources, in contrast, reversed responsibility for the Greek flight. According to Ismet Inönü, chief of staff of the Turkish Nationalist army and chief Turkish representative at the Lausanne conference, it was not fear of the Turks, but coercion by the Greek army that encouraged panic flight from Eastern Thrace.[24] At the same time, however, Turkish publicity equated ever larger groups of Greek civilians with Greek soldiers. One of the most controversial Nationalist provisions after the capture of Smyrna was the decision to treat all adult Greek men between eighteen and forty-five as prisoners of war subject to forced labor.[25] Defending the Nationalist position, a publication of the Military Staff of the Western Front argued that Greeks and Armenians who "joined hands with the Hellenes...have served the Greeks as soldiers and by forming special military organization and doing criminal acts have ultimately betrayed the Turk." It was not clear, however, that active military service was the key evidence of treason. In addition "the male population of military age, that is those between the ages of 18 to 45 who were ready at every occasion to strengthen the enemy and who were as a matter of fact soldiers in the Greek army were sent to camps as prisoners of war." Following this logic, the Turkish Nationalists regarded virtually all adult male Greeks and Armenians as enemy soldiers subject to capture as prisoners of war. As for the rest, they "were left free to go anywhere outside the Turkish country which they had betrayed so meanly."[26]

Looking back on the previous decade after the final war, Greek authorities and those sympathetic to Greece saw a systematic and

---

[23] Ernest Hemingway, *By-Line*, edited by William White (New York, 1967), 51; and Bierstadt, *The Great Betrayal*, 243.

[24] *Lausanne Conference on Near Eastern Affairs*, 627.

[25] Llewellyn Smith, *Ionian Vision*, 318.

[26] *Greek Atrocities in Asia Minor Part 4*, 21-22.

continuous effort by Turks to clear Turkey of Greeks. The Greek delegation at Lausanne charged that "since 1914 Turkey has never ceased within her territory to take all those notorious measures the effect and deliberate object of which has been to diminish her own population by 2,000,000 Greeks and Armenians."[27] Surveying the surge of refugees at the war's end, Henry Morgenthau referred to the "deliberate intention of the Turks to remove utterly all Greek population from Asia Minor," and Edward Hale Bierstadt charged that "the Turkish Nationalist Government had been carrying out a systematic policy of extermination toward the minorities ever since its inauguration...."[28]

Persecution of Turks

Even as Greek sources described a Turkish campaign of extermination against Greeks, Turkish sources and some foreign observers publicized systematic Greek persecution of Turks during the Greek campaign in Asia Minor. Controversy over Greek atrocities began with the Greek landing at Smyrna on May 15, 1919 during which Turks suffered 300 to 400 casualties (Greeks reported lower numbers) before the imposition of stricter order by the new Greek governor of Smyrna, Aristeidis Sterigiadis.[29]

Whereas Greek sources stressed the lack of abuses under Sterigiadis, accounts of persecution of Turks saw the Smyrna landing as the starting point for a brutal campaign of persecution against Turks in the entire region. A Turkish report on Greek atrocities in the Vilayet of Smyrna described a campaign of burning villages, massacres, and rape.[30] To the south of Smyrna, Greek forces moved up the Büyük Menderes/ Meandre river valley from Aydin/Aidin to Nazilli, burning towns and villages and pillaging along the way.[31] To the north, Greek forces entered Menemen, car-

---

[27] *Lausanne Conference on Near Eastern Affairs*, 680.

[28] Morgenthau, *I was sent to Athens*, 47; and Bierstadt, *The Great Betrayal*, 52.

[29] Llewellyn Smith, *Ionian Vision*, 89-92; Arnold J. Toynbee, *The Western Question in Greece and Turkey: A Study in the Contrasts of Civilizations* (London, 1922), 270-72; and Cagri Erhan, *Greek Occupation of Izmir and Adjoining Territories: Report of the Inter-Allied Commission of Inquiry* (May-September 1919) (Ankara, 1999), 58.

[30] Permanent Bureau of the Turkish Congress at Lausanne, *Greek Atrocities in the Vilayet of Smyrna (May to July 1919) in Edited Documents and Evidence of English and French officers* (Lausanne, 1919), 4-5, 60.

[31] Ibid., 6, 59-60, 66, 75-76.

rying out a massacre, and Manisa (Magnesia), killing and beating civilians.[32] The Turkish governor of Aidin reported that Greek forces advanced the terror by ordering non-Muslims to wear hats—this prevented errors during massacres—and by ordering Greek and Armenian shops to have signs in Greek, "this to prevent the molestation of the non-Turks when the town was pillaged."[33] Providing a similar overview of Greek conduct, Harold Armstrong, a British military attaché with the Inter-Allied Commission at Constantinople and supervisor of Turkish Gendarmerie, wrote, "From Smyrna the Greeks pushed out, massacring, burning, pillaging and raping as they went, in the ordinary manner of the Balkan peoples at war. Before them the Turks fled...."[34] A Turkish commander reported that Muslims fleeing the terror near Aidin went south of the Meandre or took "refuge in the mountains."[35] Closer to Constantinople, the Gemlik Yalova Peninsula on the southeast of the Sea of Marmora was another area of Greek aggression. In one of the best-known contemporary accounts of the consequences of the Greek invasion, Arnold Toynbee noted reports of the destruction of villages.[36]

How many fled the initial Greek offensives? There was little consensus on numbers. As early as 1919, the Inter-Allied Commission of Inquiry into events at Smyrna described abandoned homes and fields to the south. In the "Meander Valley," it reported, "entire villages have had to be abandoned, even if they were not destroyed by fire." However, the Commission was not able to estimate the total number of refugees.[37] Turkish sources were less cautious about providing figures. Thus, a "memorial of Greek atrocities" in Aidin and Nazilli declared that "hundreds of thousands of refugees...are now wandering in the mountains without shelter."[38]

Flatly rejecting Turkish charges of forced emigration of "the Moslem population," the Greek delegation at Lausanne declared

---

[32] Ibid., 5,8, 39-40, 51.

[33] Ibid., 63.

[34] Harold Armstrong, *Turkey in Travail: the Birth of a New Nation* (London, 1925), 83, 101.

[35] *Greek Atrocities in the Vilayet of Smyrna*, 61.

[36] Toynbee, *The Western Question*, 283-84, 311.

[37] Erhan, *Greek Occupation of Izmir*, 63, 65.

[38] *Greek Atrocities in the Vilayet of Smyrna*, 60.

that "at no time during the Greek occupation of Asia Minor was any such measure promulgated or actually applied...." If Muslim civilians were forced to move, "it can only have been under orders received from the Government and from the military authorities of Angora."[39] Such flight by civilians, of course, could also have demonstrated fear of Greek forces. The Inter-Allied Commission of 1919, for its part, was also been unable to determine whether Turkish refugees stayed away from their homes because of "mistrust of the Greeks or because Turkish irregulars are preventing them from returning for political grounds."[40]

Amidst the reports of abuses of civilians, foreign observers and Greek and Turkish authorities disputed the goals of Greek policy toward Turks. Even in the early years of the war when Greece forces held a military advantage in western Anatolia, Greece struggled to pacify and hold areas where Greeks formed a minority. This was especially true as Greek forces advanced beyond river valleys of western Anatolia. Greek and Turkish authorities bitterly disputed the precise breakdown of population, but there was little doubt that Greeks and other Christians made up a minority outside of Smyrna itself and some urban quarters. What, then, was to become of a Muslim majority under Greek rule? The past decade of violence in Turkey and the Balkans offered varied solutions ranging from full-scale ethnic cleansing or worse to more limited attacks on political and cultural leaders. Resettling Ottoman Greeks who had taken refuge in Greece during the First World War was one option, and more than 100,000 returned—if not for long, but despite the onslaught of 1919, the Greek offensives did not empty western Anatolia of Turks.[41] Where Muslims did not flee during the initial Greek advance, Greek forces more typically targeted local leaders. North, south, and west of Smyrna, in regions near Aidin, Nazilli, Kula, Alasehir, Manisa, and Salihli, local Turkish notables faced the threat of deportation. Toynbee referred to a "policy of striking at the Turkish upper class."[42] Some Turkish accounts of Greek misdeeds similarly described a pattern of targeting officials and notables. A report on Greek atrocities in Smyrna detailed killings of police, officials, and a newspaper publisher as well as killings of officials and

---

[39] *Lausanne Conference on Near Eastern Affairs*, 680.

[40] Erhan, *Greek Occupation of Izmir*, 65.

[41] Llewellyn Smith, *Ionian Vision*, p. 100.

[42] Toynbee, *The Western Question*, pp. 290-91.

notables at Aidin.[43] This terror, perhaps intended to ease Greek occupation, did not necessarily amount to full-scale ethnic cleansing.

Even though Greek forces and irregulars may not have aimed at driving out the entire Turkish population in Greek occupied areas, contemporary Turkish publicity perceived the Greek goal as nothing less than extermination. Condemning the early months of Greek occupation, the Committee for Defense of Ottoman Rights in Smyrna in June 1919 described an occupation "which one seems decided to strengthen by Hellenic measures, for a policy of rapid extermination of the enormous Musulman majority inhabiting the Vilayet of Aidin." A letter by Turkish officer Chukri Bey, commenting on occupation of the region around Aidin, similarly charged that the Greeks "have begun after a short period of calm to practice...the policy of the extermination of the Turkish element, with the object of being able to claim and annex these countries the 95% of whose population are Turks and Mussulmans."[44]

Nineteen twenty-one marked a key turning point in the war. Violence intensified as Greek and Turkish irregulars fought each other. Harold Armstrong described a country where all, whether Greeks, Armenians, or Turks, lived in fear: "Wherever I traveled I saw stark fear lay in the eyes of every man in the country."[45] Greek forces attempted to drive the Turkish Nationalists out of the war through an offensive aimed at Ankara, but the Greek advance stalled near the Sakarya River and ended in retreat. This campaign featured prominently in Turkish accounts of Greek atrocities. Pamphlets produced by the staff of the Nationalist army on the western front listed dozens of villages burned, along with their mosques and schools, near the Sakarya itself and then west toward Sivrihisar and Eskişehir, and pillage and rapes.[46]

To what extent did burning villages in the Sakarya campaign form part of a program to expel the Turkish population? Greek representatives later offered military explanations for the actions of Greek forces. Venizelos said that villages had been burnt during the Greek retreat between Sakarya and Eskisehir because of

---

[43] *Greek Atrocities in the Vilayet of Smyrna*, 12, 59; and Toynbee, *The Western Question*, 278.

[44] *Greek Atrocities in the Vilayet of Smyrna*, 10, 60.

[45] Armstrong, *Turkey in Travail*, 162.

[46] *Greek Atrocities in Asia Minor First Part*; and 2me bureau d'Etat-Major du front d'occident, *Atrocités Greques en Asie-Mineure 3me Partie* (Constantinople, 1922).

"purely military exigencies."[47] British reports, for their part, of-fered a picture of extensive devastation during the Greek retreat. An October 1921 memo by Lord Curzon on the Greek retreat from Sakarya referred to a report from September 20 by Major Johnston, Liaison Officer with the Greeks, "that the Greeks were destroying the railway west of the Sakarya by blowing up each rail, and were laying waste the country."[48] Armstrong noted Greek devastation of 200 miles during the retreat: "They tore up every mile of the permanent way of railway. They killed every Turk who was fool-ish enough to be still there, and for 200 miles behind them left desolation and the villages flat with the ground."[49]

Turkish Nationalists described an intentional Greek policy of destruction during the Sakarya campaign. First, they charged that the Greek army made particular units responsible for arson. The preface to one of the atrocity pamphlets asserted that the entire Greek 10th division "is designated by the Greeks as a division of vengeance and destruction."[50] Testimony from a Greek prisoner of war indicated that two retreating Greek divisions set up special detachments to burn villages.[51] Second, Turkish Nationalist pub-licity accused Greek commanders of approving destruction. Turkish accounts placed particular blame on the Greek com-mander Prince Andrew—today probably best known as the father of Queen Elizabeth II's husband Prince Philip—who had only re-cently returned to Greece after Venizelos lost the election of No-vember 1920. One interrogated prisoner of war said: "The com-mandant of the division Prince Andreas, ordered the soldiers to set on fire all the villages through which they were to pass," al-though another was only "inclined to believe that the Greek com-mander knew about the destruction of the villages by fire."[52] An-other lurid account presented without any specific testimony from prisoners of war reported Prince Andrew present during the pil-laging and burning of the village of Karhan, watching fires "like Nero," warning that he would come back and burn other places if

---

[47] *Lausanne Conference on Near Eastern Affairs*, 547.

[48] *Documents on British Foreign Policy 1919-1939 First Series Volume XVII*, 422.

[49] Armstrong, *Turkey in Travail*, 209.

[50] *Greek Atrocities in Asia Minor First Part*, 4.

[51] *Atrocités Greques en Asie-Mineure 3me Partie*, 106.

[52] *Greek Atrocities in Asia Minor First Part*, 10, 13.

Kemal did not make peace.[53] Were these charges accurate? They appear to have made little impression on Andrew himself, who made no mention of them in his memoir about the campaign, though he was determined to defend his military decisions from criticism. The prince's account concurred with Turkish reports only in that Andrew mentioned the Greek use of young Turkish men to dig trenches.[54]

What exactly had happened on the Sakarya? Was this a difficult military campaign in which an overextended army beat a hasty retreat in an inhospitable region as Prince Andrew suggested? Or, had the Greek army deliberately targeted the Turkish civilian population for attack as Nationalist accounts of atrocities maintained? The paucity of independent reports by eyewitnesses made it difficult to sift through the claims and counterclaims. Even Turkish accounts did not suggest that the Greeks had literally tried to expel most of the civilian population from near the Sakarya, so in one sense Greek persecution of civilians was not full scale ethnic cleansing. On the other hand, Turkish nationalists perceived something even worse. Thus the preface to one of the Nationalist pamphlets on the Sakarya campaign drew readers' attention to the Greek "policy of extermination" evident in recent years. Greek atrocities since the landing in Smyrna, "exceed all similar crimes recorded up to now in the annals of history."[55]

The last major evidence in the Turkish case against Greece before the end of the war concerned the final Greek retreat to Smyrna in 1922 after Greek defenses collapsed in August. Turkish Nationalist publicity described a retreating army destroying towns, and villages as it fled to the sea; although, without always making clear exactly how much damage had been inflicted during occupation before the Greek retreat. In any case, a report by Captain Ismail Hakki Effendi of October 29, 1922 charged that "the enemy has burnt all the villages on the railway line . . .with the exception of one or two, has looted....killed a part of their inhabitants with tortures...violated women....." Atja's population fell from 12,000 to 3,000. The Greeks left Nazilli so damaged that "the city remains hardly a village. Out of 17,000 inhabitants only 3000 remain." Aidin and Sokkia had also suffered extensive burning.[56]

---

[53] *Atrocités Greques en Asie-Mineure 3me Partie*, 31-32.

[54] Ibid., 86; and Prince Andrew of Greece, *Towards Disaster: the Greek Army in Asia Minor in 1921* (London, 1930), 281.

[55] "Policy of extermination," *Greek Atrocities in Asia Minor First Part* , 3.

[56] *Greek Atrocities in Asia Minor Part 4*, 9-11, 15, 17-18.

"Carrying out their plan of annihilation and extermination" also provided the Greeks with opportunities for extortion.[57]

The Greek delegation at Lausanne responded to such Turkish accounts by conceding misdeeds committed by the Greek army along its line of retreat. "Excesses had certainly been committed," Venizelos acknowledged, but added that "an army which was no longer willing to fight...could not have gone out of its way to ravage wide tracts of country to right and left; at the most it could only have set fire to what was on its line of retreat." Ismet, angered by this logic, countered that the Greek army, "extending over a front several hundred kilometers long" had retreated across a broad front and had also carried out "a large part of the destruction ...prior to the last offensive."[58]

Where Greek sources admitted excesses and British reports noted extensive devastation, Turkish nationalists consistently identified extermination throughout the Greek-Turkish War of 1919-1922. The key to labeling any given series of "atrocities" as extermination was the sense that each episode of violence fit into a long-term pattern of driving out Muslims and Turks, which dated back to wars in Turkey in Europe during the nineteenth century and now extended into Anatolia. Thus Turkish reports of atrocities during the occupation of Smyrna and its region in 1919, charged that Greek authorities "began immediately after the occupation, to put into force again their former ignoble system which met with such success in the Morea (Peloponnesus), in Thessaly, in Epirus, in Crete and recently also in Macedonia." The results of Greek terror could be seen in the presence throughout Turkey of immigrants from Crete, Morea, Macedonia, Epirus, and elsewhere, living evidence, that "to give over a country to Greece is to doom its inhabitants to torture, to depopulate it...."[59] The Ottoman League at the end of May 1919 noted a past of "systematic extermination of the Turks of Thessaly, of Macedonia, and the Islands...."[60] In a lengthy presentation to the Lausanne Peace Conference, Ismet presented a historical overview looking back to "troubles and massacres" in the Morea during the Greek War of Independence. Recalling a history of mass flight of Muslims from regions lost by the Ottoman state, Turkish nationalists interpreted

---

[57] *Greek Atrocities in Asia Minor Part 4*, 11.

[58] *Lausanne Conference on Near Eastern Affairs*, 53, 85.

[59] *Greek Atrocities in the Vilayet of Smyrna*, 3-5.

[60] *Greek Atrocities in the Vilayet of Smyrna*, 96.

a wide range of abuses and atrocities during the war of 1919 to 1922 as steps in a comprehensive, long-term campaign to exterminate Muslims and Turks. Indeed, this remains a major theme of Ottoman historiography to the present day.[61]

Atrocities, Extermination, and Ethnic Cleansing

A decade of conflict from the Balkan Wars to the Treaty of Lausanne brought massive suffering to the many Greeks and Turks who experienced the burning of their homes, loss of possessions, deportation, and in many cases death. Those who spoke for Greek or Turkish authorities made few apologies. They either defended persecution of civilians as military or legal necessity or, at most, conceded excesses committed either during retreat or by armies excited to vengeance after having seen evidence of other atrocities.

From a contemporary perspective, the Greek-Turkish conflict brought a wide range of abuses of human rights that in some cases included ethnic cleansing: in the Balkan Wars, in some coastal regions of Turkey during the First World War, and at the very end of the Greek-Turkish war of 1919-1922. In other cases, it is more difficult to determine whether ethnic cleansing occurred. It is probably most difficult to gauge the precise level of flight and expulsions in areas in the interior of Anatolia. Greek forces occupying western Anatolia faced enormous practical obstacles to any plan to literally expel all of a large Muslim majority in areas outside of Smyrna. Indeed, there was never a clear sense of how Greece intended to hold western Anatolia should it win the war other than by harassing and expelling local Turkish notables and political leaders.

In an era before definitions of either ethnic cleansing or genocide, however, many contemporaries, Greeks, Turks, and some foreign observers, interpreted the violence of the conflict as extermination. Even when some significant proportion of the population under assault remained in their homes, Greeks and Turks explained the violence to the world as nothing less than extermination. Extermination in this conflict could threaten either a culture, such as Hellenism, a community, such as the many villages and towns pillaged and burned, or people themselves whether Greeks or Turks.

---

[61]*Lausanne Conference on Near Eastern Affairs*, 194. For a similar interpretation see Justin McCarthy, *Death and Exile: The Ethnic Cleansing of the Ottoman Muslims 1821-1922* (Princeton, NJ, 1995).

Seeking to make a case to the world at large, Greek and Turk-ish leaders and publicists created parallel narratives of extermina-tion, in which each respective population at risk faced a long-term threat to its survival. Such narratives provided little sense of any alternatives to escalating conflict that targeted civilian popula-tions. For some foreign observers this provided more evidence of "Balkan" behavior. From a Greek or Turkish perspective, entire populations, including civilians came to be identified as combat-ants. Thus, all adult Greek men could be treated as potential pris-oners of war at the war's end, and both sides could accept the logic of ethnic cleansing if expelling civilians en masse was neces-sary to prevent extermination.

Consequences of Population Transfers: The 1923 Case of Greece and Turkey

ELENI ELEFTHERIOU

As with many ethnic conflicts, the relationship between Greece and Turkey has been characterized by a series of interrelated disputes that represent an infringement on the freedom and basic rights of one group by the other. Early relations between Greek and Turkish populations in the nineteenth century quickly evolved into a zero-sum game of control and domination. Not only was the exchange of territory that demarcated the border between the young Greek state and the declining Ottoman Empire viewed in these terms, but more importantly, the control and coercion of each other's ethnic group was played out in a "tit-for-tat" manner. The violence and abuses by one side seemed to justify violence by the other, providing both sides with the rationale and opportunity to continue committing atrocities against each other. Gaining control and power over one's own fate meant unifying one's own group, which often led to taking control and power away from another.

National unity was idealized and pursued by centralizing power and demanding a common language, culture, and religion for citizens, which provided for a strong self-identification of the state. Acquiring unity through such measures, however, resulted in the state's intolerant and repressive behavior toward those perceived as "other."[1] Due to Ottoman domination of the Greeks from the fifteenth century, and with the later rise of nationalism in the nineteenth century, the separate identities of Greeks and Turks would eventually develop into a game of conquest between "self" and "other." The formation and development of "self" became inextricably linked to the control or conquest of "other." Each successive conflict between the Greek and Turkish populations built upon the previous conflict, and then served as the foundation for each subsequent conflict. This set the foundation for a history of animosity between the two peoples that has continued, in a variety of settings, until the present day.

The international community had few methods with which to manage the ethnic conflicts that erupted with the rise of nationalism, and most of these methods were ineffective. At the turn of the

---

[1] Patrick Thornberry, *International Law and the Rights of Minorities* (Oxford, 1991), 1.

twentieth century, the international treaties that were created to protect the rights of ethnic minorities were limited and ill defined, establishing only restricted rights for minorities within a limited geographic region of mostly European states. The most common strategies for diffusing the violence between ethnic groups were to exchange populations or to attempt to redefine territorial boundaries. Both were accomplished with immeasurable human suffering, and with little success in resolving the ethnic animosities.

This following study is an examination of the consequences of the population exchange between Greece and Turkey in 1923, and how this exchange shaped the nature of relations between Turkey and Greece over the course of the twentieth century. Section I will describe the general historical events that brought about the population exchange from the time of Greek independence until 1923. Section II will then discuss the transfer of populations and the immediate impact of uprooting and relocating entire communities. Section III offers a broad account of the long-term consequences of the population transfer. The fundamental concern in this analysis is to describe the circumstance that led states to destabilize their own societies, either by disenfranchising or abusing certain minority groups or by attempting to expel or otherwise eliminate these groups entirely. The study concludes with an examination of how a commitment to the protection and advancement of minority rights can mitigate the continuing abuses against minorities in both Greece and Turkey. For the purposes of this examination, the broad definition of a minority group as a non-dominant group distinguished by a shared ethnicity, race, religion, or language will be adopted.

Greek Statehood: Defining Borders and Separating Identities

In 1821, a portion of the Greek people gained its independence after four centuries of being under Ottoman rule. The nascent Greek state, however, included less than one-third of all Greek nationals from the Ottoman Empire.[2] This reversed the irredentist policy of a minority group seeking to unite with the majority of its "motherland." In the case of the Greeks, the minority group that was independent sought to expand its territorial control to include the majority of Greeks living to their north in Thessaly, Epirus, and Macedonia, and southward to encompass the Aegean islands and Crete. For many Greeks, this was simply a policy of reclaim-

---

[2] Theodore George Tatsios, *The Megali Idea and the Greek-Turkish War of 1897: The Impact of the Cretan Problem on Greek Irredentism, 1866-1897* (New York, 1984), 3.

ing territory that had once been theirs during the Byzantine Empire. For the Ottoman Turks, however, this Megali Idea (the "Great Idea" that would unite all ethnic Greeks), was perceived as a threat to their territorial integrity.[3]

The Birth of the Megali Idea

The Protocol of 1832 established the territorial boundaries of Greece, and marked the beginning of a national struggle to unite all ethnic Greeks under one nation-state. The Greeks sought to create a modern state with a strong centralized government and a clearly defined nationality. Greek leaders espoused the concept of nationalism and advocated the need for promoting a homogeneous culture based on a common language and religion, as well as national customs and traditions.[4] The purpose of this unifying process was two-fold. First, it rallied all citizens around the national goal of the Megali Idea. Populist nationalism was developed and utilized, not only to unify the citizens of this newly created state, but also to divert attention away from the overwhelming domestic problems that plagued Greece at the time. This allowed for greater attention to be focused on the foreign policy goal of freeing all Greeks who remained under Ottoman rule and by building a pan-Greek state. Greek sociologist Constantine Tsoucalas describes the emergence of Greece's policy of unification and its first ill-fated results:

> The period from 1895 to 1907 saw a substantial setback in Greece's economic development. The new government could not solve the problem of the external debt, and a state of permanent depression set in. Faced with rising popular discontent, the government had recourse to the Megali Idea. Public opinion was aroused; in 1897, despite economic depression and a total lack of military preparations, the government was pushed into declaring war on Turkey.[5]

In 1897, the Megali Idea brought Greece and Ottoman Turkey to fight their first war against each other as independent states. Turkey succeeded in defeating the ill-equipped Greek army, but

---

[3] Ibid.

[4] Keith R. Legg and John M. Roberts, *Modern Greece: A Civilization on the Periphery* (Boulder, 1997), 19.

[5] Constantine Tsoucalas, *The Greek Tragedy* (Middlesex, UK, 1969), 25.

did not succeed in destroying the Greek dream of a unified Greater Hellas. As a consequence of the war, Greece was required to pay a large indemnity to Turkey and to give up parts of Thessaly. Greece also failed in its attempt to unite with Crete. The only solace to Greece was that the island was given autonomy from Turkey.[6] The war also brought about grave economic problems for Greece, leading an international commission to take control of its finances. The pursuit of the Megali Idea was suspended,[7] but it would rise again, bringing tragedy to both Greeks and Turks, and establishing a pattern of mutual resentment and bitterness that would mar relations between the two countries for the remainder of the century.

The second function of the Megali Idea was to give the major revolutionary figures from the Greek War of Independence the opportunity to gain domestic political support in their competition for government positions.[8] One of these dominant figures was Eleftherios Venizelos, who became Prime Minister in 1910 and led Greece through both victory and defeat in pursuit of the Megali Idea. Greece was a victor in both of the Balkan Wars of 1912-13, the first against Ottoman Turkey and the second against Bulgaria. With the 1913 Treaty of Bucharest, which ended the Balkan Wars, Greece was able to expand her territory to include Crete, Epirus, most of Macedonia, as well as many of the Aegean islands.[9] The Megali Idea seemed to be gaining momentum and ostensibly becoming a reality, and it continued to focus the attention of Greeks both at home and those remaining in Ottoman Turkey who hoped to be unified with their ethnic brethren.

World War I and the Decline of the Ottoman Empire

From the onset of the First World War until 1916, the Turks deported approximately 150,00 Ottoman Greeks to Greece and moved 50,000 or more into the interior of Anatolia. In the process of the deportations many thousands of Greeks died.[10] The Armenians suffered a worse fate at the hands of the Turks, enduring

---

[6] Legg, 95.

[7] Tatsios, 110.

[8] Legg, 21.

[9] Tsoucalas, 31.

[10] Norman M. Naimark, *Fires of Hatred: Ethnic Cleansing in Twentieth-Century Europe* (Cambridge, Mass., 2001), 28.

several massacres during this period, and ultimately, genocide. Ottoman Turkey was facing its own demise and its rulers felt threatened by the heterogeneity and pluralism represented by its ethnic minorities.[11] The Turkish government's preferred strategy for dealing with the rebellious factions within the minority groups was to expel from its territory or exterminate all members of the ethnic communities that Turkey believed posed a threat.

During the First World War, Prime Minister Venizelos succeeded in overriding the King's interest in maintaining a position of neutrality by bringing Greece into the war on the side of the Allies. Britain encouraged Greece to join the Allied powers in the war by promising to grant Greece control over Western Anatolia. Greece, albeit belatedly, accepted Britain's offer. The defeat of the Ottoman Empire and Allied victory provided Greece with the opportunity to reap its share of the spoils of war. With the support of Britain and France, Greece sought to unite the Greeks of Asia Minor into a single Hellenic state. Since Ottoman Turkey had allied itself with Germany, the West was interested in bringing about the dissolution of the Ottoman Empire and the dissemination of its territory to the victors of the war.

As part of the Paris Peace Conference, the 1920 Treaty of Sèvres gave Greece almost all of Thrace (with the exception of Constantinople), as well as the Aegean islands of Imbros and Tenedos, and the Dodecanese except for Rhodes.[12] More importantly, the treaty provided Greece with administrative control over Smyrna (Izmir) and the surrounding Anatolian provinces with large Greek populations.[13] This appeared to bring the Megali Idea into fruition. In fact, the difficulty of administering an area heavily populated by Turks would prove fatal to Greek and Western calculations. Greece's determination to unite all ethnic Greeks by expanding its territory, despite Turkish nationalist opposition, would be met with a crushing defeat of the Greek army and would mark the beginning of the end for the Megali Idea.

The Path to War

The Treaty of Sèvres remained an illusive victory for Greece as long as the Turkish nationalist forces opposed the terms of the

---

[11] Ibid., 42.

[12] Michael Llewellyn Smith, *Ionian Vision: Greece in Asia Minor, 1919-1922* (Ann Arbor, 1998), 129.

[13] Tsoucalas, 34.

treaty and resisted its enforcement.[14] Although the Treaty of Sèvres offered international recognition and a tacit acceptance of the Megali Idea, this only provided the Greeks with a false sense of security by making them think that they would receive support from the West in their effort to annex territories from the collapsing Ottoman Empire. Greece's overestimated expectation of Western backing, combined with its underestimated view of Turkish nationalism, fueled a misguided Greek military strategy for uniting all ethnic Greeks under one nation-state.

In 1919, the Greek army occupied Smyrna, committing its own atrocities against the Turkish population.[15] By 1921, the Greek army had advanced eastward and northward, attempting to weaken and expel Turkish nationalist forces.[16] This was met with severe reprisals from the nationalists under the emerging leadership of Mustafa Kemal, who would later become the founding president of modern Turkey and still later be given the name of father of his country, *Atatürk*. The anti-nationalist Turks, as well as ethnic Greeks and Armenians, were targets of Turkish nationalist attacks. Nur Bilge Criss describes the conflict between the anti-nationalists, who sought to protect the minorities, and the nationalists (also known as Kemalists) who feared that the minorities would aid the secessionist movements.

> Ankara ordered the Turkish police in Istanbul to identify acts of treason, starting with the minorities. The anti-Nationalist activists who were Turks were already well known to Ankara. Those Turks who had a serious reason to fear Nationalist retribution were leaving the country. ... Turkish police were trying to gather information about the sympathies and about direct help provided by these people to the Greek cause.[17]

Just as the anti-nationalists opposed the Kemalist attacks against ethnic minorities in Turkey, Greece's strategy for enforcing the Treaty of Sèvres also had its critics. Ioannis Metaxas, a Greek military officer and staunch anti-Venizelist, opposed the enforcement of the Treaty of Sèvres on the basis that it was neither possible nor appropriate to impose Greek rule on Turkish soil. Metaxas believed that even if Greece had succeeded in controlling Smyrna

---

[14] Smith, 133.

[15] Clogg, Richard, *A Concise History of Greece* (Cambridge, 1992), 94.

[16] Naimark, 45.

[17] Nur Bilge Criss, *Istanbul Under Allied Occupation, 1918-1923* (Boston, 1999), 145.

and enforcing the treaty, a new war would be inevitable as Turkish nationalism would rise again and new guerrilla groups would be formed to oppose Greek rule. Metaxas' views epitomized the anti-Venizelist movement. In his response to questions of why he opposed the policy of war, he argued: "Because in fact you are seeking the conquest of Asia Minor, and without preparing for it through the Hellenization of the country. It is only superficially a question of the Treaty of Sèvres. It is really a question of the dissolution of Turkey and the establishment of our state on Turkish soil."[18]

Metaxas' position, had it been accepted by the ruling powers in Greece, would have avoided the fatal policy that led to the devastating consequences of 1922. However, it represented, nonetheless, a misguided belief about minority groups. His belief in the necessity of Hellenizing those territories Greece desired to rule was indicative of the mind-set of the time, which viewed minorities as a liability to civil society rather than as an entity to be protected. As long as minority groups were not treated with equanimity, they would represent the potential for instability. However, little effort was made to provide minorities with the same rights as the majority populations. Rather, governments sought to create homogeneous populations by changing the identities of the minorities who could be converted and expelling those who refused conversion.

Both Greek and Turkish elites believed that minority groups increased the potential for conflict within a democratic society and expected that a diverse population would only lead to division and unrest. The political leadership feared that the differences among minority groups would lead to a weak government that lacked the solid support of its populace. Thus, the idea of establishing civil society by protecting the identity of minority groups often was not even a consideration. A strong central government required unity among the populace, and this unity was achieved through the homogenization of people who might otherwise represent diverse religions, languages, and customs.

Even Metaxas' desire to Hellenize non-Greeks before imposing Greek rule on them, however, was not considered seriously because the policy of the Megali Idea had become so firmly entrenched among the Greek populace that its abdication would have caused a severe political backlash. The objective to create a Greater Hellas was pursued regardless of the fact that the territories sought by Greece were inhabited mostly by Turks. The vision

---

[18] Smith, 203.

of the Megali Idea had unified the Greek people since the emergence of the Greek state, and in many ways, it had taken on a life
of its own. More importantly, the Kemalists' persecution of ethnic
minorities, including ethnic Greeks, was difficult for Greece to
ignore. So the pursuit of unification continued, and the relationship of reciprocated violence between Greeks and Turks made
war a reality.

By September 1922, the Turks succeeded in launching a massive counter-offensive against the Greek occupation and destroyed
the Greek army. The Turks then occupied Smyrna, killing an estimated 30,000 Greeks and Armenians living there, and then setting
fire to the city.[19] As chaos set in, hundreds of thousands of the
surviving ethnic Greeks of Turkey were forced to flee and sought
refuge in neighboring Greek islands or the mainland.[20] The Megali
Idea, which eluded Greece since its inception, was finally put to
rest.

Sanctioning Ethnic Cleansing

The devastating war led to a formal exchange of populations between Greece and Turkey. In 1923, Turkey's victory gave it the
necessary advantage to replace the Treaty of Sèvres which had
favored Greek interests, with the Treaty of Lausanne. Capitalizing
on their position as victors, the Turks insisted on compulsory
population exchanges as the final settlement to the war and refused to allow for provisions in the treaty that would safeguard
the rights of the minorities who remained.[21] The *Convention Concerning the Exchange of Greek and Turkish Populations* provided for
the transfer of approximately 1.4 million Anatolian Greeks to
Greece, and 356,000 Thracian Turks to Turkey.[22] The terms of the
treaty essentially had been fulfilled prior to the agreement, as all
but 290,000 of the Greeks to be transferred had already been
forced from their homes and were living in refugee camps in
Greece.[23] To argue, then, that it was the population exchange that

---

[19] Clogg, 97.

[20] Tsoucalas, 35.

[21] Naimark, 54.

[22] Ibid.

[23] Ibid.

brought about a cessation to the hostilities between Greece and Turkey is to misconstrue the course of events.

The exchange simply prohibited the Greeks from returning to their homes, which left Greece with the heavy burden of having to absorb over one million citizens. By the time the population transfer was formally accepted, the Greek army had been destroyed, the majority of the Greek population in Turkey had been almost entirely expelled or killed during the hostilities, and Greece was exhausted and demoralized by her overwhelming defeat. Furthermore, the population exchange only served to escalate rather than reduce the tension between Greece and Turkey. The idea that creating virtually homogeneous populations in Greece and Turkey would stabilize relations proved unfounded for two reasons. First, it was impossible to transfer all minority groups, and second, the exchange caused immense suffering and further destruction for both the Greeks and Turks who were forcibly expelled. Those who were not expelled included the Greeks of Constantinople who had lived there prior to 1918. As the historical residence of the Orthodox Patriarchate for the Greek Orthodox Church, Constantinople's value could not feasibly be replaced by another city. In return for this exception, the Turks of Western Thrace were also exempted from the exchange.

Without effective measures for the protection of the minorities who remained, these populations faced continued injustices from their governments. The most immediate suffering, however, was endured by "the tens of thousands of Greeks [who] perished in the process of being driven from their homes and resettling in Greece."[24] In the course of the Greeks' journey out of Anatolia, they faced violence from bandits, suffered from the diseases of infested camps, and risked suffocation, starvation, or drowning due to the overcrowded and underprovisioned ships that were meant to transport them to Greece. It was estimated that hundreds died every day in makeshift hospitals and camps. And a full year after the exchange took place, well over one million refugees in Greece remained destitute.[25] The conditions for the Turkish refugees were better only to the extent that they were fewer in number and could be more readily absorbed in the regions of Anatolia vacated by the Greeks. Like the Greek refugees, the Turks from Greece found it difficult to support themselves and find acceptance among the Anatolian Turks. Since many of the Turkish refu-

---

[24] Naimark, 55.

[25] Ibid.

gees spoke Greek and new little Turkish, they were treated as aliens from Macedonia.[26]

Patrick Thornberry describes the treaty that legitimized the population exchange as an unjust and crude expression of state power, leading to "appalling human misery."[27] At the time, and to a great extent still today, the state system has promoted international law that tends to accord primacy to states as legal actors or subjects of law. Although it is less pronounced today than during the interwar period, the consequence of limiting the conduct of international law to relations among states has been that individuals or groups within a state have found it difficult to use the international legal system to protect their rights. With the limited protection for minority rights, individuals or groups whose rights had been violated by their own governments were left to rely on another state for representation in the international legal system. Most likely, this state would be one with ethnic or religious ties to the minority group being persecuted and would be willing to make a claim against the state responsible for violating the minorities' rights. Most often, however, the minority populations were simply left to their own defenses.

Perhaps as a cruel irony of history, the idea of adopting population exchanges as a measure to mitigate the atrocities committed against minorities was popularized after the Balkan Wars in which Greece and her Balkan allies (Bulgaria, Serbia, and Montenegro) had triumphed over Ottoman Turkey. Only now, Greece itself was suffering the hardship of such a painful settlement. It is an even greater cruelty of history that the forced expulsion of populations continues to be used as a solution to, or in conjunction with, ethnic violence in the Balkans today. The unnatural goal of creating a complete identification of state with nation (or national group) continues to lead to state sanctioned ethnic cleansing, either through the expulsion or concerted destruction of minority groups.

In 1923, the League of Nations accepted the idea of the population exchange with reservation and with a sense that no other option to end the atrocities against the minorities was viable. In some respects, the Treaty of Lausanne was an attempt to codify the need to find a solution based on the reduction of fear between Greece and Turkey. This required that both states would need to recognize the borders imposed by the proposed settlement and

---

[26] Ibid.

[27] Thornberry, 51.

believe that they would not be challenged in the future. For if either state believed it could challenge the borders based on the existence of its ethnic groups living in the neighboring state, then irredentist policies would continue and war would ensue again. Under these circumstances, minorities might be perceived and treated as insurgent elements that could encourage irredentist or secessionist aspirations. Therefore, the treaty sought to make the borders permanent, to reduce the number of ethnic minorities in both states, and to provide for basic measures that would protect the minority groups that did remain.

The League's rationale underlying the support of the exchange was that it was better to provide a legal framework for transferring populations than to allow the inevitable violence against the minority groups to continue. In this vein, other states with equally intractable ethnic conflicts today occasionally invoke the Greek-Turkish population exchange as a successful approach to ending the violence and settling territorial disputes. This interpretation of the population exchange, however, is based on a misrepresentation of the history that followed the 1923 exchange.

The Post-1923 Consequences of the Population Transfer

The population exchange left behind minority groups in both states who continued to suffer at the hands of their ethnic rivals. The expulsion of populations that had lived in these areas for centuries left the emigrants with a grave sense of injustice. And this sense of injustice continues even today to be played out between Greeks and Turks. Forcing ethnic groups to leave their ancestral homes serves to inflame ethnic animosity, resulting in destructive consequences for future relations between these groups. The descendants of those forced to emigrate remember the blame and resentment that resulted from the population exchange and feel they are justified in withholding certain rights and privileges afforded to the citizens of the majority.

The lack of any substantial minority populations in Greece and Turkey left the small remaining minorities at an even greater risk of being oppressed, as their voice was significantly weakened by the exchange. The belief system that ethnic heterogeneity is harmful and that homogeneity should be achieved only perpetuates conflict and prevents these ethnic groups from prioritizing mutually beneficial goals and establishing a civil society based on equal rights for all citizens. Moreover, the active pursuit of such homogenization of populations may actually sow the seeds of future conflict.

Minority Rights Abuses in Greece

According to Helsinki Watch, human rights abuses against ethnic minorities continue in both Greece and Turkey today. In a 1990 report, Helsinki Watch describes efforts by the Greek government to assimilate the ethnic Turks of Western Thrace by discriminating against them and denying them their ethnic identity. The report indicates that the Greek government has failed to provide ethnic Turks with equal protection under the law by depriving them of certain rights of citizenship if they leave Greece; denying them the right to purchase property, establish businesses, or repair Turkish schools in Western Thrace; and restricting their freedom of expression, religion, and travel.[28]

The Turks of Greece claim that the Greek government denies them their identity by referring to them only as Moslem Greeks rather than Turkish citizens of Greece. "Greek authorities deny the existence of the Turkish minority, asserting only that there is a Moslem minority in Greece, a minority population that is homogeneous in religion, but heterogeneous in origin."[29] The Greek government's response is that the Treaty of Lausanne refers only to "the Moslems of Greece" rather than the Turks of Greece, therefore the identity of "Turk" can apply only to citizens of Turkey.[30] In this regard, the Turkish minority is denied its ethnic status and becomes a religious minority only. Greek courts have ruled that reference to Greek citizens as "Turkish" endangers public order by "openly or indirectly inciting citizens to violence or creating rifts among the population at the expense of social peace."[31]

Even though the Treaty of Lausanne refers to the Turkish minority as a religious minority, the reasoning behind the court rulings and the Greek government's policy indicates a far deeper and more fundamental problem in the relationship between Greece and Turkey. The history of violence and animosity between the two nations, and the lack of substantive measures to educate the majority population on the values of ethnic heterogeneity and minority rights serves only to propagate ethnic conflict. The nature of the relationship between Greece and Turkey inevitably affects the relationship between their own majority and minority popula-

---

[28] Ibid., 1.

[29] Lois Whitman, *Destroying Ethnic Identity: The Turks of Greece* (New York, 1990), 3.

[30] Ibid., 15.

[31] Ibid., ii.

tions; and in turn, the domestic conditions and continuing civil strife in both states cause increasing bitterness in their interstate politics. Only by implementing and abiding by laws that respect the human rights of all citizens regardless of ethnicity, can this vicious cycle of antagonistic internal politics and belligerent foreign affairs come to an end.

In 1955, Greece enacted the Greek Nationality Law, which discriminates against non-ethnic Greeks regarding their right to citizenship. Chapter B, Section VI, Article 19 states that:

> A person of non-Greek ethnic origin leaving Greece without the intention of returning may be declared as having lost Greek nationality. This also applies to a person of non-Greek ethnic origin born and domiciled abroad. His minor children living abroad may be declared as having lost Greek nationality if both their parents or the surviving parent have lost the same.[32]

Although the extent to which this law is enforced is not clear, the fact that it has not been repealed raises questions about Greece's interest in avoiding appearances of discrimination. Ethnic Turks have also complained of restrictions on their freedom of movement, claiming that their passports have been confiscated without cause, and returned a few months later without receiving any explanation.[33] The existence of discrimination in any country weakens civil society and has the potential to cause conflict and perhaps violence on some level. The importance of reporting discriminatory policy is to rectify the problems that exist, and to prevent discrimination from escalating into more serious abuses and violent conflict.

Minority Rights Abuses in Turkey

The denial of ethnic identity is by no means limited to the Greek government. While the Turks of Greece have complained openly about their experience of discrimination, the Greeks of Turkey live in fear of speaking out against the Turkish government. Helsinki Watch reported that Turkey's restrictions of the Greek minority are in violation of the Lausanne Treaty, the European Convention on Human Rights, and the minority rights agreements issued by the Conference on Security and Cooperation in Europe, including

---

[32] Ibid., 11.

[33] Ibid., 13.

the Paris Charter.[34] Since Turkey has signed all of these international documents, its violation of the terms of these agreements harms its reputation as a supporter of international law.

Greeks have lived in Constantinople, now Istanbul, since the seventh century B.C.E. Of the 100,000 to 110,000 ethnic Greeks left there after the population exchange in 1923, only approximately 2,500 remain.[35] The largest exodus of ethnic Greeks occurred in 1955 and 1964. In 1955, a bomb exploded in the Turkish consulate in the Greek city of Thessaloniki, damaging in addition the birthplace of Mustafa Kemal Atatürk. Violent attacks against the Greek minority in Turkey ensued, and riots occurred in the Greek neighborhoods of Istanbul. After these riots left fifteen people dead and caused approximately $300 million in damages, thousands of ethnic Greeks left Turkey. Six years after the attack, a Turkish court found that the Turkish Prime Minister at the time was responsible for the bombing, and that his intentions had been to create the exactly the results that actually took place in Istanbul.[36]

In 1964, while tensions between Greece and Turkey were high due to the ethnic conflict in Cyprus, the Turkish government chose to expel all Greeks who held Greek citizenship from Turkey. The thousand Greeks who were expelled were Greeks who had been born and lived in Turkey, but who simply held Greek citizenship. Many of them were given only a few hours' notice, and were prohibited from taking anything with them except for $22 and one suitcase of clothing. The official government position was that "[a]s the result of the unfriendly policy of the Greek government towards Turkey, the Turkish government is terminating the privileged treatment accorded in the past to Greek nationals."[37]

That same year, an additional 10,000 to 11,000 Greeks were expelled from Turkey when the Turkish government refused to renew the residence permits of Greeks citizens. A Turkish newspaper reported that in addition to those who were expelled, 30,000 ethnic Greeks of Turkey had left permanently. The newspaper reported that "they were not allowed to sell their houses or property or to take money from their bank accounts."[38] Thousands of

---

[34] Lois Whitman, *Denying Human Rights and Ethnic Identity: The Greeks of Turkey* (New York, 1992), 3.

[35] Ibid., 6.

[36] Ibid., 8.

[37] Ibid., 9.

[38] Ibid.

additional Greeks who feared losing their lives and property chose to leave Turkey, which, in many cases, had been their families' homeland for hundreds of years. Those who remain continue to be fearful of abuses by the government. Officials from the Turkish Human Rights Association report that "the Greek community is more fearful than the Kurds or the Armenians."[39]

Those who invoke the 1923 population exchange between Greece and Turkey as a successful strategy for solving intransigent ethnic conflicts fail to understand fully the nature of ethnic cleansing. The policy of eliminating the existence of an ethnic group from a region has immensely devastating repercussions that extend beyond the immediate need to end the fighting. The loss and suffering endured by those affected by population transfers continue even after they have relocated and begun new lives elsewhere. The suffering becomes part of their identity, and that identity includes a polarized view of "self" and "other." The adversary that drove the refugee from her or his home becomes the "other" (often the embodiment of evil) that must be opposed or controlled, for fear that the "other" will become dominant. Transferring populations does not control or manage the ethnic conflict, but rather, sets it deeper within the consciousness of the ethnic group.

Extending the Ethnic Cleansing to Cyprus

Indeed, one can argue that the population exchange between Greeks and Turks in 1923, repeated itself in Cyprus in 1964 and (more forcefully) in 1974. The ethnic ties of the Greek and Turkish communities of Cyprus to their respective "motherlands" prevented them from forging a Cypriot national identity. Greek-Cypriots had even fought in the Greek War of Independence in the 1820s, which initiated the same type of distrust among the two Cypriot communities as existed between Greece and Turkey.[40] The Greek-Cypriots had hoped that the creation of an independent Greek state would lead to union between Cyprus and Greece, freeing Cyprus from its Ottoman rulers. Although the Greek-Cypriots accounted for 80 percent of the population and the Turkish-Cypriots for only 18 percent, Ottoman control of the island created a situation in which the majority was treated as a minority group.

---

[39] Ibid., 2.

[40] Andrew Bell-Fialkoff, *Ethnic Cleansing* (New York, 1996), 138.

With its independence in 1960, a constitution was imposed on Cyprus that actually served to aggravate ethnic conflict by promoting a situation in which the two communities would act separately in their decision-making process, with each side being able to veto the decisions made by the other.[41] Inevitably, conflict erupted and today ethnic division continues to plague the island. The relationship between Greece and Turkey during the previous century was an ominous foreshadowing of the conflict between the ethnic groups in the newly independent Cyprus. The aspiration for unity based on ethnic homogeneity impacted how the ethnic communities in Cyprus viewed themselves and each other.

The lack of willingness to accept and value the ethnic differences by both communities guaranteed the escalation and perpetuation of violence on the island. Without a mutually acceptable solution available, Turkey and the Turkish-Cypriot population viewed segregation along ethnic lines as the only solution to the conflict. Turkey forced the Greek population to the south of the island and congregated the Turkish population in the north. And at the time of this writing, the Turkish military continues to enforce the division of the island and the separation of the two communities (with the ultimate goal of creating two separate states) as a final settlement. Although many scholars and policy-makers alike view the present-day division of the island as a successful solution, this interpretation of events only reinforces the idea that homogeneous states are desirable and that ethnic rivals can and should be effectively displaced or otherwise "cleansed" for the purposes of establishing peace. Solutions that justify the expulsion, elimination, or other forms of cleansing of ethnic groups can only propagate ethnic hatred and foster further instability.

Conclusion: Promoting Civil Society through Ethnic Diversity

When states believe that the international community will accept the forced transfer or expulsion of minority populations, then more constructive approaches to creating civil society will not be pursued. More significantly, if ethnic groups, of either the majority or minority populations, believe that ethnic strife will always lead to self-determination, then ethnic conflict can become the means to an end rather than an unintended consequence. Violence, then, becomes an option for bringing about homogeneity.

---

[41] David Wippman, "International Law and Ethnic Conflict on Cyprus," *Texas International Law Journal* 31 (Spring 1996).

Once homogeneity is no longer viewed as the only means to a unified or strong state, ethnic diversity can exist without posing a threat to any minority group.

Thomas Franck, a prominent legal scholar, explains that ethnically homogeneous populations are an interest of government leaders that is still pursued today. While the continuing ethnic violence in the Balkans is well known, there are other, more subtle, examples that elucidate how deep-seated the desire for ethnic purity is in the world today. In 1997, Slovakia's then Prime Minister offered the Hungarian Prime Minister 600,000 Hungarians in exchange for the ethnic Slovak population of Hungary. Franck argues that "[s]uch an approach to the survival principle tends to shade into genocide, an extreme violation of basic international treaty and customary law."[42] Contemporary ethnic conflicts that pursue the elimination of certain minority groups reveal that international treaties prohibiting such behavior are being undermined by political leaders who do not understand that ethnic diversity can promote a strong a civil society rather than detract from it.

In a cross-sectional study of 125 civil wars since 1944, Nicholas Sambanis of the World Bank found that forced partitions of ethnic populations "are less likely to occur as the degree of ethnic heterogeneity increases."[43] Therefore, the more diffuse a state's ethnic composition is, the less likely it is that any given majority can impose its will on minority ethnic groups. These results indicate that increasing ethnic diversity may, in fact, prove to be a better solution to ethnic violence than promoting homogeneity.

In 1923, international norms for the protection of minorities were greatly inadequate, and international treaties failed to safeguard the rights of these groups. Consequently, the separation of ethnic rivals may have been the only practical solution at the time. Today however, these norms, and the various international treaties that promote these norms, are beginning to have a more pronounced influence on the behavior of states. While some states still violate the rights of minority populations, most states do provide for the legal protection of minorities. The focus of ethnic conflict resolution today, therefore, should not rest on the misplaced assumptions of the past. For the failures of the 1923 population exchange have shown the long-term destruction of such policies.

---

[42] Thomas Franck, *The Empowered Self: Law and Society in the Age of Individualism* (Oxford, 1999), 249.

[43] Nicholas Sambanis, "Partition as a Solution to Ethnic War: An Empirical Critique of the Theoretical Literature," *World Politics* 52, no. 4 (2000): 459.

The irreversible damage done by ethnic cleansing (of all types) throughout this century indicates that policies that encourage homogeneity are more destructive than they are stabilizing. The policies of the new century should take into consideration these failures and explore the importance of ethnic diversity in advancing civil society.

Ethnicity in and of itself, however, is not defined by conflict. Some ethnic groups live quite peacefully among each other, while others, in apparently similar circumstances, do not. The fundamental factor in determining which ethnic groups will engage in conflict is the extent to which the leadership of these groups protects and values the rights of all individuals, regardless of ethnicity. International law has played a large role in defining the rights of minorities and in influencing state behavior to protect the rights of those citizens who are the most vulnerable targets of ethnic violence. However, laws alone are not sufficient to end ethnic conflict.

There must be a certain level of acceptance for the ethnic composition of a state at any given time. This acceptance requires that states recognize the legitimacy and the validity of all ethnic groups that comprise the state. For whenever a state understands and promotes the value of its composite parts, it is in effect valuing its own identity and enhancing its own development. Similarly, when a state seeks to undermine any single element of its population (a part of its own identity), it is sowing the seeds of conflict, promoting instability, and stifling progress. It is generally when those in power view the interests of minority groups as antagonistic to the interests of the state that abuses against minorities occur. Inevitably, a spiraling level of fear encompasses the majority, or ruling group, and any minority group. The majority group may fear that the existence or presence of ethnically or religiously diverse groups somehow threatens its own identity, and so it tries to isolate and weaken these minority groups. The minority groups then seek to protect themselves, if possible, by engaging in the same type of violence that is used against them by the majority group.

This powerplay between ethnic groups is based on the belief system that empowering one's "self" (one's own ethnic group) occurs only by weakening, subjugating, or otherwise eliminating others. In many ways, those who engage in violent behavior do so out of weakness rather than strength. This weakness stems from the group's fear that by accepting the legitimacy of an ethnically diverse people, the original group must recognize that it is not superior. It is this sense of superiority that leads to fear. For in reality no ethnic group can be superior to another. Therefore, to achieve this perception of superiority, ethnic groups must engage in rhetoric and behavior that demeans or denigrates those belonging to different groups. As each group seeks to differentiate itself and empower itself by pursuing this self-imposed sense of superiority, these groups are lead inexorably to an increasing level of violence.

Few would disagree that ethnic conflict is based on ethnic hatred. Unfortunately, the customary response given regarding the origins of ethnic hatred is that such hatred is decades or centuries old. It is as though the duration of the hatred is what defines it. After all, history is replete with examples that lend credence to the idea that populations that have fought fervently against each other and have suffered at the hands of each other can never live amicably together. Yet this superficial view offers little insight and provides a limited basis for which to understand ethnic conflict. The continuation of any type of relationship, whether it is based on mutual respect or mutual animosity, requires that a population be educated to support the views that maintain that particular relationship. In this regard, the political leadership of a state or ethnic group plays a decisive role in the nature of ethnic relations.

Political leaders often derive their own power by generating fear and fostering the enmity that leads the populace to believe that without such leaders, it will be at the mercy of a vicious adversary. The political game of "self" versus "other" begins, and the general public perceives that the identity of "self" is in danger of being influenced, or dominated, or perhaps even eliminated by those viewed as "other." These leaders are then able to use fear to garner the political support they need to carry out abuses against other ethnic groups. Without such fear, the general public has little reason to react violently to those they perceive as "other." In addition, political leaders are often able to direct the kind of public information necessary to increase ethnic animosities and to secure the loyalty they need to make the abuses appear justified.

One of the most immediate challenges today is how to deal with the proliferation of violent ethnic, racial, and religious conflicts that constitute the main human rights threats at the turn of the twenty-first century. The imperative to change the conditions that lead to ethnic violence remains as true today as it was at the turn of the last century. More importantly, doing so requires that governments put forth a concerted effort to respect and encourage ethnic diversity, to promote effective democratic institutions, to foster the growth of civil society, and to create a genuine universal human rights culture.

Ethnic Heterogeneity, Cultural Homogenization, and State Policy in the Interwar Balkans[1]

VICTOR ROUDOMETOF

The Balkan system of nation-states that emerged out of the Balkan wars of 1912-13 and the subsequent World War I (1914-1918) formed a political field, where the issues of national minorities and national homogenization shaped intrastate and interstate relations. No state was able to incorporate all of its potential nationals within its own territory. Indeed, the problems posed by the incomplete nature of "national liberation" were common throughout Eastern Europe. The resolution of these issues was the first major test for the twentieth century international system of nation-states. For many of the newly created or consolidating nation-states, completion of the nation-building process entailed centralization and homogenization.[2] Both projects were an extension of state sponsored strategies of modernization, aiming to Europeanize the Balkans. The issue of ethnic heterogeneity within the Balkans became a problem for state administrators and international committees, whose goal was to construct nationally homogeneous imagined communities.[3] One must keep in mind that the dominant international consensus during this period viewed minority peoples as "anomalies within the nation-state," and as elements that weakened and divided it.[4]

Therefore, the pursuit of centralization and homogenization entailed a clash with the local cultures of various minority groups. Rival nation–states did not hesitate to lay claim to such minorities, thereby adding fuel to the fire. In the following study, I describe the impact of state-sponsored policies of cultural homogenization upon the ethnically heterogeneous population of the Balkan nation-states.

---

[1] This paper was presented at the Conference on Ethnic Cleansing in Twentieth-Century Europe (the Thirty-Fourth Annual Duquesne University History Forum), November 16-18, 2000. It is a heavily revised and abbreviated version of Chapter Seven of the author's monograph, *Nationalism, Globalization, and Orthodoxy: The Social Origins of Ethnic Conflict in the Balkans* (Westport, Conn., 2001).

[2] Anthony Giddens, *The Nation-State and Violence* (Berkeley, 1985).

[3] Benedict Anderson, *Imagined Communities* (London, 1991).

[4] Thomas D. Musgrave, *Self-Determination and National Minorities* (Oxford, 1997), 10.

220 The Interwar Balkans

Building Ethnically Homogeneous States in Macedonia and Anatolia

From the Balkan Wars of 1912-13 up to the Greek-Turkish population exchanges of 1923, the demographic and political map of the Balkan Peninsula experienced drastic changes that altered the human geography of the region. Extensive population movements followed the Balkan Wars. The trend intensified after World War I, leading to extensive population transfers aiming to separate peoples according to ethnic, linguistic, and religious criteria. The two regions most acutely affected were Macedonia and Anatolia. Both regions experienced extensive and often forced population exchanges.

Macedonia was divided among Greece, Bulgaria, and Serbia (e.g. the post-1917 Kingdom of Serbs, Slovenes, and Croats). Because of the ethnic intermixing of peoples in the region, the division was not uniformly accepted, and interstate rivalry over the region persisted during the interwar period. Bulgaria's efforts to conquer the majority of Ottoman Macedonia were frustrated both in the Balkan wars and in World War I. In the aftermath of World War I, Greece and Bulgaria agreed on a mutual population exchange during which approximately 30,000 Greeks left Bulgaria and 53,000 Bulgarians left Greece.[5] The Macedonian separatist organization iMRO opposed the exchange because it would weaken claims to the Greek held part of Macedonia.[6] Consequently, those under iMRO influence chose to remain in Greece. By far the most important development in Greek Macedonia was the departure of the Muslim population and the resettlement of Greek Orthodox refugees from Anatolia. With the 1923 Greek-Turkish population exchange agreement, 354,647 Muslims left Greece and 339,094 Greeks arrived in Greek Macedonia from Anatolia. The result was a complete change in the ethnic composition of Greek Macedonia. By 1928 Greeks accounted for 88.8 percent of the population in Greek Macedonia, totally altering the pre-1913 mixture of different ethnic groups and nationalities.[7]

In the aftermath of the population exchanges, a total of approximately 200,000 to 400,000 Slavic speakers resided in Greek Macedonia, located almost exclusively in northwestern region. For

---

[5] D. Pentzopoulos, *The Balkan Exchange of Minorities and Its Impact upon Greece* (Paris, 1962), 60.

[6] E. Barker, *Macedonia: Its Place in Balkan Power Politics* (London, 1950), 30.

[7] Pentzopoulos, Exchange of Minorities, 69, 107, 127-37.

the Greek state, this was a "Slavophone Greek" population, that is, Slavic-speaking, but Greek in terms of national identity. This formula owed much to the desire to minimize the appeal of Bulgarian protests in international forums in the interwar period, and it influenced greatly the official Greek line on the issue. In this region, the assimilationist policy of the Greek state was compounded by the conflict between local Slavs and the Turkish-speaking Greek refugees who settled in the region. The conflict typically involved possession of homes and agricultural lands. The state's support of refugees contributed significantly to the delegitimization of the state in the eyes of the local Slavic population.[8] The state administrators' view of the situation was strongly colored by the persistent effort to thwart Bulgarian propaganda (and revisionism) among the Slavic population. Their reports differentiated among groups on the basis of their loyalty to the state, rather than on purely "ethnic" criteria.[9]

During the interwar period, the IMRO had to face up to a disastrous situation. Macedonia was now divided into three regions, each under the control of a nation-state eager to consolidate its control over the newly conquered region. The organization fragmented repeatedly during the interwar period, and two main factions developed. First, a conservative wing became a powerful player in interwar Bulgarian politics, actively participating in the 1923 coup and the assassination of Bulgarian prime minister Alexander Stamboliski.[10] From 1919 up until 1934 the organization conducted paramilitary warfare in Serb-controlled Vardar Macedonia. Up until 1934, this Bulgarian IMRO remained a potent po-

---

[8] J. Koliopoulos J. *Plundered Feelings: The Macedonian Question in Occupied Western Macedonia 1941-1944*, 2 vols. (Thessaloniki [in Greek], 1994-95), 45.

[9] Anastasia Karakasidou, "Transforming Identity, Constructing Consciousness: Coercion and Heterogeneity in Northwestern Greece," in *The Macedonian Question: Culture, Historiography, Politics*, ed. Victor Roudometof (Boulder, Col., 2000), 55-98. The plight of members of the Slavic-speaking minority is illustrated by their protests against separate educational institutions. Under the treaties' regime, the Greek government decided to issue a primer ("Abecedar") for use among the Slavic community. Minority protests against its use highlighted the minority's desire not to be considered "different" from the rest of the population. This desire was directly linked to worry that such a "difference" would intensify their unequal treatment by the authorities; I. Mihailidis, "Minority Rights and Educational Problems in Greek Interwar Macedonia: The Case of the Primer 'Abecedar,'" *Journal of Modern Greek Studies* 14 (1996): 329-44. The "Abecedar" primer is considered "hard" evidence by the Macedonian side that the Greek state recognized the existence of a Macedonian population within its boundaries.

[10] I. Banac, *The National Question in Yugoslavia: Origins, History, Politics* (Ithaca, N.Y., 1984), 321-323.

litical force in Bulgarian life and practically a state within a state in the Pirin region. Its open irredentism found a constituency among the 220,000 refugees from Macedonia and Thrace who flooded into Bulgaria in 1918.[11] In 1934, the *Zveno*, a group of military officers, overthrew the government. Virtually its first act was to arrest the iMRO leadership, putting an end to its terrorist regime.

Second, a leftwing iMRO fraction developed mainly in the Serb-held and Greek-held regions of now divided Macedonia. The consolidation of an international communist organization (i.e., the Comintern) in the 1920s led to some failed attempts by the Communists to use the Macedonian issue as a political weapon. Already in the 1920 Yugoslav parliamentary elections, 25 percent of the total Communist vote came from Macedonia. But participation was low (only 55 percent), mainly because the pro-Bulgarian iMRO organized a boycott against the elections.[12] In the following years, the Communists attempted to enlist the pro-iMRO sympathies of the population to their cause. In the context of this attempt, the Comintern recognized in 1924 a separate Macedonian nationality. Still, the Comintern's suggestion that all Balkan Communist parties adopt a platform of a "united Macedonia" was rejected by the Bulgarian and Greek Communist parties.[13] Despite this rejection, their conservative opponents within the two states did not hesitate to seize the opportunity to accuse the two parties for plotting against the nation, thereby leading to the prosecution of the two communist parties in Bulgaria and Greece. By the 1930s, Slavic-speaking Communist sympathizers were viewed as separatists by the Greek state.[14]

Throughout the interwar period, the Slavic Macedonian population in Greek Macedonia was subjected to an intense acculturation campaign by the Greek state.[15] When the Metaxas dic-

---

[11] John R. Lampe, *Yugoslavia as History: Twice There Was a Country* (Cambridge, 1996), 152-153.

[12] Ibid., 140.

[13] A. Papapanagiotou, *The Macedonian Question and the Balkan Communist Movement 1918-1939* (Athens, [in Greek], 1992).

[14] Anastasia Karakasidou, "Fellow Travelers, Separate Roads: The KKE and the Macedonian Question," *East European Quarterly* 27 (1993): 453-77.

[15] P. Carabot, "The Politics of Integration and Assimilation vis-à-vis the Slavo-Macedonian Minority of Inter-war Greece: From Parliamentary Inertia to Metaxist Repression," in *Ourselves and Others: The Development of a Greek Macedonian Identity Since 1912*, ed. Peter Mackridge and E. Yiannakakis (Oxford, 1997), 59-78; Karakasidou, "Politicizing Culture"; Karakasidou, "Fellow Travelers"; and Anastasia Karakasidou, "Transforming Identity, Constructing Consciousness: Coercion and

tatorship came into office in 1936, Greek official policy took a turn toward harsh suppression of the Slavic Macedonian language and culture. The Slavic population in Vardar Macedonia was exposed to a similar assimilation campaign by the Serb authorities. The official viewpoint declared Vardar Macedonia to be "Southern Serbia." During the initial Serb occupation (1913-15) and after the return of the region in the Serb-controlled Kingdom of Serbs, Slovenes, and Croats, Serb authorities expelled the Bulgarian Exarchist clergy and teachers, removed all Bulgarian signs and books, and dissolved all Bulgarian cultural associations. Additionally, approximately 4,200 Serb families settled in the region by 1940.[16] These populations' subsequent resentment was registered in the growing support for the Yugoslav Communists. Yet, even as late as 1941, the population of Skopje did not hesitate to celebrate the coming of Bulgarian forces to occupy this part of Yugoslavia.

The other region that experienced dramatic demographic changes was Anatolia. In this case, this outcome was produced by the collision between Greek and Turkish nationalisms in the 1910s and 1920s. In the course of this conflict, Greece and Turkey were transformed into presumably homogeneous nation-states. In the aftermath of the 1912-13 Balkan Wars, the Young Turks slowly realized that political independence was impossible to achieve without real economic independence.[17] Creating a strong nation-state involved creating a national indigenous Muslim middle class that would support Young Turk policies. Only such a class could provide support for their policies. Such a class existed but was Eastern Orthodox (Armenian and Greek) Christian, not Muslim. The Young Turk policy was to displace Orthodox peoples and replace them with Muslims. This option was strengthened by the popularization of Pan-Turkism in the 1910s. Pan-Turkists, such as Yusuf Akcura, questioned the viability of the Ottoman Empire and argued for the creation of an ethnically homogeneous Turkish state.[18]

Heterogeneity in Northwestern Greece," in *The Macedonian Question: Culture, Historiography, Politics,* ed. Victor Roudometof (Boulder, Col., 2000), 55-98; S. Pribichevich, *Macedonia: Its People and History,* (University Park, Penn., 1982); H. Poulton, *Who are the Macedonians?* (Bloomington, Ind., 1995).

[16] Banac, *The National Question,* 218-319.

[17] E. J. Zurcher, *The Unionist Factor: The Role of the Committee of Union and Progress in the Turkish National Movement, 1905-1926* (Leiden 1984), 56.

[18] For details, see J. M. Landau, *Pan-Turkism in Turkey: A Study in Irredentism* (London, 1981); D. Kitsikis, *Comparative History of Greece and Turkey in the Twentieth Century,* 2nd ed. (Athens, 1990 [in Greek]), 97-99.

The Turkish losses of the Balkan wars reinforced support for the Pan-Turk option. The combination of Pan-Turkism and statism was materialized in the changing attitude toward the minority population. The guiding principle behind the new policies was to establish control of the economy by the Turkish ethnic group.[19] During the 1908-1918 period, government officials collaborated with local merchants, traders, or rural notables in order to set up a series of new companies that were to undertake major commercial functions in Anatolia. To counter the dominant role of the Greek and Armenian communities, they strongly encouraged the "Turkish" character of commercial establishments; buying from a Turk instead of a Greek or Armenian became a matter of patriotism.

To implement these goals the Young Turks began a policy of intimidation and deportation in Anatolia, where most of the Christian population lived. In 1909, the Unionists coordinated a commercial boycott of Ottoman Greek commerce aimed at punishing Athens for supporting Crete's declaration of union with Greece. The aim of the boycott was also to raise political and national consciousness amongst the Turks; over time, this factor became more important than the original aim, for the Committee became aware of the need for a "national economy" and a Turkish bourgeoisie.[20] More pressure was applied to the Ottoman Greek community when in 1909 Greece annexed the islands of Lesbos and Chios and when Greece annexed Western Thrace after the second Balkan War (1913). By the end of 1913, more Muslim refugees flooded into the Empire. The harassment of the Muslims in the newly conquered territories made the Unionists opt for a population exchange between the Muslims of the Balkan states and the Ottoman Greeks of the Empire.[21] Treaties between Bulgaria and Turkey in 1913 and between Greece and the Ottoman Empire in 1915 attempted to bring about population exchanges; but the First World War interrupted these plans.[22]

---

[19] F. Ahmad, "Vanguard of a Nascent Bourgeoisie: The Social and Economic Policy of the Young Turks, 1908-1918," in *Social and Economic History of Turkey 1071-1920*, ed. O. Okyar and H. Inalcik (Ankara, 1980), 342-43.

[20] F. Ahmad, "Unionist Relations with the Greek, Armenian, and Jewish Communities of the Ottoman Empire 1908-1914," in *Christians and Jews in the Ottoman Empire*, ed. B. Braude and B. Lewis (Ankara, 1982), 414.

[21] D. Ergil, "A Reassessment: The Young Turks, their Politics, and Anti-Colonial Struggle," *Balkan Studies* 16 (1975): 62.

[22] Pentzopouolos, *The Balkan Exchange of Minorities*, 54-55.

During the 1914-1918 period, the harassment of the Ottoman Greek minority took the form of an open economic and paramilitary warfare.[23] A total of 481,109 persons were deported into the interior of Anatolia in the four years of the war; even after the 1919 Armistice, the expulsions continued, being directed this time against the Greeks of the Black Sea and the Trebizond area.[24] Muslim refugees formed paramilitary bands, and the reign of terror against the Ottoman Greek minority continued up to 1916. The Unionist government asked all foreign firms to fire their Greek employees and hire Turks instead. Under the leaderships of Enver Pasha and Celal Bey, a "Special Organization" was formed to "nationalize" Western Anatolia. In 1914, approximately 130,000 Greeks were forced to leave for Greece or the Aegean islands from the rich suburbs of Smyrna, the main commercial center of Western Anatolia and home to a majority of the Ottoman Greek population.[25]

The final solution of the minority question took place with the Greek Anatolian venture in the 1921-22 period. The Greek state's attempt to conquer Anatolia failed miserably resulting in the 1922 Asia Minor debacle. The Greek failure also constituted the high point of the Turkish War of national independence (1919-22), conducted under the leadership of Mustafa Kemal, who became known as Kemal Atatürk, the founder of modern Turkey. The Turkish victory led to the practical disappearance of Greek Orthodox presence in Anatolia. The overwhelming majority of Ottoman Greeks in Anatolia fled their homes after the collapse of the Greek Army in 1922. In the 1923 Lausanne Treaty, the Greek and Turkish governments agreed on the compulsory exchange of the Muslim minority in Macedonia and the remaining Ottoman Greek population of Anatolia.

Minorities and State Policy in the Southern Balkans

The post-1912 Balkan boundaries divided the Albanians, with a substantial number falling within the boundaries of the post-1917 Kingdom of Serbs, Slovenes, and Croats. The Serb (and later on

---

[23] The Armenian genocide was part of this process of nation building. The issue is not discussed here because the Armenian movement falls outside the topic of Balkan nationalism. Suffice it to say there are considerable similarities between the Armenian and the other Eastern Orthodox nationalisms.

[24] Pentzopoulos, *The Balkan Exchange of Minorities*, 57.

[25] Ergil, "A Reassessment," 63.

Yugoslav) state occupied Kosovo.[26] Still, the region was populated mostly by Albanians and had, in fact, hosted the League of Prizren (1878), the initial organization of the Albanian movement. In the 1921 census, out of 436,929 inhabitants in Kosovo, there were 280,440 Albanian speakers, of whom 72.6 percent were Muslims, 26 percent Orthodox, and 1.4 percent Catholics.[27] Consequently, the stage was set for an Albanian-Serb confrontation. The *kachak* movement, actively assisted by Italy, challenged Serb rule as early as 1918. In late 1918, the movement's leaders, Azem Bjeta and his wife Shote, commanded close to 2,000 fighters and 100,000 adherents. The movement was brought under control only in 1922 with the help of Ahmed Zogu, Albanian Minister of the Interior at the time. The Serb-controlled government employed Wrangelite White Russian troops to assist Zogu in overthrowing the Fan Noli government in 1924. In return, Zogu assisted in the suppression of the *kachak* movement in order to be able to disarm the Albanian bands of the northern part of the country and consolidate his own rule. Zogu eventually turned to Italy for help, and with Italian encouragement he declared himself "king of Albania" in 1928.

The pacification of Kosovo was completed by 1924, but this signaled only the beginning of the Serb-Albanian confrontation. The Serb intelligentsia did not acknowledge that the Albanians possessed a national identity. They viewed them as "savages" or "criminal interlopers" who lived in the wild and preyed on the innocent.[28] Simultaneously, they developed the thesis many of them were initially Serbs who had been converted to Islam. They spoke of *arnautasi* (Albanized Serbs) in order to "reclassify" the Albanians as Serbs.[29] With the return of Serbian troops in 1918 all Albanian schools were closed down; authorities tried to assimilate Albanian children by allowing only Serb education. In addi-

---

[26] Considerable fighting, killing and destruction took place during the Balkan wars and World War I; see Djordje Mikić, "The Albanians and Serbia During the Balkan Wars," in *East Central European Society and the Balkan Wars*, ed. Bela K. Kiraly and Dimitrije Djordjevic (Boulder, Col., 1987), 165-96; [26] Miranda Vickers, *Between Serb and Albanian: A History of Kosovo* (New York, 1998); and Banac, *The National Question*, 298. Approximately 16,000 Albanians fought alongside the Ottoman forces. Serb reprisals were also extensive. Approximately 20,000-25,000 Albanians were killed in Kosovo; see Noel Malcolm, *Kosovo: A Short History* (New York, 1998), 254, 257. The Serb army marched through Kosovo in its 1915 retreat to the Adriatic, and close to 100,000 of them perished in the process.

[27] Vickers, *Between Serb and Albanian*, 95.

[28] Banac, *National Question*, 293-295.

[29] Malcolm, *Kosovo*, 199.

tion to the manipulation of ethnic labels, the Serb-controlled government resettled poor Serb families into the region as a means of boosting the Serb minority. The colonization program of 1922-29 and 1933-38 included the settlement of 10,877 families on 120,672 hectares of land. Support for the colonists and the appropriation of land on behalf of the colonists naturally inflamed relations between Serbs and Albanians.

In an effort to increase the gap separating the Kosovo Albanians from those in the Albanian state, the post-1929 Serb administration deported thousands of Albanians, Turks, and Muslims from Kosovo and Macedonia into Turkey. Some estimate that some 200,000 to 300,000 were deported while others estimate the number to be between 90,000 and 150,000.[30]

While the Albanians in Kosovo were thus the targets of an assimilation campaign, the newly created Albanian state had its own minority issues in the south. The boundaries of the Albanian state were a topic of discussion since 1913 but were finalized by an international commission only in 1920.[31] In its final incarnation, the boundary line between Greece and Albania fell short of Greek claims, leaving on the Albanian side a strip of land claimed by Greece. This ill-defined territory, inhabited by a Greek Orthodox minority, was destined to become "Northern Epirus," an object of territorial dispute between Greece and Albania for most of the twentieth century.[32] When Albanian entered the League of Nations (1920) it recognized the international minority treaties' regime. However, the Albanian state recognized the existence of a mere 15,000 Greek-Orthodox Christians in the south, a figure that fell far from the Greek claims. The Greek side routinely exaggerated the Greek minorities' numerical strength by counting as Greeks practically all Orthodox Albanians and Vlachs. On the other hand, Albanian official policy sought to separate the Vlachs from the Greeks.

---

[30] Vickers, *Between Serb and Albanian*, 105-20, and Banac, *National Question*, 301-03, use the higher figures; Malcolm, *Kosovo*, 286, the lower.

[31] Vickers, *Between Serb and Albanian*, 78-93.

[32] By 1911 the Greek government's internal reports estimated that the approximately 220,000 Orthodox Albanian Christians of the provinces of Skhoder, Jannina, and Monastir had a Greek "national consciousness" (Hristina Pitouli-Kitsou, *Greek-Albanian Relations and the Northern Epirus Question 1907-1914* Athens, 1997 [in Greek]), 87-94. In 1913 a Greek insurrection to the south raised the issue of autonomy for the Greek minority. In 1920-21 further clashes between Albanian nationalists and the Greek minority raised the issue of educational and ecclesiastical protection; Vickers, *Between Serb and Albanian*, 98-99.

The constitution of the Albanian state led to efforts for the establishment of an independent state church, thereby allowing the Albanian side to reduce the influence of Greek language in the south. In 1923 legislation was introduced that excluded priests who were not Albanian citizens, did not speak the Albanian language, and were not of Albanian origin. Between the wars, the Albanian state viewed Greek education as a potential threat to its territorial integrity. Consequently, in an effort to break up the alliance of Romanian-speaking Vlachs and Greeks, the state took a negative view of Greek-language schools and a positive view of Romanian-language schools. When Albania institutionalized the prohibition against "foreign" schools in 1934, the Greek minority (with the help of the Greek state) brought the issue to the World Court of Hague, which ruled in 1935 in favor of the minority.[33]

The final boundary demarcation between Greece and Albania also left approximately 20,000 Muslim Albanians (Chams) in Greece. Many of them were local landlords and resented the post-1920 land appropriation plans. During the interwar period Italy and Albania repeatedly raised the Chams' issue. More than twenty protests were filed with the League of Nations concerning violations of the Chams' rights.[34] During World War II, the majority of Chams sided with the Axis forces and terrorized the local Greek population. This fueled resentment by the Greeks, and in the aftermath of World War II, the Chams had to flee to their ethnic "homeland," Albania.

Finally, Bulgaria's cultural homogenization policies aimed at acculturating the people into the Bulgarian nation, which in turn meant Eastern Orthodox in religion and Bulgarian-speaking in language. In some instances, like that of the Turkish-speaking Christians (Gagauzes) of the Black sea coast, the people were simply declared to be "Bulgarian." In other cases, like that of the Greek urban population, the assimilation campaign led either to immigration or to acculturation. Over the post-1880 period, the Greek population declined in all Bulgarian cities, falling from

---

[33] The court's decision allowed the minority to keep its educational institutions. However, the decision was never implemented. During World War II, Romanian-language schools were opened while Greek-language schools were closed down. See David J. Kostelancik, "Minorities and Minority Language Education in Interwar Albania," *East European Quarterly* 30 (1996): 74-96; and Lena Divani, "The Consequences of the League of Nation's Minority Regime in Greece: The Foreign Office Perspective," in *The Minority Issue in Greece: A Contribution from the Social Sciences*, ed. K. Tsikelidis and D. Christopoulos (Athens, 1997 [in Greek]), 186-189.

[34] Divani, "The Consequences of the League of Nation's Minority Regime," 181-182.

53,028 in 1880-84 to 9,601 in 1934.[35] The approximately 50,000 Bulgarian Catholics were also subject to pressure and persecution.

Of all these groups, the most numerous one was the Muslim minority. The Muslim population was (and still is) divided into four main groups: the Turks, by far the largest group; the Pomaks, converts to Islam who have retained their customs and language; the Tatars, who arrived in mid-nineteenth century as refugees from Crimea; and the Gypsies (Roma). The Turkish and Muslim population was extensively impacted by the establishment of the Bulgarian principality (1878) and its expansion into Ottoman territory. Thousands of Muslims fled during the 1877-78 Russo-Turkish war. The immigration wave persisted in the following years. Between 1878 and 1912 350,000 Turks immigrated into the Ottoman empire; 101,507 immigrated into Turkey between 1923 and 1933; and 97,181 between 1934 and 1939.[36] The departure of the Turks was facilitated by legislation enacting property taxation as well as by a 1879 law prohibiting rice cultivation, a stable of their diet. Moreover, the realization that the restoration of Ottoman power was not a viable option contributed to persistent immigration. The growth of the Bulgarian state and the influx of refugees also impacted demographic trends. The Bulgarian population increased from 2,000,000 in 1880 to 4,300,000 in 1910 to 7,000,000 in 1956. In contrast, the Muslim population (whose overwhelming majority were Turks) grew slightly from 676,212 in 1887 to 821,298 in 1934. Consequently, the Turks' proportion fell from 26 percent of the population in 1878 to 14 percent in 1900.[37]

The minority had to face the attitudes of Bulgarian nationalists, who wished to see minority schools shut down and the minority itself assimilated into mainstream Bulgarian culture--failing which they should simply leave the country. However, Turkish and Muslim minority rights were guaranteed in a series of conventions signed by the Bulgarian state in 1878, 1909, and 1919. Consequently, the state undertook specific actions to protect their rights and to reassure the continuation of their educational and religious institutions. Simultaneously, the state centralized these

---

[35] R. J. Crampton, R. J., "The Turks in Bulgaria, 1878-1944," *International Journal of Turkish Studies* 2 (1989): 43-78; Kemal H. Karpat, "Bulgaria's Methods of Nation-Building: The Annihilation of Minorities," *International Journal of Turkish Studies* 4 (1989): 3.

[36] Ilia Eminov, *Turkish and Other Muslim Minorities in Bulgaria* (New York, 1997), 79.

[37] Bilal Şimşir, *The Turks of Bulgaria (1878-1985)* (London, 1988), 5; Crampton, "The Turks in Bulgaria," 42-50.

institutions by creating a religious hierarchy with the Sofia-based *mufti* as the supreme religious leader, and by bringing private minority schools under state regulation and control. From the state's viewpoint, these actions aimed at severing the ties between Muslims and the Ottoman state (later on, Turkey), thereby affirming state sovereignty. Quite predictably, there was not always sufficient good will on both sides, and complaints about regulations and their implementation were raised by the minority.[38]

Over the interwar period, the minority gradually adjusted to its new situation, and began efforts to organize itself as a political lobby in Bulgarian politics. These efforts indicated a gradual socialization into the Bulgarian political body. However, Bulgarian political life did not favor such an accommodation. Although the short-lived Agrarian government (1919-23) was favorably disposed vis-à-vis the minority, Bulgarian nationalists continued to view the minority with suspicion. The Turkish National Congress of 1929 raised once again the key issues of schools, religious institutions, and cultural societies. In the late 1920s, the minority adopted the new Turkish (Latin) alphabet, instead of the old Arabic one. Since the new alphabet had been instituted by Kemal Ataturk in Turkey, Bulgarian nationalists viewed this action with great suspicion. The state reaction was to gradually suppress the minority's educational institutions. The number of Turkish schools declined from 1,712 in 1921-22 to 949 in 1928-29.[39] The military junta that came to power in 1934 proceeded to close down even more Turkish schools, and by the end of 1944 only 377 schools remained open. Additionally, property of the *wakf* (that is, property of Islamic religious and charitable organizations) was expropriated, Turkish newspapers closed down, and Turkish intellectuals were sent into exile.[40]

The National Question in Interwar Yugoslavia

The problems that plagued the first Yugoslavia were indeed similar to the difficulties that condemned proposals for inter-ethnic and multicultural citizenship in the nineteenth century.[41] The re-

---

[38] Eminov, *Turkish and Other Muslim Minorities*; Crampton, "The Turks in Bulgaria"; Şimşir, *The Turks of Bulgaria*.

[39] Şimşir, *The Turks of Bulgaria*, 112.

[40] Eminov, *Turkish and Other Muslim Minorities*, 49; *The Turks of Bulgaria*, 114-125.

[41] Roudometof, *Nationalism, Globalization, Orthodoxy: The Social Origins of Ethnic Conflict in the Balkans*.

vitalization of the Yugoslav movement on the eve of the twentieth century coincided with the return of Alexander Karadjordjević to the Serb throne in 1903 and the ascent of Serbia as the chief challenger to Habsburg rule. Yugoslavism emerged out of the intellectuals' efforts to find a formula for South Slav unification.[42] In 1905 a Croatian and Serb coalition was formed in Dalmatia and Croatia-Slavonia and in 1906 this coalition won the Croat elections.[43] This coalition was led by Ante Trumbić and Frano Supilo, young Croat intellectuals who were convinced that the Croats could no longer survive within the Habsburg Empire and should find their future within a South Slav state. These leaders upheld an agenda of "Croat-Serb oneness" (*narodno jedinstvo*), that is, the proposition that the Croats and Serbs were—or were becoming—one people with two names.[44]

This emerging Yugoslav movement carried with it the seeds of a future conflict. It was never clarified how the "fusion" of the South Slav "tribes" would take place, and it was easy to misinterpret Greater Serbian nationalism as simply Yugoslavism. In turn, the Habsburg Serbs were themselves divided into two groups: on the one hand, Greater Serb nationalists, and on the other those looking to Serbia as the creator of a Yugoslav state in which Serbs, Croats, and Slovenes would be separate and equal with each other.[45] For the first group, the Yugoslav solution was frequently cast in terms of fulfilling the idea of Greater Serbia. The second group's ideology was almost identical to the ideology of *narodno jedinstvo* advocated by the young Croat intelligentsia.

During World War I support for the Yugoslav unification came from the Yugoslav Committee, founded by émigré leaders (Supilo, Trumbić, and Meštrović) from the South Slav Habsburg lands.[46] The Yugoslav Committee and the Serb government did not always have amicable relations: for example, the Committee

---

[42] M. Gross, "Social Structure and National Movements Among the Yugoslav Peoples on the Eve of the First World War," *Slavic Review* 36 (1977): 628-43.

[43] Dimitrije Djordjevic, *History of Serbia 1804-1920* (Thessaloniki, 1970 [in Greek]), 322-23.

[44] Banac, *The National Question*, 96-98.

[45] D. Rusinow, "The Yugoslav Peoples," in *Eastern European Nationalism in the Twentieth Century*, ed. P. Sugar (Washington DC, 1995), 358.

[46] Gale Stokes, "The Role of the Yugoslav Committee in the Formation of Yugoslavia," in *The Creation of Yugoslavia 1914-1918*, ed. D. Djordjevic (Oxford, 1980); M. B. Petrovich, (1976) *A History of Modern Serbia 1804-1918*, 2 vols. (New York, 1976) 2: 626-49; Djordjevic, *History of Serbia*, 404-10; Banac, *The National Question*, 214-225.

desired to organize separate military units while the Serb government wanted these units to become part of the Serb army. The vexing question of implementing a power-sharing agreement that would allow for peaceful coexistence between Serbs and Croats was emerging at the time. In fact, Yugoslavism represented a diplomatic solution through which the actual problems that the unification entailed could be avoided rather than dealt with. Indeed, the diplomats were forced to proceed without having a clear-cut and substantive agreement on the nature of the new state. In the July 20, 1917, Corfu Declaration, all sides agreed to the establishment of the kingdom of Serbs, Croats, and Slovenes under the Karadjordjević dynasty. By 1918 the new state was rapidly becoming a reality, as the war was ending.

The questions that plagued the early negotiations between Serbs and Croats, however, had not been resolved. While the Serb government viewed the question of Yugoslavia as an expansion of Serbia into Habsburg lands, the Habsburg Slavs (especially Croats) viewed the issue in terms of a federation of independent and sovereign units.[47] This fundamental issue would plague the new state during the interwar period.[48] During the 1920s, it was impossible to reconcile the views of Croat movement (represented by Radić's Croat Peasant Party), who proposed decentralization or federalism with the views of the Serb government (represented by Pašić's Radical Party), who opted for centralization. After the 1920 elections, the two sides were unable to agree on a constitution. Paišć used his parliamentary majority and political skill to pass the so-called Vidovdan constitution in 1920 (proclaimed on St. Vitus Day, the Serb national holiday). The constitution centralized power in the hands of the central government in Belgrade and was immediately resisted by the non-Serb minorities, most vocally by the Croats.

This was just the beginning of the infamous "national question." In practical terms, the new state could not be governed without cooperation between Serbs and Croats. In 1921 only 39 percent of the population was Serb (including the Montenegrins). The Croats amounted to 23.9 percent, the Slovenes were 8.5 per-

---

[47] Zdenko Zlatar, "The Building of Yugoslavia: The Yugoslav Idea and the First Common State of the South Slavs," *Nationalities Papers* 25 (1997): 387-406.

[48] For extensive discussions, see Banac, *The National Question*; A. Djilas, *The Contested Country* (Cambridge, Mass., 1991); and Rusinow, "The Yugoslav Peoples." Despite the government's attitude, the majority of Belgrade intellectuals were opposed to grounding all authority in the state and the church (Lampe, *Yugoslavia as History*, 188-90).

cent, and the Bosnian Muslims were 6.3 percent, followed by smaller groups of Macedonian Slavs, German, Albanians, Hungarians, Turks, Romanians, and Italians.[49] Throughout the interwar period, inter-ethnic cooperation was not forthcoming, and this failure was the primary reason for the gradual delegitimization of the state. The expansion of a Serb-dominated bureaucracy contributed to the perception of Serb domination.[50]

The persistent failure to construct a viable political coalition plagued the state in the 1921-28 period. The failure was due to the resistance of Croats and other non-Serb groups to accept the centralized structure of the new state. Passions were greatly inflamed in the process, and eventually culminated in the 1928 shooting of the Croat Peasant Party's leader Stjepan Radić in the floor of the parliament. By 1929, King Aleksandar, frustrated with the parliament's inability to implement effective centralization, carried out a coup and assumed direct control of the government until his 1934 assassination. On this occasion, the Kingdom of Serbs, Slovenes, and Croats was renamed "Yugoslavia." New provinces were constructed, allegedly based on purely geographic criteria, as a means of bringing about Yugoslav unity. The provinces allowed Serbs a lion's share in administration and were widely viewed as a Serb ploy for domination. Furthermore, the 1931 census illustrated this policy of centralization and homogenization. The census included a single "Serbo-Croat" nationality, thereby constructing a facade of homogeneity with 77 percent of the population recorded as "Serbo-Croat."[51] In 1934 the king was assassinated by IRMO terrorists, and this act signaled the beginning of a downward spiral for the first Yugoslavia. In 1939, a belated attempt at reconciliation between Croats and Serbs created a large and autonomous Croatian province, yet, without providing the Serbs of Croatia with a similar province.[52] Perhaps this compromise could have prevented

---

[49] Lampe, *Yugoslavia as History*, 129-30.

[50] The successive cabinets and political manipulations during the interwar are too extensive and inconsequential to be reviewed here in detail. There has been a lively debate in the literature regarding who's to blame for the failure of the first Yugoslavia. See Banac, *The National Question*, for a pro-Croatian viewpoint and A. Dragnich, *The First Yugoslavia* (Stanford, Cal., 1983) for an attempt to exonerate the Serb side (and prime minister and Radical Party leader Nikola Pašić in particular) from such a blame. The interpretation adopted here is quite close to the balanced and nuanced perspective of Lampe in *Yugoslavia as History*.

[51] Raymond Pearson, *National Minorities in Eastern Europe 1848-1945* (New York, 1983), 152.

[52] Rusinow, "The Yugoslav Peoples," 377.

the disintegration of the state, but the Axis invasion of April 1941 rendered it a moot issue.

Conclusions

In the first decades of the twentieth century, two main routes were available for the political reconstruction of Southeastern Europe. The first route was that of major compulsory population transfers, a strategy applied to Macedonia and Anatolia, the heartland of the Ottoman Empire. This strategy, known as "ethnic cleansing" in the 1990s, resulted from collusion among Greek, Bulgarian, and Turkish nationalisms in the first three decades of the twentieth century. Needless to say, it failed to produce homogeneity even in Anatolia and Macedonia. In the Southern Balkans, this strategy was applied only to a limited degree, and the minority question plagued interstate relations in Bulgaria, Greece, and Albania. The application of nationhood[53] as a guiding principle for centralization and state-building contributed significantly to the marginalization of minority groups. The second main route to the problem of national homogenization was the creation of a common state to house peoples of different ethnicities, religions, and cultural backgrounds. While this option was never implemented in the Southern Balkans, it was successfully pursued in the South Slav lands. The creation of the Kingdom of Serbs, Slovenes, and Croats (1917) was a skillful attempt to manage ethnic heterogeneity within the boundaries of a single state, thereby avoiding bitter territorial disputes among them.

From the Serb-Croat confrontation to the minority issues in the Southern Balkans, politicians and administrators were not able to develop formulas conducive to successful state-building. The use of terms such as "Slavophone Greeks" (for the Slavic Macedonians in Greece) and *"arnautasi"* for the Albanians in Kosovo aptly illustrates the manner in which administrators and intellectuals attempted to make reality conform to their preconceived notions. These statist policies of cultural assimilation have to be placed within the context of the European, indeed international, trends of the interwar period. The end of World War I and the proclamation of the principle of self-determination legitimized nationalist projects throughout Eastern and Central Europe, thereby paving the way for state intervention. In the 1920 Paris Peace Conference and

---

[53] Victor Roudometof, "Nationalism, Globalization, Eastern Orthodoxy: 'Unthinking' the 'Clash of Civilizations' in Southeastern Europe," *European Journal of Social Theory* 2 (1999): 234-47.

the League of Nations, the Allies were careful in excluding their own territories from the application of the principle.

Specifically, the treaty provisions that emerged from the Paris Peace Conference were twofold. First, the enemy states of Austria, Hungary, Bulgaria, and Turkey signed minority provisions a part of the peace treaties themselves. Second, the Allied states created or enlarged by the Conference (i.e., Poland, Czechoslovakia, Romania, Yugoslavia, Greece) signed separate treaties safeguarding minority rights.[54] In 1921, as part of the international minorities' treaties regime, the League of Nations established a procedure according to which minorities could submit petition to a special commission regarding their treatment. However, the minorities did not have the right to come before the council itself, and it was up to the council to decide the merit of the individual petition. Consequently, "within time, most minorities saw the League not as a judge but as a policeman, far less concerned with justice than the maintenance of law and order."[55]

The events of the interwar period shaped public institutions, mentalities, and local traditions in the Balkan states. They cast their shadow even on present-day disputes over minority issues in the region. From the persistent desire of the Macedonian Slavs to be recognized by the Greek state to Bulgaria's ongoing troubles with its Turkish minority to Greek-Albanian disputes over the Greek minority and to the Kosovo affair, it is plain that the legacy of state-sponsored policies of cultural homogenization has contributed significantly to the intensification and politicization of ethnic differences. However, unlike the interwar period, the international system of states views such issues in quite different light today. The future will show whether the new international emphasis on human rights will pave the way for a more long term and satisfactory solution to the issue of ethnic heterogeneity in the Balkans.

---

[54] For a complete listing, see Patrick Thornberry, *International Law and the Rights of Minorities* (Oxford, 1991), 41; Musgrave, *Self-Determination and National Minorities*, 41.

[55] Pearson, *National Minorities*, 143.

II.

THE ETHNIC CLEANSING OF GERMANS DURING AND AFTER WORLD WAR II

Anglo-American Responsibility for the Expulsion of the Germans, 1944-48

ALFRED DE ZAYAS

American and British historians have not given the enforced flight and expulsion of fifteen million Germans and the end of World War II the attention that this important and tragic phenomenon deserves. In itself this deliberate avoidance of a legitimate field of historical research and publication merits our attention today, considering that the flight and expulsion of the Germans constitutes the largest mass transfer of population in history, a veritable demographic revolution in central Europe, and a form of genocide in the course of which more than two million human beings perished.

The reticence of historians is coupled by the failure of the press and other news media to fulfil their responsibility to inform the general public and to generate debate about these events. Quite to the contrary, the entire subject matter of the flight and expulsion of the Germans has been subject to taboos and remains largely ignored even to this day. Only the occurrence of "ethnic cleansing" in the former Yugoslavia during the last decade of the twentieth century allowed obvious parallels to be drawn, and a beginning of a discussion on the earlier ethnic cleansing against Germans is now emerging. Much more general information, oral history, education in the schools, and reflection by policy makers is necessary.

As we have seen in the context of the war in Bosnia-Herzegovina and Kosovo, ethnic cleansing is evil *per se*. The world is now convinced that the way to lasting peace is that of human rights, of respecting the rights and cultures of minorities and not of excluding them or expelling them. The racist and vicious political tool euphemistically called "population transfers" is now not only discredited, but it is recognized as a crime against humanity.

Let us remember, however, that during the Second World War it was the Czechoslovak President-in-exile, Eduard Beneš, who advocated ethnic cleansing of the entire German population of Czechoslovakia as a measure to guarantee peace, and tried to minimize the intrinsic barbarity of mass expulsions by pretending that the "transfers" could or would ever be carried out in an "orderly and humane manner"—a contradiction in terms, since an uprooting from one's native land could never be termed "humane," and any such transfer would necessarily constitute, at the very least, a form of cultural genocide against the targeted group.

The way to the ethnic cleansing of the Germans was thus so: first all Germans had to be defamed as Nazis, as traitors, as disloyal subjects; second, the idea of removing the Germans from their 700-year-old homelands had to be packaged as a positive measure for peace; third, the Allied peace plan set out in the "Atlantic Charter" had to be quietly abandoned, as if the Charter was valid for the victors only, but not for the vanquished; fourth, any objections of conscience had to be neutralized by postulating the unlikely scenario that the proposed transfers would really be carried out in an "orderly and humane" manner; fifth, a nexus had to be created between Nazi crimes and measures of retribution, even though the victims of the proposed expulsion had nothing to do with Nazi crimes and in many cases had themselves been victims of Nazism.

Not every observer, however, was caught in this web of lies and dishonest political manoeuvres. George Kennan, Bertrand Russell, Robert Murphy, and others warned of the madness of the scheme. And as early as 1946, the noted British publisher and human rights activist, Victor Gollancz, threw light on the reality of what had happened and was still happening, recognizing the moral implications of Allied policy: "If the conscience of men ever again becomes sensitive, these expulsions will be remembered to the undying shame of all who committed or connived at them .... The Germans were expelled, not just with an absence of over-nice consideration, but with the very maximum of brutality."[1]

But in order that the conscience of mankind should become sensitive--even half a century after the events--it is necessary to have full information, open discussion without taboos, to have freedom of expression. The phenomenon of ethnic cleansing should be analyzed from all its aspects: not just historical or legal, but also cultural, demographic, economic, sociological and psychological. In this paper, however, the author will focus primarily on the historical aspects.

The Genesis of the Idea of Expelling the Germans from Their Homelands: The Role of Eduard Beneš

On 9 February 1940 Churchill had stated: "We are opposed to any attempt from outside to break up Germany. We do not seek the humiliation or dismembermemt of your country. We wholeheart-

---

[1] Victor Gollancz, *Our Threatened Values* (London, 1946), 96.

edly desire to welcome you without delay into the peaceful collaboration of civilized nations."[2]

On 14 August 1941 at the conclusion of the Atlantic Conference, Prime Minister Churchill and President Roosevelt proclaimed the Atlantic Charter in which they renounced "aggrandizement, territorial or other" and undertook a commitment to oppose "territorial changes that do not accord with the freely expressed wishes of the peoples concerned."[3] This widely praised declaration represented an attempt to set a higher standard of international morality based on the principle of equal rights and self-determination of peoples. It was subsequently endorsed by the Soviet Union and by representatives of the governments-in-exile of Czechoslovakia and Poland.

Eduard Beneš thus had an uphill battle to sell the expulsion scheme which he had concocted following his personal humiliation at the Munich Conference of September 1938. The first step in obtaining Allied approval for his programme of large-scale spoliation of billions of dollars worth of land and private property was to put the blame on the victims. The persons targeted for expulsion had to be brandmarked as morally evil and deserving of such treatment. One of the best tools to do that was to ignore cause and effect and construct an argument whereby the German minority of Czechoslovakia was accused of being disloyal, notwithstanding the fact that for two decades the Czech political establishment had engaged in systematic discrimination against its three and a half million ethnic Germans.

In order to justify the expulsion of the Germans, Eduard Beneš again and again referred to them as traitors: "We must get rid of all those Germans who plunged a dagger in the back of the Czechoslovak State in 1938."[4] Here Beneš was referring to the Munich Agreement of 1938. He neglected to mention, however, that this Agreement was a direct result of Czechoslovakia's failure to grant effective equality to its ethnic Germans, as reflected in a consistent pattern of discrimination in all areas of economic and cultural life, constituting violations of the minority rights treaties

---

[2] Mr. Stokes quoted this statement back to Prime Minister Churchill in the course of the debate in the House of Commons on 23 February 1944; *Parliamentary Debates*, Commons, vol. 397, cols. 901-02.

[3] United States, Executive Agreement, Series 236:4; Department of State, *Bulletin*, V, p. 125. See also Louise Holborn, *War and Peace Aims of the United Nations*, 2 vols. (Boston, 1943-1948), 1: 2.

[4] Beneš in a broadcast from London, 1944, cited in Holborn, *War and Peace Aims of the United Nations*, 2: 1036.

signed at the Paris Peace Conference of 1919. More concretely, the Munich Agreement was a result of Beneš' own intransigence in the 1930s that drove many loyal Sudeten Germans into the camp of Konrad Henlein and his Sudeten German Party and in turn drove Konrad Henlein into Hitler's arms, since Beneš only offered the Germans second-class citizenship in his very Czech Czechoslovakia.

As Arnold Toynbee noted in an article in *The Economist* in 1937, following a trip to Czechoslovakia and long before the Munich Agreement:

> The truth is that even the most genuine and old-established democratic way of life is exceedingly difficult to apply when you are dealing with a minority that does not want to live under your rule. We know very well that we ourselves were never able to apply our own British brand of democracy to our attempt to govern the Irish. And in Czechoslovakia to-day the methods by which the Czechs are keeping the upper hand over the Sudetendeutsch are not democratic....[5]

After Toynbee it was Lord Runciman who travelled to Czechoslovakia and reported back to the British government:

> Czech officials and Czech police, speaking little or no German, were appointed in large numbers to purely German districts; Czech agricultural colonists were encouraged to settle on land confiscated under the Land Reform in the middle of German populations; for the children of these Czech invaders Czech schools were built on a large scale; there is a very general belief that Czech firms were favoured as against German firms in the allocation of State contracts and that the State provided work and relief for Czechs more readily than for Germans. I believe these complaints to be in the main justified. Even as late as the time of my Mission, I could find no readiness on the part of the Czechoslovak Government to remedy them on anything like an adequate scale ... the feeling among the Sudeten Germans until about three or four years ago was one of hopelessness. But the rise of Nazi Germany gave them new hope. I regard their turning for help towards their kinsmen and their eventual desire to join the Reich as a natural development in the circumstances.[6]

Thus, when Beneš still refused every compromise on autonomy for the Sudeten Germans in 1937 and 1938, he effectively pre-

---

[5] *The Economist*, 10 July 1937, p. 72.

[6] *Documents on British Foreign Policy, 1919-1939*, vol. 2, no. 3, 50.

cipitated the complete separation of the Sudeten German areas pursuant to the Munich Agreement, which both Great Britain and France found fair at the time.

In a letter to the American Secretary of State, Cordell Hull, dated 14 September 1938, the US Ambassador to France, William Bullit, reported:

> during the past few days the French newspapers have published many maps showing the racial divisions in Czechoslovakia ... public opinion has begun to develop the attitude, "Why should we annihilate all the youth of France and destroy the continent of Europe in order to maintain the domination of 7,000,000 Czechs over 3,200,000 Germans?"[7]

Bullitt concluded his letter sarcastically:

> In view of the growing belief among the French and the British that Beneš in his heart of hearts has decided to provoke general European war rather than accept complete autonomy for the subject nationalities of Czechoslovakia, intense pressure will unquestionably be brought on Praha ....[8]

A similar message was sent by Joseph Kennedy, the US Ambassador to Great Britain, to Cordell Hull, in which Kennedy quoted Prime Minister Chamberlain as saying: "I can see no rhyme nor reason in fighting for a cause which, if I went to war for it, I would have to settle after it was over in about the same way I suggest settling it now."[9] In the same sense Arnold Toynbee spoke of a prevailing "feeling of acute moral discomfort" at the prospect of "fighting for the balance of power in defiance of the principle of nationality."[10]

In the light of this and many, many incontestable testimonies and written sources, it is nothing but blatant hypocrisy on the part of the British and the French after the Second World War to disavow the Munich Agreement as being unjust or even illegal. The Munich Agreement provided in September 1938 if not the best, at

---

[7] *Foreign Relations of the United States, 1938,* vol. 1: 595.

[8] Ibid., 596.

[9] Ibid., 622.

[10] A. Toynbee, "A Turning Point in History," *Foreign Affairs* (January 1939): 316. See also *The Times,* 2 June 1938, for a similar opinion expressed by the Dean of St. Paul's; *The Times* editorial of 4 June 1938 suggested that Czechoslovakia grant plebiscites to her minorities.

least a tenable solution, which was welcomed with relief by almost everyone other than Beneš himself. By its terms some three million Germans living in the affected districts were allowed to secede from Czechoslovakia and be united with Germany, while some 500,000 Germans still remained within the borders of the reduced Czechoslovak State. This entailed, of course, an economic loss for Czechoslovakia, but it was, essentially what the Sudeten Germans had been demanding in the name of self–determination since November 1918 and corresponds to the recommendations of the American commission under Harvard Professor Archibald Coolidge, which was discussed but not adopted by the Paris Peace Conference in 1919.[11]

The Munich Agreement cost Beneš his job in Prague, and as an exiled politician in London he had started already in December 1938 to reflect on how to undo the Munich Agreement and how to keep the lands and the wealth of the Sudeten Germans without the annoyance of having to take their legitimate rights and interests into account. Thus, he launched a campaign of disinformation against the Sudeten Germans, which would have had no effect whatever but for the subsequent turn of events, for which the Sudeten Germans had no responsibility: Hitler's invasion of the rest of Czechoslovakia on 15 March 1939 and his megalomanic bid for world power in the Second World War. Thus, it was the Sudeten Germans and the Germans from the German provinces East of the Oder-Neisse who would pay the bill for Hitler's crimes.

First, of course, Beneš had to soft-pedal and gradually sell the idea of population transfers as a measure of ensuring peace after the expected defeat of Nazi Germany. His first target was the British political elite. In September 1941 he wrote, "I accept the principle of transfer of populations.... If the problem is carefully considered and wide measures are adopted in good time, the transfer can be made amicably, under international control and with international supervision."[12] On the basis of these rather uto-

---

[11] "To grant to the Czechoslovaks all the territory they demand would be not only an injustice to millions of people unwilling to come under Czech rule, but it would also be dangerous and perhaps fatal to the future of the new state ... the blood shed on March 3rd when Czech soldiers in several towns fired on German crowds ... was shed in a manner that is not easily forgiven... For the Bohemia of the future to contain within its limits great numbers of deeply discontented inhabitants who will have behind them across the border tens of millions of sympathizers of their own race will be a perilous experiment and one which can hardly promise success in the long run." *Papers Relating to the Foreign Relations of the United States--The Paris Peace Conference*, 1919, vol. 12, 273.

[12] Eduard Beneš, "The New Order in Europe," *Nineteenth Century and After* 130 (1941): 154.

pian representations, British Foreign Minister Anthony Eden informed Beneš as early as July 1942 that "his colleagues agree with the principle of transfer."[13] A decision of the British Cabinet that it had no objection to the transfer of the Sudeten Germans was shortly thereafter communicated to Beneš.[14] Soviet[15] and American[16] approval followed in June 1943. And the initial proposal of removing a limited number of German "traitors" evolved into a maximalist expulsion syndrome affecting the entire Sudeten German population, including German Social Democrats and other anti-Nazis, merely on ethnic grounds.

Thus, it is Beneš that bears the responsibility for the dynamic that eventually led to the expulsion not only of the 3.5 million Germans from Sudetenland, but also of the 10 million Germans from East Prussia, Pomerania, Silesia, East Brandenburg, from pre-war Poland, from Hungary, Romania and Yugoslavia. Once the principle of "population transfers" was accepted, the floodgates were open and the ethnic cleansing of the Germans could begin.

The Anglo-American View on Limited Transfers

Although neither Churchill nor Roosevelt were at the origin of the expulsion program, they did not reject it outright because they found aspects of it useful for political and strategic reasons.

Pursuant to the Molotov-Ribbentrop Pact of August 1939, Poland was invaded by Germany and the Soviet Union and divided along the Ribbentrop-Molotov line. This arrangement largely corresponded to the old Curzon Line, which the British Foreign Minister Lord Curzon had proposed as Poland's eastern frontier, but which Poland had succeeded in pushing eastward during its war against the Soviet Union in 1920.

Whereas the loser of the war, Germany, would obviously have to abandon its occupation of the western half of Poland, it soon became apparent that Stalin had every intention of keeping the eastern part of Poland. This situation constituted an acute embar-

---

[13] Letter by British Foreign Office to Rudolf Storch (German Social Democrat leader in London exile), *Der Sudetendeutsche*, 29 October 1955, p. 1; see also Radomir Luza, *The Transfer of the Sudeten Germans* (New York, 1964), 238.

[14] Eduard Beneš, *Memoirs of Dr. Eduard Beneš: From Munich to New War and New Victory* (London, 1954), 207.

[15] Ibid., 222.

[16] Ibid., 195, 223.

rassment to England, which had entered the war because of Poland and was now faced with the prospect of ending it by accepting the Soviet annexation of Eastern Poland.

It thus became a matter of national honor to provide Poland some form of compensation in the West--at the expense of Germany. Initially this compensation was to be proportional to the loss. There were some two and a half million Poles living East of the Curzon Line, and they should have the opportunity of being resettled in what was left of Poland. Thus, if the German province of East Prussia were to be allocated to Poland after the war and its 2.5 million Germans were to be transferred to Western Germany, there would be a solution that would allow Stalin to keep his booty, give the Poles adequate compensation, and punish the Germans for starting the war. The ominous extension of the principle of population transfers would claim larger groups--not just the German population of East Prussia, but eventually that of Pomerania, Brandenburg and Silesia as well.

Upon concluding talks with Roosevelt in Washington in July 1943, Beneš cabled to the Czechoslovak government-in-exile that Roosevelt "agrees to the transfer of the minority populations from Eastern Prussia [*sic*!], Transylvania and Czechoslovakia...."[17]

By the times the Allies met at the Teheran Conference (28 November to 1 December 1943), Stalin had decided that it was to his advantage to extend Soviet influence in the West by pushing Poland's western frontier as far as possible. Instead of negotiating hard and making Stalin understand that this would not be acceptable, Churchill and Roosevelt quite light-heartedly let it happen: "Eden said that what Poland lost in the East she might gain in the West.... I then demonstrated with the help of three matches my idea of Poland moving westward. This pleased Stalin, and on this note our group parted for the moment."[18]

The landgrab had by now advanced to include not only East Prussia, but large tracts of German lands east of the Oder River, including Danzig, West Prussia, and Upper Silesia. As Churchill pointed out: "It was industrial and it would make a much better Poland. We should like to be able to say to the Poles that the Russians were right, and to tell the Poles that they must agree that they had a fair deal. If the Poles did not accept, we could not help it."[19]

---

[17] Ibid., 195.

[18] Churchill, *Closing the Ring* (London, 1953), 362.

[19] Ibid., 396.

In summing up the results of the Teheran Conference, Churchill observed: "it is thought in principle that the home of the Polish state and nation should be between the so-called Curzon Line and the line of the Oder including for Poland East Prussia and Oppeln; but the actual tracing of the frontier line requires careful study, and possibly disentanglement of population at some points."[20]

This formulation, however, exceeded the compensation favored by the United States. Half a year after Teheran, in May 1944, the Committee on Post-War Programs in the State Department prepared a memorandum containing policy recommendations with respect to the treatment of Germany in the light of long-term United States interests. On the matter of the German-Polish frontier it recommended:

> This Government should not oppose the annexation by Poland of East Prussia, Danzig and in German Upper Silesia the industrial district and a rural hinterland to be determined primarily by ethnic considerations. The United States, however, would not be disposed to encourage the acquisition by Poland of additional German-populated territory in the trans-Oder region.[21]

On 18 December 1944 *Pravda* published a long article by Dr Stefan Jedrichowski, propaganda chief of the Lublin Committee, the Communist-led Polish provisional government at Lublin, in which Jedrichowski recommended that the western frontier of Poland should run from Stettin south along the Oder and Western or Lusatian Neisse River to the Czechoslovak border. Having read this ominous article in Moscow, where he was stationed at the time, George F. Kennan immediately reported to the American Ambassador Averell Harriman on the far-reaching implications of the new arrangement. First and foremost he noted that Poland's dependence on the Soviet Union would be immeasurably increased. In a memorandum written a full six weeks before the Yalta Conference, Kennan expressed his misgivings with a frontier arrangement which

> makes unrealistic the idea of a free and independent Poland. It establishes a border in Central Europe which can be defended only by the permanent maintenance of strong armed forces along its entire extent. Despite Churchill's unconvincing optimism as to

---

[20] Ibid., p. 403.

[21] *Foreign Relations of the United States*, 1944, vol. 1: 302-3.

the ease with which new homes can be found in Germany for six million people (I believe the figure is too low) it renders the economic and social problems of the remainder of Germany ... highly difficult of solution, and reduces radically the possibilities for stability in the area.... We may not be able to prevent the realization of this project.... But I think we are being unrealistic if we fail to recognize it for what it is and give it its proper place in our thinking about the future of Europe. Above all, I see no reason why we should have to share responsibility for the complications to which it is bound to lead.[22]

The die was cast. What followed were the unsuccessful and half-hearted attempts of the Western Allies to limit the expulsions. The United States Delegation to the Conference of Malta on 1 February 1945 proposed that "We should resist vigorously efforts to extend the Polish frontier to the Oder Line or to the Oder-Neisse Line."[23] But a few days later at Yalta, President Roosevelt caved in to Stalin's demands. Churchill had meanwhile understood what it would mean to have to house and feed millions of expelled Germans in the British zone of occupation in postwar Germany and tried to put the brakes on. So he argued that "a considerable body of British public opinion... would be shocked if it were proposed to more large numbers of Germans."[24] He therefore insisted that any transfer of populations should be "proportioned to the capacity of the Poles to handle it and the capacity of the Germans to receive them."[25] This, of course, was too little, too late.

At a parliamentary debate on 1 March 1945 the Labour MP Strauss observed:

> According to the Prime Minister some parts of Germany, certainly Upper Silesia, are to go to Poland. I hope the Government will hesitate before it finally gives its approval to a proposal of this sort, which can hold out no advantage to anybody but may be exceedingly harmful to the general prospects of a lasting European peace. On what ground is such a proposal put forward? That it is going to be some compensation to Poland. But the whole justification for the Curzon Line is that it was agreed in 1919 at Versailles. Not only was the Curzon Line, but also Po-

---

[22] George Kennan, *Memoirs*, 2 vols. (New York, 1967), 1: 214.

[23] *Foreign Relations of the United States--The Conferences at Malta and Yalta*, 510.

[24] *Foreign Relations of the United States, the Conferences at Malta and Yalta*, 717, 720.

[25] *Foreign Relations of the United States, the Conferences at Malta and Yalta*, 717.

land's Western boundary was agreed at Versailles. If one is fair to Poland, so, presumably, is the other.[26]

Five months later the entire area in question was occupied by Soviet forces, millions of Germans had already fled the Soviet onslaught, and the remaining Germans were being expelled by the Poles and the Czechs. Tens of thousands arrived exhausted and dying in Berlin and elsewhere in Brandenburg and Saxony.

As George Kennan observed in his *Memoirs*:

> The disaster that befell this area with the entry of the Soviet forces has no parallel in modern European experience. There were considerable sections of it where, to judge by all existing evidence, scarcely a man, woman, or child of the indigenous population was left alive after the initial passage of Soviet forces; and one cannot believe that they all succeeded in fleeing to the West.[27]

This was large–scale ethnic cleansing, and the Anglo-Americans were at this stage helpless to stop it. All they could do at the Potsdam Conference was to go on record objecting to the extent of the expulsions and to try to gain some control over the actual transfer. Thus emerged the "humanitarian" language of Article XIII of the Potsdam Protocol, published as a communiqué on 2 August 1945 at the end of the Conference: "The Three Governments having considered the question in all its aspects, recognize that the transfer to Germany of German populations, or elements thereof, remaining in Poland, Czechoslovakia and Hungary, will have to be undertaken. They agree that any transfers that take place should be effected in an orderly and humane manner."[28]

In a letter dated 1 August 1945 from Sir Geoffrey Harrison, the British member of the negotiating party, to John Troutbeck, head of the German Section at the Foreign Office, Harrison explained:

> The Sub-Commission met three times, taking as a basis of discussion a draft which I circulated.... the negotiations were not easy--no negotiations with the Russians ever are.... we had a great struggle, which had to be taken up to the Plenary Meeting, about including the last three and a half lines. Sobolev [the Rus-

---

[26] *Parliamentary Debates*, Commons, vol. 408, col. 1655.

[27] Kennan, *Memoirs*, vol. I, 265.

[28] *Foreign Relations of the United States, the Conference of Berlin*, vol. 2: 1495.

sian negotiator] took the view that the Polish and Czechoslovak wish to expel their German populations was the fulfilment of an historic mission which the Soviet Government were unwilling to try to impede. The view of the Soviet Government was that it was the function of the Allied Control Council in Germany to facilitate the reception of the transferred populations as rapidly as possible. Cannon [the American negotiator] and I naturally strongly opposed this view. We made it clear that we did not like the idea of mass transfers anyway. As, however, we could not prevent them, we wished to ensure that they were carried out in as orderly and humane a manner as possible and in a way that would not throw an intolerable burden on the occupying authorities in Germany. Uncle Joe finally agreed to join in requesting the Polish and Czech Governments and the Control Council for Hungary to suspend expulsions until the report of the Control Council was available. This may prevent mass expulsions for the time being, but I have no doubt that hundreds of Germans will continue to move westwards daily.[29]

The human catastrophe that ensued was predictable. On 16 August 1945 Churchill said before the House of Commons:

I am particularly concerned, at this moment, with the reports reaching us of the conditions under which the expulsion and exodus of Germans from the new Poland are being carried out.... Sparse and guarded accounts of what has happened and is happening have filtered through, but it is not impossible that tragedy on a prodigious scale is unfolding itself behind the iron curtain which at the moment divides Europe in twain.[30]

On 12 October 1945, Robert Murphy, the American political advisor to General Eisenhower, described to the State Department the Berlin refugee crisis as follows:

Knowledge that they are victims of a harsh political decision carried out with the utmost ruthlessness and disregard for the humanities does not cushion the effect. The mind reverts to other mass deportations which horrified the world and brought upon the Nazis the odium which they so deserved. Those mass deportations engineered by the Nazis provided part of the moral basis on which we waged war and which gave strength to our cause.

Now the situation is reversed. We find ourselves in the invidious position of being partners in this German enterprise and

---

[29] Public Record Office, London, Document FO 371/46811.

[30] *Parliamentary Debates*, Commons, vol. 414, cols. 83-4.

as partners inevitably sharing the responsibility. The United States does not control directly the Eastern Zone of Germany through which these helpless and bereft people march after eviction from their homes. The direct responsibility lies with the Provisional Polish Government and to a lesser extent with the Czech Government....

As helpless as the United States may be to arrest a cruel and inhuman process which is continuing, it would seem that our Government could and should make its attitude as expressed at Potsdam unmistakably clear. It would be most unfortunate were the record to indicate that we are *particeps* to methods we have often condemned in other instances.[31]

On 18 October 1945 General Eisenhower sent a telegram to Washington:

In Silesia Polish administration and methods are causing a mass exodus westward of German inhabitants.... Many unable to move are placed in camps on meagre rations and under poor sanitary conditions. Death and disease rate in camps extremely high. Germans who attempt to hold onto homes and land are terrorized into "voluntary" evacuation. Methods used by Poles definitely do not conform to Potsdam agreement.... Due to mass migration into Brandenburg and Saxony, health conditions in these regions tragically low.... Reasonable estimates predict between 2 1/2 and 3 million victims of malnutrition and disease between Oder and Elbe by next spring. Breslau death rate increased ten fold, and death rate reported to be 75% of all births. Typhoid, typhus, dysentery and diphtheria are spreading.... Attention is invited in this connection to serious danger of epidemic of such great proportion as to menace all Europe, including our troops, and to probability of mass starvation of unprecedented scale.[32]

In view of these distressing reports, Secretary of State James Byrnes sent a telegram on 30 November 1945 to the American Ambassador in Poland, Arthur Lane, instructing him to convey the American displeasure to the provisional Polish government:

US Govt has been seriously perturbed by reports of continued mass movements of German refugees who appear to have entered Germany from areas east of the Oder-Neisse line. These persons presumably have been expelled summarily from their

---

[31] *Foreign Relations of the United States*, 1945, vol. 2: 1290-2.

[32] National Archives, RG 165, Records of the War Department TS OPD Message File, Nr. S 28399 of 18 October 1945.

homes and dispossessed of all property except that which they can carry. Reports indicate that these refugees--mostly women, children and old people--have been arriving in shocking state of exhaustion, many of them ill with communicable diseases and in many instances robbed of their last few personal possessions. Such mass distress and maltreatment of weak and helpless are not in accord with Potsdam Agreement... nor in consonance with international standards of treatment of refugees.[33]

Byrnes also sent a telegram to the American Ambassador in Prague, Lawrence Steinhardt, instructing him to approach the Czech government to impress upon it the need for suspending the expulsions and for using the most humanitarian methods in effecting any future deportations. He further explained: "We recognized that certain transfers were unavoidable, but we did not intend at Potsdam to encourage or commit ourselves to transfers in cases where other means of adjustment were practicable."[34]

But in the summer and fall of 1945 neither Great Britain nor the United States had an effective way of preventing the consequences of their earlier approval of frontier changes and compulsory population transfers as a method of peace making.

In reviewing the catastrophe of the expulsion of the Germans the International Committee of the Red Cross observed:

Had it been borne in mind that the repatriation of some 1,500,000 Greeks from Asia Minor, after the first World War, had taken several years and required large-scale relief schemes, it would have been easy to foresee that the hurried transplanting of fourteen million human beings would raise a large number of problems from the humanitarian standpoint, especially in a Europe strewn with ruins and where starvation was rife.[35]

Conclusion

Although the Anglo-Americans did not originate or invent the expulsion schemes, it was their approval of the principle of population transfer that led to the catastrophe that cost over two million Germans their lives.

---

[33] *Foreign Relations of the United States*, 1945, vol. 2: 1317.

[34] *Foreign Relations of the United States*, 1945, vol. 2: 1294.

[35] ICRC, *Report on its Activities During the Second World War*, vol. 1: 673-4.

This brings us to the unaccustomed perspective whereby the Germans—or at least some Germans—suddenly appear as *victims*, and not just as that familiar caricature of boot-stamping bullies bent on conquering the world.

In the light of ethnic cleansing in the former Yugoslavia, it is worth reflecting on what happened to the Germans in 1945-48. The time is ripe for today's politicians to admit the Anglo-American responsibility for allowing a demographic catastrophe to unfold, including the savage revenge taken on all Germans, large scale murder, raping, and plunder. The time is ripe to recognize that at Teheran and Yalta both Churchill and Roosevelt connived at these expulsions that so completely negated the values for which the war had been ostensibly fought.

It is certainly time for Anglo-American politicians and historians to recognize that the racist and thoroughly inhuman proposals of Anglo-American politicians like Henry Morgenthau[36] and Lord Vansittart[37] are every bit as disgraceful as the genocidal utterings and practices of a Radovan Karadžić or Slobodan Milošević.

In the light of the expulsion and spoliation of so many innocent Germans, one must ask what happened to the noble principles of the Atlantic Charter? Perhaps no one put it as clearly as British Labour MP John Rhys-Davies when he spoke on 1 March 1945 before the House of Commons: "We started this war with great motives and high ideals. We published the Atlantic Charter and then spat on it, stomped on it and burnt it, as it were, at the stake, and now nothing is left of it."[38]

Let us hope that American and British historians will soon start to consider this much neglected subject matter as a legitimate field of research for themselves and their students. Surely thousands of dissertation themes await to be developed and assigned.

---

[36] Henry Morgenthau, *Germany is our Problem* (New York, 1945).

[37] Lord Vansittart, *Bones of Contention*, 1943.

[38] *Parliamentary Debates*, Commons, vol. 408, col. 1625.

The London Czech Government and the Origins of the Expulsion of the Sudeten Germans

CHRISTOPHER KOPPER

The expulsion of the ethnic Germans from Czechoslovakia was not an act of spontaneous anger and revenge. The Czech plans for the expulsion of Germans did not only reflect the brutalization of the German occupation regime. From the very beginning, the concept of mass expulsion was a means to solve the general ethnic conflicts that had brought down the first Czechoslovak Republic.

The traumatic impact of the Munich treaty, the forced surrender of the Czech borderlands (the Sudeten region) and the disintegration of Czechoslovakia on the Czech exile government in London cannot be overrated. The government-in-exile and the Czech resistance at home had good reasons to believe that the sizeable German majority among the more than three million Germans in Czechoslovakia had willfully served Hitler's cause of dismembering the Czechoslovakian Republic.[1]

Although the British government had already condemned the German occupation of the so-called "Protectorate Bohemia and Moravia" in spring 1939, London did not seriously question the legitimacy of the Munich treaty until 1941. Edvard Beneš had to take into account the fact that the British government would not press for a full restoration of the Czechoslovak territory. Already in late 1939, Beneš had drafted a plan to cede parts of the Sudeten region to Germany in order to achieve a greater ethnic homogeneity of the Czech lands. This first draft program looked more like a concept for an ethnic rectification. For military and ethnic reasons, some smaller parts of the northwestern and northern border regions should be transferred to Germany. Since the population of 1.4 million people in these areas was almost entirely German, the overall number of ethnic Germans in Czechoslovakia would have been cut by almost half.[2] Beneš and

---

[1] For one example for the idea of a German "Fifth Column," see Beneš's article "The Organization of Post-War Europe," *Foreign Affairs* (October 1941).

[2] Václav Kural, "Tschechen, Deutsche und die sudetendeutsche Frage während des Zweiten Weltkrieges," in *Erzwungene Trennung: Vertreibungen und Aussiedlungen in und aus der Tschechoslowakei, 1938-1947: im Vergleich mit Polen, Ungarn und Jugoslawien* (cited hereafter as *Erzwungene Trennung*), ed. Detlef Brandes, Edita Ivaničková, and Jirí Pesek (Essen, 1999), 71-94.

his fellow cabinet members believed that a sizeable reduction of the German minority would have prevented the re-emergence of a "critical mass" of Germans that might destabilize Czechoslovakia from inside again.

This first draft for ethnic postwar planning was certainly not meant to be more than a painful compromise. Beneš and the Czech exile government never ceased to demand the annulment of Munich and the restoration of the historical borders of the Czech lands. Therefore, the Czech government-in-exile would only have offered territorial concessions in case of a British-German peace treaty. Deliberations of this kind were rendered obsolete by Churchill's strong determination not to conclude any peace treaty with Germany.

Already in spring 1940—and even before Churchill's appointment—Beneš had drafted a new plan to solve the problems of ethnic tension and unrest. In contrast to his program of 1939, the historical borders of the Czech lands would not be surrendered to an ethnic rectification and to strategic expediency.[3] According to this new plan, a sizeable part of the German minority in Central Bohemia and Central and Northern Moravia would be resettled in the predominantly German border regions. Three border districts in Western and Northern Bohemia should be granted a limited form of "self government" under a German majority.[4]

At first glance, such a concept for the federalization of the border territories showed many similarities with the federalization concepts of the pro-constitutional and anti-irredentist German-Czech parties. Until Munich, the "activist" (loyal) German parties such as the Sudeten Social Democrats had campaigned in vain for a thorough federalization of Czechoslovakia to overcome the widespread sentiments of civic alienation among the German minority and to steal the thunder of the irredentist Henlein movement (the *Sudetendeutsche Partei*). The exiled leadership of the Sudeten German Social Democrats in England pressed for a thorough reform of the pre-Munich minority politics in order to establish the Germans as a third

---

[3] Detlef Brandes, "Benes, Jaksch und die Vertreibung/Aussiedlung der Deutschen," in *Erzwungene Trennung*, 95-104; Lemberg, Hans, "Die Entwicklung der Pläne für die Aussiedlung der Deutschen aus der Tschechoslowakei," in *Der Weg in die Katastrophe: Deutsch-Tschechoslowakische Beziehungen 1938-1947*, ed. Detlef Brandes and Václav Kural (Essen, 1994), 77-92.

[4] Václav Kural, "Zum tschechisch-deutschen Verhältnis in der tschechischen Politik 1938-1945," in *Der Weg in die Katastrophe*, ed. Brandes and Kural, 93-118.

"state nation" on equal status with the Czechs and Slovaks. The so-called *"Treuegemeinschaft"* of the Sudeten German Social Democrats equaled the concept of national self-determination with a federal self-government within a renewed Czechoslovak Republic.[5] The Social Democrats who represented by far the strongest anti-Nazi movement among the Sudeten Germans were even ready to accept a limited territorial separation of Germans and Czechs by dissolving the so-called German "linguistic islands" in Central Bohemia and Moravia through an exchange of population.

But in the fall of 1940, Beneš gave in to the expectations of the national resistance movement at home and adopted a tougher position towards the German minority. In a radio message to the Central Committee of the Resistance Movement (UVOD) in November 1940, Beneš indicated his intention to reduce the overall number of Germans by one million through mass expulsion of Nazis and by the resettlement of ethnic Germans within Czechoslovakia. But despite his plan to drive about one million ethnic Germans to Germany, he faced severe criticism on the home front. The military resistance in particular was strongly opposed to Beneš's plans for autonomous German regions and did not conceal their eagerness to drive out at least a sizeable minority of Germans. Beneš now faced the problem to balance between the expectations of the home front and likely British objections against a potential German irredenta. The British government did not officially declare the Munich agreement null and void until August 1942. But several talks with the Foreign Office indicated to the Czech leader that the British government was not opposed to the idea of selective expulsions of German Nazis and that it would not insist on respecting the post-Munich borders of Czechoslovakia. As the exile government claimed to be the only rightful government of the Czech people, Beneš could not make any final decisions about post-war politics without prior consultation with the resistance movement at home. In order to maintain his moral authority on the home front, he had no choice but to give in to the radicalization of the Czech resistance under the impact of the German occupation. A message of the UVOD leaders about their idea of reducing the number of Germans by revolutionary action from below did not sound like a

---

[5] "Richtlinien für die Auslandspolitik der Sudetendeutschen Sozialdemokratie," undated (November 1939), in *Češi a sudetonemecká otázka, 1939-1945: Dokumenty* [Czechs and the Sudeten–German Question, 1939–1945: Documents], ed. Jitka Vondrová, (Prague, 1994), 31-34.

rhetorical figure of speech or some kind of imaginary violence, but a serious intention.

The pressure of the home front on the government-in-exile had some serious repercussions on the relations between Beneš and the Sudeten German Social Democrats. Before he received the messages from the UVOD, Beneš had publicly announced his intention to admit representatives of the Sudeten German Social Democrats to the Czech State Council in exile. Beneš had denied the Social Democrats' request to call for a federalization of Czechoslovakia, but the unresolved problem of collective German rights in a new Czechoslovak Republic did not have to be an insurmountable obstacle for German-Czech cooperation in exile. Wenzel Jaksch (the chairman of the Sudeten German Social Democrats) surely missed a chance by not joining the Czech State Council without preconditions.

For the Sudeten German anti-Nazis, the window of opportunity only remained open for a short time. After Beneš had received the home front's criticism, he denied the existence of any negotiations about a future federalization of Czechoslovakia. According to his previous experiences, he was certain that his Sudeten German counterpart and leader of the Sudeten SPD, Jaksch, would never join the National Council if any hope for a change was categorically rejected. Beneš predicted accurately that in the course of time the alienation between Germans and Czechs would intensify. Therefore, the participation of Germans in the National Council was not only postponed, but rejected permanently.

In contrast to some later allegations, the Sudeten German *"Treuegemeinschaft"* never contested Beneš's rejection of the Munich Treaty.[6] But whereas Beneš blamed the lenience of the Czech ethnic policy for the disintegration of Czechoslovakia by the Sudeten German Nazis, the Sudeten German Social Democrats believed that only a thorough federalization of Czechoslovakia could curb irredentist unrest and anti-Czechoslovak disloyalty. The crucial point, however, was not the unconditional nullification of Munich, but the ramifications of the minority policy in the First Republic. Beneš, the exile government, and the Czech resistance tended to equate autonomy for Germans with the disintegration of Czechoslovakia through Henlein's Fifth

---

[6] See the declaration of the Sudeten German Social Democrats about the future position of the Sudeten region in a democratic and federal new order of Europe, March 10, 1940 (Holmhurst declaration), in *Češi a sudetonemecká otázka, 1939-1945*, 49-54.

Column. In this respect, the perceptions of the Sudeten German anti-Nazis and the Czech exile about the failure of the First Czechoslovak Republic were diametrically opposed.

As early as August 1941 the Central Committee of the Resistance Movement (UVOD) had radioed to the exile government its intention of expelling "the Germans" from Czechoslovakia. The meaning of the term "the Germans" was unambiguous and included the entire German population, with possible exceptions for proven German anti-Nazis. Beneš partly accepted the UVOD position, but insisted on a more realistic approach to reducing the number of Germans. Taking into account the likely objections of the British to a full-fledged ethnic cleansing, he insisted that a "certain number of Germans will remain in Czechoslovakia" after the end of the war. Above all, Roosevelt's and Churchill's Atlantic Charter stipulated the right of national self-determination and rendered a maximum expulsion program unrealistic. At this point, an indiscriminate expulsion of Germans would have met serious resistance in the British and American political elite.

Therefore, Beneš developed the idea of a "minimum program" for the reduction of the German minority. To get rid of at least one million pro-Nazi Germans, he argued in favor of selective expulsions. In an article in the October 1941 edition of *Foreign Affairs*, he carefully referred to a "transfer" of Germans in general terms as a potential means of solving the minority question by increasing the ethnic homogeneity. As a backup position towards the Western Allies, but only as a last resort, Beneš was willing to accept minor territorial concessions. Worried about the consent of the Western powers, Beneš did not really identify with the "maximum solution" of an indiscriminate expulsion.

The appointment of Reinhard Heydrich as Acting *Reichsprotektor* for Bohemia and Moravia in late September 1941 brought a substantial brutalization of German occupation policy against the Czech national resistance. Because of brutal repression against the Czech population, the Czech resistance and the exile government became less inclined to distinguish between their German oppressors and the rather invisible minority of German anti-Nazis. Apart from hundreds of executions and thousands of Czechs imprisoned in the "death mills" of German concentration camps, the Czechs suffered from the near total suppression of Czech national and cultural organizations such as the *Sokol* (the Czech gymnasts' association, a traditional center of Czech nationalism). The German *Protektorat* administration even substituted German for Czech as the main administrative language, intentionally assaulting the national pride of the Czechs.[7] Under the impact of increased repression, Beneš's London cabinet felt more and more sympathetic to the maximum concept of an indiscriminate expulsion of Germans. *Reichsdeutsche*, Germans from the incorporated territories, were prominent in the *Protektorat* administration. Sudeten German Nazis themselves held some key positions as administrators, in particular the high profile Karl Hermann Frank (Acting *Reichsprotektor* after the assassination of Heydrich) and Konrad Henlein (party *Gauleiter* of the Sudetenland). Because of their knowledge of the Czech political milieu and the Czech language, Sudeten German Gestapo officers played a crucial role in advising and implementing the suppression of Czech national resistance. In the Sudeten region, Sudeten German officials and Nazi Party functionaries played a crucial role in the politics of repression by suppressing any form of Czech cultural autonomy.

Under these circumstances, the government-in-exile was even more inclined to treat the German minority as a collective entity. Beneš and his cabinet did not draw a clear distinction between the majority of Germans that had voted for and had actively supported the Nazified *Sudetendeutsche Partei* and the still sizeable anti-Nazi minority of 10 to 15 percent of all ethnic Germans. In his speech to the Czech National Council in November 1941, Beneš indiscriminately charged the entire German bourgeoisie, the educated German elite and the "Nazified working class" of active and passive support of the Nazi occupation.[8] The identification of a great majority of Germans with the Nazi oppressors certainly paved the way for the idea for a collective punishment of a majority of Germans.

The German policy of increasing repression–which escalated still more steeply after the assassination of Heydrich–had a major impact on the radicalization of Czech expulsion concepts. The London Czech government felt even more pressed to match the rising bitterness of the Czech resistance and its expectations of a tough policy against Germans. But the rising anger against the German occupiers was not limited to the core of the Czech

---

[7] See Miroslav Karny, Jaroslava Milotova, and Margita Karna, (eds.), *Deutsche Politik im "Protektorat Boehmen und Maehren" unter Reinhard Heydrich 1941-1942: Eine Dokumentation* (Berlin, 1997).

[8] Detlef Brandes, "Das Problem der deutschen Minderheit in der Politik der Alliierten in den Jahren 1940-45: das tschechische Beispiel," in *Integration oder Ausgrenzung: Deutsche und Tschechen 1890-1945*, ed. Jan Kren, Václav Kural, and Detlef Brandes (Bremen, 1986), 105-156.

resistance. From late 1941, German officials in the Sudeten region occasionally noticed that threatening remarks against Germans had been overheard and reported by German police informers.[9] Such wisps of information must have represented only a fraction of the existing simmering vengefulness.

After the massacre of Lidice (June 1942) and other brutal reprisals were reported by the British media, the British War Cabinet voted unanimously for a full annulment of the Munich Treaty and further agreed generally to an unspecified transfer of German minorities in East Central Europe. Although the War Cabinet remained unspecific in regard to the numbers of expellees, the Foreign Office favored an indiscriminate expulsion rather than expulsions based on individual criteria of responsibility and guilt. As the Chamberlain Cabinet had involuntarily paved the way for the subversion and decomposition of the Czechoslovak Republic, the War Cabinet was caught in a morally weak position to refute Czech security demands. Beneš reviewed his own expulsion plans in the wake of Lidice. In an October 1942 radio message to the Czech resistance, he proposed a more radical "medium plan" to reduce the number of Germans in Czechoslovakia by 1.5 million, by means of transferring Czech territory occupied by 500,000 Germans to Germany and further expelling one million Germans from Czech soil entirely.[10] Only two months later, he even outlined plans to reduce the number of Germans in the Czechoslovak Republic by two million through territorial concessions and the expulsion of one to 1.5 million Germans. After the Casablanca conference demonstrated the Allies' commitment to demand nothing short of an unconditional German surrender, Beneš and his exile cabinet could feel assured that the Allies would not stop short of returning an entirely liberated Czechoslovakia to the Czechoslovak government. An internal memorandum from December 1943 did not yet demand the total expulsion of Germans, but the idea of reducing the German minority to less than one third of the entire population, even in predominantly German areas, obviously implied the expulsion of up to 66 percent of all Germans.

The exiled Sudeten Germans found themselves increasingly isolated politically from the British Foreign Office and the Czech

---

[9] Volker Zimmermann, *Die Sudetendeutschen im NS-Staat: Politik und Stimmung der Bevölkerung im Reichsgau Sudetenland (1938-1945)* (Essen, 1999), 423.

[10] Brandes, "Das Problem der deutschen Minderheit," 112-113.

exile government. In 1942, the political gap between the Sudeten German anti-Nazis and Beneš became unbridgeable. The chance for securing some kind of Sudeten German status in the postwar order of Czechoslovakia had diminished after Heydrich's appointment and practically disappeared after the escalating atrocities of Lidice. The Sudeten German Social Democrats claimed to be the true and rightful representatives of their fellow Germans who, as they believed, were suffering equally from National Socialist repression. Since the exiled Sudeten Germans were equally opposed to Nazism as Czechs were, except for Czech nationalist motivations, they considered the Czech plans for mass expulsions as a collective punishment and remedy against Nazism as absolutely unacceptable and even counterproductive for the re-establishment of democracy in the Sudeten region.[11]

Additional difficulties arose from the Sudeten Germans' need to demonstrate a significant level of German resistance at home in order to establish a moral base of equal victimization and an equal commitment for their joint anti-Nazi cause. The Sudeten German exiles could never establish a radio connection to their comrades in the Sudeten region and never obtained reliable information from the "home front." The Czech national resistance had enough troubles maintaining its clandestine organization despite the growing pressure of Gestapo persecution. Installing a network of agents in the predominantly German Sudeten area was feasible neither for Czechs nor for Sudeten Germans. Among Germans, popular support for political policing remained on a high level. Reports by the Sudeten German exiles about a steep decline of pro-Nazi allegiance in the Sudeten region sounded more like wishful thinking than confirmed facts.[12] In the fall of 1944, moreover, Nazi propaganda skillfully exploited news leaks about Czech expulsion plans. In the wake of the Soviet advance in Slovakia, more and more Sudeten Germans lost their faith in a German victory and had good reasons to ask their Czech neighbors for forgiveness for past humiliations. But for fear of Czech reprisals, the Sudeten Germans rallied behind the National

---

[11] Resolution of the Sudeten German Labor Movement in exile, June 7, 1942, in *Cesi a sudetonemecká otázka, 1939-1945*, 167-168.

[12] Memorandum of the Sudeten German Social Democrats, "Die Anwendung der Atlantik Charter auf das Sudeten-Problem," in *Cesi a sudetonemecká otázka, 1939-1945*, 258-260. On the low level of anti-Nazi resistance among the Sudeten Germans, see Volker Zimmermann, "Der 'Reichsgau Sudetenland' im letzten Kriegsjahr," in *Erzwungene Trennung*, 53-70.

Socialist policy of ultimate defense rather than rebelling against a continuation of war. On the other hand, Beneš and his fellow cabinet members had not forgotten the overwhelming support for the Nazi movement among the Sudeten Germans before and after Munich. Jaksch frequently testified that the Nazified majority of Sudeten Germans had already intimidated the anti-Nazis in spring 1938. As in the rest of the Reich, the high degree of political loyalty and compliance among the population and the close surveillance through the Gestapo rendered all resistance activities beyond very small and isolated groups extremely risky.

The Sudeten German resistance found itself in a very uncomfortable situation. Although these resistants represented the political opinions of the anti-Nazi minority among the Sudeten Germans, they claimed to act in the interest of the entire German population in the Czech lands. The German anti-Nazis claimed nothing less than a moral title for political leadership over the German minority that, from the Czech point of view, had overwhelmingly supported the Nazi cause and lost any right to collective recognition. To support their claim to be the only rightful representation of the Sudeten Germans, Wenzel Jaksch and his fellow Social Democrats had to avoid any compromise that might have looked like a treason of rightful German interests. Therefore, the Sudeten German Social Democrats were never willing to negotiate about a general transfer of Germans. But despite his opposition against expulsions, Jaksch was not generally opposed to the idea of expelling National Socialists who had been guilty of anti-Czech treason and had participated in the National Socialist suppression of Czechoslovakia. But Jaksch and his comrades preferred the plan of establishing German anti-fascist peoples' courts to bring to trial German perpetrators for their crimes against fellow Germans or to extradite them to Czech courts if they had committed crimes against Czechs.[13] In January and September 1942, Beneš submitted to Jaksch several proposals for a selective transfer of Germans tainted by Nazism. In January 1942, Jaksch refused to accept the expulsion of 600 000 to one million Sudeten Germans as a precondition for representation in the Czech National Council. But even if the Sudeten German Social Democrats had accepted these terms, the National Council would likely not have taken a milder decision in favor of their German-Czech compatriots. Furthermore, Beneš's proposal from September 1942 was not really serious, since he had already

---

[13] Wenzel Jaksch, Speech on the second conference of the "Treuegemeinschaft," 4 October 1942, in *Češi a sudetonemecká otázka, 1939-1945*, 183-185.

adopted the concept of a large-scale expulsion and abandoned the criterion of individual responsibility and guilt.

Interestingly, the Czechoslovak Communist Party (KSC) was far more reluctant than the other Czech parties in exile to adopt a concept of mass expulsions. One reason for their reluctance towards a collective expulsion of Germans was the fact that the Czechoslovak Communist Party was the only real multiethnic party in Czechoslovakia and included a sizeable group of some ten thousand German members who were regarded as equals. Only just before Beneš paid Stalin a visit in Moscow in December 1943 did the KSC politburo opt for the expulsion of "guilty" Germans. According to the still prevailing communist doctrine on fascism, Nazism was defined as the "rule of the most reactionary and aggressive fraction of capitalists." If taken at face value, this doctrine was incompatible with the idea of a collective guilt of all Germans.

At this time, Beneš already favored full-fledged expulsion. Despite its propagandistic claims after liberation of having invented the expulsion, the KSC politburo adopted the concept of a large-scale expulsion only after Stalin had officially endorsed Beneš's ethnic policy.[14] Since the KSC politburo never made any important decision without prior consultation with the Kremlin, it is clear that the Czech and Slovak Communists-in-exile changed their mind because of Stalin's decision rather than as a result of their own will.

In the summer of 1944, the Czech exile government definitively accepted the idea of driving out the Sudeten Germans by violent means; just after the military liberation of the Czech lands by Allied forces, the Czech national resistance would expel the highest possible number of Germans. In a message to the Czech resistance movement from July 1944, Beneš gave the resistance organs full power of attorney to take the expulsion into their own hands until the Czech government was reestablished in Prague.[15] Beneš and the London exile government had not designed the

---

[14] See for example the speech of the leading Czech Communist politician Václav Kopecky in the Czechoslovak Parliament, November 1948, in *Dokumente zur Sudetenfrage*, ed. Fritz Peter Habel (Munich, 1984) 271-272. Secretary Vyacheslav Molotov had already indicated Soviet encouragement for expulsion, but this was not a personal endorsement from Stalin; see Zbynek Zeman and Antonin Klimek, *The life of Edvard Benes 1884-1948* (Oxford, 1997), 184.

[15] Radio message from Beneš to the resistance groups, 14 July 1944, *Dokumente zur Sudetendeutschen Frage 1916-1967*, (Munich, 1967), 289-294.

concept of "spontaneous expulsions" themselves and only responded to the wishes of the resistance movement at home, but Beneš deliberately used the resistance at home to pursue a double strategy. On one hand, he negotiated with the British and American governments an "orderly transfer" of about two million Germans, while on the other he encouraged the resistance at home to exceed these limits and to create a *fait accompli* before a peace conference would make a decision. At this point, the Czech government-in-exile had already accepted the idea to prefacing the "orderly transfer" (as negotiated with Churchill and Roosevelt) with a "wild" and surely violent expulsion with all its horrible consequences.

But the Beneš government was still hesitant to reveal its decision for a near total expulsion to the Western Allies. In a secret memorandum to the Western powers from November 23, 1944, Beneš still seemed to be determined to keep a German minority of "not more than 800 000" citizens in the Czechoslovak Republic.[16] Apart from active anti-Nazis, a part of the skilled German workforce essential to continue production in key industries should be permitted to stay. At first sight, this proposal seemed to include a large number of politically acceptable and economically indispensable Germans and implied relatively generous exemptions and expulsion without limits. But the general denial of collective minority rights such as bilingualism in public institutions and secondary school instruction in German created a very restrictive framework for the future of Germans in Czechoslovakia.[17] The Beneš government had good reasons for assuming that a majority among the "desirable" Sudeten Germans would rather opt for a "transfer" to Germany than to apply for the Czechoslovak citizenship and keep their number far below the maximum level of 800,000. Still, Czech concerns that they should limit German rights in the future were needless since the Beneš government planned to reduce the number of Germans to less than 6 percent of the entire Czechoslovak population.

---

[16] Memorandum of the Czechoslovak Government on the problem of the German minority in Czechoslovakia, 23 November 1944 (strictly confidential), *Cesi a sudetonemecká otázka, 1939-1945*, 303-308.

[17] The denial of a collective minority status was already foreshadowed in another Beneš memorandum about the transfer of Germans from Czechoslovakia, dated December 1943, in *Dokumente zur Sudetenfrage*, ed. Habel, 273-274. This memorandum was written in Czech and not meant for submission to the US or British governments.

Therefore, the German minority would never have been able to pose a potential irredentist threat to the new Czechoslovakia.

For tactical reasons, Beneš was cautious in revealing the means and modalities of expulsion to the Western public, but he did not hide the scope and scale of the transfer. In an article for the October 1944 article in *Foreign Affairs*, he floated the idea of transferring "the greatest possible number of German inhabitants," excepting only those Germans who had been loyal to the Czechoslovak Republic before the German onslaught. Interestingly, Beneš was more outspoken towards the Soviet Foreign Secretary Vyacheslav Molotov. In a talk with Molotov in March 1945, he did not conceal the Czech intention to "transfer" as many Germans as possible. Obviously, Beneš did not have to fear any criticism from inside the Soviet administration, let alone from the highly controlled public opinion in a totalitarian dictatorship.[18]

---

[18] Zeman and Klimek, *Edvard Benes*, 231.

Escaping History: The Expulsion of the Sudeten Germans as a *Leitmotif* in German-Czech Relations

SCOTT BRUNSTETTER

Coming at the end of the Second World War, the expulsion of the Sudeten Germans capped what had been one of Europe's most violent periods.[1] Yet, the completion of the expulsions did not consign them merely to the annals of history. The emotions and the memories of the Second World War and the expulsions in particular have constituted a *leitmotif* in relations between the Germans and the Czechs over the past fifty years that, if unresolved, can only hinder the further development of amicable relations.[2] The events of the past remain alive in historical memory.[3] Conflicting views of history further complicate the situation, which makes the potential to exorcize the demons of the past increasingly difficult. The purpose of this chapter is specifically to trace the role of the Sudeten German expulsion in the bilateral relations between the Germans and the Czechs. By examining the role of historical memory, some general thoughts on the long term effects of ethnic cleansing can be offered, which can give important lessons for both current and future diplomats as well as scholars.

Unfinished Memories

In contrast to the removal of the Germans from what is today Poland, which seems to engender less debate as to its history, the expulsion of the Germans from Czechoslovakia remains an issue

---

[1] On the expulsions, see the chapters by Alfred de Zayas, Christopher Kopper, and Erich Helfert, in this volume. I will use "Germany" and "West Germany" as well as "Czechoslovakia" and "the Czech Republic" interchangeably. East Germany during the Cold War and Slovakia after its end will not be examined in this context.

[2] Neta C. Crawford, "The Passion of World Politics: Propositions on Emotion and Emotional Relationships," *International Security* 24 (Spring 2000): 150; Jerzy Jedlicki, "Historical Memory as a Source of Conflicts in Eastern Europe," *Communist and Post-Communist Studies* 32 (September 1999): 226.

[3] See Walter Becher's anecdote about the relationship between past and present: "History is the Present of the Past" in *Identität und Geschichts-bewußtsein: Ihre Bedeutung für das Selbstverständnis des Menschen und seiner Gemeinschaftsbereiche* (Geislingen, 1984), 10.

that has yet to reconciled.[4] The continuing memories and, more importantly, the different historiographical approaches to this issue set the two on opposite sides of a canyon of historical thought. The memories of the past remain unfinished. Despite their expulsion, Sudeten Germans maintain strong memories and passions for their *Heimat*.[5] Throughout the ensuing fifty years, the Sudeten Germans continued to maintain a sense of cohesion, though they were expelled not only from their homeland, but also from the very territorially based culture that defined them.[6]

Maintaining a semblance of cohesiveness and a memory of the past became the primary focus of the many Sudeten German *Landsmannschaften* that have flourished in Germany.[7] The January 1950 Detmolder Declaration that announced the formation of the *Landsmannschaften* proffered the primary goal of securing the "existence and substance" of the Sudeten German *Volksgruppe*. Added to the November 1949 Eichstätter Declaration that laid out the European political goals of the Sudeten Germans, among which was the desire to regain their lost territory, the Sudeten German desire to return to the past, despite its consequences, is quite evident.[8] Both of these proclamations sought to preserve the distinct identity of the Sudeten Germans and were, as Ferdinand Seibt notes, compensations for the lost *Heimat*.[9] Sustaining the

---

[4] After Polish attempts at reconciliation in the late 1960s and early 1970s, through offers from religious figures and the 1970 Warsaw Treaty, the National Socialist period no longer seems to define Polish-German relations. See Jedlicki, 227.

[5] Based on personal interviews with expellees.

[6] Becher, 11.

[7] The word *Landsmannschaften* does not have a direct English equivalent. It is best described as an organization of former expellees who have a distinct connection with their former homeland, in this case the Sudetenland. Moreover, it reflects the concept of *Heimat*, which represents a strong feeling toward the land on which one was born, lives, and where relatives are often buried. Typically, Germans of that era remained in a certain region, forging a distinct attachment to it. Although Germans expelled from the territory that became Poland also formed such *Landsmannschaften*, they were not as strong and focused as those for the Sudeten Germans.

[8] "Grundsätze einer sudetendeutschen Europa-Politik—Eichstätter Erklärung," and "Gründung der Sudetendeutschen Landsmannschaften—Demolier Erklärung der SL," in *Krajanské organizace sudetských Němců SRN: Studie o sudetoněmecké otázce, Část II* [Sudeten German *Heimat* Organizations in the German Federal Republic: Studies on the Sudeten German Question, Volume II] (Prague, 1998), 165-67.

[9] Becher, 14; Ferdinand Seibt, *Deutschland und die Tschechen: Geschichte einer Nachbarschaft in der Mitte Europas* (Munich, 1998), 398.

memory of what had been became the critical function of the *Landsmannschaften.*[10]

In addition to preserving the Sudeten German identity, these organizations also helped keep issues relating to the expulsion of the Sudeten Germans, such as the Munich Agreement and the Beneš Decrees, from becoming relics of the past. During the development of bilateral relations during the Cold War and more prominently after its end, the expulsion of the Sudeten Germans remained a critical issue for German as well as Czechoslovak policymakers. Moreover, the power the *Landsmannschaften* exhibited, particularly with their political strong base with the CSU in Bavaria, became an important part of the genesis of German foreign policy vis-à-vis Czechoslovakia.[11] Indeed, as will be described below, the Sudeten German issue has remained one of the most pervasive issues in German-Czech relations.

Developing in entirely different domestic situations, the view of the expulsions for the Czechs and the Germans has over time solidified into vastly different perspectives. That these divergent historiographical perspectives exist[12] can be most clearly demonstrated with the terminology both sides have used and continue to use in referring to the removal of the Germans. The Germans use the word "*Vertreibung,*" or expulsion, while the Czechs prefer "*odsun,*" or transfer. The subtle difference between them illustrates the fundamentally distinct lenses, through which the two ethnic groups perceive the same event. Likewise, it represents a sharp difference in the deep-seated moral assessment of the events.[13] *Vertreibung* denotes a harsher, forced movement, with little legitimacy or legality; contrarily, *odsun* suggests a more legitimate and legal removal of the Germans. Although the Allies'

---

[10] The annual Sudeten German days play an important role by keeping the Sudeten customs alive and teaching them to younger generations. Likewise, the Sudeten German House in Munich, through historical research and exhibitions, preserves the Sudeten German existence and identity.

[11] Peter Reichel, "Die Vertriebenenverbände als außenpolitische 'pressure group'," in *Handbuch der deutschen Außenpolitik,* ed. Hans-Peter Schwarz (Munich, 1976), 235.

[12] For an exploration of this varying historiography, see Bradley F. Abrams, "Morality, Wisdom and Revision: The Czech Opposition of the 1970s and the Expulsion of the Sudeten Germans," *East European Politics and Societies* 9(April 1995): 234-55; *Zur Geschichte der Deutsch-Tschechischen Beziehungen: Eine Sammelschrift Tschechischer Historiker aus dem Jahre 1980* (Berlin, 1985).

[13] Jan Pauer, "Moralischer Diskurs und die deutsch-tschechischen Beziehungen," *Arbeitspapiere und Materialien* (Forschungsstelle Osteuropa an der Universität Bremen), nr. 17, Juni 1998, 7, fn. 1.

August 1945 Potsdam Declaration spoke of the "transfer" of the Germans, a point the Czechs continually bring to the forefront, the brutality with which some were undertaken speaks more toward expulsions. It is this perception of difference between *Vertreibung* and *odsun* that largely colors the debate surrounding the removal of the Sudeten Germans.

One of the primary differences in the interpretation of the expulsions is the perceived connection with the National Socialist atrocities. Czechs saw the removal of the Germans as intimately connected to the events during the occupation. As the Czech historian Václav Kural noted in 1994, the primary guilt lies with the Germans for starting the destructive spiral. The Germans, however, have predominately viewed the progression with a different twist. Although there has been gradual recognition that the National Socialist atrocities in occupied Czechoslovakia precipitated the expulsions, the expulsions themselves are still seen as illegitimate, for they stemmed from a "false" principle—that of collective guilt. Even German President Richard von Weizsäcker in a December 1995 speech in Prague, agreed with this idea. For the Germans in particular, as Timothy Burcher argues, the illegitimacy of collective guilt has given a fixture upon which their outrage could focus.[14] This varied perspective has helped to perpetuate the difficult memories of the past, leaving it an issue of considerable importance within the context of a developing German-Czech relationship.

Confronting the Past

Confronting the memories of the past has always been an important part of reconciliation after a conflict. For the Germans and the Czechs, however, the strict division of Europe during the Cold War largely prohibited instances where the two could pursue cordial relations that could promote historical reconciliation. The most effective instance of diplomatic progress came within the context of Willy Brandt's *Ostpolitik* in the early 1970s. The extended negotiations surrounding the 11 December 1973 Prague Treaty, which formally established diplomatic relations between Czechoslovakia and West Germany,[15] quickly brought forth contentious legal issues.

---

[14] Pauer, 8, 18-19; Timothy Burcher, *The Sudeten German Question and Czechoslovak-German Relations since 1989* (London, 1996), 10.

[15] This treaty was the last to be completed with the Eastern states. West Germany signed the Moscow Treaty with the Soviet Union in August 1970 and the Warsaw

By far, the most difficult challenge arose over the disposition of the 1938 Munich Agreement, which began the disastrous spiral of events that culminated in the removal of the Sudeten Germans. The German government resisted pressure to see it as "void (*ungültig*) from the beginning," believing that the agreement's remission would add an element of legitimacy to the removal of the Germans.[16] For the Czechs, in addition to providing a sense of legitimacy for the expulsions, the agreement's repudiation would have maintained the historical political viability of the state, which remained an important concern for the young state. The eventual compromise allowed both to see "the *measure* of this treaty [Munich] in view of the mutual relations as invalid (*nichtig*)." Seeing the measure of the treaty as invalid allowed both states to continue to view the Munich Agreement to their own advantage. The 1973 treaty also took the first step in linking the National Socialist actions with the expulsions, recognizing that the removal of the Germans "was imposed" by the violence in occupied Czechoslovakia. This admission, however, remained short of the direct connection inferred in the concept of collective guilt the Czechs used as a catalyst.[17]

For the Sudeten Germans, this treaty was extremely problematic. In July 1973, as the German Parliament debated the treaty, the Sudeten German Council proclaimed that it left the Sudeten German question unresolved and falsely portrayed its history through an incorrect interpretation of the Munich Agreement in that it did not recognize the role of the British and the French for the original injustices surrounding the end of the First World War. Furthermore, they rejected the inviolability of the common borders stipulated in the treaty, saying it allowed neither for the right of self determination nor that of a *Heimat*.[18] Throughout, the Sudeten Germans remained dogmatically

---

Treaty with Poland in December 1970. Although discussion with Czechoslovakia began in October 1970, the discussions were prejudiced by the May 1970 Czechoslovak-Soviet Friendship Agreement in which the Munich Agreement of 1938 was declared "void (*ungültig*)." See Sudetendeutschen Rat, *Der Prager Vertrag in den Parlamenten: Eine Dokumentation* (Munich, 1975), 5.

[16] Gregor Schöllgen, *Die Außenpolitik der Bundesrepublik Deutschland: Von den Anfängen bis zur Gegenwart* (Bonn, 1999),126.

[17] Schöllgen, 126, emphasis added.

[18] Oskar Böse und Rolf-Josef Eibicht, eds., *Die Sudetendeutschen: Eine Volksgruppe im Herzen Europas von Frankfurter Paulskirche zur Bundesrepublik Deutschland* (München, 1989), 116-17.

attached to the memories of the past. Despite their protests, however, the Prague Treaty was ratified, leaving issues that could only be resolved with the end of the Cold War.

While the compromises inherent in the treaty were positive first steps between the two states, the treaty fell short of its ambitious goal of escaping the past. The political German-Czech Historical Commission of the 1990s viewed this treaty as a first step toward stability and understanding between the two states. Wilfried Fiedler, however, argues that the treaty did not solve the temperamental issue surrounding the role of Munich Agreement, nor did it address the Beneš Decrees.[19] In a more general appraisal, Ann Phillips suggests that the bilateral relations between the Germans and the Czechs during the 1970s suffered from a general lack of *détente*, in part due to the Prague Spring of 1968 and a general steadfastness of the Czechoslovak government on difficult issues. Thus, German-Czech relations as a whole had to be placed on a "new footing" after 1989.[20] The treaty left the memories of the past unfinished, awaiting a time when relations could be pursued on a more cordial basis.

The monumental changes in Central Europe after the fall of the Berlin Wall, the unification of Germany, and the collapse of the Soviet Union, brought the potential for pursuing bilateral relations outside the constraints of the US-Soviet competition and opened the door for establishing a new foundation. The chance to deal with the past and move beyond it represented the potential to move into the future of Europe. Amid feelings of euphoria as the Wall fell, leaders on both sides made immediate strides to put the past behind them. In a December 1989 letter to German President Richard von Weizsäcker, Václav Havel, the newly elected president of Czechoslovakia, wrote that he personally condemned the removal of the Germans after the war, considering it a "deeply immoral act." In his first trip abroad as President of Czechoslovakia in January 1990, Havel went to Munich and Berlin, as a "gesture" to "overcome the burden of history." While in Germany, he publicly apologized for the removal of the

---

[19] Wilfried Fiedler, "Münchener Abkommen und Prager Vertrag: Vertrage der Vergangenheit—Vertrage der Zukunft?," in *Die Sudetendeutsche Frage 1985, Eine Standortbestimmung: Tagung des Sudetendeutschen Rates in Kochel vom 29. November bis 1. Dezember 1985*, (Munich, 1986), 38.

[20] Sächsische Landeszentrale für politische Bildung, *Deutsch-tschechische Beziehungen: Arbeitstexte zur politischen Bildung* (Dresden, 1998), 33; Ann L. Phillips, "The Politics of Reconciliation: Germany in Central-East Europe," *German Politics* 7 (August 1998): 71.

Germans and reiterated the condemnation of the expulsions as an immoral act.[21] In a March 1990 speech from the Prague castle, President Weizsäcker expressed similar sentiments, speaking of both the brutalities of the Nazi regimes on the Czechoslovak people as well as the German responsibility for these actions.[22] Leaders from both sides thus appeared focused on surmounting the past.

The Sudeten German *Landsmannschaften*, after initial hopes that the normalization of relations might facilitate the return of their land, attempted to become directly involved, but soon found themselves outside the direct negotiations. In early 1990, Czech officials seemed willing to entertain some talks with the Sudeten Germans, but the latter's insistence on direct involvement in the bilateral German-Czech negotiations led to Czech resistance. After a Summer 1990 meeting between the Czechoslovak Prime Minster Marian Čalfa and the head of the Sudeten German *Landsmannschaften*, Franz Neubauer, in which the two "fought their way through three hundred years of Czech-German history," the Prague government vowed that it would no longer deal with the Sudeten Germans directly.[23] The Sudeten Germans would now be only an internal factor for Germany. Thus, as Timothy Burcher suggests, the Sudeten German question, although not a specific German state interest, was tied to a powerful interest group, centered in Bavaria, that could have considerable influence on overall German interests.[24]

Despite Sudeten German objections, negotiations between the Czechs and the Germans continued. The February 1992 Treaty of Good Neighborliness and Friendly Cooperation was the first

---

[21] Pauer, 3; Burcher, 26; "Czech Attitudes to Sudeten Germans since 1989," *CTK News Archive*, CTK Czech News Agency, 1996, [http://www.ctknews.com/archive].

[22] *Deutsch-tschechische Beziehungen: Arbeitstexte zur politischen Bildung*, 33-34.

[23] Milan Hauner, "The Czechs and the Germans: A One Thousand-Year Relationship," in *The Germans and Their Neighbors*, ed. Dirk Verheyen and Christian Søe (Boulder, Col., 1993), 269; Timothy W. Ryback, "Dateline Sudetenland: Hostages to History," *Foreign Policy* (Winter 1996/1997): 171.

[24] Burcher, 17. Bavaria, which received the majority of expelled Sudeten Germans, has been both a power center of the *Landsmannschaften* as well as a contributor to the cooperation between the two nations. Between Bavaria and the Czech Republic several exchange programs exist and trade between them is quite extensive. See Barbara Stamm, "Historische Nachbarn als europäische Partner," in *Deutschland und Tschechen*, Special issue, *Zeitschrift zur politischen Bildung—Eichhold Brief* 35 (1998), Heft 4/98, 57-61.

diplomatic result in this new environment toward overcoming history. Like the 1973 Prague Treaty, the Munich Agreement again became a major issue. This time, however, the two sides took further strides forward. The Germans still demurred at calling the Agreement void, but did take the new step of unequivocally declaring that the Czechoslovak state had never ceased to exist during the Second World War, thus stepping beyond the 1973 compromise that invalidated the "measures" of the Agreement. The most important step came with Germany's recognition that it bore the primary responsibility for the subsequent suffering.[25] Although the German stipulations did not lend complete legitimization to the removal of the Sudeten Germans, especially in the eyes of the Czechs, it did seek to explain them in a larger historical context, and thus assisted in the process of building a historical consensus. In most respects, this treaty appeared to be the diplomatic start to new and better relations.[26]

The rhetoric of the Friendship Agreement, however, soon proved less forceful than first envisioned. With the collapse of the partition separating East and West, property became a heated issue that tested the positive nature of the Friendship Agreement. This debate centered on the legality of the postwar Beneš Decrees,[27] which removed Czechoslovak citizenship from the Germans and stripped them of all their property. For the Sudeten Germans, the return of their property had been a crucial part of their platform since the 1949 Eichstätter Declaration. This goal, moreover, remained as Franz Neubauer's 1986 declaration in the *Sudetendeutsche Zeitung* that they wanted "the return of the Sudeten German territory to the Sudeten Germans, nothing more, but nothing less." In a slightly softer demand, at the May 1991 Sudeten German Day, he proclaimed that the Beneš Decrees must be declared invalid and the Czechoslovaks must make it possible for those who wished to return to do so. Even in their response to

---

[25] Phillips, 75.

[26] Seibt, 406; *Deutsch-tschechische Beziehungen: Arbeitstexte zur politischen Bildung*, 34.

[27] This law of 8 May 1946 stipulated that any act undertaken between 30 September 1938, the day of the Munich Accords, and 28 October 1945, when the law determining the confiscation of German property was passed, was "not illegal, even when such actions may otherwise be punishable by law." Removing Czechoslovak citizenship from the Germans paved the way for their eventual removal. With these decrees, the expulsion of the Sudeten Germans was given a certain quasi legality. See Appendix VIIa, in *Documents on the Expulsion of the Sudeten Germans*, ed. Wilhelm K. Turnwald, trans. Gerda Johannsen (Munich, 1953), 263.

the German-Czech 1992 treaty, the Sudeten Germans reiterated their claim to their *Heimat*.[28]

The German government was largely sympathetic to their cause. It removed a clause from the 1992 German-Czechoslovak Friendship treaty that would have annulled all Sudeten German property rights in Czechoslovakia and called for Czechoslovak compensation for the survivors.[29] At the same time, however, it has continuously avoided forcing the issue; instead Germany remains committed to developing a policy with its eastern neighbors that will lead to further European integration. The Czechs, for their part, remained adamant, claiming that the decrees were legal and they have no desire to revoke them. As Havel noted in February 1995, the Czechs have no intention of returning to the past and revoking the decrees that had existed for a considerable time and had been approved by the Parliament.[30] Indeed, the Czech's rejection of Austrian calls in 2000 for linking the Beneš Decrees to Czech membership in the EU demonstrates their continued adherence to their legality.

The property issue rose to prominence in March 1995, after a Czech citizen of Sudeten German origin attempted to regain his relatives' property by challenging the legality of the Beneš Decrees in the Czech High Court. If they were deemed unconstitutional, the potential for a legal influx of former Sudeten Germans was considerable; if declared constitutional, it would be endorsing decrees that violated international norms that had existed at the time and which was later enunciated the 1948 Universal Declaration of Human rights. The court, however, according to Timothy Ryback, made the worst decision possible. It upheld the decrees and declared the Germans "collectively responsible" for the Nazi crimes. Property confiscated before 1948, the year of the declaration on human rights and the Czechoslovak coup, was not included and was thus subject to return.[31] Thus, the removal of the Sudeten Germans had gained another semblance of legality from

---

[28] Cited in Burcher, 19; *Krajanské organizace sudetských Němců SRN*, 179.

[29] Ryback, 172.

[30] Pauer, 18-19.

[31] Ryback, 173; Eva Glauber, "Tschechen und Deutsche: Der Schwierige Weg zu Guter Nachbarschaft," in *Deutsche und Tschechen: Nachbarn im Herzen Europas; Beiträge zur Kultur und Politik*, edited by Claudio Gallio and Bernd Heidenreich (Cologne, 1995), 27. Using this date made confiscation of property during the Communist period illegal while exempting the confiscation of the Sudeten German property between 1945 and 1947.

the Czech perspective. German reaction to this decision was definitive. Foreign Minister Klaus Kinkel expressed shock over the incident and in a speech before the German Parliament shortly after the decision, blamed the Czechs for their lack of desire to confront the past honestly and with justice. With this decision, it appeared the Czech Republic was more intent on protecting the gains of the past rather than reconciling differences.

Despite this apparent setback, negotiations for the future continued, culminating in the January 1997 German-Czech Declaration. In what has been the strongest representation of the developing bilateral relationship, both sides declared their desire to develop the relationship further in a "spirit of good-neighborliness and partnership." Germany continued to recognize that the ground for removal of the Germans was "prepared by" the Nazi brutalities. The Czech government for the first time noted that it "regrets (*lituje*)" the suffering and injustices incurred by the removal of the Germans. Moreover, the Czechs also admitted, albeit not definitively, that the use of concept of collective guilt caused undue suffering for the Sudeten Germans. The Czech government also recognized with regret that the Beneš Decrees of 8 may 1946 allowed for the appearance that such actions were "not illegal." Coming to a consensual conclusion about the history was to be continued through a specific German-Czech Historical Commission that was already working.[32] Thus, both sides appeared to be moving toward a view of history based on compromise.

The language of the 1997 treaty also showed an important recognition of the perception of history on both sides. In the treaty, both sides were content to use the Czech word "*vyhánĕni*." For the Czechs, this new term removed the cruelty inherent in the Czech word "*vyhnĕní*," or expulsion. The Germans likewise accepted *vyhánĕni*, for it can also be translated as *Vertreibung*.[33] The

---

[32] *Krajanské organizace sudetských Nĕmcív SRN*, 206-08, 216-18. The historical commission's work has been very much oriented toward finding a unitary view of the past. While it has not covered all of the hard issues, it does represent a thorough continuation of coming to terms with history. In addition, there were annual meetings in the early 1990s jointly sponsored by German and Czech political foundations that brought together scholars, politicians, and others to help facilitate further cooperation. The books from these conferences were published in both Czech and German. For an example, see *Jsme Evropané?; Sind Wir Europäer?* (Prague, 1996) and *Češi a Nĕmci: Cestou Dialogu; Deutsche und Tschechen: Weg des Dialogs* (Prague, 1996).

[33] Ota Filip, "Wenn man 'odsun' und 'vyhnání' kreuzt, kommt 'vyháněni' heraus," *Frankfurter Allgemeine Zeitung*, 24 December 1996, 4. Indeed, examination of the German-Czech Declaration in Czech and German does indicate that the Germans

agreement on a single word to describe the removal of the Germans represents a pronounced step toward greater reconciliation. However, this compromise should be viewed with some caution. The emotions of both sides remain, for the Germans still see the removal as *Vetreibung* and the Czechs as *odsun*. The next step must be to jettison the different emotions in favor of a common view, which at the moment appears to be a difficult path.

Despite the intense debates surrounding these issues, the 1997 German-Czech Declaration clearly represented the forward momentum of the desire to come to terms with history. The publication of the common history by the German-Czech Historical Commission, entitled *Deutsche-tschechische Beziehungen*, stressed the shift in focus from a conflict based relationship to one of reconciliation and consensus building, noting that history should no longer be a hindrance to the future of German-Czech relations.[34] Indeed, since the end of the Cold War, despite some setbacks, such as the debate over property, German-Czech relations have focused on confronting history, attempting to heal the deep wounds of the past. Havel's March 1993 proclamation, that Czechs should "look the past in the face," and resist the temptation to build a history built on "lies and prejudices," represents the primary philosophy that will lead to German-Czech reconciliation and a future in Europe.[35] Substituting injustices in the present for the injustices of history can serve only to further undermine the future.

Since 1990, the slow and at times painful process of historical reconciliation has moved forward and in so doing has set the foundation for a strong future. Both sides have moved toward a more consensual view of history, with the Czechs accepting the suffering of the expulsions and the Germans accepting the responsibility of National Socialist actions for the expulsions. The question still remains, however, whether this view of history that

---

used *"Vertreibung,"* while the Czechs used *"vyhánĕni."* See *Krajanské organizace sudetských Nĕmcŭ SRN*, 206-08, 216-18.

[34] Hans Lemberg, "Die Arbeit der Deutsch-Tschechischen Historikerkommission," in *Deutschland und Tschechen*, Special issue, *Zeitschrift zur politischen Bildung — Eichhold -Brief*, 86; Sächsischen Landeszentrale für politische Bildung, 7, 36. While the committee was unable to reach a consensus on the number of Sudeten Germans killed in the expulsions, the political process of negotiating history cannot be downplayed.

[35] Nancy Meriwether Wingfield, "Czech-Sudeten German Relations in Light of the 'Velvet Revolution': Post Communist Interpretations," *Nationalities Papers* 24 (March 1996): 100.

still leaves some elements unfinished is enough to hinder the continuing inertia of a common view of the future.

The Past as the Future?

At the heart of this discussion of the Sudeten German expulsions as a *leitmotif* in German-Czech relations, and within the general context of ethnic cleansing, is the process of reconciliation. As John Paul Lederach argues, reconciliation is that place where concerns of the past and the future can meet. "Acknowledging" the past and "envisioning" the future provide the framework for escaping the legacies of ethnic cleansing.[36] History does not need to be the defining element of the future. Escaping its grasp and charting a new vision through reconciliation must be seen as a seminal part of that which follows ethnic cleansing. For Germany in particular, overcoming its past "appears necessary" if reconciliation can truly succeed.[37] Indeed, as the Czechoslovak Foreign Minister Jiři Dienstbier had noted as early as December 1989, only by coming to terms with the legacy of history can the two states enter a period of renewal and improvement.[38] Achieving these goals can be assisted with Track II diplomacy, which develops a consensus at lower levels though commissions and negotiation that can then be pushed to upper levels of diplomacy. The goal of such actions would be to cool the memories and emotions by direct discourse.[39] Indeed, the exchanges and the historical commission represent positive steps in this direction.

Similarly, material advances are occurring that are linking Germany and the Czech Republic under the larger umbrella of Europe. The Czech Republic's 1999 inclusion in NATO and the continued negotiations for EU membership suggests that the future is now taking precedence in German-Czech relations. Indeed, the lure of the prosperity and stability of the West, embodied in the often used phase, "Return to Europe," is the primary guiding feature of the Czech Republic's policy vis-à-vis

---

[36] John Paul Lederach, *Building Peace: Sustainable Reconciliation in Divided Societies* (Washington DC: United States Institute of Peace Press, 1997), 27.

[37] Lily Gardner Feldman, "The Principle and Practice of 'Reconciliation' on German Foreign Policy: Relations with France, Israel, Poland, and the Czech Republic," *International Affairs* 75, (April 1999): 335.

[38] Burcher, 25.

[39] Jedlicki, 230.

Germany. This potential is an important feature of the reconciliation process. Continued negotiations at the state level, in conjunction with Track II diplomacy at the local level, offers a complete mechanism for escaping history. It is conceivable, that a bright vision of a future can trump the dark memories of the past.

History still plays an important role in the developing German-Czech relationship as the Sudeten Germans continue to press their desire to return to the *Heimat*. The issue of property restitution has gained some new momentum with the recent decision of the Sudeten German *Landsmannschaften* to file a collective court action in the United States against the Czech Republic. Moreover, the Sudeten Germans have continued to link the issue of property reclamation with the potential of EU accession for the Czech Republic.[40] Yet, history could simply run out on the Sudeten Germans if the Czechs become a full member without restrictions of the EU, which would allow any German citizen to settle in the Czech Republic. Overall, however, German-Czech relations appear to be pursuing reconciliation and finishing the memories of the past.

The German-Czech lesson raises some important issues for the numerous examples of ethnic cleansing since the 1990s. First, will other groups experience similar paths of historical memory, thus keeping the past of ethnic cleansing firmly entrenched in the present? This is particularly relevant for the Muslims, Croats, and Serbs of Bosnia-Herzegovina and more generally for the unsettled Balkans. Second, is overcoming historical memory-based issues after ethnic cleansing easier than dealing with the explosive intermingling that precipitated war? If so, does this mean that humanity's violence should be condoned? Are population transfers a viable alternative if they create stability?[41] While there is likely no single answer to these questions—for the long term effects of ethnic cleansing will differ according to the conflict—one fact does remain clear. The memories of ethnic cleansing will exist until they are rectified and then only with the assistance of a

---

[40] Stefan Wolff, "From Irredentism to Constructive Reconciliation? Ethnic German Minorities and Germany's Relationship with Poland and the Czech Republic Since 1990," unpublished paper presented at the Comparative Interdisciplinary Studies Section of the International Studies Association (CISS/ISA) Conference, Washington DC, 28 August 2000, p. 11.

[41] Some scholars suggest that population transfers can actually promote stability. See Chaim D. Kaufmann, "When All Else Fails: Evaluating Population Transfers and Partition as Solutions to Ethnic Conflict," in *Civil Wars, Insecurity, and Intervention*, ed. Barbara F. Walter and Jack Snyder (New York, 1999), 221-60.

vision of a common future. If no such vision exists, there is a larger potential for past grudges to explode again in conflict.

On a final note, it is important to recognize that while it should be a goal of states to escape the grasp of history and strive for a better future, the events of the past should never be forgotten. The choice of "escaping" as opposed to "forgetting" used in the title of this chapter is purposeful. The tragedies of the ethnic cleansing of the Sudeten Germans and others, whose stories are reflected in the pages of many books, including this one, clearly speaks to a need to remember the past. Indeed, escaping the pull of history on contemporary relations while maintaining a clear cognizance of what occurred can represent very different processes which actually complement each other in the creation of a better future.

Polish-speaking Germans and the Ethnic Cleansing of Germany East of Oder-Neisse

RICHARD BLANKE

This essay deals with an admittedly secondary aspect of the ethnic cleansing of East Central Europe after 1945; one that receives hardly a passing reference in most of the historical literature, although its significance should be readily apparent: the role played by people of indeterminate nationality in this epic undertaking. After all, if one proposed to remove all the Germans from east of Oder-Neisse, that did beg the question: who or what would determine who counted *as* German? Was Germanness to be a function of objective characteristics such as language and ancestry, of formal citizenship, or of individual sentiment, as manifested either currently or in the recent past?

Judging from the wartime discussions of the ethnic-cleansing project in Western government circles, few seemed aware that such a problem even existed. Representatives of the Eastern European states that were going to take over from the expelled Germans knew better, of course, but they had their own reasons for not raising the issue, in order that they might deal with it as they chose. In any case, the German-Polish borderlands included as many as one million people who might qualify as either German or Polish, depending upon the criteria employed. They were mainly Polish by language, or the descendants of recent Polish speakers, but with a more or less developed German political consciousness. One large group lived in German Upper Silesia (and are discussed in this volume by Tomasz Kamusella). This paper focuses on the other major group of "Polish-speaking Germans:" the nearly half-million Masurians who inhabited the southeastern third of the German province of East Prussia.

Although they are hardly a household term, even in professional circles, the Masurians present the clearest and best-documented example anywhere in Eastern Europe of national consciousness developing counter to native language. Although most Masurians spoke Polish and lived adjacent to Poland, they gave every indication over quite a long time period of voluntary and virtually unanimous identification with (first) the Prussian state and (then) with the German nation. Perhaps this was because they had adopted Lutheranism while other Poles remained Catholic; perhaps it was simply because they had never known anything but Prusso-German rule. In any case, they never showed much interest in being part of the Polish national community.

Polish nationalists did make a perhaps belated, but not inconsiderable effort to win the Masurians over to their cause, especially after 1880. But they had little to show for their effort as of 1919, when Poland laid claim to Masuria, on strategic as well as ethno-linguistic grounds, at the Paris Peace Conference. Whereupon the Peace Conference, in an effort to determine the Masurians' own sentiments, arranged for a plebiscite (held July 11, 1920), the result of which must rank as the most lopsided such contest ever held: 99.3 percent of the vote in the eight Masurian counties *(Landkreise)* went to Germany.[1] Nor did the following twenty-five years offer any more indications that Poland's interest in the Masurians was in any way requited. On the contrary, Masurians increasingly neglected their mother tongue in favor of German; to the point where many of them (and the young, especially) no longer qualified as even Polish-speaking. Their desire to be full-fledged members of the German national community found particularly bizarre expression in their exceptionally high levels of support for National Socialism: in the Reichstag elections of July 1932, which saw the vote for Hitler jump to 37% nationwide, it reached 58 percent, 66 percent, and 71 percent respectively in the three Masurian districts, *i.e.*, the largest Nazi majorities in the entire country.

But such was the prevalence among Polish nationalists of the ethnic-objective notion of nationality, as a matter of inheritance rather than individual choice (and thus unlikely to change from one generation to the next), that they remained convinced that some essential Polishness must still inhere in even the most Germanized Masurians. However deluded or repressed these lost Polish souls might currently be, their national consciousness was bound, sooner or later, to realign itself with their essentially Polish ethnicity. And so, as it became clear during the closing stages of the Second World War that Poland would acquire most of East Prussia, Polish leaders were unanimous in insisting that the Masurians not be expelled along with the "real" Germans. Of course, they also had every reason to take this position, for the Masurians and the other "ethnically Polish autochtons" resident in eastern Germany provided at least some demographic justification for Poland's acquisition of these otherwise German lands.

---

[1]The vote for Germany in the Allenstein Plebiscite District as a whole, which included parts of Warmia (Ermland), was 97.8 percent.

In 1943, surviving Polish-Masurian activists from the interwar period reconstituted themselves as an underground "Union of Masurians" (*Związek Mazurów*). In November 1944, their leader, Karol Małłek, met with Communist Party boss Bolesław Bierut in Lublin and informed him that East Prussia contained as many as 700,000 people, Masurians and others, "of Polish language or background," and many others could be expected to return from western Germany. While most of these people had undergone pretty thorough Germanization, they remained candidates for Repolonization, and so should be treated as Poles and encouraged to stay. Only the "truly German" population should be expelled.[2] Małłek's memorandum on the Masurians was the subject of at least one meeting of the Lublin government (in December 1944) and was probably the source of its note to Stalin asking that the Red Army, poised to enter East Prussia, "not treat the Masurian population like the German."[3]

This request apparently failed to impress Stalin and/or Red Army commanders, however, for Masurians participated fully in the tragedy that befell East Prussia in the winter of 1944/5 (the story of which has been told elsewhere).[4] Masuria's planned evacuation had scarcely begun before it was cut off by the rapid Soviet advance. Most of its people were overtaken by Red Army units while trying to make their way westward in horse-drawn wagons or on foot. These treks were often looted and demolished; females, regardless of age, were raped , and males (regardless of

---

[2] Tadeusz Filipkowski, "Zagadnienia Prus Wschodnich w memoriałach przedłożonych Polskiemu Komitetowi Wyzwolenia Narodowego" [The Problem of East Prussia in Memoranda Submitted to the Polish Comittee of National Liberation], *Komunikaty Mazursko-Warmińskie* 26 (1980): 60ff.; Edmund Wojnowski, "Warmia i Mazury w latach 1945-1989. Społeczeństwo—gospodarka—kultura" Warmia and Masuria in the Years 1945-1989. Society–Economy–Culture], *Komunikaty Mazursko-Warmińskie* 41 (1996): 30f.; Heinrich Mrowka, "Zur masurischen Frage 1944/45," *Deutsche Ostkunde* (1984), 143ff.

[3] Leszek Belzyt, "Zum Verfahren der nationalen Verifikation in den Gebieten des ehemaligen Ostpreußen, 1945-1950," *Jahrbücher für die Geschichte Mittel- und Ostdeutschlands* 39 (1990): 251ff.; Andrzej Sakson, *Stosunki narodowościowe na Warmii i Mazurach 1945-1997* [National Relations in Warmia and Masuria, 1945-1997] (Poznań ,1998), 29ff.

[4] *E.g.*, by Alfred M. de Zayas, *Nemesis at Potsdam: The Anglo-Americans and the Expulsion of the Germans* (London, 1979); Hans von Lehndorff, *Ein Bericht aus Ost- und Westpreußen* (Düsseldorf, 1960); Edgar Lass, *Die Flucht. Ostpreußen 1944/5* (Nauheim, 1964); Hans-Ulrich Stamm, *Schicksal in 7 Jahrhunderten. Aus der leidvollen Geschichte Ostpreußens* (Hamburg, 1976); Manfred Zeidler, *Kriegsende im Osten* (Munich, 1996).

age) often shot.[5] Other would-be refugees, realizing the futility of flight, decided to stay or return home. But that too could prove fatal, for the Red Army seems to have had a *carte blanche* unparalleled in modern European history to do as it pleased with conquered civilians; and Nazi atrocities in the Soviet Union had given its members every incentive to do their worst. But there is also at least some circumstantial evidence to suggest that the Red Army's extraordinary brutality, *e.g.*, in the East Prussian village of Nemmersdorf, captured in October 1944 but then relinquished again to the Wehrmacht, was not just a spontaneous reaction to earlier German misdeeds, but was consciously designed to encourage flight, *i.e.*, "ethnic self-cleansing," from areas assigned to Poland or the USSR.[6]

Soviet soldiers apparently had little information about Masuria's ethnic make-up; they were rarely able or willing to distinguish between "real Germans" and others, including the numerous conscript workers from Nazi-occupied countries as well as ethnic Poles. Locals who claimed to be Polish made no impact on them; on the contrary, the sight of putative "German fascists" pleading for mercy in a Slavic tongue seemed only to increase their rage.[7] In the wake of the Red Army came also bands of Poles from south of the pre-war border, bent upon plunder; their long-festering resentment of the Masurians' relative affluence and traditional disdain for Poles provided the necessary motivation.[8]

---

[5] According to one estimate, one in four fleeing Masurians failed to survive the attempt; cf. Bernd Martin, *Masuren. Mythos und Geschichte* (Karlsruhe, 1998), 88. Of fourteen members of two Polish-Masurian families overtaken in Sensburg County, only one survived to tell the tale; cf. Emma Babinnek, Nov. 18, 1952, *Bundesarchiv Koblenz (hereafter, BAK)*, OstDok.2, #29-30,46; cf. also Wilhelm Zakrzewski, Jan. 18, 1953, *ibid.*, OstDok.1.I, #40-1,57; Victor von Poser, "Die Räumung des Kreises Ortelsburg,", *ibid.*, OstDok.8, #571; Paul Schmolski, "Die Räumung des Kreises Neidenburg 1945," March 20, 1953, *ibid.*, #564: Friedrich Skusa, "Bericht über die Flucht im Kreise Neidenburg," April 30, 1953, *ibid.*, #566; Lass, 87, 99ff.; Reinhold Weber, *Masuren: Geschichte - Land und Leute* (Leer, 1983), 154ff.

[6] Zeidler, 15f.

[7] For example, when Soviet soldiers entered the unresisting Masurian village of Kronau (*Kreis* Lötzen), they raped most of the females and shot virtually all the men who were on hand (*i.e.*, 52 of them), including 18 French conscript laborers; cf. Katharine Goebel, *BAK*, OstDok.10, #179; Lass, 99. Among the numerous women shot resisting rape were even some conscript laborers from the Soviet Union itself.

[8] This was not an integral part of Poland's ethnic cleansing *cum* resettlement program, for few of the looters came to Masuria to stay. Rather, they took whatever they could carry with them back to Mazovia, including "doors, windows, ovens, floors," and sometimes even "entire houses," as a result of which some villages near

In May 1945, after four months of Soviet military rule, Soviet authorities transferred civilian administration of southern East Prussia to Poland. One of the first problems facing the new administration was how to distinguish between the "ethnically Polish" Masurians, who were allowed, required even, to stay, and the Germans, who were to be expelled. And this had to be done quickly if the former were to be spared the fate that commonly befell the remaining Germans: dispossession, forced labor, and (finally) expulsion. It is not easy to ascertain exactly how many Masurians remained in East Prussia at this time. In part, this was due to shifting definitions: traditionally, the term Masurian implied at least some familiarity with the Polish-Masurian language. But many, if not most Masurians no longer spoke much Polish, and so authorities declined to make language an essential criterion of Polish-Masurian nationality. Masurians were classified officially (along with most Warmians and Upper Silesians) as "autochtons," a term meaning simply people who have been born where they currently live; but applied in this case only to people of quasi-Polish descent. The criteria for autochton status were left intentionally vague in order to produce the largest possible indigenous population. Officials were doubtless aware that most of these people had identified previously with Germany, and that many had also been fervent "Hitlerites," but they chose to overlook this in the interest of reclaiming them for Poland. In June 1945, Polish authorities issued a "registration" order that proposed very lax criteria for qualifying as an autochton: a Polish language background, Polish family name, or previously expressed sympathy for Poland usually sufficed to earn one a temporary "certificate of Polishness."[9]

Based on these generous criteria, Polish authorities anticipated that they would find as many as 500,000 ethnic Poles, most of them Masurians, in their part of East Prussia.[10] In fact, there were nowhere near that many, for Masurians had responded to the approach of the Red Army the same as other Germans: most of them had fled westward, and in roughly the same proportion. In other words, most Masurians removed themselves from Masuria

---

the border "ceased altogether to exist;" cf. Andrzej Sakson, *Mazurzy — Społeczność Pogranicza* [The Masurians—A Borderland Society] (Poznań, 1990), 76f.

[9] Władysław Wach, "Veryfikacja ludności miejscowej—na Mazurach i Warmii" [Verification of the Local Population—in Masuria and Warmia], *Strażnica Zachodnia* 15 (1946): 221; Belzyt, 255f.

[10] Belzyt, 248.

before Poland could even let them know that they should remain. There are no precise figures for the number who stayed, or who failed to make good on their escape and returned. In 1947, officials in the new Allenstein Wojewodship, which comprised most of Masuria, counted only 80,000 autochtons, not all of them Masurians.[11] This number rose to 111,300 by 1950, partly as a result of returnees from Western Germany and other places, partly as a result of a further relaxation of the definition of an authochton. Meanwhile, the immigrant Polish population of Masuria grew rapidly; by 1950, Poles from away outnumbered the autochtons by at least six to one.[12]

At first, Polish officials were lavish in their appreciation of the autochtons and celebrated them as the demographic foundation of the new Polish society that was to arise in the "recovered territories." The small group of interwar Polish-Masurian activists were feted as the nucleus of a native Masurian intelligentsia and appointed to official positions.[13] The problem was that the reasoning behind this respectful attitude was often lost on the newly arrived Poles from other regions. They found it hard to accept as fellow Poles people who spoke such bad Polish, who were normally Protestant, and who had traditionally made so little effort to hide their disdain for Poland. They were more inclined to see the Masurians as just so many defeated Nazis than as prodigal sons and daughters of Poland.[14] Of course, they also had an obvious interest in seeing them this way, for the more

---

[11] Belzyt estimates that no more than 65,000 Masurians, about 15 percent of the pre-war total, were on hand for the beginning of Polish rule; cf. *ibid.*," 253; Wojnowski, 54, 212; Sakson, *Mazurzy*, 69, 84, 150; Wróblewski, 65; *Odbudowa Ziem Odzyskanych* [The Reconstruction of the Regained Territories] *(1945-1955)*, ed. Kazimierz Piwarski *et al.* (Poznań, 1957), 433; Christian Stoll, *Die Deutschen im polnischen Herrschaftsbereich nach 1945* (Vienna, 1986), 66ff., 97; Hans Joachim von Koerber, *Die Bevölkerung der deutschen Ostgebiete unter polnischer Verwaltung* (Berlin, 1958), 38.

[12] Koerber, 84ff.; Sakson, *Mazurzy*, 72ff., 108; Belzyt, 248; Alfred Bohmann *Menschen und Grenzen* (Cologne 1969), 281; Rudolf Neumann, *Ostpreußen unter polnischer und sowjetischer Verwaltung* (Frankfurt/Main, 1955), 3.

[13] Wojnowski, 55; Sakson, *Mazurzy*, 111f.; Neumann, 33; Paweł Sowa, *Po obu stronach kordonu* [On Both Sides of the Border] (Olsztyn, 1969), 170.

[14] Stoll, 67; Neumann, 31. When schools opened in September 1945, most Masurian children were not in attendance; some because schools were not yet available for them, but most because Polish-language instruction was of little use to children who no longer learned that language at home; cf. Kazimierz Pietrzak-Pawlowski, "Repolonizacja kulturalna ziemi warmińsko-mazurskiej" [The Cultural Repolonization of the Warmian-Masurian Lands] *Przegląd Zachodni* [Western Review] 2 (1946): 694.

locals who could be branded as ex-Nazis and unreconstructed enemies of Poland, the better their own chances of acquiring one of the fine (by Polish standards) Masurian farms or homes.[15]

Thus the original idea of basing Polish rule, at least in part, on an indigenous population of ethnically Polish autochtons was soon overwhelmed by a reality in which most Masurians were treated little better than the remaining ethnic Germans. They too were stripped of civil rights, dispossessed of their farms and businesses, made to do forced labor, and subjected to periodic searches of their homes, confiscation of their possessions, and physical as well as verbal abuse by Polish newcomers. While officials did not condone this behavior, neither did they intervene very energetically to stop it.[16] On the contrary, according to most of the Masurians who recorded their experiences for the West German *Ost-Dokumentation* project in the 1950s, the looting, abuse, arbitrary arrests, and other mistreatment actually increased following the shift from Soviet to Polish rule in June 1945.[17]

One possible reason for the widespread mistreatment of autochtons during the first months of Polish rule was an ill-considered decree of March 2, 1945, which declared the property of German citizens in the newly acquired lands, specifically including "German citizens of Polish nationality," to be legally "forfeit." Intended presumably to clear the way legally for various state actions, it was interpreted instead by some Poles as a license to loot.[18] And so, deprived of property as well as civil rights, most Masurians as well as ethnic Germans were reduced to a floating class of hired hands and forced laborers; often obliged to work as

---

[15] Wojnowski, 219; Stoll, 59; Sakson, *Mazurzy*, 78ff.

[16] Remaining Germans males between the ages of 15 and 55 were also subject to deportation to the Soviet Union for longer or shorter periods of forced labor, and there was apparently no effort to distinguish between Masurians and other Germans. Of the *c.* 10,000 Masurians deported to the Soviet Union, half did not return; cf. Andrzej & Agnieszka Wróblewski, *Ausreiseerlaubnis* (Dortmund, 1996), 17; Belzyt, 250; Martin, 64; Sakson, *Mazurzy*, 80; Wojnowski, 227.

[17] August Nowinski reports that Polish assumption of authority in Magdalenz (Neidenburg Country) was followed by "a thorough plundering...until the last piece of straw had disappeared from our farms;" cf. Nowinski, Jan. 12, 1953, *BAK*, OstDok.1.I, #40-1, 305. Emma Babinnek recalled how "plundering continued day and night;" cf. Babinnek, Nov. 18, 1952, *ibid.*, OstDok.2, #29-30,46; cf. also Friedrich Junga, *ibid.*,251; Friedrich Klamaschewski, *ibid.*, OstDok.1.I, #28, 221.

[18] The decree was reversed in May 1946, but by then much of the indigenous population, including the "objectively" Polish Masurians, had been effectively dispossessed; cf. Stoll, 57.

such on what had been their own farms or workshops, now in the hands of Poles. A government commission that looked into the condition of the autochtons in late 1945 found that most of them, having been robbed of "all valuables," were living "close to starvation,...unsure of their lives [and] in constant fear" of gangs of thieves who seemed to "fear no punishment."[19] As the noted writer Jerzy Putrament later charged, Polish authorities basically "squandered the moment of liberation;" above all, by "allowing the looting to take place almost without punishment." Thus the "first appearance of [Polish] statehood [in Masuria] resembled the appearance of the Tatars in Sienkiewicz's *The Flood;*" and continued misrule threatened to accomplish what the Germans had not been able to do in centuries: eliminate the last scintilla of Polishness among the Masurians.[20]

Officials assumed, of course, that those Masurians who were eligible for Polish citizenship would also want to acquire it, if only to avoid being "ethnically cleansed." All too many Masurians, however, did not even seek certification as Poles, despite the singularly unattractive alternative of removal to a war-ravaged Germany that seemed to have neither work, housing, nor even very much room for them. Instead, officials reported a "massive flight by the autochtonous population" along with the Germans who had been ordered to leave.[21] Before long, Masurians were being compelled to stay, and some who had joined the exodus westward were stopped at the border, sent home, and forced to sign a Polish loyalty oath.[22] A 1946 survey by the Polish Western Association (Instytut Zachodni) estimated that two thirds of the autochtons in East Prussia, including half of those who had

---

[19] Belzyt, 258ff.; Neumann, 27.

[20] In *Warmia i Mazury*, June 9, 1957. Writing from exile, Jędrzej Giertych could afford to be even blunter; he compared the Masurians' fate under People's Poland to that of the Pruzzi under the Teutonic Knights; cf. Giertych *Oblicze religijno-narodowe Warmii i Mazur* [The Religious-National Aspect of Warmia and Masuria] (Rome, 1957), 163. Cf. also Wojciech Wrzesiński, "Diskusja i polemika" [Discussion and Polemic], *Komunikaty Mazursko-Warmińskie* [Warmian–Masurian Communications] 2 (1957), 174-181.

[21] Sakson, *Mazurzy*, 77ff. According to one departing Masurian, the government's renewed assurances that autochtons who opted for Polish citizenship would enjoy equal rights with Poles were no longer enough; "we were so tired of life amidst the robber-gangs (*Banden*) that we preferred to leave the homeland;" cf. Frl. Pellny, *BAK*, OstDok1.I, #28, 133.

[22] Stoll, 49; Sakson, *Mazurzy*, 145.

already received Polish citizenship, preferred to relocate to Germany.[23]

As it became clear that things were not going as planned with respect to the Masurians, whose retention remained so desirable from the standpoint of legitimizing Poland's acquisition of the region, a new program was launched (April 1946) under which remaining non-citizen autochtons would undergo a formal process of "verification" of their underlying Polishness, and then have citizenship bestowed upon them.[24] Some 55,000 residents of Allenstein Wojewodship underwent this procedure; several hundred officials were assigned to the time-consuming task of evaluating individual cases. The results were mixed, however: fewer than half (*i.e.*, 26,979) "passed." The remainder were rejected, mainly because they continued to insist that they were German and did not want Polish citizenship.[25] Officials were at a loss to explain the refusal of so many Masurians to "Repolonize" themselves, despite the heavy price that came with refusal. After all, there was still no sign of any West German "economic miracle" in 1946-1947. On the contrary, Germany was threatened with widespread starvation. So the common suspicion during the following decades that Masurians were seduced by some Golden West hardly applies to this period.[26]

---

[23] Belzyt, 268; Sakson, *Mazurzy*, 97, 147; Stoll, 71; Wrzesiński, 181.

[24] Wach, 222; Belzyt, 256; Koerber, 37, 61; Stoll, 8.

[25] Belzyt, 259; Wojnowski, 229; Sakson, *Mazurzy*, 129. When officials convened one group of about 350 (mostly female) Masurians and urged them to apply for recognition as Polish citizens, most declined on grounds that, having been effectively dispossessed by a combination of official measures and unofficial looting, they saw no prospect of satisfactory lives in Poland and preferred to await relocation to what was left of Germany; cf. Sakson, *Mazurzy*, 104. Another Masurian recalled how his fellow villagers were called together and "strongly pressured to opt for Poland....But not one was prepared to do so;" cf. August Nowinski, Jan. 12, 1953, *BAK*, OstDok1.I, #40-1, 305. A third witness related how officials, including the local schoolteacher, convened the residents of her mostly-Masurian village and "tried to convince us that we were the descendants of Poles" and should therefore become Polish citizens. But "no one from our village complied voluntarily with the summons to opt (for Poland). Never before had we felt so consciously German;" cf. Hildegard Rogalski, *BAK*, OstDok.2, #29-30, 485. In another Masurian village, residents were told that they would not be allowed to remain in Poland if they did not sign a paper testifying to their Polish nationality; all but ten residents refused, and were duly expelled along with the ethnic Germans; cf. Hermann Zyweck, *BAK*, OstDok1.I, #40-1, 197.

[26] Wojnowski, 221ff.; Sakson, *Mazurzy*,104; Koerber, 64; Stoll, 75. According to one frustrated Polish teacher, the problem was simply that "the majority of Masurians consider themselves Germans, and dream only of departure for Germany;" cf. E. Bielawski, quoted in: Sakson, *Mazurzy*, 82. Another teacher requested a transfer out

The years 1948 and 1949 saw the intensification of the Polish regime's Stalinist character, and this was accompanied by a hardening of official attitudes toward the autochtons. In Masuria, new Wojewode Mieczysław Moczar proceeded basically to force verification upon the remaining Masurians; some were even kept locked in their own basements until they agreed to sign declarations of Polishness. When the last 1300 remained adamant even in the face of this kind of pressure, frustrated officials simply announced that they were all Polish citizens now, regardless. In 1950, Poland declared that its verification campaign had been concluded and that 117,000 Masurians and Warmians had being successfully inducted into the Polish national community.[27]

But Masurians continued to have problems getting along with what were supposed to be their fellow Poles. There were continued language difficulties, differences of religion and historical memory, and a residue of bitterness remaining from the behavior of some of the new arrivals in 1945.[28] Intermarriage between Masurians and newcomers remained uncommon; only 5 percent of marriages concluded in one representative county, 1949-1954, were "mixed."[29] Indeed, closer contact with "real" Poles after 1945 only seemed to reinforce the view of many Masurians that they were fundamentally different—their German national consciousness seems rather to have strengthened than weakened under Polish rule, especially as news of West Germany's phoenix-like economic recovery from the devastation of World War II began to reach them. A confidential official survey in 1952 found that half of the Masurians in Poland, though now officially verified as Poles, still considered themselves German and hoped to relocate to Germany. Indeed, a majority of them (*i.e.*, about 35,000 people) had applied by that time for formal reaffirmation of their German citizenship , either directly from West Germany or through the United States Embassy in Warsaw.[30]

---

of the region on grounds that he was tired of the "Sisyphean task" of trying to make Poles out of the Masurians; "it is insane to try to convince these people that they are Poles;"cf. *Die Arbeiterstimme* (Wrocław), Nov. 29, 1956.

[27] Neumann, 34; Belzyt, 256ff.; Stoll, 63; Sakson, *Mazurzy*,128; Wróblewski, 34; *Nowa Kultura*, Oct. 14, 1956; Paul Syburra, Jan. 20, 1952, *BAK*, OstDok.1.I, #28, 115.

[28] Jan Gisges, in *Nowa Kultura*, Aug. 12, 1956; Sakson, *Mazurzy*, 79, 121.

[29] Sakson, *Mazurzy*, 240; Buchhofer, 27.

[30] Belzyt, 50; Sakson, *Mazurzy*, 136, 161f.; Joachim Rogall, "Die Tragödie einer Grenzlandbevölkerung—polnische Forschungen über die Masuren," *Zeitschrift für Ostmitteleuropa-Forschung* 45 (1996): 110.

In 1956, an agreement between West Germany and Poland made it easier for members of families to reunite - primarily in the former country, of course - and this was followed by the departure of most Masurians from Poland during the following several years. The long history of Masurians seeking work in western Germany meant that most of them had relatives there who could now be leveraged to secure the right to emigrate. About 30,000 Masurians, *i.e.*, about half the number currently in Poland, left for Germany from 1956 to 1959.[31] A significant amount of "quiet emigration" on other than family-reunification grounds continued during the following two decades, so that by 1980 only ten to fifteen thousand Masurians remained in Poland. Although still cited regularly as demographic justification for Poland's possession of the region, the autochtonous population was now a mere three percent of the total. And most of this remnant left during the subsequent turbulent decade, marked by Solidarity's challenge, the imposition of martial law, and the virtual collapse of the Polish economy.[32]

Aside from brief outbursts of candor in 1956 and 1980-1981, the Polish public remained largely uninformed about these developments until 1989. Thus the revelation (*e.g.*, in the 1989 reportage by Agnieszka and Andrzej Wróblewski, *Zgoda na wyjazd* [Permisssion to Leave], and in Andrzej Sakson's more scholarly study of 1990, *Mazurzy—Społeczność pogranicza* [The Masurians—A Borderland Society]) that Poland's Masurian population had shrunk by that time to about 8000, *i.e.*, less than ten percent of the 1950 figure and just two percent of what it had been a century earlier, came as quite a shock. It was suddenly clear that, while Poland had managed to gain Masuria, it had lost most of the Masurians.[33] Today, the Masurian population of (what is still called) Masuria consists of a few thousand isolated individuals who are more likely to describe themselves as Germans than as Masurians (although most of them, ironically,

---

[31] Sakson, *Mazurzy*, 162ff., 203; Ekkehard Buchhofer *Die Bevölkerungsentwicklung in den polnisch- verwalteten deutschen Ostgebieten* (Kiel, 1967), 39.

[32] Sakson, *Mazurzy*, 178ff., 204ff., 228; Korbel, 113; Leszek Belzyt, "Zur Frage des nationalen Bewußtseins der Masuren im 19. und 20. Jahrhundert," *Zeitschrift für Ostmitteleuropa-Forschung* 45 (1996): 53.

[33] Wróblewski,100; Sakson, *Mazurzy*, 64ff.; cf. also Stoll, 97; Belzyt, "Frage," 54. The main reason for the survival of even this small remnant was that most of them were married to Poles, and so more inclined to remain in Poland. (The number of mixed marriages between Masurians and Poles proper increased gradually; cf. Sakson, *Mazurzy*, 243ff.)

now speak only Polish).[34] Writing in 1985, Igor Newerly summarized the Masurian experience in People's Poland as follows: "The Masurians lived among us for a certain time and then left. They were Germans, and so they left for Germany; the Masurian question no longer exists."[35]

In conclusion, the flight of most Masurians, as well as the experience of those who remained in their homeland after 1945, is significant, first of all, as evidence of the tenacity with which they stuck to their chosen national identity, however counter-intuitive that choice might seem from the perspective of ethnic-objective concepts of national identity. Although they were exempted from the expulsion decrees, promised an equal role in the Polish national community, and given every opportunity to draw a line under their Germanophile past, it turns out that Polish East Prussia has been just about as thoroughly cleansed of its "ethnically Polish autochtons" as of its "objectively German" population; except that, in contrast to other Germans, most Masurians appear to have relinquished their homeland quasi-voluntarily. This is not an altogether fair assumption, of course, but it raises the not-entirely-facetious question (suggested also by the history of the German minority in interwar Poland): What if the Grand Alliance, instead of authorizing the ethnic cleansing of the lands east of Oder-Neisse, had merely placed this region under the rule (or misrule) of Polish Communists? Might not a decade or two of that sufficed to persuade most Germans to leave in the same quasi-voluntary manner as the Masurians, thus relieving the United States of its co-responsibility for having sanctioned and facilitated history's greatest-ever ethnic-cleansing operation?

---

[34] See, for example, the interviews with surviving Masurians in Ralph Giordano, *Ostpreußen Ade: Reise durch ein melancholisches Land* (Köln, 1994), *passim*.

[35] Quoted in Sakson, *Mazurzy*, 306. This sentiment was echoed in the resigned conclusion of historian Leszek Belzyt: "The Masurians have opted for the German state and for German culture, and in return have left their homeland, given up their community, and sacrificed their mother language"; cf. Belzyt, "Frage," 54.

Ethnic Cleansing in Upper Silesia[1], 1944-1951

TOMASZ KAMUSELLA

"There is no peace without remembrance"

—Pope John Paul II
at the mass commemorating the Fiftieth
Anniversary of the end of the Second
World War

In 1995, on the Fiftieth Anniversary of the end of World War II,
numerous semicentennials of the founding of various
institutions and enterprises in the Polish western and northern
territories[2] (i.e. former *deutsche Ostgebiete* [eastern territories of
Germany]) were celebrated—but quite unreflectively. Few

---

[1] Upper Silesia formed the territorial basis of Prussia's Oppeln (Opole)
Regierungsbezirk, a component of the Province of Silesia. This province contained
two further regencies that coincided with Lower Silesia. A further part of historic
Silesia lay across the border as Austrian, after 1918 Czech lands. The industrial
basin with coal mining and metallurgical industry was concentrated in the southeastern corner of Upper Silesia. After the plebiscite (1921) Upper Silesia was divided between Poland and Germany. Warsaw gained most of this industrial basin,
which became the autonomous Silesian Voivodeship. This voivodeship also included a section of eastern Austrian Silesia transferred to Poland from former
Austrian lands. The truncated Oppeln (Opole) *Regierungsbezirk* remaining in Germany became the only constituent of the newly established Province of Upper
Silesia. Additionally, in 1919 the southern section of the Ratibor (Racibórz) county
had been transferred to Czechoslovakia and together with most of former Austrian
Silesia formed the Kraj of Silesia before it was disbanded in 1928.
  In 1938 the Germans seized the western half of Czech Silesia and the Poles the
eastern half of this region. Next year Berlin regained the Polish part of Upper Silesia and added to it the adjacent counties of the Kielce and Krakow voivodeships.
From 1941 to 45 this "greater Upper Silesia" was organized as the Province of Upper Silesia. This wartime German province closely corresponded to Poland's postwar Silesian Voivodeship that got split into the Opole (Oppeln) and Katowice
(Kattowitz) voivodeships in 1950.
  It is necessary to add that the first Upper Silesian cleansing occurred in the
wake of the division of this land in 1922. In the years 1922-39, 190000 people left
the Silesian Voivodeship for Germany and 100,000 the Province of Upper Silesia
for Poland (F. Serafin, "Stosunki demograficzne i społeczne" [The Demographic
and Social Relations], in *Województwo śląskie (1922-1939)* [The Silesian Voivodeship
(1922-1939)] , ed. F. Serafin, (Katowice, 1996), 88; H. Weczerka, *Schlesien* (Ser.: *Handbuch der historischen Stätten* (Stuttgart, 1977), lxxxvii.

[2] The "western and northern territories" (*ziemie zachodnie i północne*)—a largely
neutral designation of this part of *deutsche Ostgebiete* that was transferred to Poland.

consciously alluded to the fact that the incorporation of the *deutsche Ostgebiete* into postwar Poland had been an unusual indemnification for the eastern lands (*Kresy*) of Poland. These lands had been lost to the Soviet Union in line with the secret Ribbentrop-Molotov Pact, an instrument that has never been found invalid as it pertains to the Soviet Union's territorial gains. Rarely is the point emphasized that the post-1945 shape of the European political map was the legacy of the failure at establishing and maintaining a German nation-state (prohibited by the Versailles Treaty) within a Europe organized according to the wishes of Pan-Germanists, as well as of Moscow's success at actualization of its Pan-Slavist empire.

Having secured the *deutsche Ostgebiete* for Poland through his usual policy of *faits accomplis*, Soviet leader Joseph Stalin got a steady leverage on Warsaw. It was clearly realized that the existence of postwar Poland depended on the Soviet Union because of the virtual impossibility of any rapprochement between the Poles and the Germans under such circumstances. The incorporation of the relatively highly developed *deutsche Ostgebiete* enabled postwar Poland to transform itself from an agricultural to agricultural-industrial country. This was not without influence on the initial improvement in the overall standard of living. Thus, the "regained territories"[3] (christened so by the Polish Communist-cum-nationalist propaganda) let the Communists consolidate the unwilling Polish society around the unwanted goal of "building socialism" under Soviet supervision. The inclusion of these territories in the new Poland also legitimized the Communists' undemocratic seizure of power.

The Onslaught of the Red Army and Deportations to the Soviet Union

The relative security of Silesia was over with the destruction of the *Heersgruppe Mitte* (Army Group Center). Next the Soviet armies reached the Vistula at the turn of June and July 1944. The first Allied bombing raid in Upper Silesia took place on 7 July 1944.[4]

---

[3] The name "regained territories" (*ziemie odzyskane*) was bestowed to "prove" the "legitimate return" of the Polish section of the *deutsche Ostgebiete* to the "rightful owner"—Poland. Another propaganda-ordained designation of the "Piast lands" (*ziemie piastowskie*) even more fortified this mythologizing trend by associating these territories with the first Polish royal House of Piast.

[4] S. Siebel-Achenbach, *Lower Silesia from Nazi Germany to Communist Poland, 1942-1949* (New York, 1994), 29.

On 19 October, the Red Army entered the territory of the Third Reich for the first time and seized two counties in East Prussia. The German troops recovered them on 5 November. It was discovered that on 20-21 October the Soviet *soldateska* had killed all the population and livestock in the village of Nemmersdorf (Mayakovskoye) and committed other atrocities.[5] In autumn 1944 and the following months, the fright evoked by the Red Army's excesses was increased by disinformation and the Reich authorities who strove to dissuade the Upper Silesian population from fleeing from their homeland. Berlin needed the population in place in order to continue the production in the industrial basin, which had to take over the destroyed Ruhr as the Reich's powerhouse.[6]

Therefore, only small segments of the Upper Silesian population had a chance to escape or to be evacuated before the rapidly advancing Red Army—20 to 50 percent of the rural populace, and sporadically larger percentages of the urban populace from the areas east of the Oder.[7] The refugees sought shelter mainly in the Sudetenland and in the Protectorate of Bohemia and Moravia.[8] On 12 January 1945 the Soviet armies launched the winter offensive. On 19 January Soviet soldiers burst into Upper Silesia near Herby and Kreuzburg (Kluczbork).[9] By the close of January almost all the Upper Silesian basin was in Soviet hands. Subsequently, using the Oder bridgeheads in Krappitz (Krapkowice) and near Cosel (Koźle), the Red Army renewed the attack. It resulted in the speedy seizure of the rest of Upper Silesia that was completed by 26 March.

In November and December 1944 the offensive had been preceded by a wide-scale indoctrination action which incited the Red Army soldiers to perpetrate indiscriminate massacres of and bloody revenge against the Germans.[10] Thus, it is not surprising

---

[5] Alfred M. de Zayas, *Nemesis at Potsdam: The Expulsions of the Germans from the East* (Lincoln, Neb., London, 1988), 61-62; P. Engel, *Die Vertreibung der deutschen Bevölkerung aus den Gebieten östlich der Oder-Neiße*, vol 1 (Augsburg, 1995), 7-8.

[6] Elizabeth Wiskemann, *Germany's Neighbours: Problems Relating to the Oder-Neisse Line and the Czech Frontier Regions* (London, 1956), 90.

[7] Michal Lis, *Ludność rodzima na Śląsku Opolskim po II wojnie światowej (1945-1993)* [The Native Population of Opole Silesia After World War II (1945-1993)] (Opole, 1993), 19.

[8] De Zayas, 77.

[9] Siebel-Achenbach, 58.

[10] De Zayas, 65.

that the Soviet troops cruelly avenged the years of National Socialist terror in Belorussia, Ukraine, and Russia. Rape, arson, pillage, murder, and carnage had tended to take place from time to time after the troops had crossed the wartime borders of the enlarged Reich. But the occurrence of all the phenomena skyrocketed when the Red Army reached Germany's prewar border.[11] The tragedy suffered at the hands of the Red Army by the Upper Silesian populace, regardless of their linguistic, ethnic, or national provenance, is daunting and defies description.[12]

When the Soviet occupation of the industrial part of Upper Silesia stabilized in February 1945, a forced population deportation was carried out. Wehrmacht soldiers and civilian men, but also women and children, were transported in unheated freight trains or marched in sub-zero temperatures into the Soviet Union's heartland. It is estimated that sixty-five thousand were deported to forced labor and concentration camps in the Soviet Union.[13] About fifty to seventy-five per cent of the Upper Silesian prisoners perished from hunger and disease in these camps,[14] and many were freed as late as 1949 and 1955.[15] Besides the Upper Silesians deported to the Soviet Union, the Soviet occupation authorities imprisoned several tens of thousands of other Upper

---

[11] J. Walczak, *Generał Jerzy Ziętek. Biografia* [General Jerzy Ziętek: A Biography] (Katowice, 1996), 149. The assaults against the civilian population were designed not only to unleash a vast refugee movement that would impede the military operations of the Wehrmacht, but also as an introduction to and the first stage of the subsequent ethnic cleansing; see Walther Hubatsch et al., eds., *The German Question* (New York, 1967), 313.

[12] I recommend those interested in this aspect to read Horst Bienek's novel *Erde und Feuer* (Munich, 1982); J. Kaps's moving *Die Tragödie Schlesiens 1945-1946 in Dokumente* (Munich, 1952); and B. Waleński's *Na płacz zabrakło łez* [There were No More Tears to Cry] (Opole, 1990) on the extermination of the village of Gottesdorf (Boguszyce) during which 200 inhabitants and 100-150 other civilians lost their lives.

[13] K.Cholewa, "Ślązy w ruskich łagrach" [Silesians in Russian Camps], *Gazeta Górnośląska/Oberschlesische Zeitung* 20 (1993), 1 and 4; N. Honka, "Ofiary nieludzkiej ziemi" [The GULAG Victims], *Gazeta Górnośląska/Oberschlesische Zeitung* 23 (1993); P. R. Magocsi, *Historical Atlas of East Central Europe* (Seattle and London, 1993), 48; Z. Woźniczka, *Z Górnego Śląska do sowieckich łagrów* [From Upper Silesia to the Soviet Camps] (Katowice, 1996).

[14] Honka, "Ofiary nieludzkiej ziemi."

[15] M. Dobrosielski, "Mniejszość niemiecka w Polsce – historia i współczesność" [The German Minority in Poland: History and the Present Day], in *Mniejszość niemiecka w Polsce – historia i teraźniejszość* [The German Minority in Poland: History and the Present Day], ed. anon. (Warsaw, 1995), 62; Wiskemann, 94.

Silesians in the so-called DP (Displaced Persons) camps which, as a matter of fact, were forced labor camps.[16]

Serious war damage sustained in Upper Silesia worsened with activities of the regional Soviet military commands. Besides establishing the occupation administration they also gathered as many "war trophies" (i.e., items of movable property left by the Germans or forcefully seized from them) as possible, and transported them to the Soviet Union.[17]

The deportations of workers and the dispatch of movable property to the Soviet Union clashed with the interests of the Polish Communists. During the talks between the delegation of the KRN (Polish National Council) and the Soviet government, in Moscow (14-21 February 1945), the KRN obtained the "right" to establish the Polish administration in the occupied territories of the Third Reich east of the Oder and Neisse, in accordance with the Soviet-PKWN[18] agreement of 27 July 1944.[19] The Polish Communist authorities strove to prevent occurrences of deportations and Soviet-approved and Soviet-executed methodical pillage already in February 1945. But only at the end of April 1945 did the deportations stop, although the planned seizure of property continued until October 1945.[20]

Ethnic Cleansing in Upper Silesia

The Polish Communists used the tactics of *faits accomplis* and as early as 5 February 1945 announced that Poland was entitled to administer the *deutsche Ostgebiete* east of the Oder-Neisse line.[21] Earlier, on 28 January, the Polish citizens of German descent from the territory of prewar Poland (i.e., also from the Silesian

---

[16] Lis, 20; A. Topol, ed., *Obozy pracy przymusowej na Górnym Śląsku* [The Forced Labor Camps in Upper Silesia] (Katowice, 1994).

[17] Norman Davies, *God's Playground: A History of Poland* 2 vols. (Oxford, 1981), 2: 481.

[18] The Polish Committee of National Liberation formed at Moscow under the Soviet tutelage. The PKWN was the kernel of the future Communist authorities of post-war Poland.

[19] Z. Kowalski, *Powrót Śląska Opolskiego do Polski* [The Return of Opole Silesia to Poland] (Opole, 1983), 37.

[20] Lis, 25 and 144.

[21] H. Marzian, *Zeittafel und Dokumente zur Oder-Neisse Linie, 1939-1952/53* (Kitzingen, 1953), 28-29.

Voivodeship) had been deprived of their Polish citizenship and simultaneously expropriated and used as forced labor or interned in labor camps.[22] On 29 January, Silesian *Voivode* (Communist regional governor), General Aleksander Zawadzki had approved seizure of all German farms and agricultural machinery.[23] The decree of 3 March stated that all Germans living in the "regained territories" would be expropriated.[24] On the same day it was announced that the "regained territories" would be populated with Polish settlers from the overpopulated areas around Warsaw and Krakow. On 5 March, the confiscation of the property of Upper Silesians, who had fled before the advancing Red Army, commenced.[25] On 14 March, the Upper Silesian Voivodeship was formed and the Dąbrowa industrial basin was included in it. On 18 March the Oppeln (Opole) *Regierungsbezirk*, already controlled by the Polish administration, was added to this voivodeship.[26] The inclusion of the ethnically and historically unrelated territory of the Dąbrowa basin in this new Silesian Voivodeship increased the percentage of Poles in line with the official policy of building an ethnically homogenous Polish nation-state.

The Polish Communists proclaimed the Slavophone "Autochthons"[27] who had resided in Upper Silesia for many generations (i.e., the Szlonzoks[28]) as Poles because Warsaw was

---

[22] Wiskemann, 96-97.

[23] Jan Misztal, *Weryfikacja narodowościowa na ziemiach odzyskanych* [The National Verification in the Recovered Lands] (Opole, 1990), 58.

[24] Th. Urban, *Deutsche in Polen: Geschichte und gegenwart einer Minderheit* (Munich, 1994), 54.

[25] Anon., "Zeittafel der Vertreibung 1945," *Deutscher Ostdienst* 2 (1995), 3.

[26] Lis, 18.

[27] The designation of `Autochthons' (*autochtoni*) Warsaw applied to the Szlonzoks and other ethnic groups of Slavic provenance in the *deutsche Ostgebiete*. Those concerned perceived this name as pejorative.

[28] The ethnic group of Szlonzoks formed most of Upper Silesia's population. They were Cathlics and spoke in their local Slavic dialect and/or Slavic-Germanic creole. In official contacts and at school they used German and Polish in church. The continued existence of this ethnic group was ensured by the mutually cancelling out endeavors of the Polish and German nationalisms to ennationalize the Szlonzoks into the Polish and German nation respectively. In the south of Upper Silesia also the ethnic group of the Morawecs lived. They enjoyed their own Moravian language printed in the Gothic type. They were Catholics too but their area of settlement was included in the Olomouc (Ölmütz) archdiocese unlike that of the Szlonzoks comprised by the Breslau (Wrocław) archdiocese (T. Kamusella, "Wyłanianie się grup narodowych i etnicznych na Śląsku w okresie 1848-1918" [The Emergence

vitally interested in retaining as many of them as possible in order to "prove" the Polishness of the prewar Oppeln (Opole) *Regierungsbezirk* and, thus, to "justify" its incorporation into post-1945 Poland.[29] The Communists did not want to depopulate the region either because that could frustrate their efforts to settle Poland's section of the *deutsche Ostgebiete* with Polish expellees from the *Kresy*. The number of the Polish expellees was considerably lower than the number of the Germans who were to be expelled from the "regained territories." Thus, in Krakow on 22 January 1945 General Zawadzki, the Polish Provisional Government's Plenipotentiary for Opole Silesia (i.e. the Oppeln [Opole] *Regierungsbezirk*)[30] saw to the establishment of the KOPŚO (Civic Committee of the Opole Silesian Poles). The KOPŚO was to work out a program of the national verification. The committee's members arrived in Katowice (Kattowitz) at the beginning of February. They decided that in order to protect those Szlonzoks, who considered themselves or were considered to be Polish, from Soviet deportations, it was necessary to ethnically cleanse Upper Silesia and especially Opole Silesia from the "German element," i.e., from those Upper Silesians who considered themselves or were considered to be German. In the memorandum of 12 February submitted to General Zawadzki, the KOPŚO members proposed the division of the Opole Silesian population into the four categories: I. Persons with full Polish consciousness; II. Persons who know Polish but do not feel any connection with the Polish nation; III. Persons who do not know Polish but have Polish surnames or are of Polish ancestry; and IV. "Indubitable Germans." This division emulated the model set by the *Deutsche Volksliste* (DVL, German National List) during the war on the

---

of the National and Ethnic Groups in Silesia During the Years 1848-1918], *Sprawy Narodowościowe* 12-13 (1998)). When, in this article, I speak of the "Upper Silesians," I mean inhabitants of Upper Silesia in general whatever their national/ethnic/religious identity might be.

[29] G. Strauchold, "II wojna światowa i jej konsekwencje. Weryfikacja i (re)polonizacja" [World War II and Its Consequences: Verification and (Repolonization], *Hoffnung* 18-19 (1995), 8-9.

[30] After the division of Upper Silesia in 1922, it became popular to refer to the Silesian Voivodeship as "Polish (Upper) Silesia" and to the Province of Upper Silesia as "German (Upper) Silesia." Warsaw, furthering its claims to this province, wished to dissociate it from Germandom and so replaced the term "German Silesia" with the new coinage of "Opole Silesia." The element "Opole" was derived from the Upper Silesian capital Oppeln (Opole). In Polish Opole Silesia is "*Śląsk Opolski*." The adjective "*opolski*" sounds almost the same as `*polski*' (Polish) which the more emphasized the postulated "Polish character" of the Province of Upper Silesia.

territory of the interwar Silesian Voivodeship. The National Socialists had used the DVL to ethnically cleanse the Upper Silesian population.[31]

The national verification[32] began on 22 March 1945 with the Silesian Voivode's decree. The MZO (Ministry of the Regained Territories) legalized this process only on 6 April 1946.[33] There was no central supervision over the national verification, which resulted in appalling irregularities. Some pro-Polish Szlonzoks were negatively verified when they were found to have nice flats, houses, and farms. Then their property as *"poniemiecki"* (ex-German) could be distributed among the authorities and their cronies. It also happened that because of family links and so as not to destroy their village communities, verification commissions composed from Szlonzokian officials positively verified persons who spoke German only, were pro-German, or declared themselves as Germans. On the whole, the verification process was quite humiliating even for those accepted as "Poles." It was just a temporary recognition that could be revoked in no time.[34]

After the front lines had moved westward, the Soviet occupation administration was formed in Upper Silesia. It was paralleled and gradually replaced by its Polish counterpart. This brought about only a modicum of stabilization but enough to attract many Upper Silesian refugees back to their homeland. At that time the Upper Silesians were not and could not be aware that at the international level it was decided to incorporate the *deutsche Ostgebiete* into postwar Poland. On the other hand, the Red Army

---

[31] Lis, 25-27.

[32] The process of national verification was applied to inhabitants of the interwar Province of Upper Silesia, i.e. German citizens who had never had any connection with the Polish state that had come into being in 1918. The policy of national rehabilitation paralleled this process on the territory of the interwar Silesian Voivodeship. Over there, the overwhelming majority of inhabitants had acquired German citizenship via the DVL during the war. Through the rehabilitation process they renounced their German citizenship and re-acquired Polish citizenship (Z. Boda-Kręzel, *Sprawa Volkslisty na Górnym Śląsku* [The Issue of the Volkslist in Upper Silesia] (, 1978); Misztal, *Weryfikacja narodowościowa*; Walczak , 144).

[33] Lis, 28-29.

[34] D. Berlińska, *Mniejszość niemiecka na Śląsku Opolskim w poszukiwaniu tożsamości* [The German Minority of Opole Silesia in Search of Its Identity] (Opole, 1999), 372; P. Madajczyk, *Przyłączenie Śląska Opolskiego do Polski, 1945-1948* [The Incorporation of Opole Silesia to Poland, 1945-1948] (Warsaw, 1996), 183-209; Cz. Osękowski, *Społeczeństwo Polski Zachodniej i Północnej w latach 1945-1956* [Society of Western and Northern Poland in the Years 1945-1956] (Zielona Góra, 1994), 111-115; Urban, 71.

was closely followed by a wave of *szabrowniks*[35] from central and southeastern Poland, and already in April 1945 the first transports with Polish expellees from the *Kresy*, began to arrive in Opole Silesia. Moreover, Polish male forced laborers coming back from Germany tended to settle in Lower and Upper Silesia and quite often found employment in the MO (Civic Militia). They frequently took vengeance on the local population for the injustices suffered at the hands of the National Socialist administration. Polish municipal organs frequently discriminated against Szlonzoks, especially in Opole Silesia where the administration was staffed mainly with officers from the Dąbrowa basin, and central and southeastern Poland. Irrespectively of positive or negative national verification, they treated all Szlonzoks as Germans because of their distinctive Slavic-Germanic creole,[36] knowledge of the German language, and attachment to German culture.[37]

Intimidation of the Upper Silesian population by the Polish administration and Polish settlers, widespread lawlessness and looting for black market profit (i.e. *szaber*), the onset of the violent struggle for power between the Communist and anti-Communist forces, and the national verification evoked the general feeling of insecurity and fear especially in Opole Silesia. Decisions of the Polish authorities worsened the state of deep anarchy. Most importantly, on 3 May 1945, Warsaw commenced Polonization of the "regained territories," which on 6 May was followed by the unilateral annulment of the autonomy of the prewar Silesian Voivodeship. On the same day the Act on Expulsion of Enemy Elements from Poland was issued.[38] It provided the Polish authorities with the legal basis to conduct so-called "unorganized," "wild" (i.e. illegal in the light of international law) expulsions of Upper Silesians even before the decisions of the Potsdam Conference. This act also "legalized" instances of expulsions prior to the date of its enactment. General Zawadzki, who was also

---

[35] Individuals who specialized in pillaging property left by Germans.

[36] Cf. T. Kamusella, "Das oberschlesische Kreol: Sprache und Nationalismus in Oberschlesien im 19. und 20. Jahrhundert," in *Die Geschichte Polens und Deutschlands im 19. und 20. Jahrhundert*, ed. M. Krzoska and P. Tokarski (Osnabrück, 1998).

[37] E. Nowak, *Cień Łambinowic. Próba rekonstrukcji dziejów obozu pracy w Łambinowicach 1945-1956* [The Shadow of Łambinowice: An Attempt at Reconstructing the History of the Labor Camp in Łambinowice] (Opole, 1991), 48 and 51; Siebel-Achenbach, 133.

[38] Urban, 55.

Voivode or governnor, wrapped up these measures in July with decisions that were to gradually shift remaining Germans (or persons the authorities considered to be Germans) to the western half of the Silesian Voivodeship formed from the interwar Oppeln (Opole) *Regierungsbezirk*. The system used to carry out this operation was the network of various camps and "German ghettoes" established in the cities.[39]

In the framework of Polonization,[40] first of all, teaching and use of the German language in public and private was forbidden at the peril of fines, imprisonment, and, eventually, incarceration in the special camp at Gliwice (Gleiwitz). Almost simultaneously with the creation of the Polish administration the Polish educational system was organized. The first Polish schools in Upper Silesia opened already in March and April 1945. Afterward, the Polish administration conducted and enforced Polonization of Upper Silesian geographical names, as well as of first names and surnames of the retained Upper Silesians. With participation of Polish settlers, German libraries and monuments were destroyed, and German inscriptions defaced or removed from signposts, shop signs, buildings, gravestones, furniture, machinery and even from table cloths and walls in private houses and flats.[41]

---

[39] B. Linek, `*Odniemczanie' województwa śląskiego w latach 1945-1950 (w świetle materiałów wojewódzkich)* ["De-Germanization" of the Silesian Voivodeship in the Years 1945-1950 (in the Light of Voivodeship Archival Documents)] (Opole, 1997), 22-63; M. Podlasek, *Wypędzenie Niemców z terenów na wschód od Odry i Nysy Łużyckiej. Relacje świadków* [The Expulsion of Germans from the Territories East of the Oder and Neisse: Stories of the Survivors] (Warsaw, 1995), 155-162; Ph. Ther, *Deutsche und polnische Vertriebene. Gesellschaft und Vertriebenenpolitik in der SBZ/DDR und in Polen 1945-1956, Kritische Studien zur Geschichtswissenschaft* 27 (Göttingen, 1998), 55-58.

[40] Under the term of Polonization I understand the official policies of "*odniemczanie*" (de-Germanization) and "*repolonizacja*" (re-Polonization). The former entailed expulsion of Germans and removing "traces of German culture" including the German language. "Re-Polonization" was to assimilate the retained population of Upper Silesia with the mainstream of the Polish nation via the Polish language and culture (Linek, 11-12).

[41] M. Kneip, *Die deutsche Sprache in Oberschlesien. Untersuchung zur politischen Rolle der deutschen Sprache als Minderheitensprache in den Jahren 1921-1998, Veröffentlichungen der Forschungsstelle Ostmitteleuropa an der Universität Dortmund,* vol. B – 62 (Dortmund, 1999), 156-175; Linek, *passim*; B. Linek, "Polonizacja imion i nazwisk w województwie śląskim (1945-1959) w swietle okólników i rozporządzeń władz wojewódzkich" [Polonization of First Names and Surnames in the Silesian Voivodeship (1945-1959) in the Light of Circular Letters and Decisions of the Voivodeship Authorities], in *Wrocławskie Studia z Historii Najnowszej* [Wrocław Studies in Recent History], ed. W. Wrzesiński, vol 4 (Wrocław, 1997); Strauchold, 8.

During the war sixty thousand Upper Silesians had lost their lives. But those who survived and found themselves in Germany were not allowed by the Polish authorities (who considered them Germans) to return to Upper Silesia, whereas their families (considered Polish) were not allowed to leave for Germany. This led to break-ups of numerous marriages. Those retained Upper Silesians who had been classified as belonging to the first and second groups of the DVL during World War II, were customarily discriminated against. It was almost impossible for them to find employment, and they were barred from any form of professional career until 1956. The few who obtained jobs more often than not carried out menial tasks (this included women and youngsters) and for two years had to give up one quarter of their earnings for the reconstruction of the Polish capital razed by the German forces in 1944.[42] Germans used as forced laborers, and holders of the first and second groups of the DVL obtained the lowest category of food rations (891 calories per day) and their family members or unemployed members of the three aforementioned groups even a lower category (604 calories per day).[43] Thus, it may be inferred that some significant albeit not openly advertised goals of Polonization were the biological eradication of Germans and the "too German" Szlonzoks, as well as the creation of an Upper Silesian population which would be thoroughly docile and indolent.

The general level of education and literacy (especially in German) among the Upper Silesians was quite high in comparison to the Polish average. The authorities strove to utilize this situation for Polonization by establishing numerous Polish libraries, and rooms and houses of culture, and also by organizing trips to central Poland which were meant to pull the Upper Silesians away from German literature and culture.[44] Another instrument of Polonization was compulsory military service in the Polish Army, not unlike the compulsory conscription of Upper Silesians into the Wehrmacht during the war, for the sake of Germanization besides the demands of warfare. In the late forties

---

[42] D. Brehmer, "Mniejszość niemiecka w Polsce jako most między narodami" [Poland's German Minority as a Bridge between the Nations], in *O problemach polsko-niemieckiego sąsiedztwa* [On the Problems of Polish-German Neighborhood] , ed. F. Pflüger and W. Lipscher (Warsaw, 1994), 422-423.

[43] Siebel-Achenbach, 170.

[44] Strauchold, 9; G. Strauchold, *Polska ludność rodzima ziem zachodnich i północnych. Opinie nie tylko publiczne lat 1944-1948* [The Polish Native Population of the Western and Northern Lands: Not Only Public Opinions from the Years 1944-1948] (Olsztyn, 1995), 95-146.

and early fifties, one of the Polonization programs was to draft young Upper Silesians into the Polish Army only to dispatch them immediately to work in Upper Silesian coal mines without compensation.[45]

Due to the influx of Polish expellees from the *Kresy* and of settlers from central and southeastern Poland, and the return of Upper Silesians of various categories, conflicts over ownership of farms, houses, and flats did arise. Often new "co-owners" were settled on farms, in houses and flats of Upper Silesians. This led to mutual acrimony and negative verification of numerous Szlonzoks, even those of Polish national leanings. In this manner they were deprived of their own property, and eventually many of them were expelled or interned in "DP" and transfer camps.[46]

May and June 1945 was the period of the "unorganized expulsions."[47] Warsaw conducted them on the basis of the unofficial Soviet permission.[48] The shift of the Silesian population to the south was terminated in mid-May 1945 with the effective enforcement of its borders by Czechoslovakia.[49] The first Polish

---

[45] K. Karwat, "Śląskie drogi. Wojna zaczęła się po wyzwoleniu" [The Silesian Fate: War Started After Liberation], *Dziennik Zachodni* 111 (1995), 8.

[46] Nowak, 48. On the appalling conditions in these camps and instances of inhumane treatment of inmates see H. Esser's *Die Hölle von Lamsdorf. Dokumentation über ein polnisches Vernichtungslager* (Dülmen, 1977), G. Gruschka's *Zgoda—ein Ort des Schreckens* [Zgoda – the Place of Fright] (Neuried, 1996), J. Ruszczewski's "Polskie obozy i miejsca odosobnienia dla ludności śląskiej i niemieckiej na Śląsku Opolskim w latach 1945-1949" [The Polish Camps and Places of Isolation for Silesian and German Population in Opole Silesia During the Years 1945-1949], *Kwartalnik Opolski* 4 (1993) and his "Nacjonalizm, szowinizm, czy syndrom odwetu i odpowiedzialności zbiorowej? (Konflikty międzygrupowe na przykładzie Śląska Opolskiego w latach 1945-1949)" [Nationalism, Chauvinism or the Syndrome of Revenge and Group Responsibility? (Intra-Group Conflicts on the Exemplar of Opole Silesia in the Years 1945-1949)], in *Fenomen nowoczesnego nacjonalizmu w Europie Środkowej* [The Phenomenon of Modern Nationalism in Central Europe], ed. B. Linek et al. (Opole, 1997), as well as J. Sack's *An Eye for an Eye: The Untold Story of Jewish Revenge Against Germans in 1945* (New York, 1993). Eponymous of these atrocities are the Łambinowice (Lamsdorf), Świętochłowice (Schwientochlowitz), and Jaworzno camps.

[47] They are often referred to as "wild expulsions."

[48] De Zayas, 104-105.

[49] Due to the lack of space, I do not consider here the phenomenon of "Upper Silesian refugees" (*hornoslezské uprychlíky*) in Czechoslovakia. They numbered about five thousand. Almost exclusively German-speaking they claimed their Morawec roots to escape to Czechoslovakia. After the "departure" of Germany beyond the Oder-Neisse line, Czechoslovakia seemed to them a better option than Poland. Cf. D. Janak's `Něklidna hranice (Slezske pohraniči v letech 1945-1957) [The Restless Border (The Silesian Borderland in the Years 1945-1957)] (parts I and II)," *Časopis*

civil servants arrived at Zgorzelec (Görlitz), on the new German-Polish border, on 23 May when the Soviet Union officially renounced its right to control Warsaw's section of the *deutsche Ostgebiete* granted to Poland. Already on 1 June the Görlitz/Zgorzelec bridge over the Oder was closed in order to limit these "unorganized expulsions," and to prevent people from leaving for Germany or returning to Silesia from Germany. In mid-June, Wrocław (Breslau), Legnica (Liegnitz), and the whole of Upper Silesia were tightly closed, barring the returning Silesian refugees from entering these areas. This caused discontent of the Soviet authorities not able to feed the swelled population inside the Soviet zone of occupation.[50]

The pinnacle of the postwar ethnic cleansing in Upper Silesia was marked by the Potsdam Conference (17 July-2 August 1945), where though not *de jure,* but *de facto,* the *deutsche Ostgebiete* were transferred to Poland. What is more, on the basis of Article XIII of the Potsdam Agreement Warsaw was allowed to expel the German population of the territories in an "orderly and humane manner."[51]

The time-consuming process of the national verification delayed expulsions from Upper Silesia. They took place later than those from Lower Silesia populated by "indubitable Germans," whom the Polish Communists, understandably, did not want to verify nationally.[52] Some Upper Silesians succeeded in returning despite the fortified border control and the danger of becoming Soviet POWs during the perilous trip. If their houses and farms had already been taken over by new Polish owners they were often sent to "DP" and transfer camps. The Soviet administration had earlier transferred these camps to its Polish counterpart. The number of these camps together with outright forced labor camps, and larger and smaller places of detention exceeded one hundred. Most of these camps and places of detention, irrespective of the name they went by, shared the characteristic of forced labor

---

Slezskeho Zěmskeho Muzea (B series) 1 and 2 (1993); P. Pałys's *Kłodzko, Racibórz i Głubczyce w stosunkach polsko-czechosłowackich w latach 1945-1947* [Kłodzko, Racibórz and Głubczyce in the Polish-Czechoslovak Relations in the Years 1945-1947] (Opole: at author's expense, 1997).

[50] Siebel-Achenbach, 121 and 125.

[51] De Zayas, 87-88.

[52] M. Całka, "Exodus. Wysiedlenia ludności niemieckiej z Polski po II wojnie świa-towej" [Exodus: The Expulsions of the German Population from Poland After World War II], *Mówią Wieki* 2 (1993), 3.

camps. But if one takes into consideration the high mortality in these camps and places of detention, they often approximated the model of extermination camp. According to sketchy data, in the years 1945-1947 forty thousand persons perished there, including women and children.[53] The authorities started to send to these camps members of the first and second DVL groups, as well as negatively verified or "inconvenient" Upper Silesians.[54] They remained there awaiting revision of their negative verifications, pardon for belonging to the first or second DVL groups, or their expulsion.

Warsaw did not fully comply with the decisions of the Potsdam Conference and continued to expel certain numbers of Upper Silesians to the Soviet occupation zone of Germany until 23 December 1945 when the Soviet Union sealed the border on the Oder-Neisse line. The subsequent legal mass expulsion lasted intermittently from February 1946 to the end of 1947.[55] The expellees were customarily robbed and intimidated. They were not furnished with appropriate food rations and were transported in freight trains largely unprepared for humans, which resulted in numerous deaths *en route.*[56] The expulsion was revived for a short time during the summer of 1948.[57] Afterward the process continued under the labels of individual departures for Germany, and of the Family Link Action in the years 1950 and 1951.[58] Meanwhile, on 6 April 1946, MZO minister Władysław Gomułka, issued the decree on the procedure of affirming Polish nationality of persons residing in the former *deutsche Ostgebiete.*[59] It was reflected in the 28 April 1946 Act on Polish Citizenship for the Autochthonous Population (i.e. original and mostly Slavophone inhabitants of the territories who could be potentially Polonized).[60] This act approved the "broad" approach to the national verification and in many cases was used to stop the expulsion of

[53] Madajczyk, 244-294.

[54] Nowak, 79.

[55] Siebel-Achenbach, 129, 139 and 144; Wiskemann, 118.

[56] Siebel-Achenbach, 147.

[57] Całka, 5.

[58] Wiskemann, 120.

[59] Strauchold, 9; Strauchold *Polska ludność...*, 75-82.

[60] Dobrosielski, 61.

Upper Silesians, and, subsequently, to make it almost impossible for them to leave for Germany.

Conclusion

After the completion of the national verification and rehabilitation in 1950, 900,000 verified and about one million rehabilitated Upper Silesians remained in Upper Silesia while 900,000 Polish expellees and settlers had moved into the region. In total, during the years 1945-50 as many as 400,000 persons were either deported to the Soviet Union[61] or expelled to the postwar Germany.[62] This number together with the similar number of those who had fled or been evacuated before the incursion of the Red Army, added up to 800,000 Upper Silesians residing in Germany in 1950.[63] The postwar ethnic cleansing in Upper Silesia continued at a low level of individual departures for West Germany until the fall of Communism in 1989.[64] During that period almost 500,000 Upper Silesians left for West Germany.[65] In 1990 when Germany reunited, over 100,000 Upper Silesians left for this state, and in 1991-2000 sixty-five thousand followed in their footsteps.[66]

The postwar ethnic cleansing in Upper Silesia had more wide-ranging repercussions for West German-Polish relations than the whole process of expelling Germans from the *deutsche Ostgebiete*. Many more Upper Silesians were retained in their homeland than expelled. Despite their intensified emigration at the turn of the 1980s and 1990s, they still add up to one-third of the Opole (Oppeln) Voivodeship's population of 1.1 million, and Upper

---

[61] Most of those hauled to the SU went to West Germany when freed in 1947-1955.

[62] Linek in K. Cordell, ed., *The Politics of Ethnicity in Central Europe* (London, 2000), 137; Lis, 31.

[63] E. Buchhofer, "Die Bevölkerung Oberschlesiens seit 1945" [The Population of Upper Silesia Since 1945], in *Oberschlesien nach dem zweiten Weltkrieg*, ed. R. Breyer (Marburg, 1975); G. Reichling, *Die deutsche Vertriebenen in Zahlen. Teil 1: Umsiedler, Verschleppte, Vertriebene, Aussiedler 1940-1985* (Bonn, 1986), 61.

[64] In this *annus mirabilis* alone a quarter of a million Upper Silesians left for West Germany.

[65] T. Kamusella, "Ethnic Cleansing in Silesia 1950-89 and the Ennationalizing Policies of Poland and Germany," *Patterns of Prejudice* 2 (1999), 70.

[66] J. Rogall, "Die Deutschen in Polen," *Informationen zur politischen Bildung* 2 (2000), 4.

Silesians number 1.2 million in the Katowice (Kattowitz)[67] Voivodeship with the population of 4.9 million.

The Upper Silesians residing in the Katowice (Kattowitz) Voivodeship are usually those who were rehabilitated and their descendants. Those of the Opole (Oppeln) Voivodeship are mainly the verified and their descendants. It was easier for members of the latter group to emigrate to West Germany. Moreover, the Polish authorities were more distrustful of them than of Upper Silesians in the Katowice (Kattowitz) Voivodeship. While in 1950 the numbers of the verified and the rehabilitated were almost equal, the number of the latter and their descendants grew a bit; the numbers of the former and their descendants plummeted by more than half, to between 350,000 and 400,000.

Polonization was more painful for the verified because they had not had any extensive familiarity with Polish language and culture prior to 1945, unlike the rehabilitated, who had lived in Poland's Silesian Voivodeship during the years 1922-1939. Those verified, due to expulsion and ongoing emigration of their kin, necessarily retained more links with their family members in West Germany than the rehabilitated. On the whole, the rehabilitated managed to enter the mainstream of the Polish nation if they chose to do so. It was more difficult for the verified, whom the authorities and Polish neighbors continued to treat as "crypto-Germans."

The continuous inflow of Upper Silesians (mainly the verified and their descendants) into West Germany during the Communist period, allowed Bonn to claim that the number of Germans in Poland was one million or more. At the same time, until 1989, Warsaw had flatly refused to acknowledge their very existence. Not surprisingly, under these conditions of forced Polonization without any respect for the regional tradition, most of the verified and their descendants began to identify rather as Germans than Poles.

Nowadays, the Opole (Oppeln) Voivodeship's Upper Silesians account for ninety per cent of Poland's German minority. Upper Silesian Germans reside in the voivodeship's eastern half and in the westernmost counties of the Katowice (Kattowitz) Voivodeship, directly bordering on the Opole (Oppeln) Voivodeship. They constitute the largest compact German group outside the German-speaking states. The situation is starkly different in the other areas of the former *deutsche Ostgebiete*, where

---

[67] After the 1999 administrative reform, the Katowice (Kattowitz) Voivodeship became known as the Silesian Voivodeship.

almost all the erstwhile inhabitants had been expelled or emigrated by the beginning of the 1950s with the exception of some tiny pockets.

Postscript

Ethnic cleansings were such a common phenomenon in the twentieth century[68] that there is a danger that they may have become entirely banal in the eyes of the common man. The sheer reality of such tragic events may be conveniently forgotten or successfully denied. For instance, to this day Ankara shies away from recognizing the genocidal nature of its ethnic cleansing of Armenians in 1915. On the other hand, the perceived "usualness" of ethnic cleansings translates into easy manipulation of facts and evidence in the postmodern age of ubiquitous mass media that "make real" anything that is broadcast. Those not heard remain at the peril of being disregarded as "non-existent."

In the course of my research I came across the most telling exemple of such an approach in the form of a photograph. It depicts a platform with a train of open freight wagons, each filled to the brim with standing people. In contrast, the platform looks desolate with few railway officers and some unidentified onlookers. It is such a "normal" sight in the midst of any ethnic cleansing conducted in the West that one hardly bothers to look twice.

The irony comes with the caption. I spotted the picture in three different publications. In G. Jochheim's *Frauenprotest in der Rosenstraße* [The Women Protest in Rosenstraße],[69] the situation is identified as a transport of Jews to a concentration camp, while T. Staněk's *Odsun Němců z Československa 1945-1947* [The Expulsion of Germans from Czechoslovakia 1945-1947][70] and A. Krzemiński's "Kompleks winy" [The Syndrome of Guilt][71] claim it to be a transport of Sudeten Germans expelled from Czechoslovakia.

---

[68] H. Orłowski and A. Sakson, eds., *Utracona ojczyzna. Przymusowe wysiedlenia, deportacje i przesiedlenia jako wspólne doświadczenie* [The Lost Fatherland: Forced Expulsions, Deportations and Resettlements as a Mutual Experience], *Studia Europejskie* [European Studies], vol. 3 (Poznań, 1996).

[69] (Berlin: Edition Hentrich, 1993), 96.

[70] (Prague: Academia-Naše voisko, 1991), photo 5.

[71] *Polityka* 13 (1995), 16.

Whatever is true, the tragedy remains the same. But meddling with evidence and relativizing the experience of ethnic cleansing shows only disdain for those who suffered and perished at a whim of an ideology. Large numbers of people and their leaders in the twentieth century succumbed to this message of hatred. In a broad sense, this hatred accompanied the gargantuan process of building nations and their nation-states over the last two centuries. It made a travesty of the basic Christian commandment, making it into an altogether novel principle of "Love thy neighbor not."[72]

---

[72] R. Kapuściński, "Nie kochaj bliźniego" [Love Thy Neighbor Not], *Gazeta Wyborcza* (18 Sept. 2000), 16-17.

Reshaping the Free City: Cleansed Memory in Danzig/Gdańsk, 1939-1952[1]

ELIZABETH MORROW CLARK

A s the conflict in the former Yugoslavia reaches its tenth anniversary, historians are facing a challenge: what do we mean by "ethnic cleansing"? When is the term appropriate or applicable? How can historians, political scientists, anthropologists, sociologists, and others refer to crimes approaching, but perhaps not including, genocide? How do we discuss crimes perpetuated against specific ethnic groups based on their ethnic identity, more specifically, how do we discuss the crimes perpetrated during and after World War II in Europe? How did we do this before the term "ethnic cleansing" emerged in general use? How do we do this after? The genie is out of the bottle. At the Duquesne History Forum Conference on Ethnic Cleansing in Twentieth-Century Europe held in November 2000, scholars wrestled with the idea that the term "ethnic cleansing" refers not only to *people* cleansed, but also to *names* and *places*. In this essay I will demonstrate that *memory* itself was also cleansed after the war, in and around the city of Danzig/Gdańsk.[2] The unique character of this place, both before and after World War II, determined its fate. Unlike other territories that changed hands after the war, the city of Danzig/Gdańsk belonged to neither Weimar Germany nor the Polish Second Republic between the wars, but existed until 1939 as the Free City of Danzig. This city's unique political and historical position in Polish history and imagination influenced its reconstruction and its new identity and new population after World War II.

Defining Ethnic Cleansing

According to Norman Naimark, "cleansing has a fundamentally dual meaning... its implications are of both external and internal purging. One cleanses a body politic or a geographical region of a

---

[1] This paper was presented as "Free City? Ethnic Cleansing in and out of Danzig/Gdańsk, 1939-1948," at the Duquesne History Forum Conference on Ethnic Cleansing in Twentieth-Century Europe in November 2000.

[2] In references to the city before the Second World War and in references that emphasize the city's German identity the term "Danzig" will be used. In general references and in postwar references, Gdańsk will be used.

people; but one also cleanses a people itself of foreign elements."[3] The conferences in Yalta and Potsdam, despite misgivings about whether Poland could handle massive movements of people and whether Germany could accommodate the influx, Poland's borders were redrawn to include the "Northern and Western Territories." This region east of the Oder-Neisse line was also called *ziemia odzyskany*, or "the recovered lands." It was believed that to prevent further attacks on Russia by Germany, a strong and independent Poland was necessary.

Both Polish governments—the government–in–exile and the Communist government—realized that, at the very least, the Eastern Territories won from the Soviets in 1920 would be lost, that the Curzon Line loomed again, and that Poles from those regions would need somewhere to live.[4] These decisions were not made lightly. The London Poles were particularly concerned about the laws governing German property and citizenship, while the Communists focused on transferring Poles from Volhynia and parts of Galicia in an attempt to reduce the effects of wartime ethnic cleansing perpetrated against Poles by Ukrainians.

Ethnic cleansing by and against Poles and Germans, influenced the picture of contemporary Gdańsk and the character of the territory of the former Free City of Danzig. Germans in and around Danzig fled or were transferred out of Poland. Many Poles from the region east of the "Curzon Line" were sent to the Baltic Coast, including urban Poles from Lwów [Lvov, Lemberg] and Vilna [Vilnius].[5] It is no accident that the parents of Jacek Rzesniowiecki, professor of modern European history at the University of Gdańsk in Poland, came to Gdańsk from Ukraine, or that his spouse, Joanna Rzesniowiecka, née Winter, traces her family to Lithuania. Stories from West Germany also reveal the tragedy of

---

[3] Norman Naimark, "Ethnic Cleansing in Twentieth-Century Europe," The Donald W. Treadgold Papers in Russian, East European, and Central Asian Studies, No 19. University of Washington, 1998, p. 8. See also Michael Burleigh and Wolfgang Wippermann, *The Racial State: Germany, 1933-1945*, (Cambridge, 1991).

[4] In fact, something like 70,000 square miles of territory was to be handed the Soviets after the war. See Alfred M. de Zayas, *Nemesis at Potsdam: The Anglo-Americans and the Expulsion of the Germans—Background, Execution and Consequences* (London, 1977), 52.

[5] For a thorough exploration of the Soviet expulsions, see Terry Martin, "The Origins of Soviet Ethnic Cleansing," *Journal of Modern History* 70 (December 1998): 813-861. Martin also refers to David Laitin's *Ethnic Cleansing, Liberal Style*, MacArthur Foundation Program in Transnational Security, Working Paper Series, no. 4 (Cambridge, Mass., 1995).

the change in the city's identity. When touring the city of Gdańsk for the first time during the summer of 1989, I first noticed the paradox of local identity upon reading grave markers lining the floor in the huge gothic St. Mary's Church and realizing they were in German *Fraktur* script, and that they dated from the late eighteenth century. It was a moment of epiphany. This Gdańsk was also Danzig, but where were the Germans? The stories about relocation are not limited to Poland. The next year, while studying in Münster, Germany, I did not know exactly how to respond to the casual comment made by the mother of a friend whose family had fled from Königsberg, admiring my interest in Gdańsk, and commenting, "you know, one day, Danzig will be German again."

Interwar Identity and Postwar Memory

The Free City of Danzig between the wars was neither fish nor fowl, and this duality continued even after the war.[6] Unlike Poznań [Posen], it did not have a discrete historical Polish identity, but it also was not as great a stretch to incorporate this port city into Poland as, for example, Wrocław [Breslau].[7] Nevertheless, twenty, even thirty, years after the end of the war there were still signs posted on the coastline west of Gdańsk to the Oder that read: "'This territory always was and always will be Polish,' as if they doubted that this was really the case."[8]

Between the wars, the Free City of Danzig was neither Polish nor German. The Free City of Danzig was established according to the Treaty of Versailles and in order to fulfill Woodrow Wilson's ideal that Poland be independent and have access to the sea. The Free City was neither fully sovereign nor a part of either Poland or Germany. With its overwhelmingly ethnic German majority and historic relationship to Polish trade, the Free City itself was a compromise between Polish economic needs and German ethnic claims. The Free City had a local government and a Constitution guaranteed by the League of Nations. Its foreign affairs were governed, ostensibly with the consent of its leaders, through Warsaw;

---

[6] According to Norman Davies, the only section of the Baltic shore with any modern relationship to the Polish interior was the Vistula delta, including Danzig. Norman Davies, *God's Playground: A History of Poland. Volume II, 1795 to the Present* (New York, 1982), 497.

[7] See Sheldon Anderson, "The Oder-Neisse Border and Polish-East German Relations, 1945-1949," *The Polish Review* 42 (No. 2, 1997): 185-199.

[8] Naimark, 34.

it belonged to a Customs Union with Poland, while Poland (and Poles) enjoyed special rights related to trade and administration within the city itself, including the right to maintain a post office within the Free City's borders. The city was ethnically, and in recent experience culturally, German. Statistics vary, but it was estimated that ninety-five percent of the city's population identified itself as German.

The city's long history of association with its Polish hinterland made its *physical* absorption into Poland after World War II unique. In the late eighteenth century, Danzigers had resisted incorporation into Prussia during the Partitions and had enjoyed Free City status for a short time during the Napoleonic Wars (the interwar Free City was sometimes referred to as the "second Free City"), but by the turn of the twentieth century, Danzig had evolved into a regional capital within Prussia, a military outpost, and a port of secondary significance to the German Reich, behind Bremen, Hamburg, even Königsberg. Danzig's historic economic significance to Poland was its strategic location and the convenience of exporting Polish agricultural goods up the Vistula to the Baltic Sea. Even during the Partition era, this natural relationship between the Polish hinterland and the Baltic port remained a constant.

It may surprise students of ethnic conflict that the Free City of Danzig pursued a vibrant period of reconciliation and cooperation with Warsaw in the twenties. Evidence of cooperative actions taken may seem exceptional, but are also relevant, as recent scholarship has demonstrated.[9] Moderate leaders in the Free City and interwar and postwar Polish politicians manipulated the Hansa image in order to emphasize the city's independence from Germany and its longstanding economic relationship with Poland. Danzig had joined the Union of the Hanseatic Cities in 1361, and one of the names initially suggested for the Free City of Danzig in 1919 was the Free and Hanseatic City of Danzig. To this day Bremen, sister-city of Gdańsk, uses the term "Hanseatic" in official titles and on the city's seal. The official city website refers to the "Freie Hansestadt Bremen."[10] It is understandable that the urge to

---

[9] The author's dissertation deals with the theme of Polish-German reconciliation in Danzig. See: Elizabeth Morrow Clark, "Poland and the Free City of Danzig, 1926-1927: Foundations for Reconciliation" (Ph.D. diss., University of Kansas, 1998). See also Franz Dwertmann, *Danzig 1944: Gespräche nach 50 Jahren.* (Gdańsk, 1994). This is a collection of interviews conducted in Gdańsk and in Germany.

[10] See "Bremen.online—Die Bremer Stadtinformanten," http://www.bremen.de/haupt.html..

link the Hansa past to an independent and partially sovereign present was strong among some key Free City leaders. Yet the Prussian era had brought many changes to the city, diluting the old merchant identity, with its fortunes also long past, incorporating Danzig's history into a broader German national identity. This form of German nationalism was imported in part with the civil servants who arrived to administer this regional political and military seat. The strongest advocates for the return of the city to Germany during the interwar era were former German civil servants who remained in the Free City and served the backbone of the conservative leadership.[11] The various Free City governments, with assistance from Berlin, did achieve one of their primary goals, the protection of the city's "German character."[12] The city's population remained overwhelmingly of German ethnicity. German civil servants and pensioners were offered financial incentives to relocate to the Free City. Polish dockworkers and day laborers were often not residents of Free City territory, and ethnically Polish children born in hospitals in Danzig were not offered Free City citizenship. The definition of "Deutschtum" in the city was subject to interpretation, but it manifested itself as a latent fear that the city's political separation from Germany would lead to eventual "Polonization." It is natural, then, that this same Prussian era, conjured up as an ideal by nationalist and conservative politicians, was rejected by the postwar Polish government. Those who sought reconciliation between Poland and the Free City, however, touted the Hansa image. Since this image emphasized a relationship that was outside the era of virulent nationalism, and also coincided with Poland's golden age, it was more amenable to Polish authorities after the war. The physical restoration of the city reflects this sentiment.

One of the most tragic consequences of World War II from both a human and from a diplomatic point of view is that the short-lived pursuit of *Verständigungspolitik* between Germans and Poles vis-à-vis the Free City was cut off by the Nazi seizure of power, the war, and the expulsions. Once Danzig was reclaimed by the Reich, reconciliation between local Germans and Poles was irrevocably damaged. Ethnic cleansings out of Danzig eliminated

---

[11] One prominent example of this was Heinrich Sahm, who served as President of the Free City's Senate. See: Heinrich Sahm, *Errinerungen aus meinem Danziger Jahren, 1919-1930* (Marburg, 1958).

[12] Horst Jablonowski, "Die Danziger Frage: Schicksal einer Freien Stadt im Widerstreit von Nationalbwusstsein und Interessenpolitik," in *Europäische Begegnungen: Beiträge zum west-östlichen Gespräch,* Heft 3 (March 1969): 115.

forever any chance or reconciliation, and the city itself was resettled by those who had also experienced ethnic cleansing. Thus we
see that the city in a very *physical* sense stands as a witness of *two*
crimes against humanity.[13]

One difficulty in contemplating the history of ethnic cleansing
in regions where different ethnic groups have traditionally lived
side by side is that to discuss it at all seems to preclude any discussion or examples of coexistence, cooperation, conciliation, or
inclusion. These traditions are shunted aside in the search for an
Urgeschichte that explains or justifies the ethnic cleansing as a result of longstanding conflict. It is a tragedy that the cooperative
efforts of the interwar era were overshadowed not only by the rise
of the Nazi Party but also by the elimination of any further reconciliation by the homogenizing of populations, and not just of
populations, but also of places, after the war.

In the year 1997, the city of Gdańsk celebrated a millennium of
history. Historians date this anniversary from a reference in *The
Life of St. Adalbert.*[14] In 999 this work described Gdańsk as a city
urbs Gddanyzc. A sincere attempt was made to reconcile the city's
divided history with its post-Communist present at a recent academic conference, "Danzig—sein Platz in Vergangenheit und
Gegenwart." This conference brought historians and politicians
from Warsaw, Bonn, Bremen, and the United States during the
city's millennial year.[15] The efforts to cleanse the public of un-
pleasant memories after World War II have been replaced with
honest discussion of the successes and failures of ethnic policy in
all eras.

The City of Danzig and German Refugees in 1945

What role did the city of Gdańsk play in the actual experiences of
ethnic German refugees? The city was best known, perhaps, as the
exit point from which Germans in the East could escape the ravages of war by sea. Approximately 1.5 million refugees and over
700,000 soldiers were evacuated from an area described in sources
as "East and West Prussia," which comprised East Prussia,
Königsberg, the territory of the Free City of Danzig, and the Polish

---

[13] See Tadeusz Piotrowski, *Poland's Holocaust: Ethnic Strife, Collaboration with
Occupying Forces and Genocide in the Second Republic, 1918-1947* (London, 1998).

[14] Adalbert, or Wojciech, Bishop of Prague, reportedly visited the area in 997.

[15] Udo Arnold, ed., *Danzig: sein Platz in Vergangenheit und Gegenwart* (Warsaw,
1998).

section of the so-called "Corridor," these last two having been claimed for Germany after the invasion of Poland on September 1, 1939. Among the accounts of the Baltic evacuation as related by Alfred de Zayas in *The German Expellees: Victims in War and Peace*, he describes the sinking of the *Wilhelm Gustloff* on January 30, 1945, by a Soviet submarine. The ship had been provided with lifesaving gear from the German Naval Service Bureau in Danzig, and had over 3,000 refugees on board, including patients from local hospitals and wounded soldiers.[16] Only 1,100 survived.[17] De Zayas reports that "in spite of disasters, the German navy and merchant marines continued their rescue missions until the last days of the war."[18] In one testimony shared by de Zayas, the fate of the Free City's last traditional residents is revealed: "The technical personnel from the Schichau Shipyards and Elbing and Danzig were able to save themselves on one of the last ships to sail from Danzig. They would later build a new existence for themselves in Bremerhaven."[19] According to some historians, most former Free City of Danzig citizens moved to West Germany, but there is no comprehensive study of their postwar lives.[20]

Refugees fled through Gdynia on passenger liners escorted by naval vessels. Oddly, the *German Expellees* refers to the Gdynia port, which until the German invasion lay entirely within Polish territory and was built into a workable port between 1926-1936, as "Gotenhafen." This is geographically and ethnically confusing for the reader, though it may simply reflect the vocabulary used by refugees. On March 30, 1945, the Free City of Danzig, though a free city only in name after the invasion of September 1, 1939, was

---

[16] Polish witnesses tell stories of German soldiers, desperate to escape in the last days of the war, trading their uniforms for civilian clothes. Those who were caught deserting were hung from trees along the park like streets by their own officers, as the stories go, sentenced on the spot for their crime. Doubtless they had shed their uniforms to hide their identity from local Poles, as well as from Soviet or German troops.

[17] Alfred M. de Zayas, *The German Expellees: Victims in War and Peace* (New York, 1993), 65.

[18] Ibid., 69.

[19] Ibid., 71.

[20] Herbert S. Levine, *Hitler's Free City: A History of the Nazi Party in Danzig, 1925-1939* (Chicago, 1973), 161. Prominent Danzig citizens also fled to foreign port cities to pursue their careers after the war, as was the case for Ludwig Noë, former President of the International Shipbuilding and Engineering Company, the joint-stock company that replaced the German Imperial Shipyards in 1920. Noë moved to the Netherlands.

captured by the Soviet Red Army. The city itself was systematically demolished, and over ninety percent of the historic quarters
were destroyed. On this same day the Gdańsk Voivodship was
announced. As one modern Polish account puts it: "New inhabitants started to arrive from the eastern lands, formerly belonging
to Poland and now cut off by the new Polish-Soviet border decided at Yalta, and to settle next to Kashubians and other native
inhabitants of Pomerania."[21]

The Selective Reconstruction of Gdańsk

The very reconstruction of Gdańsk, which had been destroyed by
Soviet artillery from the sea and by land, was influenced by the
change in both population and administration. This city, along
with Warsaw, suffered some of the most extensive wartime damage experienced on Polish territory. Over ninety percent of the city
was destroyed, and only a handful of historic buildings survived.
According to city promotional materials and tour guides, Gdańsk
was buried under two to three million cubic meters of rubble.[22]
The city literally had to be rebuilt from the ground up. The maritime history of Poland was incomplete without it, and the fate of
the city was closely bound up with Poland's fortune. As one tour
book published in 1993 put it, "Although these ties had their moments of conflict, yet Gdańsk remained established permanently
in the national consciousness."[23] The "true" history of Gdańsk,
while perhaps disputed in its interpretation between Polish and
German historians throughout the centuries, was valuable to Poles
in the immediate postwar era. There was no need to invent a history for the city, and the first projects slated for renovation were
understandably those which were at once most ancient and also
linked most closely to Polish history. The bitter negotiations after
the First World War over the future of the city of Gdańsk/Danzig
(whether the city should remain under German sovereignty,
should be given over to the Second Republic of Poland, or should
become the Free City of Danzig) found their ultimate conclusion
in 1945. The new Polish state could reconstruct the city in a particular incarnation, choosing the one that de-emphasized the Prussian era and emphasized the golden age of Polish history and

---

[21] Lech Krzyżanowski, *Gdańsk* (Warsaw, 1993), 18.

[22] *Gdańsk 2000: Bound for the Future* (Gdańsk, ND), 14; and Lech Krzyżanowski,
Gdańsk (Warsaw, 1993), 19.

[23] Krzyżanowski, 19.

grain trade. This idea is alluded to in Krzyżanowski's *Gdańsk*, when he notes "facades were to be reconstructed according to their state before the war, or even earlier."[24] He also comments on the purpose of this reconstruction, saying that it "provided a harmonious backdrop for the restored original historic monuments, witnesses of former glory."[25] In 1952, Warsaw sanctioned the reconstruction of the city and the design and construction projects already underway.

Just a cursory look through the photograph collections in the local state archives in Gdańsk reveals the transformation of certain areas of the city. It is not so much the construction or invention of Polish stylistic elements, as the absence or neglect of uniquely Prussian structures, that is notable. Many buildings from the era of German unification were not renovated or repaired at all. During the late nineteenth century, the city had developed industrial and architectural characteristics in keeping with its Prussian connection. One can see the changes in the historical and ethnic character of the city in the story of Leopold Winter (1823-1893), the first non-native Danziger/Gdańsk citizen to serve as *Oberbürgermeister* (chief mayor). The rapid industrialization of German lands in the mid to late nineteenth century can be seen in the massive restructuring and modernization projects Winter undertook.[26] During his tenure he oversaw the construction of a sewer system, comprehensive street paving, traffic regulation, primary and secondary school construction and organization, the modernization of the port, and the improvement of public transportation. Danzig was not linked by rail to its western neighbor-port Stettin/Szczecin until 1870. The main train station itself, recently renovated with great success, was built in 1900. With modernization and Germanization came a change in style and preferences, and the lovely stone entrances along the side streets of the old town were endangered. A Professor Schulz, of the School of Fine Arts, founded a society for conserving the historic style of the city at the turn of the century, and his ideals were valued into the Free City period, and the character of this port city was maintained into the interwar era.

By choosing to reconstruct or renovate selected architectural elements of the Danzig/Gdańsk landscape after the war, a "cleansing of memory" could also take place. City planners could

---

[24] Ibid.

[25] Ibid.

[26] *Gdańsk 2000*, 21.

reinvent the landscape without betraying the city's *Hansa* heritage, of which Poles are also proud, by erasing the evidence of the most reviled period of German domination—Prussian administration, while romanticizing and emphasizing the long *Hansa* history of the city, its glory days as the outlet for Polish grain and agricultural products, and its future as a leading industrial and ship-building center for Poland after World War II.

Danzig/Gdańsk claims such famous citizens as Gabriel Fahrenheit (1686-1736), who lived for several years in the city, and based the zero degree on the lowest Baltic temperature at Danzig in 1709. Danzig/Gdańsk also claims as a native son philosopher Arthur Schopenhauer (1788-1860), who was born in the city, although he grew up in Hamburg. Poles are particularly proud of astronomer Jan Hevelius (1611-1687), born of long-time Danzig citizens, the Hevelke family. He lived, worked, and made his observations in Gdańsk, and also helped with the family brewing business.

In 1807 Napoleon Bonaparte reached the city of Danzig and pronounced it a Free City, separating its administration from Prussia. For this he was hailed as a liberator. Soon after, of course, it became evident that Napoleon intended to use the city as a jumping-off point for his invasion of Russia, and the city suffered severe deprivations and loss of life. In 1814, the city was returned to Prussian control, even before the settlement at the Congress of Vienna. Nevertheless, the short era of independence has been looked upon as a model for the Free City that existed from 1920 to 1939 and also as a period of prominence for Gdańsk politicians today. In the preface to the album *Gdańsk 2000: Bound for the Future*, the first in a series of publications promoting businesses in prominent Polish cities, Tomasz Posadzki wrote, "Now that the twentieth century is closing, it is time to look to the past. Napoleon said that Gdańsk was the key to everything. What we wish today is to once again bring truth to these words." This publication touts Gdańsk as a city having liberal values and as a place that welcomed "foreigners and their ideas."[27]

The historic merchant character of the city, both today and shortly before World War II, has been emphasized in order to downplay perceptions of ethnic competition between Germans and Poles. This phenomenon makes the city of Danzig/Gdańsk an excellent model for studying how individuals, businesses, and local concerns have identified themselves in the twentieth century. Is it possible that they might place belonging to a local or city

[27] Ibid.

culture, on an equal footing with a national culture? What can be said about the loyalty and pride felt by Poles in Gdańsk today? How have they formed their link to this place, when so many came from families that were transferred there in the 1940s? These are questions inspired by this study of the reconstruction of the city, if not answered by it.

The main historic districts in Gdańsk reveal the character of the city at its height. Unlike Warsaw, where one finds the old town square reconstructed as a showpiece of Polish recovery, and unlike Krakow, which itself survived the war relatively intact, and has a large square dominated by the famous Cloth Hall, Gdańsk has no true square, but a series of parallel thoroughfares connecting market space and businesses to the Motława river, which flows on to the Baltic Sea and forms the center of all commerce and activity in the city. The city shows some Italian Renaissance influence, but visitors to Gdańsk most often compare the city to Amsterdam or Lübeck. Some monuments from the city's past remain; others, like the monument to Emperor Wilhelm I, were not replaced after the war. The monument to the Kaiser was erected in 1903 and stood before the sixteenth-century High Gate, itself emblazoned with the Prussian coat of arms, flanked by unicorns, the Polish coat of arms, flanked by eagles, and the Danzig coat of arms, flanked by lions. The interwar headquarters of the High Commissioner of the League of Nations and the Town Hall were renovated, but the parliament buildings that housed the Danzig *Volkstag* and Senate were not. The Danzig Great Synagogue was also not rebuilt. This building was constructed around 1885, but in the spring of 1939, Danzig Jews sold many of their synagogue's treasures in order to purchase their own lives, escaping before the German invasion, but not returning. The synagogues of Danzig and surrounding towns were not rebuilt.[28]

There are three categories of historical reconstruction undertaken after the war: modern structures meaningful or useful to the postwar Polish state, medieval and renaissance buildings linked to the economic and political history of the city, and monuments of great historical and cultural value. The following descriptions of

---

[28] *Danzig 1939, Treasures of a Destroyed Community* (New York, 1980). Precious objects from the Great Synagogue were sent to the Jewish Theological Seminary of America for safekeeping. The Seminary was to house the collection for fifteen years. If after that period the Danzig Jewish community was revived, the materials would be returned. Otherwise, it was to remain in the United States. A major exhibition of these objects was held at New York City's Jewish Museum in 1980. The above publication is the catalogue of the exhibition. One of the essays included in this catalogue is "What shall we tell our Children," by Günter Grass.

these buildings and monuments reveal the motivation behind the reconstruction of the city according to a postwar interpretation of the role Gdańsk has played in Polish history.

Prussian Era Buildings

*Komenda Garnizonu/Generalkommando/Żak/
Gdańsk City Council*

For many years, this building has housed the student club known as Żak. In 1996 the city announced that the Gdańsk City Council would move into this structure. The Żak building belonged to the High Commissioner of the League of Nations during the Free City era, but before that the complex, built in the years 1898-1901, formed the headquarters of the Prussian General Staff.

Above this building flutters the city's traditional emblem—a gold crown above two stylized white crosses, on a background of red. This very symbol demonstrates the blended history of the city: red and white (the Polish colors), Teutonic style "cavalry" crosses, a crown of gold dating back to 1454 and the incorporation of Danzig and Pomerania into the Polish Commonwealth by King Kasimir, son of King Jagiełło (himself of Lithuania). A pair of lions was also added to the city's coat of arms at this time but during the Prussian era were replaced in some cases with a black Prussian eagle. It was a significant decision, then, when the Free City of Danzig which emerged after World War I began again to use this symbol which bridges the city's many personalities and its unique character. Once again, the adoption of this symbol by present-day Poles, reveals a point of agreement between the ethnic German population which disappeared after the war and the generations of Polish postwar residents. Here, memory adopts symbol as well as structure, keeping the positive images of the League of Nations presence and the loyalty of medieval Danzig, although the residents of this proud city had fled or had been transferred away from their homeland.

Polish Post Office

A symbol of interwar Polish sovereignty and rights in the Free City of Danzig, and a major site of Polish resistance to the German invasion on September 1, 1939, the most famous depiction of which can be found in Günter Grass's *Blechtrommel/Tin Drum*. The defenders of the post office were executed. A modest bronze monument to the passionate defense of the post office stands behind the building. Another monument from this era can be seen to the north of the city, on the Westerplatte, commemorating the place where the first shots of the war were fired.

Politechnika Gdańska / Technische Hochschule

Established in 1904, renovated after World War II, the Technical University of Danzig / Gdańsk is perhaps the best known of the academic institutions in Gdańsk. In 1945, the university was declared a Polish state academic institution.

Buildings Representing Economic and Political Life

Długi Targ/Langer Markt/Ulica Długa /Langgasse

The main avenue at the core of the city leads to the canals, unlike Warsaw and Krakow or other cities where trade went on in a central square, Danzig / Gdańsk is structured so that all business streets lead to the canal, though this street is the largest. Along it one will also find the *Artushof/ Dwór Artusa*, or the Artus manor house, a merchant house. The street, dominated by the Artus house, built around 1380, and remodeled in the Italian Renaissance style, represents the golden era of Danzig building and architecture, and not coincidentally, the golden era of the Polish-Lithuanian Commonwealth. One house along this street features the coats of arms of Poland, Royal Prussia, and the city of Danzig. The Artus building was severely damaged during World War II, and has recently undergone continued renovation, as has the *Rathaus/Ratusz*, the town hall. Both house city administrative offices and museum exhibits.

Ratusz/Rathaus

This building was begun in the mid-fourteenth century and is an excellent example of the Gothic style. The tower was built after a fire in the late fifteenth century and includes a statue of Polish King Sigmund II August.[29] The Town Hall underwent significant renovations in the 1990s and was completely refurbished in time for the 1000-year anniversary of Gdańsk in 1997. Beside the town hall stands the late sixteenth century Neptune fountain by Abraham van den Blocke. This fountain remains a central feature in the old town.[30] Van den Blocke also designed the Golden Gate that

---

[29] *Piękny Stary Gdansk* (Bydgoszcz, 1994), 159.

[30] Van den Blocke was Flemish.

constitutes the entrance to Ulica Długa, which also forms the opening of the famous Royal Road, the road the Polish kings used when entering the city. An inscription on the eastern side of the gate reads, "Small states grow due to unity, big ones fall due to disunity." One on the western side reads, "Let these who love you fare well, let peace reign within your walls and happiness in your palaces."[31] The Polish tour book *Gdańsk* by Lech Krzyżanowski points out that in the late nineteenth century, during the Prussian era, streetcars caused such vibrations that this gate was structurally compromised, and that in the 1950s, etchings made in 1648 were used to repair the damage.[32]

Motława—Żuraw/Krantor

The property along the Motława canal is still under construction. Along this canal one can visit a medieval crane, one of the few, perhaps the only one, still standing in a former *Hansestadt*. The crane was built in the years 1442-1444[33] and unloaded the huge grain shipments that arrived in the city from the Polish hinterland. Visitors can also see that some of the hundreds of former warehouses and granaries of Danzig/Gdańsk have been reconstructed as storage, as the Central Maritime Museum, and as new townhouses.[34] After the war, only twenty of these warehouses on the so-called "Granary Island" were reconstructed. As recently as 1989, most of the area was dominated by ruins standing empty and ridden with shrapnel and bullet damage.

Buildings and Sites with Great Historical and Cultural Value

Kosciół Mariacka/Marienkirche

This immense Gothic church completed in 1502 is, according to tour guides in Gdańsk, the largest Gothic church in Europe. The entire population of the town, 20,000, fit inside easily.[35] This

---

[31] Krzyżanowski, 14.

[32] Ibid., 15.

[33] *Gdańsk 2000*, 14.

[34] In 1813, the siege of the city destroyed most of the medieval warehouses on the island.

[35] Krzyżanowski, 9.

church was lovingly reconstructed after the Second World War. While the nearby Baroque style Sobieski chapel represents a long-standing Polish influence in the city of Gdańsk, the Jan Sobieski monument (1897) that stands in the open space of the former Timber Market was moved to Gdańsk from its former home in Lviv [Lwow, Lemberg], Ukraine, in the mid-1960s. Before the war, this open space was occupied by a war memorial erected in 1904 to honor those who fell in the German wars of unification. This memorial was shaped like an obelisk and bore the dates 1864, 1866, and 1870. By erecting the Sobieski monument in its place, the city could be brought back into the Polish sphere, as a city along the Royal Road of the Polish kings.

Malbork/Marienburg

Outside the city, in the small town of Malbork rests one of the most fascinating reconstruction efforts in this region: the former fortress of the Teutonic Knights. The Knights also occupied Danzig in 1308. This site had been carefully reconstructed and restored in the nineteenth century by German architects and archeologist Conrad Steinbrecht. The German reconstruction of the site had been rooted in the romantic urge to link the newly unified Germany into a nation by conjuring up its Teutonic past. It was also closely linked to the rise of German archeologists to prominence during that era. From 1882 until 1921, Steinbrecht led the effort, and after his death the work continued into the 1940s. At about the same time that the renovations were completed in 1944, the fortress was claimed by the local German commander for use as an armory. It was subsequently bombed into dust by the Soviets, and then rebuilt by the Poles. The Polish guidebooks do emphasize the battle of Grünwald and the Polish victory against the knights in 1410, as well as a depiction of the city of Gdańsk as having been reluctant to cooperate with the Teutonic Knights, and the royal privileges offered the city after Grünwald. That the fortress, sections of which had been little more than a pile of overgrown rubble where shepherds let goats roam only a century before, was part of a reconstruction plan at all is remarkable. Almost immediately after the war was over, renovation plans were underway, and by 1956, a national committee for the reconstruction of the castle was brought into being. Parts of the fortress complex were still being restored as recently as 1997.[36]

---

[36] *Marienburg: Historischer Grundriss und Informationen über die Sehenswürdigkeiten der Marienburg* (Warsaw, 1993), 15.

Memory and Ethnic Cleansing

Timothy Snyder has said, "Territory and nationality are among the most powerful sources of bias."[37] Sometimes territory and nationality collide. The shape of Gdańsk/Danzig is emotional, as well as physical. Gdańsk itself has a new image, a post–World War II identity so closely tied with the Polish spirit through the Solidarity Movement that it is impossible to argue that the city is not intrinsic to Poland as a nation state in the year 2000.

The "citizens of Gdańsk" undertook the restoration of the city into the 1960s.[38] Youth organizations from Warsaw and around the country, as well as local residents, themselves only connected to Gdańsk for a generation, completed an impressive reconstruction of this Baltic city, complete with historical references from its genuine past, and retaining many famous street names, if in Polish translation. The names retained are descriptive, as in *Długi Targ/Langer Markt*, "Long Market Street," and *Ulica Długa /Langgasse*, "Long Street," *Kamienna Grobla/Steindamm*, "Stone dam," *Zielona/ Grüner Weg*, "Green Way." The retained names also describe the traditional use of a space or its proximity to a building or church, as in *Biskupia/Bischofsberg*, "Bishop's Hill," *Chlebnicka/Brotbänkengasse*, "Bakers' Street," *Targ Dzewny/ Holzmarkt*, "Timber Market," and *Garncarska/Töpfergasse*, "Potters' Street." In this way, many of the historic sections of the city remained unchanged in name as well as in physical shape, or the reconstruction and reinterpretation of the physical shape.

Since the fall of Communism, efforts were stepped up in the renovation of museum space and of historic sites, as well as of the reconstruction of resort space and facilities in nearby Sopot. The Polish Ministry of Culture and the Arts, the Friedrich Ebert Foundation, and the Polish-German Cooperation Foundation have all had a hand in this effort. *Gdańsk: Dwa Oblicza Miasta*, a dual-language (Polish/German) volume, was published in honor of the city's one thousand year anniversary. The volume features photographs from the turn of the century opposite present views in the city. The book was prepared, according to the authors, for

---

[37] Timothy Snyder, "'To Resolve the Ukrainian Question Once and for All': The Ethnic Cleansing of Ukrainians in Poland, 1943-1947," *Journal of Cold War Studies* 1 (Issue 2, 1999).

[38] *Gdańsk 2000*, 14.

"Gdańsk residents and Tourists, so they can see another view of the city and to increase their interest in the city and its history."[39]

Images of cities reborn, of identities disjointed, stories of suffering unacknowledged, all these cannot be ignored. The city of Gdańsk is one example of a *place* from which many ethnic Germans were forced to flee, a *place* which then absorbed a new or reinterpreted identity through its new residents, through war–torn Polish communities. The cultural identity of Gdańsk is not simply German. One cannot simply point to this city and say, "This was Danzig." The Free City legacy must be acknowledged as more than a local oddity; the link between the Polish hinterland and this port city was traditionally strong. The successes, and the failures, experienced by the city cannot be extricated from the relationship between the Poles and the Germans living at and on the Vistula delta. Why else would Gdańsk rise from the ashes of devastated Danzig, retaining the beauty and historical character valued by the postwar Polish state and its citizens? The tragedy of the highest order for the city of Gdansk is that expelled Germans and resident Poles alike must live with the paradox of a partially "cleansed" memory, signified by gravestones at the *Marienkirche*, St. Mary's Church, and in countless unmarked graves in the Baltic.

The reconstruction of the city portrays a non-threatening history. Tour books for Polish and foreign consumption today don't emphasize Germanness, but the character of the reconstruction does not deny it. One might also consider the Free City era to have been one of transition, and the need to demonstrate historic "Polishness" therefore might not have been as intense as in other areas. As recently as 1998, a group of former Danzig residents claimed to be a sort of government–in–exile, maintaining that the German invasion in 1939 not only led to the takeover of Poland, but also of the Free City, and that the fate of this semi-sovereign entity had never fully been decided.[40] Even among the *Vertriebenen* organizations at the *Haus der Heimat* in Stuttgart, Germany, the Danzigers kept their distance, maintaining office space separate from those organs that represented everyone from Sudeten Germans to Baltic Germans to Volga Germans. The *Gdańsk 2000* volume discussed in this essay had also declared the preservation of

---

[39] Robert Hirsch, Krzystof Krzempek, Piotr Popiński, *Gdańsk: Dwa Oblicza Miasta* [City of Two Faces] (Gdańsk, 1997), 7. The authors credited the completion of successful renovations in Gdańsk in part to records that the Soviets had captured during the war, and then returned in 1956.

[40] For an impressive, and alarming, statement on behalf of this group, see "Free City of Danzig," http://danzig-freestate.org/freedanzig.html.

the physical history and memories in Gdańsk by suggesting that businesses might consider investing in the restoration of historic districts: "...repair and sanitation of historical districts of the town are particularly good opportunities for business activities, mostly in such housing estates as Wrzeszcz [or] Oliwa . . . , [which] has a large stock of decapitalized buildings in ... poor technical condition. These buildings are very interesting examples of urban architecture from the turn of the century."[41] The turn of the century—was this not the very era initially avoided by those who planned the reconstruction of the city immediately after the Second World War? Here we see, in the name of business and economic well-being, an opportunity to undertake the repair of memory, the regaining of memory, to reveal what had been "cleansed," or, at the very least, willfully forgotten.

What does Gdańsk stand for? What can it stand for in such a context? Lech Wałęsa spoke before the United States Congress in 1989 and answered that famous question "Why die for Danzig?" by saying that it was no longer a question of dying for this city, but of living for it. German Danzig is gone, Polish Gdańsk lives, and the streets of this rebuilt city do speak, asking that its *complete* history be remembered.

Photos

View of Teutonic Knights' Fortress,
Malbork/Marienburg. Photo by
Author, 1997.

---

[41] *Gdańsk 2000,* 56.

Free City of Danzig Flag by *Jaume Ollé*

City of Gdansk Flag, 1996 by *Adam Kromer*[42]

---

[42] Flag images from "FOTW Flags of The World" website at http://fotw.digibel.be/flags/", accessed 27 June 2001.

Dlugi Targ patrician home façades. Photo by
Author, 1997.

Zak Student Club/League of Nations High
Commissioner former headquarters.
Photo by Author, 1997.

Polish Post Office in the Free City of Danzig. Gdansk, Poland.
Photo by Author, 1997.

Cleansed Memory: The New Polish Wrocław (Breslau) and the Expulsion of the Germans[1]

GREGOR THUM

The settlement of the Potsdam Conference in 1945 did not only move the German-Polish frontier 150 miles to the West. It also shifted the ethnic border between Poles and Germans. This act of political engineering changed dramatically the ethnic and cultural face of a whole region—a region that had been Germany's East until 1945 and that subsequently became Poland's West. Within a few years this vast experiment subjected millions of people to resettlement, thereby redrawing the ethnic geography of Central Europe. In 1939 about eight million Germans lived east of the Oder-Neisse Line. After the war they were prevented from returning to their homes. All were forced to leave their homes between 1945 and 1947. In Upper Silesia and in the Masurian region about 900,000 former German citizens with an established, complex ethnic identity were declared to be Poles by the Polish government. They were allowed to remain as so-called "autochthons," after they had successfully undergone "verification" and had accepted Polish citizenship.[2]

According to the census of 1950, about 4.5 million Poles settled in the former German territories after the war. These became the Western and Northern Territories of the newly established Polish State.[3] Two and a half million Poles came from Central Poland, from some of the most devastated regions and from cities like Warsaw and Poznań. During the privation of the postwar days, they came with hope of finding tolerable living conditions in

---

[1] This article is based on the author's dissertation, "Die fremde Stadt: Breslau nach dem Bevölkerungsaustausch," submitted to the Faculty of Cultural Science at the European University in Frankfurt (Oder) in 2001.

[2] Czesław Osękowski, *Społeczeństwo polski zachodniej i poółnocnej w latach 1945-1956: Procesy integracji i dezintegracji* [The Society of Western and Northern Poland: Processes of Integration and Disintegration] (Zielona Góra, 1994), 80-138; Michael G. Esch, *'Gesunde Verhältnisse': Deutsche und polnische Bevölkerungspolitik in Ostmitteleuropa 1939-1950* (Marburg, 1998), 299-321. As a rule, the autochthons did not feel accepted as equals within Polish society after the war. The majority declared themselves to be ethnic Germans later on and emigrated to the Federal Republic of Germany after 1950.

[3] Figures from Piotr Eberhardt, *Między Rosją a Niemcami: Przemiany narodowościowe w Europie Sprodkowo-Wschodniej w XX w.* [Between Russia and Germany: Ethnic Changes in East Central Europe in the Twentieth Century] (Warsaw, 1996), 125-126.

the new territories. One and a half million were the "repatriated" Poles, the Poles who had been expelled from Eastern Poland, the region annexed by the Soviet Union in 1944/45 and cleansed of its two million Polish inhabitants.[4] One hundred and fifty thousand Poles re-immigrated from France, Western Germany, Yugoslavia, Romania and Belgium.[5] Many of them were miners. They settled in the industrial regions of Silesia. About 200,000 Polish Jews, who had survived the Holocaust in the Soviet Union, returned to Poland after the war. A sizeable portion of them took up residence in the cities of Lower Silesia.[6] Another group consisted of 200,000 Ukrainians. In 1947 the Ukrainians were forcibly deported by the Polish government from the southeastern boarder regions and dispersed all across the newly acquired Western Territories.[7]

I will not cover in detail the enormous movement of populations. We know a great deal about the genesis of the Oder-Neisse Line and the idea of resettling all Germans from Poland. The traumatic experiences of the expellees have been documented in numerous publications in Germany.[8] In a multi-volume work, a Polish-German research team is now publishing all the important documents from Polish archives about the Germans in Poland in the years between the war and resettlement.[9] However, at the present time we know very little about the long-term cultural consequences of this Central European ethnographic shift. As in the

---

[4] Jan Czerniakiewicz, *Repatriacja ludności polskiej z ZSRR 1944-1948* [The Repatriation of the Polish Population from the USSR, 1944-1945] (Warsaw, 1987).

[5] Andreas R. Hofmann, *Die Nachkriegszeit in Schlesien: Gesellschafts- und Bevölkerungspolitik in den polnischen Siedlungsgebieten 1945-1948* (Köln, 2000), 142-157.

[6] Hofmann, *Die Nachkriegszeit*, 339-343; Bożena Szaynok, "Ludność żdowska na Dolnym Śląsku 1945-1950" [The Jewish Population of Lower Silesia], *Historia 146* (Wrocław, 2000).

[7] Timothy Snyder, "'To Resolve the Ukrainian Question Once and For All': The Ethnic Cleansing of Ukrainians in Poland 1943-1947," *Journal of Cold War Studies* 1/2 (1999): 86-120.

[8] The German literature on this issue is impressive in quantity, though not always in quality. The first systematic collection of records concerning the expulsion—with clear political intentions—was Theodor Schieder, ed., *Dokumentation der Vertreibung der Deutschen aus Ost-Mitteleuropa*, 8 vols. (Bonn, 1954-1961). For a survey on the Polish literature, see Włodzimierz Borodziej, "Historiografia polska o 'wypędzeniu' Niemców" [The Polish Historiography of the "Expulsion" of the Germans], *Polska 1944/45-1989: Studia i materiały* 2 (1997): 249-69.

[9] Włodzimierz Borodziej and Hans Lemberg, eds., *Die Deutschen östlich von Oder und Neisse 1945-1950: Dokumente aus polnischen Archiven*, vol. 1 (Marburg, 2000).

case of all other regions that were forcibly deprived of parts of their traditional population during the twentieth century, the cultural history of the Western Territories after the expulsion of the Germans remains to be written.[10]

In this paper I will discuss the cultural transformation of the region after its ethnic cleansing. I am especially interested in the psychological experiences of the Polish settlers who took over the former German towns and villages, houses, and flats in 1945 and attempted to create new homes there.[11] I have chosen the city of Wrocław/Breslau—the largest urban center of the Western Territories and economically and culturally the most important—as the case study. The problems associated with the evacuation of an entire city and its repopulation with people of another nationality raise questions that are of primary interest to me. How did the Polish settlers and their political leaders cope with the fact that they had to start living in a city so completely foreign to them? How did they come to terms with the German past of Wrocław/Breslau after the forcible resettlement of the former inhabitants? How did they manage the cultural transformation of an initially alien German city into a city which today is as Polish as any other town in Poland?

The earliest documents of the city of Wrocław/Breslau date back to the time when it became a Polish Episcopal See in the territory of the medieval Polish State in the year 1000. From then on it was the capital of the region of Silesia. As a result of long-term demographic and political changes beginning in the thirteenth

---

[10] The interesting Polish sociological literature on the Western Territories concentrates on the integration of the different Polish population groups after the war and generally does not deal with cultural questions. See Osękowski, *Społeczeństwo*; Andrzej Kwilecki, ed., *Ziemie Zachodnie w polskiej literaturze socjologicznej* [The Western Territories in the Polish Sociological Literature] (Poznań, 1990). But in the last years, we have seen the first substantial attempts to take up the complicated cultural issues linked to the population shift. See Zbigniew Mazur, ed., *Wokół niemieckiego dziedzictwa kulturowego na Ziemiach Zachodnich i Północnych. Praca zbiorowa* [Concerning the German Cultural Heritage in the Western and Northern Territories: An Anthology] (Poznań, 1997); Zbigniew Mazur, "Das deutsche Kulturerbe in den West- und Nordgebieten Polens," *Osteuropa* 47 (1997): 633-49. In the field of art history and preservation of the architectural legacy in the Western Territories there already exists a literature, going back to the fifties. See the excellent article by Adam S. Labuda, "Das deutsche Kunsterbe in Polen: Ansichten, Gemeinplätze und Meinungen nach dem Zweiten Weltkrieg," *Deutschland und seine Nachbarn: Forum für Kultur und Politik* 20 (1997): 5-23.

[11] A survey of the research on the Western Territories since 1945 is in Gregor Thum and José M. Faraldo, "Las Regiones Occidentales Polacas: Experimento social y arquitectura de las identidades," *Cuadernos de Historia Contemporanea* 22 (Madrid, 2000).

century, Silesia came to be inhabited by German-speaking settlers. Irrespective of the ethnic situation, the political affiliation of Silesia changed frequently. The region belonged successively to Poland, Bohemia, and the Habsburg Empire before it was annexed by Prussia in the middle of the eighteenth century. From 1871 on, Silesia was part of the German nation-state just like Bavaria or Saxony. On the eve of the Second World War the Silesian capital of Breslau, an urban center of over 600,000 inhabitants, was the biggest German city east of Berlin.

The 1945 decision of the Allied Powers on the future German frontiers represented a profound turning point in the history of Wrocław/Breslau. Because there was only a very small Polish minority in the city in the first half of the twentieth century, the Potsdam Agreement resulted in a complete population exchange. The inhabitants of Wrocław/Breslau, who had survived the ruthless evacuation of their city by the German authorities in January 1945, were not allowed to return to their homes after the war. Those who had remained in the city and survived the siege of "fortress" Breslau in the last three months of the Second World War, and those who had managed to return after the war—altogether about 200,000 to 300,000 people—were programmatically resettled between 1945 and 1948. During the same period, about 250,000 Poles settled in Wrocław. About 20 to 30 percent were "repatriates." Among them, the former citizens of Lwow [Lvov, Lemberg] played an important role.[12] The rest were settlers from Central Poland. The number of autochthons in Wrocław was negligible.[13] Today Wrocław has again more than 600,000 inhabitants and is the fourth largest city in Poland. With rare exceptions, hardly any present-day Wrocławians can claim

---

[12] On the composition of the postwar population in Wrocław see Elżbieta Kaszuba, *Między propagandą a rzeczywistością: Polska ludność Wrocławia w latach 1945-1947* [Between Propaganda and Reality: The Polish Population of Wrocław, 1945-1947] (Warsaw, Wrocław, 1997); and Irena Turnau, *Studia nad strukturą ludnościową polskiego Wrocławia* [Studies on the Population Structures of Polish Wrocław] (Poznań, 1960).

[13] In a document of 1949 the Wrocław authorities quoted the number of authochthons in Wrocław as 2,769 persons; see Archiwum Państwowe we Wrocławiu [State Archive in Wrocław] (hereafter, AP Wr), UWW 784, 30; Kaszuba, *Polska ludność*, 69, quotes 2954. In contrast to the autochthons of Upper Silesia or the Masurian region, these Wrocław autochthons could hardly be seen as a deeply rooted population group in the city. Most of them were immigrants, who came to Wrocław during the first half of the twentieth century from the Poznań region or Upper Silesia, or they were the husbands or wives of Polish-Geman mixed marriages and their descendants. See the report on this population group in Wrocław from 1946, AP Wr, UWW 748, 1-5.

family origins in Wrocław that reach further back than the year 1945. Everybody is a new settler. They, or their parents, or sometimes even grandparents, came to Wrocław as a result of postwar events.

Joanna Konopińska, a Polish student from Poznań, came to Wrocław in the autumn of 1945. She lived in a former German house. Some weeks after her arrival, she wrote in her diary:

> I am sitting at the desk writing down my impressions, although it would certainly be better if I started cleaning the flat instead. Cleaning and sweeping out this foreignness, this Germanness, which is peeping out from every corner . . . . At present I am stumbling at every turn onto things which belong to somebody else, which testify to another life I know nothing about, which testify to the people who built this house, who have been living here, and who now, perhaps, are not alive any more. How could one start a new life here? No, I can't imagine that I will ever be able to say: This is my house.[14]

The foreignness described by Konopińska was widely experienced among Polish settlers of the postwar era. Many of them were from Warsaw, Poznań, or Lwow. The majority came from the countryside, without any experience of urban life. For them the modern city with its German face and departed German inhabitants was a completely alien milieu.[15] Moreover, the fear that the Germans would return sooner or later, take over the land, and chase away the Poles was omnipresent.[16] All these factors created an atmosphere of instability during the postwar years, leaving impressions that the settlers owed their present status to a provisional conquest and to a temporary settlement. The administration of Wrocław was chaotic, as it was everywhere else in the new territories. Corruption and nepotism were widespread among public servants.[17] For most inhabitants, looting and black marketeering

---

[14] Joanna Konopińska, *Tamtem wrocławski rok* [A Year of Wrocław] (Wrocław, 1987), 53.

[15] Włodzimierz Kalicki, "Breslau—das Zuhause von Paweł und Małgorzata," *Transodra: Deutsch-polnisches Informationsbulletin* 17 (1997): 14-28, 18 (first published, "Dom Pawla, dom Małgorzaty," *Gazeta Wyborcza*, 8 Sept. 1995, 6-12).

[16] Hofmann, *Die Nachkriegszeit*, 118; Philipp Ther, *Deutsche und polnische Vertriebene: Gesellschaft und Vertriebenenpolitik in der SBZ/DDR und in Polen 1945-1956* (Göttingen, 1998), 273.

[17] See the inspectoral reports about the administration in Wrocław 1945-1949 in Archiwum Akt Nowych we Warszawie [Archive of New Documents in Warsaw] (AAN), MAP 2338, 32-34; MAP 2340, 19-20; AP Wr, UWW VI 750, 36-58.

were the only way to survive. Crime was four times as high as in Central Poland. The "Recovered Lands" came to be called Poland's "Wild West."[18] This negative attitude toward their new homes in Western Poland represented much more than a temporary problem in adaptation. It had far-reaching political effects in slowing down the entire settlement and the integration of the new territories into the rest of the Polish State. The Polish government was forced to take action. It had to demonstrate its ability to expand to the new frontiers, to establish law and order, to settle and integrate the new territories. To achieve these ends, one of the most needed measures would be to counteract the feeling of foreignness and instability in the region. Only when Polish settlers believed that the Western Territories were Polish and would remain Polish could they be expected to commit themselves to the future of the new West.

An extensive national pedagogical program was developed. Thousands of Polish scholars, teachers, journalists, and writers took part, along with a number of newly founded regional research institutions, with the West-Institute (Instytut Zachodni) in Poznań leading the way. Their goal was to disseminate knowledge about the Western territories and facilitate identification with them. A view of history was propagated that justified the arbitrary move of Poland 150 miles to the West. The radical break was presented as a natural culmination of a long-term historical process. One has to agree with Norman Davies that "the elaborate historical ceremonial, which the ideologists have laid on to sanctify the decision of the 1945, is almost as breathtaking as the great settlement itself."[19] A brief outline of the interpretation of the westward shift of Poland might go as follows. The Western Territories were old Polish lands. These had been occupied by imperialistic Germans in the Middle Ages and subjected to a forced Germanization policy ever since. During the whole period of foreign rule the economic and cultural ties between these territories and Poland continued to exist. A significant Polish minority preserved the Polish character of the land and successfully resisted Germanization. During World War II German imperialism culminated in the Nazi occupation of the rest of Poland, in genocide, and the scorched

---

[18] About the problematic everyday situation in Wrocław in the years after the war, see: Jędrzej Chumiński, "Czynniki destabilizujące proces osadnictwa we Wrocławiu (1945-1949)" [Factors of Destabilization during the Process of Settlement in Wrocław, 1945-1949], *Socjologia* 10 (1993): 55-78.

[19] Norman Davies, *God's Playground: A History of Poland*, 2 vols. (Oxford, 1982), 2: 518.

earth policy. After the triumph over Nazi Germany, the lost Polish territories along the River Oder and the Baltic Sea were returned to their "Motherland" (*Macierz*) as an act of historical justice. In this interpretation of history, the former German territories were called the "Recovered Lands" (*Ziemie Odzyskane*) in the official language of the People's Republic.

For the new citizens of Wrocław this meant that the history offered to them in the official propaganda concentrated on four different episodes. Upon these local Polish identity had to be built:

The medieval history of the city under the reign of the Polish dynasty of the Piasts;

the history of the Polish inhabitants during the centuries of foreign rule;

the era of the Third Reich, especially the fanatic, militarily senseless, barbaric defense of the "fortress" of Wrocław in the last weeks of World War II, during which the Germans destroyed most of the city;

the Polish recovery of Wrocław after the war and its reconstruction in the so-called "Pioneer Period."

The settlers were taught the history of the Polish minority, but did not hear much about the German majority. They learned about the negative impact of Prussian rule on the economy of Wrocław/Breslau in the eighteenth century, but nothing about its industrialization and modernization since the beginning of the nineteenth century. They knew everything about Wrocław's cultural ties to Poland, but nothing about those to Germany. They could read the shocking chronicle of the siege in 1945 written by the German priest Paul Peikert, but were not allowed to know what the same chronicler had to say about the beginnings of the Polish administration in Wrocław.[20] They should know everything about the heroism of the first settlers and their achievements during the Pioneer Period, but not about the setbacks and the chaos of the postwar era.[21]

---

[20] Paul Peikert, *Kronika dni oblężenia: Wrocław 22.I - 6.V 1945* [Chronicle of the Days of Siege: Wrocław, 22. I.—6. V. 1945], ed. K. Jonca and Alfred Konieczny (Wrocław, 1964). See the remarks of Sebastian Siebel-Achenbach, *Lower Silesia from Nazi Germany to Communist Poland* (New York, 1994), 127 fn.

[21] Considering the endless number of titles in which this view of history became manifest, only one perfect example in English language should be given here. It is

The official view of history was promoted in a flood of publications, political speeches, and exhibitions aimed at the general public during the entire existence of the People's Republic. Annual ceremonies such as the "Weeks of the Western Lands" or the "Days of Wrocław" were platforms for spreading the historiographic propaganda. During the great jubilees celebrating ten, twenty, twenty-five, and forty years after the "liberation" of the Western territories and their "return to the Motherland," the historic significance of the events in 1945 and their official interpretation were imposed on the minds of the entire society. Wrocław played the major role as the venue of all these nationwide ceremonies. One of the biggest propaganda shows the People's Republic ever organized was the "Exhibition of the Recovered Lands" in Wrocław, in the summer of 1948.[22] During these three months of the exhibition, 1.5 million visitors were informed not only of the progress in reconstruction and integration of the Western territories, but also of the Polish view of the history of the territories on the Oder and Baltic and Poland's historical claims on them.

Just as important as an instrument of building and shaping a local Polish identity were the popular products of regional studies, such as surveys of local histories or guidebooks. Most of the Poles entered the Western Territories after the war for the first time in their lives. They did not know very much about the history of the former German towns and landscapes. This was a chance to shape the settlers' and the visitors' historical knowledge of the region. The first Polish guide to the city of Wrocław was published in 1946.[23] Its author, Andrzej Jochelson, had served as an enlightened

---

Marian Orzechowski, "Wrocław in the Recent History of the Polish State and Nation," *Polish Western Affairs*, no. 13, 2 (1972): 305-33. One of the most striking interpretations of the Potsdam settlement as a final result of a long-term historical development is Gerard Labuda, *Polska Granica Zachodnia: Tysiac lat dziejów politycznych* [The Polish Western Frontier: One Thousand Years of Political History] (Poznań, 1971) (Abridged version in English: "Poland's Return to the Baltic and the Odra and Nysa in 1945: Historical and Current Conditions," *Polish Western Affairs*, no. 16, 1 (1975): 3-36. See also the late, but not less impressive, answer of Klaus Zernack, "Deutschlands Ostgrenze," in *Deutschlands Grenzen*, ed. Alexander Demandt (Munich, 1991), 140-65.

[22] Jakub Tyszkiewicz, *Sto wielkich dni Wroc Deutschlands Grenzen awia. Wystawa Ziem Odzyskanych we Wrocławiu a propaganda polityczna ziem zachodnich i północnych w latach 1945-1948* [One Hundred Great Days: The Exhibition of the Recovered Lands in Wrocław and the Political Propaganda of the Western and Northern Territories 1945-1948] (Wrocław, 1997).

[23] Andrzej Jochelson, *Przewodnik po Wrocławiu* [Guide Through Wrocław] (Kraków, 1946).

pioneer of the Polish administration in Wrocław since May 1945 and later as a leading activist of the "Society of the Lovers of Wrocław" (Towarzystwo Miłośników Wrocławia), founded in 1956. This guidebook was followed by many similar books. Without giving a detailed analysis, it is obvious that it gave its readers a version of Wrocław's history in which the Poles had been the actual citizens of Wrocław from time immemorial. By contrast, the Germans played, if any role, only that of conquerors and oppressors of the legitimate inhabitants. A remarkable periodical of this type is the "Wrocławian Calendar" (*Kalendarz Wrocławski*), edited each year since 1957 by the "Society of the Lovers of Wrocław," mentioned above. The reader finds information and articles about the past and present of the city, but at the same time, there is a real calendar which supports the official view of history with innumerable commemorative days, such as: "The 350th anniversary of the death of Fryderyk Guesau, pastor of the Polish Protestant parish in Wrocław." Or, "55 years ago members of German nationalistic organizations expelled Polish students from the lectures at the University of Wrocław." Or, "15 years ago the National Council of the city of Wrocław created the medal 'The builders of Wrocław' for those who rendered great services to our city."[24]

Beyond the dissemination of a Polonocentric view of Wrocław's history, an attempt was made to reshape public position in a way that would fit it into the official historiographic interpretation. After the resettlement of the last Germans in 1947, the Prussian-German face of Wrocław was to disappear entirely. The immediate records of the German past had to be removed and the city would be given a Polish face. The needed measures were termed "Polonization," "Repolonization," "Degermanization" and "Removing the traces of Germanness."[25] Both the main development and promotion of an authoritative view of history and its physical expression in the public position, were aimed at repressing the centuries-long presence of Germans in Wrocław in the collective memory of Polish society. It is this deliberate attempt of Degermanization which I call the cleansing of memory.[26]

---

[24] *Kalendarz Wrocławski* [Wrocław Calendar], ed. Towarzystwo Miłośników Wrocławia (Wrocław, 1974), 7-29.

[25] Bernard Linek has examined this policy in the region of Upper Silesia in '*Odniemczanie' województwa śląskiego w latach 1945-1950 w świetle materiałów wojewódzkich* [The "De–Germanization" of the Silesian Voivodship, 1945-1950, in Light of Materials from the Voivodship Archive] (Opole, 1997).

[26] This policy was partly a response to the German policy towards Wrocław after the World War I. See Gregor Thum, "Bollwerk Breslau: Vom 'Deutschen Osten' zu

The first step in this cleansing was the Polonization of tens of thousands of geographic names in the Western territories.[27] This was much more than a pragmatic adaptation of the names of towns and streets to the Polish language. The renaming represented a highly symbolic expression of claiming the former German land. Apart from the necessity to have Polish names throughout the entire territory of the Polish State, the new place names were to reflect the thesis that lost territories had returned to their "Motherland." The linguist M. Rudnicki, one of the leading specialists in this field, explained the political goal of the naming policy in the Western Territories:

> The aim of the changing of place names is to give back to the Recovered Lands their purely Polish character, which the Germans . . . endeavored to blur by introducing purely German names.[28]

At the First Onomastic Conference, opened in Szczecin [Stettin] on 11 September 1945, Rudnicki declared:

> We are not newcomers in this land, we are returning to it. We revive that which has remained, we are largely fetching back the names from the documents: we return that which was defaced completely, which was changed by chance, to the state in which it should be.[29]

The Polish government established teams of experts, led by historians and historical linguists. These naming commissions had to make sure that the newly selected names were consistent with the official interpretation of history. In the case of place names, these consisted largely of Reslavization of geographic names. These had either undergone a long-term, unintended Germanization in pronunciation and spelling, or they had been deliberately Germanized by governmental actions after the end of the nine-

Polens 'Wiedergewonnenen Gebieten,'" in *Preußens Osten—Polens Westen: das Zerbrechen einer Nachbarschaft*, ed. Helga Schultz (Berlin, 2001), 227–252.

[27] Maria Wagińska-Marzec describes the organizational aspects of the renaming in "Ustalanie nazw miejscowości na Ziemiach Zachodnich i Północnych" [The Naming of Places in the Western and Northern Territories], in *Wokół niemieckiego*, ed. Z. Mazur, 367-416.

[28] Wagińska-Marzec, "Ustalanie nazw," 407, citing Archiwum Państowe w Poznaniu (AP P), Akta UW, 1946/1947, 70.

[29] Wagińska-Marzec, "Ustalanie nazw," 380, citing AP P, Akta UW, 1945, 221.

teenth century. This process was carried out with great fervor after Hitler came to power in Berlin. Thus, the Polish naming policy after 1945 represented a partial annulment of a calculated Germanization policy of the Prussian or German authorities of the records of Slavic settlements in the eastern parts of Germany. At the same time, Polish policy also attempted to erase the historical memory of place names testifying to the centuries of German settlement. Of interest is that this policy came in conflict with the autochthons in Upper Silesia and in the Masurian region. During the time when they lived within the frontiers of Germany, these people had used their own Slavic forms instead of the official German place names. After 1945 they expected them to be adopted by the Polish authorities. As these names reflected the mixed German-Polish character of this ethnic group, they were often not in compliance with the Polish written language. With some exceptions, the authorities rejected them outright.[30]

In the case of Wrocław, the Polish form of the German "Breslau" did not, of course, have to be invented or fetched back from documents in 1945. "Wrocław" had always been used in the Polish language. "Breslau" and "Wrocław" were linguistic variants that shared a common root. But in Wrocław itself about 1,500 German names of districts, streets, squares, and bridges had to be changed.[31] Most of the neutral German names were simply translated into Polish. Of note are the nationally significant names, such as those that represented German public figures or events significant in German national history. In most of these cases a semantic reencryption was carried out by finding Polish equivalents. The German "Emperor Bridge" (Kaiserbrücke) was translated into "Grünwald Bridge" in commemoration of the Battle of Grünwald in 1410, when a Polish army defeated the Teutonic Order. The "Street of the SA," a name the Nazis introduced in place of "Kaiser Wilhelm Street," became the "Street of the Silesian Rebels" in commemoration of the Polish attempts to change the borders in Upper Silesia after the First World War. The "Tauentzien Square" was changed to "Kościuszko Square." In this renaming, a Prussian war hero of the Napoleonic era was replaced by a Polish one of the same period. The same treatment was applied to the Prussian generals Gneisenau and Blücher, who were replaced by the Polish

---

[30] See, for example, the long discussions about the "right" Polish city names in the case of Nysa/Nisa, Lignica/Legnica, or Kistrzyn/Kostzryn: AAN, MAP 580, 1-18, 162-175.

[31] See the report of Andrzej Jochelson, member of the Wrocław Street Name Commission of 1948 in AP Wr, UWW XVIII-63.

generals Bem and J. Poniatowski. "King St." (Königsstraße) was at first simply translated into Polish. But as "King St." was originally named for a Prussian king, it seemed desirable to change it into "Street of Stanislaw Leszczyński," the name of the Polish king of the eighteenth century. Remarkable is the case of the former "Junker Street." By changing it to the "Street of the Victims of Auschwitz," city officials linked the Holocaust implicitly to the bearers of the Prussian State. This move reflected the policy of the Anti-Hitler-Coalition that considered Prussia to be the main foundation of the Third Reich. It also reflected the traditional Polish aversion to Prussia. These few examples demonstrate that the re-naming had been a highly symbolic act, one that revealed a great deal about self-identity and attitudes toward Wrocław's German past.

Of course, the German monuments in Wrocław/Breslau were removed. The huge equestrian statue of Emperor William I, the biggest monument in the city, was pulled down in October 1945 with all due ceremony. Speeches were held, an orchestra played, and at the spot where the statue had stood the Polish flag was raised.[32] To this day the site is empty. Attempts to erect a statue of Bolesław Chrobry for the millennium of Wrocław in 2000 were not successful. There were many examples in which the *horror vacui* of removed German monuments were filled with Polish monuments. In the central square, the equestrian statue of the Prussian king Frederick William III was replaced by the statue of the Polish poet Aleksandr Fredro, a move that required the transfer of the statue from Lwow to Wrocław. In place of the former Tauentzien monument in the middle of modern-day Kościuszko Square, a memorial stone commemorates the Polish fighters for national and social liberation in the years 1791 to 1945. The entire city of Wrocław was outfitted with a network of monuments and commemorative plaques.[33] They taught the citizens the Polish history of the city, the suffering of the Polish nation during the German occupation in World War II, and the "recovery" of the Western territories in 1945.

The ensemble of St. Martin's chapel on Cathedral Island is a most impressive monument and an excellent example of the well thought out political iconography of Polish Wrocław. The chapel,

---

[32] See the description in the diary of Konopińska, *We Wrocławiu jest mój dom, Dziennik z lat 1946-1948* [My House is in Wrocław: Diary from the Years 1946-1948] (Wrocław, 1991), 94.

[33] Zygmunt Antkowiak, *Pomniki Wrocławia* [Monuments of Wrocław] (Wrocław, 1985).

once part of the medieval castle of the Piasts, was excavated from the ruins of the neighboring buildings and reconstructed between 1957 and 1960. A description on the outside wall of the chapel informs visitors of this early record of Polish settlement and rule in Wrocław. Nearby a commemorative plaque cites the four profoundly patriotic principles of the former Union of Poles in Germany. And, a large statue of Pope John XXIII in front of the chapel commemorates the Pope. It was unveiled in 1968. During the Second Vatican Council in 1962, the Pope publicly approved the Polish right to the Western territories by describing them as "recovered after centuries." The statue stands not only in front of St. Martin's chapel. It is displayed in front of the entire ensemble of the mediaeval churches on Cathedral Island. This Island, the seat of the bishop, remained Catholic during the Reformation and its changes that swept through the city of Wrocław/Breslau. It had remained part of the Polish Church when Breslau was a city in Prussia. To Poles this made the island the noblest part of Wrocław at least until 1990. After that the old town with its traditional center of trade and bourgeois life became more attractive.

The policy of Degermanization of public places was not limited to the renaming of streets and the removing of German monuments. Wrocław was strewn with everyday records of its German past. There were all sorts of German inscriptions on the outside walls of the houses, in the flats and corridors, on the manhole covers, or on railroad equipment. There were company signs, advertising, nameplates, technical explanations, art objects, and epitaphs at and in the churches, and in German cemeteries. In April 1948 the Wrocław authorities, along with all others in the Western Territories, received a secret circular from the Ministry of the Recovered Lands. The ministry gave orders to intensify the Repolonization process. Repolonization, as Undersecretary J. Dubiel stated, was "not producing sufficient results everywhere," since "the removal of traces of Germanness was not being brought to a close everywhere and not to a sufficient degree." For the sake "of a radical redressing of this state of affairs" the following measures were to be taken:

1. Eliminating the German language;

2. Removing the vestiges of German inscriptions;

3. Polonizing first and last names;

4. Struggling against all forms and vestiges of the
Nazi and Germanization ideology.[34]

Everything reminiscent of the former presence of Germans in
public places had to disappear. The implementation of this directive required a tremendous effort. For weeks and months groups
of workers were sent out with pails of paint, ladders and tools, to
remove or paint over the German inscriptions. They had to reach
the gables of five story houses to remove the German word *"Er-
baut"* (built), while the neutral digits of the year in which the
house was built could be left. In some cases they had to use ham-
mers and chisels to remove inscriptions carved in stone. At times
this policy produced comic results. The facade of the former
Building College, the present Department of Architecture at the
Engineering College in the B. Prus Street, had a stone relief containing German inscriptions. Because the workers did not remove
the inscriptions completely, one can still decipher the proverb
"Ohne Fleiß kein Preis" (no pain, no gain) on the front of the
building, thus demonstrating the truth of the proverb. During de-
Germanization the Wrocław authorities had to deal with sensitive
questions: What to do with gravestones with Polish names but
German inscriptions on them? Should they be preserved as evidence of the existence of Poles in former eastern Germany, or
should they be removed as embarrassing records of a successful
Germanization?[35]

The authorities had to deal with the pervasive question of
what to do with thousands of German cemeteries over the entire
Western territories. Some in the center of the cities and in church
graveyards were removed immediately after the war. The majority
remained. Since they were no longer in use, they dilapidated over
time. The graves were destroyed by grave robbers. Often the
cemeteries served as hiding places for criminals. At the end of the
1950s the Polish government decided to rectify this situation. Devastated German cemeteries made a bad impression both on Polish
society and foreign visitors. Since 1958 government decrees directed local authorities to patch them up. Plans were made to
transform them into green spaces, parks or new cemeteries--forty

---

[34] Secret circular N° 18 (26.04.1948) of the Ministry of the Recovered Territories in
AAN, MZO 496, 33; printed in Maria Rutkowska, "Kilka dokumentów lat czterdzi-
estych" [Some Documents from the Forties] *Wokół niemieckiego*, ed. Mazur, 257-300,
298-300.

[35] See the inquiries from the districts of the voivodship Wrocław: AP Wr, Urząd
Wojewódzki Wrocławski VI-749, 62, 79.

years after the last burial.[36] In Wrocław the authorities did not wait as long. From 1970 to 1972 they dissolved all remaining German cemeteries without any protests on the part of the inhabitants. The gravestones and mausoleums were used as building material for the stands of the Wrocław sports stadium, for pens in the zoo, or to reinforce the ditches in the center of the city.[37]

> Years later thousands of German gravestones were brought to the territory of the former German cemetery in Osobowice. In 1984 stonemasons with pails of paint walked between hills of German gravestones. They marked in white the slabs which they wanted to buy as raw material for their production.[38]

The only remaining cemetery is the Jewish cemetery in the south of the city, one of the largest and most impressive in Europe. It was damaged during the siege of Wrocław and decayed after the war. The Jewish cemetery escaped the fate of dissolution perhaps because it had been used by the Polish Jewish Community in Wrocław after 1945 or perhaps because Ferdinand Lassalle, the famous socialist thinker and one of the leaders of the German Labor Movement, was buried there. In 1975 the Jewish cemetery was listed as a historical monument of the city.[39] Today the only remaining cemetery from the time prior to 1945 is a tremendous monument to the history of Jews in Wrocław. As nearly all inscriptions on the gravestones are in German and reflect the advanced assimilation of German Jews at the beginning of the twentieth century, the Jewish cemetery is an excellent record of Wrocław's German past.

The effort to wipe out German documents from Wrocław's past sometimes produced extreme results. Works of art were destroyed, as in the case of the ancient epitaphs in which the German text was chiseled out. The authorities had not intended such excesses, and Polish art historians and preservationists immediately protested against such acts of cultural barbarism. The destroyed epitaphs are an eloquent record of a de-Germanization mania that spread over the Polish Western Territories in the postwar era. To-

---

[36] On the situation of the closed cemeteries in the Western Territories in 1960 and future plans in AAN, Urząd Rady Ministrów 2.2/47.

[37] Kalicki, "Breslau," 21.

[38] Ibid., 21.

[39] Maciej Lagiewski, *Stary Cmentarz Żydowski we Wrocławiu* [The Old Jewish Cemetery of Wrocław] (Wrocław, 1995).

day in Wrocław one is lucky to find a forgotten manhole cover with a German label or an old German inscription on a house wall, where postwar paint may be flaking off.

The government in Warsaw had to impose de-Germanization. Often the local authorities were not keen on searching for German inscriptions everywhere and thought they had more important things to do.[40] On the whole, it can be assumed that the principle of Degermanization met with the desire among the Polish settlers to forget about the expelled inhabitants. Joanna Konopińska wrote in her diary in January 1947:

> Almost in every newspaper one is reminded of the necessity to remove the German inscriptions on public buildings, on signs of the shops, everywhere you find them. This is completely appropriate, because if you stumble at every turn on inscriptions in a foreign language you are permanently reminded of the Germanness of the city which we want to see as ours, as a Polish one.[41]

Degermanization had its limits. Inscriptions, street names, or monuments could be removed. But the city as a whole, its architectural substance, was a record of its past, including its German history. Every building built before 1945 stood in the way of the cleansing of memory. Without razing Wrocław to the ground and spiriting the old city away, as the Nazis tried to do with Warsaw in 1944 and planned to do so with Moscow and Leningrad, the memory of the city could not be erased. Poland had no such plans. From the start it was clear that the German history of Wrocław, in view of its architectural records, could not be denied in total. To a certain degree it was possible to push the architectural records of the Prussian-German rule into the background while emphasizing all those buildings which could somehow be linked with a Polish or, at least not specifically German, past.

Poland did not take over Wrocław as the intact city it had been until the end of 1944. During the siege in 1945 Wrocław was badly destroyed. After three months of street fighting, artillery fire, and air raids, Wrocław was a field of rubble. In a strategically senseless way and without respect for human life and cultural values, the German defenders used every opportunity to delay the

---

[40] The removal of the inscriptions was a strenuous job which made repeated reminders necessary. In the case of the voivodship Wrocław, see AP Wr, UWW VI-749. See also the detailed govermental instructions of the mentioned secret circular letter N° 18.

[41] Konopińska, *We Wrocławiu*, 94.

city's capitulation. The defenders burned down and blew up entire streets and districts to delay the advance of the Soviet troops.[42] The residential areas in the south and in the west of the city center, where the street fighting took place, were destroyed up to 100 percent. The old town was also heavily damaged, in some areas irreparably so. The Polish authorities estimated that sixty-eight percent of greater Wrocław and about fifty percent of the old town were destroyed. When the advance guard of the Polish administration arrived in Wrocław in May 1945, the city lay in ruin and had to be rebuilt, much like the majority of the big towns in postwar Poland. Although this was a heavy burden for Poland in the Western Territories, it was also an opportunity. In fact, the assimilation of the former German territories seems to have been psychologically easier in places where major reconstruction had to take place, prior to the arrival of new occupants. The tremendous efforts made in rebuilding Gdańsk/Danzig, Szczecin/Stettin, or Wrocław/Breslau were well documented in numerous illustrated books. They presented the rebuilding efforts as an act of heroism during the pioneer era, as the beginning of a new period that had little in common with the preceding period of time.[43] Furthermore, the reconstruction of the destroyed cities provided an opportunity for dissemination of the official view of history through architecture as well.[44]

Emil Kaliski, one of the leading architects of the reconstruction of Wrocław, wrote in a programmatic essay in 1946:

> We are fully aware of the importance of the task placed by history into the hands of the Polish city planners. It is our task to give back to our fatherland the city that once was perhaps more Polish than Krakow.
>
> This problem is that of placing Wrocław in a special position, one different from the position of other destroyed Polish cities . . . .

---

[42] Karol Jonca, "Destruction of 'Breslau': The Final Struggle of Germans in Wrocław in 1945," *Polish Western Affairs* 2 (1961): 309-39.

[43] See, for instance, Ingnacy Rutkiewicz, *Wrocław: Gestern und heute: Touristische Informationen von Olgierd Czerner* (Warsaw, 1973).

[44] For the cases of Gdańsk and Poznań, see Jacek Friedrich, "Gdańsk 1945–1949: Oswajanie miejsca" [Gdańsk 1945–1949: Taming the Place], *Gdańsk pomnik historii* 2 (2001): 27–41; José M. Faraldo, "Medieval Socialist Artefacts: Architecture and Discourses of National Identity in Provincial Poland, 1945–1960," *Nationalities Papers* 29 (2001): 605–632.

In the case of Wrocław, they [the mediaeval architectural monuments] have a special, specific significance, because they are, as I would say, the Polish birth certificate of this city. The certificate has already faded, in some parts it is not legible any more, in others it has been destroyed. That's why we have to restore with the utmost caution everything that has miraculously survived up to our days in order to preserve it for the future. And we have to restore not only what has survived here in spite of the destruction, but also what the Germans did not preserve, which was buried over the centuries and in the end was hidden under the earth . . . .

The Polish Wrocław will become the antithesis of the German Wrocław. We will connect it with Warsaw, Poznań, Łódź, and Krakow in an indissoluble way.[45]

Kaliski was a staunch advocate of a radical "de-Prussifaction" of Wrocław. In his opinion the old town should be returned as much as possible to a mediaeval appearance that recalled the Polish beginnings of the city. The architectural sketch he drew up during the fifties for the symbolically important Cathedral Island is revealing of his conception. This island was the part of Wrocław which could most easily be linked to the Middle Ages and the medieval Polish State. To make this as visible as possible he suggested an archeological reconstruction of an architectural site which would correspond to the very first period of the Piasts' rule in Wrocław.[46] Kaliski's radical conceptions of "de-Prussification" and "Medievalization" did not gain general acceptance during the reconstruction of Wrocław. His basic idea, that the reconstruction should be used to emphasize the Polish-medieval origins of Wrocław, nevertheless exerted a strong influence on the architects and the institutions involved in the building.

The reconstruction of the old town of Wrocław did not proceed as smoothly as it had in Warsaw, Gdansk, or Poznań. The principles of reconstruction and preservation, the esthetic criteria and the general political setup changed repeatedly during the long process of rebuilding Wrocław. This process began immediately after the war. It remains incomplete to this day.[47] In general, an

---

[45] Emil Kaliski, "Wrocław wrócił do Polski" [Wrocław Returned to Poland], *Skarpa Warszawska* 9 (3 March 1946): 4-5.

[46] Jan Harasimowicz, "Architektura Wrocławia w rysunku i grafice" [Wrocławian Architecture in Drawings and Graphs], in *Atlas architektury Wrocławia*, ed. Jan Harasimowicz (Wrocław, 1998), 221–315.

[47] For a critical revaluation of the reconstruction and its shortcomings as a result of repeatedly changing political conceptions, see Małgorzata Olechnowicz, "Architektura na obszarze Wrocławskiego starego miasta po 1945 roku" [Architecture

attempt was made to restore all recorded medieval or early modern structures—i.e., everything that was built before Wrocław became a Prussian city—and to display these structures prominently in the cityscape. This meant that reconstruction efforts were concentrated on the old town and Cathedral Island.[48] Those parts of the city constructed during the rapid urban expansion during the nineteenth and twentieth centuries within the context of Prussian and German rule attracted far less attention. Special emphasis was placed on the reconstruction of the Gothic churches. Although nearly all of them were badly damaged during the war, the authorities spared no expense in restoring them. Seventy percent of St. John's Cathedral, the main church of the bishopric, had been destroyed. Its reconstruction at first seemed to be pointless. Nevertheless, the rebuilding began in 1946. Its first stage was brought to an end in 1951. Other churches followed with their own reconstruction efforts.

Apart from medieval churches, special emphasis was placed on the historical reconstruction of the market place (Rynek) and the medieval city hall at its center. The market place was reconstructed in the spirit of the "Polish preservationist school."[49] During the Warsaw uprising in 1944, German occupation forces intentionally annihilated Warsaw as an expression of the complete destruction of the Polish nation. The old town of Warsaw, left completely razed, was then historically reconstructed "as a sign of Poland's indestructibility"[50] In a manner parallel to the rebuilding of the old town of Warsaw, some of the ancient houses at the mar-

---

in the Area of the Wrocław Old Town in 1945] (Ph.D. diss., Dept. of Architecture, Polytechnikum Wrocław, 1997). The reconstruction of outstanding architectural monuments during the fifties was accompanied by the demolition of a huge number of restorable ruins, especially of bourgeois Wrocław, in order to recover building material for the reconstruction of Warsaw and other cities in Central Poland. See Jakub Tyszkiewicz, "Jak rozbierano Wrocław" [When Wrocław was Pulled Down], *Odra* 9 (1999): 17-21.

[48] See the detailed analysis of the leaders of the reconstruction in Marcin Bukowski, *Wrocław z lat 1945-1952: Zniszczenia i dzieło odbudowy* [Wrocław 1945-1952: Devastation and Reconstruction] (Warsaw, 1985). Also Edmund Małachowicz, *Stare Miasto we Wrocławiu: Rozwój urbanistyczno-architektoniczny, zniszczenia wojenne, odbudowa* [The Old Town of Wrocław: The Development of Architecture and Planning, Wartime Devastation, Reconstruction], 2nd ed. (Warszawa, Wrocław, 1985); *Wrocław na wyspach: Rozwoój urbanistyczny i architektoniczny* [Wrocław on the Island: Development of Architecture and Planning] (Wrocław, 1981).

[49] Konstanty Kalinowski, "Der Wiederaufbau der Altstädte in Polen in den Jahren 1945-1960," *Österreichische Zeitschrift für Kunst und Denkmalpflege* 32 (1978): 81-93.

[50] Kalinowski, "Der Wiederaufbau," 82.

ket place in Wrocław were completely rebuilt in the old style. There are various examples on the marketplace, where the reconstruction of buildings was used to correct the architectural transformation of the nineteenth and twentieth centuries. Polish architects often removed the neoclassical, historicist or modernist modifications that had taken place under the Prussian rule, in order to restore the style of an earlier time, when Wrocław had been a city under the reign of the Catholic Habsburgs. In this way the general effect of the Wrocław market after the postwar reconstruction came to resemble a rather baroque place, much more akin to the architecture of Vienna than that of Berlin.[51]

The history of one of its buildings, the so-called "House Under the Golden Crown" is particularly interesting. The original building, the first Renaissance house in Wrocław, was pulled down in 1904 in spite of the protests of art historians. On its site a modern storehouse was erected. In 1945 the building was damaged, but its skeleton survived the war in fairly good condition. In 1958-1960 the Renaissance House was reconstructed with a facade of the pseudo-historical copy of the "House Under the Golden Crown." The architects decided to reconstruct the portal by fitting it with the portal of another building. The restored house is now fifteen feet higher than the original had been. But the floors, too high for a sixteenth-century building, now align perfectly with those of the neighboring storehouse from the beginning of the twentieth century. This manner of reconstruction came to an end with the "House Under the Golden Crown." In the sixties modernist conceptions gained prominence.[52]

In addition to the reconstruction of the Gothic churches and the market place, the damaged main building of the university was reconstructed at great expense. The building had started as a Jesuit College. In 1811 it became the Prussian Friedrich-Wilhelm University. In its reconstruction the Catholic and Habsburg and not the Prussian traditions of Wrocław were emphasized. The Austrian reminders on the inside of the auditorium of the Aula Leopoldina, one of the most beautiful rooms in Wrocław, were not

---

[51] In the 1930s the German architect and historian of architecture Rudolf Stein criticized the disfiguring of the Wrocław market place during the nineteenth and twentieth centuries and proposed the preservation of an earlier architectural stage. The Polish reconstruction was largely based on his work. See Rudolf Stein, *Der große Ring zu Breslau* (Breslau, 1935).

[52] Olechnowicz, *Architektura* [Architecture], 47; see also Jan Harasimowicz, *Atlas architektury Wrocławia* [Architectural Atlas of Wrocław], 2 vols. (Wrocław, 1997-1999), 2 (1999): 43.

touched. Only the Polish eagle replaced the Habsburg double eagle in the entrance door to the Aula, as if it had been there all along. The eight portraits on the walls of the Aula included the Prussian King Frederick II, hung there after Silesia had become part of Prussia. On the 10th of October 1945 the National Council of Wrocław voivodship asked the president of the Wrocław University to replace this portrait in order to remove the last traces of Germanness. The president referred the Council to the curator of the voivodship, Jerzy Güttler.[53] He responded in December 1945 as follows:

> Considering the disgraceful role Frederick II played during the partition of Poland and his relation to the Polish nation, which was full of hatred and disdain, the curator of the voivodship thinks the removal of his portrait to be appropriate.[54]

There are few cases in Wrocław where entire buildings were pulled down because they were identified with the German rule. Interesting enough, buildings symbolic of Prussian rule were more likely to fall victim to this de-Germanization policy. Buildings representing the German nation state or even of the Third Reich were less affected. It is no coincidence that both the castle of the Prussian king, built in the middle of the nineteenth century, and the neo–classical Hatzfeld Palace, that had served as residence of the Prussian provincial governor since 1808, were not reconstructed. The Prussian castle burned down during the siege, but it could have been restored without problems. It was not until the 1960s that the Wrocław government decided to remove the ruins of the castle in spite of the opposition of the preservationists. Afterward, the authorities were not able to fill the gap left by the removal of the ruin. Today, the visual effect of the Square of Freedom—with an immense gap instead of the castle—is oppressive. The history of the Hatzfeld Palace is different. The destruction of the palace was much more extensive than that of the castle. The palace, built as an aristocratic residence in the eighteenth century, had been a jewel of architecture in Wrocław and all of Silesia. Nevertheless, the majority of the ruins were pulled down. Only the entrance hall was restored and connected to a modern building during the sixties.

Despite the Prussophobe/Germanophobe movement after 1945, a pragmatic approach to the "German" and "Prussian"

---

[53] AP Wr, UWW XVII-109, 153.

[54] AP Wr, UWW XVII-109, 156.

buildings prevailed over the ideological desire to remove them. Even in the old town there were so many of them that it would have been impossible to remove them all. Moreover, their modern construction had survived the war in a much better condition than the older buildings. And it seems that the monumental public buildings from the first half of the twentieth century especially met the representational demands of the People's Republic. Thus the massive buildings of the police headquarters, the court, or the railway management were used in the same way as before 1945. Even the purely Nazi-style building of the provincial government of Lower Silesia became the headquarters of the voivodship administration. The voivodship committee of the communist party (PZPR) took its residence in one of the few buildings in Wrocław erected during the Third Reich.

It is hard to say what would have happened to Wrocław if the hardliners of Degermanization had been alone in the field of reconstruction. There were many people who tried to subdue their ideological eagerness, a demonstration of the fact that even in the postwar era the official view of history was not accepted by everybody. Among art historians and preservationists there had been many who had played important roles in seeking less ideological approaches to the history of Wrocław.[55] As an example, the Wrocław curator Jerzy Güttler performed a great service during the postwar years by saving many architectural monuments that would have been lost in the anti-German atmosphere after 1945. Probably it would not have been possible for him to contradict the Degermanization in principle. However, he did manage to protect parts of the architectural legacy from the ideologically motivated demolition by declaring it was not really German. In August 1949 he wrote:

> Silesian architectural monuments as a rule do not have a German character, even if they were built during the German occupation. Exceptions to this are some buildings of the Protestant cult. But often buildings built in the second half of the nineteenth and in the twentieth century do have a real German character.... Looking for the traces of Germanness, the curator of the voivodship recommends focusing one's attention on objects of this type. They could serve as a reservoir of building material so far as

---

[55] Adam S. Labuda, "Das deutsche Kunsterbe in Polen: Ansichten, Gemeinplätze und Meinungen nach dem Zweiten Weltkrieg," *Forum für Kultur und Politik* 20 (1997): 5-23. Also "Kunst und Kunsthistoriographie im deutsch-polnischen Spannungsverhältnis: eine vernachlässigte Forschungsaufgabe," in *Deutsche Geschichte und Kultur im heutigen Polen: Fragen der Gegenstandsbestimmung und Methodologie* ed. Hans-Jürgen Karp (Marburg, 1997), 119-35.

practical points do not speak against their demolition. But architectural monuments must be protected . . . .[56]

This letter is most remarkable. It shows the way in which a building from the German period could be protected by the preservationists. Either they declared the architecture to be regional or European in style. This was true of all the buildings built before the middle of the nineteenth century. Or the building had to have a high practical value, such as those typically "German" or "Prussian" public buildings of the late nineteenth and twentieth centuries. In this manner Güttler, without contradicting the general goal of removing the traces of Germanness, supplied the authorities with arguments to preserve and reconstruct everything.

In this approach the Degermanization in the area of architecture in Wrocław did not necessarily mean the demolition of buildings or the alteration of their exterior. On the whole, these buildings were used either without speaking of and thinking much about their history, or their history was reinterpreted in such a way that they no longer represented the German past. The Wrocław guidebooks and city histories created a network of all sorts of Polish architectural monuments, especially from the Middle Ages, and other records of Polish history. The baroque buildings, which went back to the Catholic rule of the Habsburgs, were held in high regard as representatives of a multicultural period. Everything else that could not be presented as Polish, Austrian or at least as a Silesian regional style remained a blank space on the map. The tenement blocks and villas, the public buildings of the nineteenth and twentieth centuries, the storehouses, the industrial architecture, the important examples of a creative architectural avant-garde during the twenties that included the majority of the buildings surrounding the inhabitants—all these seemed to lack a specific history. Given these conditions, Wrocław remained an anonymous, alien place in the minds of most Polish citizens. At least within the framework of the official view of history it was hard to develop a living relationship with the city. Andrzej Zawada, a Polish philologist at the University of Wrocław, wrote in the nineties:

> Wrocław is a city whose memory has been amputated. I only became accustomed to it with difficulty because its debility troubled and tormented me at every turn. You could not walk down the streets of Wrocław without thinking about it . . . .

---

[56] AP Wr, UWW XVII-109, 156, 34.

That's why it is good to go for a walk through any of the European cities, where the past is not only not hidden but also where every tourist is taught about it. Wrocław's past was covered up, just like a family member's offensive origins in one of the so-called better families.[57]

Zawada's harsh judgment indicates how paradoxical the effects of the cleansing of memory in Wrocław have been. On the one hand, the intense efforts to create a visible Polish tradition in Wrocław helped to reduce the feeling of foreignness among the citizens. This approach corresponded to their interests. Many of the postwar settlers had lost their homeland through expulsion from the East or through wartime destruction, or they had been forced to leave their homes for economic reasons. Given no real alternative to living in Wrocław, they strongly wanted to feel as the legitimate inhabitants of Wrocław. It was this need that called for a view of history in which Wrocław was an old Polish city. After many centuries, it had returned to its "Motherland." In an era when expulsion, uprooting and homelessness were common experiences, the idea of a city coming home to its roots was consoling. It was probably this idea which made it possible for the Polish settlers to start living in a city that was completely alien and even hostile to them. Under these extreme circumstances the Degermanization or the "sweeping out the foreignness," as Joanna Konopińska put it in her diary, was necessary for survival. For settlers, who had experienced the arbitrary fate of warfare and had come involuntarily, this was the first step in making an alien place their new homeland.

Thus based upon an extremely selective memory and an ideological interpretation of the local history, the local identity had its shortcomings and limits. These appeared where the contradictions of the artificially constructed history became too obvious. The German past, the substantial part Germans played in the history of Wrocław, could be eliminated from public places. But its records kept on existing in the private sphere. In the houses and flats, the old furniture, the dishes, the books and other relics of the former occupants told a story which did not fit into the official view of history.[58] In this way the German past was much more present in everyday life and in the private memory than one would have expected, given the official position of the People's

---

[57] Andrzej Zawada, "Bresław," *Bresław* (Wrocław, 1996), 41-63, 52.

[58] See the literary treatment of this issue in the novel of Stefan Chwin, *Hanemann* (Gdańsk, 1996).

Republic. This tension between private experience and public position called for eventual clarification. It took two generations and the profound political changes at the end of the twentieth century for a view of history to be abandoned, a view that had long served as a psychological anchor after the traumatic population movements of the 40s.

From the time Germany accepted the Polish Western frontier and cleared the way for an amicable relationship between Poland and Germany, the identity of the Wrocławians underwent a revolution.[59] The fact that Wrocław had not always been a Polish city is no longer a political issue. Today the citizens of Wrocław are discovering the entire past of their hometown, including its German sections. The German history of Wrocław is no longer kept secret, but is increasingly seen as an important part of the local identity. The blank spots on the maps are being filled in very quickly. A new generation of urban historians, among whom art historians, again play a leading role and are completely reshaping the collective memory of Wrocław. The work they have achieved within only one decade is impressive.[60] The image of the city is changing rapidly for the better. Today every building is seen as an equally integral and valuable part of a city shaped by a dramatic history.

---

[59] On the processes of changing identity in Wrocław, see Mark Zybura, "Breslau and Wrocław," in *Erinnern, vergessen, verdrängen: polnische und deutsche Erfahrungen*, ed. Ewa Kobylinska and Andreas Lawaty (Wiesbaden, 1998), 369-380.

[60] For example, see Harasimowicz, *Atlas architektury*; and *Encyklopedia Wrocławia* [Encyclopedia of Wrocław] (Wrocław, 2000). Michał Kaczmarek et al. *Wrocław: Dziedzictwo wieków* [Wrocław: A Heritage of Centuries] (Wrocław, 1997). See also the numerous works on the architecture of the first half of the twentieth century in Wrocław by Janusz Dobesz, Jerzy Ilkosz, Wanda Kononowicz, Jerzy Rozpędowski and others.

Yugoslavia's First Ethnic Cleansing: The Expulsion of the Danubian Germans, 1944-1946

JOHN R. SCHINDLER

It is a truism surrounding the human catastrophe which has recently enveloped the Balkans that the wars of the 1990's are a refighting of the conflicts of the 1940's, a case of genocide redux. Historically this viewpoint has much to commend it, yet it ultimately falls short as an all-purpose mantra: in the former Yugoslavia, as elsewhere, it is sometimes easier to explain away than to actually explain. It is undeniably true that the crimes committed in Bosnia-Hercegovina and Kosovo in 1992-1995 and 1998-1999, respectively, bear strong resemblance in outline to crimes committed during the Second World War. Certainly there is a direct line from the virulent Greater Serbian ideology born during that war to the phenomenon of "ethnic cleansing" today.

That said, it is unfortunate that historians, as much as politicians, have frequently employed selective memory when dealing with the travails of ex-Yugoslavia. For although the concept of ethnic cleansing dates to the Second World War, only once in the former Yugoslavia has it been brought to its logical, if evil, conclusion. Although several Yugoslav peoples have been subjected to partial or incomplete ethnic cleansing—Serbs in Croatia during the Second World War, Muslims in Bosnia–Hercegovina and Albanians in Kosovo in the 1990s—only the ethnic Germans of Yugoslavia experienced complete destruction. From 1944 to 1946, Yugoslavia's half-million-strong German community was annihilated: more than ten percent were murdered, while the rest were displaced or expelled from their homeland. This underreported story, long a focus of particular historical amnesia among the Yugoslavs, deserves greater scholarly attention, not merely for the sake of the dead and deported, but because the theoretical and practical roots of the Balkan tragedies of the past decade, the origins of what the world has come to denounce as ethnic cleansing, lie in the terrible experience of Yugoslavia's ill-fated ethnic Germans.

Yugoslavia's Germans: Roots and Realities

The more than 500,000 ethnic Germans, known as Danubian Swabians (Donauschwaben), who found themselves in Yugoslavia[1] with the collapse of the Habsburg Monarchy in the fall of 1918 were in some ways a privileged minority. Germans, though only four percent of Yugoslavia's population, were by far the best-educated and most economically productive ethnic group in the country; indeed, they were the largest Yugoslav minority, and the fourth largest nationality after the founding Serbs, Croats, and Slovenes.[2] Although Germans could be found in every large town and city in the regions of Austria-Hungary which were assigned to Yugoslavia after World War I, the Donauschwaben were particularly heavily concentrated in what the Serbs termed Vojvodina, historically a region of south Hungary astride the Danube. German populations settled in this border region, particularly in the Hungarian counties of Torontál and Bács-Bodrog (also known as Banat and Bačka, respectively), in the eighteenth century, particularly during the reign of Empress Maria Theresia.

During that tumultuous period, German colonists also settled in easternmost Croatia, Srijem, and east Slavonia. For Vienna, German settlers served as a vital bulwark against Balkan instability as part of the famed Military Border (*Militärgrenze*). Throughout the Danubian region of south Hungary, ethnic Germans were productive from the outset in agriculture, commerce, and the civil service; their presence in the larger towns and cities—Novi Sad (Neusatz), Zemun (Semlin), Subotica (Maria Theresiopel), Osijek (Esseg)—was decisive.[3] The only other significant concentration of Germans was found in the Kočevje (Gottschee) region of Slovenia, actually the oldest German settlement in Yugoslavia, established in the fourteenth century.

The dissolution of the Habsburg Monarchy and the occupation of south Hungary by Serbian forces was doubtless an unwelcome change to most Donauschwaben. Serbian dislike of ethnic Germans, whom they termed "*švabi*," was profound and widespread; Donauschwaben were never trusted by the government in Belgrade, which during the 1920s restricted German schools and

---

[1] Officially the Kingdom of Serbs, Croats and Slovenes (*Kraljevina Srba, Hrvata i Slovenaca*—SHS) until 1929, when the country was formally retitled Yugoslavia ("the land of the South Slavs").

[2] According to Yugoslav statistics, which did not consider Slavic Muslims to be a separate ethnic group. Joseph Rothschild, *East Central Europe Between the Two World Wars* (Seattle, 1990), 202-4.

[3] For a survey of Donauschwaben across old Hungary see Friedrich Gottas, "Die Deutschen in Ungarn," in *Die Habsburgermonarchie 1848-1918: Band III/1.Teilband: Die Völker des Reiches*, ed. A. Wandruszka and P. Urbanitsch (Vienna, 1980).

cultural life. That said, the Germans' interwar political motto *staastreu und volkstreu* ("loyal to the state, loyal to one's people") exemplified the attitude of most Donauschwaben, whose economic status was largely unaffected by Belgrade's misgivings about its German minority. Germans were frankly less of a concern to the Serb-dominated state during most of the interwar period than the politically restive Croatian and even Hungarian populations of Yugoslavia. Additionally, the main German political party (*Partei der Deutschen*) worked wholly within the system and was represented in the federal parliament, winning eight seats (of 312) in the 1923 elections.[4] Until a royal dictatorship was proclaimed in 1929 and elections were suspended, Donauschwaben participated fully in Yugoslav politics.

Turmoil in Central Europe beginning in the mid-1930's took its toll on the position of Yugoslavia's Donauschwaben. To the Nazi regime in Berlin, ethnic German populations in the Danubian region, what it termed *Volksdeutsche*, existed to be exploited for the political benefit of the Third Reich. Nazi overtures to Yugoslavia's Donauschwaben met with at least some success, due mainly to rising hostility to the German minority emanating from Belgrade, yet in general Nazi propaganda made relatively little impact on the mainly Catholic ethnic Germans of Vojvodina; certainly they were less enticed by pan-Germanism and political offerings from Berlin than their ethnic kin in Hungary and Romania, whose *Volksdeutsche* populations were notably radicalized by the late 1930's. Nevertheless, Yugoslav authorities were increasingly dubious about the loyalty of the Donauschwaben, and in the three years leading to the Axis invasion of Yugoslavia, Belgrade's intelligence and security services considered ethnic Germans disloyal at best, "fifth columnists" at worst.[5]

The rapid collapse of the Yugoslav military under German and Italian blows in April 1941 was at least partially due to the unwillingness of some groups, principally Croats, to fight for a Serb-dominated dictatorship. Donauschwaben too were unenthusiastic soldiers in the Serb-run Yugoslav Army, but accounts of their disloyalty, much less treason, have been greatly exaggerated for propaganda purposes. Nevertheless, German military occupation authorities (the Banat was administered by the *Wehrmacht*,

---

[4] Rothschild, *East Central Europe*, 219.

[5] Uglješa Popović, *Deseti po redu: Tajna vojna obaveštajna služba bivše Jugoslovenske Vojske od 1938. do maja 1941. godine* [Rows of Ten: The Military Intelligence Service in the Yugoslavian Army from 1938 to May 1941] (Belgrade, 1976), 79-81.

while the Bačka was returned to Hungary) quickly pressed some Yugoslav Germans into military service.

For Heinrich Himmler in particular, the Donauschwaben offered excellent recruitment possibilities. Restricted by the *Wehrmacht* from adding to the ranks of his Armed SS (*Waffen-SS*) with Reich Germans, Himmler turned to the *Volksdeutsche* of southeastern Europe to flesh out his special corps. The *Waffen-SS* soon raised its first *Volksdeutsche* division, the 7th Volunteer Mountain Division *"Prinz Eugen,"* among the Vojvodina Germans, in conscious emulation of the German role in the former *Militärgrenze*.[6] When volunteers eventually dried up, the division was kept up to strength through conscription of Donauschwaben. *Prinz Eugen*, committed throughout the war to counterinsurgency operations inside Yugoslavia, gained a hard-won reputation for tenacity and brutality, particularly against the Communist Partisans. Donauschwaben were also conscripted into other *Wehrmacht* and paramilitary units to fight guerrillas, especially in Banat, where thousands of ethnic Germans were recruited into SS-run police regiments.[7] Due to this cynical exploitation of the Donauschwaben by Himmler and his SS, the German minority became inextricably linked in the minds of many Yugoslavs with war crimes and anti-guerrilla excesses. Retribution, when it came, would be ferocious.

Brotherhood and Unity (For Some)

The winners of the Yugoslav civil war which took place from 1941 to 1945, running simultaneously to the conflict between Axis and Allies, were the Communist Partisan forces led by Marshal Josip Broz, better known as Tito. The main reason for Communist victory was not military proficiency but rather political effectiveness: by appealing to all, or nearly all, of Yugoslavia's many peoples, the Communists were guaranteed a broader base than any of their rivals; certainly the Partisans possessed more widespread appeal than either Croatian or Serbian ultranationalists, Tito's main opponents. The official Communist policy of "brotherhood and

[6] Otto Kumm, *"Vorwärts Prinz Eugen!": Geschichte der 7. SS-Freiwilligen-Division "Prinz Eugen"* (Osnabrück, 1978), 12-22. The naming of the division after the most esteemed of Habsburg field marshals, Prince Eugene of Savoy, liberator of Belgrade from the Turks, was another conscious emulation of the Danube Swabian military tradition.

[7] See the brief but detailed work by Nigel Thomas and Krunoslav Mikulan, *Axis Forces in Yugoslavia 1941-45* (London, 1995).

unity" (*bratstvo i jedinstvo*), stressing the equality of Yugoslavia's nationalities, went far in securing the support of many who were otherwise unsympathetic to the Yugoslav variant of Bolshevism.

Yet there was a significant exception to the policy of "brotherhood and unity." Donauschwaben were considered by the Communists to be implacable foes of Yugoslavia, traitors and collaborators without exception; Tito and his Partisans offered ethnic Germans no concessions at any time. In this, the Donauschwaben were unique. While the Communists offered political concessions to other "enemy peoples," i.e., Hungarians and Italians, for Germans there was nothing but contempt and hatred. Indeed, Tito's Partisan army raised special ethnic units among the Hungarians of Vojvodina (principally the Sándor Petőfi Brigade) and several Italian brigades named after Garibaldi; though these were more propaganda exercises than viable combat units, they indicated the desire of the Communists to co-opt all Yugoslavia's minorities save the Donauschwaben. Even multiethnic Partisan units raised in Vojvodina were careful to exclude ethnic Germans.[8]

Communist intentions towards the Donauschwaben were clarified by a legal act accepted by the Anti-Fascist Council of the People's Liberation of Yugoslavia (AVNOJ, Tito's pseudo-parliament) on 21 November 1944. This momentous act declared that all ethnic Germans were enemies of the state, and would be deprived of Yugoslav citizenship, civil rights, and property; the principle of *Volksdeutche* collective guilt thus became official Communist policy. Although the Partisans heartily embraced collective guilt against "class enemies"—as defined rather loosely by the party—against no other ethnic group did Tito's forces apply such an inflexible standard.[9] The stage was set for the destruction of the German minority, a process already underway when AVNOJ announced its proclamation.

Communist dictates were not the sole force behind the crimes committed against the Donauschwaben. Hatred of the German minority, always notable among Serbs, grew precipitously during

---

[8] For instance, the 51ˢᵗ Vojvodina Division of Tito's army included an all-Slovak brigade. See Sveta Savić, *51. vojvodjanska divizija* [The 51ˢᵗ Voivodina Division] (Belgrade, 1974), 9-22. Tito's Partisans raised a "Free Austrian Brigade"—really hardly a battalion—for propaganda purposes, on Moscow's instructions, but this was explicitly an effort to bolster Communism in Austria and was never a "German" unit: its ranks were filled with Austrian Communists, not Donauschwaben.

[9] See Vladimir Geiger, "Germans in Ex-Yugoslavia," in *Southeastern Europe 1918-1995*, ed. A. Ravlić (Zagreb, 1998).

the war. Among the Serbian nationalist guerrillas known as Čet-niks, who were the Communists' main rivals, Germanophobia reached new and dangerous heights once Yugoslavia fell under Axis occupation. The exterminationist intent of Četnik ideology evolved during the war under the guiding hand of Stevan Mol-jević, a virulently nationalist Bosnian Serb lawyer who served as the top ideologist of the movement. Četnik plans for Yugoslavia after their hoped-for victory, envisioned in the context of an Allied triumph which would bring the deposed Karadjordjević monar-chy back to the throne in Belgrade, called for a Greater Serbia, an ethnically pure state encompassing not only Serbia (including Vojvodina and Kosovo) but also Montenegro, Macedonia, Bosnia-Hercegovina, most of Croatia, and even neighboring parts of Hungary and Romania. Moljević's concept, which he termed "Homogenous Serbia," was formalized as early as June 1941, and a very similar proposal to create an "ethnically pure" Serbia was delivered to the Yugoslav government in exile in London that summer. Achieving a monoethnic Serbia would mean expelling 2.675 million non-Serbs, including the entire half-million-strong German minority, which Moljević explained forthrightly would be accomplished by what he termed "cleansing."[10]

What would later be called ethnic cleansing was endorsed as the official Četnik plan for postwar Yugoslavia at the January 1944 congress held at Ba in western Serbia. At the Ba con-gress—significantly convened on 27 January, St. Sava day, honor-ing the patron saint of Serbia and Serbian nationalism—Četnik representatives accepted a version of Moljević's outline. A Ger-man agent at Ba reported that Draža Mihajlović, the Četnik leader, affirmed that for every Serb killed by German occupiers, a hun-dred Germans would be killed in retribution.[11]

Fortunately for the Donauschwaben and all Yugoslavia's non-Serbs, the nationalist Četniks were defeated by Tito's forces, de-ciding the fate of the Yugoslav civil war. However, for the coun-try's doomed Germans the ultimate outcome was similar, if not identical. For although the Partisans and Četniks were implacable foes, their views on the Donauschwaben were quite similar: these

---

[10] Jozo Tomasevich, *The Chetniks: War and Revolution in Yugoslavia, 1941-1945* (Stanford, CA, 1975), 166-70, 185-92; on the "Homogenous Serbia" plan, refer to Bože Čović, "Homogena Srbija" [Homogenous Serbia], in *Izvori velikosrpske agresije: Rasprave, dokumenti, kartografski prikazi* [The Sources of Great Serbian Aggression: Debates, Documents, Cartographical Evidence], ed. B. Čović, (Zagreb, 1991).

[11] Tomasevich, 399-406.

traitors and collaborators were to be punished brutally and collectively, without regard to age or status, much less personal guilt or innocence. Unfortunately for the German minority, the Serbs who dominated Tito's secret police and special security forces were no more prone to leniency towards Donauschwaben than their countrymen in Četnik ranks were. Therefore when the *Wehrmacht* began to retreat northwards from the Balkans in October 1944, two-thirds of the 510,000 Donauschwaben in Yugoslavia were forcibly evacuated by German military authorities. Ethnic Germans in Srijem and eastern Slavonia were subject to evacuation in the first week of October. By the third week of that month the process had been extended to all of Vojvodina. In this harrowing evacuation, extended columns of women, children, and the elderly—most of the men from teenage years through very late middle age were in German uniform—retreated north towards the Reich, mixed with haggard *Wehrmacht* units, subjected to frequent Allied air attack. Thousands did not survive this tragic exodus. Yet for the Donauschwaben left behind, the future was grimmer still.

Mechanisms of Destruction

The fate of the Donauschwaben was sealed with the collapse of German power in the Balkans in the late summer of 1944. With the Soviet breakthrough into Romania and the southern Balkans in August, Army Group E was forced into an epic retreat; by November the front had stabilized, and would remain stable until the spring of 1945, but what Yugoslavia's Germans would remember as the "bloody autumn" of 1944 found Vojvodina—home to most of Yugoslavia's German minority—under Communist occupation.[12] With help from the Red Army, Tito's forces liberated Belgrade by the third week of October and quickly crossed the Danube into Vojvodina; by the end of November, the Yugoslav 3rd Army under Lieutenant-General Kosta Nadj, the spearhead of Tito's over half-million-strong armed forces, had cleared *Wehrmacht* troops, and with them the ethnic German population, from nearly all of Vojvodina.[13]

---

[12] For the story of Army Group E's retreat, see Erich Schmidt-Richberg, *Der Endkampf auf dem Balkan: Die Operationen der Heeregruppe E von Griechenland bis zu den Alpen* (Heidelberg, 1955), 73-6, 91-3.

[13] Mladenko Čolić, *Pregled operacija na Jugoslovenskom ratšitu 1941-1945* [Summary of Operations in the Yugoslavian Theater of War 1941–1945] (Belgrade, 1988), 240-52, 289-97.

Although Tito's military was responsible for some atrocities against Donauschwaben, the main force behind the ethnic cleansing of Yugoslavia's Germans was the feared Yugoslav secret police, the Department for the Protection of the People, known as OZNa.[14] Founded in May 1944 as "the sword of the party," and modeled closely on the Soviet secret police (NKVD) which had long employed Tito himself, OZNa was a particularly Serbian preserve, unlike the multinational Yugoslav Army.[15] Indeed, the secret police chief, Aleksandar Ranković (who would remain in his post until 1966, serving as the Lavrenti Beria of Yugoslavia), was the top Serb in the Yugoslav Communist hierarchy, and chose his countrymen for most leading positions.[16] Together with his assistant and factotum, Svetislav Stefanović, another Serb, Ranković oversaw the destruction of the hated "*švabi,*" acting as much in Serbia's interests as in the party's.

The *corps d'élite* of the secret police was a newly raised paramilitary body, the People's Defense Corps (KNOJ), established in mid-August with the mission of destroying "fifth columnists," "quislings," and "subversives."[17] Formed into well armed detachments organized and equipped as conventional battalions and brigades, like the paramilitary forces of the NKVD on which it was modeled, KNOJ rapidly rose to a complement of nine divisions—almost one-sixth of Tito's armed forces and, relative to overall strength, a larger force than Soviet secret police field units. KNOJ units were especially charged with politically sensitive actions against state enemies (including operations against the last holdouts among the Četniks), but became even better known for attacks against civilians across Yugoslavia. Beginning in Vojvodina, KNOJ units practiced what was euphemistically termed

---

[14] *Odeljenje zaštite naroda;* among Yugoslavs, the mordant if accurate rhyme "*OZNa sve dozna*" (OZNa knows everything) quickly gained currency.

[15] On OZNa's establishment, mandate, and provenance, see Gojko Miljanić, *Rukovodjenje i komandovanje u oslobodilačkom ratu 1941-1945* [Leadership and Command Structure During the War of Liberation, 1941–1945] (Belgrade, 1990), 119-27; Milovan Dželebdžić, *Obaveštajna služba u narodnooslobodilačkom ratu 1941-1945* [Military Intelligence During the War of National Liberation, 1941–1945] (Belgrade, 1987), 13-40, 266-9.

[16] See Jovan Kesar and Pero Simić, *LEKA: Aleksandar Ranković* (Belgrade, 1990) for the full story. Ranković was nicknamed "Leka."

[17] *Korpus narodne odbrane Jugoslavije* [The People's Defense Corps of Yugoslavia]. See Dželebdžić, *Obaveštajna služba,* 208.

čišćenje terena ("cleansing the terrain"), in other words the destruction and expulsion of anti-state populations, as defined by the party and the secret police.[18]

KNOJ actions against the Donauschwaben remain obscure, like all aspects of Yugoslavia's crimes against the German minority, but the outline is clear enough. The terror inflicted against the Donauschwaben beginning in November 1944 was a model for later KNOJ "cleansing" operations against state enemies, including against Kosovar Albanians in the last months of the war, actions which were as vicious, though on a much reduced scale, as any Communist crimes against Germans.[19] The secret police's "final solution" to Yugoslavia's *Volksdeutsche* problem commenced with roundups and mass shootings of "fifth columnists" in heavily German areas of Vojvodina, including Pančevo (Pantschowa), Vršac (Werschetz), Veliki Bečkerek (Gross–Betschkerek), and Veliki Kikinda (Gross–Kikinda). Probably 10,000 Donauschwaben were shot by OZNa in the four months that followed the arrival of Communist forces. This methodically run operation, supposedly termed *Akcija Intelligencija*, succeeded in demolishing ethnic German civil society in Vojvodina.[20]

Of the approximately 200,000 Germans remaining in Yugoslavia after the *Wehrmacht*'s fall 1944 retreat, the strong majority of them women, children, and the elderly, between 60,000 and 70,000 noncombatants died at the hands of OZNa and other Communist authorities (not counting some 27,000 Donauschwaben who died serving in the German military). While at least 12,000 Germans were dispatched to slave labor in the Soviet Union in December

---

[18] KNOJ rounded up Četnik bitter-enders well into 1946; Draža Mihajlović was caught and executed after a perfunctory trial, while Stevan Moljević was apprehended and imprisoned by the Communist secret police.

[19] Yugoslav authors are now willing to admit to OZNa/KNOJ offensives, including "cleansing" operations, against ethnic Albanians in the last months of the war: see Marko Lopušina, *Ubij bližnjeg svog: Jugoslovenska tajna policija 1945-1995* [Kill Thy Neighbor: The Yugoslav Secret Police 1945–1995], 3 vols. (Belgrade, 1996–1998), 1: 147-63. Though Lopušina is very detailed about Yugoslav secret police actions (often in a disturbingly approving manner), like nearly all Belgrade sources he is almost completely silent about the destruction of the Donauschwaben.

[20] The term *Akcija Intelligencija* is widely cited in Donauschwaben sources, though documentary evidence is lacking; for example Hans Sonnenleitner, *Donauschwäbische Todesnot unter dem Tito-Stern: Tatsachenbericht einer Familie* (Munich, 1990), 91. While the name may be incorrect, that a systematic OZNa plan existed is beyond doubt, as the destruction of the *Volksdeutsche* was anything but random.

1944, of whom twenty percent died, a far larger number were sent to Tito's own labor camps, modeled on Stalin's. As a result, the entire German population remaining in Yugoslavia was shot outright or sent to highly lethal slave labor internment; by early 1946 the Donauschwaben were no more.[21]

Civilians of all ages were committed to slave labor in OZNa-run camps as soon as Communist authorities arrived in Vojvodina. Of the ten camps where Donauschwaben were sent in considerable numbers, six were in Vojvodina (Molin and Kničanin in the Banat; Bački Jarak [Jarek], Gakovo [Gakova], and Kruševlje [Kruschiwl] in Bačka; Sremska Mitrovica [Mitrowitz] in Srijem), while two were in heavily German eastern Slavonia (Krndija and Valpovo), and two smaller camps were located in Slovenia. Due to abysmally poor conditions in the camps, where overcrowding and underfeeding were ubiquitous, as were brutal work details, the death rate among the internees, many of them children and the elderly, was exceptionally high. Typhus in particular decimated the ranks of slave laborers. Exact loss rates are difficult to determine, but based on available statistics, among the over 150,000 Donauschwaben imprisoned by OZNa circa 1944-1948, some 60,000 died.[22]

As a result of Titoist policies, Yugoslavia's German minority ceased to exist. The roughly 90,000 Donauschwaben who survived slave labor, stateless and unwanted, were deported, mainly to Germany and Austria, in the early 1950s. Although individuals may have remained, they had no legal status; German was no longer a recognized nationality in Yugoslavia, even though minorities as small as the Vlach, Czech or Ukrainian (each a mere .01 percent of the country's population) continued to receive official recognition and rights. Indeed, Belgrade's nationalities policy, particularly towards small ethnic groups, was unquestionably the most lenient in the Communist world.[23] Yet Germans received no such largesse, in 1944 or after.

All Yugoslavia's peoples suffered in the aftermath of Communist victory, when Tito employed his own Red Terror to accomplish in only three years the *Gleichschaltung* which it took Sta-

---

[21] Ingomar Senz, *Die Donauschwaben* (Munich, 1994), 126-32.

[22] The most comprehensive treatment is Georg Wildmann et.al., *Verbrechen an den Deutschen in Jugoslawien 1944-1948: Die Station eines Völkermords* (Munich, 1998).

[23] Sabrina P. Ramet, *Nationalism and Federalism in Yugoslavia, 1962-1991* (Bloomington, Ind., 1992), 19-22.

lin a generation to achieve in the Soviet Union; Croatian martyrdom, in what is remembered as "the Way of the Cross,"was profound, and tens of thousands of innocent Serbs died as well. Some quarter-million Yugoslavs were murdered by OZNa and KNOJ from 1944 to 1948 (from an overall Yugoslav population of fifteen million, which had already suffered something like a million dead during the Second World War)[24], either through executions or slave labor, but no group suffered more than the Donauschwaben.[25] Certainly no other ethnic group was singled out for collective punishment and annihilation. Tellingly, although Yugoslavia's other "enemy peoples"—Hungarians and Italians—also endured the depradations of the secret police, and though the ancient Italian population in Istria was mostly expelled in the *"foibe"* of 1945, both groups retained their official recognition as protected nationalities down to the collapse of Yugoslavia in 1991 (and Vojvodina Hungarians, over 400,000 strong, accounted for almost two percent of Yugoslavia's population)[26]

The impact on Yugoslavia of destroying the country's best-educated and most productive minority was baleful. The hunger which remained pervasive throughout Yugoslavia until the early 1950's (and was only remedied by large-scale Western food and economic aid, encouraged by Tito's unplanned break with Stalin in mid-1948) was exacerbated by the thousands of Donauschwaben fields which lay fallow in Vojvodina, the region which before World War II had been Yugoslavia's breadbasket. Gradually farms and towns in Vojvodina and eastern Slavonia which were ethnically cleansed in late 1944 were resettled by Serbs and Montenegrins, colonists known as *došljaci*. Under this Communist land reform program, by the early 1950s, recently German regions were reoccupied by some 200,000 newcomers from the poorer and more war-ravaged regions of Serbia, Montenegro, and Bosnia-

---

[24] Full and fair analysis of the vexing issue of Yugoslavia's population losses 1941-1945 can be found in Vladimir Zerjavić, *Gubici stanovništva Jugoslavije u drugom svjetskom ratu* [Population Losses in Yugoslavia During the Second World War] (Zagreb, 1989).

[25] For an English-language account of this little-known story, see Borivoje A. Karapandzich, *The Bloodiest Yugoslav Spring 1945—Tito's Katyn and Gulags* (New York, 1980); n. b.—this work typically fails to account for the destruction of the Donauschwaben.

[26] Ramet, *Nationalism*, 20-2, 55.

Hercegovina.[27] Vojvodina's population, deprived of its Germans and filled with Serbs, thus resembled much more that of Serbia, of which Vojvodina remained a province (as it still is). In a bitter irony, Jovica Stanišić, the chief of Serbia's secret police, the feared RDB, from 1991 until 1998, the architect of the next round of ethnic cleansing in Yugoslavia, hailed from the once heavily German town of Bačka Palanka (Deutsch–Palanka) on the Danube. His parents were *došljači* (converts), and Stanišić, like many Serbs, grew up in a town, indeed in a house, built by ethnically cleansed Donauschwaben.[28]

Conclusions

The destruction of the Donauschwaben remains one of the most underreported stories in twentieth century Balkan history, a bout of extended historiographical amnesia which is only beginning to lift, and slowly; most Yugoslav accounts still relegate the expulsion of the country's largest ethnic minority to a non-event, or even just punishment for collaboration.[29] Yugoslav crimes against ethnic Germans at the end of the Second World War do not exist in isolation: they were part of a concerted Communist campaign across Soviet-dominated Europe to dispense with German minorities throughout the region. In Yugoslavia's case, the wholesale destruction of the Donauschwaben was merely one component, though the most violent and comprehensive one, of a Communist effort to settle scores with enemies, real and imagined.

For decades, Western silence regarding Titoist crimes against noncombatants in the mid-1940's has been pervasive and almost impenetrable. NATO countries, eager to bolster Yugoslavia, from 1948 a useful anti-Soviet bulwark in Europe, were mostly unwilling to examine closely any of the unpleasantness of the Tito regime; certainly the fate of a half-million Germans inspired neither much interest nor much sympathy in the West. In this avoiding of awkward questions the ranks of Yugoslav experts in the West

---

[27] Nikola Gačesa, *Agrarna reforma i kolonizacija u Jugoslaviji 1945-1948* [Agrarian Reform and Colonization in Yugoslavia, 1945–1948] (Novi Sad, 1984), 75-9, 347-8.

[28] *Intervju* (Belgrade), 9.5.1997; *Naša Borba* (Belgrade), 22.11.1997.

[29] One recent and otherwise commendable historical survey refers ambivalently to the fact that Donauschwaben "fled or had to leave Yugoslavia and Croatia" without mentioning massacres or slave labor. See Ivo Goldstein, *Croatia: A History* (London, 1999), 158. This is still an improvement over most Yugoslav accounts, which traditionally omit the fate of the Germans entirely.

have been regrettably complicit almost without exception. All that remains of the once-vibrant German community in Yugoslavia are the memories of survivors, now very elderly, and touching books commemorating a long and presumably irrevocably lost homeland.[30]

Yet the fate of the Donauschwaben deserves greater scholarly attention not merely because it served as the first act of totalitarianism in Yugoslavia—for the destruction of the German minority was among the Communist regime's most important initial acts, a model of much unpleasantness to come—but also because the "ethnic cleansing" visited on the Germans, an evil blend of Serbian xenophobia and Communist secret police methods, presaged the Balkan horrors of the 1990's. It is no exaggeration that the Greater Serbian ideology and secret police techniques which formed the backbone of ethnic cleansing against Bosnian Muslims and Kosovar Albanians have their roots, theoretical and practical, in the Communist campaign against all Danube Swabians at the end of the Second World War. In the Danubian city of Vukovar, cleansed of its Germans in late 1944, and cleansed of its Croats in late 1991, both times by Serb-dominated Yugoslav military and police forces, there exists a direct line from the initial Yugoslav ethnic cleansing to the next. It is thanks only to NATO intervention that Belgrade was less successful in its latter-day crusades against Bosnians and Kosovars than it was five decades earlier against Yugoslavia's ethnic Germans.

Much remains unknown about the destruction of the Donauschwaben. More hard facts are needed, yet they remain elusive by design. There can be no doubt that OZNa produced vast amounts of paperwork concerning its activities in Vojvodina circa 1944-1946, records which surely survive in the archives of the Interior Ministries of Yugoslavia and the Republic of Serbia, located in Belgrade. However, Interior Ministry archives, including as they do so many of Titoism's dark secrets, remain closed to outsiders, especially Western researchers. Despite the recent fall of the corrupt hypernationalist regime of Slobodan Milošević, the still depressingly nationalist government of Vojislav Koštunica seems ill-inclined to pursue serious examination of Serbian crimes in the 1990s, much less those perpetrated a half-century earlier. Therefore there appears scant opportunity for Western researchers to learn the full facts of the crimes committed against Yugoslavia's

---

[30] See Andreas Rödler, *Franzfeld 1792-1945: Bilder einer donauschwäbischen Großgemeinde im Banat* (Reutlingen, 1985), which is typical of books produced by survivors of the humanitarian disaster of 1944-1946 to commemorate home villages.

doomed German minority. Yet the moral mandate is clear, and at last there are some historians in the Balkans, notably in Croatia, who are willing to ask painful questions. Historians in the West must do the same; this article is at least a start.

The Expulsion of the Germans from Hungary after World War II

JÁNOS ANGI

Hungary, the Hungarians, and the various nationalities living together with the Hungarians in the Carpathian basin could not avoid becoming subject to genocide, deportation, violence, and humiliation during and after the Second World War. Among a series of tragedies the calamity of the Hungarian-Germans was one of the most shocking. The majority of Hungarians did not lay blame on their German minorities for the disaster of the war, nor did they take revenge against them as it happened in other East Central European countries. Yet Hungarian political leadership still exercised collective punishment against them for the crimes committed by their alleged *Vaterland,* the Third Reich. Though the Germans within the borders of a much reduced Hungary, sanctioned by the Treaty of Trianon of 1920, avoided persecution on the level of their cousins, the *Donauschwaben* of Yugoslavia, this fact cannot exempt the Hungarian political leaders from blame. They applied collective punishment against a national minority that had coexisted in harmony with the Hungarian majority for centuries, and many of them had lived in the Carpathian basin for a longer period of time than some of the other ethnic groups who, in the course of the centuries, became assimilated with the Hungarians. These included the Turkic Cumans, various Slavic nationalities, and many Vlachs—later called Romanians—of Transylvania.

German settlers had reached the Carpathian Basin and found new homes in the Kingdom of Hungary in two large waves. The first of these migration waves took place in the twelfth and the thirteenth centuries, when the migrant Germans settled both in Upper Hungary in the region of *Szepesség* or *Zipserland,* as well as in southeastern Transylvania. Although these Germans came from various regions of the Holy Roman Empire, the Hungarians called them "Saxons"—a term that these German settlers themselves came to adopt as their own. "Saxon" became a legal category in Hungarian constitutional developments.

These German settlers established walled and fortified towns in Hungary's northern and eastern regions, where they received special privileges and immunities under the so-called "Saxon law." The Transylvanian Saxons, known as *hospes* [guests] gained their legal charter, the so-called *Andreanum,* from King Andrew II (r.

1205-1235) in 1224. The rights and privileges granted in this document were extended to all German-speaking inhabitants of Transylvania in 1486. The Germans of the *Szepesség/Zipserland* were granted a similar charter by king Stephen V (r. 1270-1272) in 1271. In the sixteenth century, in consequence of the Protestant Reformation, most of the Hungarian "Saxons" became Lutherans, or Evangelicals as they are known in Hungary.

The expulsion of the Turks in the late seventeenth and early eighteenth century was soon followed by a second wave of migration, when the Habsburg rulers of Hungary made it their policy to repopulate the reconquered and depopulated Hungarian territories of central and south-central Hungary. Although the ranks of the new settlers included Slovaks, Serbians, and even Frenchmen, the majority of them were Catholic Germans from the Rhine region, who in Hungary came to be known as "Swabians." They settled primarily in the country's four regions: 1) the central plains, 2) southeastern Trans-Danubia, 3) Szatmár County in eastern Hungary, and 4) the southern borderlands, known as Bácska [Bachka] and Bánát [Banat], and subsequently as Délvidék [Southern Lands].

To counteract the appeal to Protestantism, the Viennese court wanted to settle only Catholic Germans in Hungary. Accordingly they selected the "Swabians" from the upper Rhine region from the area situated between Lake Boden and the upper Danube. This region had suffered greatly during the Thirty Years war, and the impact of that calamity was still being felt a century later. Therefore, its population was eager to move to new territories with more promise of a better life.[1]

Once settled, the Germans of both of these major migration waves were generally better off than the local Hungarian population. There were several interrelated reasons for this: 1) the town-dwellers among the settlers were on a higher level of urbanization than their Hungarian counterparts, and they were also more receptive to modernization and innovations; 2) the agriculturalists among them brought with them a higher level of material culture and more advanced agricultural technology and

---

[1] On the general aspects of Germans in Hungary see Béla Bellér, *A magyarországi németek rövid története*, [A Short History of Germans in Hungary] (Budapest, 1981); Claus Jürgen Hutterer, "Die Deutsche Volksgruppe in Ungarn," in *Beiträge zur Volkskunde der Ungarndeutschen*, ed. Iván Balassa (Budapest, 1975), 1133; and István Szabó's still guiding analysis: *A magyarság életrajza* [Biography of the Hungarian People] (Budapest, 1941), 42-43, 145-154. In English see G. C. Paikert, *The Danube Swabians* (The Hague, 1967). On the southern settlements see, Sue Clarkson, *The History of German Settlements in Southern Hungary*, manuscript.

better methods of cultivation; 3) they all enjoyed privileges and immunities—such as temporary exemption from taxation, followed by favorable tax regulations, living under their own laws, not being subject to county administration, etc.—that were not available to the native Hungarian population; 4) the settlers followed the principle of primogeniture, which left all the property in one piece in the hands of the first-born son, while Hungarian practice favored the division of the property among all children; 5) the Germans also appeared to be more diligent and more productive than the country's other ethnic groups, including the Hungarians. While this generally accepted view cannot be verified by hard evidence, there is no question that the Germans were more successful than their neighbors, which only increased their confidence and their drive to succeed.

By the early twentieth century the Germans in the Kingdom of Hungary numbered close to two million, or slightly over ten percent of the population. This figure, however, also included not only the "Saxons" and the "Swabians," but also those Austrians who had settled on Hungary's western frontierlands, adjacent to Austria. Although bilingual and bicultural, most of these German still retained much of their German identity.

Table

Nationalities in Hungary (1880–1910)[2]

Nationality	1890		1900		1910	
	number of people	%	number of people	%	number of people	%
Hungarian	6 404 070	46.6	8 651 520	51.4	9 944 627	54.5
German	1 870 772	13.6	1 999 060	11.9	1 903 357	10.4
Slovak	1 885 451	13.5	2 002 165	11.9	1 946 357	10.7
Romanian	2 403 041	17.5	2 798 559	16.6	2 948 186	16.1
Ruthen	353 229	2.6	424 774	2.5	464 270	2.5
Croatian	639 986*	4.6	196 781	1.2	198 700	1.1
Serb			520 440	3.1	545 833	3.0
Other	223 054	1.6	244 956	1.4	313 203	1.7
Total	13 749 603	100	16 838 255	100	18 264 533	100

* Croatian and Serb people were counted together in 1890.

[2] This table is from the study by Péter Hanák, "Magyarország társadalma a századforduló idején" [Hungarian Society at the Turn of the Century], in *Magyarország története, 1890-1918* [History of Hungary, 1890–1918], ed. Péter Hanák, 2d ed. (Budapest, 1975), 414.

Following World War I and the dismemberment of Austria-Hungary (1919-1920), there remained approximately half a million Germans in much reduced post-Trianon Hungary.[3] They comprised the only substantial ethnic minority (c. 6 to 7 percent) in a country of eight million inhabitants, The proportion of all other minorities together did not reach 2 percent.

The position of these Germans changed significantly after the *Anschluss* (1938), when Hitler annexed Austria, and for a while Hungary became the immediate neighbor of the much-enlarged and aggressive Third Reich. In 1940, when Hungary joined the Tripartite Pact, the Hungarian-Germans were given the choice to join either the Hungarian or the German armed forces. The majority of the conscripts felt sufficiently Hungarian that they chose the Hungarian Army. A significant minority, however, joined the Waffen SS, which placed them under direct German command. Although Hungary joined the war on Germany's side, after 1941 German-Hungarian relations became more and more strained. Hitler became increasingly distrustful of Regent Miklós Horthy's political sympathies and leadership in the war. This strained relationship eventually led to Hungary's German occupation (19 March 1944), and then Regent Horthy's unsuccessful attempt to leave the war altogether (15 October 1944).

The Soviet Red Army had reached Hungary's borders following the Romanian *volte-face*. Among the tens of thousands of Hungarians who fled the country with German troops, there were also some 60,000 to 70,000 Hungarian-Germans.[4] Having been classified as Germans, they were most concerned about Soviet retaliation.[5] These fears were well justified in their case, for the victorious Allies agreed to accept the principle of collective responsibility, which was incorporated into the Potsdam Agreement in 1945:

> The Allied forces have occupied the territory of Germany, and the German people will have to suffer for the crimes committed

---

[3] Concerning the Peace Treaty of Trianon the most important English–language reference is still *Total War and Peacemaking: A Case Study on Trianon*, ed. Béla K. Király, Peter Pastor, and Ivan Sanders (New York, 1982).

[4] Péter Hanák, "A magyar társadalom a két világháború között" [Hungarian Society Between the Two World Wars], in *Magyarország története, 1918-1919, 1919-1945* [History of Hungary, 1918–1945], ed. György Ránki, 3d ed. (Budapest, 1984), 765.

[5] Lóránt Tilkovszky, *Ez volt a Volksbund: A német népcsoportpolitika és Magyarország* [This was the Volksbund: The Politics of the German Ethnic Minority and Hungary] (Budapest, 1978), 353.

under the leadership of those persons whom they openly adored and whom they blindly obeyed.[6]

The Potsdam Agreement permitted Poland and Czechoslovakia—victims of Nazi Germany—to punish and to expel their German minority population. The existence of the latter within those states had been used by Hitler as a justification for the German attacks against them and for their dismemberment. Curiously, the Potsdam Decrees were also applied to Hungary, even though it had been an ally of Nazi Germany. To quote from the relevant documents:

> The conference has made the following resolutions concerning the resettlement of Germans from Poland, Czechoslovakia and Hungary: The three governments,[7] considering all aspects of the problem, concede that there is a need to take measures in order to resettle the German populace remaining in Poland, Czechoslovakia and Hungary. They agree that whatever the method of relocation, it should be conducted in an organized and humane manner... this problem should first be considered by the Allied Control Commission in Germany, which should also handle the equal distribution of Germans among the occupation zones.[8]

The Allied Control Commission set the number of Germans to be expelled at 6.5 million, half a million of whom were to come from Hungary. The Hungarian-Germans to be expelled were assigned to the American Zone.[9] Though the text of the decree only made *allowance* for the resettlement of German minorities, the Hungarian government—eager to appease the victors—interpreted it as a *request and command*. Thus, this statement was

---

[6] The text of the Potsdam Agreement in Hungarian was published by Dénes Halmosy, *Nemzetközi szerződések, 1918-1945* [International Agreements, 1918-1945] (Budapest, 1983), 649.

[7] The British, American, and Soviet governments.

[8] Halmosy, *Nemzetközi szerzdések*, 661-662. The problem is discussed with respect to the three countries by Andrea R. Süle, "A Közép- és Kelet-Európai német kisebbségek kitelepítése a második világháború után" [The Expulsion of the German Ethnic Groups of Central and Eastern Europe after World War II], in *Medvetánc*, vol. 88/489/1. 107-130. [with some archive photos].

[9] Süle, "Német kisebbségek," 113.

incorporated into the Hungarian expatriating order.[10] In consequence of this, an ethnic minority whose ancestors had lived in Hungary long before the ancestors of many other assimilated nationalities were expelled from their homeland. They also lost their Hungarian citizenship.[11]

In theory at least, the Hungarian political parties were against collective punishment, since the same principle could also be applied against the Hungarians by the Romanians in Transylvania, by the Slovaks in Upper Hungary, or by the Serbs in Bácska/Bachka, i.e., the upper half of Voivodina. Yet, it was only the Independent Smallholders' Party that resolutely opposed German expulsion. As the sole legally operating right-of-center party, it could count on the support of Hungary's German minority. The Communist Party, the Social Democratic Party, and even the Smallholders Party welcomed deportations, since they were bound to weaken the electoral base of their chief rival, which had received more than 50 percent of the votes at the 1945 elections. The expellees would also lose their land holdings, which could then be distributed among Hungarian peasants.[12]

During these expulsions Hungary was under Soviet military occupation, and until the Paris Peace Treaty of 1947 all *de facto* executive powers were in the hands of the Allied Control Commission. It was chaired by Marshal Voroshilov, who was in charge of the Soviet occupying forces in Hungary. Thus, even though the country had a Smallholder prime minister in the person of Zoltán Tildy, he was in no position to resist the will of the Hungarian Communists and the Soviet occupiers. The position of the minister of interior—by Soviet instructions—was the Communist László Rajk. It was Rajk who was responsible for organizing the deportation of the Hungarian-Germans.[13]

---

[10] The decree issued on 22 December 1945, no. 12.330/1945. M. E., was entitled: "A magyarországi német lakosságnak Németországba való kitelepítéséről" [About the Resettlement of the German Minorities of Hungary in Germany], in *Magyar Közlöny* [Hungarian Gazette], 29 December 1945. See Appendix no. 4, below.

[11] See Appendix 5, below.

[12] See Appendices 1, 2, and 3.

[13] For a summary of the deportations, see Sándor Balogh, "A német nemzetiségű lakosság kitelepítése Magyarországról a második világháború után" [The Expulsion of the German Ethnic Group from Hungary after World War II], in *Információs Szemle* [Informational Review], no. 4 (1981), 96-129 (in 1981, this periodical was available only to the Hungarian Communist apparatchiki). See also György Zielbauer, *Adatok és tények a magyarországi németség történetéből, 1945-1949* [Data and Facts on History of Germans of Hungary, 1945-1949] (Budapest, 1989); and István Fehér, *A magyarországi németek kitelepítése, 1945-1950* [The Expulsion of

The Deportation Decree used the census of 1941 as the basis for the decision as to who should be expelled, even though this policy violated the promises made by the framers of the 1941 census. The Decree stated that "those Hungarian citizens who in the last census claimed to be German-speaking or of German nationality, who reassumed their former German names in place of their Hungarianized names, and who became members of the *Volksbund* or other German armed organizations (SS) should be resettled in Germany."[14]

Only very few in the above categories were exempt from deportation. These included: 1) those who had non-German marriage partners and under-age children, or grandparents above age sixty–five living in the same household; 2) members of the so-called "democratic" parties, which, in contemporary Communist terminology, included the Communist Party, the Social Democratic Party, and the Peasants' Party; and 3) those who had suffered persecution at the hands of war-time governments.

These exemptions did not apply to name-changers, to members of German associations, and to those who had served in the German armed forces.[15]

Once the program was implemented, two major problems emerged concerning the deportation: 1) Because of previous flight of many Germans to the West, there were simply "not enough" (half a million) ethnic Germans left in Hungary to fill the quota; 2) Hungarian police forces—who remembered the deportation of the rural Jewry with the help of the Hungarian gendarmerie—in many instances refused to carry out the orders for deportation.

As it turned out, it was simply impossible to implement the orders for the deportation of 500,000 Germans. The State Security Department (ÁVO) did set up a special unit for the execution of the decree, and most of the deportations were in fact carried out by the members of this unit. By the summer of 1946 about 120,000 Germans had been deported to the American Zone of Germany.[16] But in that year—because of the turns of great-power politics and the beginning of the Cold War—deportations were stopped.

Germans from Hungary, 1945-1950] (Budapest, 1988). This is a particularly important book, because it contains many relevant and otherwise not very easily accessible, Hungarian documents (hereafter: Fehér, *Magyarországi németek*).

[14] "Decree of Deportation," in *Magyar Közlöny*, 29 December 1945, 1.§. See Appendix no. 4.

[15] *Ibid.*, 25.§. See Appendix no. 4.

[16] Fehér, *Magyarországi németek*, 115.

American authorities in Germany did not want to accept more Germans, and they alluded to hygienic deficiencies in the transport carriages, as well as to the deportation of innocent people.

Resettlement, however, did not stop at that point. After the Communist takeover in 1947, the Hungarian government deported an additional 50,000 Germans, but this time to the Soviet zone of Germany. On 1 October 1948 all deportations were suspended, and on 23 September 1949 they were officially stopped.[17] Hungarian statistical data gives the number of deported Germans as 185,655. German official statistics, on the other hand, speak of 213,196 people. The latter data, however, also includes those Germans who had fled voluntarily prior to the implementation of the deportations.[18]

The tragedy of the unjustifiable expulsion of around 200,000 Germans from Hungary was further complicated by the fact that the homes of some of the deported Trans-Danubian Germans were given to Székelys from Bukovina, who had been resettled several times before and during the war. Following the Second Vienna Arbitration Award of 1940, which returned Northern Transylvania to Hungary, and then the Hungarian reoccupation of northern Voivodina in 1941, these Hungarian-speaking people from beyond the northeastern Carpathians were first settled in Bácska/Bachka. After the war, however, with the loss of Bácska and the massacres of some 30,000 Hungarians by Tito's partisans, they fled back to Hungary and then were resettled in Trans-Danubia. Their plight has been chronicled by Sándor Sára's powerful and staggering film entitled *Sír az út előttem* [The Road Cries before Me]. This film deals with the encounters between these two humiliated and exploited ethnic groups: the Hungarian Germans and the Bukovina Székelys. Living in the same villages in Tolna county since 1945-1946, there is still a strong ill feeling between them. The passage of over half a century did not heal the wounds of the Székelys and the Germans. Their plight appears to be a never-ending story.

---

[17] Az Országos Földbirtokrendezési Tanács körlevele [Circular of the Land-Ordnance Council], 1 October, 1949; and *Magyar Közlöny*, 23-24, September 1949. See Appendices 1, 2, and 3.

[18] Fehér, *Magyarországi németek*, 163. Statistische Bundesamt, *Die Deutschen Vertreibungsverluste* (Wiesbaden, 1958), 400.

APPENDIX

1. Decree No. 600/1945 M. E. of the National Provisional Government abolishing the system of large estates and the provision of land for landless people. (Extracts)

Chapter II
The confiscation of landed estates

Paragraph 4
The landed estates of traitors, arrow-cross national-socialists, and other fascist leaders, *Volksbund*-members, and also those who committed war-crimes and crimes against the people should be confiscated in their entirety, regardless of their size.

Paragraph 5
Those Hungarian citizens who, contrary to the interest of the Hungarian people, supported the political, economic, and military interests of German fascism, who volunteered to serve in fascist military or police forces, supplied to these organizations information harmful to Hungarian interest, acted as informers, and have reassumed their German-sounding names are considered as traitors, and as perpetrators of war-crimes and crimes against the people.

> Debrecen, 15th March 1945
> In the name of the National Provisional Government
> Béla Miklós
> Prime Minister

2. Decree No. 1710/1945 M. E. of the National Provisional Government establishing the Department for Peoples' Care

The Ministry, authorized by the Provisional National Assembly on 22 December 1944 decrees the following:

Paragraph 1
For the administration of the tasks connected with persons fleeing or expatriated to Hungary, the settlement and provisioning of people who are mentioned by the armistice concluded on 20 January 1945 in Moscow ... and the expulsion of fascist Germans the ministry sets up a Department for Peoples' Care (hereafter: Department).
The headquarters of the Department are in Budapest.

Paragraph 2
The tasks of the Department include:

...

4. Taking care for those Hungarians who are returning from deportation.

5. Executing the deportation of fascist Germans.

Paragraph 7
The execution of point 5 of Par. 2 is arranged for by a separate decree.

Budapest, 4th May 1945

Béla Miklós
Prime Minister

3. Decree No. 3820/1945 M. E. concerning the persons and the extension of the authority of the Department for Peoples' Care under Point 5, Par. 2 of the Decree No. 1710/1945

Paragraph 2.
1. In those districts and their communities where a significant portion of the population had shown sympathy for Hitlerist (*Volksbundist*, fascist, arrow-cross, etc.) activities, district committees are to be set up to investigate the "allegiances to the nation" of those people who come under the provisions of Point 5, Par. 2 of Decree 1710/1945.

Paragraph 4. The committee[s] ...
1. Assert(s) and, if needed, certifie(s) that the person under investigation had a prominent role in a Hitlerist (*Volksbundist*, fascist, arrow-cross) organization. The same assertion should also be made if the person under investigation has volunteered to the armed division of the SS. In justified cases a decision could exempt the members of the family (wife and minor children living together) of the person under investigation.

2. Asserts and, if needed, certifies that the person under investigation was a member of a Hitlerist (*Volksbundist*, fascist, arrow-cross) organization. A similar assertion should also be made about those who reassumed their former German names. In justified cases a decision could exempt the members of the family (wife and minor children living together) of the person under investigation.

3. Asserts and, if needed, certifies that the person under investigation, though not a member of a Hitlerist organization, supported their programs. In the resolution it should be made clear if the above assertion also applies to the members of the family (wife and minor children living together) of the person under investigation.
...

Paragraph 9
1. All persons who are proven to have played a prominent role in a Hitlerist (*Volksbundist*, fascist, arrow-cross) organizations (Par. 4, Point 1)

should also be taken into police custody (internment) beyond the disadvantages deriving from the Decrees concerning landed estates. Members of their family (wife and minor children living together)—unless exempted—should likewise be conveyed to the place of internment.

2. Under the interning order, the internment is implemented by the police in accordance with the Department for Peoples' Care to locations selected by the Department. The internees and their family members are allowed to take 200 kgs [c. 440 pounds] in personal possessions, except when these are judged to be economic equipment or fall under public liabilities. Other personal property items and real estates are sequestrated with the exception of foodstuff and economic equipment that are to be given to new settlers in accordance with the decision of the Department.

Budapest, 28 May 1945

Béla Miklós
Prime Minister

4. Decree No. 12.330/1945 M. E. about the deportation of the German population of Hungary to Germany (proclaimed on 29 December 1945, *Magyar Közlöny* No. 211)

The Ministry and the Allied Control Commission concerning the execution of the decree arranging for the deportation of the German population of Hungary to Germany and acting under the Paragraph 15 of Act IX/1945 decrees the following:

Paragraph 1. Those Hungarian citizens who in the last census claimed to be German-speaking or of German nationality, who reassumed their former German names in place of their Hungarianized names, and who became members of the *Volksbund* or other German armed organizations (SS) should be resettled in Germany.

Paragraph 2. (1) Paragraph 1 does not apply to the cohabiting wife and minor children of a non-German (non-German-speaking) national, nor to his parents who, prior to the coming into force of the present decree, have been living with him and have turned 65 year old on 15 December 1945.

(2) The instructions of Paragraph 1 should not be enforced upon those who were active members of some of the democratic parties or, at least since 1940, were members of any of the trade unions that now are part of the Trade Unions' Council.

(3) The instructions of Paragraph 1 should also not be enforced upon those who, though German-speaking, have acknowledged themselves as Hungarian nationals, and who can officially prove that remaining faithful to the Hungarian nationality had caused them to suffer persecution.

(4) Exemptions stated under point (2) and (3) are extended to wives (widows), minor children (orphans), and parents who, preceding the coming into force of the present order, have been living with them.

(5) Exemptions stated under point (2) and (3) cannot be used in case of those who have reassumed their former German names instead of their Hungarianized ones, and those who became the members of the *Volksbund* or other fascist or military organizations.

...

Budapest, 22 December 1945

Zoltán Tildy
Prime Minister

5. Decree No. 7.970/1946 M. E. about the loss of Hungarian citizenship of those deported to Germany (proclaimed on 16 July 1945, *Magyar Közlöny* No. 158)

(1) The citizenship of all Hungarian citizens who, under the decision of the Allied Control Commission of 20 November 1945, have been deported to Germany, is considered to have lapsed on the day they have left the country.

(2) The instructions of the above paragraph also apply to those who had been deported to Germany before the coming into force of the present decree.

Paragraph 2
The present decree is coming into force on the day of its proclamation; the minister of the interior is responsible for its implementation.

Budapest, 12 July 1946

Ferenc Nagy
Prime Minister

The Deportation of the Germans from Romania to the Soviet Union, 1945-1949

NICOLAE HARSÁNYI

Besides its significance in the history of World War II, the coup of August 23, 1944, whereby Romania left the Axis camp and joined the war effort of the Allies, also opened the door for a later series of events, the study of which was prohibited until the fall of Communism in 1989. The common denominator of all these events (deportations, expropriations and evictions, forced collectivization, mass arrests) was the use of violence in efforts to attain goals of social engineering, viz., to rearrange the political, social, and ethnic fabric of Romania in the second half of the twentieth century along the line of the blueprints of Communist construction. Violence became an "integral part of the ongoing community-structuring enterprise,"[1] a quintessential expression of proletarian dictatorship. Since violence perpetrated by either a home state or an occupier brings credit or accolades to neither, the element of violence was systematically shrouded in a pall of silence. This silence was imposed on perpetrators, victims, witnesses, and the researchers of these events. Whenever such instances of violence could not be ignored altogether, they were conveniently coated in euphemistic terms. The deportation of ethnic Germans from Romania to the Soviet Union in January 1945 was such a case: it never entered official histories of Romania until 1989, while the deportees and their families found out that in fact they had participated in the "reconstruction of the Soviet Union."

Description of Facts and Events

In January 1945 the ethnic Germans from what is now Romania's territory were rounded up to be transported to the Soviet Union for reconstruction work. The large-scale operation was launched on January 10, shortly after the Christmas and New Year holiday season, a period of lull in the traditional cycle of agricultural activity. It was also the time when families were still gathered together. The choice of the date, therefore, maximized the

---

[1] Amir Weiner, "Nature, Nurture, and Memory in a Socialist Utopia: Delineating the Soviet Socio-Ethnic Body in the Age of Socialism," *American Historical Review* 104 (October 1999): 1114.

authorities' effort to round up a large number of people. Although the deportations focused on Transylvania and the Banat, the two regions which the ethnic German communities of Saxons and Swabians called their home, the measure also affected the ethnic Germans living scattered in the cities and villages of Wallachia and Moldavia. The same fate befell the Germans living in Satu Mare [Szatmár] county in northwestern Romania, who had been gathered up for the same purpose already on January 2-3, 1945. All men aged seventeen to forty-five and women eighteen to thirty years old were given two hours to pack up clothes and food that would last for fourteen days, then marched under guard to the nearest railroad station, where a brief medical examination established each individual's suitability for work. The unfit ranged from people with physical disabilities to pregnant women and women with children less than one year old; individuals in these categories were all sent back home. The rest had to embark in freight cars[2] belonging to the Romanian railroads. These box cars had previously been outfitted with bunks and a small stove. Light and air came through one or two small windows covered with an iron grating or barbed wire. Once a person got inside the car, there was no way that he or she could be recalled on the basis of some late minute reprieve. The cars, carrying about thirty persons each, were latched shut from the outside and set in motion eastward through Romanian territory. Once they reached the Soviet border (after several days of traveling) at the towns of Iaşi or Ungheni, the involuntary passengers disembarked only to be loaded again into wide gauge box cars belonging to the Soviet railroads. Larger in size, these cars accommodated forty people, well packed. Otherwise, the furnishings offered little change: the same bunks, a stove, plus a hole in the floor serving as a toilet (*ubornaya*). The space inside was so glutted with luggage that people had to sleep, sit, and stand in shifts. After a further journey of three to six weeks into Soviet territory, these trains reached their destinations in the Ukrainian and Ural industrial regions. Their human loads were transferred onto trucks and taken to camps in the neighborhood of industrial units, mines, or state farms. Another medical exam established each internee's fitness for either hard or normal labor, which also led to corresponding differentiations in the food rations. Conditions in the camps were harsh and eventually took

---

[2] "One can chisel a permanent image into the edifice of twentieth century ethnic cleansing of freight cars, overcrowded with deportees, hungry, thirsty, starving, diseased, suffocating in unhygienic and barbaric conditions." Norman M. Naimark, *Ethnic Cleansing in Twentieth Century Europe* (Seattle, 1998), 23.

their toll: housed in rundown buildings or wooden sheds, with several dozens crowded onto two and three tiered bunks in a room, working from dawn to dusk, chronically malnourished, infested with parasites, the deportees experienced a rising mortality that climbed to 15 percent.[3] Meanwhile, those who became ill and incapable of working were sent back to Romania or eastern Germany. The first returnees arrived in the fall of 1945. It was from them that relatives learned the first news about the deportees' fate and location.

Conditions started improving after 1946: a more diversified and ameliorated medical care reduced the mortality rate; with the working hours cut down to eight (provided one fulfilled her/his daily quota), the internees enjoyed more leisure time, even a weekly day off, filled, however, with supervised cultural programs. The relations with the local Russian and Ukrainian population, who upon arrival had often thrown stones at the Germans, had warmed up: little by little they befriended them, and also bought some of the Germans' clothes (as they were more fashionable than the ones available in the Soviet Union). Many German women who could sew were asked by the Russian officers to make dresses for their wives.

Finally, the years 1948-1949 marked the beginning of the end of the forced labor for the ethnic Germans from Romania. Gradually, they returned, again by trainloads, although under improved circumstances, to Romania (disembarking at Sighet [Máramarossziget]) or East Germany (end-station Frankfurt an der Oder). The last internees made the return voyage in 1951. The total figure of ethnic Germans from Romania subjected to this form of *Zwangsarbeit* (forced labor) is estimated at 75,000, of whom some 10,000 never lived to see the day of their return home.[4]

The end of deportation showed again the duplicity of the Soviet and Romanian Communist regimes. On the one hand, the Germans were sent off with speeches full of gratitude for their contribution and bands played while they embarked on the

---

[3] Theodor Schieder, ed., *Dokumentation der Vertreibung der Deutschen aus Ost-Mitteleuropa*, 8 vols. (Bonn, 1954-1961; reprint ed. 1994), vol. 3—*Das Schicksal der Deutschen in Rumänien*, 80E (hereafter cited as *Schicksal*).

[4] Both *Schicksal* and Georg Weber et al., *Die Deportation von Siebenbürger Sachsen in die Sowjetunion 1945-1949*, vol. 1 (Köln, Weimar, Wien, 1995) give these figures. According to Günter Schödl, *Land an der Donau* (Berlin, 1995), 596, the number of deportees ranged between 75,000 and 80,000 (40,000 from the Banat, 26,000 from Transylvania, and some 15,000 from Satu Mare county and the rest of Romania). Elemér Illyés in *National Minorities in Transylvania* (Boulder, 1987) gives a higher estimate, that of up to 100,000.

westbound trains decorated with fir branches. On the other hand, authorities in both countries did their best to erase the traces of material memory: diaries and sketches of scenes from their life in the USSR were confiscated. As to forms of discourse connected to the deportation, the Romanian regime took good care that censorship remained tight and that the episode did not surface; to speak of it remained a taboo. Even when Romania experienced a relative cultural thaw between 1963 and 1971, the episode of the deportations was not tackled either in scholarly or artistic works.

Until the overthrow of the Ceauşescu dictatorship in December 1989, this episode in the series of deportations that were a dreaded fact of life in the decade following the end of World War II was only dealt with in books published in the West. Many publications issued by various German émigré organizations mentioned quite briefly the deportation to the Soviet Union, usually as an integral part of the larger narrative concerning the history of the Saxons and Swabians of Romania.[5] Scholarly research in the Federal Republic of Germany has either included these deportations in the larger theme of the uprooting/cleansing (*Vertreibung*) of ethnic German populations from East Europe in the wake of World War II,[6] or else within the larger pattern of demographic changes induced by the migration of Germans in Europe within the last 1000 years.[7] The memory of the deportations was saved through oral history. The testimonies of a great number of deportees were archived in the decade after the war the Federal Ministry of Expellees, Refugees, and War Casualties (*Bundesministerium für Vertriebene, Flüchtlinge und Kriegsgeschädigte*). After 1989, when some Romanian and Soviet archives became more accessible to researchers, a comprehensive study was published in 1995 by a group of historians from

[5] Alliance of Transylvanian Saxons, *The Transylvanian Saxons: Historical Highlights* (Cleveland, Ohio, 1982); G. C. Paikert, *The Danube Swabians: German Populations in Hungary, Rumania and Yugoslavia and Hitler's impact on their Patterns* (The Hague, 1967); Johann Schmidt, ed. , *Die Donauschwaben 1944-1964: Beiträge zur Zeitgeschichte* (Munich, 1968).

[6] Wolfgang Benz, ed. , *Die Vertreibung der Deutschen aus dem Osten: Ursache, Ereignisse, Folgen* (Frankfurt am Main, 1985).

[7] Lothar Dralle, *Die Deutschen Ostmittel- und Osteuropa: ein Jahrtausend europäischer Geschichte* (Darmstadt, 1991); Andreas Baaden, *Aussiedler-Migration: historische und aktuelle Entwicklungen*, (Berlin, 1997); Hans-Ulrich Engel, ed. , *Deutsche unterwegs: von der mittelalterlichen Ostsiedlung bis zur Vertreibung im 20. Jahrhundert* (Munich, Vienna, 1983); Dieter Kraeter and Hans Georg Schneege, *Die Deutschen in Osteuropa heute* (Bielefeld, 1969).

Westfälische Wilhelms-Universität, Münster.[8] The hefty, three-volume work offers the inquisitive reader not only a detailed narrative of the events and personal testimonies, but also pictorial representations (photos and sketches) that the deportees managed to smuggle out of the camps. However, by focusing on the deportation of only the largest segment of the German population of Romania, the Saxons, this work fails to render a full picture of what happened to the Swabians or the Zipsers. In post-Communist Romania historians and anthropologists tried to recover the memory of the deportation by the methods of oral history. This may betray a need for unmediated access to the voice of those who suffered, as well as the urgency to make it heard, since many of these survivors have now reached an advanced age. However, an analysis of archival documents and of the "other side" as it were, would be crucial to the understanding of this period.

Motive

First of all, in establishing the categories of population that were deported, the question arises why these people were subject to the indemnity of forceful removal from home and relocation.

The Germans of Romania (the Saxons of Transylvania, the Swabians of the Banat and of Satu Mare [Szatmár] county) were singled out for deportation because of their ethnicity. This was the sole criterion of selection reflecting the collective guilt ascribed to this ethnic group in the wake of World War II, and which, after the end of the war, would justify the expulsion of the ethnic Germans from Czechoslovakia and Poland, and the exchanges of population between Czechoslovakia and Hungary (the infamous Beneš decrees). The Soviets, under Stalin had already applied the principle of collective guilt in order to deport whole ethnic groups to remote regions in the Asian part of the USSR. The Volga Germans were deported to Kazakhstan and other regions of inner Asia a couple of months after Germany attacked the Soviet Union on June 22, 1941. It was a prophylactic reason that the Soviet authorities provided for this forced mass relocation: the Volga Germans were viewed as potential saboteurs and traitors who might cooperate with the invaders. The deportation of Germans for labor at the end of the war, on the other hand, can be viewed as a punitive measure directed against the entire ethnic group. The Germans of Romania were not the only ones shipped away to

---

[8] Weber, *Die Deportation von Siebenbürger, vol. 1.*

labor camps in the USSR: the same fate befell the German population living in the south of Hungary, in the Polish lands east of the Oder and the Neisse, the Baltic countries, Bulgaria, Yugoslavia, and Czechoslovakia. All in all, some 500,000 German men and women participated in the reconstruction of the war-torn Soviet economy.[9] It is true that before August 23, 1944, many of the ethnic Germans from Romania had fought in the *Waffen SS*. After Romania joined the side of the Allies, some withdrew with the German army from the territory of Romania, while others stayed at home relying on their status as Romanian nationals and, consequently, thought that they would not be affected by the quick advance of the Red Army. However, among the ethnic Germans of Romania, a sizeable number of individuals had, throughout the war, held antifascist, liberal, or social democratic views which caused them to refrain from joining the *Volksgruppe* (the pro-Nazi organization of the Germans living in Romania). A few Germans were even members of the underground Romanian Communist Party. Such subtleties meant nothing to the Soviet military authorities who oversaw the deportations and who made no effort at determining guilt or innocence on a case-by-case basis. The only decisive factor remained the individual's German name.

Let us now see how the lists of deportees were drawn up. Late in August 1944, three days after August 23, the Romanian authorities registered all ethnic Germans who were Romanian citizens with a view toward later use in public work. A second, and similar, registration occurred in October 1944, whereby again the ethnic Germans were conscripted for work. It was on the basis of these lists that the Germans were rounded up in the first days of January 1945.

Agency

Who actually carried out these deportations? In the case of the Germans deported to the Soviet Union, establishing clear-cut responsibility is complicated by Romania's political status after the armistice with the Allies (the treaty was signed in Moscow, on September 12, 1944). This made Romania a "friendly country," thereby not occupied by the Red Army. The Romanian authorities retained jurisdiction and formally restricted the Red Army's freedom of initiative. In practice, however, the agreement was observed only on paper. In reality, the Romanian authorities had to comply with the orders of the Soviet representatives. When on

---

[9] Ibid., 72.

January 3, 1945, Lieutenant General Vladislav Petrovich Vinogradov, the representative of the Allied Control Commission in Bucharest, ordered the deportation of the ethnic Germans, the Romanian authorities provided the railroad cars, as well as the personnel necessary for the roundup of deportees. According to the Armistice Treaty, Romania was to help the Soviet war effort by supplying, among other things, means of rail transportation. Since the Armistice Treaty made no mention of possible forced labor, or any other kind of labor for that matter, the German deportees had to be integrated with cargo shipped from Romania to the USSR. Units of the Romanian police, army, and gendarmerie participated in the action together with Soviet troops, but the initiative, the organization and the supervision of the entire operation remained in Soviet hands. Owing to its precarious grip on power,[10] the government of General Rădescu had little choice but to fulfill the Soviet demands. Aware that the Red Army was deporting citizens of Romania, a country termed as "friendly," the Romanian government issued an official note of protest, transmitted to General Vinogradov, accusing the Soviets of breaking the provisions of the Armistice Treaty. Needless to say, this protest, as well as other similar statements made by political leaders Iuliu Maniu and Dinu Brătianu, went totally unheeded.

Legal basis

The deportation had no previous legal basis since neither international documents or agreements, nor the Armistice agreement had stipulated it by the time the deportation started. The deportation of the ethnic Germans in January 1945 was codified a month later, in February, during the Yalta conference, when the use of forced German labor came to be included in the category of war reparation. The Soviets thus created a situation and only afterwards justified it legally. Moreover, the Germans were not even told where they would be taken away. The Germans expected that they would be transported to different areas in Romania to help repair the ravages of the war, as they had previously been mobilized for a couple of days to clean streets, clear roads of snow, or repair railroads. Upon finding out that they had crossed the Romanian-Soviet border, their anticipations turned to the worst.

---

[10] Ibid., 154-163.

In Lieu of Conclusion

For the Swabians of the Banat, the deportation to the Soviet Union was not the only episode of forceful removal. In 1951, those living in the countryside were again rounded up and deported to the Bărăgan, a vast plain in the southeast of Romania. This time, the deportation was not motivated ethnically, but socially: the Romanian Communist regime, following the Soviet model, decided to deport all the owners of middle and small size farms in the Banat, irrespective of their ethnicity. The deportation to Bărăgan lasted until 1955.

Certainly, the deportation to the Soviet Union and its memory was one of the causes that led to the dramatic shrinking in size of the German community in Romania up to the present day. The massive waves of emigration to West Germany in the seventies, eighties, and early nineties had among their root causes also the fact that, although the Swabians and Saxons called Romania their *Heimat*, since the end of World War II this homeland, through its Communist and nationalist policies, had proved not to be so sheltering any more. They left behind a material culture that testifies to a rich historical past, but a culture that is now irretrievably fading from reality, to be stored only in memory.

III.

THE AFTEREFFECTS OF ETHNIC CLEANSING IN THE WAKE OF WORLD WAR II

The Isolationist as Interventionist: Senator William Langer on the Subject of Ethnic Cleansing, March 29, 1946

CHARLES M. BARBER

Republican U.S. Senator William Langer of North Dakota looked at the *Vertreibung* (the ethnic cleansing of the Germans from eastern Europe) through the eyes of a *Volksdeutscher* (ethnic German) and an anti-Communist Socialist. His father, Frank J. Langer, was an ethnic German from Michelsdorf (now Ostrava) in Moravia, and young William, born in 1886, imbibed enough of his German heritage as a child to remain close to it throughout his life. As Attorney General for North Dakota during World War I, and U.S. Senator during World War II, he had ample opportunity to downplay his German background, but he did not. He confronted the nativism that afflicted German-Americans in both World Wars and thus became a dependable source of support for those who were pronounced guilty by reason of ethnicity by their super-patriotic fellow citizens.[1]

His political career, stretching from 1914 to 1959, was spawned in the midst of the remarkable prairie populism of the North Dakota Non-Partisan League. The Bank of North Dakota, for example, is not a private bank. It is owned by the people of North Dakota. The people of the state also own their own grain mill, which has competed successfully in the American marketplace since the 1920s. When the large grain, telecommunication, and railroad corporations attempted to shut down this brand of democratic socialism, they were defeated in a unanimous 1920 Supreme Court decision.[2]

In the words of William O. Douglas: "This chapter in our history, little known even to Americans, was one I told later to try to convince foreigners that our country was not reactionary in a con-

---

[1] Raymond Lohne, *The Great Chicago Refugee Rescue* (Rockport, Maine, 1997); Charles M. Barber, "A Diamond in the Rough: William Langer Reexamined," *North Dakota History: Journal of the Northern Plains* 65, no. 4 (Fall 1998): 2-18; Theodore Pedeleski, "The German-Russian Ethnic Factor in William Langer's Campaigns, 1914-1940," *North Dakota History* 64, no. 1 (Winter 1997): 2-20; William Sherman and Playford Thorson, eds., *Plains Folk: North Dakota's Ethnic History* (Fargo, N.D., 1988), 147, 170; and Charles M. Barber, "The Nordamerikanischer Saengerbund versus the U.S. Treasury Department, 1944-46," *Yearbook of German-American Studies* 30 (1995): 73-116.

[2] William O. Douglas, *Go East, Young Man: The Early Years* (New York, 1974), 420, and Robert Morlan, *Political Prairie Fire: The Nonpartisan League, 1915-1922*, 290-292.

stitutional sense, as the propaganda made out." The socialist enterprises of North Dakota were market-driven, the farmers of that
state having little use for the Bolshevik version. As such the Republican William Langer's anti-Communism came from the left
and not the right, a source of confusion for some interpreters of
North Dakota.[3]

Langer and the Nonpartisan League excoriated the rigid
anti–Communist policies of what came to be known as McCarthyism. This was a lonely, poorly understood political stance in the
years immediately following World War II. Yet Langer could stand
up to the vilification that his left wing anti-Communism produced,
because he was a man of great civil courage. As his colleague, Russell Long of Louisiana (Democrat), pointed out in a eulogy in 1959:

> Bill Langer had great courage. Often he was the only one to vote
> against measures which at the time were very popular. Many of
> us have lived to see that his judgment on some of those occasions
> was much more correct than the public was willing to admit at
> that time...
> ...From time to time some would say that Bill Langer was un-
> predictable or perhaps inconsistent. Yet in a great many ways he
> was one of the most consistent of all Senators to serve in this
> body during the 11 years I have been here...in that he was always
> on the side of the poor man, the little man, the underprivileged,
> or those who had been neglected by society as a whole.[4]

Langer linked the plight of the neglected in postwar Europe to
the fate of the neglected farmers in his own State. He accurately
reflected North Dakota farmers' anger at their colonial status vis-à-
vis Wall Street. His sense that the people of North Dakota were
kindred to colonial peoples around the world gave him a larger
sense of moral anger. Despite the perseverance of its citizens, North
Dakota was perennially in the dilemma of a colonial relationship to
the cities to its southeast, exporting "raw fossil fuels, unfinished
farm products, and young people," and importing "finished products and capital."[5]

---

[3] See, for example, Michael Paul Rogin, *The Intellectuals and McCarthy: The Radical
Specter* (Cambridge, MA, 1967), especially the chapter "Agrarian Radicalism, Ethnic
and Economic."

[4] United States of America, 86th Congress, 2nd Session, "Memorial Services Held in
the Senate and House of Representatives of the United States Presented in Eulogy of
William Langer, Late a Senator from North Dakota," (Washington, 1960), 41.

[5] David B. Danbom, "North Dakota: The Most Midwestern State," in *Heartland:
Comparative Histories of the Midwestern States*, ed. James H. Madison (Bloomington,

The inferiority complex derived from its colonial status, which David Danbom sees as characterizing much of the State's attitudes towards itself and the rest of the country, is seen by Daniel Rylance to characterize its leading spokesman, William Langer.[6] However, a sense of inferiority on the part of the state and the man is only a partial explanation. Langer belonged to a group of Senators whose views are well represented today by Non-Governmental Organizations (NGOs) like Amnesty International and Doctors Without Borders. The older holders of these views have been described by Justus Doenecke as "Old Isolationists":

> Hostility to the Marshall Plan by no means implied that the old isolationists were indifferent to the fate of all Europe. During the debates of the late forties, veteran isolationists, almost to a man, were making one claim: a revived Germany could save Western Europe, and save it without the necessity of America's underwriting half the Continent.[7]

While not calling for military action, Langer feared a Communist takeover of Central Europe that others would later see as a reason to maintain a strong American military build up in Europe: "I want to read into the Record at this point a statement incorporating the communist program for Germany, entitled 'Strategy and Tactics Prescribed by the Seventh World Congress.' This is the resolution drawn up by the Communist Party on the anniversary of the French Socialist Party Day of October 1944." After reading this plan, Langer continued:

> ...Mr. President, if there is any doubt in the minds of my listeners that the document which I have just read constitutes the real aims of Soviet Russia, perhaps the following...will serve to dispel that doubt. On March 18 Mr. C. L. Sulzberger cabled the *New York Times* from Berlin as follows: 'The world diplomatic crisis, which unfortunately shows signs of crystallizing into a contest for ascendancy by the bloc of nations led by the Union of Soviet Socialist Republics on the one hand and by the United States and Britain on the other, is sharply reflected in Germany by the clear-cut endeavor on the part of the General Communist Party to

---

1988), 109; see also William Cronon, *Nature's Metropolis: Chicago and the Great West*, (New York, 1991), 214, 284, 376.

[6] Danbom, "North Dakota," 107-110; Daniel Rylance, "William Langer and the Themes of North Dakota History," *South Dakota History* 3 (Winter 1972): 41-62.

[7] Justus Doenecke, *Not to the Swift: The Old Isolationists in the Cold War Era*, (Lewisburg, PA, 1979), ch. 7.

force the social Democrats into a merger that would be controlled by the Communists.'[8]

In addition to seeing power/political realities as well as or better than many pundits or politicians of his time, William Langer saw and spoke about what some Americans might have been able to see, but apparently could not bring themselves to talk about at any length, so recently after the terrible secret of the Holocaust had finally been revealed.[9] He noticed and he abhorred the indifference among Christians in the United States to the remnant surviving after Hitler's killing of millions of Jews, many of whom were now caught up in and suffering in the ethnic cleansing of the Germans from Eastern Europe. As for the vast majority of the expellees (*Vertriebenen*), their only crime had been to be the wrong ethnicity in the wrong place at the wrong time. William Langer knew this, and he made sure that the rest of his colleagues in the Senate would not be able to forget it, until something was done to rectify the situation.

On March 29, 1946 William Langer rose in Senate chambers to address the twin horrors of enforced deportations (what is now called "ethnic cleansing") and the resulting starvation:

> While America sits in judgment against men responsible for enforced deportations of nationals of non-German races, we have accepted an arrangement in regard to eastern Germany which has led to the forced deportation from their homes of millions of Germans, or German-speaking persons, among them a large proportion of children. Despite the fact that the Potsdam declaration declares that the evacuation of Germans from what has suddenly, by unilateral decision of one ally, become Poland, as well as from Czechoslovakia and other eastern countries, despite the fact that the Potsdam declaration clearly states that this must be halted until it can be carried out in "an orderly and humane manner," the deportations have gone ahead until, according to the report of General Eisenhower himself some time ago, the American zone had already then had to take in half a million of these people.[10]

Langer went on to emphasize that the refugees were moving into defeated Germany,

---

[8] *Congressional Record*-Senate, 79th Congress, 2nd Session, 29 March 1946, 2801-2802.

[9] Walter Laqueur, *The Terrible Secret: Suppression of the Truth about Hitler's "Final Solution"* (New York, 1980), 26.

[10] *Congressional Record*—Senate, 79th Congress, 2nd Session, 29 March 1946, 2804-05.

...a country whose war destruction is incomprehensible to those who have not seen its completely shattered cities and communications.... Not all these refugees, or evacuees, arrive, of course, at all. Thousands of them perish on the trek from Silesia or Bohemia. Those who arrive are without ration cards, and already in an advanced state of hunger or actual starvation....[11]

Someone who had seen devastated Germany was George Orwell. As a war correspondent between February 15 and May 24, 1945, he had sent eighteen dispatches about the effects of the war on liberated Paris and occupied Cologne, Nuremburg, Stuttgart, and Salzburg back to the *Observer* and the Manchester *Evening News*.[12] Five months previous to Langer's speech he had written:

In so far as the big public in this country is responsible for the monstrous peace settlement now being forced on Germany, it is because of a failure to see in advance that punishing an enemy brings no satisfaction. We acquiesced in crimes like the expulsion of all Germans from East Prussia—crimes which in some cases we could not prevent but might at least have protested against—because the Germans had angered and frightened us, and therefore we were certain that when they were down we should feel no pity for them...

...Actually there is little acute hatred of Germany left in this country, and even less, I should expect to find, in the army of occupation. Only the minority of sadists, who must have their "atrocities" from one source or another, take a keen interest in the hunting-down of war criminals and quislings.[13]

In his speech Langer also gratuitously declassified "so-called confidential material" that paralleled Orwell's observations, like the official report on conditions in the Russian zone of the German Central Administration for Health submitted to the Berlin *Kommandanture*, a German agency created by the Russian authorities. This report, though it might have been expected to minimize conditions, was so brutal that Langer compared its descriptions of physical, spiritual, and psychological hunger to the "famous black death of

---

[11] Ibid.

[12] Jeffrey Meyers, *Orwell: Wintry Conscience of a Generation* (New York, 2000), 231-2.

[13] Sonia Orwell and Ian Angus, eds., *The Collected Essays, Journalism, and Letters of George Orwell*, vol. 4, *In Front of Your Nose, 1945-1950* (New York, 1968), chap. 5, "Revenge is Sour," 9 Nov. 1945.

the Middle Ages."[14] He thought it was "high time" that the American people were informed by their representatives in the Senate as to the true facts about conditions in Central Europe. Despite the unwillingness of the Truman administration to let those facts out, and despite what he called "almost a conspiracy of silence in the press regarding what is going on," Langer stated that "sufficient facts are available to those determined to find them," and his colleagues agreed.[15] These were the days before the creation of the C.I.A. in 1947 and before the Internal Security Act of 1950 produced a chilling effect upon the Senate, what McGeorge Bundy called "a grim and illiberal public mood."[16]

William Langer and George Orwell both favored a form of democratic, pluralistic socialism. They also both loathed imperialism in general and the British Empire in particular. Since the phenomenon of a socialist Republican, imbued with anti-Communism and anti-McCarthyism, seems to be so foreign to American commentators, I have paired the obscure Senator from the distant State of North Dakota with the author of *Homage to Catalonia* and *Animal Farm* and the creator of "Big Brother" in order to underline Langer's outspoken prescience, and to support a conclusion that his isolationist stances against the expansion of a "Pax Americana" did not preclude American intervention on behalf of underdogs abroad, as well as at home. Moreover, Langer's humanitarian position, sometimes taken lightly by most pundits at the time, and has been ignored by most historians of the Cold War afterwards, eventually became a cornerstone of Allied policy, in word and deed, once Lucius Clay, among others, saw that the policies of revenge and the politics of hunger were detrimental to interests of the United States in Europe.[17]

One article can scarcely explain fully the distinct aspects of North Dakota which helped to define the politics of William Langer, as well as the personal traits that influenced his stands. A key characteristic, however, was William Langer's unwillingness to distinguish between victims of the Holocaust, victims of the *Vertreibung* (Expulsion), or victims of persecution anywhere in the

---

[14] *Congressional Record*—Senate, 79th Congress, 2nd Session, 29 March 1946, 2805.

[15] Ibid., 2805, 2808.

[16] Eleanor Bontecou, *The Federal Loyalty-Security Program* (Ithaca, 1953), ix.

[17] See Edward N. Peterson, *The American Occupation of Germany: Retreat to Victory* (Detroit, 1978); and Kai Bird, *The Chairman: John J. McCloy, The Making of the American Establishment* (New York, 1992), esp. "German Proconsul, 1949," 308-31.

world during his tenure in the Senate from 1940 until his death in 1959. He spoke up for them all, and in some cases, like that of the Danube Swabians, was instrumental in helping their cause.[18]

Langer's speech of March 29, 1946, occupies eleven full pages in the Congressional Record and can be broken down into several categories, even as they are intertwined almost seamlessly into Langer's challenge to the American conscience:

1. Non-Partisan League Anti-Communism

The practical, socialist ventures of the Non-Partisan Leaguers immunized them against the theoretical appeals of Bolshevism to others on the left. In addition, the German-Russian communities in both Dakotas were well aware— through the medium of F.W. Sallet's *Dakota Freie Presse*, and other German-language sources—of the mass starvations of Lenin and Stalin which affected their countrymen in Ukraine.[19]

2. Anti-Imperialism/Humanitarian Interventionism

Many of the same economic conditions which affected ethnic Germans in eastern Europe also affected their more fortunate cousins in North Dakota, who were blessed with political means and the will to use them. An Old Isolationist like Langer might have been unpopular in Washington, D.C., but his views were acceptable also to the other dominant ethnic group, the Norwegians, in the western part of his state, whose sparse conditions and oppressive outside factors were all too reminiscent of the other side of the Atlantic.

3. U.S. Domestic Food and Farm Policies and the Politics of Starvation

Langer's anger at the way in which farmers were exploited by the politicians in urban states was coupled with his dismay at the unwillingness of the U.S. government to allow food providers to send food to those who so desperately needed it. Langer's target here was the Morgenthau Plan in particular, but also the cynicism of other officials in the Truman Administration who allowed such a draconian measure to continue.

4. The Refugee Situation (Ethnic Cleansing)

---

[18] Lohne, *Refugee Rescue*; Barber, "Diamond in the Rough," 2-18.

[19] Sherman and Thorson, *Plains Folk*, 148; and La Vern J. Rippley, "F. W. Sallet and the Dakota *Freie Presse*," *North Dakota History* 59, no. 4 (fall 1992): 16-17.

Langer saw Teheran, Yalta, and Potsdam as being the West's acquiescence in the face of Czech and Russian ethnic cleansing, and his public utterances have since been backed up by the classified dispatches of key government officials like George Kennan.

Non-Partisan League Anti-Communism

In the primary and November elections of 1916, "socialism with a human face" took root all over the Upper Midwest of the United States, but most tenaciously in North Dakota, one year before the more famous inhuman variant of socialism broke out in Russia. Yet the Northern Great Plains version of market-driven socialism, the Non-Partisan League, has gone unnoticed in American History beyond the scope of regional, Great Plains, and populist histories. Most of mainstream U.S. historians, and virtually all of the major media, have been mesmerized, instead, by the Bolshevik Revolution of November 1917. Since then English–language readers of history have been subjected to an unrelenting barrage of editorial opinion, sophisticated academic opinion, overt and covert McCarthyism, respectable and shameless forms of well- and ill-intended propaganda to the effect that socialism and freedom are incompatible.

Most writers and journalists in Anglo-America were content to publish this fraudulent mixture of an attack on Leninist and Stalinist Bolshevism with an attack on farmer and labor organizers in London, Detroit, New Ulm, Minnesota or Fargo, North Dakota. A short list of well known writers who were and are both consistently on guard against this deception and brave enough to point it out is distinguished, but not lengthy: George Orwell, C. L. R. James, George Steiner, I. F. Stone, and Christopher Hitchens. These writers have seen clearly that such attacks almost always were motivated by the "haves" in efforts to hoodwink or intimidate the "havenots."[20]

What seems to have escaped most American discussions of socialism, Communism, and McCarthyism, both then and now, is that socialism had been declared legal in a unanimous U.S. Supreme Court decision in 1920 (253 U.S. 233).[21] The defendant in the

---

[20] See, for example, Neil Middleton, ed., *The I. F. Stone's Weekly Reader* (New York, 1974), 32.

[21] Douglas, *Go East Young Man*, 420.

case was North Dakota. The decision has been commented on at length in Robert Morlan's superb study of the Non-Partisan League from 1915-1922,[22] but its significance at the time was grasped by only a few who embraced progressive causes and also made it as far as Washington D.C. One of them was William O. Douglas, a distinguished liberal member of the Supreme Court from the State of Washington. Justice Douglas praised the Non-Partisan League experiment in economic activism:

> In September, 1922, having decided to go to Columbia Law School, it was time to go East. After two years of teaching I had saved seventy-five dollars and, of necessity, my transportation across the country was to be by freight train.... The immensity of my country came home to me as we cleared the Rockies and entered the Great Plains. Dust storms were making up every day and beginning to cast a pall over the land,.... I was soon to find full confirmation of what my Yakima friend, O. E. Bailey, who grew up in North Dakota, had told me about the privation of the farmers who settled west of the rich Red River Valley. As a consequence of their suffering, North Dakotans had rebelled, formed unorthodox political parties, and turned socialistic under such leaders as William Langer, whom I was to meet years later in Washington, D.C., when he became a senator.... On this freight train ride that offbeat program gained new meaning in my mind. My reading of Thorstein Veblen had taught me the manner in which the Establishment was exploiting our material and human resources, and this situation suddenly came alive for me. I began to see, in the rawness of North Dakota, the stuff out of which great reforms were being created.[23]

When the Supreme Court finally struck down the challenge to the Tennessee Valley Authority's constitutionality on January 30, 1939,[24] it was merely extending to the federal level what it had already granted North Dakota. A key figure in justifying Non-Partisan League legislation was United States District Judge Charles F. Amidon. His comments are revealing:

> It is hopeless to expect a population consisting of farmers scattered over a vast territory, as the people of this state are, to create any private business system that will change the system now existing. The only means through which the people of the state have

---

[22] Morlan, *Prairie Fire*, chaps. 10-13.

[23] Douglas, *Go East Young Man*, 127-129, 420.

[24] Steve Neal, *Dark Horse: A Biography of Wendell Willkie* (Lawrence, Kans., 1984), 36.

had any experience in joint action is their state government. If they may not use that as the common agency through which to combine their capital and carry on such basic industries as elevators, mills and packing houses and so fit their products for market and market the same they must continue to deal as individuals with the vast combinations of those terminal cities and suffer the injustices that always exist where economic units so different in power have to deal the one with the other.[25]

Such was the nature of the socialism that Republican Senator Langer carried to Washington D.C. in 1940. It was decidedly anti-big business, but it was also decidedly anti-Communist, since North Dakota farmers knew well that they were entrepreneurs, however much they may have resented the manipulation of market forces from outside of the state. Professional patriots in 1917-1919, who had tried to discredit the Non-Partisan League as either too pro-German or too pro-Soviet, had succeeded in crushing the League in every other state except North Dakota, especially violently in neighboring Minnesota.[26]

During the 1920s the Leaguers were rent by many internal disputes, much to the delight of their external enemies, but when the League was needed most in North Dakota, it came back in the Depression under the leadership of William Langer.[27]

Anti-Imperialism; Humanitarian Interventionism

What about lumber, Mr. President? The farmers of the United States cannot get sufficient lumber to build a chicken coop...; But Mr. President, within the last few months, a billion feet of lumber was shipped to Mexico, and other foreign countries, and indeed, shipped all over the world.... We have here an amendment which, if adopted, will raise the price of wheat to only $2.10 a bushel, whereas in the last war, as I stated before, the farmers were getting $2.26 a bushel in Minneapolis.... Yet today, instead of our doing something to help the farmers of this country, we find Senators saying they want to refer the Russell amendment back to the committee on Agriculture and Forestry for further study.[28]

---

[25] Morlan, *Prairie Fire*, 243-244.

[26] Ibid., chap. 8, "The Reign of Terror"; Carl H. Chrislock, *Watchdog of Loyalty: The Minnesota Commission of Public Safety During World War I* (St. Paul, 1991).

[27] Rylance, "Themes," 49-53; and Agnes Geelan, *The Dakota Maverick: The Political Life of William Langer* (Fargo, 1975), 47-78.

[28] *Congressional Record*—Senate, 79th Congress, 2nd Session, March 29, 1946, 2798.

Mr. President, something terrible has happened to America. The American people have been and continue to be maneuvered into one of the most diabolic crimes ever committed against a civilized people. Unless and until we are able to compel a change and to force this mad fanaticism to break its vicious grip which is dragging the whole world down into chaos and agony, we shall earn the everlasting contempt and hatred of our fellow men and the enduring condemnation of our Creator.[29]

Langer was frustrated with colleagues who supported the policies of Agriculture Secretary Henry Wallace, which favored the British Empire and interests in Mexico over American farmers. And it was by voicing these frustrations that he began his speech of March 29, 1946. By the time he finished he had enlarged his theme to attack what he saw as an abandonment of American principles of democracy and humanitarianism. In the months following this speech he would expand his attacks on British and American imperialism to include Cold War policies that he felt were more harmful to the interests of the American taxpayers than they were to the Soviet Union. Many of these views, though not well accepted at the time, proved to be prophetic. To be called an isolationist in the 1940s sometimes was close to being called "soft on Hitler," or "soft on Stalin," depending on the accuser, or the timing of the accusation. But William Langer was not an isolationist in the sense of being a pacifist or totally withdrawn behind a "fortress America." He was also sensitive to the dominant German-speaking minority of his state, targets of nativism in World War I, and frustrated by their inability to send food, clothing, and farming implements to their brothers and sisters in Europe.[30] Langer considered these restrictions on humanitarian aid by the American government to be totalitarian in nature, and he was not impressed with the efforts that had been put forth at the time of his speech:

At this moment, to my knowledge, this administration has consented to four sops being thrown to American public opinion.

On April 1, we are now informed, it is the intention of the Army to permit the reopening of mail service to letters weighing not over 24 grams.

Secondly, after having stood in the way for many months we are permitting the International Red Cross to operate in the greater Berlin area, at the request of the Russian.. General Smirnow...

---

[29] Ibid., 2810.

[30] Barber, "Diamond in the Rough," 3-6.

The State Department has also now permitted the American Relief Societies for Foreign Service to ship 2,000 tons of relief goods into Germany per month...

This administration has finally consented to the creation of a 12-man Famine Committee under the leadership of former President Hoover to hasten the imposition of self-rationing by the American people on themselves and to find some means of putting the whole food crisis in its proper perspective.[31]

Langer characterized the policy changes as just "so much superficial window dressing in the face of the tragedy now threatening the world."[32] His call for the revision of the Morgenthau Plan, and for removal of restrictions on parcels from his and other states, show him to be an interventionist, rather than an isolationist. Langer saw food, clothing, and shelter in the form of aid as going much further than the introduction of more American soldiers or American backed soldiers. That point, after all, was moot. The American Army already occupied much of Germany. Langer saw himself as defending the top American general there, Lucius Clay, from insane policies of revenge which would only make the peace-making and peace–keeping process that much more difficult:

It is perfectly obvious, Mr. President, that General Clay himself has become a victim of these vicious policies for which the group within this administration who have been bent on the destruction of the German people are responsible.[33]

This defense of Clay's role as a peacemaker, over and above his role in Cold War maneuverings, shows that Langer was primarily an isolationist when it came to military intervention. He was not against non-military interventions grounded in humanitarian impulses and economic common sense. Like the State of North Dakota itself, this is very much an underdeveloped theme in any assessment of William Langer's views on foreign policy, and, with the exception of the work of Justus Doenecke, one which has not received much attention in the works which deal with Langer's "old isolationist" views.[34]

---

[31] *Congressional Record*—Senate, 79th Congress, 2nd Session, March 29, 1946, 2810.

[32] Ibid.

[33] Ibid., 2808.

[34] Glenn H. Smith, *Langer of North Dakota: A Study in Isolationism, 1940-1959* (New York, 1979); Larsen, *William Langer*, chap. 4; Geelan, *Dakota Maverick*, 105-120; Elwyn B. Robinson, *History of North Dakota* (Lincoln, 1966), 469-470; David A. Horowitz, "North Dakota Noninterventionists and Corporate Culture," *Heritage of the Great*

U.S. Food and Farm policies and the Politics of Starvation

> Mr. President, what has been the result of the position taken by the Department of Agriculture for the last 12 or 14 years? One of the Secretaries of Agriculture used to take great delight in quoting the Bible. Even in Biblical times people knew enough to store products so that in time of need there would be food available. But what do we find today: this morning's *Washington Post* contains a letter written by Henry C. Taylor who enclosed a letter from Karl Brandt, who is the economic adviser to the United States Army stationed in Germany, whose letter is also printed in this morning's *Washington Post*. That letter is from the one who is more expert than anyone else in the world on this subject. I read as follows under the heading 'Famine in Germany.'[35]

William Langer was expert in providing context for his arguments from the very media sources that were read and quoted by his rhetorical targets. His frustrations with the lack of genuine support for farmers from Henry Wallace and other so-called "farm-friendly" New Dealers were shared by many of his Senate colleagues in the South as well as the North and Northwest. Proper compensation for farm productivity was a permanent issue for North Dakota. But how to get this food to people in Europe before they starved, without bankrupting the American farmer? That was the question that clearly was being dodged by the Truman Administration in March, 1946. Langer quoted Karl Brandt as stating that General Lucius Clay was being let down by his own government, and that the Russians, who had benefited from massive food shipments via Lend Lease, were releasing tons of sugar and potatoes at the height of the famine in Germany, Austria, and Poland, while Britain and the U.S. were "going on record as the ones who let the Germans starve."[36]

After Senators Shipstead (Minnesota) and Eastland (Mississippi) broke in briefly with supporting remarks, Langer recalled the efforts of Senators Eastland, LaFollette (Wisconsin), and Wherry (Nebraska) to persuade the President to intervene in the food crisis.

Plains 17 (Summer 1984): 30-39; Robert P. Wilkins, "Senator William Langer and National Priorities: An Agrarian Radical's View of American Foreign Policy, 1945-1952," *North Dakota Quarterly* 42 (Fall 1974): 42-59; Eric P. Bergeson, "Wild Bill Goes to Washington: A Reassessment of the Senate Career of North Dakota's William Langer" (M.A. thesis, University of North Dakota, 1990).

[35] *Congressional Record*—Senate, 79th Congress, 2nd Session, March 29, 1946, 2798.

[36] Ibid., 2799.

He also recalled the State Department refusing the plea of Senator Vandenberg of Michigan after he had "stood upon the floor of the Senate with tears in his eyes and presented a petition signed by 8,000 citizens of Michigan, begging our State Department to permit people in this country to send food to their relatives in Germany and Poland and Rumania and Austria."[37]

Any rational person who has spent a good deal of time in a bureaucracy is always amazed at the longevity and staying power of a dumb idea. If the goal is to build a better world with people's cooperation, starving them is a dumb idea. Feeding them is a better idea, even a good idea. Those who have toiled in bureaucracies also know that a good idea, for some strange reason, suffers like Kierkegaard's "Truth." Its power is suppressed often because it is the truth or a good idea, and it only becomes worthy of acceptance when others have joined it, not because it was the right thing to do in the first place.[38] The Morgenthau Plan and its enabling directive, JCS 1067, was a dumb idea from the moment it was conceived in the conversations of Henry Morgenthau and FDR.[39] In addition, as one early postwar analysis pointed out, Morgenthau's proposals were based upon certain assumptions, regarded as facts in 1944 that did not turn out to be the case a year later.[40] Orwell said it best, however, in "Revenge is Sour." The plan was hatched when the *Wehrmacht* was still a frightening phenomenon, and the ghastly news of the death camps was fresh. By the time Germans were within the Allied grasp, the Nazi song had ended, but their fearful melody lingered on in the ears of those in power. Moreover, Morgenthau and his men were used to being actors rather than the acted-upon, and they saw no reason why their scheme should not be tried on a nation which had caused the world so much mischief.

Langer was not naive about Germany's war-making potential in a revived Europe, but as a Senator from a farm state, he under-

---

[37] Ibid.

[38] William Hubben, *Dostoevsky, Kierkegaard, Nietzsche, and Kafka*, (New York, 1952), 9; "Truth is a power. But one can see that only in rare instances, because it is suffering and must be defeated as long as it is truth. When it has become victorious others will join it. Why? Because it is truth? No, if it had been for that reason they would have joined it also when it was suffering. Therefore they do not join it because it has power. They join it after it has become a power because others had joined it."

[39] Bird, *The Chairman*, 223-227.

[40] B. U. Ratchford and William. D. Ross, *Berlin Reparations Assignment: Round One of the German Peace Settlement* (Chapel Hill, N.C., 1947), 30-33.

stood the intimate relationship between agriculture and industry, farm and city, which Morgenthau and his circle did not.[41]

In this insight he was in the company of General Charles de Gaulle and Jean Monnet, the architect of the European Coal and Steel Community of the 1950s. These men rejected the "strategy" of closing down the Ruhr Basin: they understood the importance of placing Germany's industrial base in a Western European rather than a national context.[42]

On March 29, 1946, William Langer was concerned about a judgment of history upon Americans whose ignorance could not plead the excuse of a police state:

> ...history will not judge merely the governments or military authorities of the Allied countries. History will judge the countries themselves; history will judge the people of those countries. And history will certainly judge the most august representative body of the richest of those countries—the United States.[43]

In that speech Langer had also attacked the politics of hunger by "a savage minority of bloody bitterenders" who were forcing the "brutal Morgenthau Plan" upon the Truman Administration and empowering Stalin as a result:

> ...among the crimes with which this (Nazi) leadership has been charged (at Nuremberg) is the crime of systematic and mass starvation of racial or political minorities or opponents.... Yet to our utter horror, we discover that our own policies have merely spread those same conditions...I hold in my hands absolutely authentic photographs which have been taken at the beginning of the winter in the city of Berlin. These photographs are interchangeable for horror with the photographs with which we became familiar from Dachau, Mauthausen, Buchenwald, and other extermination camps. These are photographs of children between the ages of 5 and 14....

While reiterating his advocacy of a "swift, just, and sure trial of Nazi war criminals," Langer refused to be associated with policies which echoed of those very war criminals:

> To my knowledge, America has never known any enemy children. I refuse now to indict the American conscience as being

---

[41] *Congressional Record*—Senate, 79th Congress, 2nd Session, March 29, 1946, 2803.

[42] Kai Bird, *The Chairman*, 234.

[43] *Congressional Record*—Senate, 79th Congress, 2nd Session, March 29, 1946, 2805.

anything but revolted by the actual fact that millions upon millions of the helpless, of the sick, of the innocent, from infants to the aged, are now suffering the tortures of the damned,.... Mr. President, all the terrifying admissions on the part of our Government officials—and they are admissions—concerning the desperate food crisis now confronting humanity, are steps at least in the right direction, but there has not yet been either an official or unofficial admission of the fact that this grave threat of famine is deeply rooted in policies and practices which have already been largely responsible for the world's present predicament...."[44]

Strong stuff. But less than two months before, Orwell had written the following:

Food is a political weapon, or is thought of as a political weapon.... Many people calculate that if we send more food to, say, Hungary, British or American influence in Hungary will increase; whereas if we let the Hungarians starve and the Russians feed them, they are more likely to look towards the USSR. All those who are strongly Russophile are therefore against sending extra food to Europe, while some people are probably in favour of sending food merely because they see it as a way of weakening Russian prestige....

...The folly of all such calculations lies in supposing that you can ever get good results from starvation.... This is the "realistic" view. In 1918 the "realistic" ones were also in favour of keeping up the blockade after the Armistice. We did keep up the blockade, and the children we starved then were the young men who were bombing us in 1940. No one, perhaps, could have forseen just that result, but people of goodwill could and did foresee that the results of wantonly starving Germany, and of making a vindictive peace, would be evil. So also with raising our own rations, ...while famine descends on Europe. But if we do decide to do this, at least let the issues be plainly discussed, and let the photographs of starving children be well publicised in the press, so that the people of this country may realise just what they are doing.[45]

Like Orwell, William Langer had spent his entire career identifying with the "acted-upon," within his own state from 1914 to

---

[44] Ibid., 2801.

[45] Orwell and Angus, *Collected Essays*, "The Politics of Starvation," *The Tribune*, January 18, 1946, 85; see also, Marion C. Siney, "British Official Histories of the Blockade of the Central Powers During the First World War," *American Historical Review* 68 (January 1963): 392-401.

1940, and around the rest of the world from his seat in the U.S. Senate until his death in 1959. Although he attained the powerful positions of Governor of North Dakota, and U.S. Senator, he never really changed this attitude, often referring to himself as the "messenger boy" for the people of North Dakota.[46] Like other politicians from North Dakota throughout the twentieth century, he saw the farmers of that state in an economic and political relationship to "the actors," the large corporations headquartered in Minneapolis, Chicago, and New York, not unlike that of a Third World Country.[47] As Assistant States Attorney for Morton County in 1914 he successfully sued Northern Pacific Railroad for back taxes on behalf of the City of Mandan.[48] He thus won the enduring hostility of the powerful in his state and beyond, which in the 1930s would come to include representatives of the New Deal like Harry Hopkins.[49] His election to the U.S. Senate despite the bitter opposition of his political enemies was a triumph which he always attributed to his willingness to speak and act on behalf of the powerless.[50] He was from a state that had fought big business and Washington, D.C., to a standstill from 1915 through Pearl Harbor, but whose citizens could never relax in the face of richer and more powerful outside forces.

Even more consciously than the people of his state, William Langer had a career of making enemies of the large corporations and the leaders in both the Republican and Democratic Parties. The

---

[46] Lawrence H. Larsen, "William Langer, Senator from North Dakota," (M.A. thesis, University of Wisconsin, 1955), chap. 2; see also Geelan, *Dakota Maverick*.

[47] Danbom, "North Dakota," 107-126; Rylance, "Themes," 43-44; and Robinson, *History of North Dakota*, esp. "The Character of a People."

[48] Rylance, "Themes," 47.

[49] Judge Robert Vogel, former U.S. Attorney for North Dakota under President Eisenhower, details the antagonisms that Langer's politics and policies produced in an unpublished manuscript, "Bill Langer and His Enemies: An Unequal Contest." In researching this monograph he made full use of the Harry Hopkins papers in the Franklin Delano Roosevelt Library in Hyde Park, New York. I am indebted to Judge Vogel for permission to cite his work.

[50] Ibid.; Rylance, "Themes," 53, 55; Geelan, *Dakota Maverick*, 79-104; U.S. Senate Committee on Privileges and Elections, 77th Congress, 2nd Session, January 29 (legislative day Jan. 23) 1942, S. Rept. 1010; U.S. Senate, Election, Expulsion and Censure Cases, 1793-1990 (Washington, 1995), Case 123, 368-370; and William O. Douglas, *The Court Years, 1939-1975*, 151; "In 1939 [*sic*] the Senate had tried to bar my old friend William Langer, Senator-elect of North Dakota, from a seat, and at long last the Senate voted it had no power to exclude when the candidate had the proper credentials and possessed the constitutional qualifications."

political tightrope that he gladly walked gave him a unique perspective on the plight of the "unwanted" of the earth. The relationship that he felt he owed to the powerless in his state and around
the world, regardless of ethnicity or religion, is what explains the
passion of his rhetoric in the Senate. It explains why he argued with
equal fervor for the starving in Europe as victims of Hitler, Stalin,
and the West, whether they were German Christians or Polish
Jews. It is why he attacked the U.S. policy of imprisoning hundreds
of South American citizens of German ancestry without a trial or
hearing and kept behind wire in Texas and at Bismarck, North
Dakota, for four years.[51] It also explains why he feared for powerless Jews in Europe, and around the world, because one of their
own, Henry Morgenthau, took it upon himself to be an avenging
angel, and others, like Harry Dexter White and Frank Coe, for their
own reasons, encouraged him in this folly:

> All over Europe and South America, and in the United States as
> well, the same group continues its fanatical persecution and in-
> timidation of the German people, even the most violently anti-
> Nazi, with such abandon that across the world a new wave of
> anti-Semitism is rising.[52]

Because he was a major figure, as Treasury Secretary, in the
Roosevelt administration, and FDR's friendly neighbor from New
York, Henry Morgenthau was major figure in Washington D.C.,
and it may not have occurred to him in 1944 that the acted-upon in
his campaign against the Germans would include Jews. If it did,
that is a matter of even greater shame. By 1946 it was too late for
Henry Morgenthau, whatever his realizations might have been.
The machinery of revenge was in full throttle and would only
change when the motives of those in power would begin to coin-
cide with the sentiments of those who identified with the powerless. As such a one, Lucius Clay began a retreat away from the
Morgenthau Plan—in keeping with the ideas of many of his subor-
dinates who had always resisted it—and towards the sanity that
Langer sought in U.S. policy. In a Directive of May 3, 1946 he halted
all further deliveries of reparations from the American zone to the
East, just a few months after Langer's March 29 speech.[53] Although

---

[51] *Congressional Record*—Senate, 79th Congress, 2nd Session, March 29, 1946, 2804.

[52] Ibid.

[53] George F. Kennan, *Memoirs: 1925-1950* (Boston, 1967), 259-260; Peterson, *Retreat To Victory*, chap. 2.

there is no established connection between Langer's speech and Clay's actions in Germany, Langer was, by definition, an actor in Washington, D.C., as a U.S. Senator. Yet William Langer always thought and behaved as one of the acted-upon from North Dakota, and he could see from the beginning that Morgenthau's plan of revenge was a disaster for Nazism's victims as well as the perpetrators, anti-Nazis, and the large majority who were bystanders. To make his position absolutely clear, Langer quoted the chief rabbi of Berlin, a camp survivor, who was in the U.S. to advocate gentler policies than the ones proposed by Morgenthau and at Potsdam:

> No crime, no wrong is ever committed only against a single individual; it is as well committed against oneself... the right to punish only becomes man if he is prepared to help.... To punish and to help belong together; only together do they represent justice.[54]

Unlike many of his contemporaries in Washington, William Langer did not suffer either from the "genteel anti-Semitism" which understandably infuriated Henry Morgenthau[55] and was so well described by Laura Z. Hobson in *Gentleman's Agreement*,[56] or from the nastier versions of it in other parts of America. It was a nastier version, in fact, a speech in Congress, that had inspired Laura Zametkin Hobson to write her novel two years after that first shock:

> ...a first-page story in the National Affairs section of *Time* for the week of February 14, 1944... told of Congressman John Rankin, the Mississippi Democrat, ... referring to Walter Winchell as "the little kike I was telling you about."
> ..."This was a new low in demagoguery," *Time* said, but in the entire House, no one rose to protest.'[57]

No one in the House, perhaps. But in the Senate William Langer would do battle with anti-Semitism as well as racism a few months later in a debate with Theodore Bilbo of Mississippi, before

---

[54] *Congressional Record*—Senate, 79th Congress, 2nd Session, March 29, 1946, 2810.

[55] Kai Bird, *The Chairman*, 222-223.

[56] *Gentleman's Agreement* (New York, 1947). This novel, which tore the lid off of anti-Semitism in America right after the Second World War, was made into a movie the same year by Darryl Zanuck of 20th Century Fox, with a screenplay by Moss Hart, directed by Elia Kazan and starring Gregory Peck; see Laura Z. Hobson, *Laura Z: A Life* (New York, 1983), chap. 20.

[57] Hobson, *Laura Z*, 317, 322-323.

the latter's disgrace in his own re-election campaign and when he was still in the full flower of views which were repugnant to many, but by no means all, Americans.[58]

Because of his experience with bigoted colleagues in the Senate, Langer likely knew that his concern for the sufferings of all peoples of Europe did not resonate much in an atmosphere of indifference to the sufferings of the enemy Germans and a longstanding hostility toward Jews. Nevertheless, he had been in the forefront of Senators trying to give aid and comfort to Jews in Europe and their supporters in America. At the Manhattan Center in New York on November 17, 1942, Langer had said that the Senate's job was to "continue the fight so well begun at the First World Zionist Congress at Basel, Switzerland, where in 1897 the delegates called for 'a publicly recognized, legally secured home for the Jewish people.'"[59]

In May 1943 he was one of only five Senators on the letterhead of the Emmanuel Cellar Committee to rescue the Jews of Europe, although he is incorrectly identified as being from South Dakota, a powerful metaphor for the obscurity of both States in the minds of the movers and shakers of an organization based on the East Coast.[60] On October 6, 1943, Langer noted that five months after a refugee conference in Bermuda, Sweden had taken in Jews from Denmark, but that the United States had done nothing:

> The unprecedented mass murders and deliberate starvation of European Jews by Nazis continues unabated.... I submit that by doing nothing we have acquiesced in what has taken place over there. Perhaps if a special agency were established on behalf of this Government or in combination with other governments, to deal with the saving of the Jews of Europe, something more effective would be done.[61]

---

[58] *Congressional Record*—Senate, 78th Congress, 2nd Session, May 12, 1944, 90, Pt. 4, 4385-4387, 4404-4417, 4424-25.

[59] *Congressional Record*—Senate, 77th Congress, 2nd Session, vol. 88, Pt. 10, Appendix, A4260-1; "The acknowledgement of that fact, the establishment of that nation, will come the sooner, the harder we battle....May God bless Palestine; may God bless the United States; may God bless the United Nations. Jews, rally-rally for victory-rally to take your proper place as a nation among the United Nations." When the eventual establishment of the State of Israel began to displace Arab Palestinians in 1948, Langer spoke up for these refugees as well, see *Congressional Record*-Senate, 83rd congress, 1st session, offprint, Wednesday, February 18, 1953.

[60] "Post-Presidential Papers, Individual Correspondence File [PPI]," Historical Materials, Herbert Hoover Presidential Library, West Branch, Iowa, Box 16, File 2165, Folder I.

[61] *Congressional Record*—Senate, 78th Congress, 1st Session, vol. 89, pt. 6, 8125.

As other members of the legislative and executive branches began to pay heed to the disaster, the special agency that Langer advocated became the War Refugee Board, and on December 19, 1944 he praised President Roosevelt for his role in setting it up, while castigating State Department officials like James Dunn who, in his opinion, had stood in the way if its implementation:

> About 18 months ago the American Minister in Switzerland cabled the State Department reporting in detail regarding Nazi atrocities against civilians, especially Jews.... Sumner Welles, the Under Secretary of State, immediately called in various Jewish leaders and gave them the information. He was shocked, and felt they should know the facts. Subsequently mass meetings were staged in Madison Square Garden to raise money to aid these persecuted people and to urge further steps to protect them as far as possible.... Simultaneously, Mr. Welles instructed his colleagues in the State Department to intensify their rescue efforts. But behind his back a cable was sent to the American Minister in Switzerland which told him to send no more reports regarding atrocities.... That cable was initialed by James Dunn. Mr. Welles did not see that cable. Other officials in other branches of the Government did see it and were amazed that the State Department should tell an American minister abroad to ignore these atrocities and not report on them. Subsequently Mr. Welles called in his subordinates and... countermanded the instructions. The whole matter eventually got to the White House, where the President himself intervened and ended the whole question by appointing a new and reorganized War Refugee Board.[62]

Langer's March 29, 1946 speech reiterated a similar frustration with the Treasury Department to that which he had described three years earlier in connection with the State Department. Citing the work of the Wallenbergs of the Enskilda Bank of Stockholm, who "time and again, at grave risk to their own personal safety, served as intermediaries between the German underground and the Allied government," Langer expressed his amazement that their bank's funds were frozen in the U.S. on the orders of "Mr. Frank Coe, acting as a mouthpiece of Mr. Morgenthau's cohorts."[63] He challenged the Senate and the Truman administration to rise to the occasion, end vindictive policies, and confront a "disintegration of

---

[62] *Congressional Record*—Senate, 78th Congress, 2nd Session, vol. 90, pt. 7, 9694-9695.

[63] *Congressional Record*—Senate, 79th Congress, 2nd Session, March 29, 1946, 2804.

human society" that cut across all "racial and religious lines; ...across all...distinctions between the strong and the weak, the ex-enemy and the ally, and the guilty and the innocent alike.[64]

These words predate the appearance of Victor Gollancz's famous book *Our Threatened Values* in June 1946,[65] a work in which Gollancz attacked the stated intention of Field Marshall Montgomery to keep the German rations at 1000 calories a day (Britons were getting 2,800) because "they gave the inmates of Belsen only 800." "These words," concluded Gollancz "revealed more clearly than if they had been spoken for the purpose—the moral crisis with which Western Civilisation is faced." Gollancz doubted that even a thoughtful minority in the English-speaking world realized how grave the crisis was, and in this he was echoing the words put forth by William Langer on the floor of the Senate two months earlier.[66]

The Refugee Situation—Ethnic Cleansing

> Mr. President, this great human tragedy now confronting us holds in its horrible embrace every group, race, and nationality represented in the vast cross section of the peoples of the world who now constitute the backbone of America. What many Americans cannot understand is that we are all in the same boat together. The tragedy is no respecter of persons. For instance, the minority which has suffered most under Nazi persecution is now crying aloud for redress and for help in the midst of the new catastrophe. On February 25, Mr. Bernard Baruch described the plight of European Jewry in the following words: 'Added to their physical suffering is their mental anguish, for they have become the unwanted, driven   from place to place, welcome nowhere. Constant fear presses them to move on somewhere— some-how—anywhere away from the persecutions existing even now. They do not want to go back to the countries they left because there robbery, riot, and even murder stalk the land.'[67]

Senator William Langer's wholesale championing of the downtrodden on March 29, 1946, affords us the opportunity to see a point where Holocaust and *Vertreibung* intersect—German-speaking (or even Yiddish-speaking) Jews, surviving the Holo-caust, only to be swept up with Gentiles to become victims of the

---

[64] Ibid., 2799-2800.

[65] Victor Gollancz, *Our Threatened Values* (London, 1946).

[66] Ibid., 7.

[67] Ibid., 2799.

Vertreibung, or to be put in DP camps where conditions remained deplorable until the dictates of the Morgenthau Plan and JCS 1067 were countermanded.[68] To bolster his arguments he cited several eyewitness reports from sources like Countess Helmuth von Moltke, travelling from Silesia to Berlin; F. A. Voigt, erstwhile Berlin correspondent of the *Manchester Guardian* on the policies of Russian Marshal Zhukov regarding Danzig, the Sudetenland, Mecklenburg, and Brandenburg; and an antifascist Zurich journal, the *Welt-Woche*, on the fate of Germans in the Polish zone.[69] Although the fate of the ethnic Germans in all these areas was grisly, Langer showed considerable sympathy for the Poles:

> In describing the conditions in the Polish zone one must be very careful to point out that in this instance our decisions at Potsdam only compounded the crimes which we committed against Poland at Teheran and Yalta. Poland has been stabbed in the back not only by Germany, but by Russia, England, and the United States as well.... Pandemonium reigns in Poland. There is no security left for man or beast. Hordes of desperate and hungry men swarm over the countryside. The treacherous Warsaw government, spawned in Moscow, is dividing and liquidating Poland in order to conquer the Polish people. Murder, looting, and rape are the order of the day.[70]

Interesting here is Langer's clear-eyed, public description of the consequences of Soviet motives and methods in action, only five weeks after George Kennan's now well-known, but then still confidential "Long Telegram" warning his government of the duplicities and atrocities of Soviet policies.[71] Kennan had been an eyewitness to the *Vertreibung* in 1945, and his memoirs demonstrate that Langer was right on target in his remarks from the floor of the Senate:

> Not the least depressing, to me, of the published results of the Potsdam Conference, was the sanctioning and further refinement of the earlier decisions relating to the severance of East Prussia from Germany, the partition of that province between Russia and Poland, and the cession to the Soviet Union, in particular, of the

---

[68] Ibid.

[69] *Congressional Record*—Senate, 79th Congress, 2nd Session, 29 March 1946, 2805, 2806.

[70] Ibid., 2806.

[71] Kennan, *Memoirs*, "Excerpts from Telegraphic Message from Moscow of February 22, 1946," 547-559.

administrative center and port of Koenigsberg.... These territorial changes seemed to me to be doubly pernicious, and the casual American acquiescence in them all the less forgivable, because of the fact that they served... simply to extract great productive areas from the economy of Europe and to permit the Russians, for reasons of their own military and political convenience, to deny these areas and their resources to the general purposes of European reconstruction.... The disaster that befell this area...has no parallel in modern European experience.... [the Russians] swept the native population clean in a manner that had no parallel since the days of the Asiatic hordes.... There were some, I suppose, to whom 2.5 million human inhabitants of East Prussia could be regarded as expendable; but what about the 500,000 horses, the 1.4 million head of cattle, the 1.85 million pigs that were once to be found in the place? And what about the near 4 million tons of wheat, 15 million tons of rye, and 40 million tons of potatoes it had once produced annually?... I have never been able to understand the indifference of our statesmen, and of our public, to these circumstances. It is one thing for a great country to develop a territory for its own uses.... It is another thing... to take fertile territory merely in order to turn it, because this suits one's military convenience, into a wilderness.[72]

William Langer was not like the American officials and public whose attitudes were so shocking to Kennan. The Senator had not been indifferent to these developments, and his speech predates by nine months Kennan's short telegram to the State Department advising the proper handling of the Russians,[73] and by more than a

---

[72] Ibid., 263-266.

[73] Ibid., 291-292; "Excerpt from 'The United States and Russia'[the 'short telegram'],"
  A. Don't act chummy with them';
  B. Don't assume a community of aims with them which does not really exist;
  C. Don't make fatuous gestures of good will;
  D. Make no requests of the Russians unless we are prepared to make them feel our displeasure in a practical way in case the request is not granted;
  E. Take up matters on a normal level and insist that Russians take full responsibility for their actions on that level;
  F. Do not encourage high-level exchanges of views with the Russians unless the initiative comes at least 50 percent from their side;
  G. Do not be afraid to use heavy weapons for what seem to us to be minor matters;
  H. Do not be afraid of unpleasantness and public airing of differences;
  I. Coordinate, in accordance with our established policies, all activities of our government relating to Russia and all private American activities of this sort which the government can influence;
  J. Strengthen and support our representation in Russia."

year his famous "Mister X" article in the July 1947 issue of *Foreign Affairs*, a cornerstone of containment policy vis-à-vis the Soviets.[74]

Langer's timing was closer to that of someone he viewed as the ultimate imperialist, Winston Churchill—his use of the term "iron wall" coming a few weeks after that of Churchill's "iron curtain" speech in Fulton, Missouri:

> Certainly the iron wall which Russia has erected around the Eurasian continent, within which the nitrate deposits and the agricultural areas of Germany have been imprisoned, make the threat of famine all the more grim in the years immediately ahead. With Russia stripping the countries under her control, pillaging and living off the land, and unable herself to meet her own food requirements, the prospect of famine becomes increasingly inevitable.[75]

Tacit approval of the "iron wall" was count number one in Langer's four-count indictment of American policy. Count number two was the wholesale deindustrialization of Germany called for in the Morgenthau Plan. Count number three was America's tacit or deliberate commitment to the revival and extension of slave labor. Count number four was American acquiescence and silence in face of the *Vertreibung*:

> A fourth problem, which is inseparably intertwined with the three I have mentioned, now lays the ghastly consequences of Mr. Morgenthau's policies squarely on the conscience of the world. I speak in bitter protest of the fiendish mass deportations which are making a charnel house of central and eastern Europe, and which are being carried on in another name, namely, denazification, in every country of the world.[76]

Langer's intense tone of resentment irritated one of his colleagues from the wealthier states. Towards the end of his March 29th speech, Langer's critique was called into question by Senator

---

[74] William Appleman Williams, *The Shaping of American Diplomacy: Readings and Documents in American Foreign Relations* (Chicago, 1956), 947-948, 994-996.

[75] Ibid., 947, 992-994; *Congressional Record*—Senate, 79th Congress, 2nd Session, March 29, 1946, 2803; see also Wynona H. Wilkins, "Two If By Sea: William Langer's Private War Against Winston Churchill," *North Dakota History* 41, no. 2 (Spring 1974): 20-29.

[76] *Congressional Record*—Senate, 79th Congress, 2nd Session, 29 March 1946, 2803, 2804.

William Knowland of California, in phrases that would be repeated throughout the Cold War against critics of the emerging bipartisan foreign policy:

> I would not want this afternoon to pass following the re-marks of the Senior Senator from North Dakota and have the impression remain in the Senate and throughout the country that the United States Government had pursued a policy comparable to the Nazi policy in Europe, because in my opinion such is not the case. It happened to have been my privilege to have served for a time in Germany, and I think I know first-hand some of the conditions existing there.... Mr. President I would not want to go unchallenged today any suggestion that the American Army has fallen down in its occupation job, because I do not believe that is so... it is not a fact that this Government is engaging in any con-spiracy to starve the German people....
>
> ...the dismantling of German industries...is essential to the future peace of the world...it is very easy to turn chemical facto-ries (for fertilizer) and factories which in the normal course of events are used only for peacetime purposes into a vast war ma-chine. In this Nation, one of the most peaceful nations of the world, we found that in the relatively short period of a few years we could change from a nation without an armament industry, so to speak, to the greatest armament producer on the face of the earth.... I did not mean to take the time of the Senate today to dis-cuss this subject, but I felt, in fairness to the Senate and in fair-ness to the country, that at least a different point of view should be presented.[77]

Langer's reply to Knowland was that, despite his highest re-gard for the "distinguished Senator from California," he preferred to believe the reports of Mr. Karl Brandt, the man appointed by the U.S. government to go to Berlin, and his ten-day-old report, than impressions gathered by Senator Knowland from six to eight months previously. He also pointed out that he and Senator Knowland were sitting in a comfortable, air-conditioned, Senate chamber, while Brandt was toiling amidst the wreckage of Berlin, and the immediate, visceral evidence of starving thousands, with reports coming in of starving millions.[78] He then started in on any doubts that the Senator from California might have about his patri-otism:

---

[77] *Congressional Record*—Senate, 79th Congress, 2nd Session, 29 March 1946, 2811.

[78] Ibid.

I have just as high a regard as has the distinguished Senator from California for what has been done by the Army and Navy and the Air forces in this war, if not a higher regard. I yield to no man when I say that I believe we have the finest fighting force in the entire world.[79]

Langer then challenged Senator Knowland to listen to the words of shame uttered by the American representative in Berlin:

I wish particularly to have the distinguished Senator from California listen to [Karl Brandt's] statement: "The greatest famine catastrophe of recent centuries is upon us in central Europe. Our Government is letting down our military government in the food deliveries it promised, although what General Clay, General Draper, and General Hester asked for and were promised was the barest minimum for survival of the people....

...The few buds of democracy will be burned out in the agony of death of the aged, the women, and children...It makes all the many hard-working officers of Office of Military Government, Food and Agriculture Branch, ashamed."[80]

Langer concluded by vowing never to be quiet when such injustices were being carried on in the name of the American people, "no matter how unpopular the cause may be; and when the time comes when I cannot speak for what I believe to be right, I shall walk out the door with my head up and my conscience clear."[81]

Paul Douglas confirmed in 1959 that, despite his disagreement with Langer over America's position in the world, it was performances like these during Douglas's tenure in the Senate, that caused him to hold Langer in such high regard, and to be convinced that he "was the most misunderstood, the most improperly attacked, and the one who perhaps suffered more from the hands of his opponents than any other Member of this body."[82]

Conclusion

It is, perhaps, understandable that Langer's words were sometimes too painful for his Senate colleagues to hear without at least one of

---

[79] Ibid.

[80] Ibid.

[81] Ibid.

[82] 86th Congress, 2nd Session, *Memorial Services*, 54.

them firing back as did the Senator from California. But what about posterity? What about historians? Why have they been so slow to acknowledge a prescience similar to that of his great contemporaries, George Orwell and Kennan? McCarthyism, covert murders, and wars by the C.I.A.; domestic spying on U.S. citizens by all kinds of federal, state, and local agencies; Vietnam; Watergate; Iran-Contra? These and other manifestations of an America made cynical by its victory in World War II are as much a testimony to Langer's warnings in 1946 as they are of Orwell's in 1949. Yet Langer remains obscure, except in the memory of his own state, among the ethnic Germans, and many other people he helped around the country, colleagues in the U.S. Senate, and in the pages of the Congressional Record.

In an article published in 1974 Robert Wilkins addressed Langer's anti-internationalism as a source of why he had been so misunderstood, marginalized, and ridiculed by the media of his day, and why his influence in the Senate was undercut so drastically.[83] But Wilkins also pointed out that Langer's attack, as early as 1940, on the exclusion of the American voter from foreign policy decision-making through the device of bi-partisanship, would turn out to be extremely popular soon after his death, when taken up by Senators Morse, Fulbright, McGovern, and others in the 1960s.[84]

William Langer was a consummate and canny politician, but he considered it an abomination to play politics with human misery. He is there in the Senate record, to be read, if not by the cynics who still play politics with human misery today, then at least by historians who find it interesting that a confirmed isolationist when it came to taking lives, would advocate intervention when it came to saving them.

---

[83] Robert Wilkins, "National Priorities," 42-59.

[84] Ibid., 50.

A House Divided: The Catholic Church and the Tensions between Refugees-Expellees and West Germans in the Postwar Era

FRANK BUSCHER

In November 1989 Germans in both East and West greeted the opening of the Berlin Wall with exuberance and joy. The celebratory mood, however, soon gave way to the realization that forty years of division had left its mark on German society. East and West Germans generally viewed and treated each other as strangers.[1] In the following decade—once the true economic, financial and social costs of reunification began to manifest themselves—resentment and tension replaced the feelings of estrangement. East Germans considered their Western countrymen pampered, selfish and annoying know-it-alls, so-called *Besser-Wessis*. At the same time, West Germans regarded the residents of the former GDR as lazy, slothful, and incapable of adjusting to the Federal Republic's capitalist economy.

Such feelings had precedent in recent German history. In the decades after the Second World War West Germans and the millions of refugees and expellees who came to the West in and after 1945 also resented each other. This paper will examine these social tensions and how the Catholic Church in Germany dealt with them. It will be based on primary source records from the Archdiocese of Cologne and the *Land* North Rhine-Westphalia. Both played important roles. The archbishop of Cologne, Cardinal Joseph Frings, served as special papal deputy for refugees (*Hoher Protektor für das Flüchtlingswesen*) and chairman of the Fulda Bishops Conference, the annual meeting of the German episcopate, in the postwar period. As such, Frings was continuously involved in the expellee and refugee issue. His offices received and produced numerous documents, and the files of Cologne's archdiocesan archive are thus particularly valuable. The Archdiocese of Cologne was one of four dioceses in the *Land* North Rhine-Westphalia (the others were the Archdiocese of Paderborn and the dioceses of Aachen and Münster). At the same time, among the West German *Länder* North Rhine-Westphalia became home to the largest number of expellees and refugees. By 1961 over 3.2 million had moved to the

---

[1] Mary Fulbrook, "*Ossis* and *Wessis:* the Creation of Two German Societies," in *German History Since 1800*, ed. Mary Fulbrook (London, 1997), 411-31.

Land whose industries in the Rhine and Ruhr offered abundant employment opportunities. Owing to this immense influx, the documents left by officials of the *Land* have also been of great interest to historians.

It is the aim of this study to show that then as now native West Germans were preoccupied with their own problems and largely indifferent to the needs and sensitivities of their countrymen from the East. An important contributing factor to this regrettable attitude was undoubtedly the knowledge that the refugee and expellee problem was going to be solved primarily on Western terms. (At the end of the century an almost identical viewpoint emerged during the process of reunification, once again dividing the German people.) To be sure, not all West Germans disliked or rejected the expellees and refugees. Since many residents of the cities in the Rhine and Ruhr had earlier been migrants themselves, they tended to accept and show compassion for the newcomers from the East.[2] Further, as Catholic clergymen duly noted, many natives provided material help or donated money, although the extent of their assistance and donations routinely fell short of expectations. For example, at the end of 1946 Cologne's *Caritas* chapter reported "numerous, quietly conducted charitable acts" by both clergy and laypersons.[3] The Catholic Church also conducted several special collections for expellees and refugees. In 1946 the West German dioceses collected the seemingly impressive but wholly inadequate sum of RM 2.79 million. The money was intended to secure pastoral care for an estimated 2.5 million Catholic expellees and refugees deported to the Protestant regions of North Germany and the Soviet zone.[4] Six years later, when West Germans enjoyed noticeably greater prosperity, the government of the Federal Republic asked its citizens to donate funds earmarked for refugees from the Soviet zone. Again, the natives opened their wallets, but only half-heartedly. North Rhine-Westphalia's share of the money collected was almost DM 130,000, but the *Land's* Social Ministry rightfully complained that this amount was "relatively

---

[2] Alexander von Plato, "Skizze aus dem Revier—Thesen zur Integration von Flüchtlingen und Einheimischen in die neue Zeit," in *Flüchtlinge und Vertriebene in der westdeutschen Nachkriegsgeschichte*, ed. Rainer Schulze et al. (Hildesheim, 1987), 264-68.

[3] Diözesan-Caritasverband Köln to Frings, 23.12.1948, Historisches Archiv der Erzdiözese Köln (HAEK), CR II 25.20b/8.

[4] Diaspora-Kommissariat der deutschen Bischöfe in Paderborn, Kollekte für die Ostflüchtlinge, 1.6.1946, HAEK, CR II 25.20b/5.

small,"[5] an understatement in light of the over 620,000 refugees from the Soviet zone who had migrated to the *Land* by August 1953.[6]

In the West German countryside the greatest problems were outright rejection and resentment rather than miserliness. The war had left rural areas and small towns relatively unscathed, but in the immediate postwar years they bore the brunt of the influx of expellees and refugees. In West Germany the primarily rural *Länder*—Bavaria, Schleswig-Holstein, and Lower Saxony—received the vast majority as the level of wartime destruction made most urban areas unsuitable for expellee and refugee settlement. By April 1947 10,096,000 Germans expelled from Eastern and Southeastern Europe had arrived in the four zones of occupation. Two years later, the American and British zones alone had accepted 7.5 million and in 1950 the Federal Republic counted almost 8 million. By then expellees and refugees made up 16.5 percent of the total West German population but 33 percent in Schleswig-Holstein, 27.2 percent in Lower Saxony and 21.1 percent in Bavaria. North Rhine-Westphalia initially accepted relatively few expellees and refugees. On 1 May 1947 only 941,000 lived in this *Land*.[7] During the next two years, this number increased only marginally, to 1.1 million. Almost three-quarters of the new arrivals were assigned to rural regions, a move the *Land's* government defended by pointing to the great destruction of housing in urban areas.[8]

Rural West Germans rejected the expellees and refugees for a variety of reasons. They resented having to make economic sacrifices for the new arrivals and feared *Überfremdung*, i.e. too much alien penetration of their communities.[9] Further, robbed of their material possessions and their dignity by the process of expulsion, the expellees and refugees arrived as dispirited paupers and usually lived in camps or other types of mass housing, where

---

[5] Sozialministerium NW, Vermerk betr: Verteilung der vom Bund zugewiesenen Spendenanteile an SBZ-Flüchtlinge, 7.9.1953, Hauptstaatsarchiv Düsseldorf (HSAD), NW 67/1369/191-2.

[6] Tätigkeitsbericht 1952 und 1953 der Diözesanseelsorge für die Heimatvertriebenen im Erzbistum Köln, HAEK, CR II 25.20b/25.

[7] Das Flüchtlingswesen in Nordrhein-Westfalen, 1947, HSAD, NW 7/110/97-101.

[8] Sozialministerium NW, The Refugee Problem in *Land* North Rhine/Westphalia, 19 September 1949, HSAD, NW 7/120/54-61.

[9] See Helga Grebing's and Kurt Jürgensen's brief essays on Lower Saxony and Schleswig-Holstein in *Flüchtlinge und Vertriebene*, ed. Schulze et al., 269-81.

they suffered from overcrowding and decidedly unhealthy conditions. Unemployment among male expellees and refugees was relatively high, in large part a result of the lack of employment opportunities in rural regions. In 1949 9.8 percent of the expellee and refugee workforce in the Detmold district and 10.3 percent in the Münster district did not have work.[10] The situation in Bavaria was far worse. There, the unemployment rate among expellees and refugees in 1950 was almost 30 percent, in some camps in the countryside it reached 70 percent.[11] Consequently, many West Germans in rural areas and small towns viewed and treated their new, unwelcome neighbors as losers, lazybones and thieves.[12]

Unfortunately, those who should have served as positive examples frequently failed. In his research on the churches and the expellee and refugee problem in Bavaria, Ian Connor found a divergence in the behavior of local priests and their superiors. While church leaders pushed for speedy integration to prevent the emergence of political radicalism, parish priests sided with their original flocks and rejected the newcomers, even though the latter's need for comfort and pastoral care undoubtedly exceeded that of the natives.[13] Judging by the number of complaints received by the Archdiocese Cologne, the conduct of Bavaria's local clergy was by no means unique. In general, the documentary evidence shows that parish priests behaved like the native West German population; some chose to look after the expellees and refugees and showed remarkable sensitivity in the process,[14] others preferred to ignore the problem and instead focused exclusively on the distress of local parishioners. The Catholic Church,

---

[10] Source unknown, 1949, HSAD, NW 7/110/102-03.

[11] Ian Connor, "The Churches and the Refugee Problem in Bavaria, 1945-49," *Journal of Contemporary History* 20 , no. 3 (1985): 399-421.

[12] Rainer Schulze, "Growing Discontent: Relations between Native and Refugee Populations in a Rural District in Western Germany after the Second World War," *German History* 7 (1989): 332-49; and Paul Erker, "Revolution des Dorfes? Ländliche Bevölkerung zwischen Flüchtlingszustrom und landwirtschaftlichem Strukturwandel," in *Von Stalingrad zur Währungsreform*, ed. Martin Broszat et al. (Munich, 1990), 367-425.

[13] Connor, "The Churches," 401.

[14] Among these was Pfarrer Dörenkamp in Cologne who invited expellees and refugees to his parish for a conversation about their religious and charitable needs. Their dreadful mental and economic condition convinced Dörenkamp to call for intensified pastoral care and charity. Dörenkamp to Frings, 6 June 1946, HAEK, CR II 25.20b/6.

however, needed to set a good example to persuade West Germans to open their "hearts and doors" to the expellees and refugees. Allied and German authorities as well as Catholic relief officials were convinced that the churches were the only institutions in the immediate aftermath of the war with the moral authority necessary to encourage local populations to assist the newcomers. They urged the Catholic hierarchy to remind West Germans of their "moral duties" to perform "deeds of true Christian charity."[15] The Archdiocese of Cologne reacted to such requests by ordering all diocesan priests, particularly those serving in communities with expellee and refugee populations, to act as role models for their parishioners through "words and deeds."[16]

As the influx of expellees and refugees reached its peak during 1946/47, Catholic leaders began to worry about another issue. *Christ Unterwegs*, a Catholic expellee publication, reported that 5.5 million of the roughly ten million expellees and refugees who had arrived in the four zones by April 1947 were Catholics. Seventy percent were assigned to heavily Protestant areas, where Catholic places of worship and priests were few and far between. As a consequence, the Church and its relief agencies encountered great difficulties in providing badly needed pastoral care and material assistance. Even worse, at least from the Catholic perspective, were efforts by Protestant clergy to fill in the gap. A Catholic official from the Soviet zone alleged angrily that Protestant pastors were boasting about the large numbers of Catholic expellees and refugees attending their services.[17] But priests who worked directly with the new arrivals often provided a more balanced assessment. While lamenting that many expellees, and "not even the worst," attended Protestant services, with some participating in the Eucharist, they also praised Protestant pastors and officials. Protestant churches, *Christ Unterwegs* reported, were being made available for Catholic purposes without requests for compensation.[18] In the Archdiocese of Cologne, where officials

---

[15] Deutscher Caritas-Verband to Frings, 27 September 1945, HAEK, CR II 25.20b/3 and Regierungspräsident Düsseldorf to Frings, 26 October 1945, HAEK, CR II 25.20b/3.

[16] Erzbischöfliches Generalvikariat to Pfarrämter in der Erzdiözese Köln, 25 July 1946, HAEK, CR 25.20b/6.

[17] Erzbischöflisches Geistliches Gericht Erfurt to Provinzial der Jesuiten P.W. Flosdorf, 26 April 1946, HAEK, CR II 25.20b/5.

[18] Diasporaberichte, *Christ Unterwegs*, July 1948.

emphasized their good relationship with the "other Christian religion," Catholics made use of eighteen Protestant houses of worship throughout 1947.[19] Nonetheless, Catholic leaders continued to watch the efforts of the Protestant Church with unease, caused in large part by suspicions that the latter was attempting to expand its influence at Catholic expense as well as considerable feelings of inferiority. The Protestant Church appeared to be ahead in the race to solve the expellee and refugee problem, its relief efforts seemed to be superior and its clergymen came across as better role models than their Catholic counterparts.[20]

Complaints about the indifference shown by local Catholic clergymen began to increase in 1948. One cause was the growing willingness of expellee clerics to criticize their West German colleagues. Emphasizing that the collective attitude of the native clergy depended on the cooperation and goodwill of each individual pastor,[21] expellee priests cited reports that, in contrast to the Vatican and the German episcopate, local clerics "in the villages and small towns" did not consider assisting the expellees and refugees a priority.[22] One of the most vocal critics was Monsignor Albert Büttner, the head of the *Kirchliche Hilfsstelle*, a Catholic office which looked after and sought to achieve the integration of expellees from Southeastern Europe. The new arrivals, Büttner wrote in August, were highly critical of the Catholic Church for the slowness of its effort to find a solution to the expellee and refugee problem. They particularly resented native pastors "who continue to live in palaces and have plenty of food." Büttner warned that the Protestant Church was striving to become the leading provider of care for expellees and refugees. Even worse, if local clerics persisted in ignoring their plight, the

---

[19] Bericht über das Flüchtlingsproblem in der Britischen Zone, Fragebogen B-Kirchen, 9.1.1948, HAEK, CR 25.20b/11.

[20] Dr. Püschel, Abteilung Caritas-Flüchtlingshilfe, Deutscher Caritas-Verband, to Frings, 21 April 1949, HAEK, CR II 25.20b/16; Friedrich Froehling, "Beurteilung der Konferenz der Flüchtlingsabteilung des Ökumenischen Rates der Kirchen, 3 March 1949, HAEK, CR II 25.20b/16; Sladek to Frings, 2 June 1950, HAEK, CR II 25.20b/21; and Referat über die gegenwärtige Situation, 1 August 1951, HAEK, CR II 25.20b/23.

[21] Der Diözesanseelsorger für die Ostvertriebenen im Erzbistum Köln, Jahresbericht 1947, 1 February 1948, HAEK, CR II 25.20b/11.

[22] Kirchliche Hilfstelle München to Erzbistum Köln, 1 March 1948, HAEK, CR II 25.20b/11.

desperate newcomers could become easy targets for Communist or neo-Nazi agitators. Büttner was also concerned that the expellees and refugees might become a permanent underclass in the wake of the currency reform and ending of many price controls in the Western zones in June 1948.[23]

These economic measures indeed hurt the two groups. The expellees and refugees had benefited from price controls and the lack of a stable currency which had caused many West German employers to offer them sham employment. After the introduction of the *Deutsche Mark*, such employees were laid off and replaced with native labor. Similarly, the lifting of price controls on crucial goods and services such as food and transportation negatively affected expellees and refugees. This development contributed to increasingly sharp verbal attacks by expellee clerics. During a conference in November they accused West German pastors of "frequent failure." Local priests, they charged angrily, avoided expellee and refugee camps and other types of mass housing for fear of the dreadful conditions they would encounter.[24] Hans Lukaschek, the former *Oberpräsident* of Upper Silesia who became expellee minister in Konrad Adenauer's first cabinet, adopted a more moderate tone. Nonetheless, the prominent Catholic expellee considered it necessary to bring reports about the callousness of Bavarian clergymen to the attention of the episcopate. In contrast to the generous attitude shown Catholic expellees and refugees by Protestant pastors, several Catholic parish priests had refused Protestant expellees use of Catholic places of worship, a development Lukaschek found distressing.[25]

For Cologne's diocesan refugee pastor Oskar Golombek, an expellee from Upper Silesia, 1948 also ended on a sad note. The clergy had not become "electrifying models" who inspired love of one's fellow men among the faithful, he noted in his annual report. More importantly, Golombek already envisioned the next difficult issue to drive a further wedge between natives and newcomers as well as between local and expellee clergy. That issue was the *Lastenausgleich*, the concept that those who had

---

[23] Die kirchliche Betreuung der ausgewiesenen Volksgruppen: Aufgabe und Leistung der Kirchlichen Hilfsstelle, 4 August 1948, HAEK, CR II 25.20b/13.

[24] Einige Merksätze aus der Aussprache der Diözesanflüchtlingsseelsorger, November 1948, CR II 25.20e/2.

[25] Lukaschek to Bishop Ferdinand Dirichs, 26 December 1948, HAEK, CR II 25.20b/15.

suffered serious material losses as a result of the war deserved compensation. Although the *Lastenausgleich* did not become federal law until August 1952, the debate concerning the necessity of such an equalization of burdens began shortly after the war. Church leaders, who viewed the uneven distribution of property as a major source of social problems, supported the idea from the beginning.[26] Germany's Catholic bishops were eager to protect private property, but unwilling to prevent its concentration in the hands of the few.[27] Unfortunately, lower-ranking clerics once again failed to act in accordance with the hierarchy's wishes. Cologne's Golombek noted bitterly the absence of a *Lastenausgleich* within the clergy itself as West German churchmen were unwilling to share with their impoverished colleagues from the East.[28] But Catholic expellees were not alone in their criticism. Some West German pastors also disapproved of their colleagues' tendency to side with their original parishioners and ignore the expellees' and refugees' plight.[29] For the most part, however, local clerics and church officials were remarkably insensitive, accusing the expellees and refugees of living at the local population's expense and attempting to elude their share of responsibility for the Nazi past.[30]

The German episcopate remained concerned about the attitude of local clerics well into the 1950s. In August 1953 the Fulda Bishops Conference announced that it expected a more intense effort from the clergy providing pastoral care and charity to the expellees and refugees. The bishops reminded clergymen of continuing tensions between local populations and the new arrivals. They also emphasized that the *Lastenausgleich* had yet to alleviate the latter's suffering.[31] West German churchmen by and large did not respond positively to such appeals. In the Advent of 1953 the Catholic pastors assigned to the refugee camps

---

[26] Frings to Prälat Paul Ramatschi, 24 March 1947, HAEK, CR II 25.20b/8; also see *Christ Unterwegs*, 1 February 1947.

[27] Hirtenbrief der in Fulda versammelten deutschen Bischöfe, *Christ Unterwegs*, 1 August 1947.

[28] Golombek to Frings, 19 February 1949, HAEK, CR II 25.20b/15.

[29] Pfarrer Magnani to Erzbistum Köln, 27 December 1949, HAEK, CR II 25.20b/19.

[30] Kläre Schwarzer to Frings, 3 April 1950, HAEK, CR II 25.20b/20.

[31] Protokoll der Plenarkonferenz der Bischöfe der Diözesen Deutschlands in Fulda vom 18. bis 20. August 1953, HAEK, CR II 25.20b/25.

Friedland, Sandbostel and Uelzen appealed to the faithful in the Archdiocese Cologne to share a portion of their Christmas presents with their Eastern co-religionists now residing in the camps.[32] The local clergy's response was particularly disappointing, as 210 of the diocesan churchmen donated only DM 3,200. The camp pastors criticized their colleagues for failing to recognize the importance of the Church's work in the camps, while Cardinal Frings launched a renewed request for donations.[33]

Since Catholic clergymen were by and large poor role models, it should not be surprising that West German Catholics as a whole also failed the millions of expellees and refugees. To be sure, the natives had their own problems in the postwar years. However, their reaction to the newcomers went beyond the mere determination to focus on one's own troubles. Instead, the historical record reveals that from the beginning the locals' response was frequently governed by outright and bitter resentment. In North Rhine-Westphalia the representatives of expellees and refugees used the first meeting of the *Landtag's* refugee committee in February 1947 to complain bitterly about the lack of compassion on the part of officials and the *Land's* population. For many ordinary West Germans, the newcomers were unwelcome for a variety of reasons. In the popular imagination, the expellees and refugees were responsible for the lost war, competed with the natives for scarce food and resources, and received better treatment and greater privileges.[34] Already at the height of the expulsion in 1946/47, some locals began to refer to expellee children and youths as "Polacks" and "Russians."[35] The most senior of the refugee clergy, Maximilian Kaller, the former bishop of Ermland and special papal deputy for German refugees from June 1946 until his death in July 1947, complained that West Germans treated expellees and refugees like "undesirable intruders and beggars."[36] A top official from Hesse,

[32] Lagerpfarrer von Friedland, Sandbostel und Uelzen to Confratres, December 1953, HAEK, CR II 25.20b/25.

[33] Krahe to Frings, 18 January 1954, HAEK, CR II 25.20b/25.

[34] Volker Ackermann, *Der "echte" Flüchtling: Deutsche Vertriebene und Flüchtlinge aus der DDR 1945-1961* (Osnabrück, 1995), 66.

[35] Der Diözesanseelsorger für die Ostvertriebenen im Erzbistum Köln, Jahresbericht 1947, 1 February 1948, HAEK, CR II 25.20b/11.

[36] *Christ Unterwegs*, 1 February 1947.

Peter Paul Nahm, noted with amazement the inexplicable callousness displayed by the locals in light of the suffering they witnessed daily.[37] The seemingly deteriorating relationship between the two groups also drew the attention of the Catholic Refugee Council, a body of leading expellee and native clergymen who served as advisors to the German episcopate and Cardinal Frings. At its first meeting in October 1948 the council determined that the chasm between locals and newcomers had deepened and called on the Catholic Church to provide ecclesiastical leadership for the expellees and refugees.[38] A meeting of diocesan refugee pastors came to a similar conclusion and reported that most expellees were now surviving solely on their hope for a return to their homes in the East.[39] The heads of Germany's leading Catholic relief organizations saw the problem as a largely rural phenomenon and decided to step up their charitable activities in villages and small towns to create a better understanding between the locals and their new, unwanted neighbors.[40]

In 1949 growing native-expellee tensions began to impact the *Länder* of the French zone. Until then, among the Allies occupying Germany, the French had accepted by far the smallest number of expellees and refugees for their zone.[41] But this changed with the start of several major relocation programs during which millions of expellees and refugees were resettled from Bavaria, Schleswig-Holstein and Lower Saxony to the other *Länder* in West Germany. Initially, the *Länder* of the French zone balked, however, demanding financial compensation for accepting their quota of expellees and refugees. These demands caused deep disappointment among Catholic expellees who had hoped for a show of solidarity on the part of their "Catholic brothers and sisters" in Southwestern Germany. Concerned about the image of local Catholics, the leading churchmen in the French zone lobbied

---

[37] Nahm to Frings, 11 December 1946, HAEK, CR II 25.20b/8.

[38] Kurzprotokoll der ersten Sitzung des Katholischen Flüchtlingsrates, 10 October 1948, HAEK, CR II 25.20b/18.

[39] II. Konferenz der Diözesan-Heimatvertriebenen-Seelsorger, 24 and 25 February 1948, HAEK, CR II 25.20e/2.

[40] Kurzprotokoll der Arbeitsgemeinschaft für katholische Flüchtlingshilfe in Limburg-Lahn vom 20.10.1948, 4 November 1948, HAEK, CR II 25.20b/14.

[41] By August 1948 the French zone had accepted 68,000 expellees and refugees. This compares to 3.1 million for the British zone, 3.5 million for the American and 4.4 million for the Soviet.

their *Länder* governments for a more lenient approach.[42] Unfortunately, the native population showed little sympathy for the expellees' and refugees' physical and mental suffering. This disturbed both state and ecclesiastical authorities. In November 1949 the Adenauer government ordered the *Land* Rheinland-Pfalz to accept 90,000 expellees during 1950. To prepare for the influx, the *Land's* Social Ministry asked for and both churches agreed to publicize pastoral letters urging the locals to do their duty as humans and receive the new arrivals with open hearts. In those parts of the *Land* belonging to the Archdiocese of Cologne Catholics were called upon to offer special assistance to youths, the unemployed and young mothers. While admitting to the faithful that social tensions were unavoidable, the Archdiocese nonetheless instructed them to treat the new members of their communities with goodwill and kindness.[43]

Both churches hinted that the mixing of denominations throughout Western Germany, which had resulted from the authorities' expellee and refugee settlement policies, made it more difficult to inspire parishioners. In the spring of 1949 the British military government asked the *Länder* of its zone about changes in their denominational structures. According to the North Rhine-Westphalian Social Ministry, the influx of expellees and refugees had caused the number of Protestants to increase by 16-17 percent (to about 41 percent of the population) and that of Catholics by only 6-7 percent (to roughly 55 percent).[44] Both churches bemoaned that religious preferences had not been taken into account during the settlement process. As a result, most Protestant expellees and refugees were settled in predominantly Catholic areas, whereas their Catholic fellow-sufferers ended up in the largely Protestant regions of Northern Germany. In short, the expulsion had led to the emergence of numerous diaspora communities throughout the four zones. Both churches complained about the difficulties this situation had created. Pastoral care and charity were problematic due to a lack of clergy and the great distances clergymen had to cover in the diaspora to reach their co-religionists. More significantly, in many places the sudden mixing of denominations had undone historical tradition

---

[42] Prälat Kreutz to Frings, 11 July 1949, HAEK, CR II 25.20b/17.

[43] Sozialministerium Rheinland-Pfalz to Frings, 13 April 1950, HAEK, CR II 25.20b/20.

[44] Sozialministerium Nordrhein-Westfalen to Public Health Department, HQ Military Government, 9 March 1949, HSAD, NW 7/120/53.

as villages and towns which had been exclusively Catholic for two or three hundred years had to accept large numbers of Protestants and vice versa. Yet, neither church found that confessional differences impacted the relationship between the locals and newcomers more than other factors. Both described relations between the groups as ranging from good to bad.[45] Nonetheless, at least Catholic leaders preferred to return to greater religious segregation. The Archdiocese of Paderborn argued that Catholic expellees and refugees would find it easier to come to terms with their fate if they were settled in areas offering them a place of worship and adequate pastoral care, i.e. predominantly Catholic districts.[46] The Archdiocese of Cologne reiterated its commitment to working towards the integration of expellees and refugees but insisted that the "problem of mixed marriages has become a heavy cross for both Catholic and non-Catholic expellees."[47]

While religious differences mattered less than the churches had anticipated, the negative consequences of the currency reform and the lifting of price controls considerably strained the relationship between the locals and newcomers. As unemployment and business failures increased among expellees and refugees during 1949/50, locals tended to blame the victims.[48] West German officials worried that the two groups were drifting even farther apart.[49] Catholic expellee clerics blamed a lack of knowledge of Eastern Germany and its people on the part of West Germans. The natives, they complained, tended to hold the expellees and refugees responsible for their misfortune because they could not "understand that innocent people can be expelled from their homeland."[50] To Catholic leaders, West Germans

---

[45] Evangelische Kirche im Rheinland to Sozialministerium Nordrhein-Westfalen, 3 August 1949, HSAD, NW 7/120/34-5; Erzbischöfliches Generalvikariat Paderborn to Sozialministerium Nordrhein-Westfalen, 22 April 1949, HSAD, NW 7/120/42.

[46] Ibid.

[47] Erzbistum Köln to Sozialministerium Nordrhein-Westfalen, 29 March 1949, HSAD, NW 7/120/44-7.

[48] Uwe Kleinert, *Flüchtlinge und Wirtschaft in Nordrhein-Westfalen 1945-1961* (Düsseldorf: Schwann, 1988), 28-9.

[49] Lukaschek to Frings, 4 November 1950, HAEK, CR II 25.20b/22.

[50] Paulus Sladek, "Die religiöse Lage der Heimatvertriebenen," *Stimmen der Zeit*, 6 March 1949.

appeared psychologically unprepared not only for the influx of large numbers of strangers into their communities but, more importantly, for the sacrifices needed to achieve their speedy economic and social integration. The debate concerning a future *Lastenausgleich* was thus bound to lead to even greater tensions, causing the Catholic Refugee Council to make working on a solution to this problem a top priority.[51] Compensating the large number of Germans who had experienced material and financial losses due to war and currency reform was by necessity expensive and thus controversial. The losses caused by the expulsion alone were estimated at RM 62 billion. But the expellees and refugees were not the only ones who had suffered losses. All in all, the wealth of one in every three Germans had been negatively affected by the war and its consequences.[52]

The predecessor of the *Lastenausgleich* was the 1949 Law for the Alleviation of Social Distress, popularly called the *Soforthilfegesetz* (immediate relief law). This law entitled Germans harmed by the expulsion, currency reform, political persecution and wartime damage to monthly support payments from the federal government. As an entitlement, the measure was also intended to remove the stigma of living on public assistance from the members of these groups. But it failed to bring the native population and expellees and refugees closer together. A meeting of diocesan refugee pastors noted with concern that the law and the discussion regarding the *Lastenausgleich* had widened the chasm between these groups.[53] Other expellee clerics accused the German people of lacking a sense of solidarity. During a meeting of the Catholic Refugee Council in April 1951, Cardinal Frings repeated the Catholic Church's support for an equalization of burdens guided by the "Christian notion of private property."[54] But the hierarchy's position appeared to have little impact among ordinary Germans, including Catholics. A report presented to the Fulda Bishops Conference in August criticized expellees and

---

[51] Protokoll der Tagung des Katholischen Flüchtlingsrates am 28.9.1950, HAEK, CR II 25.20b/22.

[52] Werner Abelshauser, "Der Lastenausgleich und die Eingliederung der Vertriebenen und Flüchtlinge—Eine Skizze," in *Flüchtlinge und Vertriebene,* ed. Schulze et al., 229-38.

[53] Protokoll der VI. Konferenz der Diözesanflüchtlingsseelsorger, April 1950, HAEK, CR 25.20e/3.

[54] Protokoll der Tagung des Katholischen Flüchtlingsrates am 25. April in Unkel a. Rhein, HAEK, CR 25.20b/23.

refugees for frequently making exaggerated demands, causing negative reactions among the natives. The latter, on the other hand, stood accused of lacking the necessary awareness regarding the immense significance of the issue. West Germans, the bishops were told, opposed not only the concept of compensation but also the very idea of integrating the expellees and refugees for fear of change. The natives' opposition thus threatened the key policy pursued by the federal and *Länder* governments, both churches and most political parties to provide millions of new arrivals from the East with a new home in the West. Eliminating the negative attitudes each group held vis-à-vis the other was crucial. It was portrayed as the "greatest educational undertaking" ever faced by the German people, a process in which the churches would serve as decisive participants.[55]

At the beginning of the 1950s Catholic churchmen were alarmed by two developments related to the *Lastenausgleich*. One area of concern were the demands put forth by expellee and refugee political parties such as the *Bund der Heimatvertriebenen und Entrechten* (Federation of Expellees and Dispossessed: BHE) and organizations, especially the *Zentralverband deutscher Vertriebener* (Central Association of German Expellees: ZdV);[56] the other was the settlement of expelled farmers on land in West Germany. With respect to the first issue, even expelled clerics accused the leaders of such interest groups of asking too much and using improper tactics. Their criticism was justified. The most prominent expellee functionary at the time was Linus Kather, the leader of the ZdV-BvD and a CDU *Bundestag* deputy. Kather also headed an influential committee, the *Unkeler Kreis*, which for two years offered the ZdV-BvD, the parties of Adenauer's coalition and the federal government the opportunity to iron out their differences with respect to the approaching equalization of burdens. Moderates considered Kather a radical. The latter was not afraid to resort to political blackmail to pressure Adenauer and his government. Among others, he threatened to leave his party together with other expellees and join the BHE. He also made his and the ZdV-BvD's support of West German rearmament dependent upon a *Lastenausgleich* he considered

---

[55] Referat über die gegenwärtige Situation im deutschen Flüchtlingsproblem, notwendige Massnahmen und Bitte an die Hochwürdigste Bischofskonferenz 1951, 1 August 1951, HAEK, CR II 25.20b, 23.

[56] In 1954 the ZdV was renamed *Bund der vertriebenen Deutschen* (Federation of expelled Germans: BvD).

satisfactory.[57] Catholic clergymen charged Kather with leading the Federal Republic toward social revolution and intensifying the contrast between natives and newcomers. His conduct and a perceived increase in right wing radicalism among other expellee organizations led them to call for a greater Catholic effort to influence the expellees and refugees.[58] For this purpose the Catholic Church did establish a number of organizations corresponding to the different expellee groups, among them the *Ackermanngemeinde* for Sudeten Germans, the *Eichendorffgilde* for Silesians, the *Ermlandjugend* for young East Prussians and the *Danziger und Südostdeutsche Jugend* for young expellees from the Danzig area and Southeastern Europe. Their leaderships acknowledged that the expellees and refugees had made greater sacrifices than native West Germans. At the same time, the former were urged to overcome their resentment and bitterness and to cease making demands of and launching accusations against the local population. The expellees and refugees also needed to be realistic. The upcoming *Lastenausgleich* would not compensate them for all their material losses or spare them the difficult struggle to establish a new existence in the West.[59]

The impact of such efforts is difficult to measure. It must have been rather limited, however, as Catholic clergymen remained apprehensive during 1951 and 1952. The *Lastenausgleich* debate, one leading expellee cleric warned, had already sown many seeds of discontent among the natives and expellees and refugees. Further troubles lay ahead unless the Catholic Church stepped up its efforts to enlighten both sectors of the population regarding the moral and social necessity of the *Lastenausgleich*.[60] Yet, the Church's own expectations were bound to be controversial and thus unsuited to bring natives and newcomers closer together. A November 1951 meeting of fifty-one leading Catholic personalities, including Bundestag deputies, moral theologians, government officials and businessmen, set very high standards

---

[57] Reinhold Schillinger, "Der Lastenausgleich," in *Die Vertreibung der Deutschen aus dem Osten,* ed. Wolfgang Benz (Frankfurt/Rhine, 1985), 183-92; and Arnulf Baring, *Außenpolitik in Adenauers Kanzlerdemokratie* (Munich, 1969), 155-157.

[58] Kirchliche Hilfsstelle, "Um die Zukunft der kirchlichen Volksgruppenarbeit in München," May 1950, HAEK, CR 25.20b/21.

[59] Leitsätze der Ackermanngemeinde zur Meisterung des Flüchtlingsschicksals, undated, HAEK, CR II 25.20b/22.

[60] Hartz to Frings, 8 January 1951, HAEK, CR II 25.20b/23.

bound to be unpopular with native West Germans. The German people, the participants concluded, not only had to accept the equalization of burdens as a moral duty, but would also have to make sacrifices "to the limits of what can be endured."[61]

The *Lastenausgleich* continued to preoccupy the Catholic clergy throughout 1952. In May the *Bundestag* approved the law's final version, which went into effect in mid-August. Hard bargaining and controversy marked the crucial months before its final passage. The expellees and their parties and organizations considered the legislation inadequate. After the Adenauer government agreed to spend an additional DM 2 billion on the *Lastenausgleich*, it obtained the support of leading expellee functionaries, including the difficult Linus Kather, and had more than enough votes to secure legislature's approval. In the heady months before the May vote some in the Catholic hierarchy were willing to abandon the Church's moderate tone. Noting that the Protestant Church had taken the side of expellees and refugees by forcefully criticizing the earlier, less painful versions of the law, and that the West German press viewed the Catholic Church as absent from the debate, Prelate Franz Hartz, the leading Catholic expellee clergyman by 1952, urged Cardinal Frings to be more aggressive. He proposed to make Catholic support for West German rearmament dependent upon the solution of the expellee problem. He also expressed his support for an October 1951 resolution by CDU lawmakers to carry out the equalization of burdens "with the goal of restoring destroyed and lost property within the limits of economic possibilities."[62] Frings disagreed on all points and requested that Catholic officials remain committed to moderation. The Cardinal also took exception with expellee and refugee demands for restoration of their former property. In his opinion, the state's duty to assist the members of these groups was limited to helping them make a fresh start.[63]

Frings' response resulted in large part from his concern about the effect of the *Lastenausgleich* debate on the already strained relationship between the natives and the newcomers. Most West Germans considered the demands of the latter excessive if not radical. The aggressive tactics of expellee political parties, organizations and functionaries made matters worse. Catholic

---

[61] Kirchlich-politisches Gespräch zum Lastenausgleich, 15 November 1951, HAEK, CR II 25.20b/23.

[62] Hartz to Frings, 4 February 1952, HAEK, CR II 25.20b/24.

[63] Frings to Hartz, 22 February 1952, HAEK, CR II 25.20b/24.

leaders worried that even six years after the expulsion had reached its peak expellee-native relations appeared to deteriorate rather than improve. But their response was not particularly imaginative or inspiring. They agreed to revive a four year old but now defunct umbrella organization for the Catholic relief groups involved in the effort to aid expellees and refugees. When the leaders of these groups met in September 1952, they agreed in general to increase Catholic relief activities "because the expellee problem requires the cooperation of all social strata."[64] The meeting demonstrates that the Catholic Church continued to prefer a moderate tone and goals it considered reasonable. Undoubtedly, many Catholic expellees must have been disappointed by the refusal of their church to join the more radical voices, but at the same time the continued "quiet" approach appeared to be least offensive to the native population.

In addition to the tensions brought on by the *Lastenausgleich*, Catholic leaders also worried about friction created by plans to appropriate land to expellee and refugee farmers. Expellee Minister Lukaschek identified farmers as particularly strong opponents of the *Lastenausgleich*. Their opposition, Lukaschek asserted, was causing despair and bitterness among expellee farmers, who had hoped for more solidarity from their Western colleagues.[65] The latter, however, considered expellee farmers primarily unwanted competitors for already scarce agricultural land.[66] Many objected to the establishment of agricultural settlements to integrate expellee farmers in West German villages. In September 1952 the Catholic Refugee Council described the degree of integration "completely inadequate" and called on the Adenauer government to support plans to settle some 130,000 expellee farmers on wastelands.[67] The natives objected to this and other endeavors, including the provisions of the 1953 Federal Expellee Law regulating the economic integration of expellee

---

[64] Hartz Tagebuch Nr. 985, 12 August 1952, HAEK, CR II 25.20b/24; and Kurzprotokoll aus der Tagung der "Arbeitsgemeinschaft für katholische Flüchtlingshilfe" am 10.9.1952, HAEK, CR II 25.20b/24.

[65] Lukaschek to Frings, 10 August 1950, HAEK, CR II 25.20b/21.

[66] Herbert Kötter, "Vorschläge und Hypothesen für die Erforschung der Rolle der Vertriebenen bei den Veränderungsprozessen in der Landwirtschaft und den ländlichen Gebieten der Bundesrepublik Deutschland seit 1946," in Schulze et al., eds., *Flüchtlinge und Vertriebene*, 239-44.

[67] Kurzprotokoll der Tagung des Katholischen Flüchtlingsrates in Königstein vom 11.9.1952, HAEK, CR II 25.20b/24.

farmers, calling them "unreasonable burdens."[68] In the months before final passage of the law Catholic leaders worked on a compromise solution with the West German farm lobby, insisting that native farmers would benefit themselves from their assistance to their desperate and impoverished colleagues. Fearing that differences between the two groups threatened to "split the Catholic camp,"[69] they also urged the members of the parliamentary committee reporting the legislation to show "courageous generosity."[70] It is difficult to determine the impact of such efforts, but in the end a majority of deputies voted in favor of the law.

The social divisions discussed in this essay did not prevent the integration of expellees and refugees in West German society. As the Federal Republic's economy recovered from the war and experienced a period of impressive growth during the 1950s, its demand for labor was immense. Millions of expellees and refugees entered the workforce, a process strongly encouraged by *Länder* governments like North Rhine-Westphalia's and the federal administration to ensure that they would not become a permanent underclass and political radicals. Historians agree that the workplace in fact became the primary agent for integration. Catholic leaders also sought to integrate the newcomers as speedily as possible, but the lower ranks of the clergy often did not share this goal. As a result, the Catholic Church unfortunately sent mixed messages and had much less influence in this area than the hierarchy had hoped to achieve.

---

[68] Lukaschek to Hartz, 23 December 1952, HAEK, CR II 25.20b/24.

[69] Golombek to Frings, 29 December 1952, HAEK, CR II 25.20b/24.

[70] Hartz to Bundestagsausschuß für Ernährung, Landwirtschaft und Forsten, 30 December 1952, HAEK, CR II 25.20b/24.

The United States and the Refusal to Feed German Civilians after World War II

RICHARD DOMINIC WIGGERS

A t least seven million Germans perished during the Second World War, and 25 to 50 percent of that country's housing and transportation systems were destroyed by wartime bombing and shelling. When the fighting ended, ten million internal refugees and expellees from the eastern territories were already crowded into the devastated towns and cities to join twenty million other homeless Germans.[1] Worse was still to come. The first Allied assessments of the food situation indicated that there were acute shortages, and that starvation was almost certain to occur within occupied Germany later in the year.[2] The ensuing famine continued for nearly three years, but the few scholars who have examined the issue have concluded that the German famine was simply a tragic by-product of world-wide food shortages.[3]

The situation in defeated Germany was unique, however, because the feeding of a civilian population ruled under conditions

---

[1] Alan Kramer, *The West German Economy, 1945-1955* (New York, 1991), 11-7; Michael Ermarth, ed., *America and the Shaping of German Society, 1945-1955* (Providence, 1993), 5; Eileen Egan and Elizabeth Clark Reiss, *Transfigured Night: The CRALOG Experience* (Philadelphia, 1964), ix, 6-11, 160.

[2] John J. McCloy, ASW, to President, 26 April 1945, National Archives (NA)/RG107/E180/B29; Press conference of Secretary of War and ASW John J. McCloy, 26 April 1945, NA/RG107/E180/B29; CCS 844, United States Chiefs of Staff, "Employment of German Prisoners of War in European Industry," 26 April 1945, NA/JCS, Strat/R12; SHAEF Forward, Eisenhower, to AGWAR and Combined Chiefs of Staff, 6 June 1945, NA/RG332/ETO,SGS/B57.

[3] John H. Backer, *Priming the German Economy: American Occupational Policies 1945-1948* (Durham, 1971), 50, 52, 200; James F. Tent, "Food Shortages in Germany and Europe, 1945-1948," in Guenter Bischof and Stephen E. Ambrose, *Eisenhower and the German POWs: Facts Against Falsehood* (Baton Rouge, 1992), 97, 100; Edith Hirsch, *Food Supplies: in the Aftermath of World War II* (New York, 1993). See also Douglas Botting, *From the Ruins of the Reich: Germany 1945-1949* (New York, 1985), 137-257; Eugene Davidson, *The Death and Life of Germany: An Account of the American Occupation* (New York, 1961), 127-61; Franklin M. Davis, Jr., *Come as a Conqueror: The United States Army's Occupation of Germany 1945-1949* (New York, 1967), 135-61; Josue de Castro, *The Geopolitics of Hunger* (New York, 1977), 425-39; Guenter J. Trittel, *Hunger und Politik: Die Ernaehrungskrise in der Bizone* (Frankfurt, 1990); Harold Zink, *The United States in Germany 1944-1955* (Princeton, 1957), 293-303. A published account by Canadian novelist James Bacque has alleged recently that as many as nine million civilians died of starvation and mulnutrition in postwar Germany: *Crimes and Mercies: The Fate of German Civilians Allied Occupation, 1944-1950* (New York, 1997).

of belligerent occupation was considered by most contemporary experts to be an obligation under international law. According to Article 43 of The Hague Rules of Land Warfare, "The authority of the legitimate power having in fact passed into the hands of the occupant, the latter shall take all the measures in his power to restore, and ensure, as far as possible, public order and safety."[4] Although there was no explicit obligation contained within that clause that required the occupying power to adequately feed the enemy civilian population that had fallen under its control, that was certainly the interpretation that was accepted at the time.

More than a year after the war ended, during special Senate hearings held in Germany, U.S. military government officials affirmed that it was generally recognized that under international law, the conquering nation does have "an obligation, as far as possible, to prevent epidemics and pestilences."[5] During Senate hearings in Washington in 1946, another witness testified that

> The Hague conventions are generally recognized as laying down the law which has to be followed by an occupying power. They are based on the assumptions that when a country has been defeated and occupied, the occupier or occupiers have become responsible for the orderly government of the people in their power. They must safeguard the basic rights of the local population and see to it that their basic needs are met just as if they were the national government of that country. Willfully to

---

[4] Major General J.H. Hilldring, Director, Civil Affairs Division, War Department, Office of the Chief of Staff, for Mr. McCloy, "Comments on Mr. Warburg's proposals re surrender and post-surrender policy toward Germany," 23 March 1944, NA/RG107/E180/B38. See also Major General O.P. Echols, Civil Affairs Division, testifying before the U.S. Senate, Judiciary, 20 June 1946; Eyal Benvenisti, *The International Law of Occupation* (Princeton, 1993), 7-18; Brigadier General C.W. Wickersham, United States Army Commandant, School of Military Government, "The School of Military Government," also printed in *Military Review*, 22 1944, Public Archives of Canada (PAC)/MG42/8176/NSC1812; F.S.V. Donnison, *Civil Affairs and Military Government North-West Europe 1944-1946* (London, 1961), 173.

[5] U.S. Senate Special Committee Investigating the National Defense Program, Hearings on *Investigation of Military Government*, 79th Cong., 26 May 1946 (Washington, 1946). According to a memorandum produced for the U.S. President's Famine Emergency Committee in 1946 entitled "The Disease Potential in Germany," Hoover Institution Archive, Stanford University (HIA)/USPFEC/B25, a military force has a "definite responsibility for the citizens of a conquered nation under the rules of 'Land Warfare' and the Geneva Convention. This responsibility devolved upon the Supreme Commander of the Allied Forces and was in part delegated by him to the Military Government chiefs of the respective armies." See also James F. Byrnes to the President, "Subject: Responsibilities for Relief and Supply in Occupied Areas," 1 November 1946, Harry S. Truman Library (HST)/WHCF/B38.

deny them the necessities of life is a violation of international law.[6]

Most legal scholars agreed that ensuring adequate feeding of civilian populations under their control and care was "not simply an act of charity or generosity but the fulfilment of a duty of international law, which is part of the general duty of an occupant, even a belligerent one, to restore and maintain law and order in the occupied territory."[7]

The best evidence that the Allied governments recognized a legal obligation to feed the civilian populations living under their control came from their own earlier practices and statements. During the war, the U.S. Army followed a "disease and unrest" formula of civilian feeding based on a 2000 calorie level requirement for the average adult.[8] If insufficient food was obtained from indigenous sources to maintain that minimum, the shortfall was imported at the expense of the occupying armies. In postwar Germany and Japan, the U.S. Army financed the most urgent food imports by citing obligations under Article 43 of The Hague Rules of Land Warfare. This practise continued during the drafting of the 1946-47 budget when they created a new appropriation known as the Government and Relief in Occupied Areas (GARIOA) fund. That year and in subsequent ones, whenever this portion of the budget was being considered by Congress, U.S. Defense Department officials argued that they were obligated by international law to import food at their own expense to prevent "disease and unrest" in the occupied territories under their direct military control.[9]

---

[6] Alexander Boker, "Human Events," U.S. Senate Committee on the Judiciary, Hearings on *A Bill to Amend the Trading with the Enemy Act, as Amended, to Permit the Shipment of Relief Supplies*, 79th Congress, 2nd Session, 25 April 1946 (Washington, 1946).

[7] Max Rheinstein, "The Legal Status of Occupied Germany," *Michigan Law Review* 47 (November 1948): 28-31; Josef L. Kunz, "The Status of Occupied Germany Under International Law: A Legal Dilemma," *Western Political Quarterly* 3 (December 1950): 561-2; Louis M. Gosorn, "The Army and Foreign Civilian Supply," *Military Review* 32 (May 1952): 28; Marjorie M. Whiteman, *Digest of International Law* 10 (Washington, 1968): 979.

[8] Major General J.H. Hilldring, Director, Civil Affairs Division, the War Department, to the Secretary of War, "War Department Responsibility for Procurement and Financing of Civilian Supplies in Occupied Countries," 8 October 1945, NA/RG107 E106/B3; President Truman to Secretary of War Robert P. Patterson, 1946, HST/WHCF/B38.

[9] Ben Hill Brown, Oral History Interview, 24 May 1975, HST/OH/Brown; Gunther Harkort, Oral History Interview, 12 November 1970, HST/OH/Harkort; J.W. Brab-

There was also an agreed standard for the calorie level required to maintain what Article 43 of The Hague Rules of Land Warfare referred to as "public order and safety," and what the Allies referred to during the war as "disease and unrest." During World War II, the Food and Agriculture Organization (FAO) concluded that levels of 2200 calories or lower should be temporary, and that 1700 should be considered the "upper limit of the 'semi-starvation level.'"[10] The U.S. Department of Agriculture used 2000 calories as the standard to estimate postwar global food needs.[11] At the outbreak of the war, Great Britain also established a baseline "disease and unrest" formula of 2000 calories for civilian populations. In 1945, that country's Standing Committee on Medical and Nutritional Problems still recognized 2000 calories "as the minimum necessary to prevent serious loss of life from the spread of epidemic."[12] The U.S. National Research Council also determined that the emergency food subsistence level for the average Western European or North American should be 2000 to 2200 calories.[13] Twenty-five years later, in 1970, the same body reported that even 2000 calories were inadequate to maintain health in the average adult.[14]

---

ner-Smith, "Concluding the War—The Peace Settlement and Congressional Powers," *Virginia Law Review* 34 (July 1948): 553, 555, 568.

[10] "Memorandum on Food Consumption and Related Matters," Appendix on "Classification of Food Consumption Levels on the Basis of Their Relations to Health, Well-being, and Capacity for Work"; "Attachment to Appendix: Excerpts from the report of a special joint committee (published April 1944) on Food Consumption Levels in the United States, Canada, and the United Kingdom, U.S. Edition, pages 30-33," HIA/Becker/B2.

[11] United States Department of Agriculture, Office of Foreign Agricultural Relations, Washington, D.C., "The Food Situation and Outlook in Continental Europe, the Mediterranean Area, and the Soviet Union," 16 April 1945, Library of Congress (LOC)/Harriman/B178.

[12] John E. Farquharson, "Hilfe für den Feind: die britische Debatte um Nahrungsmittellieferungen an Deutschland 1944/45," *Vierteljahrshefte Fuer Zeitgeschichte* 37 (April 1989): 254; Harry L. Coles and Albert K. Weinberg, *Civil Affairs: Soldiers Become Governors* (Washington, 1964), 150-1.

[13] National Research Council, Food and Nutrition Board, "Memorandum on Questions Submitted by the Cabinet Committee on World Food Programs, Part I: Calorie Consumption Levels and Their Relation to Health, Well-Being, and Capacity for Work," 13 December 1946, HST/WHCF/B8.

[14] "Ration Scale Established by Supreme Headquarters, European Forces (SHAEF) in January 1945 for the West German Population and the Type of Food Supplied," 1970, HIA/Becker/B1.

Not surprisingly, the Allied governments protested at wartime reports indicating that civilians in German-occupied Belgium, France and Holland were being forced to subsist on ration levels of 1100-1600 calories when "the absolute minimum necessary to sustain the life and health of a sedentary adult male is estimated by experts as 2500."[15] Prosecutors at Nuremberg accused German defendants of committing a war crime when they conspired to force down the ration levels in occupied France below 2000 calories.[16] Allied officials also complained when they discovered that POWs liberated from German captivity were subjected to "starvation rations" of 1800-2000 calories during the last phase of the war, and were forced to rely on Red Cross packages to supplement their diet.[17] Despite its imperfections and imprecision, the generally accepted 2000 calorie level provided a realistic benchmark for what the Allied armies referred to during World War II as their "disease and unrest" formula. Most importantly, civil affairs officials employed it to determine the requisitions and import requirements for civilian relief supplies in liberated and occupied territories.[18]

But this benchmark was deliberately ignored when it came to planning for the occupation of Germany. The first step towards a retributive food policy was taken during the fall of 1944 in response to a draft SHAEF handbook that suggested that German civilians be guaranteed a base ration of 2000 calories after the war.[19] While the authors of the handbook were naturally preoccu-

---

[15] Spencer Coxe, "Relief and Reconstruction In Western Europe," *New Europe and World Reconstruction* (May 1943): 3.

[16] Adolf Arndt, "Status and Development of Constitutional Law in Germany," *The Annals of The American Academy of Political and Social Science* (November 1948): 4; Montgomery Belgion, *Victors' Justice: A Letter Intended to Have Been Sent To a Friend Recently in Germany* (Hinsdale, 1949), 97.

[17] George W. Wunderlich, Office of the General Counsel, to Mr. Alfred E. Davidson, General Counsel, "Subject: Food Rations of German Prisoners of War in American Prison Camps," 13 April 1945, HST/Rosenman/B10.

[18] Major General J.H. Hilldring, Director, Civil Affairs Division, the War Department, to the Secretary of War through The Deputy Chief of Staff, "War Department Responsibility for Procurement and Financing of Civilian Supplies in Occupied Countries," 8 October 1945, NA/RG107/E106/B3; Military Government of Germany, "Monthly Report of the Military Governor, U.S. Zone", No. 5, "Monthly Report of the Military Governor," 20 December 1945, NA/RG94/OpBr/B1175.

[19] David B. Woolner, "Coming to Grips with the 'German Problem': Roosevelt, Churchill, and the Morgenthau Plan at the Second Quebec Conference," in David B.

pied with the postwar rehabilitation of the European economy, this particular undertaking struck officials in Washington as being far too soft and constructive.[20] President Roosevelt proposed a much harsher food policy in its place: the Germans "should have simply a subsistence level of food—as he put it, soup kitchens would be ample to sustain life—that otherwise they should be stripped clean and should not have a level of subsistence above the lowest level of the people they had conquered."[21] Soviet officials also proposed that the Allies limit grain and food production in postwar Germany, and U.S. Secretary of the Treasury Henry Morgenthau conceived of a similar program that would limit the civilian population "to a subsistence level."[22]

Even State Department planners believed Germany should be guaranteed only bare subsistence for a period of at least several years after the war and not immediately brought up to the level of the other European states that had been the victims of her wartime aggression and occupation.[23] Secretary of State Cordell Hull concluded that "it is of the highest importance that the standard of living of the German people in the early years be such as to bring home to them that they have lost the war and to impress on them that they must abandon their pretentious theories that they are a superior race created to govern the world. Through lack of luxuries we may teach them that war does not pay."[24] By the end of 1944, British and U.S. Civil Affairs officers being trained in London were informed that despite official "disease and unrest" targets of

---

Woolner, ed., *The Second Quebec Conference Revisited: Waging War, Formulating Peace: Canada, Great Britain, and the United States in 1944-1945* (New York, 1998), 72.

[20] President Roosevelt to the Secretary of War, 26 August 1944, U.S. Department of State, *Foreign Relations of the United States, Diplomatic Papers 1944*, I (Washington, 1966), 544.

[21] Forrest C. Pogue, *George C. Marshall: Organizer of Victory 1943-1945* (New York, 1973), 467-8; Walter Millis, ed., *The Forrestal Diaries* (New York, 1951), 10.

[22] Canadian Ambassador in Washington to the Secretary of State for External Affairs, 21 August 1944, PAC/RG25/F7-E-2[8]); Secretary of War Stimson, "Suggested Recommendations on Treatment of Germany from the Cabinet Committee for the President," 5 September 1944, Franklin D. Roosevelt Library (FDR)/RG24/B333.

[23] Eleanor Lansing Dulles, interview #8, 28 June 1963, Dwight D. Eisenhower Library (DDE)/OH/Dulles.

[24] Cordell Hull, *The Memoirs of Cordell Hull*, II (New York, 1948), 1619; Davidson, *The Death and Life*, 8-10.

2000 calories, many inhabitants of liberated Europe were still receiving only 1600 calories per day. In apparent retaliation, they were to ensure that after the war the average German adult "will receive 1500 calories as a maximum although there is no assurance that he will get that much; that is all he can have during our occupation." The implications were clear:

> As for supplying the Germans with food, it will only be as a last resort. We are going to treat Germany as a defeated country. We have to make them realize they are defeated and they are not a liberated country. We expect to put out food to the German people only when there is no other food available... The food problem will probably cause more trouble from a public safety angle than any other one. But we have to be strict with them and we have to watch the food now because later we will have to feed them if supplies become exhausted. We do not want circumstances to force us to import food for Germans.[25]

Months before Germany surrendered, Allied officials seemed to have agreed on a revised "disease and unrest" formula for the occupied enemy states that was significantly lower than the generally accepted wartime standard. Recommended calorie levels for the average adult were set well below the generally accepted 2000, and the Allied occupying armies were not obligated to furnish supplies to meet even those levels.[26] The determination to inflict postwar punishment on the German population had clearly overcome concerns about obligations to Article 43 of the 1907 Hague Rules of Land Warfare. On May 10, only days after the German surrender, the following orders were issued to Dwight D. Eisenhower, Supreme Commander of SHAEF forces in occupied Germany:

> You will estimate requirements of supplies necessary to prevent starvation or widespread disease or such civil unrest as would

---

[25] Lieutenant General John C.H. Lee, The Disarmament School, *Disarmament School Lectures*, Second Course, Vol. 1, London (December 1944), Brig. General F.J. McSherry, "Civil Affairs as it Pertains to Disarmament and Control Machinery," 23 November 1944.

[26] "The Treatment of Germany," and attached proposals for "Economic Policies Toward Germany," 12 January 1945 , FDR/RG24/B337; SHAEF to Headquarters, 21 Army Group, and commanders of the Twelfth and Sixth Army Groups, "Control of Distribution and Rationing of Food in Germany," 25 January 1945, DDE/Smith/WWII/B37; AGWAR and CCS to SHAEF Main and Eisenhower, 19 April 1945, NA/RG332/ETO,SGS/B57; "Germans to Get only Bare Needs," *Evening Standard*, 17 February 1945, NA/RG218/E102/B13.

endanger the occupying forces. Such estimates will be based upon a program whereby the Germans are made responsible for providing for themselves, out of their own work and resources. You will take all practicable economic and police measures to assure that German resources are fully utilized and consumption held to the minimum in order that imports may be strictly limited and that surpluses may be made available for the occupying forces and displaced persons and United Nations prisoners of war, and for reparation. You will take no action that would tend to support basic living standards in Germany on a higher level than that existing in any one of the neighboring United Nations and you will take appropriate measures to ensure that basic living standards of the German people are not higher than those existing in any one of the neighboring United Nations when such measures will contribute to raising the standards of any such nation.[27]

Paragraph 5 of Joint Chiefs of Staff (JCS) 1067—the operational guidelines for the U.S. occupation—also ordered Military Government officials to restrict themselves to promoting the production and maintenance of only those indigenous goods and services "required to prevent starvation or such disease and unrest as would endanger the occupying forces."[28] The U.S. Deputy Military Governor, Lucius Clay, confided that "I feel that the Germans should suffer from hunger and from cold as I believe such suffering is necessary to make them realize the consequences of a war which they caused," but also warned officials in Washington that "this type of suffering should not extend to the point where it results in mass starvation and sickness."[29]

---

[27] "Directive to Commander in Chief of U.S. Forces of Occupation Regarding the Military Government of Germany," 10 May 1945, in Dennis Merrill, ed., *Documentary History of the Truman Presidency, Volume 3: Unconditional Surrender and Policy in Occupied Germany after World War II* (University Publications of America: 1995), 7-8; See IPCOG 2/1, paragraph 3(d), as cited in February 1950, Russell Fessenden, Foreign Policy Studies Branch, Division of Historical Policy Research, Office of Public Affairs, Department of State, Research Project No. 143, "Negotiations Concerning German Reparations, Part I - Yalta Through Potsdam," HST/PSF/B179.

[28] SHAEF Food and Agriculture Section, Economic Control Agency, G-5 Division, "The Food Position in Western Germany as of 1 June 1945", 3 July 1945, NA/RG332/ETO,SGS/B57; Backer, *Priming the German Economy*, 37; John H. Backer, *Die deutschen Jahre des Generals Clay* (Muenchen, 1983); John H. Backer, "From Morgenthau Plan to Marshall Plan," in Robert Wolfe, ed., *Americans as Proconsuls: United States Military Government in Germany and Japan, 1944-1952* (Carbondale, 1984), 157.

[29] Lucius D. Clay, Deputy MG for Germany, to John J. McCloy, ASW, 29 June 1945, NA/RG107/E180/B29; Clay to McCloy, 29 June 1945, in Jean Edward Smith, *The*

German POWs were among the first to feel the pinch when their Geneva Convention protection and treatment was removed and they were transformed from POW into Disarmed Enemy Force/Surrendered Enemy Personnel (DEF/SEP) status immediately after the war. SHAEF officials set the calorie level for non-working POWs at 1500, though they understood at the time that it was well below their own suggested standards.[30] The ration for a normal adult civilian consumer, meanwhile, was set slightly higher at a maximum of 1550 calories.[31] Several months later, the Level of Industry Committee concocted a formula that would permit the Allies to reduce ration scales in Germany below the European average.[32] First, the German Standard of Living Board calculated that during the years of relative hardship that preceded the outbreak of World War II, the average German adult was consuming 2900 calories per day, 10% above the European average at the time and higher in overall quality and fat content. They then recommended that the future German ration level be reduced to 2150 calories, equivalent to the level that prevailed in 1932, the

---

Papers of General Lucius D. Clay, Volume I: Germany 1945-1949 (Bloomington, 1974), 24, 42.

[30] Archer L. Lerch, Major General, Office of the Provost Marshal General, to Commanding General of Army Services Forces in Washington, "Inspection of Concentration Camps and Other Internment Camps in the European Theater of Operations," 9 June 1945, NA/RG160/B331; Brigadier General Frank J. McSherry, Deputy Assistant Chief of Staff, G-5, to SHAEF G-5 Division, Chief of Staff, "Food Situation in Western Germany," 15 June 1945, DDE/Smith/WWII/B37; John Dos Passos, *Tour of Duty* (Boston, 1946), 252.

[31] G-5 Division, SHAEF, to Chief, Food and Agriculture Section, 22 June 1945, NA/RG332/ETO, SGS/B57; for Chief of Staff, British Zone, "Imports of Wheat for Consumption by German Civilians in the British National Zone," August 1945, Public Records Office, Great Britain (PRO)/DBPO/S1/V5/F8; Colonel O.W. Hermann (US), F. Hollins (UK), and Colonel P. Dessus (France), Combined Resources and Allocations Board, Combined Food and Agriculture Committee, "Food Import Requirements for British, American, and French Zones of Germany for the 1945-46 Consumption Year," 6 August 1945, PRO/DBPO/S1/V5/F7; Colonel T.W. Hammond, Brigadier T.N. Grazebrook, Lt. Colonel M.P.F. DuPont, and Major General N.T. Sidorov, Coordinating Committee, note by Allied Secretariat, "Nutrition of the German Civil Population," 24 August 1945, PRO/DBPO/S1/V5/F14.

[32] EC (S)(45) 44, Some Random Notes on the Reparations Discussions in Berlin, September-November 1945, by Mr. G.D.A. MacDougall, "Economic Section of the Cabinet Secretariat," 29 November 1945, in M.E. Pelly, H.J. Yasamee, and G. Bennett, *Documents on British Policy Overseas* 1 (London, 1990): 519-30; Mr. A.K. Cairncross, Economic Advisory Panel, Allied Commission on Reparations, U.K. Delegation, to Sir Percy Mills, "Future German Population," 29 November 1945, PRO/DBPO/S1/V5/F33.

worst year of the prewar depression.[33] What is most striking is that even this planned "standard of living" formula, harsh as it was, was only a long term goal. According to a State Department release issued at the end of 1945, the even lower "disease and unrest" formula, with a 1500-1550 calorie ceiling, would continue to be enforced during at least the first two years of the occupation while reparation removals were carried out.[34]

The shortage of indigenous food sources was further exacerbated by other Four Power policies. First was the influx of millions of expellees into the increasingly overpopulated western zones with their devastated water supplies, dwellings, and hospital facilities. Second was the disruption of the 1945 planting season combined with the isolation of the food producing lands in the eastern part of the country. Third was the decision to give priority to the housing and feeding of millions of non-German DPs and liberated Allied nationals.

SHAEF planners soon realized that food imports would be needed to sustain even a minimum standard of 2000 calories for the DP population.[35] They also acted quickly to encourage the

---

[33] E. Lewin, British Secretary, Allied Control Authority, Directorate of Economics, Level of Industry Commitee, "A Minimum German Standard of Living in Relation to the Level of Industry," and "Explanatory Notes to Table IV, V, and VI," 17 September 1945, NA/RG107/E106/B3; Military Government of Germany, "Monthly Report of the Military Governor, U.S. Zone", No. 2, 20 September 1945, NA/RG94/OpBr/B1175; Major General J.H. Hilldring, Director, Civil Affairs Division, the War Department, Civil Affairs Division, to the Secretary of War through the Deputy Chief of Staff, "Comments on Preliminary Report by the Working Staff of the German Standard of Living Board," 9 October 1945, NA/RG107/E106/B3; "Draft First Report on Reparation by Economic Advisory Panel," 3 November 1945, PRO/DBPO/S1/V5/F33; OMGUS, Office of the Legal Adviser, to Director, Economics Division, "Average of Standards of Living of European Countries," 8 February 1946, NA/RG260/LD/B55; Lewis H. Brown, *A Report on Germany* (New York, 1947), 247; John Gimbel, *The American Occupation of Germany: Politics and the Military, 1945-1949* (Stanford, 1968), 20-1; John Gimbel, "Governing the American Zone of Germany," in Wolfe, *Americans as Proconsuls*, 93; B.U. Ratchford and Wm. D. Ross, *Berlin Reparations Assignment: Round One of the German Peace Settlement* (Chapel Hill, 1947), 71.

[34] "Reparation Settlement and Peacetime Economy of Germany," December, 1945, LOC/Harriman/B184.

[35] Major General F.F. Scowden, Chief, SHAEF, Supply and Economics Branch, G-5 Division, to CCAC, "Justification of food import requirements for SHAEF Zone of Germany during June, July and August 1945," 12 March 1945, NA/RG218/E88/B65; CCS 551/15, Combined Civil Affairs Committee for consideration by the Combined Chiefs of Staff, "Control and Distribution and Rationing of Food in Germany," 9 April 1945, NA/JCS/1/R11; SHAEF Forward, signed Eisenhower, to AGWAR and Combined Chiefs of Staff, 16 May 1945, NA/RG332/ETO,SGS/B57.

planting of home gardens, revive production of farm machinery, and resume the operations of the German fishing fleet. By 1948, in fact, the western occupation zones had attained 95% of prewar domestic food production. But Germany had always been a net importer of foodstuffs, and the 25% increase in the population of the western zones due to the influx of refugees and expellees meant that domestic production could provide barely half of total needs.[36]

Not surprisingly, the average daily ration level in the western occupation zones during the summer of 1945 fluctuated between 700-1190 calories, far below not only the generally accepted minimum of 2000, but also the substandard ceiling of 1550 calories established by the Allies through their revised "disease and unrest" formula.[37] While non-German refugees living within DP camps were soon receiving 2300 calories thanks to emergency food imports and Red Cross supplements, German civilians living in the U.S. and British zones were authorized to receive just over half that amount (1354), and were in fact believed to be obtaining only 1250 on average.[38] Conditions appeared to be only marginally better in the Soviet zone, which contained most of Germany's best agricultural lands, and slightly worse in the French zone. In most dire need were the millions of ethnic Germans being expelled from their homes in eastern Europe.[39]

Even Allied officials began to protest the conditions. In an October 1945 letter to the Assistant Secretary of War, U.S. Deputy Military Governor Lucius Clay reported that "undoubtedly a large number of refugees have already died of starvation, exposure and disease.... The death rate in many places has increased several fold, and infant mortality is approaching 65 percent in many places. By

---

[36] Backer, *Priming the German Economy*, 37, 46.

[37] SHAEF G-5 Division, to Chiefs of Staff , SHAEF, 26 June 1945, NA/RG332/ETO,SGS/B57.

[38] Military Government of Germany, "Monthly Report of the Military Governor, U.S. Zone," No. 4, "Monthly Report of the Military Governor," 20 November 1945, NA/RG94/OpBr/B1174.

[39] German League of free Welfare Associations, Central Committee for the Inner-Mission of the German Protestant Church, Caritas Association for the Roman Catholic Germany, and the German Red Cross, to the Allied Control Council, 5 September 1945, HIA/Lochner/B2; Committee Against Mass Expulsion, *The Land of the Dead: Study of the Deportations From Eastern Germany* (New York, 1946); Egan and Reiss, *Transfigured Night*, 21-30, 54-8; Alfred-Maurice de Zayas, *The German Expellees: Victims in War and Peace* (New York, 1993), 149-50.

the spring of 1946, German observers expect that epidemics and malnutrition will claim 2.5 to 3 million victims between the Oder and Elbe."[40] A British report warned that the 1150 calorie levels prevailing in places like the Ruhr would almost certainly lead to "Belsen" conditions, and reminded readers that "2000 calories is considered the minimum necessary to keep body and soul together," and that by comparison the British civilian ration was 3000 calories at the time.[41] At a Cabinet meeting in London in early October, the participants acknowledged that the overall death rate among German civilians had already climbed to four times the prewar normal, while the mortality rate for children had risen tenfold.[42]

During the fall of 1945, the Combined Nutrition Committee completed a detailed survey of the food situation in the three western zones of occupation. They reported prevailing ration levels that ranged from 840-1400 calories, and warned of the probable side effects: reduced worker efficiency, an expanding black market, retarded growth in children, and increasing symptoms of malnutrition in children and pregnant women. Although they did not observe any "unusual incidence of disorders arising from malnutrition among the few children that we examined," and only a few cases of famine oedema, the authors of the report warned about the progressive loss of weight documented among civilians of all ages. Their study also noted that "the sole justification for a policy involving partial starvation for the German people is that the needs of the liberated Allied countries must come first."[43]

Later that fall, another inquiry into conditions within Germany was completed by Byron Price. He reported to U.S. President

---

[40] Lieutenant General Lucius D. Clay, Office of the Deputy Military Governor to John J. McCloy, Assistant Secretary of War, War Department, 5 October 1945, NA/RG107/E180/B26; Hans W. Schoenberg, *Germans from the East: A Study of their Migration, Resettlement, and Subsequent Group History since 1945* (The Hague, 1970), 32, 38.

[41] Lt.Colonel. J.H.B. Lowe, "Report on the Visit to the British Zone of Germany and British Sector in Berlin," September 1945, PRO/DBPO/S1/V5/F7.

[42] Cabinet, Gen 93 (Ministerial)/1, Minister of State, "Control of Epidemics," 5 October 1945, PRO/DBPO/S1/V5/F27.

[43] "Combined Nutrition Committee (US/BR/FR) Report," July 30-August 8, 1945, HIA/Becker/B1; Military Government of Germany, "Monthly Report of the Military Governor, U.S. Zone," No. 1, "Monthly Report of the Military Governor," 20 August 1945, NA/RG94/OpBr/B1175; "Die Lebensmittel - Rationierung in Deutschland 1939-1946," 1946, HIA/Becker/B1.

Truman that he knew of "no competent medical authority who would regard a ration of 1550 calories as satisfactory, or who considers that present rationing in Germany is adequate for a people who are expected to work, and who have no heat at home and no way to reach their places of employment except by walking." According to Price, a growing body of medical evidence was showing a widespread and dangerous loss of weight and an alarming loss in resilience to disease, and he warned that "epidemics and rioting will not be far behind." Even the generally accepted minimum ration of 2000 calories to prevent starvation would not permit the "bombed-out, freezing, pedestrian Germans to live anything like as well as the European average." Price reported to Truman that there could be no question that

> the vengeance of Nature's God lies heavily on the German people. They are paying in kind for the unparalleled miseries and cruelties for which they are responsible. As cold weather begins, millions find themselves housed against the raw climate in rubble heaps and caves, without fuel for heating, and with a food supply rated by medical standards well below the level of subsistence. Just now these people are quiescent, and lawlessness is negligible, although epidemics begin to threaten the health of western Europe.[44]

U.S. Secretary of War Robert Patterson wanted the Price Report made public, and agreed with his conclusion that the official calorie ceiling of 1550—let alone the lower ration amounts that were actually being administered in postwar Germany—was inadequate to maintain the health of the German civilian population.[45] Months earlier, the U.S. Catholic Bishops had already spoken out against the restrictions on food imports into occupied Germany, and warned that "it is unworthy of the victors to revenge injustices by violating human rights and heaping insults on human dignity. As things are now, future generations may well charge the victors with guilt of inhumanities which are reminiscent of Nazism and Fascism."[46] Other Americans also began to complain that the fail-

---

[44] Byron Price to President Truman, 9 November 1945, HST/B-File/Germany/F1; J.J. McCloy, War Department, Office of Assistant Secretary, to Judge Patterson, 24 November 1945, NA/RG107/E106/B4.

[45] Secretary of War Robert P. Patterson, War Department, to President Truman, 27 November 1945, HST/B-File/Germany/F1.

[46] "Bishops Discuss World Peace," *Catholic Action* 27 (May 1945).

ure to restore mail service to Germany was preventing them from exercising their rights as U.S. citizens to communicate with or send personal packages to friends and family members.[47]

At first, President Truman turned a deaf ear to the growing volume of protests. He explained to one U.S. Senator that though all Germans might not be guilty for the war, it would be too difficult to try to single out for better treatment those who had nothing to do with the Nazi regime and its crimes:

> While we have no desire to be unduly cruel to Germany, I cannot feel any great sympathy for those who caused the death of so many human beings by starvation, disease and outright murder, in addition to all the regular destruction and death of war. Perhaps eventually a decent government can be established in Germany so that Germany can again take its place in the family of nations. I think that in the meantime no one should be called upon to pay Germany's misfortunes except Germany itself. Until the misfortunes of those whom Germany oppressed and tortured are obliviated, it does not seem right to divert our efforts to Germany itself.[48]

Several months later, however, the first chink appeared in the armor of harsh Allied food policy. Throughout 1945, the Allied occupation armies centralized relief efforts to ensure that any international aid flowing into occupied Germany went exclusively to liberated Allied POWs, concentration camp survivors, and non-German DPs awaiting repatriation or resettlement.[49] Only a few licensed international relief agencies such as the United Nations Relief and Rehabilitation Administration (UNRRA) and several Papal relief missions were even permitted to operate in the U.S. zone. To ensure that they assisted only non-German nationals, the

---

[47] Karl Brandt, Economist, the Food Research Institute, Stanford University, to Burton K. Wheeler, Senate Office Building, 14 September 1945; Milton S. Young, U.S. Senator, Committee on Agriculture and Forestry, to the President, 3 December 1945, HST/B-File/Germany/F1; W.O. Lewis, Baptist World Alliance, to the President, 10 May 1946, HST/B-File/DP/F2; Merrill, *Documentary History* vol. 3,114.

[48] Harry S. Truman to Burton K. Wheeler, U.S. Senate, 21 December 1945, HST/B-File/Germany/F2; Harry S. Truman to Burton K. Wheeler, United States Senate, 6 October 1945, HST/B-File/Germany/F1; Mr. Philip E. Ryan to Mr. William H.G. Giblin, 15 October 1945, NA/RG200/B1016.

[49] Pogue, *George C. Marshall*, 458; Edward N. Peterson, *The American Occupation of Germany: Retreat to Victory* (Detroit, 1977), 118; L.P. (45) 182, Cabinet, Lord President's Committee, "World Food Outlook: Memorandum by the Minister of Food," 1 October 1945, PRO/DBPO/S1/V3.

U.S. Military Government controlled all supplies, transportation and travel permits.[50] Only indigenous organizations such as the *Innere Mission* and *Caritas Verband* were permitted to help the German people, but the national Red Cross was dissolved and its remaining activities severely curtailed, and none of the agencies were permitted to obtain outside supplies.[51] In their determination to ensure that international relief was denied to enemy civilians, the State Department even prohibited efforts by the Vatican and the German community in Chile to transmit food supplies to infants living in the western zones of occupation.[52]

By the beginning of 1946, the tide of public opinion was beginning to turn against the official policy of harshness. In January, thirty-four U.S. Senators signed a petition urging that Germany and Austria be opened to private relief organizations. In particular, they expressed concern about the desperate food situation "which presents a picture of such frightful horror as to stagger the imagination, evidence which increasingly marks the United States as an accomplice in a terrible crime against humanity."[53] Even Lucius Clay, by now the U.S. Military Governor in Germany, was warning officials in Washington about the dangers of allowing hunger to persist in postwar Germany: "there is no choice between becoming a Communist on 1500 calories and a believer in democracy

---

[50] Magda Kelber, "Patterns of Relief Work in Germany," *The Year Book of World Affairs 1951* (London, 1951), 10-11.

[51] Rose B. Dolan to Mr. George Kulp, Supervisor of 7th Army ARC-CWR, "The German Red Cross," 1 July 1945; N. de Rouge, League of Red Cross Societies to Francis B. James, 11 July 1945 and 8 June 1945; Office of the Legal Adviser, USGCC, "Status of the German Red Cross," 23 August 1945; Office of the Legal Adviser, USGCC, 20 August 1945, NA/RG200/B1016; Headquarters USFET, G-5 Division, to Commanders of Third and Seventh US Armies, "Control of Benevolent German Welfare Organizations," August 1945; Captain Anton J. Vlcek, Public Welfare Branch, Headquarters, USGCC (Germany), Public Health and Welfare Division, Public Welfare Branch, to Major General Stayer, "Preliminary Report on Welfare Activities in Bavaria," 10 August 1945; Fred S. Reese, Legal Adviser, Public Health and Welfare Division, USGCC (Germany), Public Health and Welfare Division, Office of Legal Adviser, to Mr. Charles Fahy, Director of Legal Division, "Opinion on Status of German Red Cross," 20 August 1945; Fred S. Reese, Legal Adviser, Public Health and Welfare Division, USGCC (Germany), Public Health and Welfare Division, Office of Legal Adviser, to Lt. Col. William G. Downs, Public Welfare Branch, "Status of (Deutsches Rotes Kreuz) German Red Cross," 20 August 1945, NA/RG260/LD/B60.

[52] Secretary of State to U.S. Embassy, 14 September 1945, LOC/Harriman/B182.

[53] Kenneth S. Wherry, United States Senate, Committee on Appropriations, to the President, 4 January 1946, HST/WHOF/B1272.

on 1000 calories. It is my sincere belief that our proposed ration allowance in Germany will not only defeat our objectives in middle Europe but will pave the road to a Communist Germany."[54]

Responding to growing pressure from Congress and public opinion, President Truman permitted representatives of seven U.S. relief organizations to survey the situation in occupied Germany, and their final report was critical of the conditions that prevailed.[55] On February 19, 1946, he decided to approve the creation of a Council of Relief Agencies Licensed for Operation in Germany (C.R.A.L.O.G.), an umbrella organization which would operate under the direction of the U.S. Military Government. Several months later, relief organizations were permitted to send humanitarian aid to starving German children for the first time, and during the summer they expanded their operations to include other age groups and the British and French zones.[56]

As outside observers began to pour into occupied Germany during the spring and summer of 1946, reports about the terrible conditions began to reach a broader audience. U.S. relief workers and journalists were equally critical of the famine conditions that they witnessed. A group of editors and publishers travelled throughout Germany to survey the situation, and almost all concluded in later editorials and articles that more food aid was needed.[57] During Senate hearings in June 1946, one U.S. Army official testified that given the prevailing ration scales of 1180-1225 calories, if it was not starvation, "it is very close to it."[58] Another witness reported on the high infant mortality rates prevailing in postwar Germany, and asserted that both food and shipping space were available if only sufficient political will could be found to

---

[54] Clay to Echols and Petersen, March 27, 1946, in Smith, *The Papers*, 184.

[55] Joseph Buttinger, International Rescue Relief Committee, and Abram Becker, Acting Executive Secretary of the International Rescue Relief Committee, 25 April 1946, U.S. Senate, Judiciary, *A Bill to Amend the Trading with the Enemy Act*.

[56] Egan and Reiss, *Transfigured Night*, 64; James F. Tent, "Simple Gifts: The American Friends Service Committee and the Establishment of Neighborhood Centers in Post-1945 Germany," *Kirchliche Zeitgeschichte* (Spring 1989): 66-69; Wallace J. Campbell, *The History of CARE: A Personal Account* (Westport, 1990), 8, 40, 49.

[57] *New York Times* , 4 April 1946; OMGUS Berlin to War Department, 14 April 1946; "The Food Situation in Germany's American Zone," *Baltimore Sun* , 16 April 1946, NA/RG107/E187/B1.

[58] Major General O.P. Echols, Civil Affairs Division, War Department, U.S. Senate, Judiciary, *A Bill to Amend the Trading with the Enemy Act*, 20 June 1946, 50.

help the vanquished foe.[59] One relief worker tried to describe circumstances that most Americans could barely imagine:

> Starvation is not the dramatic thing one so often reads and imagines... of people in mobs crying for food and falling over in the streets. The starving... those who are dying never say anything and one rarely sees them. They first become listless and weak, they react quickly to cold and chills, they sit staring in their rooms or lie listlessly in their beds... one day they just die. The doctor usually diagnoses malnutrition and complications resulting therefrom. Old women and kids usually die first because they are weak and are unable to get out and scrounge for the extra food it takes to live. It is pretty hard for an American who has lacked enough food to become ravenously hungry perhaps only once or twice in a lifetime to understand what real starvation is.[60]

No individual did more to inform the world about the situation in postwar Germany and win support for an expanded civilian feeding program than former U.S. President Herbert Hoover. He was already a veteran of two relief missions to Europe during and after World War I, and pioneered the use of food as a tool of diplomacy. Hoover was widely respected by the public and members of both parties in Congress, and at Truman's invitation, he began assembling a Famine Emergency Committee in February 1946.[61] For the next few months, they visited both food producing and food deficit countries. In Germany, they discovered that urban and industrial areas continued to be the worst hit by wartime damage and postwar famine, with most civilians relying on official rations of 1,000 calories or less. Available food stocks were quickly running out, the quality of the ration was generally poor, and the condition of children was particularly tragic.[62] While typhus and

[59] Mr. James M. Reed, Friends Committee on National Legislation, U.S. Senate, Judiciary, *A Bill to Amend the Trading with the Enemy Act*, 18 June 1946.

[60] HST/Andrews/30; Testimony of Mr. G.V. Gaevernitz, U.S. Senate, Judiciary, *A Bill to Amend the Trading with the Enemy Act*, 18 June 1946.

[61] Gary Dean Best, *Herbert Hoover: The Postpresidential Years, 1933-1964, Volume Two: 1946-1964* (Stanford, 1983), 284-86; Richard Norton Smith, *An Uncommon Man: The Triumph of Herbert Hoover* (New York, 1984), 351-3; Merle Miller, *Plain Speaking: An Oral Biography of Harry S. Truman* (New York, 1973), 219-20; David McCullough, *Truman* (New York, 1992), 389-90.

[62] Herbert Hoover's "Address Before the Emergency Conference on European Cereal Supplies," London, 5 April 1946, HIA/Hoover/B108; "Some Notes on Trip with the Hoover Mission (March-June 1946)," 1 August 1946, HIA/Tuck/B7.

diptheria were already evident in 1945, by the spring of 1946 there was growing evidence of weight loss, rickets, nutritional edema and vitamin deficiency. The infant mortality rate in Germany was double the prewar rate in the U.S., and in the population as a whole deaths continued to exceed births. As the report concluded, "The collapsed Germany of 1945 presented a situation almost without parallel in the annals of modern warfare... Their disintegration was not only physical but psychological as well."[63]

Many historians view the second winter of the occupation as an important period of transition in U.S. and Allied policy towards postwar Germany. With relief agencies finally permitted to operate, C.R.A.L.O.G. shipped 10,000 tons of private relief supplies in the form of food and clothing by the end of 1946, and C.A.R.E. sent another 550,000 packages. Private parcels also flooded in, reaching a total of 17 million pounds per month by December. In three years, a total of 441 million pounds of goods valued at 200 million dollars were sent to German families and individuals. Furthermore, the country received one third of all U.S. foreign aid during the period 1945-1949, making the United States by far the largest single contributor.[64] Years later, German President Konrad Adenauer referred to the "great psychological effects" of this relief work: "It was not so much the material assistance that helped us as the connection with the outside world, the hope for reconciliation, a ray of light pointing to a brighter future—all these were awakened in Germany by these actions."[65]

Official policy was also undergoing a radical transformation. In September 1946, Secretary of State James Byrnes made his famous Stuttgart speech, and the following summer the operational directive guiding U.S. policies until then—JCS 1067—was

---

[63] U.S. President's Famine Emergency Committee, "The Disease Potential in Germany," 1946, HIA/USPFEC/B25.

[64] "Report of Conference of Senator W.F. Knowland and OMGUS Officials Held at Headquarters, Office of Military Government for Germany (US)," Directors Building, Berlin, Colonel Lenzner, Deputy Director of Internal Affairs and Communications Division, Dr. R.T. Alexander, Chief, Education & Religious Affairs Branch, Internal Affairs & Communications Division, OMGUS, 8 October 1947, HST/Rockwell/B31; Gabriele Stueber, *Der Kampf gegen den Hunger 1945-1950: Die Ernaehrunglage in der britischen Zone Deutschlands, insbesdondere in Schleswig-Holstein und Hamburg* (Neumuenster, 1984), 475-519; Edward McSweeney, *American Voluntary Aid for Germany 1945-1950* (Freiburg, 1950), 22-85.

[65] Konrad Adenauer and Beate Ruhm von Oppen, *Konrad Adenauer Memoirs 1945-53*, trans. (Chicago, 1966), 59-60; Louis P. Lochner, *Herbert Hoover and Germany* (New York, 1960), 203-35.

scrapped and replaced with JCS 1779, and a "stable and productive Germany" was proclaimed to be the new economic priority.[66] By the fall of 1946, more than two-thirds of the entire military goverment budget was being spent on civilian relief costs, U.S. officials stationed in Germany were trying to launch a limited export program, and more effort was being made to monitor civilian health.[67]

But even the influx of international relief and the apparent transformation of official U.S. policy during 1946 were failing to bring adequate relief to German civilians, and they remained at the end of the world food line. Although infants were doing well overall and their death rate was finally declining, there continued to be a disastrous decline in body weights in all other consumer groups. The situation was worst among those over 70, whose death rate had increased to 34% of the total in May and 45% of the total in October. The rate of suicide in that age group had also increased.[68] A report by the Combined Nutrition Committee completed just before Christmas 1946 concluded that fewer than 1500 calories had been made officially available to most adult German civilians during the past year, even though 2000 was still considered "the minimum amount of food upon which normal consumers could subsist in reasonable health for a limited period of time." Even assuming that some were able to supplement their diets with as many as 200-500 calories of non-ration food, the variety and quality of the rations remained poor, which explained why the number of hunger oedema cases was increasing.[69]

---

[66] Earl F. Ziemke, "The Formulation and Initial Implementation of U.S. Occupation Policy in Germany," Hans A. Schmitt, ed., *U.S. Occupation in Europe After World War II: Papers and Reminiscences from the April 23-24, 1976, Conference Held at the George C. Marshall Research Foundation, Lexington, Virginia* (Lawrence, 1978), 39.

[67] Office of Military Government for Germany (U.S.), Public Relations Office, Press Conference with General Clay, Mr. Petersen, and Mr. Allen, 15 October 1946, NA/RG107/E187/B1.

[68] Dr. Wetzel, Regierungspraesident Darmstadt, Department I, General and Inner Administration, Public Health, "Report on the Sanitary Problems within the area of the U.S. Zone in Germany," 20 November 1946, HIA/USPFEC/B25.

[69] Combined Nutrition Committee (Br./Fr./U.S.), "Report of the Sixth Combined Nutrition Survey of settled areas in the British, French and U.S. Zones of Germany made during the period 1 to 12 December 1946," 12 December 1946, HIA/Germany (ACC)/B33; "Summary of a Meeting of the Regional Commissioners of the British Zone with the Hon. H. Hoover," 10 February 1947, HIA/USPFEC/B11; Chief, Public Health Branch, OMGUS, "Nutrition Summary Report, U.S. Zone, May 1947," 20 June 1947, HIA/Becker/B1.

According to one White House official who recalled the situation years later, the food supply situation reached "rock bottom" during the bitterly cold and seemingly endless winter of 1946-47:

> [it] is stuck in the memory of the Germans, who lived through it, as the time of the shortest and most meager supply ever: 1,000-1,500 calories per day, very little heating fuel, worse, in that aspect, than in any winter of the war and postwar years. The war had been over for one and one half years; nevertheless privation and misery were unequalled, with no chance of improvement in sight. The urban masses were too much in need of rest, physically too weak, too resigned to rise in rebellion.[70]

All of this suffering was taking place during a period when the average U.S. calorie intake for an adult was 3200-3300 calories (2900 in Great Britain), and the normal U.S. Army ration was 4000.[71] Economic output in the western zones had barely reached one-third of its 1938 levels, and in January 1947 a delegate sent to Germany by the Secretary of War observed no improvement in the overall situation.[72] Six months later, Assistant Secretary of War Howard Peterson reported that ration scales for the average adult still ranged from 900-1200 calories, and that the recent bad winter had only worsened an already deplorable food situation.[73] The former U.S. Ambassador to Russia and current Secretary of Commerce, Averell Harriman, observed after a six day stay in Germany that "probably, the strongest impression I carry back from my week's stay is hunger and the hopelessness of the people."[74]

---

[70] Oral History Interview with Gunther Harkort, 12 November 1970, HST/OH/Harkort.

[71] U.S. House of Representatives, Subcommittee of Committee on Appropriations, Hearings on *First Deficiency Appropriations Bill for 1947*, 80th Cong., 1st Sess. (Washington, 1947); Secretary of War Robert P. Patterson to George M. Marshall, Secretary of State, 13 June 1947, NA/RG107/E106/B4.

[72] Forrest Davis to Senator Robert A. Taft, 29 January 1947, HIA/USPFEC/B21; Alan S. Milward, *The Reconstruction of Western Europe 1945-51* (Berkeley, 1984), 13.

[73] War Department Public Information Division, Press Section, statement by Secretary of War Howard C. Petersen, 13 June 1947, HIA/USPFEC/B24; War Department, Public Information Division, Press Section, "Statement by the Assistant Secretary of War Howard C. Petersen," 13 June 1947, DDE/Fitzgerald/B1.

[74] Averell Harriman to Executive Committee of the National Grange, 16 June 1947, LOC/Harriman/B256; W.A. Harriman, Secretary of Commerce, to the President, 12 August 1947, HST/PSF/B178.

Recognizing that responsibility to avert famine "follows the flag," the War Department finally announced that feeding civilians in Germany and other territories around the world occupied by US forces would henceforth assume first priority.[75] The most important weapon in the public relations campaign to help postwar Germany was Herbert Hoover, who was recruited in February 1947 for a second food mission directed exclusively at Germany and Austria.[76] Hoover worked closely with U.S. military government officials to persuade Congress and the American public of the need for an expanded food program.[77] As historian Jean Edward Smith explained, while U.S. Military Governor Lucius Clay had not been able to spur the government bureaucracy in Washington into action to increase food exports to Germany, "Hoover rolled over it."[78]

In his final report, the former President complained about the fact that the U.S. government was spending $600 million per year to prevent starvation in Germany. On purely practical grounds, Hoover felt that this costly and stopgap effort had to be stopped, and a new and more effective policy of German and European reconstruction pursued: "These conclusions are not the product of sentiment nor of feeling toward a nation which has brought such misery upon the whole earth. They are not given in condonement of the enormity of her crimes. They are the result of a desire to see the world look forward, get into production and establish a lasting peace."[79] Besides reducing the costs to the U.S. taxpayer of the occupation, and preserving the safety and health of Allied troops stationed in postwar Germany, Hoover was concerned that conditions there had sunk "to the lowest level known in a hundred years of Western history. If Western Civilization is to survive in Europe, it must also survive in Germany. And it must be built into a cooperative member of that civilization. That indeed is the hope

---

[75] Secretary of War Robert P. Patterson to the Under Secretary of War, 15 December 1946, NA/RG107/E106/B4.

[76] Earl Harrison of the Agriculture Department traveled on a similar mission to Japan and China.

[77] Oral History Interview with Dennis A. FitzGerald, 21 June 1971, HST/OH/FitzGerald; Lochner, *Herbert Hoover and Germany*, 182-91.

[78] Jean Edward Smith, *Lucius D. Clay: An American Life* (New York, 1990), 339-40, 361-3.

[79] Hoover quoted in Gustav Stolper, *German Realities* (New York, 1948), 301.

of any lasting peace. After all, our flag flies over these people. That flag means something besides military power."[80]

As the third winter of the occupation began, public reaction against the prolonged famine was intensifying. Even German citizens were becoming increasingly outspoken and critical of the Allies for depriving them "of the first of all human rights, the right to keep on living."[81] A group of German medical doctors submitted a report to the Second U.N. Conference on Food and Agriculture complaining about the disastrous food situation in their country.[82] There were also massive food strikes in some cities in the British zone.[83]

U.S. officials payed close attention to the growing voices of protest both at home and abroad. Secretary of War Robert Patterson warned that if outright famine broke out in Germany, "Such a calamity would be a damaging blow to our foreign policy, to say nothing of considerations of humanity and the unfavorable reaction with our own people."[84] By the beginning of 1948, the Secretary of the Army, Kenneth Royall, told an audience at a Denver Rotary Club that the current adult ration of 1425 calories was equivalent to little more than a "hearty American breakfast," and that the Allies would soon have to choose between three options as described by a former War Department official: "starve 'em, shoot 'em, or feed 'em."[85] A report issued by the U.S. Military Governor the following year acknowledged that until the end of 1948, Ger-

---

[80] Herbert Hoover, *An American Epic, Volume IV: The Guns Cease Killing and the Saving of Life from Famine Begin 1939-1963* (Chicago, 1964), 242-3; Stolper, *German Realities*, 67.

[81] Dr. F.H. Rein, University of Goettingen, "The Hunger Problem" (June 1947), HIA/Becker/B1.

[82] Nutrition Board of the German Medical Profession, "The German Medical Profession on The State of Nutrition in Germany," July 1947, HIA/Becker/B1; De Castro, *The Geopolitics of Hunger*, 429-30. See also Manfred J. Enssle, "The Harsh Discipline of Food Scarcity in Postwar Stuttgart, 1945-1948," *German Studies Review* 10 (October 1987): 488, 492, 495, 500-1.

[83] Kramer, *The West German Economy*, 83-4.

[84] Secretary of War Robert P. Patterson to Mr. Petersen, 9 May 1947, NA/RG107/E106/B4.

[85] Department of the Army, Public Information Division, Press Section, "Address by the Honorable Kenneth C. Royall, Secretary of the Army, before the Denver Rotary Club, Denver, Colorado," 8 January 1948, HST/Rockwell/B31.

many's infant mortality rate was double that of other Western European nations, and its birth rate remained the lowest in Europe.[86] Even a Military Government official who believed that the shortages of food in Germany were exaggerated, and that the average adult was actually receiving 2000-2100 calories per day thanks to unreported production and black market purchases, complained that

> It seems to me we are wasting our time arguing over 50 or 100 calories when we know that to get the German population back to any degree of normal recovery we will have to increase the ration level somewhere near their pre-war level, which for non-self-suppliers was 2870 calories daily. I am sure at least it would take another 500 calories to really be significant as far as industrial recovery is concerned.[87]

During May 1948, yet another combined U.S.- British Special Commission surveyed the situation in Germany. Their final report indicated that conditions had improved during the past year, and that the current official rate of 1500 calories was now being supplemented by an additional "spread" of as many as 400 calories through nutrition programs for special groups, and an average of 300 calories of "off the ration" intake. But even these more optimistic estimates remained far below the level required to assure full work capacity for the average adult, and had not prevented an additional loss of 10-15% of body weight in German adults during the past year. To restore productivity and morale in postwar Germany, the commission recommended that the revised "disease and unrest" formula (with a ceiling of 1550 calories) and the "standard of living" guidelines (the eventual goal of which was a ceiling of

---

[86] Office of Military Government for Germany (U.S.), "Statistical Annex, Issue No. 23", Report of the Military Governor, No. 43, January 1949, HIA/Germany (OMGUS)/B13; Office of Military Government for Germany (U.S.), "Statistical Annex, Issue No. 28," Report of the Military Governor, No. 48, June 1949, HIA/Germany (OMGUS)/B13; Dr. L. Bachmann, Technical Section (Medical), U.S. Naval Forces, Germany, "Infectious Diseases in Germany Prior to and After World War II," 20 April 1951 HIA/Becker/B6; Harold Zink, *American Military Government in Germany* (New York, 1947), 123; Hans Schlange-Schoeningen, *Im Schatten des Hungers: Dokumentarisches zur Ernaehrungspolitik und Ernaehrungswirtschaft in den Jahren 1945-1949* (Hamburg, 1955), 292.

[87] L.J. Stahler, Chief Food Rationing Branch, OMGUS to J.C. Ebbs, Nutrition Advisor, Office of the Quartermaster General, Department of the Army, 19 February 1948, HIA/Becker/B2.

2100 calories) should both be scrapped and replaced with a new base ration of at least 2540 calories.[88]

As it turned out, however, the end of the famine was already at hand. Thanks to an unusually good harvest during the summer of 1948, the onset of Cold War tensions and the accompanying launch of the Berlin airlift and Marshall Plan, and the unification of the western zones of occupation, economic conditions finally began to improve across Western Europe. The following year, food was no longer a source of concern for most Germans living in the newly created Federal Republic of Germany, and in 1950 the rationing system was abandoned entirely.[89]

In conclusion, the Allied governments did pursue a stern food policy towards the German people in 1945, portions of which persisted into the third year of military occupation. Most of the terrible conditions that prevailed were brought on by the chaos and destruction of war, but some—including the reduction in rations for POWs and civilians and the initial denial of international relief to both—were at least partly the result of a determination to ensure that this time the Germans would feel the sting of defeat and pay reparations for the damage that their nation had wrought across the continent.

The inadequate feeding of enemy civilians began with the wartime decision to scale back the original "disease and unrest" formula from a minimum of 2000 to a maximum of 1550 calories, as well as the decision at Potsdam to reduce Germany to a standard of living below the average for Europe. Depending on the time of year and the region, the actual supply of rations for civilians fluctuated between 1000-1300 calories. Added to that were the problems of unfair distribution and poor quality. While it is true that many Germans were able to supplement their official diet with "off-ration" supplies, usually obtained on the black market, these averaged anywhere from 200-300 additional calories, and the majority obtained less or nothing at all to supplement their official rations. The effects of malnutrition were made worse by accompanying shortages of clothing and shelter, the disruption or contamination of water supplies, shortages of medical personnel, hospital space, and medicines, and the influx of millions of additional DPs and refugees from the East, most of whom were in even worse shape to begin with.

---

[88] "Report of the Special Commission Appointed by Secretary of the Army Royall to Study Nutrition in Bizonal Germany," 27 May 1948, HST/B-File/Germany/F1.

[89] Zink, *The United States in Germany*, 298.

For reasons largely of geography, conditions tended to be best in the U.S. zone and worst in the French and Soviet zones, despite the fact that the latter was traditionally a food surplus region. Urban centers tended to be hardest hit as well, with inhabitants of Berlin and the Ruhr suffering most of all. Small children, university students, and older people were hardest hit by the malnutrition. Though deaths resulting directly from starvation remained rare even during the worst phases of the postwar occupation, there was an increase in the rate of deaths from suicides and diseases like tuberculosis, typhus, diptheria, and influenza in which malnutrition likely played at least some contributing role. There was also an undeniable increase in the overall death rate in postwar Germany throughout this period. In 1946, when famine was still widespread in Europe and elsewhere, it was double the prewar figure. By 1948, when conditions had improved in the remainder of Europe, it was still 30 percent higher than the prewar level, and about 35% higher than in the U.S.

Clearly, the Allies' own interpretations of Article 43 of The 1907 Hague Rules of Land Warfare were not followed when it came to the postwar occupation of Germany. When they insisted in 1945 on imposing unconditional surrender and supreme authority on the defeated enemy, most officials in London and Washington believed they had found a way to circumvent their legal obligations to feed and otherwise care for civilians living in the occupied territories. As late as 1953, a State Department legal adviser noted that the laws relating to civilian feeding were vague, and that "There is no provision in The Hague rules, requiring the military occupant to furnish subsistence to the inhabitants of the occupied territory." The same official argued that the occupation of Japan—and by implication that of Germany as well—as a result of unconditional surrender "placed the United States in an entirely different position from the occupant of enemy territory during hostilities and left it free to makes its own rules of occupation, subject to the dictates of conscience and humanity."[90] The British delegates to the conference that negotiated the expanded 1949 Geneva Conventions also tried to argue that Article 43 of The Hague Rules of Land Warfare did not apply in the case of the Allied occupation of Germany because of the special circumstances of unconditional surrender and the accompanying assumption of su-

---

[90] Yingling to Mr. Robertson, "Did the United States as the Military Occupant of Japan Have an Obligation to Furnish Subsistence for the Japanese People?," 23 October 1953, NA/RG59/E684/Hiss/B1.

preme authority over the German state by the victors.[91] In truth, all four of the Allied powers violated both the spirit and probably the letter of The Hague Rules of Land Warfare when it came to the feeding of enemy civilians.

In the end, tens of millions of Germans lived through at least several years of malnutrition and deprivation in the wake of the 1945 surrender. It is unlikely that any historian will ever be able to calculate how many civilian deaths can be attributed—either directly or indirectly—to the prolonged suffering that prevailed in postwar Germany. What is certain is that many more POWs and civilians suffered and perished than needed to in the aftermath of World War II, and that the victorious Allies were guided at least partly by a spirit of postwar vengeance in creating the circumstances that contributed to those deaths. Having returned from a tour of devastated Germany in 1947, British socialist and writer Victor Gollancz attempted to put the best face possible on these and other Allied actions:

> I have criticised in this essay our treatment of Germany. It cannot be criticised too strongly: for these policies for which we have been jointly or solely responsible—annexations, expulsions, spoliation, economic enslavement, non-fraternization and starvation—are more in the spirit of the Hitler we fought than in that of the western liberalism for which we fought him. But to go on to suggest that all distinction has vanished, and that we have been utterly corrupted by the thing we have been fighting—this would be to exaggerate, and grossly. We have alienated great territories of the enemy: Hitler would have annexed all Europe, and eventually the whole world. We non-fraternised with the Germans: Hitler murdered six million Jews. We are starving the people in our charge, not deliberately but because to feed them as we ought would be to lower our own standards: Hitler would have starved, and did starve, anyone it might suit him to starve, with complete deliberation and even, God forgive him, as a matter of preference. These are vast differences, and we must cling to the thought of them if we are to retain our self-respect.[92]

---

[91] W.H. Gardner, Chairman, Committee on the Revision of the Geneva Conventions, Deputy Leader, United Kingdom Delegation, War Office "Report on the Work of the War Office Members of the United Kingdom Delegation to the Diplomatic Conference for the Establishment of International Conventions for the Protection of War Victims at Geneva 21st April-12th August 1949," October 1949, PAC/MG42/DO35/3359/ ReelB-6141.

[92] Victor Gollancz, *Our Threatened Values* (Hinsdale, 1948), 215. See also the following publications by the same author: *Leaving Them to Their Fate: The Ethics of Starvations* (London, 1946) and *In Darkest Germany* (Hinsdale, 1947).

The German Expellees and European Values

EMIL NAGENGAST

The history of the German expellees is both tragic and bitterly ironic. After suffering flight and expulsion from East Europe in 1945 the millions of expellees were pushed to the fringes of German politics throughout the following half century. Their political interests were dismissed as a set of revisionist demands that ran directly counter to the German Federal Republic's forward-looking, post-nationalist, "European" priorities. Ironically, in the 1990s Europe's international organizations and the full spectrum of German politics embraced many of the same principles that the expellees themselves decades earlier had defined as key components of a future pan-European peace order. The final tragedy for the expellees today is the failure of German and other European politicians to acknowledge the fact that in some respects the expellees were more progressive than the national governments that treated them with much disdain.

This essay presents an overview of the tension since 1945 between, on the one hand, the efforts to marginalize the expellees from the FRG policymaking process, and, on the other hand, the determination of the expellee organizations to turn their priorities into German and European priorities. The eagerness to confine the tragedy of the expellees to the history books is widespread, but it is unjustified. Over the past fifty-five years the expellees have stood for a wide range of demands. But at the core of the expellees' political efforts, often hidden by their zealous Cold War territorial revisionism, there has been a set of principles that today are commonly associated with "European values," namely: minority ethnic group rights, the right to a homeland, the increasing permeability of national borders, and the renunciation of ethnic cleansing.

Antje Vollmer, Vice President of the Bundestag, declared in a June 2000 Bundestag debate on the EU's eastern expansion: "The topic of the [German] expulsion belongs in the Museum of German History."[1] But some prominent figures have acknowledged the mistake of ignoring the progressive aspects of the expellees' demands. In 1995, Ayalo Lasso, the UN High Commissioner for Human Rights, stated: "I am convinced that if the states had thought more since the end of the Second World War about the

---

[1] *Stenographischer Bericht*, Deutscher Bundestag—109. Sitzung (9 June 2000), p. 10298.

implications of the flight, expulsion and resettlement of the Germans the current demographic catastrophes, which are labelled as ethnic cleansing, perhaps would not have occurred on the same scale."[2] In 1999 German Interior Minister Otto Schily offered an apology to the expellees: "The political left has, in the past— unfortunately, it cannot be denied—shut its eyes to the crimes of the expulsion, to the million-fold suffering that was inflicted on the expellees; whether out of disinterest, or out of fear of the accusation of being labelled a Revanchist.... Contrary to many prejudices, the vast majority of expellees have taken an active part in the reconciliation of European nations, and they do this still today." But in the same speech Schily echoed the arguments of Antje Vollmer and Chancellor Gerhard Schroeder: "We must draw a line across this part of our history."[3] International law scholar Robert Hayden draws a parallel between the ways in which the German expellees and the Serbs have been depicted as unworthy of sympathy:

> At this point, a number of legal arguments could be made, but we will leave them aside for the moment. What is more important is that almost no one would bother to make them. The brutal expulsion of ten million Germans, as Germans, from east central Europe, the destruction of their culture, their elimination as groups as such, is, and was, of no interest to anyone not German. Indeed, I am quite aware that in raising the issue, I am doing something unseemly. The Germans, it is usually said, deserved it.[4]

The expellee organizations carry most of the responsibility for the fact that the majority of German policy-makers (and of the German public) have come to view the expellees as anachronisms and as a threat to Germany's European identity. By focusing so fiercely and, at times, exclusively on territorial demands throughout the Cold War era the leaders of the expellee organizations portrayed themselves as aggressive nationalists. In the 1990s, even after these organizations dropped their legal complaints about lost German territory, no one has been willing to acknowledge the European character of the expellees' other political demands. Outside Germany, Austria, Poland, and Czechoslovakia (and later the

---

[2] Ayala Lasso (1995) UN High Commissioner for Human Rights, cited in Alfred-Maurice de Zayas, "Zur Aktualität des Rechts auf die Heimat," in Alfred-Maurice de Zayas and Christian Hillgruber, *Gerechtigkeit Schafft Frieden* (Bonn, 1997), 22.

[3] Speech by Otto Schily at the *Tag der Heimatvertriebenen* in Berlin, 29 May 1999.

[4] Robert M. Hayden, "Schindler's Fate: Genocide, Ethnic Cleansing, and Population Transfers," *Slavic Review* 55 (Winter 1996): 730.

Czech Republic), the expellees have been ignored. Even in the post-Cold War era, in East Europe the expellees have been vehemently condemned as a dangerous, nationalist group.[5] In short, the expellees encounter deep mistrust and hatred in East Europe, but condescension and apathy in Germany.

The political history of the German expellees since 1945 provides an interesting perspective on two broader aspects of European studies: first, the evolution of German *Ostpolitik* from a policy grounded in the rejection of the postwar settlement, to a policy aimed at "drawing a line across history"; second, the evolution of certain principles from self-interested expellee demands in the 1950s to widely accepted components of the European *Wertegemeinschaft* (community of values) in the 1990s.

The return of ethnic cleansing to Europe in the 1990s made it impossible for political leaders to pretend that the expellees belong to a distant past. European politicians cannot deny that the expellees were among the earliest to make the argument that a lasting peace in Europe depends upon an international commitment to certain values. In Pristina in 1999, Tony Blair pronounced: "We fought in this conflict for a cause and that cause was justice. We fought for an end to ethnic cleansing, we fought for peace and security for all people in Kosovo.... We know that justice must apply to all people whatever their race... whatever their background."[6] The same words that were once dismissed as destabilizing, revisionist rhetoric (from the German expellees) are now used by contemporary European leaders as the justification for devastating multilateral military action against a state that violates moral and legal norms. The history of the German expellees has been one of a failed attempt to make their demand for legal and moral justice a priority of German and European politics.

The Expulsion of Germans from East Europe

Approximately fifteen million Germans either fled or were expelled from East Europe as a result of World War II.[7] In addition

---

[5] For example, "Polens Regierungschef attackiert Vertriebene: Buzek bezeichnet Verband als 'extreme Gruppe,'" *Die Welt*, 8 Sept. 1998; and Hans-Helmuth Knutter, "Das Bild der Vertriebenen in der Agitation und Propaganda der Sowjetunion und der DDR," *Verständigung der deutschen Vertriebenen mit den östlichen Nachbarn*, ed. Christof Dahm (Bonn, 1992).

[6] Speech by UK Prime Minister Tony Blair in Pristina, Kosovo , 31 July 1999.

[7] Alfred M. de Zayas, *Nemesis at Potsdam: The Expulsion of the Germans from the East* (Lincoln, Neb., 1988), 25.

to shifting the Polish-German border, the Potsdam Protocol announced the Allies' agreement with the demands of regional governments for a solution to the "German problem" in their respective lands. The Allies gave their support to the "orderly and humane" expulsion of the Germans living in Poland, Hungary and Czechoslovakia."[8] By providing humanitarian aid to only a small part of the expelled Germans when they reached the territory of what became East Germany, the Soviets forced millions of expellees to continue their westward trek through the Soviet occupation zone and into what became the Federal Republic of Germany (the FRG–West Germany). Ironically, these expellees later assisted the US greatly by becoming the strongest proponents in Europe of American anti-Soviet policies. Stalin's intention was to destabilize western German society by overloading it with poor, disgruntled refugees.[9] As proclaimed in their 1950 Charter (see below), however, the expellee organizations rejected the path of foreign policy militancy and played a vital role in the economic reconstruction and political stabilization of the FRG.

In 1945 Czechoslovak President Eduard Beneš issued a set of laws known as the "Beneš Decrees." These laws announced, first, the immediate confiscation of all property belonging to the "Sudeten" Germans, and second, the immediate expulsion of all Germans from Czechoslovakia. According to Beneš: "We must get rid of all those Germans who plunged a dagger in the back of the Czechoslovak state in 1938."[10] The Beneš Decrees resulted in the expulsion of approximately three million Germans from Czechoslovakia. Between 40,000 and 300,000 Germans were killed during the expulsion.[11] Throughout the Cold War, FRG governments repeatedly condemned the legality and morality of these actions, but

---

[8] "Potsdam Protocol, Article XIII, 87.

[9] Alfred M. de Zayas, *A Terrible Revenge: The Ethnic Cleansing of the East European Germans, 1944-1950* (New York, 1986), 132; Günter Böddeker, *Die Flüchtlinge: Die Vertreibung der deutschen im Osten* (Berlin, 1980), 471.

[10] De Zayas, *Nemesis at Potsdam*, 17.

[11] German sources place the number between 200,000 and 300,000. According to the *Sudetendeutscher Atlas*, 302,000 died during the expulsions; E. Meynen, ed. (Munich, 1954). In contrast, Czech historians have argued that the number was no more than 40,000. For examples see Radomir Luza, *The Transfer of the Sudeten Germans* (New York, 1964); Jaroslav Kucera, *Odsun nebo vyhnání?: Sudetstí Nemci v Ceskoslovensku, v letech 1945-1946* [Transfer or Expulsion?: Sudeten Germans in Czechoslovakia, 1945–1946] (Prague, 1992), cited in Ferdinand Seibt, *Deutschland und die Tschechen* (Munich, 1993), 462-463.

until 1990 Czechoslovak leaders refused even to acknowledge the FRG's efforts to represent Sudeten German interests.

Between 1945 and 1950 eight million expellees and refugees moved to the FRG from East Europe. Prior to the closing of the inner-German border in 1961, 3.6 million Germans fled from the GDR to the FRG—a large portion of them having previously moved from elsewhere in East Europe to the GDR. From 1950 to 1989 1.2 million more Germans left the territory of Poland.[12]

Initial attempts in the 1940s to establish formal expellee organizations were prohibited by the Western occupation powers.[13] This prohibition resulted in formal ties between the expellees and the major political parties in the FRG. The mainstream parties created internal committees (*Flüchtlingsausschüsse*) as a means to address expellee concerns and to bind the expellees to the parties. Against the vehement complaints of the Polish government, the Western powers ended the prohibition on expellee organizations in April 1950. Despite the efforts of the dominant parties to incorporate them, in the early 1950s the expellees established a few of their own political parties. The largest of these, the *Bund der Heimatvertriebenen und Entrechteten* (BHE), won 5.9 percent of the vote (1.6 million voters) in the 1953 national election and was part of the ruling coalition in Bonn until it sank below the five percent hurdle in 1957.[14] The first President of the BdV, Linus Kather, claimed that in 1951 Konrad Adenauer offered him a cabinet position in exchange for Kather's efforts to destroy the BHE.[15]

The Charter of the German Expellees

On 5 August, 1950, delegates from the thirty leading expellee organizations issued the "Charter of the German Expellees" (*Charta der deutschen Heimatvertriebenen*) in Stuttgart. The purpose of the charter was to "establish the duties and rights that the German expellees view as their own statute and as the unavoidable pre-

---

[12] Edmund Spevack, "Ethnic Germans from the East: Aussiedler in Germany, 1970-1994," *German Politics and Society* 13 (Winter 1995): 74.

[13] Dennis Bark and David Gress, *From Shadow to Substance 1945-1963* (Oxford, 1993), 308; Peter Reichel, "Die Vertriebenenverbände als aussenpolitische 'Pressure Groups,'" cited in *Handbuch der deutschen Aussenpolitik*, ed. Hans-Peter Schwarz (Munich, 1975), 233.

[14] Johannes-Dieter Steinert *Vertriebenenverbände in Nordrhein-Westfalen 1945-1954* (Düsseldorf, 1986), 25-91.

[15] Cited in ibid., 158.

condition for the creation of a free and unified Europe." The Charter makes three central points. First, "We expellees renounce revenge and retaliation." Second, "We will support with all of our strength every initiative that aims to build a unified Europe..." Third, "...we feel we are called upon to demand that the Right to a Homeland, as one of the God-given rights of humanity, be recognized and realized." The Charter concludes: "The nations must recognize that the fate of the German expellees, like that of all refugees, is a global problem, the solution of which commands the highest moral responsibility and commitment to powerful achievement."

Despite both their numerical strength within the newly created FRG and the strength of their legal arguments (as codified in the FRG Basic Law, the Potsdam Agreements, the UN Charter, and the Universal Declaration of Human Rights, among other international documents), the expellees took an early stance against a violent resolution of their demands. With the Charter the expellees linked their specific concerns with the broader idealistic aim of promoting pan-European unity based on the emergence of a European *Wertegemeinschaft* (community of values). Throughout the Cold War the expellees repeatedly stressed their devotion to a united Europe. The 1955 Berlin resolution of the *Verband der Landsmannschaften* declares: "The right to a homeland must be recognized and realized by all nations as a human right.... We pledge ourselves to a united Europe. This united Europe can only grow out of a community of free nations."[16] In 1964 the *Ostdeutschen Landesvertretungen* declared: "The goal of a united, free Europe is achievable only on the basis of a lawful ordering of states and nations that protects the right to self-determination and the right to a homeland, and not on the basis of the tolerance of, or even the support for, unlawful violence."[17]

The significance of these proclamations is clear when we compare the priorities of the expellees with those of other displaced peoples. Tilman Mayer points out that the expellees did not burden the whole nation with their own militancy, as did the IRA, ETA, and PLO. There was never a "Palestinization" of the German question, because the expellees renounced this path at the founding of the FRG.[18] The expellees linked the resolution of their legal

---

[16] "Berliner Entschliessung des Verbandes der Landsmannschaften," cited in *BdV-Erklarungen zur Deutschlandpolitik I (1949-1972)* (Bonn, 1984), 49.

[17] Ibid., 105.

[18] Tilman Mayer, "Die Vertriebenen im heutigen Deutschland—Schrittmacher der Verständigung mit den östlichen Nachbarn?" in *Die Bundesrepublik und die Ver-*

and moral demands with the creation of a pan-European union that was defined by moral values and legal principles. Throughout the Cold War the highest priority of the expellees was to secure a central position in the formulation and implementation of FRG foreign policy priorities.

Bund der Vertriebenen

The *Bund der Vertriebenen* (the "Federation of Expellees") was founded in 1951. By the late 1950s the BdV was the national umbrella interest group for the 13,470 expellee organizations that had sprung up across West Germany.[19] Since 1951 the BdV has maintained a national office in Bonn representing the 21 different territorial associations (*Landsmannschaften*) such as the Silesian Germans, the East Prussian Germans, and the Sudeten Germans. All BdV Presidents have served simultaneously as parliamentarians in the German *Bundestag*.

It has always been difficult to determine the exact number of BdV members. The organization is intentionally vague about such statistics, partly because it is extremely difficult for the national office to tally the membership of the BdV's numerous suborganizations, but also because the BdV prefers to give the impression that it speaks for all German expellees—not just for BdV members. In the 1950s the BdV reported a total membership of 1.7 million.[20] The BdV estimated its membership at the beginning of the 1960s to be three million West German citizens. In the 1960s all German expellees and refugees (that is, BdV members and non-members) comprised over 20 percent of West Germany's total population.[21] In 1983 the FRG Interior Ministry reported that 25 percent (16 million) of the FRG population were expellees. In more recent decades such numbers are misleading, however, because the offspring of the original expellees have the legal status of "expellee"

---

triebenen, ed. Christof Dahm and Hans-Jakob Tebarth (Bonn, 2000), 169; Mayer draws a contrast between the character of the "Charter of the German Expellees," in which violence is renounced, and the 1968 "Charter of the PLO," in which "armed struggle" is proclaimed as the "only way" to regain lost territory from Israel.

[19] Walter Bradatsch, *Neue Heimat in Niedersachsen* (Hannover, 1979), 74-88; see also "Wir informieren über uns," *BdV Broschüre* (1993).

[20] Bradatsch, 88.

[21] Timothy Garten Ash, *In Europe's Name* (New York, 1993), 30; Uwe Anderson and Wichard Woyke, *Handwörterbuch des politischen Systems der BRD* (Bonn, 1995), 309.

and are typically included in such statistics.[22] According to one BdV official, in the early 1990s about two million East and West Germans were members of one of the BdV's component organizations.[23]

Throughout the Cold War the BdV played the role of foreign policy watchdog of the FRG constitution. Their central aim was to uphold the legal claims on the lost territories with all available political means.[24] The BdV mobilized its resources against any official who compromised the constitutionally mandated priority of *full* German unification. In 1950 (when 18 percent of the parliamentarians in the Bundestag were expellees[25]), US High Commissioner John McCloy feared that the expellees formed "a group which might readily be swayed by political extremists who offer a plausible solution to their problems. In Germany's external relations they exert constant irredentist pressure."[26] In the 1950s both Konrad Adenauer and Kurt Schumacher argued that no German government would ever be in the position to recognize the Oder-Neisse border.[27] In 1966 BdV President Wenzel Jaksch summarized his organization's priority of a pacifist, but legally oriented, foreign policy:

> The spirit of presumption that hung over us during the twelve years of Hitler came to a terrifying end. And yet the spirit of self-abnegation of national interests vocalized by an audible minority is just as ominous. We cannot continue to drive the youth of Germany from one extreme to the other... Our people have unanimously repudiated a policy of force, leaving but two alternatives: the enforcement of our legal position or the acceptance of Germany's partition. A third alternative does not exist.[28]

---

[22] Kurt Hirsch, *Rechts von der Union* (Munich, 1989), 173.

[23] Interview with Markus Leuschner, Head of Youth Affairs of the BdV (interviewed at the BdV national office in Bonn on 13 July 1994).

[24] Hirsch, 173.

[25] Reichel, 237.

[26] Bark and Gress, 307.

[27] Herbert Czaja, "Parteien sollen ihre Vertriebenenpolitik überdenken," *Deutscher Ostdienst* (27 May 1994).

[28] "Expellees to Demonstrate for Right to the Homeland," *Central Europe Journal* (May 1966): 174-175.

Adenauer's "Holding Policy"

Adenauer's foreign policy between 1949 and 1963 is best described as one that sought "in the West a lasting peace but in the East merely an armistice."[29] Adenauer pursued cooperation and integration within West Europe, but he held firmly to legal and moral arguments that the post-war settlement on Germany's eastern border was unacceptable. Hanrieder describes this conflict of priorities as "looking forward and backward in history at the same time."[30] His "holding policy" concerning East Europe served an important electoral purpose. This policy allowed Adenauer to placate the expellees by presenting himself as the staunch representative of German legal claims in East Europe, while at the same time making important steps toward supranational cooperation in West Europe.[31]

Adenauer's government took several additional steps to conciliate the expellees. The "Federal Expellee and Refugee Law" guaranteed the BdV federal funds designed to promote the expellees' integration into West German society. The 1952 "Equalization of Burdens Law" (*Lastenausgleichgesetz*) imposed a tax on those Germans who retained property through the war and provided grants to the expellees and others who had lost everything. By 1994 over DM140 billion had been redistributed through this law.[32] A *London Times* article from June 1956 reports with dismay that the FRG government granted DM4.5 million for an inquiry into the fate of the German expellees. The article notes that: "...the refugee organizations have emerged as the spokesmen of a politically powerful minority. Most are nationalists, and a few are outright Nazis, and last week the Social Democratic Party was required under their pressure to dissociate itself from one of its parliamentary representatives who said that the Sudeten territory [which Hitler annexed in 1938] was Czech and should remain Czech."[33]

Peter Reichel estimates that the BdV's yearly income (from dues, donations and publications) in the 1960s reached DM14 mil-

[29] Wolfram Hanrieder, *Germany, America and Europe: Forty Years of German Foreign Policy* (New Haven, 1989), 147.

[30] Ibid., 149.

[31] William Griffith, *The Ostpolitik of the FRG* (Cambridge, 1978), 45-53.

[32] "Jahresbericht der Bundesregierung 1994," *Bundesministerium der Finanzen*.

[33] "Inquiry into a German Exodus," *London Times*, 2 June 1956.

lion, which was still insufficient to cover the costs of the 1200 expellee functionaries. To help cover the expellees' remaining expenses, federal, state, and local governments contributed a total of DM32 million in the 1960s.[34] In addition to substantial financial support, the expellees were given a cabinet position (the Federal Expellee Minister) in Adenauer's government. This ministry lasted until 1969 when Brandt's government subordinated "expellee affairs" under the Interior Ministry (at the time headed by Hans-Dietrich Genscher).

The expellees were especially influential within the Christian Social Union (CSU), the Bavarian sister party of Adenauer's Christian Democratic Union (CDU). In contrast to the other German expellees who settled throughout the FRG, the expelled Sudeten Germans settled primarily across the Czech border in Bavaria. The 1.1 million Sudeten Germans became what former Bavarian CSU Minister-President Franz Josef Strauss called "Bavaria's fourth tribe."[35]

In the late 1950s and early 1960s public opinion indicated decreasing belief in German unity as a reality, and a growing desire for reconciliation with Germany's eastern neighbors.[36] The CDU's gradual "policy of movement" in the 1960s concerning relations with East Europe met with fierce opposition from the CSU, but was overrun in 1969 by Willy Brandt's dramatic move to take on the expellees and move toward cooperation with Moscow and the East European Communist regimes.

From Brandt's *Ostpolitik* to the End of the Cold War

In 1969 the SPD, in coalition with the FDP, took control of the federal government under the Chancellorship of Willy Brandt (SPD). In the 1950s the FDP had advocated the restoration of Germany's 1937 borders. Similarly, in the 1950s the SPD had claimed to be the FRG's leading voice of revisionism.[37] By 1969, however, the FDP shared Brandt's view that strict adherence to Adenauer's "holding policy" would not allow Bonn to construct a more pragmatic

---

[34] Reichel, 234-235.

[35] This term implies that the Sudeten Germans were thus accepted as equal partners of Bavaria's three original ethnic German "tribes." Franz Josef Strauss, *Errinerungen* (Berlin, 1989), 66.

[36] Clay Clemens, *Reluctant Realists: The Christian Democrats and West German Ostpolitik* (Durham, 1989), 47.

[37] Hans Georg Lehmann, *Der Oder-Neisse Konflikt* (Munich, 1979), 114.

European framework which might ease the effects of Germany's division. According to FDP Foreign Minister Walter Sheel, *Ostpolitik* was an "expression of the identity of our interests with the interest of Europe."[38] In concrete terms, Brandt signed a series of bilateral treaties with the East European Communist governments which acknowledged the territorial status quo in Central Europe. Brandt's SPD successor, Chancellor Helmut Schmidt, signed the CSCE Final Act in 1975 which was a further German "acknowledgment" of the status quo. Brandt received the Nobel Peace Prize for his policy of cooperation and reconciliation with East Europe, but the leadership of the BdV viewed Brandt's pragmatic *Ostpolitik* as a legitimization of authoritarian regimes and as a "total surrender of Germany's natural rights" in the East.[39]

The SPD aimed at marginalizing the BdV from the policy process by restricting expellee activities to "protecting the cultural legacy of the previously German regions in East Europe."[40] West German public opinion had also turned against a revision of the territorial status quo.[41] Likewise, a growing number of expellees preferred a foreign policy based on pragmatic reconciliation, as opposed to the stubborn advocacy of moral and legal principles vis-à-vis East Europe.[42] Kurt Sontheimer, like many observers in the 1970s, argued that the expellees' opposition to Brandt's *Ostpolitik* would be the final significant political engagement by the expellee organizations.[43]

Helmut Kohl replaced Schmidt as Chancellor in 1982 with a CDU/CSU/FDP coalition, but promised to continue Brandt's and Schmidt's *Ostpolitik*. Foreign Minister Genscher and his FDP had already affirmed this continuity by making adherence to the Brandt/Schmidt *Ostpolitik* a precondition for abandoning the SPD for the CDU/CSU in 1982.[44]

Until his death in 1992, Willy Brandt and most of his SPD col-

---

[38] Garton Ash, 19-20.

[39] Herbert Czaja, *Ausgleich mit Osteuropa* (Stuttgart, 1970), 36; see also Heinrich Windelen, *SOS für Europa* (Stuttgart, 1972).

[40] Reichel, 238.

[41] "Immer Mehr Vertriebene finden sich mit der Oder-Neisse Grenze ab," *Neue Rhein–Zeitung*, 14 April 1972.

[42] Ibid.

[43] Kurt Sontheimer, *Grundzüge des politischen Systems der BRD* (Munich, 1971), 136.

[44] Hanrieder, 196.

leagues attacked every CDU or CSU concession to the expellees as a contradiction of Germany's European identity. Genscher made it clear that he was determined to "draw a line across history" and ignore the BdV's demands: "The [1970] Warsaw Treaty is binding for all and makes respect for the territorial integrity and sovereignty of all European states in their current borders the fundamental precondition for peace in Europe. This applies to everyone, and no one has an interest in turning back the wheel of history. Europe is our only hope."[45] The CDU was split between those members who saw the expellees as an obstacle to Germany's European priorities and those who either feared the loss of expellee votes or were sympathetic with the expellees' legal demands.

CSU Chairman and Federal Finance Minister Theo Waigel assured the Sudeten Germans that "We stand up for the expellees without ifs and buts,"[46] and that his party was the "lawyer for your legitimate petitions."[47] Between 25 and 30 percent of Bavaria's population of twelve million claim Sudeten German origin.[48] Their electoral strength in Bavaria, coupled with the CSU's position in the ruling coalition in Bonn, gave the Sudeten Germans a stronger voice than the other expellee associations in FRG foreign policy.

As the East European Communist regimes began to collapse in 1989 Kohl's government had to choose between Genscher's determination to "draw a line across history," on the one hand, and fulfilling the forty years of CDU/CSU promises to the expellees to defend their legal demands, on the other hand. Writing in the 1980s, Clay Clemens remarked that a "tectonic shift in European politics" would almost certainly shake the foundations of the Kohl/Genscher foreign policy consensus.[49] At every possible occasion Genscher emphasized: "We reject national unilateral actions, instead, we embed our fate in Europe's fate—namely European

---

[45] "Genscher zu Waigel," *Die Welt*, 5 July 1989.

[46] "Unterstützung für Polen nur wenn sie den Menschen hilft," *Rheinische Post*, 3 July 1989.

[47] "Vehemente Rufe nach dem 'Recht auf Heimat,'" *Süddeutsche Zeitung*, 9 June 1992.

[48] Jan Obrman, "Sudeten German Controversy in the Czech Republic," *RFE/RL Research Report*, 14 Jan. 1994, p. 12.

[49] Clemens, 313.

convergence."[50] According to Otto Graf Lambsdorff (FDP): "It is a difficult conclusion for every [expellee], but whoever questions the post-1945 borders also questions the peace in Europe." Against this emphasis on "European interests" Herbert Czaja, President of the BdV, declared that: "A Foreign Minister who is disloyal to all of Germany and to the most fundamental commandment of the constitution is not tolerable in a constitutionally-founded FRG."[51]

The BdV and German Unification

In October 1989 Kohl delivered a speech to a gathering of BdV members in which he made clear his view of German relations with post-Communist Poland: "Without German-French friend-ship there would have been no process of unification in the free part of our continent; and without German-Polish understanding and friendship there is no prospect of a Europe without walls and barbed wire, without hate and animosity."[52] At the same time, however, Kohl tried to appease the expellees (thereby provoking outrage in Germany and abroad) by supporting the legal view of the Polish-German border as a *de facto* acknowledged, but *de jure* provisional, frontier. Kohl maintained that neither he nor the FRG would ever call into question the Oder-Neisse border, but he added, "I cannot speak for [unified] Germany."[53]

In the spring of 1990 Kohl made it clear that he would no longer allow the expellees to hinder reconciliation with Poland. Through his initiative, both the FRG and GDR parliaments passed resolutions in June, 1990, that confirmed the permanence of the Polish-German border and thereby removed any remaining legal or rhetorical hindrances to Polish-German cooperation. Most sig-nificantly, Kohl's government (with opposition support) removed the last sentence of Article 23 (which read: "After their accession, [the Constitution] becomes binding for other parts of Germany") from the FRG Constitution. These steps meant that Germany had formally renounced all legal claims on lost territory.[54]

---

[50] "Interview with Genscher on RIAS," (26 Nov., 1989) *Kommentarübersicht* (BPA), 27 November, 1989; see also "Konsultationen Genschers in Washington," *Neue Zürcher Zeitung*, 24 Nov. 1989.

[51] *BdV Pressemitteilung Nr. 122* (15 Nov. 1989).

[52] "Kanzler Kohl lehnt Grenzdiskussion ab," *Bonner Rundschau*, 23 Oct. 1989.

[53] "Interview with Kohl," *Kommentarübersicht (BPA)*, 15 Nov. 1989.

[54] "Polens Westgrenze soll offiziell bestätigt werden," *Suddeutsche Zeitung* (5 April 1990).

In one of the most important speeches of the entire unification process, Kohl spoke at a BdV meeting about the need to move beyond past disputes and lay the groundwork for post-Cold War European stability and cooperation. At this meeting on August 5, 1990, marking the fortieth anniversary of the expellees' founding charter, Kohl was determined to win the support of the expellees with an appeal to the overriding goal of fulfilling Germany's "European responsibility."[55] "[German unity] has nothing to do–as some maintain—with nationalist dreams of a dominating German role on the European continent. The Polish people should know: a free and united Germany wants to be a good neighbor and a reliable partner on the road to Europe—to a community of free peoples which does not end at the Oder-Neisse."[56]

On 17 October, 1991 the *Bundestag* of the newly unified German state ratified the Border and Friendship Treaties with Poland. According to one expellee writer, this "new Versailles Treaty" would henceforth "become a memorial to the rape of millions of Germans" who live or once lived beyond the Oder-Neisse.[57] The most memorable speech from the ratification debates was that of CDU parliamentarian Ottfried Hennig, who had served eleven years as the Speaker of the East Prussian *Landsmannschaft*. Despite the difficult implications of these treaties for the expellees, Hennig explained, it is now the duty of Germans and Poles "to build bridges to a common European future."[58] In the end, only four parliamentarians voted against the friendship treaty. On the border treaty there were only thirteen no votes and ten abstentions.[59]

The Sudeten Germans and German-Czech Relations

In December, 1989 Bavarian Prime Minister Max Streibl (CSU) called on the new post-Communist Czechoslovak government to apologize to the Sudeten Germans for the 1945 expulsions, "just as

---

[55] "Der Kanzler bleibt vor den Vertriebenen standfest," *Stuttgarter Zeitung* (6 August 1990).

[56] "Helmut Kohl's Speech in Bad-Cannstatt on August 5, 1990," *Bulletin*, 17 Aug. 1990.

[57] Johanna Grund, "Vertrag des 17. Juni—ein neues Versailles?" *Der Schlesier* (10 June 1991).

[58] "Bundestag ratifiziert mit grosser Mehrheit das deutsch-polnische Vertragswerk," *Süddeutsche Zeitung* (18 Oct. 1991).

[59] Miszczak, 456.

Germany had recognized the injustices perpetrated by the Nazis."
Streibl also suggested that the Prague government accept the
Sudeten Germans as partners in dialogue in the future development of German-Czechoslovak relations.[60] Czechoslovak President Vaclav Havel provided the apology (independently of
Streibl's request), but due to continued Sudeten German demands
and due to the inability of the FRG government to marginalize this
expellee association, Bonn's relations with Prague became "Germany's most difficult in East Europe."[61]

The various demands within the Sudeten German community
ranged from an official acknowledgement by the Czechs of injustices committed in 1945 against the Sudeten Germans, to full restitution of all property lost by Germans through the Beneš Decrees. At the extreme right, a few Sudeten Germans called for the
annexation of the Sudetenland (as established in the 1938 Munich
Agreement). As noted above, in the 1950s not even the SPD would
renounce unequivocally this legal claim to the Sudetenland. But
by the early-1990s, no mainstream German politician took such
territorial demands seriously. The Sudeten Gemans are the politically strongest of the expellee groups, not simply because they
have a patron in the CSU. More important, perhaps, is the fact that
the political demands of the Sudeten Germans cannot be dismissed as dangerous revisionism. The Sudetern German *Landsmannschaft* has not sought the return of territory, but the granting
of rights which are clearly codified in European and UN institutions. The strongest and most widely held demands of the Sudeten
Germans were for the legal renunciation of the Beneš Decrees, for
some form of restitution from Prague for the property lost in 1945
and for the "right to the homeland" (*Recht auf die Heimat*). For the
Sudeten Germans this tricky phrase most often describes their
right to resettle in the Sudetenland.[62]

In December 1996 the German and Czech governments signed
a Statement of Reconciliation which had been delayed for years
due to domestic disputes on both sides. The Sudeten Germans
managed first to delay the signing, then succeeded in scuttling a
provision that would have dropped German demands for property restitution. Germany's most influential newsmagazine, *Der
Spiegel*, has been a harsh critic of the Sudeten Germans' impact on

---

[60] "Streibl erwartet von Prag Bedauern," *Frankfurter Allgemeine Zeitung* (14 Dec. 1989).

[61] Interview with Schäfers.

[62] "Genutze Chancen," *Bayern Kurier*, 5 May 1990.

relations with Prague. Through their influence on the CSU, and thereby on the federal government, *Der Spiegel* reports, "the Sudeten German *Landsmannschaft* is an entirely overvalued interest group with disproportionately great influence, which has almost single-handedly determined the progress of discussions and the condition of relations with our neighbor the Czech Republic."[63]

The BdV's Cultural Agenda

In a fundamental policy shift, after 1991 the BdV dropped territorial claims from their list of demands on the government.[64] The BdV announced that it was dedicated to fulfilling the spirit of the "federal expellee law" (BVFG) which provided DM40 million annually to the expellee associations for cultural and charitable activities. According to the law: "The federal government fosters through financial allocations to the territorial associations... expellee organizations... [the preservation of] the cultural heritage of the regions of German expulsion among the expellees and refugees, among the entire German people and abroad...."[65] In 1992 the federal government created an additional program to foster "measures undertaken by the expellees to promote peaceful cooperation with the peoples of East Europe."[66]

The BdV recommitted itself to "prohibition of expulsion" in Europe, and the "preservation of [German] cultural heritage."[67] The SPD, however, harshly criticized the BVFG as a generous payoff from the ruling coalition to the expellees in exchange for the BdV's shift away from revisionist principles. SPD critics contended that the expellee organizations were compensated with a substantial increase of state financial support as a means to soften

---

[63] "Tu Oma den Gefallen," *Der Spiegel*, no. 21, 1996; see also "Beim Hintern des Kanzlers," *Der Spiegel*, no. 5, 1997.

[64] "Wir informieren über uns," *Bund der Vertriebenen* (1993 Informational brochure).

[65] "Paragraph 96 BVFG: Jahresbericht der Bundesregierung, 1995," *Bundesministerium des Innern*.

[66] "Jahresbericht der Bundesregierung, 1994," *Bundesministerium des Innern*; In the 1994 Jahresbericht, the government described its goals concerning the expellees as: 1) improving the lives of Germans in East Europe through development aid; 2) assuring the acceptance and integration of German immigrants from East Europe; 3) protecting and promoting German culture in historical German regions of East Europe.

[67] Franz Wittmann, "Aufgaben und Ziele des Bundes der Vertriebenen," *Deutscher Ostdienst*, 20 May 1994.

the blow to the expellees of the conservatives' failure to fulfil their Cold War promises. What was especially irksome to the SPD was that the government funds went only to the traditional expellee organizations, which have been a perennial CDU/CSU constituency, and not to some of the newer and moderate expellee organizations, which include SPD members and defectors from the BdV. The SPD also objected to the provision of the law that excluded funding for "non-expellee" German cultural organizations which pursue the cultural goals spelled out in the BVFG.[68]

Under the SPD-Green government, oversight of the BdV's federal subsidies was transferred from the Interior Minister to the Deputy of the Federal Government for Culture and Media (Michael Naumann). Naumann slashed the federal subsidies for the BdV's cultural work from DM52 million in 1998 to DM34 million in 2004, resulting in the closure of the BdV's two main cultural branches: the *Kulturstiftung der deutschen Vertriebenen* and the *Stiftung Ostdeutscher Kulturrat*. Naumann declared that it is time to "modernize" these subsidies: "In the future there will be no more courses on cross-stitching Silesian folk costumes."[69]

This was a bitter step for the BdV, who have focused their political energy even more intensively on two issues: first, gaining international support for a universal (or at least, European) condemnation of ethnic cleansing; second, assuring the consistent application of international law (and especially European Union law) to all EU members and candidate members. The conflict in Yugoslavia and the eagerness of the Czech Republic and Poland to gain EU membership have brought both of these BdV priorities back into mainstream German political debates.

Yugoslavia and the Return of Ethnic Cleansing

According to T. Meron, "International humanitarian law has developed faster since the beginning of the atrocities in the former Yugoslavia than in the four-and-a-half decades since the Nuremberg Tribunals and the adoption of the Geneva Conventions for the Protection of Victims of War of August 12, 1949."[70] When the EU, followed by the UN, NATO, and the United States, took steps

---

[68] Ibid; see also Walter Stratmann, "Die SPD und die grenzüberschreitende Kulturarbeit im östlichen Europa," *Deutscher Ostdienst*, 13 May 1994.

[69] "Naumann: Nein zu Vertriebenen-Staette," *Die Welt*, 21 Sept. 2000.

[70] T. Meron, "War Crimes Law Comes of Age," *American Journal of International Law* 92 (1998): 462, cited in *Deportation, Vertreibung 'Ethnische Säuberung'*, ed. Wilfried Fiedler (Bonn, 1999).

to intervene in the Yugoslav crisis, the German expellees were both embittered and heartened. On the one hand, the expellees saw that after refusing for forty-five years even to acknowledge the legal legitimacy of the BdV's appeals to international law concerning ethnic cleansing, the international community was outraged by the ethnic cleansing of Muslims (among others) and willing even to take military action against the perpetrators of ethnic cleansing. On the other hand, the German expellees took some satisfaction in seeing that the EU, and the German government, had finally recognized the need to take a strong multilateral stance against ethnic cleansing in Europe and to defend the principle of minority group rights.

When the EU national governments finally became outraged over the ethnic cleansing taking place during the Croatian War of Independence and after it became clear that the breakup of Yugoslavia was unavoidable, the EU spelled out the criteria for the recognition of any new states. At the center of these demands on the former republics (and primarily on Croatia) was the protection of the rights of minority national groups. Through the four years of fighting in Bosnia-Herzegovina the EU, NATO, and the UN maintained the firm stance that ethnic cleansing could not be justified in any way. The 1992 UN London Conference on Yugoslavia issued a set of principles, among which were the "total condemnation of forcible expulsions, illegal detentions and attempts to change the ethnic composition of populations" and the "effective promotion... of the safe return to their homes of all persons displaced by the hostilities who wish this."[71]

NATO launched the massive military assault against Yugoslavia in 1999 with the primary purpose of halting ethnic cleansing in Kosovo. Nicole Fontaine, President of the European Parliament, provided justification for the military campaign against the Serbs: "The images of the abomination that is systemic ethnic cleansing, which might once have been considered to be expunged for good, have awakened the moral conscience of Europeans in all parts of the Union."[72] Parallel to the nearly universal condemnation of ethnic cleansing was the demand, "...on the Serb and Yugoslav authorities ... to create the conditions for the refugees to be able to return safely to their homes."[73] The German expellees recognized

---

[71] Cited in Fiedler, *Deportation, Vertreibung 'Ethnische Säuberung'*, 22.

[72] Speech by Nicole Fontaine, President of the European Parliament, at the University of Pristina, 21 Sept. 1999.

[73] From the "Resolution on the Situation in Kosovo," issued by the European Parliament (15 April 1999).

this demand as the expression of their own (ignored) demand since 1945 that the governments in Prague and Warsaw grant the Germans the right to return to their former homes (*Recht auf Heimat*).

Another paradox of the Yugoslav conflict for the German expellees was the fact that, on the one hand, the Serbs have been punished severely for their acts of ethnic cleansing, but, on the other hand, the international community has shown little sympathy for those Serbs who were themselves the victims of ethnic cleansing. When US Secretary of State Madeleine Albright pressured Croatian President Tudjman to allow the hundreds of thousands of expelled Serbs to return to their homes in the newly independent Croatia, Tudjman responded: "Who is supporting the return of the Sudeten Germans?"[74] Likewise, after NATO troops made Kosovo safe from the ethnic cleansing of Albanians by the Serbs, the Albanians immediately began their own campaign of ethnic cleansing against the Serbs.

The BdV has tried to build on this new-found international willingness to renounce ethnic cleansing. The motto of the June 2000 Sudeten German Annual Conference was "Condemn Expulsion Worldwide" (*Vertreibung weltweit ächten*). Also in June, 2000 the BdV announced their plans to work for the creation of a "Center against Expulsions" (*Zentrum gegen Vertreibungen*) in Berlin. Peter Glotz has been a supporter of this proposal: "It deals with expulsion not as an isolated problem of German history, but as a general problem of the twentieth century."[75] Cultural State Minister (*Kulturstaatsminister*) Michael Naumann and Chancellor Schroeder, however, have rejected this proposal.

EU Expansion and the BdV

The issue that has brought the expellees back into the German foreign policy debates has been the controversies surrounding the expansion of the EU to include Poland and the Czech Republic (among other candidates). In April 1999 the European Parliament passed a resolution that was applauded by the German expellees: "The European Parliament... calls upon the Czech government, in the same spirit of the reconciliatory statements made by President Havel, to repeal the surviving laws and decrees from 1945 and

---

[74] Reported to the author in an interview with Professor Dieter Blumenwitz, University of Würzburg (11 June 2000).

[75] "Naumann: Nein zu Vertriebenen-Stätte," *Die Welt*, 21 Sept. 2000.

1946 insofar as they concern the expulsion of individual national groups from the former Czechoslovakia."[76] Although not aimed specifically at the mass expulsions of Germans (and Magyars) from Poland and Czechoslovakia in 1945, in the 1990s the EU produced several similar resolutions and statements that confirmed the long held stance of the BdV that Europe must uphold certain principles. Paragraph 10 of the 1997 "Resolution on respect for human rights in the European Union" states that the European Parliament: "Stresses that European Union accession is out of the question for states which do not respect fundamental human rights, and calls on the Commission and Council to lay particular stress on the rights of minorities (ethnic, linguistic, ...) at the time of enlargement negotiations."

Hartmut Nassauer, the chairman of the CDU/CSU group in the European Parliament, called it "absurd and hypocritical" that the Czech Republic is not even made aware of its violations against international law and against Article 6 of the EU Treaty, while the Austrian government, which violated "in the smallest way" the spirit of Article 6 of the EU Treaty (which concerns the rights of ethnic minorities) was punished with sanctions. According to Nassauer, if an EU member state upheld laws similar to the Beneš Decrees, the EU would be required by Articles 6 and 7 of the EU Treaty to activate sanction mechanisms.[77]

As a result of the sanctions against Austria the EU member states were forced to adopt a mechanism for measuring a government's commitment to European values. In September, 2000, the report of the "Three Wise Men" (Ahtisaari, Frowein, Oreja) allowed the EU to lift the sanctions against Austria, and it provided a clearer codification of the international legal framework for defining "common European values." In many ways, the report supports Nassauer's (and the expellees') arguments concerning the EU's inconsistent defense of European values. The report lists the various international agreements (including specific references to EU law) that forbid discrimination against national minority groups. The report concludes: "We strongly recommend the development of a mechanism within the EU to monitor and evaluate

---

[76] "Resolution on the Regular Report from the Commission on Czech Republic's progress towards accession," European Parliament, 15 April 1999.

[77] "Prag soll sich von volkerrechtswidrigen Dekreten verbindlich trennen," *Deutscher Ostdienst*, 19 May 2000, 5-6.

the commitment and performance of individual member states with respect to the common European values. "[78]

In the summer and fall of 2000 the BdV focused much of its energy on making certain principles part of the forthcoming EU "Charter of Basic Rights." According to Erika Steinbach, President of the BdV: "What is urgent for a Europe of peace is the conscious and desired elimination of laws that violate human rights. What is urgent for a Europe of peace is a ban on expulsion and the right to a homeland to be included in the future Charter of Basic Rights of the EU."[79] The initial draft of the EU Charter (released in September 2000) has reinforced several of the BdV's long-standing political and legal demands. Article 21 asserts: "Collective expulsion of aliens is prohibited"; Article 22: "Any discrimination based on sex, race, colour or ethnic or social origin, language, religion or belief, political opinion, association with a national minority... shall be prohibited"; Article 30: "Every citizen of the Union has the right to move freely and reside freely within the territory of the Member States."

Conclusions

In the 1950s no mainstream West German political party criticized the expellee associations as an obstacle to the formulation and pursuit of the FRG's national interest. Between the 1950s and 1990s, however, the FRG's foreign policy priorities underwent a fundamental ideological change. Ironically, the expellees were viewed as incompatible with Germany's determination to define national interests in terms of European interests. At the time of reunification, terms such as "forward looking policies," "European identity," and "drawing a line across history" permeated West German political discourse. During the Cold War the expellees contributed significantly to their marginalization by refusing to abandon their legal claims to lost German territory in the East. But the political parties are also to blame for the fact that no one was willing to consider the possible compatibility of the expellees' *full* range of legal and political demands, on the one hand, and the FRG's determination to assert a European identity, on the other hand. As noted above, Otto Schily chastised the political left in the FRG for creating an unfair stereotype of the expellees as a group

---

[78] Report by Martti Ahtisaari, Jocehn Frowein, Marcelino Oreja, adopted in Paris on 8 Sept. 2000.

[79] "50 Jahre Charta der deutschen Heimatvertriebenen," speech by Erika Steinbach at the *Tag der Heimat* in Berlin, 3 Sept. 2000.

of dangerous nationalists. On the right, the CDU and CSU, eager to appease their BdV clientele, maintained "legal-symbolic paper castles"[80] for the expellees during the 1980s, and thereby encouraged the expellee associations to maintain their revisionist political agenda.

When we consider the degree to which the expellees, since 1950, have been increasingly marginalized with either platitudes or slanders, it is remarkable that the millions of German expellees never became a destabilizing nationalist pressure group. The September 2000 report of the "Three Wise Men" makes the argument that the EU sanctions against Austria "have already stirred up nationalist feelings in the country, as they have in some cases been wrongly understood as sanctions directed against Austrian citizens."[81] It would be easy to apply this same argument to the German expellees, namely, that over the past fifty years nationalism has been stirred up by the intentional marginalization of the expellees as a political voice within Germany, and by the determination of the international community to apply the principle of collective (Nazi) guilt to the expellees. BdV President Steinbach makes this same point: "The still unresolved Palestinian question makes it clear how easily masses of refugees can be politically manipulated. The expellees have contributed to the fact that no anti-democratic or radical movements ever developed [in the FRG]. Extremism and anti-democratic attitudes have never found a home in the associations of the expellees."[82] In the 1990s there was a dramatic increase in the amount of attention paid to the European principles advocated by the German expellees in their 1950 Charter. The reemergence of ethnic cleansing, the calls for an emphasis on the rights of national minority groups, the determination of the Czech Republic and Poland to gain EU membership, the drafting of an EU Charter of Basic Rights have all been recent developments in European politics that should have brought the German expellees on to center stage. The final tragedy for the expellees, however, is that after waiting decades for pan-European integration to become a real possibility they have been abandoned—even by their own government.

---

[80] Garton Ash, 227.

[81] Report by Martti Ahtisaari, Jocehn Frowein, Marcelino Oreja, adopted in Paris on 8 Sept., 2000.

[82] Erika Steinbach, "50 Jahre Charta der deutschen Heimatvertriebenen" (4 Aug. 2000), comments posted on the BdV web page (www.bund-der-vertriebenen.de).

Ethnic Cleansing and Collective Punishment: Soviet Policy Towards Prisoners of War and Civilian Internees in the Carpathian Basin

TAMÁS STARK

In the wake of World War II, Soviet politics and propaganda achieved its greatest success in convincing the outside world that the foreigners in their forced labor camps were prisoners of war and war criminals. Since German armies and those of her allies had inflicted enormous damage during their retreat, Western public opinion was not much interested in what happened to these soldiers. Naturally, the Hungarian and German public felt quite differently.

Indeed, the fate of the prisoners and the push for and organization of their return were in the forefront of Hungarian domestic policy in the years after the war. Up until this time the Hungarian press and official documents had only mentioned "prisoners of war," even though everybody in the country knew that the Soviet armed forces had abducted not only POWs but also a great many civilians. Following the Communist takeover, even the euphemistic "POW" question became a taboo. Only after the fall of Communism in 1989/90 did the fate of several hundred thousand Hungarian prisoners of the Soviet Union again come into focus, and it did so both for political reasons and for historical research.

Today all history textbooks make reference to the fact that it was not only Hungarian soldiers who fell into Soviet captivity at the end of World War II, that a large number of civilians were captured and imprisoned or deported as well. Yet the few sentences generally devoted to the subject are not enough to reveal what actually happened to the captive Hungarians, and the entire topic gets lost within the broader treatment of the conclusion of the war and the rebuilding of the country. Even among Hungarians who consider themselves educated, there are many who believe that only soldiers were captured, and that whatever happened to them was really just part of the natural course of the war. The decades of collective amnesia have, it seems, left their mark, leaving scholars with quite a task if they are to uncover the once widely known facts.

Immediately after 1989/90 a number of survivors of postwar Soviet captivity were still alive and could be interviewed. Writer Ilona Szebeni published a set of interviews about forty-four women abducted from the Bodrogköz region.[1] Péter Rózsa spoke to survivors deported from the Nyírség area. Zoltán Szente recorded the troubles of innocently convicted young people.[2] Mihály Herczeg edited a book from the recollections of one-time "Levente" paramilitary youth organization members in Hódmezővásárhely.[3] Both Sándor Sára and the Gulyás brothers made documentary movies about captured women and men. Dozens of real POWs, civilians taken to forced labor, and former internees sentenced under false charges have written their memoirs, some of which were published. The output of historians is somewhat more modest. György Dupka and former KGB colonel Alexei Korsun have published contemporary documents about the deportations from Carpatho-Ruthenia/Subcarpathia.[4] Ferenc Dobos has written about the deportations in eastern Slovakia, and the Romanian-Hungarian Democratic Alliance (RMDSZ) described in a White Book the tragedies suffered by individuals from Transylvania.[5] György Zielbauer has discussed the story of members of Hungary's German population. [6] Comprehensive treatments about the fate of the prisoners have been published by Miklós Füzes, Zalán Bognár, József Domokos, and myself. [7]

---

[1] Ilona Szebeni, *Merre van a magyar hazám? Kényszermunkán a Szovjetunióban, 1944-1949* [Where is My Homeland? In Forced Labor Camps in the Soviet Union, 1944-1949] (Budapest, 1991).

[2] Péter Rózsa, *Ha túléled, halgass!* [If you Survived, Be Quiet!] (Budapest, 1989); Zoltán Szente, *Magyarok a GULAG szigeteken* [Hungarians in the GULAG Archipelagos] (Szeged, 1989).

[3] Mihály Herczeg, ed., *A vásárhelyi leventék háborús kálváriája* [Sufferings of Leventes from Vásárhely in the War] (Szeged, 1990).

[4] György Dupka and Alekszej Korszun, *A "Malenykij Robot" a dokumentumokban* [Documents on the Deportations of Hungarians from Carpatho-Ruthenia] (Budapest, 1997).

[5] Ferenc Dobos, "Magyarok a történelem senkiföldjén" [Hungarians in the No Man's Land of History], *Régió* 1992/3, 126-129; *Fehérkönyv az 1944. Őszi magyarellenes atrocitásokról* [White Book about the Atrocities Committed against Hungarians in the Fall 1944] (Kolozsvár, 1995).

[6] György Zielbauer, *Adatok és tények a magyarországi németség történetéből, 1945-1949,* [Facts and Figures on the History of the German Ethnic Group in Hungary, 1945-1949] (Budapest, 1989); György Zielbauer, "Magyar polgári lakosok deportálása és hadifogsága, 1945-1948" [Deportation and Captivity of Hungarian Civilians, 1945-1948], *Történelmi Szemle* 3, no. 4 (1989): 270-292.

[7] Miklós Füzes, *Modern rabszolgaság: Magyar állampolgárok a Szovjetúnió munkatáboraiban* [Modern Slavery: Hungarian Citizens in Soviet Labor Camps] (Budapest, 1990); Miklós Füzes, *Forgószél, Be- és kitelepítések Délkelet-Dunántúlon 1944-1948*

Although together with local histories there are over a dozen works on the subject of Soviet captivity, a large number of blank spots remains in the history of POWs, civilian internees, and political prisoners. In the last few years several important documentary volumes have been published in Russia, but the original Soviet sources are themselves frequently contradictory or contain information that cannot be interpreted. However, the original documents and the international literature clearly indicate that everything that happened in Hungary was not unique but an integral part of Soviet policy regarding the occupied territories. The point was not even just to punish the people they had conquered; collective responsibility and ethnic cleansing were at the very heart of the Soviet system.

The Soviet Union was not a national state, and its leaders never wanted it to become one. In the early 1920s Soviet policy towards its nationalities supported the strengthening of the various ethnic groups in order to stabilize the regime. But in the second half of the decade the policy changed fundamentally. Following the famine and destitution which resulted from the forced collectivization of agriculture, Poles, Finns, Lithuanians, Estonians, Latvians, and Bulgarians living along the borders fled in droves to their mother countries. Because of the troubled atmosphere along the borders and the great number of escape attempts, the border zone was expanded, and massive relocations were carried out to weaken the ethnic groups' cross-border ties. Ethnic cleansing *per se* began in 1930.

The story of the massive deportations, relocations, and internments that took place in the following two decades is not easy to unravel. The great ethnic make-over in the Soviet Union took place at the same time as the Gulag empire was developed, while specific groups of society were completely or partially annihilated. Each of these actions, such as the deportations of "kulaks" in Belarus, Ukraine, and former Polish territories in the spring of 1930, served both aims at once. Some ethnic groups were "merely" relo-

között, Tanulmány és interjúkötet [Whirlwind: Population Transfers in the Southeastern Part of Transdanubia in Hungary. Essays and Interviews] (Pécs, 1990); Miklós Füzes, *Embervásár Európában, hadifogoly magyarok a második világháborúban* [Hungarian Prisoners of War in World War II] (Pécs, 1994); Zalán Bognár, "Hazatérés" [Repatriation], *Ármádia*, September 2000, 10-12; Tamás Stark, "Hungarian Prisoners in the Soviet Union (1941-1955)," *Bulletin du Comité international d'histoire de la Deuxiéme Guerre Mondiale* (Montreal, 1995), 202-213; Tamás Stark "Genocide or Genocidal? The Case of the Hungarian Prisoners in the Soviet Union," *Human Rights Review* 1 (April-June 2000): 109-120.

cated, while others were forced into labor camps. Some of the ethnic groups were designated as potential enemies and deported for internal security reasons, while during the war collective retribution was the dominant motive. To uncover what actually happened and to carry out the investigative work is further complicated by the fact that POWs and the civilian deportees from the Soviet-occupied territories were assigned to the same network of camps and suffered the same fate.

In the spring of 1935, under the rubric of a "preventive" strike for internal security reasons, 45,000 persons from the area of Kiev and Vinnica were relocated to East Ukraine. Many of the forced settlers were ethnic Germans and Poles. That same year and in the spring of 1936, 29,000 Finns were taken from the border zone of Leningrad to Siberia.[8] In 1936/37, 171,000 Koreans were removed from the border zone in the far east.[9]

With the outbreak of World War II ethnic and political cleansing gained a new impetus. As the Soviet Union's western borders expanded considerably and the Baltic states were annexed, a large number of prisoners of war and "other hostile elements" came under Soviet domination. In order to control the masses of "potential enemies," the Soviet leadership set up a new camp system, namely the GUPVI—*Glavnoe Upravlenie NKVD SSSR po delam voennoplennyich i internirovannych* (Main Administration for POW and Internee Affairs).

The existence of the GULAG camp administration has been known in the Western Hemisphere since the mid-fifties. The eventual access to ex-Soviet archives made it possible for scholars to discover another "archipelago," namely the GUPVI, which was built for POWs and internees. The establishment of the GUPVI goes back to the time immediately following the outbreak of the war. L. P. Beria issued a general directive on September 19, 1939, to govern the administration of POWs and internees.[10] The directive proves that from the very beginning of the war the Soviet administration did not make any distinction between soldiers and the civilian population.

---

[8] Between 1928 and 1936 altogether 45,000 to 60,000 Ingrian Finns were removed. See Michael Gelb, "The Western Finnic Minorities and the Origins of the Stalinist Nationalities Deportations," *Nationalities Papers* 24 (1996): 243.

[9] Michael Gelb, "An Early Soviet Deportation: Far Eastern Koreans," *The Russian Review* 54 (July 1995): 389-412.

[10] On the organization of "archipelago GUPVI" see Stefan Karner, *Im Archipel GUPVI. Kriegsgefangenschaft und Internierung in der Sowjetunion 1941-1956* (Vienna, Munich, 1995), 55.

Some 250,000 Polish soldiers were captured during the Soviet attack on Poland. After a while some of the privates were set free, but about 40,000 prisoners were taken to forced labor camps. Roughly 14,000 of the captured officers were executed in summer 1940. Of those who had escaped from the German-occupied territories, 145,000 civilians were taken to a GUPVI camp, and around ten thousand ended up in the Gulag. The civil servants who had worked in the Polish administration were also considered "socially dangerous elements" and were arrested and sentenced. Several tens of thousands of undesirable Poles were able to escape POW and internment camps and the Gulag, but were deported instead. For security reasons, the convicts' families, roughly 60,000, were taken to Kazakhstan. Also for the purpose of strengthening the Soviet regime, Polish settlers who had arrived in the early 1920s were deported to western Siberia. This operation affected some 140,000 people. In the spring of 1941, 86,000 Poles were relocated to the central Soviet Union to "cleanse" the border zone.[11] A similar wave of arrests and deportations took place in the Baltics and Bessarabia. The actions in these territories in 1939 and 1940 served both to change the ethnic face of these areas and to utterly transform—to decapitate—the local population.

Internal security and collective retribution were both factors in the deportation of Germans in the Soviet Union. By June 1942, 1.2 million Germans had been taken to Central Asia and Siberia, and tens of thousands to the GUPVI and Gulag camps.

The charge—in fact unfounded—of collaboration with the Germans, based really on the notion of collective responsibility, was the reason for the deportation of Karachays, Kalmyks, Ingush, Chechens, Balkars, and Crimean Tartars, altogether some 900,000.[12] In November 1944 the Meskethians, Kurds, and Hemchins living along the border with Turkey were taken to Siberia, also on ethnic grounds, as well as to protect the border zone, for internal security reasons.

The practice of charging collective responsibility and cleansing on ethnic and social-class grounds continued in the Eastern European areas liberated from German occupation just where they had left off in 1941. The special units of the NKVD (the People's Commis-

---

[11] On the tragedy of Polish prisoners I used the figures of Nicholas Werth, "Ein Staat gegen sein Volk," in *Das Schwarze Book des Kommunismus* (Munich, 1998), 230.

[12] M. Guboglo and N. Kusnetsov, *Deportatsii naradov SSSR, 1930-je—1950-je godu* [A Compilation of Documents] (Moscow, 1992); see also N. Bugai, *L. Berija—J. Stalinu, "Soglasno vashemu ukazaniu"* (Moscow, 1994), 27-55.

sariat of Internal Affairs) arrested and deported the families of the Ukrainian National Union (*Organisatia Ukrainsky Nationalistov*), who had fought against both the German and Soviet armies, and the Ukrainian Insurgent Army's (*Ukrainska Povstanska Armia*) partisans, some 100,000 people in all.[13] They arrested over a hundred thousand army deserters and collaborators, and persons who were considered possible enemies of the Soviet regime. Ethnic and political cleansing continued in the Baltics and the recaptured Polish territories.

Soviet policy toward Hungary, as far as collective retributions and ethnic and political cleansing are concerned, was in many respects the same as Soviet policy toward all the occupied areas, but with the state of war that existed between the two countries there were also major differences.

The captured soldiers and the civilians abducted from Hungary for a variety of reasons met a similar fate as their counterparts elsewhere. Until autumn 1944 relatively few, around 70,000, members of the Hungarian army were taken prisoner.[14] Soviet troops entered Hungary in September 1944, and at that point there were close to 800,000 soldiers in the Hungarian Army. By January 1945 their number had diminished to half a million, while by late February only 200,000 soldiers remained. The majority of the quickly dwindling number of servicemen, on recognizing that the war was senseless, left their divisions without leave and fled home. There were relatively few who fell into captivity during a military operation. Most became prisoners of war because they had believed in what the Soviet fliers had proclaimed and had surrendered or gone over to the other side at the front in the hope of a quick release.

But civilians suffered the same extent as ordinary soldiers. On the basis of contemporary documents of Hungarian localities and eyewitness accounts, I have come to the conclusion that civilians were arrested in two waves.

The first wave took place two or three days after the Soviet takeover of a given larger settlement. The occupational forces rounded up civilians to do communal-reparation work. In most cases the unsuspecting civilians were gathered into concentration camps and then deported to the Soviet Union via Romania. We have only occasional information on the actual number of civilians

---

[13] *Deportáltak, Nyíregyháza, 1944-1948* [Deportees, Nyíregyháza 1944-1948] (Nyíregyháza, 1989), 9.

[14] Archives of Military History, Budapest, HM 1945 eln. 29055, Report to the Allied Control Commission, June 21, 1945, Budapest.

who were arrested just after the end of hostilities. On October 28, 1944, six days after the occupation of Hajdúböszörmény, 300 civilians were rounded up and deported. More than 2,000 civilians were deported from Nyíregyháza on November 2.[15] In mid-November 300 civilians were rounded up in Hajdunánás.[16] The approximate figure in Kolozsvár/Cluj-Napoca, the center of Transylvania, is three to five thousand.[17] On February 13, 1945, the victory announcement of the Red Army proclaimed that the Soviet forces had taken 110,000 POWs during the fifty days' battle for Budapest, the capital of Hungary. Out of these 110,000 prisoners, 40,000 were civilians.[18]

In general, the second wave of arrests took place a month to a month and a half after the arrival of Soviet troops. Unlike the first wave, it was a carefully planned operation directed against the ethnic Germans in general and also against the Hungarian population in the territories Hungary had regained between 1938 and 1941. This second wave of deportations was the instrument of collective punishment.

There are two documents which indicate that the Soviet leadership intended to punish not only Germany but to a lesser extent Hungary as well. V. M. Molotov wrote in a June 7, 1943, letter to Archibald Clark Kerr, the British ambassador in Moscow: "The Soviet government believes that it is not only the Hungarian government that must bear responsibility for Hungary's armed support to Germany ... but also to a certain degree the Hungarian people."[19] A few months later, on December 14, 1942, Molotov reacted to Eduard Beneš's anti-Hungarian invective by again exclaiming, "The Hungarians must be punished!"[20]

Soviet policy, however, was not driven solely by its notion of collective responsibility. During the war and in the early postwar period, the Soviet Union was certainly desperately short of labor.

---

[15] *Deportáltak, Nyíregyháza,* 1944-1948, 9.

[16] MOL (*Hungarian National Archives*) KÜM Szu tük. XIX-J-1-j IV-48229. Box 25044/45.

[17] *Fehérkönyv az 1944. Őszi magyarellenes atrocitásokról,* [White Book about the atrocities committed against Hungarians in the Fall 1944], 27-30.

[18] On the numbers of captured Hungarians see Bognár, *Hazatérés,* 11; and Krisztián Ungváry, *Budapest ostroma* [The Battle for Budapest] (Budapest, 1998), 293.

[19] Gyula Juhász, *Magyar-brit titkos tárgyalások 1943-ban* [Hungarian-British Secret Negotiations in 1943] (Budapest, 1978), 158.

[20] Péter Gosztonyi, *Háború az háború!* [War is War] (Budapest, 1989), 26.

Hence, forced labor from occupied territories seemed a logical measure to relieve the shortage. Changing the ethnic structure of the occupied and annexed lands was another element of Soviet policy. A further aim of the security forces was the elimination of potential political enemies. Deportation of civilians was also a part of a "mind control" tactic of the Soviet occupational forces. The Soviets wanted the civilians to remain in a state of fear. Individuals could sense that deportation might be hanging over them, but they never knew if they would be deported or not, or why or for how long.

Although we have limited access to ex-Soviet archives we are able to document some Soviet directives which were aimed at the concentration and the deportation of ethnic Germans and Hungarians. The 4th Ukrainian Front released its Resolution No. 0036 on November 12, 1944, which states that in Carpatho-Ruthenia, "Hungarian and German service-age nationals are living in numerous localities who, like enemy soldiers, must be arrested and sent to prison camp!"[21] In a report from December 17, 1944, the commander of the NKVD troops securing the resolution of the 4th Ukrainian Front, Major General Fagayev, wrote that "Between November 18 and December 17, NKVD details arrested altogether 22, 951 individuals in the area of Carpatho-Ruthenia and transferred them to prison camps.... The purging operations continue on the home front."[22] According to local historians the total number of deportees, however, reached 40,000.[23] The deportations continued outside the territory of Subcarpathia. The male population was deported in the whole upper Tisza River region, an area next to Subcarpathia.

The punishment character of the deportations is obvious in southeastern Slovakia, the region which belonged to Hungary during the war. In this region (between the eastern border of Slovakia and Rozsnyó/Rožňava) Hungarian males were rounded up and deported the same way as in Subcarpathia. The mass deportations stopped in early February, and the Hungarian population in the central and the southwestern part of Slovakia escaped deportation at this time. The next year, in 1946, however, about 100,000

---

[21] György Dupka and Alekszei Korszun, A „Malenykij Robot" a dokumentumokban, 15.

[22] Ibid., 28.

[23] Emlékkönyv a sztálinizmus kárpátaljai áldozatairól [Memorial Book on the Victims of Stalin in Carpatho-Ruthenia] (Budapest, 1993), 200.

Hungarians were transferred to Hungary within the framework of the Czechoslovak-Hungarian population exchange agreement. The Czechoslovak attempt to dissolve its Hungarian minority is a subject beyond the subject of this study. Here I can only suggest the reasons why the deportations in the Rozsnyó/Rožňava area were halted. For one, the strip of land in southwest Slovakia, which had belonged to Hungary during the war, was not part of the 4th Ukrainian Front's operational zone; therefore the November 12 resolution could not be applied here. Another possibility is that the Hungarian armistice of January 20 was the reason for halting the massive deportations.

Whatever the case, the armistice did not affect the deportations in the territory left to Hungary. In most of this area, organized deportations "officially" affected only German nationals. On December 16 the High Command of the Soviet Army issued a directive for the 2nd, 3rd, and 4th Ukrainian Fronts. This directive dealt with the "mobilization" and deportation of German males between seventeen and forty-five years of age, and women between eighteen and thirty, in the territories of Czechoslovakia, Hungary, Romania, Yugoslavia, and Bulgaria.[24] The directive was signed by Stalin and targeted ethnic Germans. In reality it also struck Hungarians.

In the present-day territory of Hungary the second wave of mass internment began in early January 1945. On the basis of contemporary documents and memoirs it seems that internment of civilians was carried out on the basis of lists and quotas set up presumably by the GUPVI-camp administration. Since the local organs of the NKVD had to fulfill the claimed contingents, not only ethnic Germans, but Hungarians with German names and Hungarians with Hungarian names were also deported. We still do not have a clear picture of the whole process of deportation. In some regions the total male population was "mobilized." There were dozens of villages where Hungarian females were deported too. Other regions, however, were not affected at all by the directives on "mobilization."

As is clear from NKVD chief Lavrenti Beria's January 5, 1945, letter to the Bucharest headquarters of the NKVD, the job of compiling the list of designated internees was assigned to the local authorities on the basis of registration certificates and police records.[25] An indication that the deportations were centrally orga-

---

[24] Karner, *Im Archipel GUPVI*, 25.

[25] *Transilvansky Vopros, Vengero-Rumunsky Territorialnuy Spor i SSSR 1940-1946, Dokumenty* [The Transylvanian Question, Hungarian-Romanian Rivalry for Transylvania and the Soviet Union, 1940-1946] (Moscow 2000), 288-289.

nized is that unsuspecting people were rounded up in much the same way everywhere. In villages all men between eighteen and fifty years old would—to a drum beat—be called on to do compulsory communal work. Those who appeared would be locked into schools, movie houses, or larger cellars, and after a wait of several hours or days they would be driven in a forced march to reception camps twenty-five to thirty kilometers from the front. During several days of marching the captives would get nothing to eat, and the guards, NKVD men, would shoot anyone who tried to escape. On Hungarian territory there were some eighty reception camps. Linked to these was a network of concentration camps. On Hungarian territory there were ten such concentration camps, in which over 20,000 prisoners were held. They were located in Baja, Debrecen, Gödöllő, Jászberény, Székesfehérvár, Vác, Cegléd, Szeged, Kecskemét, and Békéscsaba.

The Provisional Hungarian Government intervened with the Allied Control Commission in the interest of the mobilized Hungarians and Germans who were not members of Nazi organizations. The interventions had no direct result, but finally the Allied Control Commission gave its consent to set up lists of deportees for each locality. The work was done by local authorities in late 1944. These once top secret registers are held in the Military Archives in Budapest. According to the still incomplete collection of documents, the number of deported civilians from within today's Hungarian borders ranges from 120,000 to 140,000.[26]

In northern Transylvania, under the pretext of "mobilizing" Germans, mainly Hungarians were rounded up and handed over to the Soviet authorities. The tally for deported Hungarians ranges between forty and fifty thousand. If we consider the wartime territory of Hungary, twice the country's current size, the total number of civilian internees ranges between 180,000 and 200,000.

In addition to the real prisoners of war and civilian internees, there was a third set of prisoners. This group included those Hungarians who were arrested and sentenced by Soviet military tribunals on charges of "anti-Soviet" activity. Who belonged to this group of victims? Soldiers who served in the occupational forces

---

[26] *Magyarország a második világháborúban, Lexikon* [Hungary in the Second World War, Encyclopaedia] ed. Péter Sipos-István Ravasz (Budapest, 1997), 498. This is a statistic on the number of those Hungarians who were reported by local governments to be captured by the Soviet Army. The final figure is 94,778. Since most of the captured civilians from Budapest were not reported, the released figure is not accurate. On the numbers of the captured civilians see also *Tájékoztató gyorsfelvétel a a községek és városok közérdekű viszonyairól* [An Informative Quick Survey of the Public Relations of Villages and Towns], *Magyar Statisztikai Szemle* 1 (1946): 12-13.

on Soviet territory were arrested automatically, regardless of what they had actually done there. Members of the "Levente" paramilitary youth organization were also suspicious. The members of this organization were teenage boys who had to serve in auxiliary forces in the final months of the war. Most of these formations were evacuated to Germany during the great retreat in early 1945. Those boys who were able to avoid evacuation and remained in Hungary were considered by the NKVD as potential partisans and were treated as such. High-ranking officials and representatives of non-Communist political groups and parties were also arrested and sentenced to death or twenty-five years of imprisonment.

The fate of the convicts was different from that of POWs and civilian internees. While these groups belonged to the GUPVI camp administration, the convicts were sent to the Gulag. In their case deportation was an instrument of political cleansing. The convicts who were still alive could return to Hungary after Stalin's death. In the sixties, the Soviet Attorney General annulled the sentences of most of the surviving convicts, if they applied for this. Thirty-five hundred former convicts were exonerated in this way, although they never received any compensation from the Soviet Union. Since only survivors were rehabilitated, the actual original number of convicts must have been around 10,000, according to "SZORAKÉSZ," the central organization of Hungarian GULAG survivors.[27]

According to the postwar statistics of the Ministry of Defense, within Hungary and outside Hungary about 450,000 soldiers fell into Soviet captivity. Together with civilian prisoners, the total number of Hungarians and ethnically German Hungarians in Soviet custody must have been 600,000 to 640,000. Some newly available Soviet documents confirm these estimates, which are based largely on Hungarian sources. We know now that the Soviet camp authority registered 526,000 Hungarian captives.[28] These camp

---

[27] Tamás Stark, "A szovjet fogolyszedés néhány kérdése a magyar állampolgárok körében" [On the Soviet policy of capturing Hungarian civilians], in *Magyar kényszermunkások és politikai rabok a Szovjetunióban a II. világháború után* [Hungarian Forced Laborers and Political Convicts in the Soviet Union after World War II] (Budapest, 2000), 51. On the Hungarian political convicts in the Soviet Union see also: Gusztáv Menczer, "A Szovjet hadbíróságok által magyar állampolgárok politikai okból történt elitélése és e tény jogosságának néhány kérdése" [Hungarian Citizens, Sentenced by Soviet Military Tribunals], in *Magyar kényszermunkások és politikai rabok a Szovjetunióban a II. világháború után*, 15-33.; János Rózsás, *GULÁG Lexikon* [GULAG Encyclopaedia] (Budapest, 2000).

[28] Gosudarstvennaia archivnaia sluzba Rossiiskoi Federacii, Centr hranenia, istorikodokumentalnih kollekcii [State Archives of the Russian Federation, Center for preservation of historical collection], Moscow, MVD, fond: 1/n opis: 01e delo: 81.

statistics were compiled well after the arrival of prisoners. Consequently, the figure of 526,000 does not include the number of those who died in transit camps or during transit. With these added, the Soviet figure of 526,000 coincides with my estimate of 600,000 to 640,000.

From the tales of survivors it becomes evident that life in the camps, the prisoners' quarters, and working conditions varied greatly. Camps were usually surrounded by a triple wire fence. Day and night there were guards in the watchtowers and along the fences. Initially, inmates stayed in bunkers dug into the ground, later in communal wooden barracks. The latter had three tiers of bunks, with a sixty-centimeter-wide space per prisoner. In most places the prisoners slept on wooden boards, but elsewhere there were straw bags for mattresses. In contrast to transit camps, barracks could usually be deloused, but cockroaches continued to plague the prisoners. Inmates worked ten to fourteen-hour days. In theory, Sundays they were exempt from work, but the commanding officers would find work for them on these days too. There were places where prisoners got unsubstantial wages, but not everywhere. Those had the best chances for survival who were able to work as skilled laborers in some factory or farm. They came into contact with the local population who, as a form of barter, would give them some food or simple articles. But the majority of prisoners worked in mines, forest clearing, or road and railway construction. The worst conditions were in the Gulag camps, where condemned prisoners were held. The three most notorious camp districts (Vorkuta, Norilsk and Kolyma) were located north of the Arctic Circle. Survivors remember winter temperatures often reaching -60° C, and even worse was the constant wind. In the far east, the Taiset camp deserves mention, where the prisoners began to build the Baikal-Amur railway (BAM), hailed as the construction project of the century.

The minimal and unbalanced nourishment, abominable living conditions, and overstraining work caused quick deterioration in the physical state of the prisoners. Medical care was practically nonexistent. Although there was a "dispensary" in most camps, it was run by prisoner doctors without equipment or drugs. If someone had a fever he had a chance to get to a hospital. With no supplies there either, actual treatment was out of the question, but at least the prisoners were left in peace for a while. It was really the

---

See also V. Galicki, "Vengerskie voennoplennie v SSSR" [Hungarian POW's in the Soviet Union], *Voenno Istoriceskii Jurnal* [Review of Military History] 10 (1991): 45.

infections that caused massive deaths. Many prisoners suffered chronic diarrhea. Most of these lost more and more weight and wasted away until they succumbed. Malaria and typhus were rampant.

A great many prisoners fell victim to the abuse of their guards. Even more died from work-related "accidents," such as mine cave-ins or explosions, and freezing to death. The dead were usually buried in unmarked mass graves, often without being registered. No one officially notified the family of the death of their relative. Such news usually came, if at all, much later from a fellow prisoner who had survived.

Although the statistics of survivors are not absolutely precise, I will draw the conclusion that out of 600,000 prisoners, fewer than 400,000 were repatriated to their original countries.[29] Most of the survivors returned by the end of 1948. A total of 200,000 and 240,000 prisoners perished, a tally much higher than the complete casualty number of the Hungarian army during the war. These prisoners became the martyrs of Soviet forced labor camps.

Referring to the tragic fate of the prisoners, some historians use the terms "Holocaust," "Hungarian holocaust," or "genocide."[30] In my judgment, the fate of most of these prisoners, in general, does not meet the criteria of genocide, as established by the Genocide Convention of the United Nations in 1948. These deportations did not take place with the intent to destroy the Hungarian or German nation as a whole. Apart from exceptional cases, the Russian guards did not kill any inmates deliberately. On the other hand, due to epidemics and malnutrition, massive casualties were an integral part of life in the forced labor camps.

Contrary to the Soviet role in the fate of Volga Germans, Karachay, Kalmyks, Chechen, Ingush, Balkars, Crimean Tartars, and Meskethians, the Soviet government had no program to transfer the whole Hungarian nation, or even the entire German ethnic group in Hungary. Soviet policy in Hungary followed the Baltic, Polish, and Romanian scenarios of 1939-1941 and 1944/45. Moreover, to eliminate the real or potential critics of the Soviet regime, the Soviets punished these nations "only" by deporting a part of their manpower to use as forced labor for an unspecified period of time.

---

[29] See details on the repatriation in Stark, "Hungarian Prisoners in the Soviet Union (1941-1955)," 211-213.

[30] Füzesi, *Modern rabszolgaság*, 48.

My earlier comment about genocide referred to the fate of the Hungarian prisoners in general. My conclusion is different, however, if I limit my evaluation to the fate of Hungarians in Subcarpathia and Eastern Slovakia. Although here only the male population was deported, there is an undeniable similarity with the fate of above mentioned nations, the Karachai, Kalmyks, Chechen, Ingush, Balkars, Crimean Tatars and Meskethians. In other words, these two Hungarian groups were deported solely because of their Hungarian ethnicity. Consequently their story is an example of ethnic cleansing, and as such should become part of the international literature on genocide.

Forgotten Victims of World War II: Hungarian Women in Soviet Forced Labor Camps

AGNES HUSZÁR VÁRDY

The countless number of lives lost during World War II and the displacement of millions from their native lands are among the most tragic events in the history of the twentieth century. According to experts, the past century proved to be the most violent and the bloodiest one hundred years in human history, an assertion convincingly substantiated by scholars, researchers, and journalists who have published scores of books and articles about civilian and military victims of both world wars.[1] Special emphasis has been put on victims of the Second World War, especially Hitler's crusade against the Jewish population in the countries occupied by the Third Reich. Efforts to inform the general public about the horrors of the Jewish Holocaust have been especially successful and have led to widespread knowledge about these events among practically all the nations of the world.[2]

This level of historical awareness does not exist for most other ethnic groups and nationalities whose lives were adversely affected by World War II. Millions of non-Jews were forced to endure previously unheard of deprivation and hardship, before, during, and after the war. Compared to the extensive investigation of the Jewish Holocaust, historical research has paid little attention to the lot of other victimized groups. Relatively little has been

---

[1] For an excellent summary of this mass cruelty, see Norman M. Naimark, *Fires of Hatred: Ethnic Cleansing in Twentieth-Century Europe* (Cambridge, Mass., 2001).

[2] One of the most authoritative standard works (among many hundreds) on the Jewish Holocaust is still Raul Hilberg's *The Destruction of the European Jewry* (Chicago, 1961). See also the highly regarded work on the largest death camp, Debórah Dwork and Robert Jan van Pelt, *Auschwitz: 1270 to the Present* (New York, 1996). On the Hungarian segment of the Jewish Holocaust, see Randolph L. Graham's massive synthesis, *The Politics of Genocide: The Holocaust in Hungary*, 2 vols. (New York, 1981); and Randolph L. Braham and Béla Vágó, eds., *The Holocaust in Hungary Forty Years Later* (New York, 1985). A classic Hungarian-language summary is Jenő Lévai, *Zsidósors Magyarországon* [Jewish Fate in Hungary] (Budapest, 1948). See also the documentary collection: *Vádirat a nácizmus ellen. Dokumentumok a magyarországi zsidóüldözések történetéhez* [Indictment of Nazism. Documents on the History of the Persecution of the Jews in Hungary], ed. Ilona Benoschofsky and Elek Karsai, 3 vols. (Budapest, 1958-1967).

written about them, and as a result, these events have failed to become common knowledge. The fate of those who fled their native lands in Eastern and Central Europe in fear of the invading Soviet Army, and later settled in Western Europe, or immigrated to North or South America, has not been adequately researched. Not much is known about the victims of forced population expulsions and ethnic cleansing, such as the sixteen million Germans who were expelled from Poland and Czechoslovakia (from former East and West Prussia, Pomerania, Silesia and the Sudetenland). In the same vein, with the exception of Hungarian scholars, the general public is not aware of the retribution suffered by about 120,00 ethnic Hungarians who were driven across the Danube from Slovakia to Hungary as a consequence of the Beneš Decrees.[3] Furthermore, knowledge about Stalin's extermination of close to fifty million of his own "compatriots," including Russians and many other ethnic groups and nationalities, is limited. His victims included Poles, Ukrainians, Baltic peoples, Crimean Tatars, and many others who were exterminated by the millions.[4] Yet, knowledge about the fate of these peoples is not widespread, at least not to the extent to which the horrors of the Jewish Holocaust have permeated the historical consciousness of the broader public. It is evident that the victims of war and prejudice, misery, and extermination mentioned above have not attracted the same degree of interest of experts, researchers, and journalists. As a result, a gigantic gap of public ignorance and misinformation exists regarding these shocking events, both in Europe and the United States.

---

[3] On the Beneš Decrees, as they relate to the Hungarians, see Róbert Barta, "The Hungarian-Slovak Population Exchange and Forced Resettlement in 1947," and Edward Chászár, "Ethnic Cleansing in Slovakia: The Plight of the Hungarian Minority," in the present volume, which also contains a list of the Beneš Decrees in the Appendix. According to Barta, originally 73,187 Hungarians were slated for expulsion on the basis of the parity list. An additional 106,398 were to be expelled as "major war criminals" and 1,927 as "minor war criminals." Had this been implemented a total of 181,512 would have been expelled. But the Czechoslovak government was not satisfied even with these numbers. It turned to the Peace Conference and demanded approval for expulsion of an additional 200,000 Hungarians. These goals, however, were not allowed to be implemented, and thus by April 10, 1948, only 68,407 Hungarians were officially resettled in Hungary. Continued illegal expulsions, however, almost doubled this number, for according to the Hungarian census of 1949 at least 119,000 of these expellees were living in Hungary. Cf. Ignác Romsics, *Magyarország története a XX. században* [Hungary's History in the Twentieth Century] (Budapest, 1999), 302.

[4] See the relevant studies in the present volume by Alexander V. Prusin on the Poles, and Brian Blyn Williams on the Crimean Tatars.

The lives and fate of foreign victims of Soviet forced labor camps also falls into the category of "the little known." Insofar as historically aware individuals in Western Europe and North America have given the whole issue any thought, most have assumed that only POWs and members of the armed forces of the defeated nations were deported to the Soviet Union after World War II. This assumption could not be further from the truth. In the case of Hungary, it was only after the demise of Communism in 1989-1990 that researchers, journalists, and political leaders could begin to focus on the fate of the several hundred-thousand Hungarian civilians, including innocent women and children, who were deported to forced labor camps after the Soviet invasion of Hungary in September 1944.

It is virtually impossible to uncover the full details about these deportations, but the publication of several studies—based on memoirs, diaries, and interviews of survivors—provides a vast amount of information that contributes substantially to our knowledge of these tragic events.[5] These include published works by Tamás Stark, György Dupka, Péter Rózsa, János Rózsás, Zoltán Szente, Mihály Herczeg, and Zsolt Csalog; the memoirs of Imre Badzey and Mrs. Sándor Mészáros; and the documentary films of Sándor Sára and the Gulyás brothers.[6] Two volumes of interviews collected by Ilona Szebeni and Valéria Kormos document the fate of innocent men and women.[7] These victims included young girls

---

[5] According to Tamás Stark, since the collapse of the Communist regime in 1989-1990, about two dozen memoirs and collections of memoirs have appeared in print. See Tamás Stark, "Magyarok szovjet kényszermunkatáborokban' [Hungarians in Soviet Forced Labor Camps], in *Kortárs* [Contemporary], vol. 46, nos. 2-3 (February-March 2002), 70.

[6] These volumes are: Tamás Stark, *Magyarország második világháború emberveszteśége* [Hungary's Population Loss during World War II] (Budapest, 1989); Péter Rózsa, *Ha Túléled, halgass!* [If You Survived, Be Quiet!] (Budapest, 1989); János Rózsás, *Keserű ifjúság* [Bitter Youth], 2 vol. (Budapest, 1989); Zoltán Szente, *Magyarok a Gulag-szigeteken* [Hungarians on the Gulag Archipelago] (Szeged, 1989); Mihály Herczeg, ed., *A vásárhelyi leventék háborús kálváriája* [The War Sufferings of Leventes [Paramilitary Youth] from Vásárhely] (Szeged, 1990); György Dupka and Alekszei Korszun, *A "Malenkij Robot" dokumentumokban* [Malenkij Robot/Little Work in Documents] (Budapest, 1997); Mészáros Sándorné, *Elrabolt éveim a Gulágon* [My Stolen Years on the Gulag] (Ungvár-Budapest, 2000); Illés Zsunyi, *Nehéz idők* [Difficult Times] (Budapest, 2001); Zsolt Csalog, *M. Lajos, 42 éves* [Lajos M., 42 Years Old] (Budapest, n.d.); Imre Badzey, ed., *A haláltáborból: Badzey Pál szolyvai lágernaplója* [From the Death Camp: Pál Badzey's Szolyva Camp Diary] (Budapest-Ungvár, n.d.). On the documentary films, see Tamás Stark, "Ethnic Cleansing and Collective Punishment," in the present volume.

[7] The volumes in question are: Ilona Szebeni, ed. *Merre van a magyar hazám? Kényszermunkán a Szovjetunióban, 1944-1949* [Where is my Homeland? Forced Labor in the

and boys, who were forcibly taken to the Soviet Gulag to work from three to five years under the most primitive and excruciating circumstances. In spite of the attention given to these catastrophic events since the early 1990's, it will take many more years before most of the facts will be uncovered. It will probably take even longer for this story to work its way into the realm of "general knowledge," and for elementary and secondary level textbooks to do justice to this tragic segment of Hungarian history.

In order to gain a clear understanding of the deportation of hundreds of thousands, it is necessary to examine briefly the nature of Soviet policy toward Hungary after World War II. Since Hungary was at war with the Soviet Union, POWs and abducted civilians were treated somewhat differently from deportees of other occupied territories such as Poland and Czechoslovakia. Soviet intentions regarding Hungary emerged as early as June 1943, when V.M. Molotov, the future Soviet foreign minister, outlinted Soviet policy in a letter to Sir Archibald Clark Kerr, the British Ambassador to Moscow. He wrote that because Hungary was providing armed support to Germany, not only the government, but the the entire Hungarian nation must be held responsible.[8] In December of the same year, Molotov reemphasized this view when he reacted to Eduard Beneš's anti-Hungarian invective. He emphasized that no matter what, "the Hungarians must be punished."[9]

The status of deportations in Hungary was not affected by the armistice, as was the case in Slovakia. In December 1944, the Soviet High Command, in a decree directed to the 2nd, 3rd and 4th Ukrainian Fronts and signed by Joseph Stalin, proclaimed that all German males between the ages of 17 and 45, and all German females between the ages of 18 and 30 must be deported.[10] These deportations were to be carried out on the territories of Czechoslovakia, Hungary, Romania, and Yugoslavia. Although the major targets of the proclamation were the ethnic Germans in the states mentioned, in reality it struck a heavy blow against other nation-

---

Soviet Union, 1944-1949] (Budapest, 1991); Valéria Kormos, ed. *A végtelen foglyai: Magyar nők szovjet rabságban, 1945-1947* [Prisoners Forever: Hungarian Women in Soviet Captivity, 1945-1947] (Budapest, 2001).

[8] See Gyula Juhász, ed., *Magyar-brit titkos tárgyalások 1943-ban* [Hungarian-British Negotiations in 1943] (Budapest, 1989), 158.

[9] Péter Gosztonyi, *Háború van, háború!* [There is War, There is War!] (Budapest, 1989), 26. See also Tamás Stark in "Ethnic Cleansing and Collective Punishment," in the present volume.

[10] Dupka and Korszun, *A "Malenkij Robot" dokumentumokban*, 33-34.

ality groups as well, especially the Hungarians. If the quota could not be filled with Germans and with Hungarians with German surnames, they took any Hungarian off the streets, even if they "did not speak a single word of German."[11]

Soviet policy toward Hungarians was motivated by the concept of collective responsibility and collective retribution.[12] Consequently, unsuspecting civilians suffered the same fate as the Hungarian POWs, or those who believed Soviet propaganda, and in the hope of quick release, surrendered to enemy forces. Civilians were transported to the same network of forced labor camps and had to endure the same dreadful circumstances as the military personnel. The Association of Hungarian Veterans, an emigré organization based in Germany in the postwar years, found that as late as 1951 there were still 3,500 forced labor camps in the Soviet Union holding Hungarians, as well as other nationals such as Germans, Poles, Romanians, Japanese, Spaniards, Finns, Chinese, Ukrainians, and many others.[13]

It has been estimated that over 600,000 Hungarians—both military and civilian—were abducted by the Red Army to work in coal and lead mines, railway and road construction projects, and on collective farms.[14] According to eyewitness accounts and contemporary official documents, civilians were generally arrested in two waves.[15]

---

[11] This process of collecting people for forced labor is described in detail in the documentary collection *Moszkvának jelentjük. Titkos dokumentumok* [We Report To Moscow. Secret Documents], ed. Miklós Kun and Lajos Izsák (Budapest, 1994), 35 ff.

[12] On the brutal treatment of civilians by the "liberating" Soviet Armed Forces, see Cecil D. Eby, *Hungary at War: Civilians and Soldiers in World War II* (University Park, PA, 1998), 229-323. For the treatment and abuse of women specifically, see pp. 249-281.

[13] For a list of 3500 prisoner–of–war and slave labor camps in post-war Soviet Union, see *Fehér könyv a Szovjetunióba elhurcolt hadifoglyok és polgári deportáltak helyzetéről* [White Book on the Condition of POWs and Civil Deportees in the Soviet Union] (Bad Wörishofen, Germany, 1950), 67-100. Also included is a location map of these camps.

[14] This was the estimate of the Hungarian Central Statistical Office in 1946. But because this figure does not include those Hungarians who had been taken from Romanian-controlled Transylvania and newly Soviet-controlled Carpatho-Ruthenia [Sub-Carpathia], the actual figure may be significantly higher. Cf. Tamás Stark's assessment in "Magyarok szovjet kényszermunkatáborokban," 75-76; and in Szebeni, *Merre van a magyar hazám?*, 302-310.

[15] This is discussed by Tamás Stark in his "Magyarok szovjet kényszermunkatáborokban," 72-73.

The first wave of deportations took place primarily in north-eastern Hungary, from regions that were in the path of the invading Soviet Army. There is no accurate record on the actual number of civilians who were deported during this time, but we know that the first wave of arrests usually took place a few days after the Soviet occupation of a given settlement. The arrests were executed with the help of Hungarian collaborators popularly called *"policáj."* The Soviet Army rounded up civilians under the pretense of asking young able-bodied men and women to participate in short cleanup operations popularly dubbed "malenkij robot." or "little work."[16] Unsuspecting civilians were told to assemble in schools, movie theaters, and public buildings so as to perform a few days or weeks of communal work. However, they were not permitted to return home after the work was done. Rather, they were forced to walk twenty, thirty, or even fifty kilometers to reception centers in such cities as Debrecen, Miskolc, and Szerencs. From there, they were loaded into cattle cars, with between forty to sixty people to a wagon, and taken to one of the Soviet forced labor camps in the Trans-Ural Region.

The second wave of mass internments began in January 1945, when all of Hungary was affected. Memoirs and contemporary documents reveal that the deportations of civilians were carried out on the basis of quotas and lists set by the Soviet Secret Police, the NKVD, the People's Commissariat of Internal Affairs. The Secret Police controlled and administered all the forced labor camps in the Soviet Union. Since the local organs of the NKVD had to fulfill the numerical quotas, the collection process extended—as was mentioned above—beyond the ethnic Germans to Hungarians with German names, and to many who simply happened to be at the wrong place at the wrong time.[17]

The random nature of deportations is illustrated by the recollections of one of the deportees who related an incident that occurred during her long journey to the Soviet Union:

> I witnessed a dreadful incident near a train station. We had not reached the village yet and our train was standing at a railroad crossing. A farm wagon pulled by two horses, transporting tobacco leaves, stood on the other side of the rail gate, waiting to cross. The driver must have been about thirty-six or thirty-eight,

---

[16] On this topic, see the already cited work, Dupka-Korszun, *A "Malenkij Robot" dokumentumokban.*

[17] See note #11 above, and Stark, "Magyarok szovjet kényszermunkatáborokban," 72-73.

his son about thirteen or fourteen. A Soviet soldier ran over to them, yanked them both off the wagon, and shoved them into one of the cattle cars. It was terrible to listen to the hysterical cries of the man who screamed, "Take me anywhere you want, I don't care, but let the boy go so that he can drive the wagon home. My wife will never know what happened to us." They [the Russian soldiers] did not listen. They took them anyway. The train started, and as I looked back as long as I could, I saw the two horses standing there stock still, without their master. They did not move at all. The wife would have to wait in vain. Except for us, there were no other eyewitnesses.[18]

The compilation of lists of designated deportees was assigned to the authorities of each locality. These lists were partially drawn up on the basis of registration certificates, but they were also motivated by personal likes and dislikes. As a result, ethnic Hungarians in Romania and Slovakia were routinely selected for deportation by Romanian and Slovak authorities just because they happened to be Hungarians. It also happened that in Hungary itself, in a given village with a pure Hungarian population, lists were compiled by the local authorities motivated by revenge or jealousy.[19]

The full details of the process of deportation are unclear even today. For example, in some regions only able-bodied males were mobilized and deported, while in dozens of other villages authorities concentrated on the deportation of women only. In still other settlements, members of the Soviet Army simply took anyone to fill the quotas.

There were thousands of young women among the deportees. Their exact number is unknown, for many perished either on their way to the Soviet Union or as a result of the inhuman working conditions in the camps. The ratio of men and women internees varied from region to region. It is known, however, that from among those who were deported from the Upper Tisza Region, 60 percent were women. Most of them were between the ages of sixteen and twenty, and since legally they were still not adults, their deportation also violated the laws on the protection of minors. In this region, 42 percent of the women deportees were between twenty and thirty years, while 5 percent were between the ages of

---

[18] Ilona Vinnai (Vojtó Ferencné), in Szebeni, *Merre van a magyar hazám?*, 137-141, quotation from p. 138.

[19] See Ilona Szebeni's introductory essay to her documentary compilation, *Merre van a magyar hazám?*, 9-10.

thirty and forty. Women over forty were generally not considered for deportation.[20]

In most settlements, young, healthy, and able-bodied girls and women were put on the lists and were deported along with young boys, young men, and men in their forties. To fill the quotas, Soviet soldiers and their accomplices arrested and deported anyone who fit the age categories, regardless of family status. Survivors told of women three, four, or five months pregnant being dragged out of their beds and taken to the gathering places. They did not receive any special treatment, and were forced to march along with the other detainees twenty to fifty kilometers to the reception centers. At the time of their arrest, the majority were forced to leave without proper clothing and food supplies, and even if they were permitted to take along some food, it could serve only as a temporary solution to their minimum daily sustenance. Their clothing and footwear proved to be totally inadequate for the extreme weather conditions of the Siberian winters. As a result, after years or at times only months of excruciating hard labor, thousands perished by freezing to death.

Ilona Vinnai (Vojtó Ferencné), a young newly married woman in the village of Gávavencsellő in Szabolcs County, Northeastern Hungary, is a typical example of the countless young women who had to endure forced labor in the Soviet Union.[21] Seized in January 1945, she survived three years of harsh labor on a collective farm and in the coal mines under horrendous circumstances. Like many young women in her village, she happened to be on the list of internees but refused to assemble at the beckoning of the village drummer. Hungarian collaborators, the *"policáj"* quickly found, seized, and escorted her to the school where other detainees were held. She remembered how these *policáj* were often worse than the members of the Soviet occupational forces. They showed no mercy, were eager to search the homes of the villagers, and even pulled people from under their beds where they were hiding. Ilona Vinnai painted a vivid picture of the hardships in these camps: "In the winter we suffered from -40, -45 Centigrade temperatures, while during the summer we had to endure the scorching rays of the sun. We could barely move our limbs. In the winter tears froze on our cheeks; we cried from the cold and the pain. During the summer we fainted from the intense heat. But who paid any attention to this? They did not diagnose illness there

---

[20] Ibid., 13.

[21] Ilona Vinnai (Vojtó Ferencné), in Szebeni, *Merre van a magyar hazám?*, 137-141.

as they do back home. A person without fever was considered sick only after he or she collapsed. Our physical strength was waning, and because of uncertainty, fear, and constant dread, our spiritual strength likewise."[22] She recalled the agony survivors suffered when their fellow workers perished one after the other, especially those who had lost their fathers, sisters, brothers or husbands. Ilona continued: "But we never abandoned faith in God. When our despair was greatest, we turned to Him, and we continued to believe from one hour to the next that our captivity will end, and that we will see our loved ones again."[23]

Living conditions in the forced labor camps were inhuman. Proper nourishment was nonexistent. Watery cabbage soup, or something similar, and black bread made up their daily food. The bread was often so coarse that the prisoners suffered constant severe stomach pains. Those who worked on collective farms learned to smuggle vegetables for themselves and their fellow deportees. In most of the camps, upon their arrival, the detainees lived in underground bunkers. And even later, when housed in barracks, they had to sleep on bare wooden planks. They were plagued by lice and cockroaches. Although medical care was provided at least symbolically by doctors who were prisoners themselves, there were no drugs or medications available for treatment. Thousands died in accidents suffered at the workplace, but the majority of deaths were the result of infections and diseases. Malaria, typhus, and diarrhea were rampant, and because of the lack of medical treatment and adequate nourishment, prisoners succumbed easily. In addition, the internees were constantly mistreated—screamed at, pushed, kicked, and shoved. The guards and camp administrators forced them to work even on Sundays, notwithstanding the fact that one day of the week was designated as a day of rest.

The working conditions in the mines were horrendous and completely unsafe. Margit Krechl, a native of the village of Sajóbábony, was deported at the age of sixteen, along with her younger sister and older brother. Her story is similar to those of thousands of unsuspecting young girls who suffered similar fates.[24] As was customary, the Krechl siblings were asked to assemble at the school for questioning. Their trusting father, a black-

---

[22] Ibid., 139.

[23] Ibid.

[24] Reminiscences of Margit Krechl (Kürti Sándorné), in Valéria Kormos, *A végtelen foglyai*, 7-12. See note #7 above.

smith, who had not even been drafted into the Hungarian army because of poor health, urged his children to obey the authorities. The family was told that the Soviet liberators needed some help, and that those taken would be allowed to return home in a few weeks. The three siblings walked eighteen kilometers to the gathering center in Miskolc. Once they reached their destination, their fate was sealed. "No one said a word to us," Margit Krechl recalled. "They were screaming at us left and right, as they drove us into the cattle cars. There must have been thousands like us."[25] The train took them to the Donets Valley in Ukraine, to a village called Voroshilovka, where they were incarcerated and forced to do heavy labor in the nearby mines.

The experience of working in the mines made Margit Krechl and those with her feel like hell had been unleashed upon them. This hell affected the internees mentally, psychologically, as well as physically. Even decades after her repatriation, Margit Krechl still has deep scars on her head and legs. "You know, these are the permanent marks caused by the mine that collapsed on top of us," she explained. "They were even stingy with the proper timbering of the shafts. We had to worm our way through narrow corridors, crawling on all fours, like moles. My task was to shovel the coal into the mine car below. Many perished when the mine caved in. Only those young people survived who had enough lifeblood in them to crawl to the surface. Even today [in the late 1990s] I have nightmares of having to crawl in the dark, while something is constantly pulling me back."[26]

Since most camps were surrounded by double- or triple-wire fences and closely scrutinized by guards perched in watch towers, escape from the camps was virtually impossible. Those who were caught were severely punished and tortured. Mária Melik, one of the young women who was abducted from Rakamaz in Northeastern Hungary, related that of one of the cruelest punishments for escapees consisted of lowering them into a bunker enclosed by concrete walls, and filling the bunker with ice cold water. The detainees were forced to stay in the bunker until they froze to death.[27]

The prisoners were subjected to constant chaos, uncertainty, and disarray. Having been transported enormous distances on seemingly endless roads, being dragged from one labor camp to

---

[25] Ibid., 7.

[26] Ibid., 7-8.

[27] Reminiscences of Mária Melik (Tilki Jánosné) in Kormos, *A végtelen foglyai*, 33-39.

another, and not knowing whether they would ever be released, frightened even the most courageous young men, let alone young girls. But seventeen year old Gizella Csatlós of Balkány, Szabolcs County, thought she had no other choice but to escape.[28] This is part of her story in her own words:

> We had been outside only for a few days. The winter weather was becoming milder. As the snow began to melt, on our way to and from work, only a couple of meters from us, we saw the arms and legs of the dead sticking out from the ground. In the evenings in the barracks everyone was whispering that we should try to escape. People usually set out in pairs. My cousin said we should go too, but the guards were already bringing back prisoners who had been caught. There were even some who surrendered voluntarily because they got lost and simply circled around on the immense prairie. We were forced to watch the punishment they received. They had to strip practically naked, and were beaten until they collapsed, unconscious.[29]

Gizella and her cousin decided to escape nonetheless: "We were scared to death of the punishment, but I was plagued by an even stronger emotion. It wasn't even fear, but horror. It happened that next to me on the berth a girl from my village, Margit Krakomperger, was dying. She was exactly seventeen years old like me. I kept telling myself: 'This is certain death, I don't want to end up like she has. My cousin was very encouraging; he claimed that he could make his way by following the stars."[30]

After months of vicissitudes, narrow escapes, and hardships, while passing through several clearing camps in Odessa, Kishinev, Chernovitz, they reached the largest reception camp in a place called Bedyichev, which was the gathering place of those who were to be repatriated. Unrecognized as escapees from another camp, in early September 1945 they were told to gather their meager belongings and to go to the railroad station the next morning because they would be going home. "In a week we arrived in Máramarossziget [a former Hungarian city in Romania]. We were sobbing and laughing at the same time. We kissed the ground in joy. Apparently ours was the first train that brought back deportees from Russia. This was probably true, because other unfortu-

---

[28] Reminiscences of Gizella Csatlós (Réti Béláné) in Kormos, *A végtelen foglyai*, 19-26.

[29] Ibid., 19.

[30] Ibid., 20.

nate prisoners did not receive the kind of treatment and supplies we received. We were given canned foods, fruit, and candy. At the border a local leader even made a speech, but was cut short. Within moments a huge crowd descended on the station. Where did you come from? Who are you? Did you meet my son, my daughter, my father? Everyone was searching for his/her loved ones. But we had to reembark, because Budapest was designated as our final destination."[31] Their train passed through Gizella's native village, where someone from the crowd yelled that her cousin had jumped off the train at Bodrogszegi, and that he was already safe at home in Balkány. Gizella felt betrayed and abandoned because she feared that she would be deceived again, and then transported somewhere else. But her cousin notified Gizella's mother that she was on Hungarian soil. Her mother immediately took the next train to Budapest. When she spotted her daughter at the train station she was so shocked by her changed appearance that she fainted, even though Gizella had endured only seven months of forced labor. Compared to the other deportees detained for three to five years, her internment was relatively short because of her successful escape. Those who were forced to stay longer were in much worse shape than she was.

When Gizella and her mother returned to Balkány, a large crowd assembled in front of their house. Everyone was looking for news about their loved ones. Gizella was frightened. "What should she tell them? That their relatives are treated like beasts? That many of them had perished?"[32] Gizella recalled: "I just uttered a few sentences about where they were, and that they were working in coal mines. The rest they could read from my eyes."[33]

Few were as fortunate as Gizella Csatlós and a select few who succeeded in escaping from the camps. There are no reliable statistics regarding the exact number of Hungarian civilians and military personnel who were incarcerated and eventually permitted to return home. Official records were not always kept, and when the prisoners died of starvation, disease, freezing temperatures, or in an accident at the workplace, they were simply shoved into mass graves without their names being recorded. Of the 600,000 deportees, approximately one third—200,000 men, women

---

[31] Ibid., 23.

[32] Ibid., 24.

[33] Ibid.

and children—never made it back.[34] They died a miserable death under the most excruciating and inhuman conditions imaginable.

Mass repatriation from the forced labor camps back to Hungary began in the fall of 1947. Those who survived were marked for life, psychologically, spiritually, and physically. Many lost limbs, contracted incurable diseases, or suffered serious injuries that plagued them for life. The reigning Communist regime warned them to keep quiet and threatened them with retaliation from the moment they reached the Hungarian border. They could not count on anyone to appreciate their plight, and the local and state governments repeatedly rejected their requests for financial assistance. They received no help for further training or for the completion of their studies, and the seriously ill and disabled were denied sick benefits and disability allowance. They were given no compensation for their financial losses, and if they were, the sums were minimal. In Debrecen, for example, deportees were given 5 forints, and later 20 forints as final reparation.[35]

The dreadful effects of these deportations affected not only the internees themselves, but also their loved ones who had been left behind. Wives who lost their husbands received no pensions without producing death certificates. But these were often nonexistent because camp administrators failed to keep records of the dead. Mrs. Gyula Kéky, who did everything humanly possible to free her husband and son, pleading with authorities in Debrecen and elsewhere, eventually had to resettle in the town of Fót because her house in her native village was confiscated. Her forty-two year old husband and seventeen-year old son were deported from their native town of Hajdúböszörmény in October 1944, and she never saw or heard from them again.[36]

Her words, over four decades later, at the age of eighty-five, when she was interviewed by Ilona Szebeni, describe poignantly these tragic events that touched the lives of so many blameless, unsuspecting civilians: "Why did they take them? Why? My God, but why? They were innocent! Innocent! It was a terribly cruel

---

[34] This estimate by the Hungarian Central Statistical Office is cited by Tamás Stark in Szebeni, *Merre van a magyar hazám?*, 310.

[35] See Ilona Szebeni's introductory essay in her book *Merre van a magyar hazám?*, 14. Five Hungarian forints in 1947 were worth less than one U.S. dollar. For more details about the treatment of survivors of forced labor camps by Hungary's Communist government, see Kormos, *A végtelen foglyai*, 45-53.

[36] The reminiscences of Kéky Gyuláné are recorded in ibid., 16-22.

world in those days. Why did they do this to us? It's horrible, horrible, even today."[37]

Perhaps with time, as historical research makes greater effort to shed light on these injustices that befell humankind in the twentieth century, the deportation and internment of innocent Hungarian men, women, and children to Soviet forced labor camps will also receive the attention it deserves.

---

[37] Ibid., 16.

Revolution and Ethnic Cleansing in Western Ukraine: The OUN-UPA Assault against Polish Settlements in Volhynia and Eastern Galicia, 1943-1944

ALEXANDER V. PRUSIN

T he outbreak of the Soviet-German war in June 1941 heralded the most tragic period in the modern history of Polish and Ukrainian communities in the densely populated eastern border areas of the interwar Polish Republic. The enormous loss of life and property brought about by German and Soviet occupations can hardly be calculated. To make matters worse, in the midst of the war, longstanding animosities between Poles and Ukrainians reached their boiling point. In 1943-1944, the radical Organization of Ukrainian Nationalists (OUN) and its military arm, the Ukrainian Insurgent Army (UPA), carried out a brutal assault against Polish settlements in the regions of Volhynia and Eastern Galicia. As a result, thousands of Poles lost their lives, while many more were forced to flee their homes. In spite of the magnitude and tremendous demographic losses, this tragedy remains largely unknown outside of Poland, while the existing Ukrainian and Polish accounts on the subject are often permeated by embarrassing partiality or outright distortion.[1]

---

[1] Ukrainian nationalist authors maintain that the anti-Polish campaign was retaliation for Polish attacks on Ukrainian villages and political activists. See, for example, Mykola Lebed', *UPA, Ukrains'ka Povstans'ka Armiia: ii heneza, rist i dii v vyzvol'nii borot'bi ukrains'koho narodu za Ukrains'ku Samostiinu Sobornu Derzhavu* [UPA, The Ukrainian Insurgent Army: Its Origin, Evolution, and Activities in the Liberation Struggle of the Ukrainian People for the Independent and United Ukraine] (n.p.: Vydannia Presovoho Biura UHVR, 1946); Petro R. Sodol, *UPA, They Fought Hitler and Stalin: a Brief Overview of Military Aspects From the History of the Ukrainian Insurgent Army, 1942-1952* (New York, 1987). As unreliable are accounts offered by Ukrainian authors in post-independent Ukraine. See, for example, S.M. Mel'nychuk, *U vyri voennoho lykholittia: OUN i UPA v borot'bi z hitlerivs'kymy okupantamy* [In the Whirlpool of War Calamity: OUN and UPA in the Struggle against the Hitlerite Occupiers] (L'viv, 1992); V.I. Kucher, *OUN-UPA v boro'tbi za nezalezhnu Ukrainu* [OUN-UPA in the Struggle for Independent Ukraine] (Kiev, 1997); Yu. Makar, V. Strutyns'kyi, "Pol's'ko-ukrains'ki stosunky pid chas drugoi svitovoi viiny ta po ii zakincheni: sproba uzahal'nennia," [The Polish-Ukrainian Relations during World War II and its Aftermath: General Conclusions], and K. Smiian, "Volyn' u vidnosynakh z Pol'shcheiu u 1941-1945 rokakh," [Volhynia's Relations with Poland, 1941-1945] 49-50, all in *Volyn' v druhii svitovii viini ta pershi povoyenni roky: materialy naukovoi istoryko-krayeznavchoi konferentsii,*

The aim of this essay is to draw attention to this much-over-looked theme and propose tentative assumptions on the background causes and mechanisms of the "forgotten ethnic cleansing" in Western Ukraine. In particular, this study focuses on the revolutionary ideology of the perpetrators, the wartime functional exigencies that made the cleansing possible, and connection between the OUN-UPA state-building objectives and the violent campaign against Polish settlements.

This essay begins with a brief outline of the activities and ideology of the Organization of Ukrainian Nationalists in the inter-war Poland. The second section of this essay examines Polish-Ukrainian relations after the outbreak of World War II in Volhynia and Eastern Galicia. The third section focuses on the evolution of the OUN-UPA ethnic cleansing in 1943-1944.

Ukrainian national aspirations in the border regions of the former Austro-Hungarian and Russian empires gained momentum in the two decades prior to World War I and culminated in the wars of independence of 1918-1920. Although the Ukrainians managed to establish two independent states, in Eastern Galicia and Ukraine, they were eventually defeated by Polish and Soviet armies. The Ukrainian leaders, however, never gave up their aspirations for independence, and in early 1920s former Ukrainian officers organized clandestine groups in Eastern Galicia and western Volhynia, the regions incorporated into the new Polish state. The main Ukrainian underground group, the Ukrainian Military Organization (UVO), started its activities immediately after its inception, carrying out an anti-Polish propaganda campaign, acts of sabotage, and assassinations of Polish officials and Ukrainians suspected of pro-Polish sympathies. In 1929, the UVO was fused

---

prysviachenoi 50i richnytsi peremohy nad fashyzmom [Volhynia in World War II and the First Post-War Years: The Materials of the Historical-Ethnographic Conference Dedicated to the 50th Anniversary of the Victory over Fascism] (Luts'k, 1995). An exception is a valuable collection of documents pertaining to the resettlement of Ukrainians and OUN-UPA activities in Poland in 1944-1947, *Akcja "Wisła": Dokumenty,* [The Action "Vistula": Documents], ed. Eugeniusz Misilo, (Warsaw, 1993). On the other hand, Polish sources in general ignore national aspirations of Ukrainian nationalists and label them merely as the "hirelings" of Nazi Germany. See, for example, Edward Prus, *Atamania UPA: tragedia kresów* [The UPA Henchmen: The Tragedy of the Eastern Lands] (Warsaw, 1985), also his *Herosi spod znaku tryzuba: Konowalec-Bandera-Szuchewycz* [The Heroes of the Trident: Konovalets'-Bandera-Shukhevych] (Warsaw, 1985). A standard Polish book on the subject remains well-researched, but still tendentious *Droga do nikąd: działalność Organizacji Ukraińskich Nacjonalistów i jej likwidacja w Polsce* [The Road to Nowhere: The Activities of the Organisation of Ukrainian Nationalists and its Liquidation in Poland] (Warsaw, 1973), by Antoni B. Szcześniak and Wiesław Z. Szota.

with several other nationalist groups and formed the Organization of Ukrainian Nationalists (OUN) which continued anti-state propaganda and terrorist activities.[2]

The OUN ideology reflected the predominantly agricultural character of Ukrainian society and represented an "integral nationalist" current similar to the "agrarian" right-wing independence-seeking organizations such the Croatian Ustaše, the Slovak Hlinka Party, and the Macedonian VMRO. Since the mid-1920s, the influence of Italian fascist ideology on the OUN hadalso strengthened. The organization aimed at national and social revolution and displayed visible tendencies towards a strong authoritarian rule with syndicalist or corporatist foundation. Militarization, strict hierarchy, blind obedience of the OUN rank and file to its leadership, and OUN symbols constituted clear-cut emulation of the fascist movements in Western Europe. Nevertheless, the OUN ideology and violence displayed more separatist and nationalist rather than social tendencies. The main goal of the organization remained national independence. The socioeconomic reconstruction of the state, a main corollary of Italian fascists and German national-socialists, was to be achieved only at a later stage of national state-building.[3]

---

[2]For the Ukrainian war of independence of 1918-1919 in Western Ukraine, see Maciej Kozłowski, *Między Sanem a Zbruczem: walki o Lwów i Galicję wschodnią, 1918-1919* [Between the San and Zbrucz Rivers: The Fighting for Lwów and Eastern Galicia, 1918-1919] (Kraków, 1990). For UVO-OUN activities in the 1920s-1930s, see Volodymyr Martynets', *Ukrains'ke pidpillia vid UVO do OUN* [The Ukrainian Underground from UVO to OUN] (Winnipeg, 1949) 25, 41-50; Petro Mirchuk, *Narys istorii Orhanisatsii Ukrains'kykh Natsionalistiv* [Sketch about the History of the Organisation of Ukrainian Nationalists] (Munich, 1952), 281, 338, 371; Alexander Motyl, "The Ukrainian Political Violence in Inter-War Poland, 1921-1939," *East European Quarterly*, XIX, 1 (1985), 45-55. Among the chief victims of the OUN were the advocate of the Polish-Ukrainian rapprochement Tadeusz Hołówko and Polish minister of the Interior Bronisław Pieracki.

[3] *OUN v svitli postanov Velykykh Zboriv, Konferentsii ta inshykh dokumentiv z borot'by 1929-1955 r.* [OUN in the Light of the Resolutions of the Great Congresses, Conferences, and Other Documents pertaining to the Struggle of 1929-1955] (n.p, 1955), 10-12; Myroslav Prokop, "V sorokovi rokovyny protynimets'koi borot'by" [In the 40th Anniversary of the Anti-German Struggle], *Suchasnist'* 10 (1981), 57; *Tsentral'nyi Derzhavnyi Arkhiv Vyshchykh Orhaniv Vlady i Upravlinnia Ukrainy*, [The Central State Archive of the Supreme State Institutions and Management of Ukraine] (*TsDAVOViUU*), fond 3833, opis' 1, delo 36, lists 9, 14, 16. Henceforth Ukrainian and Russian archival sources will be cited in the following order: archive, fond/opis'/delo, list. See also, *Tsentral'nyi Derzhavnyi Arkhiv Gromads'kykh Orhanizatsii Ukrainy* [The Central State Archive of the Civil Institutions of Ukraine (*TsDAGOU*), 1/23/926, l. 37; *Nash klych* [Our Watchward], 9 July 1938. For a typology of violence, see Fred R. von der Mehden, *Comparative Political Violence* (Englewood Cliffs, N.J., 1973), 7-9.

The guiding principle of OUN ethnic policies was the removal of non-Ukrainians, especially Poles and Jews, from the socio-economic spheres of the future Ukrainian state. Such a "zoological" approach (as defined by the most influential ideologist of Ukrainian integral nationalism Dmytro Dontsov), was to alter economic disparities between the Ukrainian majority and non-Ukrainian minorities. If necessary, argued Dontsov and other nationalist ideologists, physical destruction of potential enemies was not to be excluded in the struggle for national liberation.[4]

To achieve its main aim—a Ukrainian state within ethnographic borders—the OUN established links with and gained assistance from the German, Czechoslovak, and Lithuanian intelligence services. Until the late 1930s, the organization remained a monolithic, disciplined, and highly centralized organization. However, after the 1938 assassination of its leader, Evhen Konovalets, by a Soviet agent, the OUN was rent by an internal split. War veterans of 1918-1920 grouped themselves around Colonel Andrii Mel'nyk and formed the OUN-M, while radical young nationalists in Eastern Galicia and Vohlynia led by the charismatic Stepan Bandera formed the OUN-B. Mutual hostility between the two groups mounted with time and culminated in denunciations to the Polish police and physical assaults. Yet both groups looked to the Germans for financial support and military training.[5]

In the interwar period the repressive anti-Ukrainian policies of the Polish government and the socio-economic distribution in the Eastern Borderlands—*Kresy Wschodnie*—exacerbated ethnic animosities. Although ethnic Ukrainians comprised a numerical majority in Volhynia and Eastern Galicia, Polish and Jewish minorities in the countryside and urban areas dominated local economies. The administration, police, and school system were almost entirely in the hands of ethnic Poles. (See Table, below)

---

[4] Dmytro Dontsov, *Patriotyzm* [Patriotism] (L'viv: Kvartal'nyk Visnyka, 1936), 40-41; M. Stsibors'kyi, "Problemy hospodars'koi vlasnosty," [The Problems of Household Property] 13-14, and Mykhailo Podoliak, " Suspil'nyi zmist natsional-ismu," [The Social Context of Nationalism] 37, 39, all in *Na sluzhbi natsii* [In the Service of Nation] (Paris, 1938), 13-14.

[5] *Tsentral'noe Khranilishche Istoriko-Dokumental'nykh Kollektsii* [The Central Depository of Historical-Documentary Collections] (*TsKhIDK*) 308/19/157, ll. 12-15, 59; *TsDAGOU*, 57/4/430, 3-6; *Rossiiskii Gosudarstvennyi Voennyi Arkhiv* [The Russian State Military Archive] (*RGVA*), 4/19/74c, l. 55; Roman Il'nyts'kyi, *Deutschland und die Ukraine 1934-1945: Tatsachen europäischer Ostpolitik*, 2 vols. (Munich, 1955), 1: 270-273.

In the 1930s, the Polish government pressure on the Orthodox church in Volhynia accorded OUN hundreds of new recruits. Not surprisingly, the majority of OUN members welcomed the outbreak of the German-Polish war. Nationalist units ambushed Polish columns and served as guides for the advancing German troops. Although many Ukrainians fought in the ranks of the Polish army, OUN sabotage and murder of individual Poles created a powerful popular image of the Ukrainian "fifth column" in Poland. In retaliation, armed Polish groups terrorized Ukrainian villagers. In the second half of September 1939, after the Red Army invaded Poland, Soviet propaganda further fuelled Polish-Ukrainian antagonism by inciting Ukrainians and Belorussians to rise up against "Polish oppressors."[6]

Ethnic Distribution in the Eastern Borderlands--1939 (in 1000s)[7]

Region	Greek-Orthodox	Eastern Orthodox	Roman-Catholics
Polesye (south-eastern part of Belorussia)	1,800 (Belorussians)	875,800 (Belorussians and Ukrainians)	125,200 (mostly Poles)
Vohlynia	11,100 (10,100 in countryside) (Ukrainians)	1,455,900 (1,396,000 in countryside) (largely Ukrainians)	327,900-343,250 (264,100 in countryside) (Poles)
Three provinces of Eastern Galicia (Lwów, Stanisławów, Tarnopol)	3,256,300 (2,944,700 in countryside) (Ukrainians)	10,900 (7,500 in countryside) (Ukrainians)	2,280,3000 (1,719,600 in countryside) (Poles)

Source: *Mały Rocznik Statystyczny* [*The Small Statistical Almanac*] (Warsaw, 1939), 22,24,26; *The Polish and Non-Polish Populations of Poland: Results of the Population Census of 1931* (Warsaw: The Institute for the Study of Minority Problems, n.d.),36, 41-42.

After the demise of the Polish state in September 1939, according to the Nazi-Soviet Pact, the eastern Polish provinces -- Eastern Galicia, Western Volhynia (after 1921 the eastern part of

[6] Józef Anczarski, *Kronikarskie zapisy z lat cierpień i grozy w Małopolsce Wshodniej 1939-1946* [Chronicle Notes from the Years of Suffering and Horror in Eastern Małopolska, 1939-1946] (Kraków, 1996), 29; *RGVA*, 25880/4/35, l. 1316; Ibid., 25880/4/35 (13), ll. 1125, 1168-1169; Ibid., 25880/4/35, l. 1315; Louis de Jong, *The German Fifth Column in the Second World War* (Chicago, 1956), 153, 155.

[7] These statistics artificially inflated the numbers of Poles, and should be taken with caution. I did not include in this table other minorities such as Jews, Germans, and Czechs.

Vohlynia remained within the borders of the Soviet Union), and Western Belorussia fell under Soviet rule. Sovietization of Eastern Galicia and Volhynia included the nationalization of land, mines, banks, and educational institutions. Although in theory the regime encouraged the growth of local national cultures and languages, in reality Soviet policies were motivated by a "class approach" to the ethnic question. Therefore, potentially pernicious individuals and groups were rendered harmless by arrests and deportations. As the former nationals of a "bourgeois-landowner" state, Poles found themselves under increasing pressure. Polish cultural institutions and combatant organizations were closed, and civil employees were replaced by Ukrainians, Belorussians, and Jews. In the winter of 1940 the "class approach," however, transcended ethnic lines as the first large-scale deportations affected all ethnic communities. Thousands of Poles, Ukrainians, Belorussians, and Jews were arrested and shipped off to remote areas in the Soviet Union. Poles comprised the majority of the deported--approximately 210,000 or 63 percent (including the southeastern regions of Lithuania). Nevertheless, mutual suspicions and competition for positions in socio-economic sphere permeated the relations between Poles and Ukrainians throughout the Soviet occupation.[8]

Meanwhile, the OUN made necessary adjustments to the new political order in Eastern Europe. In April 1941, the Second Congress of the OUN-B in German-occupied Poland outlined immediate objectives for the national revolution. Economic and social life of the former Polish state were to be purged of "aliens" – Russians, Poles, and Jews. In coordination with the German military intelligence, OUN-B diversionary units were to carry out sabotage in the rear of the Red Army. The nationalist leadership also planned to take over the administration in Ukraine and to present itself as an "equal" partner to Germany rather than a mere executor of German orders. If the invading army proved hostile to the Ukrainian statehood, the resolutions of the Second OUN-B Congress implied that the national revolution would enter a "new stage." Thus, the collision with imperial policies of Nazi Germany was not excluded.[9]

---

[8] Andrzej Sowa, *Stosunki polsko-ukraińskie 1939-1947: zaryz problematyki* [The Polish-Ukrainian Relations, 1939-1947: Outline of Problems] (Kraków, 1998), 120; Anczarski, *Kronikarskie zapisy*, 38, 51, 163; Halina Czarnocka, et al., *Armia Krajowa w Dokumentach, 1939-1945,* [The Home Army in Documents, 1939-1945], 6 vols. (London, 1971-1978), 1: 252-253.

[9] *OUN v svitli postanov*, 36, 48-57; Heinz Höhne, *Kanaris: Patriot im Zwielicht* (Munich, 1976), 343, 439.

With the German invasion of the Soviet Union in June 1941, OUN-B attacked retreating Soviet troops and carried out acts of sabotage on communications. OUN propaganda in the country-side and towns incited Ukrainians to a national struggle, and hurled the full vent of ethnic vitriol against Poles and "Judeo-Bolsheviks." The OUN-organized Ukrainian militia staged anti-Jewish pogroms, and alongside the German security units participated in arrests and executions of the Polish intelligentsia and Jewish civilians.[10]

However, soon after the German invasion of the Soviet Union, the OUN-B was in full-blown conflict with the Germans. At the end of June 1941 in L'viv, the OUN-B announced the resurrection of Ukrainian statehood. This action infuriated Hitler, who had envisioned for Ukraine an entirely different place in his world order--as a breadbasket and a source of slave labor. The German response was swift, and hundreds of nationalists were arrested or executed. OUN militarized formations were disbanded and fused into police battalions deployed in anti-partisan struggle in Ukraine and Belorussia. The surviving OUN-B leadership went underground.[11]

In 1941-1942 the Nazi policies in Ukraine contributed to growing anti-German hostility. Contrary to Ukrainian national aspirations, the entire territory of Ukraine was divided by the occupying powers. Eastern Galicia was incorporated into the so-called General-Government (carved out of central and southern Poland), while Volhynia and Polesie were included in the Reichskommissariat Ukraine. Parts of south and southwestern Ukraine were accorded to Germany's ally, Romania. While the Nazi regime in Eastern Galicia was comparatively mild, in the Reichskommissariat Ukraine the population was subjected to the most brutal economic exploitation, forced labor, and frequent requisi-

---

[10] *RGVA*, 25880/4/35, l. 1316; de Jong, *The German Fifth Column*, 153, 155; L. Jurewicz, "Niepotrzebny," [Useless], *Zeszyty Historyczne* 15 (1969), 161; Anczarski, *Kronikarskie zapisy*, 180, 186; Józef Żeliński, "Stanisławów 1941," *Biuletyń-Koło Lwowian*, n. 43 (1982), 3-7.

[11] *United States National Archives Microfilms*, T-175, roll 233, frame 721,406; Ibid., T-175, roll 124, frame 599,082-599,084; *Archiwum Ministerstwa Spraw Zagranicznych*, [The Archive of Foreign Affairs], oddział 10, sygnatura 213, teczka 24, kartki 346, 352. Henceforth Polish archival sources will be quoted in the following order: archive, sygnatura/sprawa, kartka. Thus, *Archiwum Akt Nowych* [The Archives of New Files], (*AAN*), 202/III/134, vol. 1, kk. 252, 260; *Rossiiskii Tsentr Khraneniia i Izuchenia Dokumentov Noveishei Istorii* [The Russian Center of Preservation and Study of the Documents of Modern History] (*RTsKhIDNI*) 69/1/1027, l. 127; Ibid., 69/1/563, l. 119.

tions of grain and cattle.[12] Anti-German sentiments soon were widespread, and partisan units grew in numbers. In this atmosphere, in the spring of 1942, the OUN-B leadership decided to create a national army which would seize power at the moment Germany and the Soviet Union had exhausted each other in battle. To this effect, several OUN leaders took to Volhynian forests and began organizing armed units. The marshy and densely forested terrain in the region provided a suitable environment for guerrilla struggle, and had become a base for other Ukrainian formations (independent of the OUN) which attacked German police stations and warehouses in the spring and summer of 1942.[13]

According to the OUN-B planning the national revolution would unfold in several stages. The crucial step would be the elimination of "foreign elements" in socio-economic sphere in Ukraine. While the "Jewish question" was effectively "solved" by the Nazis, it was the Poles who were primarily targeted by the OUN-B.[14] Between April and August 1942, the OUN-B intensified its anti-Polish propaganda. Posters and leaflets incited the Ukrainian population to murder Poles and "Judeo-Moscovites." Ukrainian blue-yellow national flags adorned village headquarters, and marching songs of Ukrainian policemen contained undisguised death threats to the Polish community. Attacks on Polish officials in German service also intensified.[15]

Through the winter and early spring of 1943, the OUN was occupied with building its military formations. Organized territorially, OUN units utilized abandoned Soviet military warehouses as main supply sources. A number of OUN commanders had had military or police experience in the Polish, Czech, and Hungarian armies or in German-sponsored police battalions. The OUN leadership and its security service watched after discipline, and in-

---

[12] *AAN*, 202/III-134, vol. 1, k. 252.

[13] Taras Bul'ba-Borovets', *Armiia bez derzhavy: slava i trahedia ukrains'koho povstans'koho rukhu* [The Army without State: Glory and Tragedy of the Ukrainian Insurgent Movement] (Winnipeg, 1981), 147-148, 168-170, 185; *RTsKhIDNI*, 69/1/747, l. 165; Ibid, 69/1/19, l. 145.

[14] Mykola Lebed', "Orhanizatsiia protyvnimets'koho oporu OUN 1941-1943," [The Organisation of the Anti-German Resistance, 1941-1943] *Suchasnist'* 1/2 (1983): 152-153; *OUN v svitli postanov*, 61-70; *RTsKhIDNI*, 69/1/1027, ll. 127-128; *AAN*, 202/III/131, k. 45; Ibid., 202/III-134, vol.2, kk. 56-57. The OUN-M program also promoted "Ukraine for Ukrainians." *TsDAVOVU*, 3833/1/36, ll. 19-19zv.

[15] *AAN*, 202/III-134, vol.2, kk. 56-57; Borys Levyts'kyi, "Einsatzgruppen," *Kultura* 15 (1969), 161.

fractions by the rank and file and officers were severely punished. By May 1943, the OUN units, which received the name of the Ukrainian Insurgent Army – the UPA – had been forged into a relatively well-armed, disciplined, and mobile force.

While the UPA buildup was in progress, contingency plans for a national uprising were laid down. In February 1943, taking into account the possibility of Germany's defeat, the Third Conference of the OUN-B decided that anti-Polish action had to begin shortly. The logic of the OUN leadership stemmed from two factors. First, the nationalists feared that the long-standing Polish-Ukrainian conflict before and during the war would compel Poles to gravitate towards an alliance with the Soviets. OUN concerns about potential Polish-Soviet collaboration were not entirely unfounded. The Soviets indeed planned to use pro-Soviet Poles in establishing communist rule in the Eastern Borderlands. Thus, in August 1942 Soviet political circles contemplated the employment of Polish intelligentsia, workers, and peasant activists in the administration of liberated western provinces. In January 1943, the Soviet Staff of Partisan Movement also made plans to "foment" a more active partisan movement in Poland to prepare the ground for Sovietization of the country.[16] Therefore, the OUN leadership counted that a preventive strike would eliminate a potential Soviet power-base in Western Ukraine.

Second, after the removal of Poles, the entire socio-economic sector would be taken over by Ukrainians. The OUN leadership reasoned that the border settlement would be most likely decided by the victorious Allies. Facing a new ethnic distribution in the Eastern Borderlands, the Allies would perforce have to accept it as a fait accompli. An OUN resolution, therefore, provided for the "strengthening of Ukrainian character" of Western Ukrainian lands by all means "depending on political possibilities."[17]

In the late winter 1943, the OUN turned to the implementation of its plans. First random attacks on Polish settlements established a pattern that was to be followed throughout the war.[18] After a

---

[16] *Deportatsii, zakhidni zemli Ukrainy kintsia 30-kh - pochatku 50-kh rr.: dokumenty, materialy, spohady u triokh tomakh* [Deportations, the Western Lands of Ukraine between the 1930s and the Beginning of the 1950s: Documents, Materials, Memoirs in Three Volumes], 3 vols., (L'viv, 1996), 1: 167-174, 205-206.

[17] Bul'ba-Borovets', *Armiia bez derzhavy*, 272; *OUN v svitli postanov*, 84, 85; *AAN*, 202/III/134, vol. 1, k. 242.

[18] Instances of mass murder of Poles were recorded as early as November 1942. Władysław Filar, "Zbrodnicza działalność OUN-UPA przeciwko ludności polskiej na Wołyniu w latach 1942-1944," [The Criminal Activities of the OUN-UPA

thorough reconnaissance, armed OUN formations backed by peasant self-defense detachments surrounded Polish villages at night or in the early morning. Guards were posted at the village outskirts and crossroads to prevent possible escape, and then the attackers would move from house to house, butchering all inhabitants regardless of age or sex. Bullets were often spared in favor of axes, knives, and pitchforks. After the murder, Polish houses were thoroughly looted and set on fire. In the course of February and March the attacks grew in intensity, and more than 800 Poles were murdered in the districts of Sarny and Kostopil' of the Rivne province.[19] In the spring, the assault intensified as the nationalist forces were strengthened by the influx of 6,000 deserters from the Ukrainian police. By May 1943, the UPA comprised about 8,000 armed men ready for action.[20]

The magnitude of the OUN-UPA assault on Polish villages was reflected in the haste with which the Germans increased their garrisons in Volhynia. For example, in a main regional railroad crossroads Kovel', the garrison grew from 300 soldiers in January to 4,000 in April. German garrisons and police forces were also reinforced in other urban areas. However, while German units controlled towns and cities, they were unable to restore order in the countryside, where the OUN-UPA violence rapidly gained momentum. Most Polish settlements were caught unaware and perished without much resistance. The main Polish underground organization, the AK (the Home Army), was unable to stem the assault due to the lack of arms and sufficient manpower.[21] Only a few Polish strongholds stood their ground. For example, in Przebraże in the Sarny district, a thousand Polish civilians, in cooperation with Soviet partisans, fought off numerous UPA units.[22]

On the other hand, the OUN-UPA leadership thoroughly planned and organized its attacks. By rapid concentration and

against the Polish population in Volhynia in the years 1942-1944], *Semper Fidelis* 5/28 (1995), 3.

[19] *RTsKhIDNI*, 69/1/1032, ll. 75-76; Ibid., 69/1/1033, l. 5.

[20] Bul'ba-Borovets', *Armiia bez derzhavy*, 261; AAN, 202/III/131, kk. 74-75; TsDA-HOU, 1/23/527, ll. 2-3; RTsKhIDNI, 69/1/563, l. 110; Władysław Filar, *Eksterminacja ludności polskiej na Wołyniu w Drugiej Wojnie Światowej* [The Extermination of the Polish Population in Volhynia during World War II] (Warsaw, 1999), 9.

[21] *TsDAHOU*, 1/23/523, ll. 44-46, 91; RTsKhIDNI, 69/1/25, l. 200; Filar, *Eksterminacja*, 31.

[22] Edmund Łoziński, "Obrona Przebraża: 30 sierpnia 1943 r." [The Defense of Przebraże: 30 August 1943], *Wojskowy Przegląd Historyczny*, 31 (1964), 293-299.

swift convergence upon a targeted village, the nationalists almost always had the element of surprise and superiority in numbers. After a village was overrun, mass murder, pillage, and burning of houses followed in quick succession. All vestiges of Polish existence were eradicated, and a few hours after the initial assault only smoking ashes, charred corpses, and crude mass graves bore witness to former Polish presence. Some villages were given short notice—normally forty-eight hours—to leave. Evacuated villages were still burned to the ground. More often, however, attacks came without warning. The death toll grew dramatically, and in April-May 1943 only in the Rivne province alone 3,000 Poles lost their lives. According to Soviet partisan reports to Moscow, all Polish villages within twenty miles of Rivne were annihilated.[23] The OUN-UPA made deliberate efforts to generate a mass flight of the Polish population by beheading its victims and mutilating corpses. Apparently, rape of Polish women also took place during the massacres. However, since wartime documents have no references to such occurrences, it must be surmised that murder and burning obliterated all traces of sexual violence.[24]

The Germans tried to stem the spread of terror by enlisting Polish refugees into police units, which were dispatched to pacify Ukrainian villages. These expeditions, as well as widespread banditry in the region (often committed by Soviet partisan bands with no clear political affiliation) added to the chaos. The economic situation of the region deteriorated further when mass influxes of Poles overcrowded towns and cities. Hunger and disease spread rapidly, and the Germans used the opportunity to initiate a mass inducement of Poles in the labor force in the Reich. With little alternative, many Poles volunteered to be sent to Germany in labor detachments. The OUN-UPA terror and the evacuation of Poles impeded German communication and supply routes. The Germans therefore made all efforts to vilify the nationalists. The German-sponsored Ukrainian press in Volhynia launched a vehement press campaign against the nationalists, labeling them "bandits on the Soviet pay" and accusing them of hindering German war efforts.[25]

---

[23] *TsDAGOU*, 1/23/530, l. 12; Ibid, 1/23/527, l. 6; Anczarski, *Kronikarskie zapisy*, 307; *Biuletyń-Koło Lwowian*, n. 43 (June 1982), 8; Filar, *Eksterminacja*, 35; *RTsKhIDNI*, 69/1/708, ll. 94, 118.

[24] *RTsKhIDNI*, 69/1/709, ll. 27, 66; Ibid., 69/1/581, l. 9.

[25] *AAN*, 202/II-7, kk. 8-9; Ibid., 202/III/128, k. 2; Ibid., 202/1-35, k. 14; *TsDAHOU*, 1/23/530, l.12; Ibid., 1/23/527, ll. 6-7; *RTsKhIDNI*, 69/1/25, ll. 36-38; Ibid, 69/1/563, l. 111.

The ethnic cleansing reached its peak in July and August 1943. The prime stimulus for the escalation of terror was the great Soviet offensive at the Kursk salient. A directive issued by a senior UPA commander in Vohlynia Dmytro Klachkivs'kyi stipulated the extermination of the entire Polish population between sixteen and sixty years of age "at the appropriate moment of German retreat."[26] Given the centralized character of the OUN-UPA, it stands to reason that Klachkivs'kyi would have not issued the order of extermination without instructions or consent of the OUN-UPA leadership.

In mid-summer the ethnic cleansing spread into the previously quiet south-western districts of Volhynia. Combined UPA and self-defense units launched simultaneous attacks on a dozen Polish settlements. During the last five days of August UPA attacks resulted in the death of approximately 15,000 Poles. Reports by Soviet partisans conveyed horrifying pictures of total destruction and catastrophic loss of life. In several districts two-thirds of the prewar Polish population was annihilated.[27] At night the sky lit up by burning Polish villages served as a grim omen for the Poles of what was soon to come. A letter by a Polish contemporary mirrors the despair of villagers facing an impending attack:

> ...What is happening here and now, the slaughter and torment of Polish families, defies all words...Almost all the Poles in the villages of the Krzemieniec county have been slaughtered...Almost all the Polish parishioners from Oleksiniec parish were robbed and killed...daily one can see fires; they are burning Polish setlements and murdering in the most bestial manner those who do not escape.[28]

The OUN-UPA aim to "cleanse" the area seemed to bear fruit as thousands of Polish refugees inundated towns and cities. By the end of August huge temporary camps, containing up to 10,000 men, women, and children, had been set up by the German administration in Sarny. The waves of Polish refugees also crossed over into Eastern Galicia, where Polish relief committees set up camps and kitchens in many localities.[29] The entire region

---

[26] Filar, *Eksterminacja*, 36-37.

[27] *TsDAGOU*, 1/23/892, ll. 81.

[28] Quoted in Tadeusz Piotrowski, *Poland's Holocaust: Ethnic Strife, Collaboration with Occupying Forces and Genocide in the Second Republic, 1918-1947* (Jefferson, N.C., 1998), 243-244.

[29] *AAN*, 202/III/128, k. 7.

in chaos, the Germans, desperate to alter the situation, committed warplanes to bomb UPA-controlled areas around Rivne.[30]

How extensive was the participation of the Ukrainian population in the murder? Unfortunately, archival documents do not provide a clear answer to this question. From the extant material, it seems that attitudes in Ukrainian villages varied from place to place and depended on the local situation. The OUN-UPA used various methods to involve the Ukrainian population in ethnic cleansing, and undoubtedly the opportunity to loot served as a powerful incentive. OUN-UPA propagandists also skillfully played on Polish-Ukrainian animosities and collective fears of Polish retaliation. The looming presence of the brutal OUN security service often sufficed for enlistment into self-defense units. Still, some Ukrainians were reluctant to join the assault, while others provided shelter and help to their Polish neighbors. A number of Polish survivors mention that they received warnings of an impending attack from their Ukrainian neighbors.[31]

Several Ukrainian political groups voiced their objections to the mass murder. In August 1943, the main Ukrainian institution in Poland and Western Ukraine, the German-sponsored Central Committee, appealed to Ukrainians to stop the massacres. At the same time, the head of the Greek-Catholic Church in Eastern Galicia, Andrej Sheptyts'kyi warned the Ukrainian population against "shedding innocent blood." The Ukrainian bishops of Stanisławów and Przemyśl issued appeals to the same effect, and similar appeals were published by several Ukrainian organizations and unions.[32]

In the fall of 1943, the Soviet offensive on the right-bank of Dnieper-river introduced new elements into the tragedy unfolding in Western Ukraine. The prospect of the inevitable encounter with the Soviet army forced the OUN leadership to reconsider its anti-Polish policies. Indeed, there are indications that there was a disagreement among OUN-B leaders in regards to anti-Polish ac-

---

[30] Ibid, 202/III/128, k. 5; Ibid., 202/1-35, k. 15; *Armia Krajowa w Dokumentach*, vol. III, 18; *United States National Archives Microfilms*, T-175, roll 124, frame 599,082-599,084; *RTsKhIDNI*, 69/1/709, ll. 81, 178.

[31] Anczarski, *Kronikarskie zapisy*, 387; Henryk Komański, "Zagłada polskiej wsi Jeziorany Szlacheckie" [The Destruction of the Polish Village Jeziorany Szlacheckie], *Semper Fidelis* 16 (1993, no. 3), 19.

[32] *AAN*, 202/III/131, kk. 61-62; Ibid., 202/1-35, k. 20; Ibid., 202/III/128, k. 9; Ibid., 202/1-35, k. 249.

tion.[33] The pace of ethnic cleansing clearly slowed down in November, and OUN leaflets promised equality of *all* citizens in a future Ukrainian state. However, while it is conceivable that more moderate OUN-B leaders tried to mitigate anti-Polish violence, at the time of their issuance these reassuring appeals seemed nothing short of a mere gesture. Mass murder and wanton destruction of property had inflicted an irreparable damage on the Polish community. In addition, at the time when OUN propaganda called for Polish-Ukrainian rapprochement, attacks still went on. Thus, in October-November 1943, the UPA units wiped out five Polish villages--430 households--in the Ludvipol' district.[34]

Similarly, high-level Polish-Ukrainian contacts failed to bring an end to the massacres. Throughout the war, meetings between Polish and Ukrainian representatives revealed irreconcilable differences in their approaches to the question of the Eastern Borderlands. While each side solemnly promised equality and justice for ethnic minorities after the war, both passionately claimed Western Ukraine as a crucial component of their respective national states. For the Poles, Eastern Galicia and Volhynia represented a historical bastion of Polish civilization in the east. For Ukrainians, they were inseparable parts of ethnic Ukraine. The credibility of Polish promises was also seriously undermined by the fact that Poland had violated an international agreement of 1923 which guaranteed Eastern Galicia autonomous status.[35] On the other hand, Polish radical circles in London considered the resettlement of Ukrainians (possibly to Polish western territories) as the only viable ethnic solution in the Eastern Borderlands. Mutual intransigence, therefore, rendered Polish-Ukrainian talks useless and had no impact on the situation in Western Ukraine. Hence, a leading Ukrainian nationalist, Yaroslav Horbovyi, pes-

---

[33] In March 1943 at the meeting between Polish and Ukrainian representatives, a high-ranking OUN member, Ivan Hryniokh, indicated that the OUN leadership was incapable of stopping mass murder due to a split within its ranks. Ryszard Torzecki, "Kontakty polsko-ukraińskie na tle problemu ukraińskiego w polityce polskiego rządu emigracyjnego i podziemia, 1939-1944" [The Polish-Ukrainian Contacts in the Context of the Ukrainian Problem in the Policies of the Polish Émigré Government and the Underground], *Dzieje Najnowsze* 1-2 (1981), 342. It is possible that Hryniokh simply tried to exonerate the OUN as the instigator of anti-Polish violence.

[34] *TsDAGOU*, 1/23/523, ll. 13, 190; Ibid., 1/23/530, l. 12; *AAN*, 202/III/131, k. 72.

[35] Torzecki, "Kontakty polsko-ukraińskie," 320-321, 323, 327-328, 330-332, 340; *AAN*, 202/III/131, kk. 71, 92-94.

simistically predicted that the future of the Eastern Borderlands would be solved by force alone.[36]

The end of the summer 1943 also witnessed the escalation of anti-Polish violence in Eastern Galicia, which until then had remained relatively quiet. The quick pace of the Soviet offense was mirrored by the OUN drive to cleanse the area before the arrival of the Soviet army. By the end of August 1943, approximately 300 Poles had been murdered in the districts of Czortków, Brzeżany, Brody, and Podhajce. In the fall the number of attacks rapidly increased, and Polish villages in Tarnopol province sustained heavy casualties. The violence followed the pattern established in Volhynia. The OUN-UPA units surrounded Polish villages, murdered all the residents, and pillaged and burned houses and barns. Corpses were often mutilated and defiled. The Polish population in some localities was given forty-eight hours to leave the premises, while the nationalist agitators called for a mass extermination of all Poles to achieve Ukrainian national liberation-- "for each [dead] Pole a tract of free Ukraine."[37]

Yet, several factors mitigated the destructive potential of the OUN-UPA in Eastern Galicia. The Polish underground here was better armed and organized. More importantly, the Germans were determined to preserve stability in this important strategic oil-producing region. In the summer 1943, after a raid by a Soviet partisan formation, the German military and police in the province were put on high alert. In addition, except districts adjacent to the Carpathian mountains, a flat terrain rendered guerrilla warfare much more difficult than in Vohlynia. German propaganda called for united Polish-Ukrainian-German efforts to stem the Soviet offensive, and in December 1943 the Galician governor, Otto Wächter, warned Ukrainians against any actions that might disrupt the functioning of the district. To prevent the spread of panic among the Poles, the Germans forbade publishing obituaries for victims of the OUN-UPA.[38]

Eventually, however, German efforts to keep Galicia quiet failed. Inadequate German manpower–all available reserves were siphoned off by the Eastern Front–allowed the OUN-UPA to attack remote Polish settlements at will. In January and February

---

[36] *Deportatsii, zakhidni zemli Ukrainy,* 214-230; AAN, 202/III-134, vol. 1, k. 247.

[37] Anczarski, *Kronikarskie zapisy,* 293-294, 305; *AAN,* 202/1-41, k. 21; Ibid., 202/III/128, k. 2; Ibid., 202/1-35, k. 108; Ibid., 202/III/129, k. 75. See a photocopy of an OUN-UPA warning in *Biuletyń Koło-Lwowian,* n. 43 (1982), 8.

[38] *TsKhIDK,* 1447/1/379, ll. 3-4; *AAN,* 202/1-35, k. 244.

1944 UPA units, augmented by self-defense detachments, wiped out several Polish villages in Tarnopol province and murdered more than 600 men, women, and children. The Polish underground retaliated by burning Ukrainian villages and killing their residents. The situation in the region began to deteriorate rapidly as the German police and gendarmerie entered the fray on one side or the other depending on a local situation. For example, according to OUN reports, in some instances German reprisals were carried out with utmost brutality. Ukrainian hostages were forced to lie down in ditches and then were sprayed with bullets, "much like Jews." Masses of refugees, blazing fires over Polish villages, and the sounds of Soviet guns in the distance further underscored the prevailing chaos. In May and June up to 100,000 Poles fled Galicia for central Poland.[39]

In view of the OUN-UPA terror, it was no surprise that the majority of Poles enthusiastically welcomed the Soviet Army. A number of Poles enlisted in anti-insurgent battalions set up by the Soviets to fight the Ukrainian underground and took positions in civil offices. The Soviet military administration and police initially encouraged Polish participation in local administration, while soldiers of Polish background harassed Ukrainians and forcibly requisitioned food and cattle in Ukrainian villages. Instances of Polish-Soviet collaboration prompted a renewal of OUN-UPA attacks, and in winter and spring of 1945, UPA units attacked several Polish villages in Buczacz district killing off all the inhabitants.[40]

The last stage of ethnic cleansing in Western Ukraine was completed by the Soviet-Polish agreement signed on September 9, 1944, which stipulated a "voluntary" population exchange between Ukraine and the newly established Polish republic. In the same month, Polish government units in Lublin and Rzeszów provinces were deployed to speed up the resettlement of Ukrainians. The repatriation of ethnic Poles from Eastern Galicia and

---

[39] Anczarski, *Kronikarskie zapisy*, 321-322, 327; *TsDAGOU*, 1/23/927, l. 6; Ibid., 3833/1/74, ll. 48, 84-85; Ibid., 1/23/927, ll. 7-8; Ibid., 1/23/892, ll. 6-7; *Deportatsii, zakhidni zemli Ukrainy*, 246-247; *AAN*, 202/III-134, vol. 1, k. 31; Ibid., 202/III/127, kk. 22, 36; Ibid., 202/III-134, vol. 1, kk. 33-34; Ibid., 202/III-134, vol. 1, no page; Ibid., 202/III-134, vol. 1, kk. 15-26; Ibid., 202/III/129, kk. 75, 82; Ibid., 202/III/129, k. 75; Ibid., 202/III/131, kk. 102-104, 107-110.

[40] *TsDAGOU*, 1/23/892, ll. 56-57, 92-93; *AAN*, 202/III/129, k. 96; Aleksandr Korman, "Polscy 'istrebitiele' z lat 1944-1945 w Małopolsce Wschodniej i Wołyniu" [The Polish Members of the 'Destruction Battalions' in 1944-1945 in Eastern Małopolska and Volhynia], *Semper Fidelis* 3(26) (1995), 7-11; Anczarski, *Kronikarskie zapisy*, 471-474.

Vohlynia also grew in numbers, and by early 1945 reached 117,114.[41]

The Polish-Ukrainian conflict continued in Poland. In Kraków, Lublin, and Rzeszów provinces the Polish army and security service waged an anti-guerrilla campaign against the OUN-UPA until the fall of 1947. Brutalities were common on both sides, and memories of the massacres in Western Ukraine often turned anti-guerrilla actions into punitive expeditions as Polish units burned to the ground Ukrainian villages and shot the villagers on the spot.[42] Hard-pressed by Polish troops and deprived of its bases and supplies by the mass resettlement of the Ukrainian population, the remnants of the OUN-UPA dispersed through the mountains to Slovakia and then to Austria. Some cut their way through to the forests of Western Ukraine.

On the surface, the OUN-UPA anti-Polish campaign in Western Ukraine seemed to have achieved its long-desired objectives. Mass murder and flight of Polish refugees irreversibly changed the ethnic composition in Volhynia and Eastern Galicia in favor of ethnic Ukrainians. In reality, however, it was a Pyrrhic victory for the nationalists. Forced to deploy its forces against Poles, Germans, and in the spring of 1944 against the Red Army, the OUN-UPA was bound to crack under the enormous pressure of fighting simultaneously too many enemies. The major political beneficiaries of the ethnic cleansing were the Soviet authorities, for whom it provided a suitable opportunity to claim and annex Volhynia and Eastern Galicia as "primordial Ukrainian lands."

The demographic implications of the massacres are difficult to assess with certainty since it is often impossible to estimate how many Poles died at the hands of the OUN-UPA as opposed to those killed by German or Soviet occupiers. In addition, these

---

[41] *Ukrain'ska RSR na mizhnarodnii areni: zbirnyk dokumentiv i materialiv* [The Ukrainian SSR in the International Arena: Collection of Documents and Materials] (Kiev, 1963), 413; Lubomyr Luciuk, ed., *Anglo-American Perspectives on the Ukrainain Question, 1938-1955* (Ontario, 1987), 191-193.

[42]*TsDAGOU*, 1/23/1475, l. 5; Ivan Bilas, *Represyvno-karal'na systema v Ukraiini, 1917-1953: suspil'no-politychnyi ta istoryko-pravovyi analiz,* [The System of Repression and Punishment in Ukraine, 1917-1953: the Socio-Political and Historical-Legal Analysis], 2 vols. (Kiev, 1994), 1: 224-225, 229; Władysław Nowacki, "Organizacja i działalność wojsk wewnętrznych w latach 1945-1946" [The Organisation and Activities of the Internal Security Troops in 1945-1946], in *Z walk przeciwko zbrojnemu podziemiu, 1944-1947* [Of the Combat against the Armed Underground, 1944-1947], ed. Maria Turlejska(Warsaw, 1966), 111-112; Jan Łukaszów, "Walki polsko-ukraińskie" [The Polish-Ukrainian Struggle], *Zeszyty Historyczne* 90 (1989), 190-192.

statistics are further complicated by the fact that historians and eyewitnesses tend to include both individual victims and entire families in the roster of dead, and also counted the displaced persons and refugees as victims of the OUN-UPA terror.[43]

Taking into consideration the prewar demographic distribution of the Polish population in Volhynia and Eastern Galicia, the time frame of the OUN-UPA assault (with its peak in July-August of 1943), its destructive capacities (the numbers and armament of the nationalist units), it can be surmised that the death toll of the OUN-UPA killing campaign in Polish settlements did not exceed 50,000 in more than 900 localities. The numbers of Poles who were forced to leave their homes can be estimated around 350,000.[44] Volhynia alone lost more than a half of its prewar Polish population. Several factors such as German presence in cities and towns and Soviet partisans' activities in the countryside (especially in Vohlynia) mitigated the genocidal potential of the OUN-UPA. The most important mitigating factor, however, was that the OUN-UPA did not posses a state-machinery that could be committed to mass murder--in comparison, for example, to the ethnic cleansing perpetrated by the Croatian Ustaše against the Serbian population in Croatia and Bosnia.

The OUN-UPA assault on the Poles of Volhynia and Eastern Galicia represented one of the most horrible ethnic tragedies of World War II.It exhibited two traits which have since become common for contemporary ethnic conflicts in Europe, Asia, and Africa: first, a revolutionary vision for remodeling a multiethnic state is often implemented through the physical removal of the members of an "undesirable" or allegedly dangerous ethnic group; second, wartime or socio-political crisis creates the most propitious atmosphere for the escalation of violence and reduces the moral and psychological restraints among the perpetrators.

---

[43] See, for example, Filar, *Eksterminacja*, 12, and Alexander Korman, "Osobowe i materialne strary Polaków wynikłe z działalności terrorystów OUN-UPA" [Personal and Material Polish Losses Caused by the Activities of the OUN-UPA Terrorists], *Semper Fidelis*, 2/15 (1993), 18-19; *AAN*, 202/III/131, k. 111; Ibid., 202/III-134, vol. 1, kk. 12, 21, 25. The German administration accounted for 100,000 Polish victims of the OUN-UPA terror. *AAN*, 202/I-41, k. 27.

[44] Computed on the basis of *Mały Rocznik Statystyczny*, 22,24,26; *The Polish and Non-Polish Populations of Poland*, p36, 41-42; *AAN*, 202/I-41, k. 10. These numbers also include those who fled the Soviet offensive in Western Ukraine in the summer of 1944. See also, Jan Czerniakiewicz, "Przemieszczenia Polaków i Żydów na Kresach Wschodnich II Rzeczypospolitej i w ZSRR, 1939-1959" [The Movement of Poles and Jews in the Eastern Lands of the Second Polish Republic and in the USSR, 1939-1959], *Studia i materiały CBR* 21 (1991): 9; also his, *Repatriacja ludności polskiej z ZSRR, 1944-1948* [The Repatriation of the Polish Population from the USSR, 1944-1948] (Warsaw, 1987), 133-134, 154, 174.

In 1943-1944 the ethnic reconstruction of Ukraine became the most essential element of the OUN-UPA revolutionary struggle. Poised to build a national state on the ethnographic Ukrainian territories, the OUN-UPA leadership viewed Poles as deadly rivals in the socio-economic and political spheres, while its ideological principles justified and promoted the physical destruction of an enemy defined in ethnic and national terms. In this respect, the revolutionary vision and murderous practices of the OUN-UPA were completely in tune with the racial war Nazi Germany waged in the East.

The Deportation and Ethnic Cleansing of the Crimean Tatars

BRIAN GLYN WILLIAMS

It has been said that one must know a nation's tragedies and the way its people commemorate them to know its soul. To understand the Russians one has to visit Russia's memorials to the millions of members of that nation who gave their lives fighting in the "Great Patriotic War" against Nazi Germany. To understand Serbian aggression in 1999 against the Kosovar Albanians one has to visit the sacred monasteries of Kosovo commemorating that people's defeat at Kosovo in 1389 at the hands of the Ottomans. The Armenians cannot be understood today without understanding the role of the collective memory of the 1915 genocidal assault on their community by the Ottoman government. The Palestinians are defined by their trans-generational narratives of their expulsion from Israel in 1948 known as *al-naqbah* (the disaster). The Jews of today, regardless of their level of religiosity, are shaped by the collective memory of the *Shoah*, the Holocaust.

The Deportation of the Crimean Tatars

The defining event in twentieth–century Crimean Tatar history is the brutal deportation and exile of this small Turkic-Muslim people from their peninsular homeland on the Black Sea (Ukraine) to the deserts of Soviet Central Asia and Siberia in the closing days of World War II. On May 18, 1944, the *entire* Crimean Tatar people, men, women, children, the elderly, unarmed civilians and those fighting for the Soviet *Rodina* (Homeland) in the ranks of the Red Army were arbitrarily accused of "mass treason" by Soviet leader Josef Stalin and deported from their villages located in the Crimea's southern Yaila mountains and on the warm southern shore of the Crimea. The official explanation for this total ethnic cleansing was announced at a later date in the Soviet paper *Izvestiia*, which declared:

> During the Great Patriotic War when the people of the USSR were heroically defending the honour and independence of the Fatherland in the struggle against the German-Fascist invaders, many Chechens and Crimean Tatars, at the instigation of German agents joined volunteer units organized by the Germans and together with German troops engaged in armed

struggle against units of the Red Army...meanwhile the main mass of the population of the Chechen Ingush and Crimean ASSRs took no counteraction against these betrayers of the Fatherland.[1]

The cleansing of the Crimean Tatars was actually part of a larger program described as Operation Deportation. Stalin took advantage of the wartime mobilization of Soviet troops and general distrust of non-Slavic minorities in many echelons of the Kremlin to eradicate several ethnic groups deemed to be untrustworthy by the Soviet regime. In addition to the Crimean Tatars, several other small distrusted nationalities living on the Soviet Union's southern borderlands (the Chechens and related highlander Ingush, the Turkic pastoralists known as the Karachai and related Balkars, the Buddhist Mongol Kalmyks, the Meshketian Turk mountain farmers, and the Volga Germans) were targeted for deportation to Siberia. Not surprisingly, the sudden "disappearance" of these ancient ethnic groups, most of whom were Muslims living in the Caucasus vicinity, went largely unnoticed in the West during the general conflagration of World War II.

It was only with the collapse of the USSR that Western scholars could begin to probe once off limits KGB documents on this tragedy and interview the survivors of the deportations. It soon became clear that in all more than 1.5 million Soviet citizens belonging to targeted ethnic groups were forcefully deported during the war years in one of the best hidden cases of mass ethnic cleansing in 20th century history. All the targeted ethnic groups were accused of "betraying the homeland" during the German invasion, and their rights as Soviet citizens were subsequently taken away.

There are some grounds for Stalin's sweeping accusation against the Crimean Tatar people. As many as 20,000 Crimean Tatars did serve in the *Wehrmacht* in varying capacities as *Hiwis* (the German acronmym for "volunteers"), but most of these were prisoners of war captured by the German army as it surrounded and captured whole Soviet armies in 1941 and 1942. Most of those captured by the Germans were used as cannon-fodder in their costly engagements with the Red Army. Others were used in vil-

---

[1] For an in-depth analysis of the role of the Crimean Tatars in the Wehrmacht and Red Army during World War II and their subsequent deportation to and adaption to Uzbekistan, see Brian Glyn Williams, *The Crimean Tatars: The Diaspora Experience and the Forging of a Nation* (Leiden, Boston, 2001); see also "Ob Utverzhdenii Ukazov Prezidiuma Verkhovnogo Soveta RSFSR," [On the Ratifying Decree of the Presidium of the Supreme Soviet], *Izvestiia*, June 26, 1942, p. 2.

lage defense brigades within the Crimea itself, and their loyalty was more to their village than to the Nazis (who initially called for the eradication of the Tatars and other "Asiatic inferiors").

It should be noticed, however, that 20,000 Crimean Tatars actually fought for the Soviet homeland in the Soviet Army. Others fought in the ranks of the partisans who launched guerrilla raids on the German occupying forces during the war. The eyewitness testimonies of Russian officers offer us an invaluable account of the anti-Nazi guerrilla activities of a Crimean Tatar partisan brigade:

> The Commissar of the Eastern formation was named captain Refat Mustafaev (prior to the war he was secretary of the Crimean regional party). Here is one episode of the military actions of his formation. In the end of the 1943 the divisions of the second and third brigades destroyed the fascist garrison in Stary Krym (Eski Kirim) destroying on that occasion two tanks, 16 vehicles with gasoline and ammunition. The partisans occupied the building of the commander of the city police and threw grenades into the restaurant where the Hitlerites banqueted. One of the group seized the Gestapo jail and freed 46 Soviet patriots.[2]

As the Crimean Tatars joined the partisans their villages suffered heavily from German reprisals. The following account is typical:

> Dozens of Crimean Tatars were shot in Alushta on the banks of the Demerci, in the foothills of the Kastel in dozens in the villages of Ulu-Sala, Kizil Tash, Degirmen Koy, Tav-Bodrak, Saly and many others.
> In July 1988 the country learned from information in Tass that in the partisan regions in the mountainous part of the Crimea all villages were burnt and a "dead zone" was created. Yes, it actually happened. More than 70 villages were destroyed. In them dwelt more than 25 percent of the Tatar population of the Crimea. In these villages, in remote woodlands, in the mountains lived only Tatars.[3]

Seen in this light, the official charges levelled against the Crimean Tatars of "mass treason" are obviously spurious. The real reason for the deportation may in fact lie in Stalin's plans to in-

---

[2] Ibid., p. 36.

[3] Svetlana Alieva, *Tak Eto Bylo. Natsional'noe Repressi v SSSR* [Thus it Was: National Repression in the USSR], vol. 3 (Moscow, 1993).

vade Turkey at this time. In particular, as the Red Army moved into a collapsing Germany, Stalin contemplated the annexation of the Turkish *vilayets* (provinces) of Kars and Ardahan on Turkey's north-eastern border with the USSR (these had been lost to Russia during World War I). The Soviets commenced a broad propaganda campaign at this time designed to lead to an Armenian uprising in this region, and Turkey in return planned a full mobilization.[4] As Stalin prepared for this operation, he, as a Georgian, must have been keenly aware of the existence of several Muslim, traditionally pro-Turkish ethnic groups located on the invasion route through the Caucasus. The "Crimean Turks," as the Tatars of the Crimea were often known, occupied the USSR's main naval base facing Turkey across the Black Sea. Other small, distrusted ethnic groups, such as the Karachai, Balkars, Chechens, Ingush and the Meshketian Turks, occupied the frontier with Turkey or the two main highways running to Turkey--the Georgian military highway and the coastal highway.

All these suspect Muslim groups were deported after having been accused of blanket treason against the *Rodina* (Homeland) during the German invasion, except for the Meshketian Turks, who were never officially accused of mass betrayal. The homeland of this small conglomerate ethnic group, made up of Turkic Karapapakhs, Muslim Armenians (Khemshils), Turkicized Kurds, and the Meshketian Turks proper was located far to the south of Georgia on the Turkish border and had never been close to the scene of combat. The fact that this patently innocent ethnic group was chosen for deportation lends the strongest credence to the claim that the deportation of the Crimean and Caucasian Muslims had more to do with Soviet foreign policy priorities than any real crimes of "universal treason" committed by these groups. As Mehmet Tutuncu surmised, "The only thing all of these peoples have in common is religion and that they inhabit areas that would be sensitive in an invasion of Turkey. And this seems the only reason for the collective punishment of all these people."[5]

Regardless of the justification, the results of the deportations were terrifying for the targeted nationalities. Just as the sanitized term "ethnic cleansing" fails to capture the true horror of rape

---

[4] Galia Golan, *Soviet Policies in the Middle East from World War Two to Gorbachev* (Cambridge, 1990), 32; George Harris, "The Soviet Union and Turkey," in *The Soviet Union and the Middle East: The Post-World War II Era*, ed. Ivo J. Lederer and Wayne S. Vucinich (Stanford, 1974), 55-78.

[5] Mehmet Tutuncu, "Why the Crimean Tatars Were Deported," *Bitig* 2, no. 5 (Dec. 1992), 16.

camps, mass slaughter, brutal expulsion, and destruction of homes, welfare, and culture, the term "deportation" fails to capture the true horror of this fate which befell the Crimean Tatars and several other small nations during the final days of World War II.

Tens of thousands of NKVD (the progenitor of the KGB) troops surrounded the Crimean Tatar hamlets in the Crimean Autonomous Soviet Socialist Republic (Crimean ASSR) and began to expel their startled inhabitants on the evening of May 18, 1944. Thousands deemed guilty of collaboration with the Germans, who had occupied the Crimea during the war, were summarily shot on the spot; those who resisted were beaten or shot. Traditionally tightknit Crimean Tatar families and villages were divided as the well–armed troops gathered them and drove them to local railheads for deportation in various directions. When families did not get on the same train together, family members were likely to be scattered throughout Central Asia and Siberia, many of them never seeing each other again. In many cases the men were separated from their families and shipped to lumber and gas camps in Siberia, where they were forced to do physical labor. The death rate in the harsh conditions in these camps deprived the Crimean Tatar community of able bodied men who might have helped their families re-adjust to life in exile.

The deportees remember with particular horror the weeks spent on the eastward-moving trains in cattle cars whose only modification for human inhabitation was a pipe fitted in the corner for defecating. For efficiency's sake the deportees had been crammed into train cars, which were then locked, and the packed, unhealthy conditions led to outbreaks of disease such as typhus, which swept away many, especially the young and the old. A survivor of the deportation recalls:

> The doors of the wagons were usually opened in stations where the train stopped for a few minutes. The panting people gulped fresh air, and they gave way to the sick who were unable to crawl to the exit to breath it. But along the length of the wagon one officer in a blue hat hastily strolled with soldiers and, glancing into the wagon, asked the same question. "Any bodies? Any bodies?" If this was the case, they pulled them out of the wagon; they were mainly children and the old. There and then, three meters from the rail embankment (the bodies) were thrown into hollows with dirt and refuse.[6]

---

[6] Alieva, *Tak Eto Bylo*, 79.

The trains carrying the bulk of the Crimean Tatar population (civilians and the wounded) trundled across the hot plains of the northern Caucasus and Kazakhstan, and, after two weeks, most reached Tashkent, the capital of the dry Central Asian republic of Uzbekistan.

According to N. F. Bugai, a specialist on the deportations, a maximum of 191,088 Crimean Tatars were deported from the Crimean Autonomous Republic in May of 1944. Another account based on conflicting NKVD sources from 1944 claims that only 187, 859 Crimean Tatars were deported from the Crimea.[7] Of these, Bugai claims 151, 604 were sent to the Uzbek SSR and 8,597 to the Udmurt and Mari Automous *Oblasts* (Ural mountain region, part of the Russian Federated Socialist Republic).[8] B. Broshevan and P. Tygliiants support this claim and reference a telegram sent from Beria to Stalin which proudly proclaims that "all the Tatars have arrived in the places of resettlement and 151,604 people have been resettled in the *oblasts* (districts) of the Uzbek SSR and 31,551 in the *oblasts* of the RSFSR (Russia)."[9] Although Soviet documents do not record the "resettlement" of Crimean Tatars in Kyrgyzstan, Kazakhstan, and Tajikistan, several thousand were eventually transferred or migrated to these regions. The Khojent (Leninbad) region in Tajikistan, in particular, saw considerable settlement according to the overwhelming testimony of those I interviewed in Uzbekistan in the Spring of 1996. Approximately 7,900 Crimean Tatars died during the actual deportation process.

Tashkent served as the main dispersion center for the majority of the Crimean Tatars who were sent to Uzbekistan (other deported groups, such as the Chechens and Ingush, were sent to Alma Ata, the capital of the Kazakh SSR and were then scattered throughout Eastern Uzbekistan, from the Fergana valley in the north to the deserts of the barren Kashga Darya *Oblast* in the south.)[10] According to records sent to NKVD head Lavrenti Beria in June of 1944, the Crimean Tatars were settled in Uzbekistan in

---

[7] V. M. Broshevan and P. Tygliiants, *Izgnanie i Vozvrashchenie* [Expulsion and Return] (Simferopol, 1994), 45.

[8] Nikolai Fedrovich Bugai, "K Voprosy o Deportatsii Narodov SSR v 30-40-x godax" [On the Question of the Deportation of Peoples in the USSR in the 30's and 40's], *Istoriia SSSR*, no. 6 (Nov.-Dec. 1989), 135-144.

[9] Broshevan and Tygliiants, *Izgnanie i Vozvrashchenie*, 45.

[10] Nikolai Fedrovich Bugai, "Pravda o Deportatsii Chechenskogo i Ingushkogo Narodov" [The Truth About the Deportation of the Chechen and Ingush Nations], *Voprosy Istorii*, no. 7 (1990): 32-44.

the following *oblasts*: Tashkent 56,632, Samarkand 31,540, Andijan 19,630, Fergana 16,039, Namangan 13,804, Kashga Darya 10,171, and Bukhara 3,983.[11] Few or no preparations had been made in advance for the arrivees, and most were forced to live in barracks outside factories, in dugouts, or in primitive earthen huts. The death rate continued to rise at this time. As many as a third of the Crimean Tatars may have died during the resettlement period in special camps in Central Asia.

The Crimean Tatar men who were still fighting for the Soviet homeland on the front (and had thus avoided deportation) were demobilized after the fall of Berlin and joined by the Tatar males deported from the Crimea in labor brigades in Siberia and the Urals region. Many Soviet military commanders, however, hid the identity of the Crimean Tatar soldiers with whom they had served during the war to protect their trusted comrades from the NKVD.

Adjusting to Life in Exile

From my own interviews with survivors of the deportation it appears that most deportees who were deposited in Kazakhstan were well treated by the indigenous populations. Those who were exiled in the Mari Republic (Siberia) found that many of the local inhabitants were themselves deported *kulaks* (a persecuted class of wealthy peasants) and political prisoners from the 1920s and 1930s and these were quick to offer assistance. Most accounts, however, stressed the hostility of the Uzbeks towards the deportees in the first year or two in Uzbekistan. The NKVD had been active in the region prior to the deportations, spreading anti-Tatar propaganda against this "nation of traitors," and it seems to have been particularly effective among the simple Uzbek *kolkhozniks* who had a xenophobic distrust of outsiders. According to the testimony of one deportee, in some instances the Uzbeks stoned the already stricken Tatars when they arrived in the comparatively backward countryside. The Crimean Tatar physicist and dissident, Rollan Kadiyev, claimed: "I personally recall how we were met by the local inhabitants, who had been poisoned by Stalin's propaganda. One of the rocks hit me. I was still only a boy." [12]

The Crimean Tatar dissident, Reshat Dyhemilev, wrote: "People were dying in droves every day, from hunger, exhaustion, and

---

[11] Broshevan and Tygliiants, *Izgnanie i Vozvrashchenie*, 46.

[12] The Herzen Foundation, *Tashkentskii Protsess* [The Tashkent Legal Case] (Amsterdam, 1976), 590.

the unaccustomed climate, but no one would help them bury their dead." According to Dzhemilev: "People died from the sharp changes in the climate and the unbearable work, from dystrophy and other illnesses, from cold and malnutrition in the absence of medical care, from nostalgia and from grief over the lost members of their family."[13] All Crimean Tatar families have stories of lost family members that recall the horrible conditions their people encountered in their first two years in Central Asia. The following account given by one deportee is sadly typical:

> My niece, Menube Seyhislamova, with ten children, was deported with us. Her husband, who had been in the Soviet Army from the first day of the war had been killed. And the family of this fallen soldier perished of hunger in exile in Uzbekistan. Only one little girl, Pera, remained alive, but she became a cripple as a result of the horror and hunger she had experienced.
>
> Our men folk were at the front and there was no one to bury the dead. Corpses would lie for several days among the living. Adshigulsim Adzhimambetova's husband had been captured by the Fascists. Three children, a little girl and two boys, remained with her. This family was also starving just as we were. No one gave either material or moral help. As a result, first of all, the little girl died of hunger, then in one day, both the boys. Their mother could not move from starvation. Then the owner of the house threw the two children's bodies onto the street, onto the side of the irrigation canal. Then some children, the Crimean Tatars, dug little graves and buried the poor little boys.
>
> Can one really tell it all? I have such a weight on my heart that it is difficult to remember all. Tell me why did they allow such horrors to happen?[14]

Crimean Tatar survivors of the deportation claim that the local Uzbeks did eventually come to the aid of the outsiders who had been dumped in their midst after the first year or two. In interviews I conducted in Tashkent with elderly deportees, they stressed the fact that the Uzbeks accepted the Crimean Tatars when the latter made a point of stressing their shared Islamic beliefs and traditions. The exiled Crimean Tatars in fact made a point of emphasizing the Muslim aspects of their culture and identity to open a dialogue with the local Uzbeks who had maintained much of the traditional, conservative religious traditions lost by the less religious, Europeanized Crimean Tatar population.

---

[13] Reshat Dzhemilev, Musa Mahmut: Human Torch (New York, 1986).

[14] *The Crimean Review.* Vol. 1 no. I May 18, 1986. p. 10.

Islam, in effect, provided a common language of idioms, symbols and shared cultural norms that bridged the differences between these two different peoples.

Several older Crimean Tatar interviewees also claimed that the local Uzbeks were taken aback when they discovered that the vast majority of the "traitors to the homeland" dumped in their midst were actually the elderly, women and children with many wounded Red Army officers in their midst. Many Uzbek villagers were, according to my informants, ashamed to discover that they had been so initially harsh to women and children who hardly looked like hardened Nazi collaborators.

Soviet statistics back up the Crimean Tatars' claims that the majority of those transported on the terrible journey from the Crimean peninsula to Uzbekistan were indeed women and children. Of the 151,529 Crimean Tatars deposited in Uzbekistan an astounding 68, 287 were children, 55, 684 women and a mere 27, 558 men according to a letter sent to Beria.[15] A full 82 percent of the Crimean Tatar "collaborators" brutally deported in 1944 to Uzbekistan then were actually women and children, and the majority of the men included in this number were, in all probability, war invalids or the elderly. The abundance of children was a pleasant surprise for those involved in the deportation, for they could squeeze more deportees in a wagon due to their smaller size.

In paintings depicting "The Deportation" that now hang in art exhibits presented by the Crimean Tatars in the post-Soviet Crimea and Uzbekistan the author noticed a common theme. Invariably the Crimean Tatar artists portrayed the horror stricken victims of the *"echelons"* (cattle transport carts) as weeping women, children and the elderly. Young men never appear in these works. To this day the Crimean Tatars reserve particular revulsion towards the Soviet regime for its treatment of this non-combatant segment of their population who were left defenseless while thousands of their husbands and fathers were fighting on the front against the German invaders in the ranks of the Red Army.

The desperate situation of the Crimean Tatar elderly, women and children in Central Asia improved significantly when the war ended and many (although not all) Tatar soldiers were allowed to search out their families in the various places of exile between 1945 and 1948. The Crimean Tatars have a distinct genre of stories which speak of the anguish of Crimean Tatar soldiers who were discharged from the Red Army only to return to a Crimea that had been emptied of their families, villages, and entire people. Those

---

[15] Broshevan and Tygliiants, *Izgnanie i Vozvrashchenie*, 45.

who did make their way with great difficulty across the war torn Soviet Union to their families in their special settlement camps in distant Central Asia were automatically declared *spetspereselenets* ("special resettlers") along with their relatives and confined to the special settlement regime. Soviet sources recorded the arrival of approximately 9,000 demobilized Crimean Tatar soldiers to the *spetsposelenets* (special settlement) camps after the war. Most interestingly, Soviet sources mention that 524 of these veterans who automatically became "traitors to the homeland" were Soviet officers and 1,392 sergeants in the Red Army.[16]

With the arrival of many of their fathers, sons, and brothers in 1946, this largely defenseless population had thousands of hardened war veterans to protect them from the abuse of MVD (Ministry of Internal Affairs) *"kommandants"* and help them rebuild their lives in their places of exile. Several older Crimean Tatar interviewees recalled the rare feelings of joy their community felt when the Crimean Tatar men came back in waves from the front to be reunited with their families. One Crimean Tatar recalled:

> In the first months after arrival in Uzbekistan, more than 40,000 Crimean Tatars perished. A primary role in this was played by the circumstance that the local population received the exiles as their personal enemies. Anti-Tatar propaganda was spread among the peoples of Central Asia and the Crimean Tatars were pictured as traitors who had betrayed Central Asian men who were fighting for the Soviet *Rodina* on the front.
>
> A short time passed then the local population began to understand. Dozens of disabled soldiers without arms or legs, with medals clinking on their chests returned from the front and searched for their mothers, wives, and children but they were no longer in this world...And then the Uzbeks understood that a monstrous injustice had taken place and they began to share their last scrap of *lepishka* (scone), their last handful of *kishmish* (raisins) or nuts.[17]

The establishment of a rapport with the indigenous Uzbek population certainly eased the resettlement process for the deported Crimean Tatars. According to first hand accounts, some Crimean Tatar widows initially married Uzbek men who were Turkic Sunni Muslims like themselves (the war and labor camps had decimated the Tatar male population), and Crimean Tatar orphans were adopted by the local Uzbeks. If one believes Soviet

---

[16] Ibid., 103.

[17] Alieva, *Tak Eto Bylo*, 93.

mythology, this tradition of adopting war orphans was in fact an Uzbek national characteristic. One Uzbek of the period, Sham Akhmudov, was reputed to have adopted fifteen war orphans, and a massive statue to this socialist hero still dominates the square in front of Tashkent's Palace of the Friendship of Peoples.

The Special Settlement Regime

Establishing good relations with the indigenous Central Asian populations was not, however, the deportees' only concern. Upon arrival in Central Asia, the Crimean Tatars, who were considered to be traitors to the homeland by the state and its officials, were forced to live under a punitive regime, in the so called *spetsposelenie* settlements (special settlement camps). These informal camps surrounded by barbed wire, which were run by the *otdel spetsposelenii* (special settlement department) of the MVD, are remembered with particular repugnance by the Tatars who lived in them. The heads of Crimean Tatar households were required to report to the *spetskommandants* every three days for a *spetsial'nyi uchet* (special accounting report on their family deaths, births, work progress, etc.), and those who left their assigned region were arrested and sentenced to five years hard labor. In these camps Crimean Tatars report that the "commandants were God and Tsar."[18]

In interviews I held in Uzbekistan, Crimean Tatars told of being awakened before dawn for twelve-hour workdays in the fields and factories, of Crimean Tatars who were sentenced to the camps for five years for leaving their restricted areas to visit family members in other camps, and of the cruelty of the hated camp *kommandants*.[19] Living conditions in the settlements were abysmal. Most deportees lived in barracks constructed next to factories, dug outs, or simple huts hastily built of unbaked dried mud bricks during the *spetsposelenie* years.

As "enemies of the people" the Crimean Tatars had no rights as Soviet citizens during this period, and their group aspirations were reduced to one basic objective, communal survival. One Crimean Tatar whose mother died in the settlement camps remembers her last words, "continue the race" (*prodolzhit rod*), and Crimean Tatars seem to have fought to keep their nation alive almost as a national mission.[20]

[18] Ibid., 95.

[19] Broshevan and Tygliiants, *Izgnanie i Vozvrashchenie*, 103.

[20] Adam Smith Albion, "Crimean Diary," *Institute of Current World Affairs* (July 20, 1995), 3.

This task was made all the more difficult by the Crimean Tatars' difficulties in adjusting to their new surroundings. The natural environment of Uzbekistan--with its blistering dry summers, droughts, and desert oasis conditions (except in the high Fergana valley)--differed markedly from that of the coastal Black Sea home of the Crimean Tatars. Uzbek medical facilities were filled during this period with Crimean Tatars who began to die off in large numbers due to their lack of immunity to local diseases, such as malaria, dysentery, dystrophy, yellow fever, and other intestinal illnesses which were not found in the Crimean peninsula where the water was purer. Women and children died in the greatest numbers. The majority of the Crimean Tatars had previously of course lived in the valleys and foothills of the peninsula's Yaila mountains or the Yaliboyu coast and were unaccustomed to the conditions they found in the arid lands of Uzbekistan.

In addition, the majority of the deportees were from the Crimean countryside; NKVD sources indicate that a mere 18,983 of the exiles were actually deported from cities in the Crimea.[21] Few Crimean Tatar farmers could acquire fields in the land-starved Uzbek oases and overpopulated Fergana valley, and most of these village peasants were forced to find work in mines or factories (the only jobs available due to the Uzbeks' loathing of such work) located for the most part in large cities such as Tashkent. One source records that during the first few years in Uzbekistan, "it was characteristic that the *spetspereselenets* from the Crimean Tatars were frequently assigned to the most trying and heaviest construction enterprises."[22]

Crimean Tatars who were settled in the Tashkent vicinity in such towns as Chircik, Angren, Gulistan, and Yangi Yul, or in the Fergana valley towns of Marghilan, Andijan, Namangan, and Fergana were forced to labor as menial workers in the many factories that had been evacuated to this region from the Nazi occupied west. In an order of May 1944, Stalin clearly directed Uzbek officials to settle the "special settlers" from the Crimea in *sovkhozes* (state farms), *kolkhozes* (collective farms) and factory settlements for "utilization" in village agriculture and industry.[23] According to one source, "The Crimean Tatars, to a considerable degree, satisfied the need for the speedy development of industry in the re-

---

[21] Broshevan and Tygliiants, *Izgnanie i Vozvrashchenie*, 44, 49.

[22] Ibid., 100.

[23] Nicolai Fedorovich Bugai, *Iosif Stalin—Lavrentiiu Berii* [Joseph Stalin to Lavrentii Beria] (Moscow, 1992), 136.

publics of Central Asia."[24] In their work on the Crimean Tatars, M. Guboglo and S. Chervonnaia write:

> In the places of "special settlement" the Crimean Tatars were subjected to a special regime, the aim of which was the destruction of the traditional modes of production, which had been forged over the centuries by systems of life security among the Crimean Tatars. Prior to the war, in the Crimea, they were primarily involved in village production and were especially famous for their skill in gardening, in wine producing, and tobacco growing. In their new regions of inhabitation they were settled in barracks, communal housing, hurriedly constructed temporary shelters, and annexes located by factories. The Crimean Tatars, regardless of their previous means of occupation, were transferred to heavy labor in various spheres of industry. The roots of national distinction were cut to the root, permanently.[25]

The cutting of the Crimean Tatars' "roots" in the soil of the Crimea was to be permanent, and few of the Crimean Tatars' traditional agricultural skills were to survive this disruption. In the post-Soviet Crimea of today the repatriated Crimean Tatars suffer from this sundering of their agrarian ties to the Crimea.

In the southern Uzbekistan region of Kashga Darya and Bukhara another form of forced labor prevailed among the Crimean Tatars. Crimean Tatar farmers who had worked for centuries maintaining the specialized mountain irrigation canals of their forefathers, were now forced to work twelve-hour days under the hot sun in Uzbekistan's "cotton *Gulag*." Moscow had turned much of the deserts of Central Asia into a vast, artificially irrigated cotton field and, with the arrival of the Crimean Tatar deportees, a class of *helots* had been provided to develop this region. Many Crimean Tatars suffered subsequent health problems from working in the pesticide coated cotton fields or as menial laborers in the unhealthy conditions of Uzbekistan's factories.

Commemorating "The Deportation" in Central Asia

More than any other event in their history, the removal of this small nation from a land it had come to define as its *natsional'naia rodina* (national homeland) under the first two decades of Soviet

---

[24] B.L. Finogeev et. al., *Krymskotatarskie Zhenshchiny: Tryd, Byt, Traditsii* [Crimean Tatar Women: Their Work, Customs and Traditions] (Simferopol, 1994), 15.

[25] M. Guboglo and S. Chervonnaia, *Krymskotatarskoe Natsional'noe Dvizhenie* [The Crimean Tatar National Movement], vol. 1 (Moscow, 1992), 76.

rule and its *atavatan* (fatherland) on the eve of the Russian revolution has shaped this people's contemporary national identity. For several generations the Crimean Tatar people worked in the factories, mines, and industrial centers of a Central Asian landscape that was in every way different from their peninsular homeland on the Black Sea, and this experience has shaped this victimized community's collective memory.

From 1944 to 1957, the Crimean Tatars worked in Central Asia's cotton *gulag* or served as a helot class working in the many factories transported to Central Asia from European Russia to put them beyond the reach of the invading Germans. In 1957 the new Soviet leader Nikita Krushchev allowed the Crimean Tatars and other deported peoples to leave their hated *Spetskommandantskii* (Special Commandant) camps, and he exculpated the deported nations on the false charges of mass treason. In addition, Khrushchev allowed several of the deported peoples from the Caucasus to return to their reconstituted homeland-republics, but three groups were omitted from Krushchev's decree. The stunned Volga Germans, Meskhetian Turks and Crimean Tatars learned that their exile was to be permanent. While the deported Chechens, Ingush, Kalmyks, Karachais and Balkars were thus allowed to return, the Crimean Tatars, Meshketian Turks, and Volga Germans were not allowed to return to their natal territories for reasons that probably had to do with the value of their former homelands.

The devastated Crimean Tatars were forbidden from returning to their republic; those who did were to be arrested. In response to this decree the scattered Crimean Tatars began to unite and mobilize their communities for a struggle to earn the right of repatriation. For the next thirty years they mounted the Soviet Union's first ethnically-based frontal challenge to Moscow's authority, demanding the right to return to their homeland. During the long exile years the Crimean Tatars began to commemorate the *Deportatsiia* on May 18, and, symbolically, they commemorated Lenin's birth date (Lenin was the founder of the Crimean ASSR and was considered much more tolerant of displays of ethno-national identity than his successor Stalin). The exiled Crimean Tatars used these commemorative events as an opportunity to demand the right to return to their homeland. Wreaths were laid at the foot of statues of Lenin; banners were carried demanding the right to return to the Crimean peninsula (which had been demoted to a regular province in the Russian Federation and Slavicized during the Crimean Tatars' absence). The MVD (Ministry of Internal Affairs), militia, and KGB often broke up these commemorative rallies, with the most spectacular attack on Crimean Tatars happening in the year 1967 in the city of Chirchik. On that occasion hundreds of Crimean Tatars were arrested, attacked by club wielding troops, or sprayed with acidic substances. This widely reported clash was in fact one of the first instances of ethnic unrest in modern Soviet history.

In addition to these outward commemorative acts, Crimean Tatar parents and grandparents kept the memory of the deportation alive in the minds of new generations who were raised on stories of this tragedy. As a whole generation grew up in Central Asia with no firsthand memories of the Crimean homeland or the deportation, the Crimean Tatar mantra became: "Nothing is forgotten, nothing will be forgotten." Rather than peacefully assimilate in their places of exile, the Crimean Tatars (even those who had never been to the Crimea) actively fought to keep their identity alive and to make sure that new generations remembered their people's tragedy.

These trans-generational transfers of grievance are in many ways similar to the narratives of the Palestinians who, like the Crimean Tatars, were expelled from their homeland in the 1940s (more than three quarters of a million Palestinians were expelled from Israel to Jordan, Egypt, the Gaza, Lebanon and the West Bank in 1948). Whole generations of Palestinians growing up in squalid refugee camps in the Middle East considered their real home to be Palestine. Unlike the Palestinians who gave up the real dream of regaining their lost lands by the 1980s, the Crimean Tatars continued to think of their people's total repatriation in a real sense. The stories of the deportation served as one of the primary vehicles for keeping this dream alive among all members of the community during the tragic exile years.

Lilia Bujorova, perhaps the most famous Crimean Tatar writer and poetess to emerge from the exile, has had her poems of the Crimean homeland published throughout the former Soviet Union. She provides the following poem entitled "Speak" (*Govori*), which captures her experience growing up in Central Asia with stories of the deportation and her lost homeland.

> Speak father speak,
> Speak until the dusk!
> Speak of the cruel war,
> Speak of the terrible day,
> In my veins let the tragedy flow,
> How salty is the sea water,
> Don't spare me, don't spare anything,
> Go again out of your native home,
> Again lose your relatives on the wagons
> Again count who remains among the living!

I want to know about everything,
So that I can tell it to your grandchildren,
Your pain cries to me,
I will bring every moment to life in them!
It will also become a homeland for them
The word "Homeland" and the word "Crimea"!
Speak father speak,
Speak father until the dusk![26]

Return to the Homeland

It was only in 1989 that a decree was published in the Soviet newspaper *Izvestiia* allowing the Crimean Tatars to return to their homeland. Since that date roughly 250,000 of the Former Soviet Union's 500,000 Crimean Tatars have returned to a homeland that most, who grew up in Central Asia, have never seen. The return migration of convoys of Crimean Tatar families from the deserts of Central Asia to the dreamed of homeland on the distant Black Sea had all the drama of the Jews' return to Israel.

While the jubilant Crimean Tatar repatriates grew up on stories of the romanticized *Yeshil Ada* (Green Island) of the Crimea during the exile years, these idealized notions of the homeland were crushed by the bitter realities of life in the post-Soviet Crimea. The Crimean Tatars' return to the Crimea in the early 1990s was strongly resisted by the local Communist *nomenklatura* (entrenched Soviet-era bureaucratic elite), which destroyed Crimean Tatar *samozakvat* (self-seized) settlements, refused to allow the Crimean Tatars to settle on their cherished southern coast (known as the *Yaliboyu*), and culturally, economically and politically marginalized the destitute returnees.

Most Crimean Tatar repatriates have thus been forced to live in what can best be described as squatter camps outside the cities of the Crimea. All Crimean cities are surrounded by distinctive Crimean Tatar settlements made up of simple rough hewn houses, with corrugated tin roofs usually lacking running water, often with no electricity. Dirt roads link them to the highway, at least until they become impassable in the winter months. The Crimean Tatar returnees, many of whom overcame the obstacles against them in Central Asia and became white collar professionals during the Soviet era, cannot find jobs in the Crimea (now a part of

---

[26] Lilia Bujurova, "Govori" *Tak eto Bylo. Natsional'nye Repressi v SSSR 1919-1952 gody* [Thus it Was. National Repression in the USSR 1919-1952], vol. 3 (Moscow, 1993), 122.

independent Ukraine). Over 80,000 Crimean Tatars were refused Ukrainian citizenship until 1999 due to bureaucratic hurdles placed in their way (one needs citizenship in order to receive a housing permit, job permit, to use hospitals and to send one's children to school). Most importantly, when land was privatized in the Crimea in 1999, the Crimean Tatars, who did not belong to collective farms from the Soviet era, were left out, and most are now landless.

Not surprisingly, the Crimean Tatar repatriates have once again begun to use commemorations of "The Deportation" as a forum for not only keeping the memory of their nation's tragedy alive in the minds of new generations, but for stating their current socio-political grievances. Every May 18th, a day known as the *Kara Gün* (Black Day) thousands of Crimean Tatars from the settlements throughout the Crimea converge on two simple monuments erected in the early 1990s in the Crimean capital, Simferopol. Those from the southern Crimea gather at a monument erected on the banks of the Crimea's main river, the Salgir (which flows through Simferopol), while those from the north gather at a monument erected opposite Simferopol's main train station.

I lived with a Crimean Tatar family in 1997 in a *samostroi* (self-built house) in the settlement of Marino just outside Simferopol, and the father of this household, Nuri Shevkiev, gave the following answer as to why he takes his family to this commemorative event every year:

> Every May 18th when I was a child growing up in Uzbekistan far from the Crimea my parents, grandparents, and aunts and uncles used to tell stories of our family losses suffered during the deportation. I know everything about those who were lost at this time, I know the name of all my father's friends killed in the deportation. Now that we have returned to the Crimea and have begun to rebuild our lives there is a danger my boy and girl will not remember our national tragedy. That they will forget those who died on the trains or in the special settlements in Uzbekistan. By taking my children to the monument on May 18th I am reminding them of the deportation and reminding them of who they are.[27]

This modern Crimean Tatar described the commemorative gatherings of May 18th as the most important annual event in the year for Crimean Tatars. Prayers are said for the *shehitler* (victims) of the deportation, and commemorative speeches are given by top Crimean Tatar political and religious leaders. These might include

---

[27] Interview with Nuri Shevkiev, Marino Crimea, Nov. 1998.

such individuals as Mustafa Dzhemilev Kirimoglu (the Crimean Tatar "Mandela" who spent seventeen years in the *gulag* for his anti-Soviet struggle to return his people their homeland) and the *Mufti* (Chief Islamic cleric) of the Crimea. At noon the two groups march from the monuments to the central square in Simferopol carrying banners, singing traditional Crimean Tatar songs, and showing their unity in the face of the militia which guards the march path. In the Central Square thousands of Crimean Tatars listen to prayers and speeches demanding more rights for their people. The fiftieth anniversary of the deportation on May 18 1994 saw a particularly large turnout as tens of thousands of Crimean Tatars converged on Simferopol celebrating their new found political assertiveness.

I visited these two memorials to the Crimean Tatars' suffering and found these stone edifices to be powerful in their simplicity. Both are about 6 feet in height with plaques mounted on them which read in Tatar and Russian "On this spot a monument will be erected to the victims of the genocide against the Crimean Tatar people." While I was visiting one of the memorials, a Crimean Tatar Red Army veteran pointed out to me that vandals had spray painted swastikas and anti-Tatar graffiti on this modest monument. The saddened veteran informed me that he had lost a brother to the Nazis, had fought for the Soviet Union, and was now called a "Nazi" by xenophobic Russians in the Crimea. Crimean Tatar cemeteries in the Crimea, often ancient sites which play an important role in Tatar Islamic celebrations and holidays, are also routinely defaced with Nazi graffiti. Long after the Soviet Union has ended and World War II has been largely forgotten by most of Europe, the Crimean Tatars of the twenty-first century continue to be burdened with the stigma of *izmeniky rodiny* (traitors to the homeland) by their detractors and those who wish to see them disenfranchized in their own homeland.

Not surprisingly, tension runs high in the largely Slavic Crimea (the Crimean Tatars now make up between 10 and 11 percent of the Crimea's population of 2.6 million), and the landless, workless, and politically voiceless Crimean Tatars, who are excluded from the Crimean parliament, have responded with mass protests. The commemoration of the deportation in 1998 turned violent as frustrated Crimean Tatars clashed with militia troops and demanded citizenship and governmental assistance to assist in the repatriation of the roughly 180,000 Crimean Tatars still languishing in exile in Central Asia (most families are divided between the Crimea and Central Asia). In 1999 the commemoration of the deportation began on April 8th and opened by marking the Russian annexation of the Crimean Khanate (the independent

Crimean Tatar state which existed from 1440 until April 8, 1783). Commemorative events continued right through to May 18th. A march which began in the eastern Crimean city of Kerch wound its way through the Crimea and called for greater rights for the Crimean Tatars. On May 18, 2000, Crimean Tatar protesters set up a tent city in the central square of the Crimean capital, Simferopol, and demanded the redistribution of land that has been given to the Russian and Ukrainian population. As their camp was surrounded by the Crimean militia the protesters chanted "nothing is forgotten, nothing will be forgotten."

The anniversary of the deportation has considerable emotional symbolism, for the generation who remember the actual event are dying off. These living memorials to this tragedy will soon disappear, and it will be left to those who grew up on the stories of the deportation to make sure that the memory of this collective tragedy is not forgotten by new generations.

Interestingly enough, the commemorations of the deportation are not limited to the Crimean Republic. During the eighteenth and nineteenth centuries, close to half a million Crimean Tatars fled from their Russian-dominated homeland to the Ottoman Empire, and today Crimean Tatar activists claim that there are five million descendants of these emigrations living in the former Ottoman provinces of Romania, Bulgaria, and, most importantly, Turkey. On past commemorations of the "Black Day," prayers have been said in Istanbul's Fetih Mehemed Cami (Mohammed the Conqueror Mosque) for those killed in the deportation. Mustafa Dzhemilev Kirimoglu, the head of the Crimean Tatars' parallel parliament known as the *Mejlis*, has also met with Turkish president Suleiman Demirel for commemorative events, and commemorative ceremonies have been held in Ankara, Eski Shehir, and smaller towns with Crimean Tatar-Turk populations.

The small Crimean Tatar enclaves found in the Dobruca [Dobruja] region of Bulgaria and Romania (the coastal strip on the Black Sea of these countries extending from the Danube to Varna) also commemorate the *Kara Gün*-Black Day. As these small diaspora enclaves become increasingly aware of their Crimean Tatar identities in the post-Communist setting, this commemoration serves as a catalyst for rediscovering and transmitting a sense of Crimean Tatarness to new generations experiencing assimilative trends (there are about 40,000 Crimean Tatars in Romania and only 5,000 left in Bulgaria). The small Crimean Tatar community of the USA, located mainly in the New York area, consists of approximately 5,000 post-World War II forced émigrés from Displacement Camps. In their commemorative ceremonies, the American Crimean Tatars hear speeches from leaders of their community, enjoy traditional Crimean Tatar cooking (such as that delightful representative of Crimean Tatar cuisine, *chiborek* pastries), and offer prayers for relatives killed in the deportation, making the effort to keep the importance of this day alive for a new generation of American Crimean Tatars immersed in American culture.

Interestingly enough, the most important monument to the deportation was actually built in Long Island, New York by Crimean Tatar architect, Fikret Yurter. This large marble edifice is located in the center of Crimean Tatar Muslim cemetery in the town of Comack and consists of a nine-foot-tall marker in the shape of the *tarak tamgha*. The *tarak tamgha*, originally the dynastic seal of the Crimean Khans of the Giray lineage descended from Chingis Khan, was adopted by early Crimean Tatar nationalists as the symbol of this people's new found national identity during the early twentieth century. It is singularly sad that the largest monument to the event that saw the Crimean Tatars scattered from their homeland by Stalin lies not in the Crimea itself but a world away on one of the many shores this diasporic people have found themselves. While visiting this monument in 1999, the author was told that most of those who were born in the Crimea have begun to die out and that generation with direct memories of the "Green Isle" of the Crimea in America must pass on memories of this homeland to those who have never seen it.

As the victimized Crimean Tatars commemorate their people's national tragedy and use it as an opportunity to gain the world's attention, it is hoped that the world will not only remember this people's long history of expulsion, ethnocide, and oppression during the Soviet period, but that they will also become aware that the "Crimean Tatar problem" has still not been solved. Half the Crimean Tatar nation is still living in the *pitmegin surgun* (the unfinished exile). Those who have returned find themselves in truly stark economic conditions in their "Zion." Both the Ukrainian authorities and local Crimean Republican authorities continue to display a shocking lack of concern (one might even call it antagonism) to the Crimean Tatars' plight in the Crimea.

The world has witnessed the spectacle of the return of a long-suffering exiled people to their traditional homeland, but the struggle for the Crimean Tatars is far from finished. As Mustafa Dzhemilev Kirimoglu, the political head of the Crimean Tatar community, told the author during a 1998 interview in the *Mejlis* (Parliament) building in Simferopol: "Our people were forced from their homes once before and for fifty years we have been discriminated against. While we are a pacifist people, even we have a breaking point. If we continue to be arrested, attacked by the Crimean mafia, and discriminated against by the local authorities in our own homeland we are not going to take it lying down anymore. The fight for true rehabilitation from Stalin's lie still goes on." [28]

As a postscript it should be mentioned that the office where this interview was held was bombed by unknown assailants in January of 1999. In addition, on several occasions the headquarters of the Crimean Tatar *Mejlis* has been raided by Crimean authorities. There have also been several mosque burnings, and Crimean Tatars have been killed in clashes with the Crimean mafia, which is known to be linked to the Crimean police. While inter-ethnic violence has not appeared on any large scale in the Crimea, one has but to look at the toppled minarets of Bosnia, the war blackened villages of Kosovo, and the bombed ruins of Grozny-Djohar to see examples of the danger to Muslim communities situated on the always uncertain fault line between the Islamic and Orthodox Christian worlds. It is to be hoped that inter-ethnic violence of the sort found in neighbouring lands divided between Christians and Muslims (such as the secessionist Georgian territory of Abkhazia, the Armenian dominated Nagorno-Karabagh enclave in Azerbaijan, Chechnya, Dagestan, Bosnia, and Kosovo) will not appear in the Crimea, and that the Crimean Tatars, who suffered so much in the twentieth century, can rebuild their culture in a democratizing Ukraine.

---

[28] Interview with Mustafa Dzhemilev Kirimoglu, Bahcesaray, Nov. 1998.

Ethnic Cleansing in Slovakia:
The Plight of the Hungarian Minority

EDWARD CHÁSZÁR

I f the term "ethnic cleansing" is equated with the elimination of
a national or ethnic minority in a given country, the Hungarian
minority in Slovakia has been subjected to systematic "ethnic
cleansing" since the end of World War II by various means. These
included forced population transfer, deportation, forced labor,
deprivation of citizenship, confiscation of property, forced assimilation, cultural oppression, and other, more subtle, means
short of the use of force. For historical background one should
keep in mind that Hungarians and Slovaks coexisted peacefully
for centuries during a stormy history of repeated invasions and
oppression by Tartars, Turks, and Austrians. Tensions developed
only in the second half of the nineteenth century, when Slovak
nationalist leaders formulated demands for the Slovak people in
the Kingdom of Hungary. Since these demands were not fully
met, another framework, namely the creation of a Czechoslovak
Republic, at first seemed to meet Slovak aspirations. At the end of
World War I not only all of Slovak-inhabited Upper Hungary, but
also areas well below the ethnic boundary between the Slovak and
Hungarian people, were annexed by the newly created Czechoslovak Republic. As a result, close to one million Hungarians
found themselves turned over to Czechoslovakia.

In 1938 the Hungarian-inhabited southern areas reverted back
to Hungary by virtue of the Vienna Arbitral Award, while in 1939,
on the ruins of Czechoslovakia, a Slovak State began its brief existence as Hitler's satellite. Czechoslovakia having been reestablished after World War II within its old borders (except Carpatho-
Ruthenia, ceded to the Soviet Union), the Slovaks again found
themselves in a framework not entirely to their liking, and many
of them continued to resist the idea of "Czechoslovakism," as opposed to a State of Czechs and Slovaks, which was more to their
liking. Eventually even the desired federal structure (established
during the "Prague Spring" of 1968 and reshaped after the "Velvet
Revolution" of 1989) failed to satisfy the aspirations of a certain
group of nationalist political leaders. Their efforts, spearheaded by
Vladimír Mečiar and Jan Slota, led to the "Velvet Divorce," (January 1, 1993) that is, the birth of a new, independent Slovak Republic. For the Hungarian minority this new burst of Slovak nationalism held little promise. Still, the difficulties faced by Hungarians
in the new Slovak Republic are dwarfed by those faced during the

years 1945-1948 in the "Second" Czechoslovak Republic, when the very existence of the Hungarian minority was at stake.

The Years of Statelessness

The architects of the second Czechoslovak Republic, chief among them Eduard Beneš in London and Klement Gottwald in Moscow, were firmly committed to the idea of creating a purely Slavic nation-state, that is, one without any non-Slavic minority, by means of the wholesale expulsion of the German and Hungarian populations. For the realization of their plan they secured the support of Stalin, then sought the same from the Western Allies. But they succeeded only partially. The Potsdam Conference in 1945 approved the expulsion of the Germans, but—due to the objection of the United States—not that of the Hungarians. For the removal of the latter the Czechoslovaks were able to attain in the end approval for a ·"population exchange," reluctantly accepted by Soviet-occupied Hungary. Under the terms of an agreement signed in February 1946, Slovaks and Czechs having permanent domicile in Hungary could voluntarily apply for transfer to Czechoslovakia. But the Czechoslovak Government was to select an equal number of Hungarians of Slovakia for transfer to Hungary regardless of their wish. (This turned out to be 74,000 in round figures. The Czechoslovak-Hungarian "population exchange" was the subject of a separate presentation at the Conference. See the paper of Robert Barta, "The Hungarian–Slovak Population Exchange," below.)

The Košice Program and the Beneš Decrees

Long before the tedious process of "population exchange" had begun, the process of getting rid of the Hungarians was already under way. The first step in this was the expulsion of 31,780 Hungarians, who took up residence in Southern Slovakia between 1938 and 1945, when this territory was under Hungarian control. The groundwork for this measure, and for others to follow, was laid down by the program of the new Czechoslovak Government, worked out by Klement Gottwald in Moscow and announced in Košice/Kassa on April 5, 1945.

The Košice Program implemented through a series of presidential decrees, so called "Beneš Decrees," contained harsh measures against the German and Hungarian minorities, declaring them collectively guilty, confiscating their property, and dismissing them from state employment. The latter entailed the dismissal of all Hungarian school teachers, as a result of which Hungarian schools ceased to function for approximately four years, when the newly created Communist government eased up some of the restrictions placed on Hungarians, restoring to them among others the right to form cultural, social, and sports associations. Exceptions to these harsh measures were granted only to those individuals who could prove that they actively fought in the defense of Czechoslovakia.

Deportations for "Public Work"

Part of the scheme for reducing the number of Hungarians in Southern Slovakia was their dispersal in the Czech-Moravian parts of the country. In November 1946 the Settlement Office for Slovakia issued a confidential order concerning the "Regrouping of Hungarians in Slovakia." According to the order all the Hungarians in twenty–three localities in Slovakia, who would not be transferred to Hungary under the "population exchange" agreement, were to be transported to the depopulated Sudeten German districts under armed escort. The order stated that the transfer was compulsory and it was to be carried out by the applications of Presidential Decree No. 88/1945 on "public labor." Subsequently the people in the designated localities were rounded up by the military and transported to their new destination in freight trains under rather inhuman conditions. The deportations undertaken between October 19, 1946, and February 26, 1947, affected 43,546 Hungarians, 5,422 of them less than six years of age. The action was also a warning to Hungary to speed up the "population exchange" process in order to avoid the possible further relocation of Hungarians within Czechoslovakia. The property of the deportees was confiscated and later distributed to Slovak colonists.

"Reslovakization"

In order to change the multinational character of Czechoslovakia to that of a homogeneous Slav state, a program called "Reslovakization" was intended to play the most important role. Acting on the assumption that much of South Slovakia's population consisted of "Magyarized" Slovaks, the program made it possible for Hungarians to apply for a change in nationality. More than half of the Hungarians, frightened and deprived of their rights, especially those in towns, in ethnically mixed villages or who were scattered, applied to call themselves Slovaks. This meant being granted citizenship and remaining in the homeland. Between June 1946 and January 1948 a total of 381,995 applications were submitted, of which 282,594 were approved by the Commission on Reslovakization. The rest were rejected on various grounds, such as lack of command of the Slovak language, or "racial deficiencies."

Consequences

Following all of the events referred to above, the ethnic composition of Southern Slovakia underwent a profound change between the censuses of 1941 (administered by Hungary at the time) and the census of 1950. The number of Hungarian native speakers is estimated to have fallen from 729,000 to 451,000. There was a marked southward movement of the Hungarian-Slovak ethnic boundary, especially in Eastern Slovakia.

To sum up: the Czechoslovak state, in spite of all the anti-Hungarian measures taken, did not achieve the goal of eliminating the Hungarian majority in the south. However, the previously uniform Hungarian character of the region was broken by Slovak resettlement and colonization, making it ethnically mixed.

Toward International Protection of the Rights of Minorities

As the tragic events of the "years of statelessness" faded, an increasing number of formerly scared and "Reslovakized" Hungarians reverted to their Hungarian ethnicity. In the 1970 census there were already 600,249 persons declaring Hungarian as their mother tongue (but only 552,006 claiming Hungarian ethnicity). After the political changes of 1948 the situation of Hungarians gradually improved. Schools were reestablished, associations reemerged, and cultural life was reborn. Still facing many adversities, Hungarians were able to preserve their identity during the communist period. In this they were helped by the emergence of human rights and the rights of minorities in the international arena.

As post-Communist Czechoslovakia and now the independent Slovak Republic have started moving toward integration with the West, joining European institutions like the Council of Europe, OECD, etc., they have acceded to a number of international agreements, and declarations protecting the rights of minorities. Moving now toward membership in the European Union, in which state borders are supposed to fade away and freedom of movement is supposed to be envisioned for all people, the postwar policy of Czechoslovakia which has sought the establishment of a homogeneous Slavic state, appears to be anachronistic.

Reconciliation and Coexistence

Winds of change are blowing at the turn of the century and the beginning of the new millennium. The Hungarian Coalition Party today forms part of the new Slovak government. The bridge between Slovakia and Hungary, blown up by the German army and in ruins for fifty-five years, is being rebuilt by the two countries jointly, with Western European help. A reappraisal of the past is taking place. Hopefully, it will lead to reconciliation and coexistence.

BIBLIOGRAPHIC ANNOTATIONS

There is a profusion of works dealing with the plight of Hungarians in Czechoslovakia (later Slovakia). Most of these are written by Hungarians who witnessed the events, described them, and supplemented their observations by documentation. In the preparation of my paper I relied mostly on the following works.

Hungarians in Czechoslovakia. New York: Research Institute for Minority Studies, 1959. Pages 11-38 of this book contain a study by Dr. Francis S. Wagner, who was Hungarian Consul General in Bratislava between 1946-1949.

Kálmán Janics. *Czechoslovak Policy and the Hungarian Minority, 1945 1948.* New York: Social Science Monographs, distributed by Columbia University Press, 1982. Janics, a Hungarian physician, spent several years in internal exile in Slovakia.

Sándor A. Kostya. *Northern Hungary.* Toronto: Associated Hungarian Teachers, 1992. A native of Košice/Kassa, Kostya left Czechoslovakia for Hungary, eventually immigrating to Canada as a teacher.

Károly Kocsis and László Szarka. *Changing Ethnic and Politic[al] Patterns on the Present Territory of Slovakia.* Budapest : G.R.I., 1999. The authors are population geographers in the Geographical Research Institute and Minority Studies Programme of the Hungarian Academy of Sciences.

Miklós Duray. *Önrendelkezési kiserleteink* [Our Attempts for Self-Determination]. Somorja/Samorin (Slovakia): Mery Ratio, 1999. The author is a leading Hungarian politician in Slovakia.

Károly Vigh. *A szlovákiai magyarság kálváriája, 1945-1948* [The Calvary of the Hungarians of Slovakia, 1945-1948]. Budapest: Püski, 1998. Vigh is a well-known Hungarian historian, originally from Slovakia.

The Hungarian-Slovak Population Exchange and Forced Resettlement in 1947

RÓBERT BARTA

Czechoslovak-Hungarian relations following the Second World War became very problematic. This was true both because these two countries had entered the war as belligerents on the opposite side, and also because the issue of the Hungarian minorities living in Slovakia[1] complicated the official diplomatic relations between the two countries. This paper will deal with the antecedents, the process, and the consequences of the forced Slovak-Hungarian population exchange in 1947. It is my contention that the Slovak-Hungarian population exchange of 1947 satisfied neither the population nor the political leaders of these two countries. Moreover, it did not solve the national minority question, nor any of the related problems.

In the given historical circumstances, the formally co-equal negotiating partners were really unequal. As a member of the victorious Western alliance, the Czechoslovak government was in a much stronger position On April 5, 1945, only two days after it had been established, it launched a program—the so-called Košice (or Kassa) Program—whose goal was to transform Czechoslovakia into a "pure Slavic state."[2] According to the regulations of Section #5 of this Program, counties and districts with "unreliable" Hungarian and German majority populations were deprived of the right to send elected delegates to the national legislature. They had to be represented by governmentally appointed state commissioners. Czechoslovak citizenship was granted only to those Hungarians and Germans who were able to prove that they had been militarily involved in the fight against Fascism, and that as such they had defended the interests of the Czechoslovak Republic. Persons who opposed the alleged "policy of Magyarization" under Hungarian rule between 1938 and 1945, and who at the same time had suffered persecution for their pro-Czechoslovak sentiments, were also exempt from forced relocation. Those who failed to demonstrate such an anti-

---

[1] On the Hungarian minority see Kálmán Janics, *Czechoslovak Policy and the Hungarian Minority 1945-48*, ed. Stephen Borsody, *Atlantic Studies on Society in Change* vol. 9, no. 18 (New York, 1982); originally published as *A hontalanság évei: A szlovákiai magyar kisebbség a második világháború után 1945-48* (Budapest, 1989).

[2] Vadkerty Katalin, *A reszlovakizáció* [Reslovakization] (Pozsony, 1993), 14.

Hungarian and anti-German stance were deprived of their citizenship and were expelled from the republic.

The agricultural lands and related properties of most non-Slovaks were immediately nationalized and made into state property. Between 1945 and 1948 roughly 236,891 acres were confiscated under this so-called "land reform" in the territory of former Upper Hungary. Inhabited mainly by Hungarian and German smallholders, this strip of territory had been lost after World War I and then returned to Hungary by the First Vienna Arbitral Award in November 1938. The anti-minority goals of this land reform were obvious.[3] Section #15 of the Košice Program ordered the immediate closing of all schools whose language of instruction was either Hungarian or German. This was done to strengthen the Slavic character of Czechoslovakia. With Soviet support, in January 1946 the Czechoslovak Ministry of Interior proposed that the Czechoslovak delegation to the Paris Peace Conference should argue for a policy that would permit the stationing of Allied Forces in Hungary to facilitate this mass population transfer.[4] But the Hungarians pointed out that as a loser, Hungary would be unable to accept and integrate into Hungarian society hundreds of thousands of penniless deportees. Yet within two months after this date, over 12,000 Hungarians were deported from their native land on the southern fringes of Slovakia, and then driven across the Danube river to Hungary. And this was only the start of the mass deportations.

The policy of mass expulsion had its roots in the decree issued by President Edward Beneš on August 2, 1945 (33:1945), which deprived Czechoslovakia's Hungarian and German populations of all of their civil and political rights, while at the same time also preparing the ground for their *en masse* expulsion as well as for the expropriation of all their property. The essence of this and other so-called Beneš Decrees was accepted and confirmed by the May 14, 1946, resolution of the Slovak National Council (65:1946), according to which all the landed and agricultural properties of the Hungarians had to be confiscated. By that time state employees of Hungarian nationality had already been dismissed from their places of employment.[5]

---

[3] Ibid., 11-14.

[4] Ibid., 15.

[5] On Beneš's activities and his presidential decrees see János Kövesdi, ed., *Edvard Beneš elnöki dekrétumai, avagy a magyarok és a németek jogfosztása* [Edward Beneš's Presidential Decrees, or the Disenfranchisement of the Hungarians and the Ger-

One of the most tragic chapters in the history of Hungarians whose families had been living in that region for a whole millennium was their deportation and forced expulsion, which allegedly—but not in reality—was based on a parity exchange of population. President Beneš declared that Czechs and Slovaks were no longer willing to share their state with Hungarians and Germans. Later Slovak pronouncements also supported this stance. The Czechoslovak government and most Slovak political leaders considered the unilateral deportation of the Hungarians in Slovakia (*všeobecný odsun mad'arov*) as the only acceptable solution, and they were seeking international backing for this undertaking. As one of the victors of the war, Prague could count on the sympathy of the international public opinion, as well as on the effective assistance of the Allied states.

The Hungarian position was much weaker. Because Hungary was a defeated state, it found little sympathy among the victors. All Hungary could do was hope that the Czechoslovak government would follow a policy of toleration based on the views of the highly regarded founding president, Thomas G. Masaryk. Budapest also hoped that the democratization of postwar Hungary would create a favorable atmosphere for the reconciliation of ethnic conflicts, and that it would open the door to a humanitarian solution of the national minority problems. That was not to be, for the Czechs and the Slovaks opted for the mass expulsion of the Hungarians.

Soon after the start of the negotiations it became evident that the conflicting opinions on ethnic questions hindered the friendly settlement of the dispute. This is clearly evident from President Beneš's speech delivered on 9 May 1945, in which he outlined the guidelines of the official governmental policy toward German and Hungarian minorities. According to Beneš, Czechoslovakia had no choice but to get rid of all non-Slavic peoples living on its territory. This was particularly true for the Germans and Hungarians, which nationalities, in his opinion, played a crucial role in the disintegration of Czechoslovakia in 1938-39. Beneš also emphasized that the Slavic character of the country had to be reinforced, and for this reason, all non-Slavs had to be resettled to their "home countries."

This policy had already been formulated by Beneš in London during the war when he headed the Czechoslovak government-in-exile in London. For this reason, the intended population

---

mans] (Pozsony, 1996); published in German as *Edward Benesch Präsidentendekrete oder die Rechtsberaubung der Ungarn und Deutschen* (Pressburg, 1993).

exchange between Hungary and Czechoslovakia was mentioned ever more frequently in official as well as semi-official reports coming out of Prague. In this deteriorating situation, while preparing for the official Peace Conference, both governments were working desperately to gain international support for their respective views. Thus, already at the Potsdam Summit in July-August 1945, Czechoslovakia had tried to secure the approval of the great powers for the expulsion of the Hungarian population. The conference, however, gave its consent only to the resettlement of the Germans from Czechoslovakia and from other German-inhabited eastern territories.[6]

The Hungarian government informed the great powers repeatedly about the deteriorating situation in Czechoslovakia and the ever more frequent incidents of illegal expulsions. Although unable to do anything about these illegal deportations, Budapest repeatedly rejected the Czech plan for population exchange. The Hungarian argument was that there were many fewer Slovaks in Hungary than Hungarians in Slovakia, and that therefore a parity exchange was really impossible beyond the small number of Slovaks who wished to leave Hungary. This argument was borne out by official population statistics. According to the 1930 census the number of Slovaks in Hungary was 104,819, while according the 1941 census, their number had been reduced to 75,877. In contrast, the Hungarians living in Czechoslovakia in 1930 numbered 592,337, while by 1941 their number rose to 761,434[7] (see Tables 1 and 2 below).

But above and beyond this, the Slovaks of Hungary were strongly split over the question of resettlement. There were more who wanted to stay than those who wanted to leave. The strongest desire to move was among the Slovak people living in the Great Hungarian Plain. They were motivated by problems of livelihood and probably also by the memories of some of the indignities that they had suffered in the period of Hungary's dismemberment following World War I. According to contemporary estimates only 14,000 wanted to leave. This, however, was

---

[6] On the Potsdam resolutions concerning the Germans, see *The UN Yearbook 1946*, 46; and *A Decade of American Foreign Policy: Basic Documents 1941-49* (Washington, 1950), 34; see also *Foreign Relations of the United States—The Postdam Conference 1945*, 2 vols. (Washington, D.C., 1960).

[7] Based on census data. Also see Szabó A. Ferenc, "Demográfiai érvek és tények a második világháború utáni magyar-szlovák lakosságcserével kapcsolatban," [Demographic Facts and Arguments Concerning the Hungarian-Slovak Population Exchange after the Second World War], *Múltunk* [Our Past] 42 (1997): 143-45.

significantly below the number of those Hungarians whom the Slovaks wished to expel from Czechoslovakia.

With tension increasing every day, serious bilateral negotiations were indispensable. In October 1945 Prague put forward a semi-official initiative, which Hungary accepted. Serious negotiations began in Prague in December 1945, which were carried out in two phases. The Hungarian delegation was led by János Gyöngyösi,[8] the Hungarian Foreign Minister, while the Czechoslovak delegation was led by Vladimir Clementis,[9] the Deputy Minister of Foreign Affairs. The first meeting on December 3rd through 6th ended without much success, because the Hungarian delegation would only agree to voluntary population exchange, rejecting the idea of forced resettlement. The Czechoslovak delegation would not even give assurance that the civil and minority rights of the Hungarians staying in the Czechoslovak territories would be guaranteed.

The second round of the talks was held between February 6th and 10th, again in Prague. The negotiations resulted in a compromise, and the resulting agreement was finally signed by both delegations.[10] The essence of the compromise was that although Budapest did not agree to the resettlement of all Hungarians from Slovakia to Hungary, the Hungarians permitted the Czechoslovak Resettlement Committee to make propaganda in favor of resettlement in those territories. Prague agreed that it would expel only as many Hungarians as the number of Slovaks who volunteered to resettle in Czechoslovakia. In the protocol attached to the agreement, both partners stated that the exchange of population was but a temporary solution, as there were seven times as many Hungarians living in Czechoslovakia as Slovaks in

---

[8] János Gyöngyösi (1893-1951), was a leading politician in the ruling Smallholder's Party, originally a teacher of Latin and History, then bookseller and journalist. Between December 1944 and May 1947 he was Hungary's Minister for Foreign Affairs, and in 1946-47 he was also the leader of the Hungarian delegation to the Paris Peace Conference.

[9] Vladimir Clementis (1902-1952) was a lawyer, publicist, and Communist politician. Before 1938 he was the attorney of the Hungarian-Slovak Communists. During the war he lived in London, where he wrote several articles on Hungarian-Slovak coexistence, the best known of which is *The Czechoslovak Magyar Relationship* (London, 1943). Between 1945 and 1948 he was the Under-Secretary of State, then the Secretary of State between 1948 and 1950. He was captured and arrested in 1951, and convicted in a show trial and sentenced to death.

[10] The text of the accord can be read in Ferenc Bacsó, ed., *Két év hatályos jogszabályai 1945-46* [The Operative Laws for Two Years 1945-46] (Budapest, 1947), 80-85.

Hungary.[11] Section 5 of the agreement designated 73,187 Hungarians for resettlement on the basis of the parity list. An additional 106,398 persons as "major war criminals" were added to this. In addition, 1,927 so-called "minor war criminals" were included as well. Thus the number of those slated for deportation reached a total of 181,512 persons. This meant that almost one-third (31.9 percent) of the Hungarians living in Slovakia were chosen to be expelled from thirty different districts.

Hoping for a more comprehensive solution, the negotiating partners agreed to continue their discussions at a later date. They also agreed that if their talks should prove to be successful, they would seek assistance from the Peace Conference to carry out the terms of the agreement. The discussions, however, were never renewed. Without consulting its Hungarian counterpart, the Czechoslovak government simply turned to the Peace Conference and declared that an additional 200,000 Hungarians would also be subject to resettlement. The Peace Conference rejected these clearly illegal Czechoslovak plans, yet it could not prevent the forcible expulsion of tens of thousands of Hungarians in direct violation of prior agreements. These mass expulsions began in the spring of 1947 and continued for several months.

The first trainload of Slovaks left for Czechoslovakia on April 11, 1947, exactly at the time when the first trainload of Hungarians left for Hungary. The Czechoslovak and Hungarian resettlement agencies began this population exchange on the basis of a so-called "Gemini-list," according to which Hungarian and Slovak families with properties of approximately equal size and value were paired up. This pairing up process, however, could only be applied to a small percentage of the expellees, for in reality the Slovaks in Hungary owned much less property than Hungarians in Slovakia. Moreover, well-to-do Slovaks generally declined to leave Hungary for an uncertain future in Czechoslovakia.

In Slovakia, it was the affluent Hungarian families who were the first to be selected for resettlement. The Czechoslovak government wanted them off their properties so as to prepare the ground for the region's Slovakization. The already mentioned land reform played a crucial role in this effort. What became Southern Slovakia after 1920, and then also after 1945 was really 4,600

---

[11] The protocol of these discussions is housed in Magyar Országos Levéltár [Hungarian National Archives] (MOL) Budapest (XIX-J-1-a), entry IV-66, box 43. The protocols are also published by György Lázár, "Csehszlovák magyar tárgyalások a lakosságcseréről 1945 decemberében" [Czechoslovak-Hungarian Negotiations about Population Exchanges in December 1945] *Múltunk* [Our Past] 43 (1998): 120-165.

square miles (12,000 square kilometers) of Hungarian-inhabited territory, 86.5 percent of whose one million inhabitants were Hungarians. It was precisely for this reason that the First Vienna Award of November 2, 1938, had opted to return this territory to Hungary after nearly two decades within the newly created Czechoslovakia.

Almost all the agricultural lands (126,000 acres) formerly owned by Hungarian and German farmers were now nationalized and parceled out to Slovaks--whether local peasants or settlers from the highlands (111,000 acres). Hungarian landowners were accused of having been fascists and of having worked for the destruction of Czechoslovakia before the war.[12] This accusation was particularly unfair in light of the fact that Msgr. Joseph Tiso's "Independent Slovakia" had been one of Hitler's most subservient allies and satellites.[13] Moreover, many of the Hungarians of Upper Hungary (now Southern Slovakia) had in fact taken an active role in the anti-fascist Slovak national uprising of 1944.

The expelled Hungarians left their homeland in what was now Slovakia again, without having the faintest idea where they were going. Usually they had only a few hours to pack their personal belongings, in many cases they did not even even know the exact date and time of departure. Their fate was in the hands of Local Resettlement Agencies, which were supervised by the Resettlement Agency in the capital city of Bratislava (formerly Pozsony).

During this period, the Hungarians in Czechoslovakia, who collectively were declared war criminals, lived in fear and uncertainty. They tried to adjust to the inevitability of resettlement, but it was still beyond their comprehension. People were in despair, especially in villages, as they could not fathom why they had to leave their ancestral homes and move to another place.

Some expellees later wrote memoirs about their experiences, in which they detailed their trials and tribulations. Others incorporated these experiences into literary works, while still others wrote sociological studies about the frightful experiences of

---

[12] On the history of Czechoslovakia between 1918 and 1948 see Victor S. Mamatey and Radomir Luza, eds., *A History of the Czechoslovak Republic, 1918-1948* (Princeton, 1973).

[13] See Yeshayahu A. Jelinek, *The Parish Republic: Hlinka's Slovak People's Party, 1939-1945* (New York, 1976). For a view more sympathetic to Tiso and independent Slovakia, see Anthony X. Sutherland, *Dr. Jósef Tiso and Modern Slovakia* (Cleveland, Ohio, 1978).

the expelled Hungarians. Most of these books, however, were in the category of forbidden literature and were not permitted to be read before the collapse of Communism in 1989. In fact, there is still no comprehensive scholarly study of the expulsion of the Hungarians from Slovakia. A rare exception—particularly on the Slovakian side—is Stefan Sutaj's recent study of this process of Reslovakization, put out in 150 copies as an internal publication of the Institute of Sociology of the Slovak National Academy in Košice in 1991.[14]

In Hungary, the Hungarian Governmental Resettlement Committee controlled the process of population exchange. This Committee was responsible both for taking account of and registering the Slovaks of Hungary, as well as for representing the interests of the Hungarians in Slovakia who were selected for resettlement. Sections 5 and 8 of the Czechoslovak-Hungarian Agreement on Population Exchange selected 181,512 Hungarians for resettlement in Hungary.

The actual resettlement was a slow and painful process. As Hungarian organizations in charge of resettling Hungarian families did not have enough trucks, vans, and other vehicles to carry the expelled families and their meager belongings, it often took several days for the expellees to make the trip from their villages to the railroad stations. Following their arrival in Hungary, they faced problems of assimilation into a "new world." This was a painful and often tragic process—especially for the members of the older generation to whom "home" could mean nothing less than their ancestral village. Initially they cherished the hope that this was only a temporary situation and that within a reasonable time they would be able to return to their homes. That hope, however, never materialized.

The Slovaks of Hungary suffered from similar problems. The only difference was that in contrast to the Hungarians of Slovakia, they were not compelled to relocate unless they wished to do so. Those who chose to resettle in Slovakia did so voluntarily. The Hungarians of Slovakia, however, had no choice. They were forced to resettle in Hungary—whether they wanted to or not.

According to official data—which does not reflect the whole truth, because of the many illegal forcible expulsions by Slovak authorities—by 10 April 1948, 73,273 Slovaks had left Hungary, and 68,407 Hungarians had resettled from Slovakia to Hungary.

---

[14] Stefan Sutaj, *Reslovkizacia: Znema narodnosti časti obyvatelstva Slovenska po II. svetovej vojne* [Reslovakization: The Ethnic Transformation of a Section of Slovakia after the Second World War] (Košice, 1991).

At that time the representatives of the two governments declared the population exchange officially over. This declaration was later formalized in the Csorba Lake Agreement in 1949.[15]

The impact of the Beneš Decrees, and the resulting policy of national discrimination, followed by forced expulsions, inflicted tremendous pain and suffering upon the Hungarian population of Czechoslovakia, and to some degree even upon the Slovaks of Hungary. Tens of thousands were forcibly uprooted, manhandled, and compelled to leave their homes permanently. The Czechoslovak Resettlement Agency even employed members of the Slovak militia to make its will known. It also pressured unwilling Slovaks, who did not wish to leave Hungary, to agree to resettlement in Czechoslovakia. The forced disintegration of ancient village communities, extended families, family relationships, and age-old friendships was a major emotional shock to the deportees, which caused lasting psychological scars, and in many cases led to personal tragedies. And this was true not only for the persecuted Hungarians of Czechoslovakia, but also for the Slovaks who had allegedly left Hungary voluntarily.[16] All these factors illustrate the extent to which any "voluntary" elements in this exchange of populations was purely sham. The whole episode must be regarded as ethnic cleansing, pure and simple.

Table 1. The Change of Numbers of Slovak Population between 1910 and 1949

Year	Hungary		
	Total Population	Mother Tongue (Slovak)	
		Number	Percentage of total population.—%
1910	7,612,114	165,317	2.2
1930	8,685,109	104,786	1.2
1941	9,316,074	75,877	0.8
1949	9,204,799	25,988	0.3

Source: Based on the census of Hungary from 1910 until 1949.

---

[15] Lázár, "Csehszlovák magyar tárgyalások," 122.

[16] Janics, Czechoslovak Policy and the Hungarian Minority, 191-205; Janics, A hontalanság évei, 277-306; on immediate postwar Hungarian-Slovak relations, see Lázár, "Csehszlovák magyar tárgyalások," 123.

Table 2. The National Composition of Czechoslovakia and Slovakia

Year	Czechoslovakia				Slovakia			
	Total population	Slovak	Czech	Hungarian	Total population	Slovak	Czech	Hungarian
		Number Percentage of the total pop. %				Number Percentage of the total pop. %		
1921	13,374,364	1,967,870 14.7	6,831,120 51.1	745,431 5.6	3,000,870	1,952,866 65.1	72,137 2.4	650,597 21.7
1930	14,479,565	2,282,277 17.1	7,406,493 55.4	367,923 5.2	3,324,111	2,250,616 67.7	121,696 3.7	592,337 17.6
1950	12,388,450	3,240,549 26.3	8,383,923 67.9	367,733 3.0	3,442,317	2,982,524 86.6	40,365 1.2	354,532 10.3
1960	13,745,577	3,836,213 26.3	9,069,222 67.9	533,934 3.9	4,174,046	3,560,216 85.3	45,721 1.1	518,782 12.4

Source: Based on the census of Czechoslovakia from 1921 until 1960.

The Fate of Hungarians in Yugoslavia: Genocide, Ethnocide, or Ethnic Cleansing?

ANDREW LUDÁNYI

The collapse of Yugoslavia witnessed a no-holds barred struggle for the territory of this former state. The world's attention has been focused on the major competitors for turf: the Serbs, Croats, Slovenians, and Bosnians. Consequently, the fate of Albanians, Hungarians, and other non-Slavic peoples had been either ignored or mentioned only as an aside to the overall conflict. The Albanians became newsworthy only after the Serbs began their massive expulsion from Kosovo in 1998-99.

The fate of Hungarians in Vojvodina, has been a "blind spot" in American media treatment. The reason for this is the relatively less visible process in Vojvodina, with less dramatic confrontations and less bloodshed. Yet, Vojvodina has been the scene of "ethnic cleansing" just the same, not just within the context of the present Yugoslav crisis, but on a number of occasions in its past history. This "ethnic cleansing" has completely transformed the ethnic/nationality profile of the region during the past 80 years. What we are witnessing at the present time may be the last chapter in the Serbianization of this region.

By ethnic cleansing I mean the elimination of one or more ethnic/nationality community by the dominant ethnic/nationality community, to assure its own population of unchallenged domination/control over territory previously shared with the "cleansed" community or communities. This term has been coined to describe the Serb-inspired policies which have been used to expel or chase away the population of Bosnian Muslims or Catholic Croats from lands coveted by Orthodox Serbs in the lands of the former Yugoslavia. This "cleansing" follows a set scenario with armed Serbs entering a non-Serb settlement and intimidating the population with calculated acts of violence—threats, beatings, rapes, and killings. These acts are preceded and followed by warnings of dire consequences if the non-Serbs refuse to leave within a specified time period (usually within a period of one or two hours up to a window of twenty-four to forty-eight hours).

In Bosnia-Herzegovina, Kosovo, Eastern Slavonia, and the Baranja (Baranya) triangle, this policy has led to a massive expulsion of Bosnian Muslims, Kosovar Albanians, Croats, and in the last two cases even Hungarians. In Vojvodina, which has *not* been a theatre of war, the "cleansing" has taken in a more subtle form un-

until recently. The threats and acts of intimidation have taken place, but not on the massive scale with which they were visited upon the above mentioned regions. Still, it has led to the continuation of a process, which is the objective of "ethnic cleansing", *i.e.*, the reduction or elimination of the non-Serb population and its replacement with a Serbian majority.

Some Key Concepts

At this point, I would like to distinguish between the concepts of ethnic cleansing, genocide, and ethnocide. Genocide, or attempted genocide, is the effort to kill an entire people by destroying its "gene pool." The classic examples of attempted genocide were the policies carried out against the Jews by Nazi Germany during World War II and by the Ottoman Turkish Empire against the Armenians preceding and during World War I. The objective here was to destroy an entire people: men, women, and children. No one is spared when people are driven into a desert on a forced march or when people are herded into a gas chamber for mass extermination.

As opposed to genocide, "ethnic cleansing" is a policy motivated by the removal and not necessarily the extermination, of a people or ethnic community. Ethnic cleansing occurs when an effort is made to dislodge a people from a coveted living space, whether it is called Kosovo, certain parts of Bosnia-Herzegovina, Vojvodina, or Southern Slovakia. In the context of what we have witnessed recently in the former lands of Yugoslavia, when ethnic cleansing is undertaken, its main objective is the eviction of a people from their traditional settlements. This eviction depends on brutal coercive tactics, intimidation, and threats of extermination. If the unwanted population can be scared into leaving then it is not necessary to kill them *all*. Thus, "ethnic cleansing" separates men of fighting age from the rest of the population and massacres them to convince the remainder to leave. Women are raped to instill fear, to humiliate, and again to convince a people that if they do not leave they too will be raped.

Ethnocide, unlike genocide, does not physically eliminate or abuse an ethnic community. It achieves its objectives by depriving a community of its culture, language, and distinct identity. Ethnocide is achieved through an aggressive policy of assimilation. Ethnocide, however, requires more time to succeed and probably generational change. It depends on a systematic process of depriving the minority's children of educational and cultural opportunities that help preserve their ethnic identity. Examples are the French attempts to eliminate Basque or Corsican distinctiveness via administrative centralization and unitary educational programs in a "French only" context.

The Land

Vojvodina was a part of historic Hungary until the end of World War I. However, the Serbian alliance with the Entente transformed this region into the northeastern corner of the newly created Kingdom of Serbs, Croats, and Slovenes in 1918, which was renamed Yugoslavia in 1929.

Geographically Vojvodina is an extension of the great Hungarian *puszta* (lowlands) which lies at the center of the Carpathian Basin. More precisely, it is the southernmost extension of these lowlands. Because it is part of this great plain, its history has usually been closely intertwined with the population and culture of the Carpathian Basin as a whole. Unlike Transylvania, it does not possess mountain barriers for frontiers. Both toward Romania and Hungary it is an open plain. It possesses natural frontiers only in the south and the west, there the Danube performs this role.[1]

In every respect, Vojvodina is characterized more by accessibility than by isolation or seclusion. It is "a region of wide valley basins, alluvial plains, sandy dune areas and crystalline hills covered with fertile loess."[2] The entire region is dominated by the large rivers which are the most conspicuous features of the landscape. These rivers divide Vojvodina into its three component parts.[3] Farthest to the west, the Darda triangle (Baranja or Baranya) is wedged in between the Drave and the Danube Rivers. In the center, Bačka (Bácska) has the Danube as its western and southern boundary and the Tisa (Tisza) River as its eastern boundary. In the east, across the Tisza, is the Banat (Bánság) region.

Vojvodina is predominantly an agricultural area. In Banat a great deal of land has been reclaimed through reforestation during

---

[1] For a good geographic description see *Jugoslavia: Physical Geography* (B.R. 493, Geographical Handbook Series, Great Britain: Naval Intelligence Division, 1944), 1: 35-42.

[2] George W. Hoffman and Fred Warner Neal, *Yugoslavia and the New Communism* (New York, 1962), 16.

[3] Although the administrative area of Vojvodina until 1988 included Srem and excluded Baranja, interwar Vojvodina included Baranja but did not include Srem. See maps A and B for the regional changes between the pre-and post-World War II boundaries of the region.

the past 200 years. In Bačka a great deal has been reclaimed by draining the marshes and by building canals. As a whole, the area possesses only limited resources for industrial growth. Consequently, most of the industry that exists is geared to the processing of agricultural goods. This includes mills, distilleries, and processing plants, which are concerned primarily with canning, sugar refining, alcohol making, and flour milling. Maize and wheat are the principle cereals of the area, but sunflower and beets are also important crops. Animal husbandry and fishing is also widespread. In short, Vojvodina became Yugoslavia's chief food-producing region, just as it had been Hungary's prior to World War I.[4]

The People

Ethnically Vojvodina has been, and is, Yugoslavia's most diverse region. Table 1 indicates the ethnic composition of Vojvodina. The diversity presented by this table does not give a complete picture of the entangled nature of the area's ethnic settlement. As C. A. Macartney noted in 1937:

> No words can, unfortunately, do justice to the distribution of the population. The Rumanians are mostly to be found in the east, the Magyars are strongest in the north, the Serbs in the south; but the three intermingle hopelessly, a wedge of Serbian settlements pushing in one place far northward, while Magyar advanced posts run to its right and left well to the south, and outlying Magyar islets are found, even in the countryside, in the extreme south, as well as in all the towns. The Sokac and Bunyevac ř[Bunjevci] settlements are near the northern frontier, islands in a non-Slavonic sea, the Slovaks and Ruthenes are rather farther south. The Germans are everywhere. The distribution can be appreciated, if at all, only from the map, and the reason for it can be learnt only from history.[5]

Stefan Possony has pointed out that the demographic make-up of East-Central Europe reflects the political rise and decline of certain nationalities. This is especially true for the present northern parts of Yugoslavia, where vast demographic changes have signaled the

---

[4] C. A. Macartney, *Hungary and Her Successors* (London, 1937), 380-81.

[5] Ibid., 381. This quotation applies to the region's interwar profile. Although the Germans have been "cleansed," for the remaining peoples of the region the description of the overall distribution is still accurate.

rise and fall of peoples. It can be said that Vojvodina's present ethnic composition is the consequence of the dominant role played there by Hungarians (Magyars), Turks, Serbs, and Austrians during the past three hundred years. Its composition and demographic profile, however, is presently in the hands of Serbia (see appendix maps A, B, & C).

TABLE 1
THE POPULATION OF VOJVODINA
ACCORDING TO NATIONALITY (IN THOUSANDS)[a]

Nationality[b]	1910***	1921***	1931***	1948	1953	1961	1971	1981	1991
				.... The Vojvodina[c]					
Serbs	382		462	841	874	1,018	1,107	1,107	1,143
Croats	7	502	16	134	128	145	138	119	98
Bunjevci & Sokci	63		68	---	---	---	---	---	---
Slovenes	---	7	8	7	6	6	---	---	---
Macedonians	---	---	---	9	12	15	---	---	---
Montenegrins	---	---	---	31	31	35	36	43	44
Muslims	---	---	---	1	---	2	---	---	---
Unspecified Yugoslavs	---[d]	---	---	---	11	3	47	167	174
Germans	301	317	317	32	35	---	7	3	3
Albanians	---	---	---	0	1	2	---	---	---
Hungarians	422	376	386	429	435	443	424	385	339
Turks	---	---	---	0	0	1	---	---	---
Slovaks	58[d]	65[d]	71[d]	72	73	74	73	69	63
Italians	---	---	---	0	0	0	---	---	---
Rumanians	76	70	72	59	57	57	52	47	38
Bulgarians	---	---	---	4	4	4	---	---	---
Czechs	---[d]	---	---	4	3	3	---	---	---
Others	12	9	16	40	41	48	84	91	107
Total	1,320	1,347	1,416	1,663	1,713	1,855	1,952	2,034	2,013

[a]This Table has been compiled on the basis of data obtained from *Jugoslavia: History, Peoples and Administration,* 76; Schieder (ed.), *Das Schicksal Der Deutschen in Jugoslawien in Dokumentation Der Vertreibung Der Deutschen Aus Ost-Mitteleuropa* Band V, 11E; Hoffman, *Yugoslavia and the New Communism,* Table 3-1, 29, Elemer Homonnay, "A délmagyarországi területek nemzetiségi megoszlása az 1948-as jugoszláv népszámlálás adatai szerint" [The Distribution of Nationalities in the Region of (Former) Southern Hungary on the Basis of the 1948 Yugoslav Census], *Lármafa* [Alarm Tree] X, no. 3 (1963), 19-40; The Mid-European Research Institute,ed., "Statistical Studies on the Last Hundred Years in Central Europe: 1867-1967' [unpublished manuscript]; *Jugoslavia 1945-1964: Statistički Pregled,* Table 3-13, 45; and Kocsis, "Adalékok a magyarság etnikai földrajzához a mai vajdaság területén," in *Tér-gazdaság-társadalom: Huszonkét tanulmány Berényi Istvánnak,* ed. Dövenyi, 348 (see fn. 9 below).

[b]In this Table "nationality"means either the declared nationality or the mother tongue of the respondent. The two have not been separated, since some of the censuses were based solely on declared nationality, while others have been based solely on mother tongue.

[c]The 1910, 1921, 1931 statistics of Vojvodina pertain to Bačka, Banat, and Baranja, whereas the postwar statistics of 1948, 1953, 1961, 1971, 1981, and 1991 pertain to Bačka, Banat, and the Srem. This different territorial basis of the pre- and postwar statistics, accounts, in part, for the doubling of the Serbian population.

[d]In the 1910, 1921, and 1931 statistics for the Vojvodina, Czechs and Slovaks were enumerated together. All "unspecified" Slavs were enumerated together with the Serbs, Croats or the Slovaks. Under "Other," the present Table includes mainly Gypsies.

[e]Since Yugoslavia is only a post-World War I creation, it has not been possible to ascertain its ethnic composition for 1910. The data for 1921 and 1931 refer to its interwar area, while the 1948, 1953, 1961, 1971, 1981, and 1991 statistics refer to the enlarged area of Yugoslavia until 1991.

[f]In the 1921 and 1931 censuses for Yugoslavia,Rumanians and Vlachs were enumerated together. Post-World War II censuses have enumerated them separately.

Until the Battle of Mohács in 1526, the area's population was predominantly Magyar. Even Belgrade (called Nándorféhervár by the Magyars) was for a long time a Hungarian fortress. But the Turkish victory over the Hungarians at Mohács led to a drastic ethnic change in what was then southern Hungary. Turkish depredations depopulated and devastated the area. Only after the ascendancy of Habsburg Austria did the area regain some of its population. However, the ethnic make-up of this new population was no longer predominantly Hungarian. It had become mainly Serbian and German.

Habsburg policy for this area was motivated by considerations of defense and consolidation of the Empire. It involved a recolonization scheme that would provide an effective defense against the Turks, while at the same time strengthening Austrian hegemony within the empire. Habsburg policy favored Serbian and German colonists rather than Hungarians due, it seems, to the historical memory which the Hungarians were likely to maintain and hence the seeds of challenge to Austrian hegemony.[6] Consequently, the population of the present-day Vojvodina became a patchwork of different nationalities. However, by the end of the 18th century the Hungarians again began to repopulate the area. They filled up especially those areas which had recently been reclaimed through the drainage of swamps. Thus, when the Treaty of Trianon (1920) dismembered historic Hungary, Vojvodina reflected a rough parity within the population—of South Slavs, Germans, and Hungarians.[7]

Ethnic Cleansing

The hegemony of the Serbs within the new South Slav state led to immediate efforts to weaken the non-Slav nationalities. The "ethnic cleansing" of Vojvodina by the Serbs has been carried out in four major phases. The first phase was achieved in the interwar years, mainly through a discriminatory land reform which rewarded Serbs and victimized non-Serbs. This was mainly put into effect right after World War I in the years 1918 to 1923. The second phase was instituted almost simultaneously, with discriminatory social, economic, cultural, and educational policies which made second rate citizens out of non-Serbs and encouraged their "vol-

---

[6] Ibid., 384, points out that for a time the Magyars (Hungarians) were even officially banned from settling in Vojvodina.

[7] See Table 1.

untary" departure from the region. This was followed by the third phase which liquidated the Swabian-German population and part of the Hungarian population at the end of World War II, mainly in the period between 1944 and 1948. Finally the last phase is underway at the present time as a consequence of the inter-ethnic strife between Serbs and Croats, Serbs, and Bosnian Muslims, and Serbs and Kosovar Albanians. This conflict has pushed Serbian refugees into Vojvodina at an unprecedented rate from 1991 to 1993 and since August, 1995. A large number of them are being settled near or in Hungarian towns and villages.

The first phase following World War I involved a colonization of Serbs within the designated new border zone with Hungary. In this fifty-kilometer border strip, Hungarians were not to be given any land. Only Serb *dobrovoljci* (war volunteers) were allowed to settle in this "security" zone. A total of 88,000 Serb families received land in this way, of which about 20,000 were brought in from elsewhere to be settled in Northern Vojvodina. This paralleled the emigration of nearly 20,000 Hungarians (*optants*) out of the region. Although this did not change the demographic balance completely, it began the erosion of the Hungarian population.[8] The major mechanism in this process was the "land reform" that took land away from the non-Serbs and redistributed it among the Serbs. In spite of the fact that in Bačka Hungarians constituted 41.4 percent of the landless peasants, the Hungarians were excluded from this "land reform." Most of the redistributed land (48.6 percent) belonged to Hungarians and (36.3 percent) to Germans, Jews, or Italians.[9]

Equally calculated were the policies instituted to put the Hungarian and other non-Slav nationalities at a permanent economic, cultural, and social disadvantage relative to the new Serbian *Staatsvolk*. This was the second phase achieved via discriminatory educational policies, restriction on the use of the Hungarian language in public discourse, and almost total exclusion of the Hungarians from government representation and public employment. For example, in 1928 in the entire Vojvodina there were only ten German and six Hungarian village notaries, while the Serbs and

---

[8] Macartney, *Hungary and Her Successors*, 401.

[9] Károly Kocsis, "Adalékok a magyarság etnikai földrajzához a mai vajdaság területén" [Data on the Ethnic Geography of Hungarians in Today's Vojvodina], in *Tér-gazdaság-társadalom: Huszonkét tanulmány Berényi Istvánnak* [Region-Economy-Society: Twenty-Two Studies for István Berényi], ed. Zoltán Dövényi (Budapest, 1996), 351-352.

other Slavs had a total of 114 notaries.[10] A similar pattern existed in city council representation, employment in the postal service, and the police force. The objective was to isolate and impoverish the minorities so that they would pack their bags and leave the country. All in all, interwar Yugoslavia followed an intolerant policy toward minorities which made existence barely possible, but did not carry out widespread atrocities as those recently resorted to against the Muslims in present-day Bosnia or the Albanians of Kosovo.

World War II changed all this. It brought to the surface brutal leaders, brutal feelings, and abuses which set the stage for the beginning of the first massive "ethnic cleansing" in the region. However, two conditions had to precede this actual process: first, the psychological preparation of the Serbian population for such a drastic policy; second, a pretext to justify expulsion as punishment for each minority's "misdeeds."

The psychological preparation began before World War II in the work of the Serbian Culture Club (Association). This organization, under the leadership of Slobodan Jovanović formulated a strategy for Serbian national revival already at its first public meeting on February 4, 1937.[11] This included a critique of the Yugoslav monarchy's "Yugoslavism" and the government's neglect of Serbian cultural values and ultimately its failure to assert Serb control over the territory that was "theirs" historically. The latter, necessitated a concerted plan to strengthen the Serbs demographically, economically, culturally, socially, and politically, so that they could take their rightful place as the ruling people of the land. Already in this statement Jovanović targeted the non-Serb populations of Sandjak, Kosovo, Vojvodina, and Bosnia-Herzegovina. He mapped out a strategy for highway construction to link together Serb settlements, a concerted effort to buy up the lands of "foreigners," and an educational policy that would give the Serbs the self-confidence and dynamism of a ruling people that can act decisively if the need arises.[12] In this context the strengthening of the Serb population and the weakening of the non-Serb populations was openly discussed. The transfer of the

---

[10] Macartney, *Hungary and Her Successors*, 411.

[11] Ljubodrag Dimić, "A szerb kultúrklub a kultúra és a politika között" [Serbian Culture Club between Culture and Politics], trans. Agnes Özer, *Hid* 59, no.9 (September 1995): 576.

[12] Ibid., 577-580.

Krajina Serbs was contemplated as a solution for consolidating Serb control of Vojvodina. At the same time, the "dispersal" of the Albanian population in the south and the Hungarian and German populations in the north was also contemplated and recommended.[13]

The plans of the Serbian Culture Club received active support from some well-known intellectuals. The historian Vasa Čubrilović was the individual who provided some of the most specific recommendations for cleansing Yugoslavia from the "foreign" Germans, Hungarians, Albanians, Italians, and Romanians. He wrote a position paper entitled the "Minority Question in the New Yugoslavia," which he submitted to the Partisan authorities on November 3, 1944.[14] In this study he recounts the history of the Serbs and how they have suffered because of "disloyal" minorities residing in their midst. Čubrilović presents the treason of the minority nationality communities from the outbreak of World War II. He then argues that the war provides an ideal opportunity to rid the state of these undesirable non-Slavic peoples. He argues that the war creates the right atmosphere for massive population transfers. Both the "German aggressors" and the "Soviet allies" have popularized this practice. He feels that Yugoslavia can depend on its allies, particularly the Soviet Union, to approve of and assist in the deportation of these enemies. Finally, he feels that resistance among the minorities will be minimal, because they are aware that they deserve deportation.[15]

In the second section of his paper he outlines how each of these minorities deserves somewhat different treatment for both tactical and political reasons. The Romanians can be taken care of most easily via an agreement between the two governments to exchange the Romanians in Vojvodina for the Serbs in the Romanian Banat. The Italians are also a simple problem, since their numbers are limited and scattered along the Adriatic coast. Istria's annexation to Yugoslavia and the deportation of most Italians who were settled in after December 1, 1918 would solve this problem. The deportation of Germans, Hungarians, and Albanians is a more complex process because they are all very numerous.

---

[13] Ibid., 579-582.

[14] Vasa Čubrilović, "A kisebbségi kérdés az Új Jugoszláviában" [Minority Question in New Yugoslavia], trans. Nándor Kartag in "Dokumentum," *Hid* 60, No. 12 (December 1996): 1042-1060.

[15] Ibid., 1043-1051.

However, because the Germans were "the enemy," they deserve no quarter: they must be expelled *en masse* from Slovenia, Bačka, and Banat. The same fate should await the Hungarians and the Albanians of Vojvodina and Kosovo Metohija respectively.[16]

Čubrilović concludes his paper by outlining how the vacated properties should be administered until South Slav colonists replace the former owners. He discusses past interwar colonization efforts and why they were not always successful. He adds that he will gladly offer his services to help with the planning and implementation of these expulsions.[17] On the basis of the massive German deportations, we can conclude that at least part of Čubrilović's proposal was adopted by the Partisan authorities. The non-implementation of the remainder is probably related to the reasons why he was not asked to help implement the proposal. After all Čubrilović was a Bosnian Serb, while Tito was Croatian and Slovenian in background. Čubrilović was thinking in terms of Serb control, while Tito was already looking forward to maintaining some sort of balance of power among the nationalities of Yugoslavia. For Slobodan Milošević four decades later, this latter consideration did not act as a restraint.

The pretext for implementing this "solution" was the "war guilt" of the minority peoples. In the case of the Germans this was an open and shut case. "Their" leaders unleashed the war on Yugoslavia in 1941 and were responsible for the subsequent bloodletting. Albanians, Hungarians, and others were only relatively less guilty, since "their" leaders also participated in Yugoslavia's dismemberment.[18] In the case of the Hungarians, the situation was also made more serious by the Novi Sad (Ujvidék) massacre of January, 1942.[19] The Partisan revenge came in October

---

[16] Ibid., 1051-1054.

[17] Ibid., 1056-1059.

[18] György Schöpflin, "Hatalom, etnikum és kommunizmus Jugoszláviában" [Power, Ethnicity, and Communism in Yugoslavia], trans. Judit Nanay, *Politikatudományi Szemle* [Political Science Review] no. 2 (1992): 86-87; Lajos Arday "Hungarians in Serb-Yugoslav Vojvodina since 1944," *Nationalities Papers* 24 (September 1996): 469. Also see my study on "Titoist Integration of Yugoslavia: The Partisan Myth and the Hungarians of Vojvodina, 1945-1975," *Polity* 12, no. 2 (Winter 1979), 230-243.

[19] Tibor Cseres, *Titoist Atrocities in Vojvodina 1944-1945* (Buffalo, N.Y., 1993), 16. The rogue commanders responsible for this massacre were court-martialed and stripped of their command. After the war they were handed over to Yugoslav authorities who had them executed.

1944, when the German army was pushed out of the region and the Hungarian authorities evacuated the area, leaving the indigenous Hungarian and German inhabitants to their fate.

Documentation of the Partisan revenge was only possible after the collapse of Communist regimes between 1989-1991.[20] It provided the "window of opportunity" to do some research on formerly taboo subjects, including the fate of Hungarians in Yugoslavia during the fall and winter of 1944. The revelations that came to light during these years, led to the conclusion that anywhere between 25,000 and 40,000 ethnic Hungarians were massacred by Tito's Partisan units during these months. While this is now supported by two thorough compilations of witness testimonies (Cseres, Matuska) and a documentary film (Siflis), it has not been dealt with in Serbian writings on either an official or unofficial level. On both levels we encounter a wall of silence. At the same time, a great deal is still said about the Novi Sad massacres carried out by Hungarian occupation forces in 1942.[21]

The latter event was seen by some as the justification for the Autumn 1944 massacres. What happened in Novi Sad in January 1942 was an overreaction by the Hungarian Military to subversive activity, including guerrilla activity targeting the Hungarian occupation forces. Two commanders in particular, Major General Ferenc Feketehalmy-Czeydner and Brigadier General József Grassy, took steps to counter this activity by rounding up all individuals considered to be sympathetic to Communism or to the guerrilla resistance. This roundup led to the summary conviction and execution of 3,309 people, mostly Serbs, but also many Jews and some known Communist activists. Of those executed, 879 people were from Novi Sad (hence the association of the massacre with that city).[22] Two things must be noted: (1) the executions

---

[20] Cseres, *Titoist Atrocities in Vojvodina*; Márton Matuska, *A megtorlás napjai* [Days of Retribution] (Novi Sad, 1990).

[21] Examples of this "denial syndrome" can be found in both scholarly and popular publications. In the Serb military historical review *Vojno Istorijski Glasnik* 64 (1995) a whole issue is devoted to the "genocidal" attempts of Croats, Hungarians, Germans, and others, without even a mention that Serbs were responsible for "ethnic cleansing" either during the two World Wars or in the recent conflicts that have swept through the lands of former Yugoslavia. This at least is understandable from the "defensive" perspective of Serb scholarship. It is unexcusable when Western scholars do the same. For an example of the latter see Andrew Bell-Fialkoff's "A Brief History of Ethnic Cleansing," *Foreign Affairs* 72 (Summer 1993).

[22] Cseres, *Titoist Atrocities in Vojvodina*, 16-17; Peter Gosztonyi, ed., *Szombathelyi Ferenc visszaemlékezései* [Memoirs of Ferenc Szombathelyi] (Washington, D.C., 1980), 26 -33, 48-53; Enikő A. Sajti, *Délvidék, 1941-1944: A magyar kormányok delszláv*

were carried out on an individual basis against individuals accused of abetting resistance to the occupation; (2) the officers in charge of this travesty were relieved of their command and were court-martialed by the Hungarian military for their abuse of power. They escaped to Germany to avoid being held accountable for their crimes, but were eventually captured and handed over to Yugoslav authorities after the War. They were then executed near the scene of their crime in 1946.[23]

The events that occurred in October-November, 1944, were quite a bit different. As the German and Hungarian military were pushed out of the region by the Soviet Red Army, Yugoslav Partisan units followed on the heels of the Red Army. Unlike many other "liberated" parts of Yugoslavia, Vojvodina was not allowed to be administered by civilian "people's councils." Instead, the entire region came under "military administration" (i.e., martial law) in line with Josif Broz Tito's decree of October 17, 1944.[24] What this meant was that Partisan units and units of the "security arm" of the Communist Party were given free rein over the entire region. This legalized the terror which was now inflicted on the minority populations. Although neither the indigenous Hungarian or German civilian population offered armed resistance of any kind to the victors, perhaps as many as 200,000 Germans and 25,000-40,000 Hungarians were massacred during the course of October-December, 1944.[25] This was done in a well-planned and systematic way to intimidate the remaining population, to liquidate the native intelligentsia, and to wreak revenge on the "guilty" nationalities. The rampage of killing was followed by the expulsion of almost all the remaining Germans and about 30,000 Hungarians.[26] In their place, Serbs from Bosnia and the Krajina region were brought in to bolster the local Serb population.

---

politikája [Southern Region, 1941–1944: The South Slavic Policies of the Hungarian Governments] (Budapest, 1987), 152-168, 174-187. Also see Sajti, "Az elmulni nem akaró mult" [The Past that Refuses to Go Away], *Könyvszemle* [Book Review], 3, no. 2 (1991), 238-239; Béla Csorba, ed., *Források a Délvidék történetéhez* [Sources on the History of the Southern Region], 3 (Budapest, 1999): 163-170.

[23] Sajti, *Délvidék*, 244; Cseres, *Titoist Atrocities in Vojvodina*, 158-160.

[24] Matuska, *A megtorlás napjai*, 383-386; Sajti, *Délvidék*, 244-246.

[25] Matuska, *A megtorlás napjai*, 380-381; Cseres, *Titoist Atrocities*, 137-141.

[26] Arday, "Hungarians in Serb-Yugoslav Vojvodina," 2; "Ethnic Hungarians in Ex-Yugoslavia," ed. S.O.S.—Geneva Committee, 1993, 16.

The method used by the Partisans was similar to what became the pattern in Bosnia-Herzegovina in 1993-95. The Partisan unit would surround and enter the targeted community. Next, groups of Partisans would go house to house demanding that the menfolk come out to the streets, squares, or soccer fields. Here they would be separated into two groups. Children and the very old men would be allowed to go home. All the others would be taken to collection points where groups of ten would be bound together. They would then be herded into a larger building, barn, or other facility where they would be tortured and sometimes executed right there. In most other instances the executions would be carried out either along the shores of the Tisza or Danube rivers or in fields at the outskirts of the settlements. The latter was usually preferred, because the rivers were less dependable in hiding the evidence. Thus, the prisoners dug their own graves. The next group would cover them up before they dug their own future graves. In this way for example, in the town of Bačko Petrovo Selo (Péterréve) some 360 men were executed and buried during the course of a few nights.[27] What was the objective of this callous brutality? In all the recollections, the witnesses are in agreement that teachers, priests, and "rich" people were prime targets for "special treatment." This meant torture and certain execution. All the other executions seemed to be done on the basis of random selection, just to fill an expected quota of executions.

While in the case of the German minority this led to their "ethnic cleansing" from Yugoslavia, in the case of the Hungarians the main objective seems to have been twofold: to intimidate them and to deprive them of all their leaders or potential leaders. The Hungarians were not to be "cleansed" at this time. The loss of the productive German population was enough to cope with through the resettlement of abandoned German homes with Bosnians or other South Slavs. Perhaps the Hungarians survived because there would not have been enough Yugoslavs to work the abandoned farmsteads. War-weary Yugoslavia could not afford to lose the agricultural production of Vojvodina and the agricultural know-how of the Hungarian inhabitants of this region. Reconstruction at this point was a higher priority than Serbianization via "ethnic cleansing." This is also borne out by the fact that Tito did not demand additional deportations in his negotiations with Hungary.[28]

---

[27] Matuska, *A megtorlás napjai*, 174-182.

[28] Ibid. 383-386; Ludányi, "Titoist Integration of Yugoslavia," 237-238; Cseres, *Titoist Atrocities*, 19-21; Peter Kaslik, "A Song For a City," *Szivárvány* [Rainbow] 14, no. 3 (1993), 136; Theodor Schieder, ed., *Dokumentation der Vertreibung der Deutschen*

What then can we say was the main objective of the 1944 massacres? Probably a demonstrative revenge for the Novi Sad tragedy of 1942, combined with the objective of breaking down all possible resistance to the reincorporation of Vojvodina into the recreated Yugoslavia. Finally, there may have been still another consideration. Perhaps Tito did not want to create a purely Serbian nation-state within the Yugoslav Federation. The "Autonomous Provinces" of Vojvodina and Kosova could act as a brake on Serbian nationalism. At least with the interwar Croat-Serb rivalry very much a live memory, as a Slovene-Croat in ethnic background, Tito could hardly have avoided considering these issues. We can at least assume on the basis of his AVNOJ program enunciated at Jajce in 1943, and his subsequent perpetuation of a transethnic "Partisan Myth," that he was constantly thinking in terms of bolstering the weaker elements of the Yugoslav federation against its strongest Serbian component.

Vojvodina's population, according to the 1910 census, was 30.2 percent Hungarian, 25.2 percent Serb, 23 percent Swabian German, 10 percent other south Slavs, and 10 percent "others." According to the 1921 census Hungarians and Germans together still constituted 48 percent of the total population. However, after World War II, according to the 1948 census 55 percent were now Serb, 22 percent Hungarian, and 23 percent "other." This demographic transformation of the region is also apparent in the ethnic profile of the settlements/cities of northern and central Vojvodina. Thus a city like Subotica (Szabadka), which had a 58.7 percent Hungarian majority in 1910, was now reduced to only 46.5 percent Hungarian. Novi Sad which was 39.7 percent Hungarian in 1910 was now only 29.6 percent of that nationality. Senta (Zenta) with 91.7 percentage of Hungarians was now reduced to 82.7 percent of the settlement's total population.[29]

The resuscitation of Yugoslavia following World War II was possible because Tito did not succumb to the intolerant nationalism of the Serbian Culture Club or Vasa Čubrilović's position paper. Instead he was focused on consolidating the power of the League of Communists of Yugoslavia and the Partisans. This objective, plus the support of the major Allied powers, enabled Tito to move away from the Serb-dominated system of the interwar years that focused on the critical role of the Serb *Staatsvolk*.

---

aus Ost-Mitteleuropa : 5—*Das Schicksal Der Deutschen in Jugoslawien* (Bonn, 1961): 119E-132E; Csorba, *Források a delvidék történetéhez III*, 222-234.

[29] Kocsis, "Adalékok a magyarság etnikai földrajzához," 355.

However, he too seems to have wanted at least to assimilate the non-Slavic minorities into the dominant South Slavic peoples. In lieu of such assimilation he was not wont to encourage emigration for the non-Slavic peoples as *Gastarbeiter* to Western Europe. From the 1950's to the late 1980's a constant stream of Hungarian immigrants or "guest workers" left for Germany, Canada, Brazil, Australia, and even the United States.

Tito was able to make the system work for almost four decades, because he imposed multi-ethnic tolerance within this patchwork state through his dictatorial control of the League of Communists of Yugoslavia, his personal charisma and cult of personality, and a clever balancing politics wherein he kept all the different peoples of the country satisfied by recognizing their right to exist, and by giving them a symbolic share in power within their own "Republics" and "Autonomous Provinces." However, with the passing of Tito, not only his dominant personality which provided the glue, but the sense of legitimacy engendered by "proletarian internationalism" and the unifying "Partisan myth" evaporated. Parallel to this development, but as a delayed reaction, the rest of the Communist world also suffered from incompetent new leaders and a loss of missionary zeal and/or lacked a pragmatic outlook that would enable them to come to grips with the persistent old problems and the new challenges.

The collapse of the Soviet bloc in 1989 and of the Soviet Union in 1991 set the stage for dramatic changes in Yugoslavia as well. In fact, already in 1989, the resurfacing nationalisms of all the peoples of Yugoslavia began to call into question the right of the Communists to exercise a monopoly of control over the state.

Just prior to the collapse of Yugoslavia in 1991, Hungarians still constituted about 17 percent of Vojvodina's total population. Today, just nine years later, they barely reach 13 percent.[30] Still they compose a relatively compact cluster in the north-central part of Vojvodina along the Tisa (Tisza) River. It is this settlement cluster that is coming under Serb attack at the present time.

This most recent effort to "cleanse" Vojvodina of non-Serbs was preceded by three important domestic developments. The first was the restatement and legitimization of the ideals expressed by the Serbian Culture Club for a Greater Serbia. This was done by the publication of the infamous "Memorandum" of the Serbian

---

[30] Estimates provided by András Ágoston, President of VMDK (Democratic Community of Hungarians in Vojvodina).

590 The Fate of the Hungarians in Yugoslavia

Academy of Sciences during the summer of 1986.[31] Its thesis was simple: Serbia should encompass all territories inhabited by Serbs. In other words, the present Serbia had the right and the obligation to "reincorporate" all lands settled by Serbs. The internal boundaries of Yugoslavia created by Tito and the League of Communists of Yugoslavia (LCY) were "unjust" and had victimized Serbs by forcing them under the jurisdiction of other Republics or Autonomous Provinces "dominated" by other peoples. Dobrica Cošić and the other authors of this document fueled the mystical martyr complex of the Serbs and played on their chauvinism to be prepared for the coming day of reckoning.[32]

The meteoric rise of Slobodan Milošević in the LCY followed closely on the Academy's approval of the "Memorandum." He became leader of the LCY in Serbia in 1986. He wasted no time in developing a cult of personality by linking his political career to the rising star of Serb nationalism stoked by his ideological patron, Dobrica Cošić.[33]

Finally, the deadlock in the collective presidency of Yugoslavia in December 1987 created a constitutional crisis which enabled Milošević to consolidate his hold over Serbia and to begin his quest to establish a Greater Serbia. This was done via a carefully orchestrated plan that first targeted Serbia's two autonomous provinces. Playing on the fear of secessionist sentiments in these provinces and the charge of harassment and persecution of Serbs, he initiated a press campaign for the replacement of "bureaucratic" and insensitive leaders. This was followed by the bussing in of thousands of Serb demonstrators demanding resignations.[34] This so-called "Yogurt Revolution" of October 1988 toppled the Party leadership in Vojvodina and led to its replacement by Milošević loyalists. This was followed by the "reintegration" of Vojvodina into Serbia proper. Using the exact same tactics, supplemented

---

[31] Ferenc Kisimre, *Joghurtforradalomtól a polgárháboruig* [From the Yogurt Revolution to the Civil War] (Szeged, 1993), 95.

[32] Ibid.; Laura Silber and Allan Little, *Yugoslavia: Death of a Nation* (New York, 1996), 31-36; Olivera Milošavljević, "The Abuse of the Authority of Science," in *The Road to War in Serbia: Trauma and Catharsis*, ed. Nebojša Popov (Budapest, 2000), 274-300.

[33] Ibid., 284-297; Kisimre, *Joghurtforradalom*, 95-96.

[34] Ibid., 87-88.

with heavy military and police reinforcements, the autonomy of the Albanians in Kosovo was also eliminated in March, 1989.[35]

The next two years witnessed Slovenia's and Croatia's efforts to break their ties with the centralizing, Serb-dominated Yugoslavia. Their efforts were crowned with success: independence. However, the Yugoslav armed forces backed Milošević in his efforts to retain those territories which had Serb populations. This led directly into the fratricidal war which began in the summer of 1991 and raced through Croatia and later Bosnia-Herzegovina. The struggle's side-effect was a constant stream of refugees in all directions. However, many of the Serb refugees saw Vojvodina as their preselected destination. The homes of evicted Croatians and Hungarians accused of "draft-dodging," were immediately offered to them. The evictions of the latter were accomplished through intimidation or military draft. In Vojvodina and in other minority inhabited areas, induction into the military has become an instrument of "ethnic cleansing" for the Milošević administration. According to Tibor Várady, one of the most astute observers of developments in Yugoslavia, the draft has been used to pressure ethnic Hungarians to leave the country. He points out that 40,000 ethnic Hungarians, mostly of military age, had left Vojvodina in the five years preceding the Dayton Agreement. "Mobilizations in Vojvodina have been pursued with more zeal than elsewhere in Serbia. In a number of Hungarian villages, police blocked the streets during the night while draft-calls were delivered. Many were taken to service forcibly, in disregard of existing regulations. Numerous cases of harassment and beating were also reported."[36]

The magnitude and extent of these forced conscriptions is evident from the appeal addressed to Lord Owen and Cyrus Vance (Co-chairmen of the Conference on Yugoslavia in Geneva) by András Ágoston, the president of the VMDK (Democratic Community of Hungarians in Vojvodina). This letter dated December 7, 1992, indicates that: "The government mobilized 2.5 times more

---

[35] Ibid., 93-94, 100-101; Silber and Little, *Yugoslavia*, 58-68.

[36] Tibor Várady, "Vojvodina - The Predicament of Minorities and Possible Solutions," paper submitted to the Helsinki Commission, May 5, 1994, 3. Also see Mark Schapiro, "Serbia's Lost Generation," *Mother Jones* 24, no. 5, (September-October, 1999): 48-53; although Schapiro does not reflect on the "ethnic cleansing" aspects of the draft, his figures (p. 50) dramatically confirm this when he points out that from March 24 to June 20, 1999, there were 2,315 "Yugoslav" applications for asylum in Hungary, and these included 825 ethnic Albanians, 787 ethnic Hungarians and *only* 506 ethnic Serbs.

Hungarians into front-line battalions than the Hungarians' percentage in Serbia's population would warrant. The forced mobilization of Hungarian reservists - barring a few brief pauses continues to the present day."[37] Drafting disproportionately high numbers of Hungarians also had a second objective: to break the opposition to the war, or at least to punish those who had been in the forefront of the opposition. This is evident from the HHRF Alert which published the full text of a protest statement issued January 28, 1992 by the leadership of VMDK. The statement maintained: "The latest large-scale mobilization campaign is having a particularly adverse effect on the representative organ of the Hungarian minority, the VMDK. Previously, draft notices had already been received by one quarter of the organization's central leadership, several leaders of tWestern Bácska (Bač) regional chapter and others. This time notices were sent to two Vice Presidents of the regional chapter in Szabadka (Subotica) and the Vice President of the Ada chapter, among others."[38]

The empty or partially inhabited homes were targeted for takeover by the Serbs. In the 1991-93 conflict we have record of only one settlement being ethnically cleansed of Croats and Hungarians in Vojvodina; this was the town of Herkóca in the Western part of the province. However in 1995 the barbarities of "cleansing" became more widespread. The first instances of such attacks took place in Sombor, Svilojevo (Szilágyi), Sonta, and Apatin.[39] The new cleansing technique in these Croat and Hungarian settlements followed the pattern that had been established in Bosnia and Eastern Slavonia. Instead of incrementally encroaching on these settlements, the Serbs resorted to a more aggressive policy, using the cover of the war to achieve their objective. The Milošević government used this new influx of Krajina "refugees," i.e., dis-

---

[37] András Ágoston's letter to Lord Owen and Cyrus Vance, Co-Chairman of the Conference on Yugoslavia, December 7, 1992, 1.

[38] Democratic Alliance of Hungarians in Vojvodina, "Statement," January 28, 1992, translated and published by Hungarian Human Rights Foundation, New York, January 30, 1992, 1-2.

[39] Béla Csorba, "Minden eddiginél nagyobb veszélyben a vajdasági magyarok" [The Vojvodian Hungarians are in Greater Danger than Ever Before]*VMDK Hirmondó* [VMDK Messenger], no. 65 (August 18, 1995), 2-3. Also see A. Ludányi, "The Domino Effect of Ethnic Cleansing in the Former Yugoslavia," *Analysis of Current Events* 7, no. 5 (January, 1996): 2-4, and "After Kosovo, Is Vojvodina Next?" *Analysis of Current Events* 11, nos. 5-6 (May/June, 1999), 6-7.

placed Serbs, to dilute even further the Hungarian inhabited parts of Vojvodina.[40]

The leaders of VMDK (*Vajdasági Magyarok Demokratikus Közössége*/Democratic Hungarian Association of Vojvodina) also contend that the Croat victory in Krajina was so rapidly concluded (in three days) because the Serb military already knew that they could not win. Instead of fighting, the Serb military units literally drove out their own people from their homes in a well-planned "population transfer." The Croats even allowed the military personnel to "escape" in this exodus if they were willing to abandon or surrender their heavy weapons. Thus the arriving refugees in Vojvodina had concealed small arms and military-led organizations. Their systematic dispersal in the region also demonstrated planning.[41]

This newest wave came as a follow-up to the settlement of Serbs in 1991-93. At that time the Serb refugee or displaced person population from Croatia and Bosnia was also resettled in different parts of the reduced Yugoslavia. While in the Serb and Montenegrin parts of the country this did not change the cultural complexion of the land, in Vojvodina it has led to more and more Serbianization. For example, the town of Čoka (Csóka) had a total population of 5,234 inhabitants in 1991. Of these, 3,229 were Hungarians. Then in 1993 it absorbed 2,315 new Serb displaced persons making the Serbs the new majority in town. Depending on its size, every town in Vojvodina was allocated anywhere from 2,000 to 3,500 new Serb inhabitants.[42] Table 2 reflects the extent of the "cleansing" in major Hungarian settlements to 1991. This preceded the loss of ca. 30-40,000 military-age young Hungarians who chose refugee status in Hungary rather than participation in Yugoslavia's fratricidal war. This also preceded the influx of ca. 200,000 Serbs displaced by the war who have settled in Vojvodina since 1991. As András Agoston observed in his letter to Lord Owen and Cyrus Vance, quoted earlier, "the greatest threat to the Hungarians of Vojvodina remains the forced resettlement process known as the 'third colonization.'"[43] As Art Chimes reports in a back-

---

[40] Csorba, "Minden eddiginél nagyobb veszélyben," 2; Michael Jordan, "Krajina's Refugees Run Into Trouble," *Transition*, December 13, 1995, 52-54.

[41] Csorba, "Minden eddiginél nagyobb veszélyben," 2-4.

[42] Ludányi, "The Domino Effect of Ethnic Cleansing," 4; "After Kosovo, Is Vojvodina Next?" 6-7.

[43] András Ágoston's letter to Lord Owen, 2.

ground report obtained by e-mail on November 2, 1997, "Vojvodina province has one of the largest refugee burdens in Europe. More than 200,000 refugees crowd into the area—about one refugee for every ten permanent residents."[44]

Béla Csorba, a Hungarian representative in the lower chamber of the Yugoslav legislature, wrote an appeal to defend the historical Hungarian settlements in Vojvodina. In this appeal, written after the arrival of the Krajina Serbs in August, 1995, he described the cleansing of the small Hungarian town of Svilojevo (Szilágyi).[45] In this instance, the Serbs--many of them armed--cruised up and down the streets of Apatin, Sombor, and Svilojevo to find vacant or temporarily empty homes. These were immediately taken over, even if the owners were only temporarily away for the weekend. This Hungarian town had 1,250 inhabitants. On August 11th, some local Serb nationalists led the displaced persons from Krajina to some homes which were confiscated from the owners. By the time the representative of VMDK, the Hungarian interest group, went to investigate the next day, the town was almost completely abandoned by its original inhabitants. Already eight homes were occupied by force. The following day forty homes were occupied by the Krajina Serbs.[46]

When the Hungarians appealed for intervention by the law enforcement authorities, they were told that this was only a peripheral case and that it was too late to do anything about it. In other words, the Milošević administration was in collusion with those who carried out these forced confiscations. The pattern that emerges is that the local population is intimidated, they request protection, and to get it, begin to collaborate with the police by providing some homes or parts of homes for the refugees. The direct consequence of this was that most of these settlements gained a Serb majority or a significant Serb plurality.

---

[44] H. A. "A nyugat-bácskai körzetben huszezernél is több a menekült" [More than Twenty-Thousand Refugees in the Western Bachka Region], *Magyar Nemzet,* [Hungarian Nation],August 15, 1995, 3; "Szerbek, települnek a Vajdaságba" [Serbians are Settling in Vojvodina], *Magyar Szó* [Hungarian Word], August 11, 1995; "Hányan érkeznek a vajdaságba" [How Many Are Coming to Vojvodina?], *Népszabadság* [People's Freedom] August 9, 1995, 2; Art Chimes, "Yugoslav Chickens: Stara Pazova," *Voice of America Background Report*, No. 5-37961 (November 2, 1997).

[45] Csorba, "Minden eddiginél nagyobb veszélyben," 2-3.

[46] Ibid., 2-3.

Consequences

On the basis of this review we can show the rate of "cleansing" by correlating the three major waves, post-World War I, post-World War II, and 1991-1995, with the declining percentage of the Hungarian population in census reports and current estimates of population changes in comparison to the population ratios of 1910 (the last census prior to the beginning of Serbianization.) As Table II shows, not only in total numbers within the population of Vojvodina, but also in their historic settlements within the region, the proportion of the Hungarian population has been drastically reduced in comparison to the ever-increasing Serb population.

TABLE 2
ETHNIC HUNGARIAN PROFILE OF SOME LARGER POPULATION CENTERS IN THE VOJVODINA

Cities	1910 pop. %	1931 pop. %	1941 pop. %	1948 pop. %	1961 pop. %	1971 pop. %	1981 pop. %	1991 pop. %
Novi Sad (Ujvidék)	13,343 39.7	17,000 30.0	31,130 50.4	20,523 29.6	23,812 23.2	22,998 16.3	19,262 11.3	15,778 8.8
Subotica (Szabadka)	55,587 58.8	41,401 41.4	61,581 59.9	46,706 46.5	37,529 50.0	43,068 48.5	44,065 43.8	39,749 39.6
Senta (Zenta)	27,221 91.8	25,924 81.1	29,463 91.7	20,898 82.7	20,980 83.7	20,548 83.1	18,863 79.6	17,888 78.4
Kanjiža (Magyarkanizsa)	16,655 97.9	19,108 07.4	18,849 97.5	10,149 87.4	9,797 91.4	10,177 90.5	10,466 89.0	10,183 88.2
Bečej (Obecse)	12,488 64.5	12,459 60.7	14,576 68.8	14,701 62.4	15,537 62.2	15,815 59.2	14,772 54.5	13,463 50.6
Bačka Topola (Bácstopolya)	12,339 98.9	12,839 85.3	13,420 95.0	12,706 91.3	12,969 86.0	13,112 82.0	12,617 74.1	11,176 66.9
Sombor (Zombor)	10,078 32.9	5,852 18.1	11,502 35.8	7,296 21.7	7,474 19.8	7,115 16.1	5,857 12.1	4,736 9.7
Zrenjanin (Nagybecskerek)	12,395 42.1	12,249 33.7	---"	15,583 40.4	18,083 32.5	18,521 25.9	17,085 21.0	14,312 17.6

'The data for this table is taken from Kocsis, "Adalékok a magyarsag etnikai földrajzához a mai vajdaság területén," in *Tér, gazdaság, társadalom: Huszonkét tanulmány Berényi Istvánnak* ed. Dövényi, 355.

"The data for 1910 and 1941 is based on Hungarian censuses based on "mother tongue." The 1931 data is based on a Yugoslav census based on "mother tongue," while the data from 1948 to 1991 is based on Yugoslav censuses based on declared "nationality." The 1941 data is missing for Zrenjanin, because it was not under Hungarian jurisdiction at the time of that Hungarian census.

The demographic change has been paralleled with a reduction in Hungarian cultural and educational opportunities. As the new Serb inhabitants have eclipsed the Hungarian population in numbers, they have been provided with Serbian language instruction in the schools. This Serbianization of the school systems has squeezed out Hungarian language instruction in many instances. In his thorough analysis of this problem, Károly Mirnics points out that not only have the total number of Hungarian pupils declined from 54,763 in 1961-62 to 27,584 in 1996-97, but that of these totals in 1961-62, 45,127 attended schools with bilingual instruction, whereas in 1996-97 only 21,848 Hungarian students were attending schools with bilingual instruction. Furthermore, that instruction in these schools, without regard to the language of instruction, was immersed in the glorification of a "Greater Serbia and Serbian greatness."[47] Although this analyst does not have exact figures on this process, personal testimonies of refugees in Hungary and the protests and testimony of VMDK leaders indicates that this is now a general trend.[48]

The Hungarian community's response to this continued erosion of their institutional support-system has been to strengthen their interest organization, to expand their contacts with the outside world, and to press for more self-government at the local and regional levels.

At the organizational level they have been able to achieve some successes. The existence and work of VMDK with 30,000 members is probably the most significant result in this area.[49] It has enabled them to run candidates for public office both at the local and regional level. They have even elected eleven representatives to the lower chamber of Yugoslavia's legislature.[50] However, personality conflicts among the leaders has led to the fragmentation of the united front that VMDK was able to project in its

---

[47] Károly Mirnics, "Asszimilációs tényezők és asszimilációs politika Jugoszláviában (Vajdaságban)" [Assimilationist Factors and Assimilationist Politics in Yugoslavia (Vojvodina)], MS., Subotica, August, 1997, 19.

[48] "Asszimilációs Önkormányzati Politika" [Assimilationist Local Politics], *VMDK Hirmondó*, No. 60 (June 3, 1995): 9; Robert Aspeslagh, "Trianon Dissolved: The Status of Vojvodina Reconsidered?" *Yearbook of European Studies*, no. 5 (1992), 131-133.

[49] Aspeslagh, "Trianon Dissolved," 134.

[50] Ibid., 135.

first few years of existence. Presently, the Hungarian community's organizational efforts have witnessed the emergence of a competing organization (VMSK). This has complicated and in some ways undermined the prospects for a united stance on critical issues vis-à-vis the Milošević administration. However, the leadership role of József Kasza and László Józsa has continued to provide a strong Hungarian presence in Vojvodina politics.

The prospects of the Hungarians in Vojvodina are grim. The outlook will only improve if their quest for local self-government and cultural autonomy receives support at the international level. While on a short-term basis this is not likely to take place, because the international community is interested mainly in the implementation of the Dayton Accord, if the Albanians *and* Hungarians continue to be abused, and "ethnic cleansing" (albeit in more subdued fashion) is continued in their settlements, the prospects for regional peace in the long-run will be nonexistent. This the international community—particularly the NATO alliance—cannot afford.

The dramatic events of October, 2000 in Belgrade, Yugoslavia provide a window of opportunity for real change. If Vojislav Kostunica, the successor to Slobodan Milošević, takes advantage of the grace period that the world is willing to grant him, he may be able to help stabilize a part of the world that witnessed the start of two world wars, as well as the recent devastations of 1991-95 and 1998-99. But Kostunica has much to overcome.

He needs to distance himself from the failed policies of Milošević. First and foremost this means that he must normalize Serbian relations with *all* the neighboring peoples of the region. This means making amends to the Albanians in Kosovo for the destruction wrought by the Serb-Yugoslav military in their lands. A similar normalization of relations is needed in the north in Vojvodina, and in the Sandjak region of western Serbia. Local self-governing institutions must be provided for the 350,000 Hungarians in Vojvodina. Similar guarantees should be provided for the Muslims of the Sandjak. The educational, cultural and social institutions which were destroyed by Serbianization, must now be reestablished and given financial and political support. Compensation and restitution must take place before normalization of relations is possible.

Map Appendix

A.   Interwar Vojvodina

B.   Vojvodina from World War II to 1988

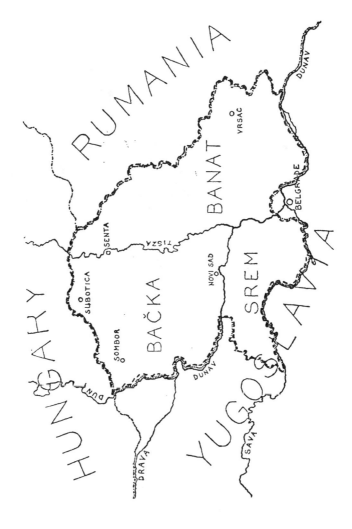

C. Hungarian Settlements in Yugoslavia

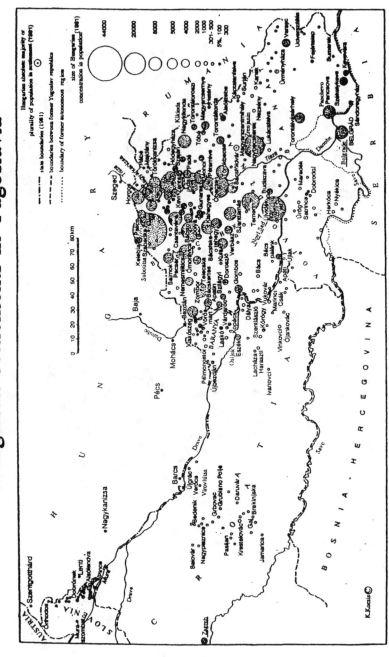

IV.

SURVIVAL AND MEMORY:

VERTREIBUNG

"1945 In Memory": A Survivor's Report

KARL HAUSNER

Almost daily, politicians, the media, and even prominent Christian leaders refer to Nazi crimes and the Holocaust, without even mentioning such crimes against humanity committed by the Bolsheviks before, during, and after World War II and the millions who were tortured and killed in China, Korea, and Indochina. There is also total silence about the Morgenthau Plan, the Allied War against civilians and the starvation camps throughout Europe and Asia during and even after World War II.

It is for this reason that I would like to contribute to the understanding and harmony between nations by presenting to those interested in the whole truth, the experience of my people in the former Sudetenland after World War II.

My Personal Experience After World War II

I was born and raised on a family farm in Schwansdorf (Svatonovice), a village with a prewar population of about 550, near the city of Troppau (Opava), formerly Sudetenschlesien (Sudeten Silesia) and now the Czech Republic. I was fifteen as the last year of the war began. The war was extremely hard on our people with a high number of casualties. The worst came when we were "liberated" by the Soviet troops and Czech partisans.

On May 4, 1945, after weeks of military and refugee movements, the German troops quickly disappeared. At about 10:00 A.M., the village was bombarded by the Soviets with grenades, and later combat soldiers appeared. There was no fighting. The Soviet troops searched every house, primarily for German soldiers and valuables such as watches and jewelry. The combat troops then moved on.

Next the Soviet support troops arrived, removed all the horses from the barns, harnessed them, hitched them to the best wagons available, took some feed, and departed. Every mature male who was located in this process had to go along. At our farm this meant five men taken away: Ernst Krebs, Fritz Krebs, Johann Kuntscher, Emil Kaimer and Franz Rohm. Of the five, only one, Franz Rohm, ever returned, and this would be after two years in a Soviet labor camp.

Other Soviet soldiers started to look for young women they could sexually abuse. Most of the population had fled or were hiding in stone quarries or woods; most who remained did so to take care of their cattle. For example, Mrs. Jahn, a fifty-year-old woman, was approached for rape. She ran out of the house and the Soviet soldiers shot her to death.

Franz Frei, during World War I, was in the Austrian Army and was captured by the Russians. He spent years with and against the Red Army. In 1923, he returned from Russia via China; thus, he had learned to speak Russian. He was in his house when Soviet soldiers entered it. They saw a family picture on the wall with his four daughters. They demanded that he produce these girls. Since he knew some Russian, he explained that they were not there. After a beating, Mr. Frei was forced to kneel down and then he was "executed."

Emma Bischof, our neighbor, about thirty-five years of age, was about to be raped. She took her two children and ran out of her farm as two soldiers ran after her. She jumped with her two children into a water reservoir. The soldiers pulled her and the children out, but her little boy Walter had already drowned. She was raped while her seven year old daughter watched.

At about 3:00 P.M. a group of Soviet soldiers entered the farm house of the Emil Kaimer family. Emil Kaimer had already been captured when they had taken his horses. Mrs. Kaimer, age thirty-eight, her three children, ages ten, six, and two, and her seventy-year-old mother-in-law were in the kitchen. Mrs. Kaimer had to lie on the floor where she was raped by dozens of Soviet soldiers, one after another. When blood flowed from her vagina, one Soviet soldier, instead of raping her, took his bayonet and stuck it into her vagina, pulled it out, and disappeared. Mrs. Kaimer was still alive for another hour or so, until a Soviet officer shot her in the head. All this was in the presence of her small children and mother-in-law.

Julius Dohmes, age sixty, hanged himself in the hay barn when they took his only horse. He was a small farmer and obviously could not handle the loss and circumstances. Hans Sommer, about fifty-five, had a small farm and Gasthaus (Inn), known as Schles, which was on the road between our village and Bautsch (Budisov). He was found shot to death near his estate. No one knows the circumstances. Most likely he fled and failed to stop and was shot.

During the next few days rape and robbery were committed by the Soviet troops whenever and wherever possible. All of the cows and the other cattle, except one for each family, were removed from the various farms and driven to the Rosmanith farm, where they were milked, awaiting their transportation to Russia. All of the young women who could be found were forced to do the chores for a period of about four weeks, during which time they were raped numerous times, in many cases daily.

At the end of May, the Soviet troops had discovered that some refugees were hiding in a stone quarry, about a mile away from the village surrounded by large wooded areas. The family of our teacher, Karl Wolny, had a hunting lodge nearby where they were hiding. On May 26, 1945, Karl Wolny (who was seventy-four years of age), his wife, his sister-in-law Mrs. Muehr, his son Oskar, with his wife Anna and her sister, were brutally slain. The two young women were raped, even though Anna Wolny was pregnant and close to giving birth. They were thrown into a mass grave without a funeral because our villagers were scared to attend. In September 1998, fifty-three years later, a stone was placed on that mass grave in memory of them.

Within about four weeks, in early June 1945, most of the Soviet troops had disappeared from the villages, and the Czech partisans took over. The situation worsened.

I personally was in hiding with a Polish-speaking family who had been on our farm prior to the end of the war. Since they spoke Polish, they could communicate with the Russians to some extent, and during more critical moments, I was hidden under sacks of feed for the horses, with clothing and the small children on top of the stack. We made our way to their hometown, Weihendorf (Wojnowice) near Ratibor, which had been claimed by the Polish Militia. Finally, at the end of June, I ventured to return to my hometown about forty miles away, naturally on foot, avoiding towns, highways and people in general. Upon my return home, all of us between the ages of fifteen and sixty-five, were to report for work, harvesting and thrashing, whatever was left in the fields. In early September (I was just about sixteen), two young men from our village, I and Ernst Frei (nineteen), who had returned from the military, were sent to the industrial and coal mining region between Ostrau (Ostrava) and Oderberg. The labor camp was built during World War II, where Soviet prisoners of war were housed and had to work in the coal mines. This was a typical labor camp with barracks, primitive sanitary facilities, and a kitchen. The camp was surrounded with barbwire fences and watchtowers for the guards.

Upon our arrival in the camp, our civilian clothing was taken away and we got a prison uniform which included wooden shoes and a helmet for use in the mine. We received shears and had to cut each others' hair as short as possible in order to reduce the habitat for lice and make us more readily detected in case we fled. This was certainly not something new, but a common practice in all labor, concentration, or prison camps.

We were housed in these barracks, sixty to eighty men in one room. In the morning we got a pot of "coffee" (roasted grain and boiled). After the shift we got about a quart of soup without fat or meat and one small loaf of bread for five days. Most of us could eat the bread during one meal; some did, and this quickly led to serious health problems. When we arrived at the camp, between the drive and the walkways there was grass. Within weeks, all of the grass was pulled out and consumed, including the roots, which further led to digestive problems, severe diarrhea and often death.

In the mine, we worked eight hours daily. Initially, we could handle the work, but within weeks, many lost strength or got injured, while others simply dehydrated due to diarrhea and died.

We were not permitted to have any reading or writing material. Thus, our parents did not know where we were. The camp I was in was within the town of Dombrau (Dombrowa) near Karwin, not far away from Oderberg.

Within weeks, I developed not just diarrhea and other health problems, but also an eye infection. Since there was no medical care and the coal dust aggravated the condition, I got to the point where I could not work in the mine under ground. Shortly after the New Year of 1946, about forty men from this camp were collected, put in a railroad car, and sent away. The train ride ended in Troppau (Opava), our county seat, about twenty miles away from my hometown. From the Troppau railroad station, we walked, naturally under guard, to Graetz (Hradez), where we were put into the castle of the huge Feudal Estate of Prince von Lichnowsky. We were to cut timber for the mine. The forest we were assigned to work in was one of the last battlegrounds between the German and Soviet Army, in late April of 1945. Most of the trees were scratched or even filled with shrapnel and in the bunkers there were still the remains of German soldiers. We cleared these woods and closed the bunkers. The equipment we used was of American manufacture: an International Harvester tractor (Farmal M) and even a few American made power saws.

In this camp things improved for us. It was much smaller and less strictly guarded, and some Czech people would slip us some food, even though it was prohibited.

In March 1946, four of us from the group were asked whether we knew how to handle horses and thus, we were transferred to the farm, where we worked with the horses, hauled wood, and later made hay.

A young Czech, who worked at the dam of the small electric power plant, found a hand grenade and played with it, and it exploded. It tore off his hand and injured him severely. We heard the blast and ran to the area and found him lying in the water. We pulled him out and carried him to the farm, from where he was taken to the hospital.

Since we saved his life, the farm manager and the other Czech people working on the farm, gave us special privileges, such as more food and more freedom. I was there until June 1946, when my family was scheduled for expulsion.

In June 1946 I received word from our guards that we would be released, sent home to our families, and then transferred to the expulsion camp in Wigstadtl (Vitkov), a town about five miles from our Village of Schwansdorf. My parents, my twelve-year-old sister, and I, along with ninety other inhabitants of our village, were to pack up and get ready to be transported to the camp. We were permitted to take with us sixty kilograms (130 lbs.) of used clothing, shoes, bedding, or utensils—no money, no jewelry, and nothing else valuable. All of this stuff was inspected by the guards of the expulsion camp in Wigstadtl. There we stayed for about five days until a complete train of about thirty box cars was assembled. Our "possessions" were loaded in railroad box cars, along with thirty people to one car. The camp was heavily guarded as was the train during the whole trip. We were not told where we were going, but within the first day of transit and waiting, we realized that we were going westwards. We hoped and prayed that this direction would be maintained, because prior to this, many German people were sent into forced labor camps to the Soviet Union.

After about four days of very slow travel and waiting, we arrived at the border crossing of Czechoslovakia and Bavaria at Furth im Walde. During the trip we were permitted, at specified locations, to leave the box car and empty the pail of human waste or use the open latrines. Occasionally, we got food and water. After the train crossed the border, the guards quickly left and we realized that we were in Bavaria, in the American Zone of Occupation. There the Red Cross and some other voluntary organizations gave us food. Then, we were ordered into barracks, where we were individually deloused by DDT powder. All of this was under the United States Military Command. We still did not know where we were to be sent next. After another day of travel, the whole train was separated into different groups and three box cars, which included us, ended up in Landshut, Bavaria. There, the railroad station had been totally destroyed by bombs, and only a small barrack housed the railroad office. All of our possessions were unloaded and put on trucks for further transportation to an unknown destination.

After about a one hour truck ride, we arrived in a remote village of about a dozen farmers. In Huettenkofen, the truck was unloaded beneath a shade tree, and that was our final destination. Within an hour, the appointed Mayor, Mr. Stelzenberger, walked with each family to another farm and told the farmer that he had to clear one room for those of us who had been expelled. The farmer gave us an ox cart, with which we transported our "valuables" to the farm and our new "home." Our new home was a twelve-foot by twelve-foot room for four persons.

The following gives an idea of the local dimensions of the expulsion in our area:[1]

> May 23, 1946—First transport, 1204 persons to Goeppingen (Schwaben).
>
> June 10, 1946—Second transport, 1204 persons to Munich (Bavaria). (This was our transport, from which three box cars were removed from the train at Landshut.)
>
> June 26, 1946—Third transport, 1108 persons to Dachau (Bavaria).
>
> July 4, 1946—Fourth transport, 1155 persons to Augsburg (Bavaria).
>
> July 18, 1946—Fifth transport, 1204 persons to Regensburg (Bavaria).
>
> August 14, 1946—Sixth transport, 1203 persons to Wuerzburg (Bavaria), another major part from our Village of Schwansdorf.
>
> August 23, 1946—Seventh transport, 934 persons to Wuerzburg (Bavaria).
>
> October 21, 1946–Eighth transport, 298 persons to Kitzingen (Bavaria). With this transport, almost every German from our district, about ten villages, and the town of Wigstadl, had been expelled.

While the end of World War II brought great relief for millions, for many other millions, hell broke loose. The crimes and the brutalities against millions of East Europeans have been kept secret and even today, very few know, or even want to know about it. Justice in the world cannot be promoted, if justice is not provided to all. A crime is a crime, whether committed by the Nazis, the Communists, or the Allies.

Other Atrocities

By comparison, my experience after World War II at the hands of the Soviet Army and the Czech Partisans, was fairly pleasant—even though, it eventually resulted in the loss of my eyesight.

What happened on May 17, 1945 in Landskron—the hometown of my wife, Hermine (Schwab)—Ober-Johnsdorf, and Kreis Landskron, is reported vividly in her account elsewhere in this

---

[1] Expulsion data from Wigstadl comes from the *Troppauer Heimat-Chronik*, January 1996.

volume and even more dramatically in the collection entitled *Documents on the Expulsion of the Sudeten Germans,* published in Germany in 1953.[2] During this massacre, her father, her uncle, a great number of local Germans, and a few German soldiers were tortured to death.

In the collection mentioned above, many other atrocities are described, such as the Death March in early May of 1945 from Bruenn (Brno) to the Austrian border, whereby, about 6,000 persons were tortured to death or shot and thrown into mass graves.

The torture and beastly killing of over 150 Germans and a few Czech "collaborators" in the Hanke Lager in Ostrau (Ostrava), was initially investigated in 1947 by the Czech Government, but the report was never released until after the collapse of the Communist Regime in 1990. Dr. Stanek, a journalist and historian, published the complete file in the Czech language in an Ostrava paper and has now made his research available to a broader audience.[3]

Mr. Franz Jenschke, who was born and raised near Grulich (Kraliki), finally made it to West Germany after the war, lived for decades in Bremen, and now resides in Berlin, reported as follows:

> A few days later, (May 20, 1945) a "trial," similar to the one in Landskron, was held in Grulich and neighboring towns including Zoellnei and Wichstadtl, towns about twenty miles from there. After the beating, torturing and killing, the previous Mayor, Mr. Grund of Zoellnei, was singled out. He was hung by his feet until he was unconscious, then he was dropped to the ground and cold water was poured over his head, until he regained consciousness. This torture was repeated a number of times and then he had to crawl on his knees and hands to the cemetery. During this "trip," Mr. Grund was beaten, kicked in his testes and forced to salute "Heil Hitler," while the survivors had to follow and watch. At the outside wall of the cemetery he had to dig a shallow grave, crawl into it, raise his right hand and say "Heil Hitler," while some of the survivors had to shovel dirt on him, until he was silent, his hand still extending out of the dirt. His grave is still there and a simple cross was recently put up.

---

[2] Wilhelm Karl Turnwald, *Documents on the Expulsion of the Sudeten Germans* (Munich, 1953). See especially Julius Friedel, "Landskron: Massacre on May 17th, 1945," 31-36.

[3] Tomáš Staněk, *Perzekuce 1945: Perzekuce tzv. státně nespolehlivého obyvatelstva v českých zemích (mimo tábory a věznice) v květnu — srpnu 1945* [Persecution 1945: The Persecution of the So-called Officially Unreliable Population in the Bohemian Lands (Outside Camps and Prisons) Between May and August 1945] (Prague, 1996). This important work has been translated into German as *Verfolgung 1945: Die Stellung der Deutschen in Böhmen, Mähren und Schlesien (außerhalb der Lager und Gefängnisse),* trans. Otfrid Pustejovsky and Walter Reichel (Böhlau, 2002)

Franz Jenschke, a devoted Christian, decided in 1988, when he visited his hometown Grulich, to restore the almost totally destroyed monastery, especially the chapels and the *Pilgerhaus*. Between 1988 and 2000, he collected over DM 2.6 million and almost finished the restoration of the *Muttergottesberg* (Hill of the Blessed Mother of God) shrine and monastery.

The brutal assassination of the *Karpaten-Deutsche* (refugees from the Carpathian region) and the blood bath in Prague (Praha) are well documented in various books (see references).

Historical Commentary[4]

In 1948, the Beneš Government, which had ordered the expulsion immediately after the war (Churchill, Roosevelt, Truman and Stalin agreed to it in Yalta and Potsdam), was overthrown by the Communists.

We expelled Sudeten-Germans had put down new roots in Germany, and many, like me, were now in foreign countries. The Czech people were tortured by their own leaders. The property was confiscated, the clergy was thrown into concentration camps, and our homeland became a land of destruction. The majority of the buildings collapsed, the land eroded, and the nation fell into poverty and atheism.

After forty years of a Communist paradise, the Marxist regimes in Eastern Europe and the Soviet Union collapsed, due to a misconceived, atheistic philosophy, bureaucracy and corruption.

Although the Czech Republic has now a democratically elected government, no attempt has been made to rectify the crimes committed after World War II and return the property and the land to us Sudeten-Germans. The Beneš Decrees of 1945/46, which permitted the killing, without trial, of Sudeten-Germans and "collaborators," the torture of virtually millions, the confiscation of all private property, and the "law" enabling expulsion of all Sudeten-Germans and even some Hungarians, remains in effect until now, the end of 2000. Even though the Czech Government has filed application to join the European Union and NATO, these unthinkable laws have not been removed or demanded by the Allies as a condition to join the European Union and NATO, except for Resolution No. 562 of October 13, 1998 by the U.S. House of Representatives.

At this time, over 120,000 churches, chapels and monasteries are in desperate need of repair, not even to mention restoration.

---

[4] The following material was prepared for the video documentary, *Brothers in the Storm* (see references below).

Thousands of such structures have been purposely destroyed or simply fell in decay beyond repair. Practically all of the farm buildings, small factory structures, and hundreds of thousands of homes in the former Sudetenland are gone or beyond repair.

Prior to the annexation of the Sudetenland to Germany in 1938, over 60 percent of the tax income for the whole Czechoslovak Republic, with a total population of fifteen million, came from the 3.5 million Sudeten-Germans. Money alone cannot and will not bring prosperity to these depressed regions. They will need people with high standards and a work ethic.

The expelled Sudeten-Germans, who came to West Germany, now (2000), own 1.5 homes per family, while the Czech Nation has a home ownership of 0.5 homes per family. The State of Bavaria honored the Sudeten-Germans by designating them as the fourth tribe in the state besides the Bavarians, Frankens and Schwabens.

Let us hope that the Czech people will find a just solution.

The Conspiracy of Silence

As pointed out before, the world knows all about the crimes committed by the Nazi Regime. Many Nazi leaders were justly punished. The German people are reminded daily about these atrocities by the media worldwide.

Where were the Western journalists when our women were raped and our people were tortured to death? While the Nazis committed their crimes behind heavily guarded concentration camp fences, the Soviet troops and the Czech Partisans committed even greater brutalities publicly in every village. Today, over fifty years later, not one of these criminals has been brought to trial, due to the Beneš Decrees.

When Tito, who also slaughtered hundreds of thousands of people in Yugoslavia, came to visit the United States, he was celebrated as a hero, and so it was when the Soviet leaders came.

The Western World and the United States will have to submit to the truth and discontinue the double standards. The Soviet Union was allied with the Western Powers and thus, the Western Powers of Britain, France, and the United States must share the responsibility for what happened in Eastern Europe after World War II.

The purpose of these memories and this commentary is not to cultivate hate, but to contribute to the understanding between nations, because truth is the foundation of all relations. Let us pray that God may bring wisdom to our leaders, so that they will return to the principles of the Constitution of the United States of America and the almost two millennia of Biblical teaching.

Documentation and References

Association for the Protection of the Sudeten German Interests. *Sudetendeutscher Atlas.* Munich, 1954.

Gauglitz, Franz J. C. *Landskroner Not und Tod.* Wiesentheid, 1997.

Hausner, Karl. *Brothers in the Storm: Sudetenland Documentary* (Video). Truman State University, Kirksville, Missouri, 2000.

_____*Heimat Zwischen Oder und Mohra.* Hinsdale, Ill., 1998.

Heimatkreis Mies-Pilsen e.V. (Dinkelsbuehl). *Miroeschau/ Mirošov oestlich von Pilsen—ein tschechisches Todeslager nach dem Krieg,* n. d.

Nawratil, Heinz. *Schwarzbuch der Vertreibung: Das letzte Kapitel unbewältigter Vergangenheit.* Fourth ed., Munich, 1999.

Sustek, F. *Hvězda pod Rosutící.* Morowský Beroun, 1997.

Staněk, Tomáš. *Perzekuce 1945: Perzekuce tzv. státně nespolehlivého obyvatelstva v českých zemích (mimo tábory a věznice) v květnu—srpnu 1945* [Persecution 1945: The Persecution of the So-called Officially Unreliable Population in the Bohemian Lands (Outside Camps and Prisons) Between May and August 1945]. Prague, 1996 . Recently translated into German as*Verfolgung 1945: Die Stellung der Deutschen in Böhmen, Mähren und Schlesien (außerhalb der Lager und Gefängnisse).* Trans. Otfrid Pustejovsky and Walter Reichel. Böhlau, 2002.

Turnwald, Wilhelm Karl. *Dokumente zur Austreibung der Sudetendeutschen.* Munich, 1951.

_____*Documents on the Expulsion of the Sudeten Germans,* Munich, 1953.

Sudeten German Council. *The Sudeten Question, Brief Exposition and Documentation.* Munich, 1984.

Sudetendeutscher Rat e.V. *Dokumente zur Vertreibung der Sudetendeutschen.* Munich, 1992.

Kuhn, Ekkehard, and Ludek Pachmann. *Tschechen und Deutsche—Böhmen und Mähren im Herzen Europas* (Video). ZDP, 1997 (aired, Südwestfunk).

United States House of Representatives. Resolutions No. 557 (October 9, 1998) and Resolution No. 562 (October 13, 1998).

Zayas, Alfred de. *Anmerkungen zur Vertreibung der Deutschen aus dem Osten.* Stuttgart, 1986.

Zway, Inglende. *The Crime of being German.* Lewes, UK, 1998.

May 17, 1945: The Day I Will Never Forget

HERMINE HAUSNER

I was, at the end of the war, eleven years old, my sister Gerlinde was seven, and my mother--Hermine Schwab--was seven months pregnant. For about one week the Soviet troops had been in our village of Ober-Johnsdorf, near Landskron, in the Sudetenland. Since my grandparents, Julius and Hermine Kreuziger, owned not only a farm and a guesthouse, but also a butcher shop, my grandfather had to butcher cattle for the troops. All of the young women were in hiding, including my mother, because the Soviet troops were still raping women. Thus we children were in our grandparents'· house. My grandmother was severely handicapped, crippled by arthritis, and thus the soldiers did not bother her.

My father, Robert Schwab, was not drafted during the war due to a problem with his legs. He worked in Landskron in the City Hall. My uncle, Reinhard Schwab, had finished his engineering education and worked in a factory also in Landskron. Our families did not feel in any way guilty of having harmed our Czech neighbors. That is why we did not flee before the Soviets and Czech partisans arrived.

On May 17, the situation had somewhat normalized and thus my father and uncle went to Landskron to work. Later in the morning a few truckloads of Czech partisans arrived in order to conduct a People's Court. Of course this was not known except where it was happening. The Czech partisans went to the nearby villages and collected all men between sixteen and sixty and even older and drove them on foot to Landskron. During the journey they were beaten and rifle shots were fired over their heads to prepare them for the tribunal. My grandfather, Julius Kreuziger, who was at that time sixty-five, was also among those who had to go to that court.

By early afternoon, hundreds of men were at the city square and the tribunal started. My father and uncle were among these. As they too appeared before the tribunal, they were beaten with rifles and they had to salute "Heil Hitler." Others had to kneel down in front of these judges, and Czech partisans would kick them in their genitals and knock them to the ground. My father was so severely beaten with rifles that his eyes were knocked out of his head. Half dead, he was then hanged on a lantern in the city square. My uncle

Reinhard was equally beaten and then, half dead, thrown in the fountain, where he drowned.

During the late afternoon, the tribunal resumed. Over forty men lay dead on the square or were hanging from the lanterns. The German men who were not killed were ordered into custody overnight, and the tribunal continued the next day. On May 19 all these dead bodies were thrown on wagons and hauled to the cemetery. Among those who came to view the tribunal were many Czech persons, who either wanted to see "justice" served or felt sorry for these men. My uncle, Emil Pelzl, was also among those at the City Square. Since my grandfather and Uncle Emil were known by many Czech farmers due to their cattle trading, they were both, though separately and unknown to each other, taken by the Czech farmers, removed from the square, sent home, and told to hide during the next few days until all this terror ended. At the cemetery, the other German men had to dig a mass grave. The dead bodies were thrown into it with a very ugly disrespect by the Czech partisans, who urinated on them.

Before my grandfather came home, we had heard of the terrible crimes and massacres which were committed at Landskron. My grandfather, in total frustration, decided to destroy his whole family, as he told us years later. During that night, my grandfather wanted, while we were sleeping, first to shoot us children, then the rest of the family, and after that himself. My grandmother, obviously suspicious of this, did not rest and kept us children awake. Thus, one hour after another went by. As morning broke, my grandfather gave up his plan. Terrible days and nights followed this massacre at Landskron.

On August 2, my sister Marlies was born, and thus my mother and grandparents had new responsibilities.

In the spring of 1946, our family, my mother, we three children, my grandparents, and Aunt Anni Kreuziger among others were expelled from our home, put in freight cars, and shipped to Germany. We arrived in Kaufbeuren, Bavaria, at the expellee camp, and a new life started.

In the summer of 1964, my husband and I traveled for the first time to Czechoslovakia on the occasion of a medical conference in Prague. We were already American citizens and hoped to be safe. During this trip, we also visited Landskron and the cemetery. Near the wall of the cemetery, where the mass grave was, we saw a pile of dirt and weeds of all kinds covering it. In this mass grave, where my father and uncle, along with the other men were buried, nothing—like a plaque or monument—had ever been placed over the grave. When we revisited that gravesite in the spring of 1992, we could not find the mass grave. The dirt had been leveled, and the area had been seeded with grass. Thus nobody knew any longer that men had been buried here in a mass grave.

At that time, not one of the criminals who had been involved in the murders had been brought to trial, and the Czech Government, even now, under a so-called democracy, has never found it necessary to investigate and punish those responsible. Some of these Czech criminals are still alive and still protected by the Beneš Decrees.

On September 17, 1995, we dedicated a chapel at our farm in Sauk City, Wisconsin, in memory of all the expellees. In this chapel a plaque was installed in memory of my slain father and uncle and all of the others, who suffered at the hands of these brutal criminals. The Memorial is under the motto: "O GOD, FORGIVE THEM AS THEY DID NOT KNOW WHAT THEY WERE DOING."

Exceptional Bonds: Revenge and Reconciliation in Potulice [Potulitz], Poland, 1945 and 1998[1]

MARTHA KENT

The problem I have tried to understand for most of my life is that I came out of captivity so well but found freedom so overwhelmingly difficult. I understand this issue best as a biological process of adaptation to the environment. From this perspective I will consider affectionate bonds as my most adaptive approach to the past situations of revenge and reconciliation.

The concentration camp Potulice [Potulitz] was located in western Poland near the city of Bydgoszsc, known in German as Brom- berg, in a region that had alternately belonged to Poland and to Germany. After 150 years of German dominion, it was ceded to Póland at the end of World War I. Many Germans remained in the region. When World War II broke out, the Nazi security forces were especially violent in subduing the former German West Prussia and Posen. Poland faced an overwhelming threat from the German Army. Polish authorities arrested 50,000 ethnic Germans, with the plan to evacuate them from the western region. In the ensuing outrage over the war, revenge, and suspicion of collaboration with the invader, 4,000 to 5,000 German-Poles were killed. During the worst outburst 1,000 people of German background were killed in the city of Bromberg on "Bloody Sunday" of September 3, 1939. The Nazi security forces and special commandos, or *Einsatzgruppen*, took a terrible revenge in Poland and killed 10,000 Poles in the following weeks.[2]

In 1940 the Nazi regime established the concentration camp of Potulice near Bromberg, to avenge and control acts of resistance. A compound with a capacity to hold 10,000 prisoners was built for the imprisonment of Polish civilians. In January 1945 the camp was dissolved, as the German guards and officials fled from the advancing Soviet army. In February 1945 Potulice became a concentration camp for German civilians who still found

---

[1] The main themes of this paper are treated in detail in my book. which will soon appear in German with Scherz Verlag in Bern, *Eine Porzellanscherbe im Graben: Über Gefangenschaft und wie man lernt, frei zu sein* (2003).

[2] Arno J. Mayer, *Why Did the Heavens not Darken?* (New York, 1988), 181. The figures given by Mayer of the number of ethnic Germans killed in Bydgoszcz during the event of "Bloody Sunday" vary from author to author and with the prevailing climate of opinion.

themselves in the western region of newly redrawn Poland. Polish former prisoners served as guards and commandants under the newly established Soviet power in Poland. In time, Potulice functioned as the central administrative camp for a network of camps and places of internment that operated in Poland until 1950 and later.[3]

That winter of 1945 all roads out of Poland were jammed by a huge westward escape and exodus of ethnic Germans living in and beyond Poland's new borders. My family joined the treks with two loaded horse-drawn wagons. I had barely turned five and had no understanding of the upheavals around me. Things just were that way. The running and hiding and shooting were normal. I had no other reality or perspective. My brother Gustav was ten years old then. I lately discovered that he had quite a cosmopolitan view. "That's a Ukrainian harness," he said when he visited me in Phoenix and we looked at photographs in Günter Böddeker's book on the great exodus of ethnic Germans in 1945.[4] I was impressed. For the first time I recognized how massive the dislocation had been—fifteen to seventeen million people. The wagons escaping with us had come from as far away as Ukraine.

What did I see then, when I was five years old? I saw basic properties of the things around me. A porcelain shard in a ditch had pretty blue flowers painted on it. Honey could be transparent, depending on whether there was a lot or a little in the can on our wagon. I saw mainly what was in front of my nose. The bricks in front of my face were pitted and rough. They were as rough as my father's whiskers when he hadn't shaved. "We're not leaving without him. You can shoot us," my mother told the militiamen who wanted to keep my father. We knew we wouldn't see him again. We knew they would kill him. At that brick wall we waited to be shot—seven children, my mother, my grandmother, and a Polish worker who persisted in staying with us. The militiamen let my father go. I saw the hollow stems of straw in front of my face, the ends cut clean, when we hid in a granary and soldiers searched for us, stomping and poking the straw above us. They

---

[3] There is one historical treatment of Potulice: Hugo Rasmus, *Schattenjahre in Potulitz 1945* (Münster, Westf., 1995). A contemporary journalist's perspective is Helga Hirsch, *Die Rache der Opfer* (Berlin, 1998). An old source with invaluable survivor testimony about Potulice is Theodor Schieder, ed., *Documentation der Vertreibung der Deutschen aus Ostmitteleuropa*—I., *Die Vetreibung der deutschen Bevölkerung aus den Gebieten östlich der Oder–Neiße* (Munich, 1984; reprint 1960 ed.), 2: 578-606.

[4] Günter Böddeker, *Die Flüchtlinge: Die Vertreibung der Deutschen im Osten* (Munich, 1980).

ransacked our wagons and left an old woman naked, raped, and dead. In the midst of the violence, I saw how somebody bigger held somebody smaller. I didn't see Ukrainian harnesses, but I could judge the harm around me from what I saw in people's faces—my mother's pained face when our heads were shaved in Potulice, my father's face after his first beating, a gray face in the black night of the open door.

We were imprisoned from March 1945 until July 1949. Captivity formed my early memory. I had no memories of a better childhood or of better places. Captivity was my place of origin, my native land. And Potulice was my hometown. Captivity is the measure by which I still judge important matters of life and death.

For the first two years we were prisoners on a large farm, along with several other families. Our daily routine: heavy labor, not enough food, and beatings. Then my father, my two older sisters, and my grandmother were sent to a labor camp in Bromberg. Shortly thereafter all the small children on the farm and my mother were sent to Potulice. This included me, then aged seven; my little brother aged five; and my little sister who was three years old. Gustav and an older sister stayed behind. My family was now scattered in three different places.

That spring of 1947 Potulice became our life. "*Potulice, nasza matka*—Potulice, our mother," the prisoners said of the camp in Polish. "*Zum Leben zu wenig. Zum Sterben zu viel.*" "Too little for living. Plenty for dying," they said of the camp in German. For the next two years I would see nothing but Potulice. A tall embankment with rows of barbed-wire fences encircled the camp. On the walkway on top of the embankment guards paced back and forth between watchtowers. Each gray barracks was fenced off from the next one. Nothing grew on the ground covered with cinders.[5]

In the children's barracks we children lined up in the morning and evening for a tin of black barley coffee and a slice of coarse bread. At noon we had a tin of watery soup—cabbage soup, a soup of unprocessed buckwheat that looked like bedbugs. We called it "bedbug soup." And there was "UNRRA soup," the worst soup, made from rotten meal that came in sacks stamped with the

---

[5] Rasmus, 88. Rasmus describes well the extent to which Potulice was walled off from the outside world. He calls it *hermetisch abgeschlossen*, hermetically sealed off. During my stay in Potulice, the embankment enclosing the camp gave the impression that the camp was sunk below the surface of the surrounding region, none of which could be seen from inside. It left me with quite a sensory deprivation that made the earth appear extraordinarily magnificent when we were freed.

letters UNRRA.[6] On some days we marched and sang Polish songs in our tiny patch of yard. With shaved heads and gray prison clothes we sang "*Miała baba koguta, koguta, koguta*—a little old woman had a rooster, a rooster, a rooster." On Sundays I was allowed to visit my mother for an hour. Once a month we left our barracks for a shower in the big kitchen building. From our fenced-off patch we children looked into the main yard. We saw prisoners harnessed to loaded wagons. They pulled them across the yard. My mother pulled such wagons. For punishment prisoners repeatedly ran and fell down. My mother did too.

In the sea of cruelty the few small gestures of kindness made my world good and whole. They stood out, were figure, while the rest was ground.

I remember the time my mother stole some cooking grease. She had wrapped it in a scrap of paper and had stuffed it into the tip of her shoe. She fed it to my little sister who was always on the brink of dying from starvation. My mother told my little sister the fat was a gift from a parcel one of the women had received. The truth, if found out, would have killed my mother and Elfie. How dangerous these small acts of kindness had been. Reaching through barbed wires to touch my head could have cost my mother a beating, the bunker, and standing for days in the bunker's water. With such few and fleeting acts I was cherished and could cherish in turn. Even when I didn't see my mother for a long time, I could have a sense of preciousness for everyone around me. I could look at death and tell its coming, a talent I have to this day when I walk through a hospital and recognize the dying.

One day my mother and little sister disappeared. I thought they were dead. People who disappeared were simply dead. Many months later my mother reappeared. She had been sent to a

---

[6] The "bedbug soup" was actually cooked unprocessed buckwheat. The disk-shaped kernels, the size of lentils, looked like the flat brown bedbugs in our barracks. UNRRA soup was made from a meal that came in brown paper sacks stamped with the letters UNRRA. Many years after our release from Potulice I discovered that UNRRA was the United Nations Relief and Rehabilitation Administration. But why were we fed the rotten corn meal when we were starving? It was bitter, acrid, and very unpalatable. I am indebted to Paul Boytinck for directing me to relevant Senate debates in the *Congressional Record*. UNRRA was actually prohibited by its own constitution from providing food to any German national. It did provide relief to all other countries, including Italy, an enemy country of the Allies during World War II. In the case of the rotten UNRRA soup, the Poles most likely received the rotten corn meal from UNRRA, found it unpalatable, and fed it to the prisoners in Potulice. See Kenneth S. Wherry, "Investigation of Starvation Conditions in Europe," *Congressional Record—Senate*, January 29, 1946, 509-520.

hard-labor prison because she would not give up any of her children. After my mother's return early in 1949, we were sold for a short while to a farmer for labor. I have since learned that it was called a "rent." The farmer paid the camp a rent for our use.

Potulice changed that spring of 1949. Everybody had hair. We could walk around in the main yard. One Sunday I saw a man wearing sandwich boards. The words on his chest and back said, "I have stolen from my comrades two pairs of socks, cigarettes, a spoon, a pencil...." On another Sunday a woman with sandwich boards stood on a table. The boards said "Thief." This was the man's and the woman's punishment for petty theft, something Potulice hadn't punished among prisoners before. In freedom theft wouldn't be allowed. We had to look more *normal* for when we were free, my mother explained. It was a terrible punishment, I thought. I'd rather have had a whipping.

On July 3, 1949, we walked through the gate of Potulice to freedom. I didn't know what freedom was. I couldn't imagine it. I thought it would transform me in some big way. Perhaps the change would feel like falling rain. The earth looked unbelievably beautiful and green. I had hardly seen any of it for over two years. The train took us to East Germany. From there we escaped to West Germany, to a small settlement of barracks in Hessen, a former *Stalag* called Trutzhain. I revelled in the magnificence of the earth, its creatures, flowers, and trees. The idea of greetings, of saying "Good Morning" and shaking hands, thrilled me. It was so superior to beatings and yelling.

I had left captivity with an intense sense of the preciousness of people. For years I puzzled over why I had this feeling of wanting to cherish everybody around me. I came to the conclusion that I had this inclination because it had been good for me. It had helped me live. It had let me grow up and flourish in very harsh places. There were ways of being in extreme danger without experiencing overwhelming stress. Recently I looked for similar situations. My favorite example is a man who spent six years in captivity. He had seen his wife and five children executed. He determined that he would not hate, but love and be helpful to everyone he met. After six years, he looked remarkably well, even though his rations had been the same as those of all other prisoners.[7]

Threat intensifies bonds. We see this in natural disasters. Every life is precious, and we save whomever we can. This strategy is very adaptive in disasters. It was adaptive for us in

---

[7] George G. Ritchie with Elizabeth Sherrill, *Return from Tomorrow* (Grand Rapids, 1978), 114-116.

captivity. Even my little sister instinctively knew it when she asked my mother to be nursed. She was four years old, long weaned, long past breast feeding. My mother let her nurse. Nursing increases the neurotransmitter oxytocin. It is calming, is incompatible with the stress response, and enhances the immune response. Nursing was good for my sister and for my mother. The rag doll my mother made for my sister also came with this calming parasympathetic response. One could say that I had grown up in a hurricane that had lasted four years, a time that made every life precious.[8]

We had intense caring bonds in captivity. I lost them in freedom. In August 1949 I started school in Trutzhain. I had had no schooling, no concept of school, and only a few reading lessons from my mother. With great timidity I sat in that first classroom in Trutzhain and went along with whatever was required of me. I glanced at the arm of the girl sitting beside me. The teacher told me not to turn around. Then he called me to the front of the room. He hit my hands and made me kneel in front of the class. This was worse than anything I had known in captivity; a whipping and a humiliation when I expected bonds to govern all relations. The cruelty in captivity had been the "weather;" the ordinary, usual background of my life. Indeed, we called the guards the "Weather" to warn of their approach. We protected ourselves against cruelty with bonds.

We immigrated to Canada in 1952. My parents had been farmers in Poland. They took readily to farming in our new country. I found our immigration unsettling. The loss of language was yet another dimension that took away my moorings. However, there were more powerful imperceptible forces outside language and culture that eroded the basic foundation of my life. These were part of my nature and physical makeup, of how the cells in my body responded to harm and to bonds. They were part of my manner of being in the world. My experience of bonds had not included harm. All harm came from events and people external to our family, from guards, soldiers, and our captors. The guards were a force of nature to be feared like the weather. In freedom harm could come from anyone around me. But I had no

---

[8] For a current review of the neurobiology of bonds see C. Sue Carter, "Neuroendocrine Perspective on Social Attachment and Love," *Psychoneuroendocrinology* 23 (1998): 779-818. For a conceptualization of parasympathetic autonomic nervous system responses in calming threat and their evolutionary significance see Stephen W. Porges, "Love: An Emergent Property of the Mammalian Autonomic Nervous System," *Psychoneuroendocrinology* 23 (1998): 837-861.

expectations of such harm, could not recognize it, and had no responses to it or ways of defending myself against it. Instead, I expected bonds, affiliation, appreciation, and happiness over being with others and seeing how they flourished. A child hitting another child on the playground in Hessen or in Canada seemed very strange to me. I had seen only guards, soldiers, and our masters use physical force.

When we immigrated to Canada, curiosity was often the first response we encountered. To the good people of Alberta, Canada, curiosity was simply a view of novelty. To me being conspicuous, standing out from the rest of the crowd, had killed people in captivity. Something harmless could appear very threatening to me. Conversely, something seen as harmful by people in freedom could make no impression on me. I expected bonds in various situations while freedom offered indifference and sometimes the destruction of bonds through hate and violence that overwhelmed me.

My family changed. Older siblings pursued their own interests. My parents had been so capable in taking care of us in captivity. In freedom they could do little to help us with the new struggles in school, with anything English, or Canadian. It seemed that people in freedom often stood by and let harm happen, as in the classroom in Trutzhain. In Alberta some people had "salvation," while others were "sinners." The idea of people as "sinners," as "bad," was more strange than English. It was an alien concept and an alien landscape, with nothing my mind could grasp. I developed recurrent nightmares of running over a charred hollow earth shot full of holes, dragging my little brother and sister along. If we lost our footing, we'd slip into a tear in the earth and fall through endless clouds. I had this dream already in eighth or ninth grade. I would have it for decades.

In 1960 I came from Canada to the United States to attend the University of Michigan. The violence against civil rights protesters shocked me. I had known a web of connectedness. Racism was a web dedicated to harming people. Films of Holocaust camps shocked me even more. The films showed a whole social system created for the deliberate destruction of people. This was a most extreme destruction of bonds, the extreme opposite of the sense of preciousness I had known in captivity. Then, while in graduate school, a professor greeted me with a mocking "*Sieg Heil!*" I could hardly mention my childhood in captivity. On rare occasions when the subject came up, people said that Potulice was "nothing." The mere mention of my captivity called forth statements of distress and grief over the atrocities of the Nazi era. That was the harm I should have experienced, some people said, the silence said, the photographs and film images said. Indeed, I got off lightly. "Whose fault is it, after all?" a German writer wrote of our captivity in Poland.[9] That young writer with a broken heart only said what others easily thought. My apprehension that I would perish grew with each year and each decade. Freedom became the fear that I would die.

By 1985 I found that I had lost all language for myself. I couldn't speak or write anything about myself. Such loss of language is documented in some ancient and interesting sources. Nearly four thousand years ago a Hittite king lost his speech during a severe storm. The Seventy-seventh Psalm says: "I'm so troubled that I cannot speak." John Krystal, a psychiatrist treating Holocaust survivors, found that his patients had great difficulty labeling emotions and speaking about the past. This state was called alexithymia.[10]

I didn't know of these examples of lost speech in 1985. I only knew that what I had was an acquired physiological response, and that I hadn't had it in Potulice. If it was acquired, there had to be a way out of it, I reasoned, without knowing what the way out could be.

Since I couldn't speak, perhaps I could write about my life. Perhaps the written word would lead to speech. I tried to write. To my complete dismay I found that I couldn't write either. Like a newly paralyzed person who might test his muscles for the slightest movement, I probed for what I could write. I started to describe how I learned to read from the Old Testament at our first place of captivity. Reading about the Covenant had thrilled me. We could do things with such agreements. If we had agreements, we wouldn't need yelling, whipping, and killing. In my writing I polished that scene for months. Eventually I turned to *Plattdeutsch*. It was my parents' private language, the language they spoke to each other. I had never heard anything harsh in it. I checked out a *Plattdeutsch* dictionary from the library and read *Plattdeutsch* stories. In time I found words for Potulice. Eventually I found words for the school scene in Trutzhain and the "*Sieg Heil!*" in graduate school. I imagined and wrote how I stood on a small table, as the woman thief had stood in Potulice. I imagined the teacher in Trutzhain hitting my hands, the professor shouting his "*Sieg Heil!*" and the whole graduate class yelling and throwing

---

[9] Martin Grzimek, *Trutzhain: Ein Dorf* (Munich, 1984), 107.

[10] Henry Krystal, *Integration and Self-Healing: Affect, Trauma, Alexithymia* (Hillsdale, N.J., 1988).

things at me, even though they hadn't done so. I experienced myself in that scene with the empathy I had felt for the woman thief. I was there with the neurobiology of a caring response. Then I imagined and wrote about how I stood in line and waited for death in places of great atrocity. The bricks in front of my face had the feel of my father's whiskers, as they did when we waited to be shot for him. I stood in scenes of great atrocity with a soothing response. The words I had found at the start of my writing, the words that let me return to great harm encountered in freedom, these were words that I had experienced in a certain neurobiological state.

Recent imaging studies show that cerebral blood flow decreases in the left frontal speech area of the brain when subjects are exposed to their own taped traumatic scenes.[11] Most likely my return to past harm used language that had normal cerebral blood flow and the physical state that came with those calm and unstressed experiences. How fantastic! I used to say facetiously that my writing was improving the circulation in my left middle cerebral artery, the artery that supplies circulation to the speech area. I may well have done so.

In the examples of empathy for the woman thief, the feel of my father's face, and *Plattdeutsch* I was engaging a physiologically ancient emotional template. Survival requires a successful response to threat. Since Walter Cannon's work of the 1920s, the fight-or-flight response has dominated all thinking to the near exclusion of other possible responses to threat.[12] Cannon's conception does not fit my family's situation in captivity. We could not escape or fight our very threatening circumstances. Some of our responses are better described as care, rescue, and empathy for one another. Rather than fight-or-flight, this is a soothing or calming response to threat. It is the neurobiology of bonds, of connection, and social affiliation that had allowed me to flourish. It comes with a lower metabolic rate, requires less food, and restores the body in an efficient manner, with an efficient immune response. If you had to live in extreme situations, you would want to do it with bonds and their neurochemistry of

---

[11] Scott L. Rauch, Bessel A. van der Kolk, Rita E. Fisler, Nathaniel M. Alpert, Scott. P. Orr, Cary R. Savage, Alan J. Fischman, Michael A. Jenike, Roger K. Pitman, "A Symptom Provocation Study of Posttraumatic Stress Disorder Using Positron Emission Tomography and Script-Driven Imagery," *Archives of General Psychiatry* 53 (1996): 380-387.

[12] Robert Sapolsky provides an entertaining and accessible review of the autonomic nervous system and the stress response in his book *Why Zebras Don't Get Ulcers: A Guide to Stress, Stress-Related Diseases, and Coping* (New York, 1994).

oxytocin and related hormonal and cellular functions. These block the noradrenaline, cortisol, and related responses of stress. Undoubtedly several emotion sybsystems are involved in this adaptation.

Bonds are as old as the caring response of any mammal to the distress call of its young. Indeed, bonds made possible the very life and evolution of mammals. I believe I engaged this ancient emotional template in my personal return to harm in the past fifteen years of my writing. The words that I found at the start of my writing were part of my experiences of bonds in captivity. I was able to use these to dispel the physiological hold of the stress response in freedom. This is my reconciliation. I'm using old bond responses from captivity to quiet the entrenched distress in freedom, a kind of "delayed calming." We are made in astonishing ways.

In 1998 I returned to Poland for a reconciliation with the historic past. For years Potulice had had a memorial for Polish prisoners who had perished under the Nazis. Now a group of German survivors formed an Initiative Group to mark a mass grave of German prisoners who had perished in the camp and to undertake a reconciliation between Polish and German survivors. I joined the efforts of the Initiative Group to support the concept of reconciliation and to meet people from Potulice. Since 1949 I had not heard of Potulice or met anyone from that time. Above all, I simply wanted to celebrate the astonishing acts of compassion and love I had seen—the Russian soldier who handed us a Bible and asked us to pray for him when he found us in our hiding place, the Polish worker who would not leave us, my mother who bartered a jacket for my father's life, my father who begged for food to save his children and took beatings. I had nothing to reconcile with Potulice, with the Polish people, or with Poland. When I left captivity, I felt I was a cherished child and not at all abused.

Thus, on an early September day a tour bus picked up Potulice survivors across Germany and headed for Bydgoszsc, Poland. On the day of our celebration we assembled at the cemetery of Potulice. To our surprise a huge crowd gathered, perhaps a thousand people, many more than we had expected. The parish priest of Potulice, Stanisław Zymuła, and the Protestant pastor, Klaus Zimmermann, conducted an ecumenical mass and service. A number of people spoke, including Gustave Bekker, our organizer, who had been in Potulice as a boy. Stanisław Gapiński spoke for Polish survivors. He had been imprisoned in Potulice as a boy under the Nazis. I read the Twenty-third Psalm for children who hadn't known freedom, for snatches of love that had been as important as a crust of bread, for the Russian soldier, the Polish worker, and the militiamen who hadn't shot us. Afterward our group walked around the walls of Potulice, now an ordinary prison in Poland for ordinary criminals. Only the watch towers reminded me of the past. The embankment had been replaced by tall masonry walls. To our surprise the officials let us into the prison. Inside the compound all barracks had been replaced by two-story masonry structures. The kitchen building still stood in the same central space. A guard handed out two loaves of bread. We broke off pieces and ate. An astonishing moment. Potulice still was a place of intense preciousness to me.

After leaving Poland, I returned to Germany to visit my old teacher, Rudolf Filtz, in Trutzhain. The former *Stalag* had become a village; the sturdy barracks stood preserved as homes and shops. We walked down the main street, now paved. We stopped at the cemetery. At the grave of my first teacher, I pulled out the undergrowth that covered his grave. This was my reconciliation, a reconciliation with events after Potulice, with silence, with a distorted reality that contained nothing of my life or of the bonds I had known. These aspects of freedom had harmed my sense of being in the world much more profoundly than Potulice ever had. At the same time, this has been an astonishing life project. I am exhilarated that I have found my way with it. Freedom is the choice of love. With it the deepest dungeons crumble. Without it freedom can be a jail and we the prisoners and guards.

Reconciliation to me starts as an individual endeavor and with our own experiences. It is first and foremost a restoring of bonds and connectedness within ourselves. In restoring our own relatedness, we reclaim our bonds with all human kind. I welcome the break in silence created by this conference for myself, for people with no language, for people with unspeakable events in their lives. There is a biological route to speech, to human connectedness, and to being cherished members of creation.

Addendum

Gustav Bekker reports on the developments since that first memorial celebration September 8, 1998 when Polish former prisoners and German former prisoners met and marked a grave site with a bilingual inscription: "For the German Victims of the Lager Potulice, 1945-1950, from the survivors." The contacts established on that day did not end with the memorial. Official representatives of both countries, who had attended the event, initiated discussions on the establishment of partnerships. One year later, Lechosław Draeger, the representative of the district Nakło nad Notecią, and Walter Kroker, the representative of the district Elbe-Elster, signed a partnership agreement. The Polish city of Nakło nad Notecią and the German city of Elsterwerda also entered into a written agreement. The aim of these agreements was to establish and support cultural, economic, and personal contacts between the people of both countries.

Dr. Bekker reports on the rapid developments that followed the agreements. Lively exchange between youth orchestras and sports groups came into being. Official delegations of district representatives and city parliaments exchanged visits.

Particularly important are the exchanges between youth groups. There are multiple contacts between the students of the Gymnasium at Elsterwerde and the Lyceum at Nakło nad Notecią These students are working on a joint history project to research the tragic events of this region during the Second World War. Survivors from Germany, Poland, and the U.S.A. are being interviewed. Students in the city of Nakło nad Notecią learn English and German. Students in Elsterwerda learn Polish. Dr. Bekker reiterates the goal of these efforts: it is to learn from personal tragic histories so that children and grandchildren can have a peaceful and happy future in a united Europe.

A Polish high school student who visited one of the resulting reconciliation meetings in Potulice became the writer of one of the first high school honors essays on the concentration camp Potulice. The writer was Jakub Leszscyński, who completed his paper at the International Baccalaureate School in Gdynia in the spring of 2001. In his essay Mr. Leszscyński asks one central question: Was Potulice a "displacement" camp or was it a "labor" camp. He comments on the lack of knowledge about Potulice in Poland. The existence of a Polish camp for German prisoners from 1945 to 1950 was a secret, one supposedly guarded by the Communist government. Moreover, there was even very little knowledge of the fact that Potulice had been a camp for the Polish people themselves, imprisoned by the Nazi regime from 1941 to 1945. Mr. Leszscyński found very few newspaper articles on the imprisonment of Poles in Potulice.

To answer his question, Mr. Leszscyński interviewed survivors of Potulice. He found that Polish former guards of Potulice described Potulice as a "displacement" camp. When Mr. Leszscyński interviewed German survivors, he found that they said something very different of Potulice. To them it was a hard-labor camp. The question concerning the nature of the camp at Potulice is treated in detail in six parts of the essay.

According to the *Encyclopedia Britannica*, displacement camps accommodate temporarily a large number of displaced persons that require resettlement. Concentration camps confine political prisoners or members of national or ethnic groups for state security, exploitation, or punishment without trial. In work camps prisoners are forced to work under very harsh conditions, without remuneration, and for benefit of the camp authorities. Armed with these definition, Mr. Leszscyński reviews the information he has collected.

The essay details how the German people were forcibly taken into captivity. These people carried out forced labor on farms and in the workshops in Potulice. They were "rented" out to whoever needed labor and could pay the "rent" to Potulice. On the farms and in the camp the prisoners were not fed well and they were beaten. One of Mr. Leszscyński's informants, Ms. Elen Guse, developed night blindness from malnutrition. At the end Mr. Leszscynski states, "I believe that Potulice was a hard-labor and not a displacement-camp." He concludes that the graves of Potulice are a reminder of the harm and tragedy produced by revenge.

This is a courageous essay. It asks an important and difficult question and, in the process, uncovers important truth, suffering, and moral culpability on a scale that touches the historical experiences of two peoples. May the young students working on this subject find new affirmations of life in a common tragedy.

Remarks by a Survivor

ERICH A. HELFERT[1]

Reflecting on the many excellent scholarly presentations we heard over the past three days, I would like to suggest that, while there is no denying the importance of carefully documented research and detached analysis of historical facts, statistics, and trends, we must remember that history is, in essence, the accumulation of human actions and experiences by leaders, followers, and ordinary people. Let us always be aware that factual narrative and interpretation of history is really rooted in a never-ending sequence of human striving and failure, of triumph and tragedy, of victors and victims. Tonight I would like to provide a little of this human perspective from personal experience.

Surviving, as a young boy, an upheaval on the scale of the Sudetenland ethnic cleansing in 1945 to 1947, causes deep emotional scarring, which persists throughout adult life. It serves a first-hand lesson in history as well, because the question of "why so much cruelty and mindless hatred from formerly peaceful neighbors" begs for some explanation and understanding. Writing a book about these experiences after fifty years, in the form of highly visual and personal vignettes within a larger historical frame, was both an emotional trauma and a welcome release. The catharsis of writing the story enabled me to view my personal fate as part of a larger drama, and gave me the perspective that allowed me to understand and even to forgive, without altering the fact that great injustice and crime reigned at the time.

I wrote the book to show the human impact of disastrous geopolitical machinations on ordinary people. I believe the main fault lies with political leaders whose virulent nationalism encourages followers to act out their darkest instincts. I also wrote the book to tell about an ignored and forgotten chapter of European history, when 3.5 million Sudeten German people were expropriated and displaced in a spasm of nationalistic retribution, despite hundreds of years of peaceful coexistence, while the victorious Allies stood by and did nothing to prevent it. Finally, I wrote the book to demonstrate, from my own experience, that there are good people on all sides, even in the darkest of times. Humanity is never fully suspended—even when incited terror reigns.

My family and I lived through the Russian conquest of our

---

[1] Dr. Erich A. Helfert is author of *Valley of the Shadow*, an autobiographical account of the expulsion of Germans from the Sudetenland.

city, Aussig on the Elbe, in May 1945, with some close encounters that luckily caused no harm to us, while others suffered badly. Soon thereafter the newly installed Czechoslovak government started waves of expropriation and primitive expulsion of the residents in whole segments of the city. Early expulsions were so cruel and disorganized and caused so many casualties that the Russian area commander forced the Czechs into a temporary halt. We then learned that my barely seventeen-year-old brother had been killed in a tragic accident during the last days of the fighting. Facing the rumored prospect of being shipped to labor camps in Russia, the three of us—my mother, my father, and I—after long deliberations, decided to commit suicide. We were about to turn on the gas and drink the poison, when we were saved by a cousin from Prague, who unexpectedly appeared and whose mixed ethnicity now placed him on the Czech side. He and my aunt behaved greedily, trying to save belongings and scheming to hang on to the villa, not caring what would happen to us. Despite their machinations our villa was soon expropriated by well-connected bureaucrats. We were marched off with fifty pounds of belongings each, in bundles that had first been picked over by them, and we were temporarily placed in a workers' flat in the industrial district. There my father, an executive previously under great pressure from the Nazi war machine, and now from Czech officials working with him on the takeover of the company, required urgent medical treatment. He died a terrible death in the now Czech-run hospital, through incompetence—or likely worse—from a botched stomach operation.

My mother and I lived for another eight months in the workers' flat, and as a young male I had to work prior to being expelled. I was lucky to have a menial job first in a German bakery, and then in a Czech bakery, where I was very humanely treated, and was able to bring home some scraps of food. Working there I could see through a back window our villa way up on the hill, which represented another life—now remote and almost unreal.

During this time we witnessed many expulsion scenes and dangerous altercations, and had several close encounters ourselves. Once I almost wound up in the worst concentration camp run by the Czechs because my white armband, which we were forced to wear to be known as Germans, had shriveled in the rain from the required width of 5 inches. I was grabbed by a gun-toting soldier and only a miraculous and humbling intervention by my mother saved me from what was likely death under inhuman conditions. Another time my mother was caught in an extortion scheme run by Czech guards of German prisoners working to clear bomb damage. Her papers were confiscated and she had to report to the concentration camp in person. She went there as if in a nightmarish dream, knowing she would never return alive, and again was miraculously saved when the commander, a known chauvinist, inexplicably let her go. A third event was the massive explosion of a large munitions dump, which I witnessed from afar, and which became the signal for sudden brutalization and murder of white armbanded German civilians by rampaging mobs of mostly young toughs in the streets, on the river bridge, and in the city squares, with people beaten, drowned and thrown into the river, and many others marched off to the camps. I barely escaped, wildly riding my bike through parts of the upheaval. By the way, it is now understood that this affair was a planned act of terror by a radical Czech group.

All this time my mother, suffering from chronic neck pains, went from bureaucrat to bureaucrat, under often dangerous conditions, to obtain a permit for us to travel about one hundred miles on the train. We sought to join her sister and family to the west in the American sector of the Sudetenland, and to be expelled together from there into Western Germany. After months of grueling queues and many disappointments she succeeded, and we joined the family. Three months later we were placed in a primitive holding camp, waiting for the next train of boxcars to transport us into Germany. A minor incident at the camp led to a break in family relations, and we were separately transported into war-ravaged Western Germany. There the difficult life of a homeless refugee began, with hardships, discrimination, and crowding into other people's homes. But a new life gradually took root. I was able to continue my schooling, and four years later I won a scholarship in the United States. My mother followed later, and I was able to work to achieve an excellent education and pursue a multi-faceted career.

What are the insights and conclusions a survivor can draw from these experiences? Apart from the wider geopolitical and human rights issues so well presented and discussed in this conference, there are several personal points I would like to make. The most important is that whatever the circumstance, one must never give up one's humanity. We learned that it is usually the "simple" people with little to gain or lose who show the greatest empathy and understanding. Also, the power of faith is immense in desperate circumstances, and we experienced a series of minor miracles that enabled us to go on.

I am often asked whether I carry feelings of bitterness and revenge. I do admit to sadness, because mankind seems unable to learn from these experiences and appears condemned to repeating such follies again and again. But I do not believe in hatred or revenge, because both are self-defeating and merely ensure the continuation of a vicious cycle of crime and retribution. I do believe, however, that the law should be applied swiftly and firmly against those leaders and individuals who incited or committed such crimes, and that widespread public condemnation should be made of injustices done. Unfortunately, in the case of the Sudeten ethnic cleansing crimes a general amnesty granted to all Czechs soon after these events has never been repudiated, and the Czech government to this day is not admitting to the illegality and human rights violations involved in the indiscriminate mass expulsions. Moreover, the infamous official decrees by the then Czechoslovak president Beneš, which "legalized" the maltreatment, expropriation and expulsion of the indigenous Sudeten population still remain on the books.

I am also often asked whether I have been back to my home town. My answer is no, because there is no one left there to see, and I do not wish to deal with the emotional impact of re-visiting past experiences that have taken so long to heal. If I picture walking up to our villa and meeting strangers there, I have only a sense of futility. There would merely be negative emotions on both sides that are better left buried. The situation is no longer reversible at this point, especially since a new generation is now living in our properties, who had no direct involvement in the illegalities of fifty plus years ago. The Sudetenland has joined so many other parts of the world where ethnic cleansing has wiped out most traces of cultures and traditions many centuries old, and has permanently displaced whole peoples from their rightful place in their own corner of the world.

The main issue remaining for all nations is unqualified recognition and condemnation of the illegality and injustice of such actions, and taking a firm stand against similar transgressions that are evolving in our present time. Such recognition should also include past actions of this kind, held up to international law, and some effort for at least a token compensation should be made. Dealing with the past in this way clearly becomes increasingly difficult, the more time has passed, but such actions would serve to tangibly underline the notion that inhumanity cannot be ignored.

I strongly feel that as individuals we must do all we can to practice tolerance and forgiveness, because we simply cannot afford to feed renewed waves of hatred and retribution. This is a difficult challenge indeed, but is there really another option for humanity? I should add that the Sudeten organizations formed in Germany right after the expulsion openly and immediately renounced any form of violence and retribution, and all along steadfastly encouraged peaceful means to obtain recognition and some form of restitution. This attitude never changed during all these decades, despite the indifference surrounding them. As it stands, however, a full and true recognition of the Sudeten tragedy has not been achieved in the postwar political climate of Europe, especially after the fall of the Soviet Empire which brought the renewal of relations between east and west. Thus another chapter in man's inhumanity to man fades into history.

Recapturing the Spirit of Nuremberg: Published and Unpublished Sources on the Danube Swabians of Yugoslavia

RAYMOND LOHNE

In a recent study of human rights in the modern era, William Korey entitled one of his chapters, "Recapturing the Spirit of Nuremberg." In that chapter we find his 1997 interview with Antonio Cassese, an Italian Jurist and scholar, who talks about the "mission" of the International Criminal Tribunal.

According to Cassese, the Tribunal's most important mission was: "to hear and record for posterity the stories of those who have suffered in the camps and killing fields..., and to dispense justice on that account in the name of the international community. The worst nightmares of the victims at Auschwitz and at the other Nazi concentration camps was that once they were free, people would not listen to the horrors they had suffered or would be indifferent or disbelieving. Like the Nuremberg Tribunal..., the International Criminal Tribunal forever would document and illuminate the horrors so that all can see and know."[1] The commission also had a second mission, namely to achieve "justice's cathartic effects." Cassese noted that witnesses before the Tribunal at The Hague reported "great relief" after their testimony. In his view, this catharsis promises "hope for recovery and reconciliation" both for the individual victims and for human society.

While the victims of Nazi concentration camps have received much attention, no international commission has ever investigated the fate of the Danube Swabians after the Second World War. Indeed, the German victims of ethnic cleansing in the postwar period have found little "relief" of any sort. They were never given the chance to speak about their trials and tribulations, even though well over fifty years had elapsed since these atrocious events. With this essay I hope to recapture something of the Spirit of Nuremberg for the Danubian Swabians (*Donauschwaben*). In doing this, I hope to aid the reconstruction of a history that is virtually lost, including the history of those who emigrated, and those who were left behind in Tito's Yugoslavia. In the following, I

---

[1] William Korey, *NGOs and the Universal Declaration of Human Rights:A Curious Grapevine* (New York, 1998). Unfortunately, this important book does not mention Yugoslavia and the Danube Swabians.

will attempt to give an overview of the published sources indispensable for reconstructing Danube Swabian history, and also suggest other directions of research needed for this reconstruction.

It is important to point out already at the outset that until recently the history of the ethnic cleansing of the *Donauschwaben* has been uniformly neglected. Not even the most recent studies on human rights mention the Danube Swabians. These include Paul Gordon Lauren's *The Evolution of International Human Rights*, Winston E. Langley's *Encyclopedia of Human Rights Issues Since 1945*, and Daniel Chirot's *Modern Tyrants*, none of which touch upon Tito's Yugoslavia.[2] This also holds true for Time-Life Books' photo-history, *Partisans and Guerrillas*, in which free-lance journalist Ronald H. Bailey gives the Yugoslav wartime background in a popular history setting. In his otherwise excellent survey, he too misses the "Schwobs"—as Danube Swabian call themselves in their own special German dialect.[3]

Important, if brief, discussions of the Danube Swabians can be found in the *Harvard Encyclopedia of American Ethnic Groups*, as well as in an essay by Victor Greene in *The Encyclopedia of American Political History*. The relevant essays in the former were authored by William S. Bernard and Frederick Luebke, while the pertinent study in the latter was written by the well-known immigration historian Victor Greene.[4]

Older European scholarship is more thorough. A good example is the *Handwörterbuch das Grenz und Auslands Deutschtum*, published in 1936, with approximately 800 contributors. This book is still a trove of information and a standard reference work for the history of the diaspora Germans in the pre-expulsion period. But it leaves off in the mid-thirties, and thus has nothing to offer on the expulsions and ethnic cleansings.[5] Anton Scherer's three relevant

---

[2] Paul Gordon Lauren, *The Evolution of International Human Rights: Visions Seen* (Philadelphia, 1998); Winston E. Langley, *Encyclopedia of Human Rights Issues Since 1945* (Westport, CT, 1999); and Daniel Chirot, *The Power of Prevalence of Evil in Our Age* (New York, 1994).

[3] Ronald H. Bailey, *et al.*, *Partisans and Guerrillas in World War II* (Alexandria, VA, 1978).

[4] See Frederick Luebke, "Austrians," and William S. Bernard, "Immigration: History of U.S. Policy," in *Harvard Encyclopedia of American Ethnic Groups*, ed. Stephan Thernstrom (Cambridge, MA, 1980), 164-171, 486-496; Victor Greene, "Immigration Policy," in *Encyclopedia of American Political History*, ed. Jack P. Greene, vol. 2 (New York, 1984).

[5] Carl Petersen, Otto Scheel, Paul Hermann Ruth, and Hans Schwalm, eds., *Das Handwörterbuch das Grenz und Auslands Deutschtum* (Breslau, 1936). I must thank Leo Schelbert for bringing this work to my attention.

works, his *Donauschwäbische Bibliographie 1935-1955*, his *Südosteuropa Dissertationen 1918-1960,* and his *Die Nicht Sterben Wollten*, contain much information about the older scholarly literature on the *Donauschwaben*.[6]

One of the best relevant recent studies from Europe is *Leidensweg der Deutschen im Kommunistischen Jugoslawien*, especially the first volume of this work co-authored by Josef Beer, Georg Wildmann, Valentin Oberkersch, Ingomar Senz, Hans Sonnleitner and Hermann Rakusch.[7] This work is built upon oral testimonies, which—as we shall see below—are vital sources of historical information.

The standard work of personal testimonies is the five-volume *Dokumentation der Vertreibung der Deutschen aus Ost-Mitteleuropa*, which is based on the enormous *"Ost-Dokumentation Project "* carried out by German historians after World War II. The goal was to elicit testimonies from as many survivors of the expulsions as possible. The resulting affidavits and protocols were placed into the German Federal Archives, where they remain even today as invaluable sources for the largest ethnic cleansing in human history. The leaders of the research project, including Hans Rothfels and Theodor Schieder, made selections from the massive collection and published them in the above-mentioned five-volumes. The volume devoted to the expulsion of Germans from Yugoslavia was published in 1961 under the title *Das Schicksal der Deutschen in Jugoslawien*. It includes a hundred-page introductory study that establishes the background for the mass expulsion of the Danube Swabians. The personal views recorded in the documents and affidavits, and published in this volume, are as important as they are horrifying.[8]

Géza Charles Paikert wrote several works on the Danube Swabians, and two of which are especially helpful. They are *The German Exodus* (1962), and *The Danube Swabians: German Populations in Hungary, Rumania and Yugoslavia* (1967).[9] Paikert was a Ca-

---

[6] Anton Scherer, *Donauschwäbische Bibliographie 1935-1955* (Munich, 1966); *Südosteuropa Dissertationen 1918-1960* (Graz, 1968); and *Die Nicht Sterben Wollten: Donauschwäbische Literatur von Lenau bis zur Gegenwart* (Graz, 1885). I would like to thank Melvin Holli for his help about these sources.

[7] Joseph Beer, *et al., Leidensweg der Deutschen im Kommunistischen Jugoslawien*, vol. 1 (Munich-Sindelfingen, 1992).

[8] Theodor Schieder, ed., *Dokumentation der Vertreibung der Deutschen aus Ost-Mitteleuropa*, 5 vols. (Bonn, 1953-1961).

[9] Géza Charles Paikert, *The German Exodus: A Selective Study on the Post-World War II Expulsion of German Population and Its Effects* (The Hague, 1962); and *The Danube*

nadian-born Hungarian bureaucrat—his parents had spent some years in Canada at the turn-of-the-century—who studied at the Austro-Hungarian Naval Academy, and then earned a law degree at the University of Budapest. Subsequently he headed the Division of Cultural Interchanges of the Hungarian Ministry of Education, which post he occupied from 1938 until Hitler's invasion of Hungary on March 19, 1944. He was then forced to take refuge with the Swedish Red Cross in Budapest, where he stayed until Soviet troops entered the city in early 1945. Paikert left Hungary soon after World War II, and then emigrated to the United States. In 1947 he was appointed professor of history and political science at Le Moyne College in Syracuse, New York. His personal experience in these matters lends special weight to what he has to say, though at the end of the brutal twentieth century, one must say that Paikert's assessment of the aftermath of the ethnic cleansing of Danube Swabians is both unnecessarily cavalier and historically unsound. I refer especially to a passage in the book's conclusion: "Merciless surgical operation and flagrant denial of human rights as the eviction (both *de jure* and *de facto*) of the Swabians certainly was, it had yet some features which—as is often the case with the most ruinous disasters—proved to be ultimately positive. It reduced ethnic tensions in exactly that part of Europe which has been most susceptible to such problems in recent history. Hungary, Rumania and chiefly Yugoslavia were rid (at however debatable a stroke) of an ethnic minority group considered too great a hazard to their national security and too disruptive to their ideological uniformity."[10] The idea that positive steps arose regionally from such a dreadful episode might well be challenged, but Paikert still deserves a prominent place in any bibliography on the *Donauschwaben;* and this is so even if one has reservations about some of his conclusions.

A general survey that is very helpful on this topic is Kurt Glaser's and Stefan T. Possony's Victims *of Politics: The State of Human Rights*. These authors note that Yugoslavia, reorganized as a "peoples' federation" in 1946, was itself the product of ethnic strife within the Habsburg Monarchy. Born out of the convulsions following World War I, and conceived as a "Greater Serbia," Yugoslavia too was a multinational state that was always rent by ethnic cleavages and nationality conflicts. From an ethnic point of

---

Swabians: German Populations in Hungary, Rumania and Yugoslavia and Hitler's impact on Their Patterns (The Hague, 1967).

[10] Paikert, *The German Exodus*, 302.

view there never was such a person as a "Yugoslav." The state contained five distinct, officially recognized major Slavic nationalities: Serbs, Croats, Slovenes, Macedonians, and Montenegrians. They were complemented by eight smaller ethnic groups, among them Germans, Hungarians, and 750,000 Muslim Slavs, all of whom maintain a strong sense of separate ethnic identities. The most significant divisive issue was animosity between the four-million strong Croats and the seven-million strong Serbians, which was rooted in their distinct national traditions, and the resulting deep-seated urge for greater ethnic self-determination. This animosity and distrust, for example, caused the Croats to take the German side in World War II. Not even a national hero like Marshal Tito—whose past injustices and atrocities, like those of his mentors in the Kremlin, tend to be forgotten—could afford to ignore these ethnic divisions.[11] These observations—written in the present tense a decade and a half before the collapse and disappearance of Old Yugoslavia—are still valid today.

An even more prophetic source may be found in Leopold Rohrbacher's journal *Neuland*, published for some decades after 1948. This periodical, especially the run from 1948 through 1958, is filled with information about the *Donauschwaben* and their situation in Germany, Yugoslavia, the United States, the Soviet Union, Brazil, and Argentina, as well as about their perception of the United Nations and the International Refugee Organization. Most instructive is also Rohrbacher's dramatic history of his people, *Ein Volk Ausgelöscht: Die Aussrottung des Donauschwabentums in Jugoslawien 1944-1948*, published only four years after the war.[12] In writing this synthesis, he drew on a wide array of sources, many of which have remained unmined ever since. Not even Paikert has used them in writing his own history of the Danube Swabians.

In addition to the above-mentioned published works, there are also various unpublished sources for the historical reconstruction of the history of the Danube Swabians and for their ethnic cleansing in Yugoslavia. One unused source consists of the extensive records kept on the postwar emigration of the Donauschwaben to the United States. The heating up of the Cold War resulted in an increased interest in Danube Swabian immigrants. Their knowledge of Tito—until 1948 Stalin's servile favorite—and of his dictatorship was of special interest to the American intelligence

---

[11] Kurt Glaser and Stefan T. Possony, *Victims of Politics: The State of Human Rights* (New York, 1979), 159.

[12] Leopold Rohrbacher, *Ein Volk Ausgelöscht: Die Aussrottung des Donauschwabentums in Jugoslawien 1944-1948* (Salzburg, 1949).

community. Hence, access to the secret files of the FBI, the OSS/CIA, the Counter-Intelligence Corps of the U.S. Army, the INS, and the Justice Department's "special investigations" could produce much additional information on the Danube Swabians, on post-World War II Yugoslavia, and on the connection between the Danube Swabians and the German-American community. Future studies on the history of the Danube Swabians will also require access to the "secret" work of the Displaced Persons Commission, as well as of the intelligence files of the State Department.[13]

Yet perhaps even these sources will not prove as vital as the personal testimony ("oral history") of the survivors of ethnic cleansing after World War II. There simply is no substitute for the testimony of people who were there. The first task of researchers into the Danube Swabians must therefore be to record the stories of the survivors, most of which remain unrecorded. Alfred de Zayas has been a pioneer in this regard. In his ground breaking studies on the expulsions, he has made available many survivors' accounts derived from interviews.[14] But much more work remains to be done, and a large number of the living "sources" reside in the United States. In my own case, reading the works of de Zayas spurred me to make contact with survivors in the Chicago area, and from my first interview (with Joseph Stein), I understood the extent to which simply telling his story was both emotionally draining and cathartic for him. It has become evident to me that the Danube Swabians have lived the "worst nightmare" of survivors for over half a century. They deserve the catharsis described by Cassese. Moreover, relating their story contributes to the available historical sources in an area marked by neglect since the original collections for the *Ost-Dokumentation* in the early fifties.

I was especially gratified to be able to interview Susanna Tschurtz, a German-American artist, who came to the United States from Yugoslavia after World War II. Her art reflects an attempt to come to terms with the worst horrors of the experience of the Danube Swabians. Her story is full of meaning for the kind of

---

[13] At the Duquesne University Conference on Ethnic Cleansing in Twentieth Century Europe, John Schindler of the Department of Defense in Washington, DC, was of the opinion in a private discussion that the Department of Defense's intelligence records may not be of much help in reconstructing the historical period in question.

[14] See Alfred Maurice de Zayas, *A Terrible Revenge: The Ethnic Cleansing of the East European Germans, 1944-1950* (New York, 1994); and his *Nemesis at Potsdam: The Expulsion of the Germans from the East*, 4th ed. (Rockport, Maine, 1998).

catharsis Cassese describes. Susanna Tschurtz was born and raised in a German farm village, which since 1920 has been part of Romania. Her village was occupied by the Red Army in the autumn of 1944, after which she came to experience war and flight at a tender age. Forced to live a refugee existence for eight years, she emigrated to America under the provisions of the Displaced Persons Act of 1948. In the United States she rebuilt her life, married, raised three sons, and graduated from Northwestern University. Her impressive *War Series* paintings initially left me speechless, for in her art was the essence of testimony. Humbled by the history, she had painted scenes of her own life and the lives of her people. I am convinced that there is an urgent need for a complete study of all of her work, starting with *Illustrations on the Theme Annihilation Camp Rudolfsgnad*, which was published in Maria Tenz Horwath's *The Innocent Must Pay*.[15]

Susanna Tschurtz's art can point the way to another source for the reconstruction of the Danubian Swabian expulsion, namely artistic manifestations. Among the many expellees, more than a few have left, or still produce, bodies of art that can help us visualize the range of experiences during and after the expulsions. Sebastian Leicht's work, for example, is highly useful in this regard. His *Weg der Donauschwaben: Dreihundert Jahre Kolonistenschicksal* is a collection of art, which includes both his *Im Todeslager* and *Die Verlorenen*. These two pieces make very clear the "mechanics" of ethnic cleansing among Yugoslavia's Germans. They are powerful enough to help us understand also the postwar American response to the refugees. The vision that emerges from the art work of Tschurtz and Leicht is both a magnificent saga of German culture in Southeastern Europe, as well as a sad and sordid tale of misguided politics, war, and inhumanity. At the same time it can also be viewed as a story of redemption, both of the individual and of an entire people.

Naturally, there are many other Danube Swabian artists and sculptors I could discuss in a more detailed and intensive study of this kind. Space limitations do not permit me to do so. At the same time I must mention the names of painters such as Erna Moser, Josef de Ponte, Hertha Karasek-Strzygowski, Oscar Sommerfeld, Franz Ferch, Konstanz Frohm, Hans Roch, and Hans Rastorfer, as well as the names of sculptors such as Hermann Zettlitzer and Wilhelm Gösser. There is also the life and work of Peter Kraemer,

---

[15] Maria Tenz Horwath, *The Innocents Must Pay: Memoirs of a Danube German Girl in a Yugoslavian Death Camp, 1944-1948* (Bismarck, N.D., 1991).

another survivor of Rudolfsgnad, which is as yet completely unexplored.

In addition to interviewing survivors and reviewing the secret files mentioned above, invaluable evidence needs to be gathered on location, in Yugoslavia, where once concentration campus stood and where the bodies of thousands of their dead are buried. In particular, an expert analysis of the topography of certain villages in Voivodina is called for. Using the analytical power generated by Geomatic research—such as that done by Scott Madry and Carole Crumley in a section of Burgundy in France—much relevant information could be gathered. These scientists tell us that the integration of advanced remote sensing within the context of a GIS [Geographic Information Systems] has significant potential for regional archaeological, cultural, and environmental research applications. Various researchers around the world are using these tools today. Some of these applications include the location of new archaeological features, such as sites, road segments, and field patterns, as well as determining current land use and land cover. SPOT, Landsat, RADARSAT, ARIES airborne scanner, and traditional aerial photography and aerial and field survey data have been collected and analyzed over a period of twenty years for this project. Traditional photo analysis and photogrammetry have been conducted, as well as advanced digital imaging processing of data.[16]

This research is especially exciting because "visualization and simulation technologies have advanced rapidly, and extend our ability to view and understand the research area, even when we are not able to be there in person."[17] Thus this tool can be used to locate and mark the Danube Swabians' mass graves, using the technology of Global Positioning Satellites.

Once located and marked, these sites should be excavated using the techniques of forensic archaeology as practiced by William Haglund. This should be done only after ground-penetrating radar has mapped the site prior to opening. This is the proper physical evidence research agenda for documenting what happened to the Danube Swabians of Yugoslavia. After the research, legal claims might be made as indicated by the evidence. Both historical and personal "closure" might be achieved through this process.

---

[16] Carole Crumley and Scott Madry, "GIS and Remote Sensing for Archeology: Burgundy, France." Taken from WWW.informatics.org/france/rsgis.htlm on December 2, 2000. My thanks to Emily Franconia for her assistance with most of my technical and intelligence-related questions.

[17] Ibid.

I recognize that we could not really "Recapture the Spirit of Nuremberg" at the Conference on Ethnic Cleansing at Duquesne University, because mere recognition of a crime is only the beginning of justice, but the ethos of the conference was justice. Indeed, the search for truths that lead to justice seems to be a component of the larger spirit of the age. In countless ways, war crimes tribunal activities, discussions of reparations payments, and quiet reconciliations can be found in the news media over the past few years. Many nations have attempted the difficult task of confronting their history, exploring past crimes, paying their dues, finding closure, and moving on to new social, political, and economic futures. All this amounts to the expression of hope. Perhaps this hope can come to the *Donauschwaben* as well.

Ethnic Cleansing and the Capathian-Germans of Slovakia

ANDREAS ROLAND WESSERLE

One of the most fateful meetings in many centuries of European history took place in the city of Pittsburgh toward the end of the First World War. On May 30, 1918, twenty-nine spokesmen for Czech and Slovak émigrés—including the well-known Thomas G. Masaryk—signed the Pittsburgh Agreement, which called for the establishment of a common Czechoslovak state, to be carved out of Austria-Hungary, which was to be dismembered.[1] The territories in question had been part of the Austrian and Hungarian realms since the late ninth century—in the case of Hungary for over a thousand years, and in the case of the Austrian realm since 1526, when the Czech nobility elected Ferdinand of Habsburg their king and thus opted to become part of the Habsburg dynastic state.

A still more radical departure for the past was the fact that the new Czechoslovak state was to be created on the basis of alleged Slavic ties. It rested on the ideology Pan-Slavism, an ideology that emphasized ethnic factors. It had been loudly proclaimed at the Pan-Slav Congress of Prague in the revolutionary year of 1848, when Prague was still largely German in language and culture. Masaryk and his disciple and close confidant, Eduard Beneš, overlooked the fact that even according to a census of 1921, the Czechs were a minority of 6.7 million in this new state, for they constituted only 49.3 percent its population. At the same time the Germans numbered 3.2 million (23.6 %), the Slovaks 2.03 million (14.9 %), and the Hungarians 800,000 (5.8 % even after large-scale expulsions). Moreover, there were also 465,000 Rusyns or Ruthenians (later arbitrarily reclassified as Carpatho-Ukrainians), and

---

[1] Popular conviction holds that the Pittsburgh Agreement—also known as the Pittsburgh Pact *[Česko-Slovenská Dohoda]* was signed by Thomas Masaryk and Eduard Beneš for the Czechs, and Gen. Milan Štefanik for the Slovaks. But the text of the Agreement clearly shows that of these three, only Masaryk was present at the signing. The signatories included fifteen Slovaks, representing the Slovak League in America, and fourteen Czechs (among them Masaryk), representing the Czech National Federation and the Czech Catholic Federation. The Pittsburgh Agreement had no legal status in international law, but it did serve as a source of inspiration for the creation of a Czechoslovak state. The Pittsburgh Agreement has been reproduced photographically many times, including in the following work: Lajos Mérey, *A pittsburghi szerződés és autonómia* [The Pittsburgh Agreement and Autonomy] (Nyitra, 1929), 2.

76,000 (0.56%) Poles.[2] Since the Pittsburgh Pact was concluded under the benevolent aegis of President Woodrow Wilson and his Fourteen Points of January 8, 1918, the signatories would have done well to heed point no. 10, which states the following: "The peoples of Austria-Hungary, whose place among the nations we wish to see safeguarded and assured, should be accorded the freest opportunity of autonomous development."[3] But Masaryk and Beneš were not satisfied with autonomy. They wanted a fully independent Czechoslovak state, even at the expense of violating the principle of national self-determination, in the name of which the new country was created. Moreover, they even overrode the wishes of some of their Czech compatriots who wanted to remain part of the Habsburg Empire. The latter included sizable factions of the Social Democrats, the Agrarians, and the Clerical parties.

The Slovaks, who hoped to be equal partners in this new dualistic state, were also soon disappointed. Having lived apart for a whole millennium, they were very different from their Czech neighbors to the west. This was manifest in their history, culture, general outlook, and even language. They soon felt just as unappreciated in the new state as did other ethnic groups who differed from Masaryk and Beneš in language or ideology. Their fate seems to have been foreshadowed when in 1919, their main spokesman, General Milan Štefanik, died in an airplane accident.[4] The results were fatal not only to the Slovak-born French general, but also to the general amity in the new country.

With time, the movement for greater Slovak autonomy gained ground despite Czech attempts to suppress it. It was spearheaded by the Slovak Catholic priest Father Andrei Hlinka, who before World War I had been close to Archduke Franz Ferdinand's so-called "Belvedere Circle." After the war Hlinka became the primary spokesman of the Slovaks, for which he suffered repeated political persecutions at the hands of the Czech bureaucracy.[5]

---

[2] Emil Franzel, *Der Donauraum im Zeitalter des Nationalitätenprinzips* (Bern, 1958), Anhang. By 1930 the demographic situation had changed in favor of the Czechs and the Slovaks as follows: Czechs—50.9%; Slovaks—6.2%; Germans—22.3%; Magyars or Hungarians—4.8%; Rusyns or Ruthenians—3.8%; others—2.0%.

[3] René Albrecht-Carrié, *A Diplomatic History of Europe Since the Congress of Vienna* (New York, 1958), 354.

[4] Some claim that Gen. Štefanik's airplane was purposely shot down by Czech anti-aircraft fire, so as to eliminate the man who could have spoken up for Slovak national interests. This claim has never been conclusively proven.

[5] I am proud to report that a cousin of my father, Deacon Toni Wässerle, the well-known *Heimatforscher* and cultural leader of the Western Slovak *Hauerland* Ger-

Masaryk's goal of creating another "Switzerland" in Central Europe remained elusive. The centrifugal pressures exerted by the cultural and political ambitions of the five major nations within Czechoslovakia proved to be too much. Attempts by a small elite group of radical Czech politicians, headed by Eduard Beneš, to channel or to suppress these efforts, and to align the country with powers outside Central Europe—such as France, the Soviet Union and her "Little Entente" partners (Romania and Yugoslavia)—only served to accelerate these problems. Finally, after the Second World War, the same radical Czech politicians joined Stalin—and other politicians of Central and Southeastern Europe—in "solving the problem" by expelling the Germans and the Hungarians from their traditional homelands, even though they have been living there for at least eight and as many as eleven centuries.[6]

To put these developments into a nutshell, the Carpathian Germans may be said to have been hit with a "quadruple whammy," a brutal ethnic cleansing and extermination in four stages:

1) The Pan-Slav Congress in Prague in 1848, where extremists appeared with a fully elaborated program of ethnic cleansing. East of the line running from Lübeck (the Hanseatic city on the western shores of the Baltic) to Trieste on the Adriatic—presumably to the Bering Straits—"re-Slavization" would be put into effect. Recalcitrant national groups like the Germans and the Hungarians would be given short shrift. Interestingly, in 1945 Titoist students of Yugoslavia echoed the same fantastic claims: *"Za Jadrana do Japana,"*—"From the Adriatic to Japan!"—a program of domination over Eurasia. This program was in fact realized by Stalin during and after World War II. He was following the footsteps of the Czarist Empire, which in 1914 sought to extend its sway from East Prussia, Silesia, and Galicia to Manchuria.[7]

2) The creation of ethnocentric new states of a "New Order" in 1918/1920—from Poland in the north to Yugoslavia, Greece and

---

mans, shared the same prison with Father Hlinka in 1919. Sadly, he was tortured to death by the Slovak partisans in September 1944. ʾ

[6] Wenzel Jaksch, *Europe's Road to Potsdam*, trans. and ed. Kurt Glaser (New York, 1958), 212; Carlton J. H. Hayes, *A Political and Cultural History of Modern Europe* (New York, 1939, 1951), 2: 613-745.

[7] Here the reference is to the Sazonov-Paléologue dialogues. Cf. Hayes, *Modern Europe*, above; and Albrecht-Carrié, *A Diplomatic History*, 69, 169. For a collection of studies on the Slav Congress of 1848 by a group of international scholars, see Horst Haselsteiner, ed., *The Prague Slav Congress 1848: Slavic Identities* (New York, 2000).

Turkey in the South. This was accompanied by a frightful cacoph-
ony of mid-level imperialism, aggression, wars, and ethnic
cleansing in those very same states. In the cases of Greece, Turkey,
and the entire Near East, French and British imperialism also came
into play.

3) The bloody Partisan Revolt of September 1, 1944, that embraced the western and central parts of Slovakia.

4) The conquest of the region in the spring of 1945 by large
and brutal military force consisting of units of the Russian and the
Romanian armed forces.

Carpathian-German History

Before World War I Slovakia had been part and parcel of the
Kingdom of Hungary for over a thousand years. The twelfth,
thirteenth, and fourteenth centuries witnessed great population
movements from western and northern regions of the German
Holy Roman Empire—including the Low Countries—towards the
east into Bohemia, Silesia, Prussia, Galicia, Hungary, and Transylvania. The two great foci of the religious-cultural movement connected with this mass migration and with the spread of the German legal systems to the east and southeast were the city of
Magdeburg on the northern Elbe River, and the city of Lübeck on
the Baltic Sea. Magdeburg had been elevated by Emperor Otto the
Great to the rank of an Archbishopric in 967 A.D., at which time it
was put in charge of converting the heathen nations living to the
east and southeast of the Empire. One of the outstanding alumni
of its cathedral school was St. Adalbert, the first Czech bishop of
Prague (973 A.D.), who in 994 baptized the Hungarian prince
Wayk [Vajk], subsequently known as King St. Stephen of Hungary. St. Adalbert—who is equally revered by Czechs, Hungarians, and Poles—was martyred by the heathen (non-German) Prussians while trying to convert them to Christianity.

Magdeburg was already a thriving political and ecclesiastical
center when Lübeck was founded in 1143 on the shores of the Baltic Sea. Though emerging as a center of influence two centuries
later, Lübeck soon extended its sway over much of Scandinavia
and the Baltic region. To a somewhat lesser extent, Freiburg im
Breisgau and Vienna in the southern German lands (Austria) became the legal model for the founding of new cities in Bohemia,
Moravia, Hungary, and Transylvania, and elsewhere. Most of
these city foundations took place in the twelfth through the fourteenth centuries, and thereby began the urbanization of East Central Europe.

After centuries of religious struggle, dynastic warfare, and Ottoman conquest, the end of the seventeenth and early eighteenth centuries saw the expulsion of the Turks from the Kingdom of Hungary (including Slovakia). Because of the tremendous population loss during the Turkish occupation, much of the country had to be repopulated. This took the form of internal migrations (e.g., Slovaks and Rusyns moving down to the lowlands), as well as external immigrations. The latter included Vlachs (future Romanians) from Wallachia and Moldavia, Serbians from the Balkans, and great many Germans from the southwestern German states. This new German migration was the famous *Schwabenzug* [Swabian Trail] that had settled hundreds of thousands of Rhineland Germans in the depopulated regions of Southern Hungary. Coming from southern Hesse, the Palatinate, and Alsace, these migrants were directed mainly to such devastated but fertile regions as Batschka [Bácska, Bačka] and Bánát [Banat], which after World War I were attached to Yugoslavia and Romania. Some of these new migrants also managed to reach Upper Hungary (Slovakia), where they left fascinating linguistic traces among the earlier German, Hungarian, and Slovak settlers.[8]

All in all, the Carpathian-Germans had put an ineradicable stamp upon Upper Hungary (later to become Slovakia). They first came in the twelfth century, but then their number was further increased after the devastating Mongol invasion of 1242. Hungary's rulers—among them Stephen III, Andrew II, Béla IV, and Ladislaus [László] IV—wished to populate, develop, and urbanize their country, and in trying to fulfill these goals they relied heavily on German immigrants. The settlers were given extensive privileges, with the understanding that they would build a series of walled cities and introduce manufacturing, commerce, and improved forms of agriculture. The new subjects of the Crown of St. Stephen hailed from various divergent regions of the Holy Roman Empire. They came from the Frankish Mosel River region, from Flanders and from old Saxony (today Lower Saxony). In a giant arc they migrated eastward through eastern Saxony, Moravia and Silesia, and then turned south into Hungary. Some moved on into the kingdom's eastern province, Transylvania, where they founded such great cities as Hermannstadt [Szeben, Sibiu], Kronstadt [Brassó, Brasov], and Klausenburg [Kolozsvár, Cluj].

In former Upper Hungary (today's Slovakia), they settled along the mountainous frontiers to guard the passes into Silesia

---

[8] Winfried Gold, *Das Zeitalter Max Emanuels und die Türkenkrieg in Europa* (München, 1976), reprint of Arminus Vámbéry, 89-158.

and Poland. Soon they were joined by miners coming up from Salzburg, Styria [Steiermark], and Carinthia [Kärnten] who wanted to take advantage of the ample deposits of silver, gold, nickel, and lead in the Fátra and the Tátra Mountains. True to the precedents set by Magdeburg City Law, all of the German-founded cities—or the cities built under the German Law—were surrounded by scores of native peasant villages. There were dozens of them,[9] including the city of Pressburg [Pozsony, Bratislava], which for over three centuries (1541-1848) served as Hungary's capital.

One of the best known of the German-inhabited regions in Upper Hungary was *Hauerland,* the encompassing the cities Neusohl, Kremnitz, Sillein, and Deutsch-Proben. It was known as *Hauerland* because many of the villages surrounding the cities ended in the suffix "hau"—Schmiedshau, Glaserhau, Drechslerhau, Krickerhau, etc.—which signified that they had all been "hewn" out of the mountain wilderness of the Carpathians. These villages were very similar to villages in the "hau" regions of southeastern Saxony and Lower Silesia.

Since many German dialects—including the "Bairisch" [Bayerisch] from southern Germany, Silesian and Mosel-Frankish from Central Germany, and some Low German—blended in this territory into a new German mix, the *Hauerland* became one of the prime areas for the development of the New High German. Today's linguists—among them Prof. Piirainen of the University of Münster—are involved in tracing the etymology of spoken German, including the Carpathian-German dialect.

Many of the German-founded villages surrounding the walled cities in Upper Hungary were East-German *Waldhufendörfer* ["single-street-villages"]. They were located in the center of a valley or a hollow, with each house or family farm situated on that single long street. Each of the homesteads had a narrow strip of field and orchard behind it, which was perpendicular to the street, running uphill right up to the edge of the woods. This beautiful spatial ar-

---

[9] Some of the most prominent of these medieval walled cities were Bartfeld [Bártfa, Bardejov] and Kaschau [Kassa, Košice] in eastern Slovakia; Leutschau [Lőcse, Levoča], Zipser Neudorf [Igló, Spišská Nová Ves], Käsmarkt [Késmárk, Kežmarok], Deutschendorf [Poprád, Poprad] in the Zips [Szepes, Spiš] region; Königsberg [Újbánya, Nová Baňa], Karpfen [Korpona, Krupina], Schemnitz [Selmecbánya, Banská Štiavnica], Altsohl [Zólyom, Zvolen], Neusohl [Besztercebánya, Banská Bistrica], Kremnitz [Körmöcbánya, Karemnica], Sillein [Zsolna, Žilina], Rosenberg [Rózsahegy, Ružomberok], Deutsch-Proben [Németpróna, Nitrianske Pravno], and Freistadt [Galgóc, Hlohovec] in West Central Slovakia.

rangement remained in place until the forced collectivization of the farms by the Communists after 1947.

These medieval German-founded cities generally remained small. Soon after their foundation in the twelfth through the fourteenth centuries, the region's gold and nickel deposits were depleted. These cities were also repeatedly ravaged by wars, among them by the brutal Hussite invasions of the early fifteenth century, and the long-drawn-out wars of the Reformation and Counter-Reformation in the sixteenth and seventeenth centuries. Likewise destructive were the wars connected with the Turkish conquest and Habsburg liberation of Hungary, the Habsburg dynasty's struggles with Prussia and Napoleonic France, the revolutions of 1848-1849 when Russian troops had helped to squash the Hungarian Revolution, and the ever worsening violence and global wars of the twentieth century. Despite these depredations, the cities retained their German characteristics and layout until the very end in 1945. The houses were grouped around a large central square, the so-called *Ring*. There were also a few subsidiary plazas, arcaded burgher's houses, straight radial streets, as well as one or two long boulevards, known as the *Lange Gassen*. As a final irony of history, Neusohl [Besztercebánya, Banská Bistrica], one of *Hauerland*'s outstanding cities that used to be a brilliant center of the Fugger banking family, was used as the headquarters of the Slovak Partisan Uprising of September of 1944.[10]

Ethnic Cleansing at the End of World War I

It is fair to say that by 1918 Russian expansionism was already viewed by Carpathian–Germans as the main threat to their ordered life. This was particularly true in light of the Russian breakthroughs at Przemysl and Gorlice, which were not very many miles away from the Carpathian-German homeland. The powerful Russian bombardments of the battles sounded to them like ominous thunderstorms through the long nights. (The Russians would attempt similar maneuvers in the Second World War.)

Despite the deep split among the left-wing Socialists, Agrarians and Clerical Conservative Germans, there can be little doubt that Carpathian-Germans were fiercely loyal to Hungary, a coun-

---

[10] Peter Kalus, *Die Fugger in der Slowakei. Materialien zur Geschichte der Fugger* (Augsburg, 1999), 11- 295; Heinrich Zillich, *Siebenbürgen. Ein abendländisches Schicksal* (Königstein im Taunus, 1976); Richard Suchenwirth, *Der Deutsche Osten* (Berg am Starnberger See, 1978); Julius Gretzmacher, ed., *Zipser Kunst. Karpatenland* (München, Wien, 1987), 5-128.

try they had protected with their lifeblood for eight centuries, even prior to the stepped-up campaign of Magyarization in the late nineteenth and early twentieth centuries. Thus, when the Peace Treaties of Saint Germain (1919) and Trianon (1920)—and the international secret deals preceding them—dismembered the centuries-old Habsburg Empire and stripped the Realm of St. Stephen of two-thirds of its territory, the anguish among the Carpathian Germans was deep and widespread. Many veterans volunteered to fight for the new Magyar republic. Others—once the days of initial turmoil and plunder in November of 1918 had been over—decided to demonstrate against the military occupation and martial law that had been imposed by the rulers of the new Czechoslovak state.[11] The entire region groaned under the fist of the unwelcome new masters. The first wave of Czech occupation was particularly savage. According to local traditions it was carried out by Czech Legionnaires who had just returned from the civil war in revolutionary Russia, where they—so the story went—had absconded with the Czarist Russian state treasury, shipping it to Prague, where it became the core of the Czech state treasury.

Although the fury of the first months of Czech occupation eased later, there can be no question that Czech imperialism continued unabated throughout the interwar period. Those desiring to advance professionally could do so only by becoming Czech in name and language. This was particularly true for the civil service, which like an all-embracing web reached from village to village, town to town, and city to city. The state owned all railroads, controlled the educational system, and oversaw the legal and police administration, as well as all state hospitals and health services. In addition, the Czech-controlled national state pursued an active policy of colonization. It would dispatch to live and work in areas that formerly had been purely German-inhabited groups of Czechs thought particularly fanatical, who would carry out their anti-German programs with the full backing of police and admin-

---

[11] My Father, for instance—at the time a high school student from the town of Deutsch-Proben in the *Hauerland*—and with him over a hundred other German *gymnasium* and university students and their friends, staged a protest rally in favor of Hungary at the county government in a neighboring town of Priwitz [Privigye, Prievidza]. They advanced with shouts of: *"Es lebe die ungarische Räterrepublik!"* —"Long live the Hungarian Socialist Republic!" In front of the government building they were met with the powerful symbols of "self-determination and development": two heavy machine guns posted by the Czechs. Prudently, they dispersed. If they had not, no doubt they would have been leveled to the ground, just as scores of civilians in the Sudetenland at the time were killed by the Czech military.

istrative authorities. Unfortunately, such practices have continued throughout our enlightened era, stretching into the late twentieth century.[12]

The newly formed "successor states" created fully or partially out of dismembered Austria-Hungary—Czechoslovakia, Romania and Yugoslavia—established a powerful modern-day precedent for ethnic cleansing by expelling hundreds of thousands of Hungarians from their homelands. Reestablished Poland and the governments of the various Balkan states and Turkey—formed out of the former Ottoman Empire—exceeded even that number with their own expulsions. Thereby, they fired the starting shots in the process of mass murder and ethnic cleansing that was practiced by Stalin after 1931 and then reached monumental proportions in the period between 1944 and 1950.

In light of the above, President Masaryk's alleged goal of creating a "New Switzerland" on the ruins of the Habsburg Empire never reached fruition, but this idea was not even allowed to germinate. A farfetched corollary of these ideas was Masaryk's plan to turn the *Burgenland* (Hungary's mostly German-inhabited western region that was transferred to Austria after World War I) into a "Slav corridor" linking Slovakia and Croatia. This would have created a buffer area cutting off new rump Hungary of 1919 from transport and communication with Austria and Western Europe. Within this true "people's prison" (a phrase coined by the French Revolution of 1789) the "Prussians of the Balkans"—as the Hungarians are known to their detractors—would have been starved into submission and swiftly Slavicized, reflecting shades of the Pan-Slav program of 1848.[13]

The Partisan Revolt of September 1, 1944

Bearing in mind the background of ethnic cleansing, highhanded elitism and opposition to national self-determination shown by Czechoslovakia, one should see the year 1943 as crucial to Slovakia. With the victories of the Red Army at Stalingrad and Kursk, and the "achievements" of the Teheran Conference (November 28-December 1)—where Germany's dismemberment was first

---

[12] Conversations with Andreas Wesserle II and Anna Wesserle, 4257 North 52nd Street, Milwaukee, Wisconsin 53216; Dr. Alfred Jüttner, Syndikus, *Hochschule für Politik München,* conversations 1986-1999 (Dr. Jüttner died in 1999).

[13] Conversation with Prof. Dr.Dr. Macha, Emperor Charles University Prague, Munich, 1997.

broached[14] against Churchill's grave objections—ex-President Beneš, now the leader of the Czech Government-in-Exile in London, hurried off to Moscow to sign a new treaty of alliance with the Soviet Union. Stalin decided to recognize Czechoslovakia, but without its eastern appendix, Carpatho-Ruthenia, which was to be transferred to the Soviet Union. This small region had been part of Hungary for a whole millennium and then part of Czechoslovakia for two decades (1919-1939), but its transfer was now eulogized by the Soviet Foreign Minister V. Molotov as the return of Carpatho-Ruthenia to its Russian homeland after a thousand years.

When Romania defected to the Soviet Union (August 23, 1944)—Italy having switched sides already in July 1943—it was high time for some Slovak patriots to decide whether to descend into defeat with Germany, or to face reality and establish a "working relationship" with the victorious Soviet Empire. The conspirators could not count on Monsignor Joseph Tiso, the President of independent Slovakia after March 1939 (not recognized by the United Kingdom), nor on his Foreign Minister Šano Mach, nor on their numerous followers, for none of the latter were willing to change their views about the Soviet Union as an "evil empire." In this regard they were forty years ahead of their time. Thus, the conspirators created an underground coalition, consisting of intellectuals and Lutheran clergymen and Communist activists, who had remained in touch with Moscow. Based on prior agreements, terrorist agents were parachuted into Slovakia from the USSR to help initiate the planned uprising.

The plan was to convince the Slovak Army, or rather its officer corps, to join a general Slovakia-wide uprising on or about September 1, 1944, to overthrow the Slovak Government in Bratislava [Pressburg, Pozsony] and to link up with a Soviet offensive into Eastern Slovakia. The grand design of this offensive was to capture the Dukla Pass in the east with the participation of the Red Army's "Czechoslovak Legion," to cross the Carpathians and to pour forth into the Hungarian Plain. These achievements were to be followed by the capture of Budapest and the Danube Valley, and then dash north along the Neutra [Nyitra, Nitra] and Waag [Vág, Vah] Rivers, smashing the German front in Galicia and Silesia from the rear.

This stupendous plan came to naught. First, the Dukla Pass was not captured as planned, because small "fighting units" of the

---

[14] The decision to dismember Germany and the details of its dismemberment were discussed and decided at the Yalta (February 7-12, 1945) and Potsdam (July 17-August 2, 1945) Conferences.

German armed forces put up a spirited and protracted resistance, which seems to have resulted in 80,000 casualties for the Czechoslovak Legion. Second, most units of the Slovak Army, except those in the *Hauerland* of Western Slovakia, refused to join the partisans. Third, despite wholesale atrocities carried out against the Carpathian-Germans and their Slovak compatriots, a significant number of the Slovak populace declined to support the insurrectionists. Nonetheless the human cost was frightful.

Personal Reminiscences about ethnic Cleansing in Slovakia

Now permit me to interject some personal reminiscences about the events that were overtaking us during those fateful final months of World War II. In those days I was a small boy spending my vacations with my mother, grandparents, and other relatives in the tiny town of Deutsch-Proben [Németpróna, Nitrianske Pravno] in the *Hauerland*. The day after my birthday on August 30, 1944, we were set to catch the train to Prague so that I could return to school. We were waiting for the train that was to leave about 4:30 A.M. when unexpectedly ragged-looking partisans emerged from the woods, in civilian clothes, rifles strung from their shoulders. They prevented us from leaving because martial law had been declared. They announced that all persons leaving their houses without their permission, or caught talking outdoors with someone else, would be shot on the spot. I was gripped by overwhelming fear, which I can still recall vividly after nearly six decades. All "intellectuals" in town—architects, businessmen, teachers, writers, almost seventy in all—who failed to hide in the nearby impenetrable woods around the city, were arrested and trucked away to Neusohl [Besztercebánya, Bánska Bistrica] some twenty kilometers (thirteen miles) away. With the exception of one, they were all massacred there.

My mother went to the City Hall—called the *Weinschtebä* or "little wine room" in our dialect—to petition the partisan chief for permission to return to Prague. As she spoke Czech excellently, because she had lived in Prague for thirteen years, the leader took her to be a Czech and outlined to her the partisans' plans "to round up all the Germans, shoot them, and divide their property among ourselves." There were about 2,200 Germans in town and only about 300 Slovaks. He rapidly changed his mind when he saw her passport, which clearly identified her as being German. Yet he was reluctant to have her killed on the spot for fear of giving away their plans and because some of the non-Communist underground leaders were not ready to commit mass murder at that point in time.

Three weeks later conditions changed for the worse, for at that time close to seventy German community leaders—among them some Socialists and even fanatical anti-Fascists—were slaughtered, some of them after unspeakable tortures. Among them was one of my father's cousins, Toni Wässerle, who was one of the local community's cultural leaders. Other towns and villages fared worse. Krickerhau [Nyitrabánya, Handlová], a well-known coal mining community nearby, lost over a hundred of its citizens. In Glaserhau [Túrócnémeti, Sklené], a poverty-stricken mountain village, all men over the age of sixteen had to dig their mass grave into which they were then machine-gunned. The village priest was the only survivor. Wounded, he was covered by numerous corpses and feigned death. Later, he crawled out of the ditch and escaped. After three weeks of terror we in Deutsch-Proben were elated to hear that a small detachment of German soldiers was approaching from the south. They had received news that all of us were going to be murdered, so they decided to rescue us.

The partisans and some of the Slovak military units associated with them had established a defensive position about four kilometers (2.5 miles) south of the town at Maut, a former toll booth near the Neutra [Nyitra, Nitra] River. The fortification was "softened up" by two German *Stuka* dive bombers which looked very puny, compared to the mighty American air fortresses that roared across Slovakia from Italy. The latter were on their way to reduce the industrial cities of German Upper Silesia, just a few miles to the north of us, to rubble. But the *Stukas* got the job done. Two self-propelled artillery pieces—whose seventy-five millimeter guns were useless because they had no ammunition, but which did possess working machineguns.

Our uncles had dug an earth-and-beam bunker in the large orchard behind grandmother's house and she took us there, her three grandsons, just before the ra-ta-tat of the machine guns. I was unable to contain my curiosity, and popped out of the bunker to see the fleeing guerrillas. As they hopped across our neighbors' five and six-foot fences, like fleeing rabbits, they must have broken all existing and future Olympic records in the high hurdles. They were proficient in torturing defenseless civilians, but they lacked the guts to face a real fight. A few of them were caught by the German police units but were treated leniently.

Once the *Hauerland* and some adjoining areas were liberated by the German units, we received numerous nuisance visits from Soviet fighter-bombers. Actually these Soviet planes belonged to the "Czechoslovak Legion" that was involved in a futile attempt to capture the Dukla Pass in the east. They strafed houses and dropped fifty-pound bombs, trying to hit the railway station. Once they almost got me and my cousin while we were were helping ourselves to his father's pears and apples in their orchard. The strategic value of these bombing missions was zero, but they did manage to kill a few civilians.

The Coming of the Red Army and the End of the War

In the mild autumn of 1944, and during the winter and early spring of 1945 the Carpathian-Germans were granted a few more months of relative tranquility. With the slow advance of the Soviet and the Romanian troops from Hungary in the south, where they perpetrated unspeakable horrors, *Hauerland's* women and children were shipped off to "safer" places in the Sudetenland, Bohemia, and elsewhere. My grandmother, aunt, and several cousins arrived in our apartment in Prague just before Palm Sunday in 1945. My uncle followed soon afterwards. They had traveled in drafty cattle cars to reach us in the Czech capital. The journey of about 300 km (180 miles) took them a whole week. They were still very grateful, for they knew that the advancing Red Army was spreading a swath of devastation comparable in its ferocity to the Mongol invasion of 1242. They were particularly harsh on women of all ages—ten to ninety—most of whom were raped if they were found.

Not all of Deutsch-Proben's German citizens had fled. Some stayed at home, like the two Wagner brothers who ran a tavern at the edge of town. When the Soviet Army arrived, they stormed out into the street with pitchfork and scythe. The Russians seized them, doused them with gasoline and burned them alive.

One of my uncles, Father Josef Petruch, was the town's Catholic priest and the Dean of the clergy in the Upper Neutra [Nyitra, Nitra] River Valley. He was well known for his open dislike of National Socialism. Nonetheless, the Soviet soldiers took him and his housekeeper, Etelka Dubrovská, to a creek outside of town, where she was repeatedly raped. Father Petruch thought he would be shot, but he was simply forced to watch, with a pistol at his head, while seventeen Russians raped her within a period of two hours.

Although Father Petruch stayed alive, the Soviets confiscated his rectory. They occupied the second floor of the huge building, from where they threw his furniture, his harmonium (a musical instrument), and his priceless Meissen china out the window. They slept in the middle of the room, rolled up in their greatcoats. They defecated in the corners of the room for the duration of their stay in town, for they were unable to make use of the water toilet (W.C.). They thought it was for peeling and washing potatoes.

Three of my grandparents were already dead in 1945, while my maternal grandmother joined us in Prague. This saved her from the brutal treatment meted out to our town's female citizens. All of her neighbors who had stayed behind were raped or killed. One of them, an old woman in her eighties (living on *Priwitzer Gasse*) died soon from the virulent, so-called Siberian gonorrhea she contracted while being gang-raped by the Russians. The mother-in-law of another uncle of mine, the late veterinarian Dr. Ladislaus Petruch, contracted a purple skin cancer which in the course of time had eaten away half of her face. Even though she was over seventy years old, she had clambered up into the hayloft, pulling the ladder up behind her, hoping that the Russians would be unable to find and follow her. However, climbing the ten-foot height proved no obstacle to these sons of the collective *kolkhozes*. She was found and gang-raped, like most of her neighbors, irrespective of age. The same thing happened to most other women who failed to flee their native towns and villages.

The barbaric devastations inflicted by the Soviet Red Army was soon followed by various anti-German laws and decrees. The orchards and fields of the Carpathian-Germans were confiscated, and the families were taken to concentration camps, where many of them died of starvation. The main camp in the *Hauerland* camp was at Novák [Nováky], about thirty kilometers south of Deutsch-Proben. The initial weeks were especially horrendous, because the camp guards liked to drench the newcomers with gasoline and set them afire, burning them alive. This was exactly what the Czechs did to their German compatriots in Prague's St. Wenceslas Square during the May 1945 revolt.

My uncle, the town priest Father Petruch, escaped imprisonment initially because of his well-known anti-Fascist past. Also, the Soviets had a well-thought-out policy of seizing power step-by-step in their newly won satellite states. Prior to 1947 it was considered imprudent by the local authorities to alarm the Western Powers by the wholesale imprisonment of well-known conservative and non-Communist public figures. Thus, on several occasions he was able to journey to the concentration camp by train and bring food to his starving flock. But these so-called "pro-German" activities and his anti-Communist views made him unpopular with his Church superiors. He was transferred to the poor Slovak village of Lazány [Lazany], but not before all his furniture and furnishings, including his bedding, had been confiscated. Unbent and undeterred, he was soon arrested and incarcerated in the ancient, sleepy fortress of Leopoldstadt [Lipótvár, Leopold] that had been built by Emperor Leopold, centuries ago, to contain Turkish incursions. There, my uncle's diabetes deteriorated rapidly. He was released a skeleton two and half years later, on Christmas Eve 1952. At first he tried to visit his former chaplain, Father Pös, then the village priest of Oberstuben [Felsőstubnya, Horná-Štubňa]. But Father Pös, who was deathly afraid of the Communist police, denied him admittance. Thus, my uncle was forced to trudge home to Deutsch-Proben through twelve miles of snow and sleet. There, cooped up in a single room, in ill health, but still unbroken, he died in 1961. He was one of those special people who openly dared to take a stand against the evil ideologies and tendencies of the twentieth century, and he paid the price for it.

Father Petruch's passing also signified the end of the Carpathian-Germans. By 1939 this thriving community of several hundred villages, towns, and cities that had extended from the borders of Silesia in the north to Hungary and Transylvania in the south and southeast, had shrunk in number to a mere 150,000. They fell victim to those who were ruthless and determined in achieving what they regarded as their own ends. Depleted by expulsion, mass murder, and oppression, according to the most recent census, today only 5,413 inhabitants of Slovakia identify themselves as Germans. They include 682 in the town of Priwitz [Privigye, Prievidza], 137 in Deutsch-Proben [Németpróna, Nitrianske Pravno], and 134 in Krickerhau [Nyitrabánya, Handlová]. According to estimates, another 10,000 to 15,000, may have escaped expulsion by hiding their German identity. Although the results were not as gruesome as in the neighboring Czech Republic, ethnic cleansing had prevailed and had been the order of the day in Slovakia.[15] Yet, despite these temporarily discouraging results, one can still hope that after all the crises and catastrophes in the world, mankind will come to its senses and millions of expellees will be able to return to their homelands in the foreseeable future. For, as Alfred Lord Tennyson predicted hopefully a century and a half ago:

> The parliament of man...
> will hold a fretful realm in awe,
> And the kindly earth will slumber,
> Lap't in universal Law.

---

[15] See the following articles: "T.G.Masaryk," and "Aus der neuesten Volkszählung in der Slowakei," in *Die Karpatenpost*, August 2000 (Stuttgart, Schlosstrasse 92/II).

V.

ETHNIC CLEANSING AND ITS

BROADER IMPLICATIONS IN

THE LAST THIRD OF THE

TWENTIETH CENTURY

Systematic Policies of Forced Assimilation Against Rumania's Hungarian Minority, 1965-1989

LÁSZLÓ HÁMOS

Of the many terms which could be used to characterize the Rumanian government's mistreatment of its more than two million-strong Hungarian minority between 1965 and 1989—"attempts at ethnic cleansing," "cultural genocide," "ethnocide," and others—I opted for an accurate expression, contemporaneous with the events: *Systematic Policies of Forced Assimilation*. "Systematic" because the phenomena to be described are not isolated incidents, but reveal identifiable, willful patterns of conduct. "Policies" because I will focus on official government courses of action during the twenty-four-year reign of Nicolae Ceauşescu. "Forced" covers the broad range of coercive techniques employed, from subtle psychological pressures, and an intricate web of interlocking legal and administrative measures, to outright physical persecution and terror. "Assimilation," or the eradication—as opposed to voluntary relinquishing—of the unique cultural, linguistic and traditional features of individual and community ethnic identity, whether through absorption of the minority, or its departure to other countries, clearly became the ultimate objective of the Ceauşescu regime.

At this timely conference on ethnic cleansing, precision and clarity are important when applying this term to phenomena, which predate the abhorrent practice. Let me clearly distinguish at the outset that the violent techniques, which characterized the Milošević-style ethnic cleansing were not evident to the same scale and intensity in Ceauşescu's Rumania. Thankfully, the dictator was overthrown before he could have employed such drastic devices—for which, incidentally, he *did* share a propensity, as evidenced by a massive village destruction he was about to carry out prior to his execution.

But an important question needs to be posed here. Was it only the drama of the violence in the former Yugoslavia, which captured world attention, and brought in the NATO bombers? Or was world reaction motivated against the objective of eradicating *minority identity*, to which the violence was only a means?

To the simple answer most people would give ("both"), I argue that it is vital to clearly identify, and prioritize as original sin not so much the *means*, but the *ends* of ethnic cleansing in their own right. If recognized as the fundamental evil, it follows that the objective of ethnic cleansing, by whatever means it is sought, must be resisted and combated in a timely and effective manner, by domestic forces and the international community alike.

The central tenet of this paper is that, although less visible and dramatic, the Ceauşescu regime's systematic efforts to forcibly assimilate a centuries-old indigenous population were no less condemnable than the ethnic cleansing practiced by Slobodan Milošević. There are many ways to skin a cat. Ceauşescu chose one, Milošević another. But, to twist another expression into present application: a thorn by any other name is still a thorn.

If we identify as primordial evil the objective of ethnic cleansing (and not only the later, violent means used to apply it), then where was the indignant, moral outrage? Indeed, the reaction was late, if ever, forthcoming. Even acknowledgement, much less punishment commensurate with the crime, never materialized. Within Rumania at the time, the majority population took little notice of the reprehensible practices being employed in their midst. Since the 1989 fall of the Ceauşescu dictatorship, none of the successive governments has visibly recognized, much less apologized for and remedied past transgressions. In fact, in a number of areas vital to the maintenance of minority ethnic identity, the evidence shows an actual continuation of Ceauşescu-era practices.

International disregard left (and leaves) even more to be desired. Let me cite only the area to which I was an intimate witness, advocating attention to Rumania's human rights abuses for two decades. It remains an open, unanswered question to this day: how could the United States, in all its Kissingerian wisdom, instead of condemning the growing persecution exercised by the Ceauşescu dictatorship, actually select this so-called "maverick" of the Eastern Bloc for special reward by granting Most-Favored Nation status in 1975? Despite clear warnings, why did the U.S. continue to consistently lend economic, political and moral support during exactly the period, 1976-1987, when domestic repression reached unprecedented intensity? How did it happen that America regarded as its best friend and closest ally in the region the only Eastern European dictator who was subsequently executed in 1989 by his own compatriots for his misconduct?

Policy recommendations are beyond the scope of this paper. This much, however, needs to be emphasized: The prolonged misery of the Hungarian minority under the Ceauşescu regime—as is undoubtedly true for other victimized communities—highlights the importance of a proactive policy. It is essential to recognize the evidence of nascent ethnic cleansing, and to introduce timely, effective, and peaceful political countermeasures to avert the escalation of tensions and outbreak of violence. The Ceauşescu-era experience of the Hungarians of Rumania may also be instructive in helping to identify seemingly obscure, insignificant and unrelated actions as markers deserving timely attention and redress.

It is impossible to convey accurately or comprehensively in a single article the myriad anti-minority policies and practices instituted under the Ceauşescu regime. Instead, I have selected four elements critical to the maintenance of ethnic identity: education, culture and language, human contacts, and physical community. I propose to describe only by way of illustrative example some of the destructive measures visited upon this population, focusing on those data, which reveal a pattern of injurious conduct.

Instead of reciting the totality of facts and phenomena, the paper will also serve as an abbreviated guide to the excellent published resources available on this subject. The author of some of the best works is also a contributor to this volume. Professor Andrew Ludányi wrote his doctoral dissertation on this subject,[1] and later co-edited a seminal publication[2] and two further analytical volumes.[3] The New York-based Committee for Human Rights in Rumania (CHRR), formed in 1976 to call attention to the intensifying pressures against Rumania's Hungarian minority, published a 1979 compendium of primary source material[4]; and a study based on these and other first-hand reports for the 1980 Helsinki Review Meeting in Madrid.[5]

---

[1] Andrew Ludányi, "Hungarians in Rumania and Yugoslavia: a Comparative Study of Communist Nationality Policies," (Ph. D. diss., Louisiana State University, 1971).

[2] John F. Cadzow, Andrew Ludányi, and Louis J. Éltetö, *Transylvania: The Roots of Ethnic Conflict* (Kent, Ohio, 1983).

[3] Rudolf Joó and Andrew Ludányi, eds., *The Hungarian Minority's Situation in Ceauşescu's Rumania* (Boulder, Col., 1994); Andrew Ludányi, ed., *Hungary and the Hungarian Minorities* [Special Topical Issue of *Nationalities Papers*] (New York, 1996).

[4] American Transylvania Federation: Committee for Human Rights in Rumania, *Witnesses to Cultural Genocide: First-Hand Reports on Rumania's Minority Policies Today* (New York, 1979).

[5] *Rumania's Violations of Helsinki Final Act Provisions Protecting the Rights of National, Religious and Linguistic Minorities: Study Prepared by the Committee for Human Rights in Rumania for the Conference on Security and Cooperation in Europe, Madrid, 1980-83* (New York, 1980). Some other important works are: Robert R. King, *Minorities under Communism: Nationalities as a Source of Tension among Balkan Communist States* (Cambridge, Mass., 1973); George Schöpflin, *The Hungarians of Rumania* (London, 1978), and a later update, George Schöpflin and Hugh Poulton, *Romania's Ethnic Hungarians* (London, 1990); Stephen Borsody, ed., *The Hungarians: A Divided Nation*

A detailed chronology of developments, contemporary with the events described, can also be found in over 1,000 pages of published transcripts of CHRR Congressional testimony.[6] Our organization took great care to research the facts and to base its assertions only on verified reports of human rights abuse. This purpose was well served by three waves of credible, underground (*"samizdat"*) sources,

(1) starting with a series of protest documents (1977-80) by Károly Király, a former alternate member of the Rumanian Politburo, and other disaffected, one-time members of the Communist Party or state administrative apparatus;

(2) continuing with *Ellenpontok* (Counterpoints), an underground journal edited by three dissident intellectuals in their late 20's who published twelve issues (1981-82) of anecdotal, detailed accounts of persecution; and

(3) ·ending with the *Hungarian Press of Transylvania*, a journalistically professional news service relying on a network of clandestine correspondents throughout Transylvania, which issued 589 releases (1983-89) translated into English and disseminated by CHRR.

(New Haven, Conn., 1988); *Destroying Ethnic Identity: The Hungarians of Rumania* (New York, 1989), and a second report: *Struggling for Ethnic Identity: Ethnic Hungarians in Post- Ceauşescu Romania* (New York, 1993); Freedom House and Hungarian Human Rights Foundation, *Rumania: A Case of "Dynastic Communism"* (New York, 1989); and László G. Antal, *The Situation of Ethnic Hungarians in Rumania* (Budapest, 1990).

[6] Between 1976 and 1987, I testified a total of 25 times before the Senate and House committees which deliberated each summer whether to continue for another year the Most-Favored Nation status originally granted to Rumania in 1975. The legislative basis for the decision was the country's human rights performance over the prior twelve months. My oral testimony, accompanied in each case by a 40-100 page written submission, documented the growing persecution against the Hungarian minority exactly during those years in which it escalated to unprecedented levels. The printed transcripts of this oral and written testimony, amounting to over 1,000 pages, before the House Ways and Means Committee Subcommittee on Trade and the Senate Finance Committee Subcommittee on International Trade can be found for each year (citations are to U.S. Government Printing Office Document Numbers): **1976** (Doc. No. 78-433 and Doc. No. 78-421-0); **1977** (Doc. No. 91-710, Doc. No. 92-972 and Doc. No. 94-837-0); **1978** (Doc. No. 31-078-0 and Doc. No. 32-633); **1979** (Doc. No. 50-500 and Doc. No. 50-437-0); **1980** (Doc. No. 66-890-0 and Doc. No. 68-772-0); **1981** (Doc. No. 83-055-0 and Doc. No. 84-209-0); **1982** (Doc. No. 99-135-0 and Doc. No. 99-400-0); **1983** (Doc. No. 24-420-0 and Doc. No. 26-235-0); **1984** (Doc. No. 40-844-0); **1985** (Doc. No. 49-957-0 and Doc. No. 52-704-0); **1986** (Doc. No. 60-795-0, Doc. No. 65-921-0 and Doc. No. 65-139-0); and **1987** (Doc. No. 81-755).

The fact that many individuals from varying backgrounds risked personal security, over an extended period of two decades, to enable their voices to be heard certainly added credence to the unambiguous content of their message: that the Ceauşescu regime was indeed waging a systematic campaign of forced assimilation against their ethnic community.

Anti-Hungarianism Elevated to State Ideology

"We live as second-class citizens . . . The state authorities treat us as if we were the enemies within."[7]

Although constitutionally defined as a "unitary" (read: ethnically pure) state, Rumania is still a multi-ethnic country, with a large minority population: more than two million Hungarians, and sizeable numbers of Roma, Germans, Ukrainians, Jews, Serbs, Greeks, Turks and others. The Hungarians alone constitute the largest national minority in Central Europe. Most live in Transylvania, which was annexed to Rumania from Hungary under the terms of the Treaty of Trianon in 1920.[8]

Although the post–World War I Rumanian governments already introduced pressures to absorb the territory and its population (limited to the Southern portion of the province when Northern Transylvania was re-annexed to Hungary, 1940-44), it was the Communist system, granting totalitarian control over all aspects of public and private life, which provided the tools to conduct an assimilationist drive with greatest impact.

During the first years after the introduction of Communism, conditions were actually favorable for Rumania's minorities. In 1952, though largely symbolic, a Hungarian Autonomous Region was created at the center of the country to include the most

---

[7] The motto here, and in each of the four sections which follow, is from *Ellenpontok* [Counterpoints], no. 8, dated October 1982. This issue consisted of a "Memorandum" and "Program Proposal" addressed to the participants of the Madrid Review Conference of the Helsinki Final Act, which met between 1980 and 1983. The document was translated into English and distributed at Madrid by the New York-based Hungarian Human Rights Foundation (the successor organization to CHRR). Reaction was swift. On November 6 and 7, 1982, Rumanian secret police unleashed a campaign of terror against Hungarians in Transylvania: more than a dozen individuals were arrested, interrogated and tortured in addition to the three editors of the samizdat, Géza Szücs, Attila Ara-Kovács and Károly Tóth, who were ultimately exiled from Rumania. The Memorandum and Program Proposal were reprinted in Stephen Borsody, ed., *The Hungarians: a Divided Nation*, 148-158.

[8] In accordance with contemporary usage, the regions of Pártium, Bánság, and Transylvania are referred to collectively as Transylvania.

densely Hungarian-populated counties. The autonomous region, however, fell victim to growing Rumanian nationalist designs. In 1960, its name and boundaries were changed to dilute its Hungarian character, and in 1967, after the advent to power of Nicolae Ceauşescu in 1965, the Region was eliminated altogether.[9] (Similar to Milošević's little noticed 1989 elimination of the autonomous status of Kosovo and Voivodina, Ceauşescu's act was a harbinger of the severely repressive measures to follow.)

Ceauşescu was the young Communist Party functionary sent to Cluj [Kolozsvár][10] in 1959 to execute the forced merger of the independent Hungarian Bolyai University with the Rumanian Babeş University. Assistant to the act was Ion Iliescu, a student union leader at the time (later President of Rumania, 1990-1996, and currently candidate for election[11]). Sanctioned by Moscow in retaliation for the 1956 revolt in Hungary, the forced merger signaled the onslaught of the campaign of forced assimilation to be waged in earnest by Ceauşescu. Three Hungarian professors, including the celebrated writer László Szabédi, committed suicide out of despair at this arbitrary act.[12]

Why did Ceauşescu craft a coordinated, interlocking campaign to destroy the ethnic identity of a centuries-old indigenous

---

[9] For a detailed description, with demographic data, of the evolution of the Hungarian Autonomous Region, see King, *Minorities under Communism*, 148-164.

[10] The Hungarian designation of placenames is included, following the Rumanian, throughout this paper.

[11] Shortly after the delivery of this paper at Duquesne University, Mr. Iliescu was elected president, defeating the government of university professor Emil Constantinescu, which had for four years included the Democratic Alliance of Hungarians in Rumania (DAHR) among its coalition partners. The retrograde sentiment of Rumanian voters was strikingly evident in the 33.2 percent of votes received in the second round by the overtly anti-Hungarian and anti-Semitic Corneliu Vadim Tudor, and by the high number of seats won by his ultra-nationalist Greater Rumania Party (37 of 140 in the Senate and 84 of 345 in the Chamber of Deputies, or 26.5 and 24.5 percentages, respectively). To thwart the specter of a Tudor presidential victory at run-off elections on December 10, 2000, the DAHR found itself in the ironic position of mildly endorsing Mr. Iliescu, a longtime adversary and opponent of democratic reform.

[12] A chilling, eyewitness account of Szabédi's public humiliation by Ceauşescu, and the impact of the merger on the Hungarian minority, can be found in "Methods of Rumanianization Employed in Transylvania," by Anonymous Napocensis, in *Witnesses to Cultural Genocide*, 66-69. The importance of the Hungarian university, its academic achievements, and the violence of its destruction are chronicled in a collection of essays, mostly by onetime Hungarian professors: *A Kolozsvári Bolyai Tudományegyetem, 1945-1959* [The Bolyai University of Sciences, 1945-1950] (Budapest, 1999).

population? What motivated his regime to antagonize this minority—otherwise proud to continue living in its ancestral homeland—to such an extent that it triggered the 1989 overthrow of the dictator?

Ion Mihai Pacepa, a onetime Ceauşescu intimate and former chief of counter-espionage who defected to the United States in 1978, recently described the rationale in a TV interview aired in Hungary:[13] "Ceauşescu was a master at using nationalism, that weapon of the emotions that has been wielded by so many dictators. A political chameleon who repeatedly changed sides and loyalties, he owed most of his rise to power and his durability to the masterly way in which he succeeded in manipulating the historic rivalries between Romanians and the principal minority groups living within Romania's borders.... One of Ceauşescu's dreams was to go down in history as the leader who Romanianized the country and to homogenize its population into a Communist melting pot. A few years after he came to power, Ceauşescu decided to create *România-Românilor*, a Rumania for the Rumanians, without ethnic diversity. As a fanatic seeking absolute power, Ceauşescu hoped thereby to kill two birds with one stone: to arouse broad popular acclaim for himself, and to eliminate all possible cause for outside interference in his reign, under the pretext of lending support to Romania's ethnic minorities."

A primary legitimizing principle of the Ceauşescu regime was the conscious promotion of anti-Hungarian sentiment, directed not only against the minority living within the country, but also against Hungary proper. During the 1980's, faced with popular discontent resulting from sharply deteriorating economic conditions, the Ceauşescu regime intensified appeals to chauvinistic sentiment. Instead of instituting long overdue reforms, the government actively propounded the myth of Rumanian cultural, historical and political superiority, hoping in this way to deflect criticism and salvage some measure of national cohesion. Minorities served as convenient scapegoats for the country's severe economic decline.

The methods of this anti-minority campaign ranged from publishing a large amount of pseudo-scholarly works, which sought to prove that Rumanians were the first to settle in the region, suffering the invasion of the primitive, barbaric Hungarians to crude, hand-lettered slogans which openly called on "Ruma-

---

[13] Transcript furnished by the interviewer, Imre Szabó. The interview aired on Hungary's Duna TV, March 21, 2001.

nian brothers" to "beat and tear asunder the traitorous Hungarians."

Anti-Hungarianism in Ceauşescu's Rumania played a twofold
purpose. On the one hand, it provided the regime with a means of
suppressing independent expression on the part of a sizeable "different" segment of the population. But perhaps more importantly,
the manipulation of ethnic sentiments and hostilities were viewed
by the regime as a valid means of justifying its own continued existence. With Rumania isolated internationally, its economy in
shambles, and Ceauşescu's once highly touted "maverick" stance
toward the Soviet Union amounting to little more than a staunch
refusal to institute Gorbachev-style reforms, heightened anti-
Hungarianism became Ceauşescu's option of choice in seeking to
salvage some measure of domestic acceptance and legitimacy. As
conditions worsened further, the Rumanian leadership seemed
increasingly imbued with a kind of siege mentality, or the active
propagation of a sense of encirclement and infiltration by the
looming "enemy," ever more frequently named as "Hungary," or
"the Hungarians." No matter how contrived or artificial, this ideology produced very real victims: the country's national minorities, whose conditions and status deteriorated radically.

Discrimination in Education

> *"The Hungarian-language school system is gradually being de-*
> *stroyed...."*

The Ceauşescu regime's prime target was, not surprisingly, the
area most critical to the survival of a minority's ethnic identity:
education in the native language. It was in the dismantling of a
once wide-ranging network of independent Hungarian elementary and secondary schools and higher education institutions
throughout Transylvania that the official Rumanian policies were
devastatingly effective. The government placed obstacles to
studying in a minority language at every step of the educational
process, and those proportionally few students able to obtain an
education in a minority language faced employment obstacles in
the Rumanian-only workplace.

Starting in 1959, individual Hungarian schools were systematically attached to Rumanian schools as mere sections, which, in
turn, were gradually phased out. At the lower levels, the openly
discriminatory Decree/Law 278, enacted in 1973, speeded the
widescale elimination of Hungarian schools by requiring a minimum of twenty–five elementary or thirty–six high school students
(both raised from fifteen) to maintain or establish a class in one of

the minority languages. Rumanian sections and classes however, were specifically exempted from the provisions of this Decree: a Rumanian section had to be maintained regardless of demand—that is, even for one Rumanian student in a village school where the remaining remaining twenty–four were Hungarian.[14]

Even in the remaining Hungarian schools and sections the subjects of literature, geography, and history, for example, had to be taught in Rumanian. The scant Hungarian-language textbooks glorified the majority, while denigrating others as barbaric invaders, as illustrated by the following quote from the textbook *Hazánk története* [History of Our Homeland][15]: "Our ancestors are the Dacians. They were characterized by powerful facial features, valor, diligence and eagerness to act, honor and courage to do battle. They cultivated wheat, forage, fruit and grapes. They built great cities. The nomadic barbarians, with bow and arrow, or spear in hand, formed plunderous gangs to search for spoils. If news of more promising riches to plunder reached them, they moved on. They built no homes, but hid out in tents. They therefore led a far more backward way of life than the Daco-Romans."[16] In any event, by 1986 all Hungarian secondary school textbooks were removed from print, not to reappear until after the fall of Communism.[17]

Whereas the Rumanian regime's initial strategy was one of *numerus clausus*—to limit the number of students studying in a minority language—a critical juncture occurred in 1976, when Rumania's entire educational system was reorganized. Technical and vocational training were emphasized, at the expense of lyceums (liberal arts high schools), where minority-language schools were strongest. *Numerus clausus* became a policy of *numerus nullus*: Hungarian-language liberal arts schools with a 300-400 year-

---

[14] *Rumania's Violations of Helsinki Final Act Provisions*, 21-23.

[15] Bucharest: Editura Didactica si Pedagogica, 1968.

[16] Quoted in Károly Nagy, *Magyar szigetvilágban ma és holnap* (New York, 1984), 146. The original Hungarian reads as follows: "A dákok a mi őseink. Erőteljes arcvonások, vitézség, szorgalom és tettvágy, becsület és harckészség jellemezte őket. Búzát, kölest, gyümölcsöt és szőlőt termesztettek. Nagy városokat építettek. A népvándorló barbárok íjjal, vagy lándzsával a kezükben, csapatokba verődve indultak zsákmányt szerezni. Ha jobb, kifosztásra alkalmasabb területről hallottak hírt, tovább vonultak. Házakat nem építettek, hanem sátrakban húzódtak meg. Tehát jóval elmaradottabb életmódot folytattak, mint a dákó-románok."

[17] "Publishers Stop Printing Secondary Textbooks in Hungarian," in *Hungarian Press of Transylvania*, Release No. 30/1986, dated May 28, 1986, No. 7/1986, dated January 27, 1986.

old past—in Nagyvárad [Oradea], Kolozsvár [Cluj], Marosvásárhely [Tirgu Mureş], Székelyudvarhely [Odorheiu-Secuiesc] and Kézdivásárhely [Tirgu Secuiesc]—were completely eliminated, while the promised minority-language vocational schools were not opened.[18] By the 1980s, all remaining Hungarian schools in the 80-90 percent Hungarian-populated counties of Hargita [Harghita] and Kovászna [Covasna] were also shut down.

As a direct consequence of these policies, official Rumanian statistics indicate that in 1955 the number of students allowed to attend Hungarian classes was roughly proportionate to the size of the Hungarian population, but by 1978 a sharp decline was evident. The proportion of those allowed to be educated in Hungarian preschools dropped by over 50 percent from 14.4 percent in 1956 to 6.3 percent in 1978; from 9.5 percent to 5.4 percent in elementary schools; and from 8.0 percent to 3.5 percent at the high school level, resulting in an overall decline from 10.0 percent to 5.3 percent.[19]

Hungarian-Language Education at the
Elementary and High School Levels*

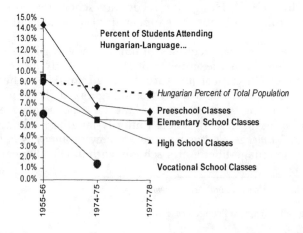

*Sources: *The Hungarian Nationality in Romania* (Bucharest: Meridiane Publishing House, 1976), pp. 8, 15-17; *A Living Reality in Romania: Full Harmony and Equality Between the Romanian People and the Coinhabiting Nationalities* (Bucharest, 1978), p. 15.

Higher education has a great historic tradition in Transylvania. The Bolyai University in Kolozsvár/Cluj, for example, can be

---

[18] *Rumania's Violations of Helsinki Final Act Provisions*, 25.

[19] *Rumania's Violations of Helsinki Final Act Provisions*, 20-21.

traced to the Jesuit academy founded by the Hungarian prince István Báthory in 1581. The forced merger in 1959 of this institution with the Rumanian Babeş University signaled the Rumanian government's clear intention to eliminate minority-language higher education as well. Following the merger, the number of Hungarian students admitted, the number of Hungarian faculty and the number of courses offered in Hungarian, were systematically reduced at this institution and its affiliated medical school, the Institute of Medicine and Pharmacology at Marosvásárhely [Tirgu Mureş].

The disappearance of opportunities to study in Hungarian was paralleled by a major drop in the number of ethnic Hungarians attending any institution of higher learning, a trend which continues to the present day. While the number of all students in the period from 1957-75 in higher education more than doubled (from 51,094 to 108,750), the number of Hungarian students rose by only 600 (from 5,500 to 6,188), or barely 10 percent.[20] In the 1986/87 academic year, despite an official proportion of Hungarians to general populace of 7.9 percent, the admission of ethnic Hungarian students to institutions of higher education was limited to a maximum of two percent of the given institution's total student population.[21] During the decade since the fall of Communism this gap has not been filled: while according to official statistics 7.1 percent of the country's population is ethnic Hungarian, in the 1999/2000 academic year Hungarians represent only 3.95 percent of all students enrolled in institutions of higher education.[22]

In the employment of university graduates, the Ceauşescu regime imposed a policy with the deliberate purpose of altering the ethnic composition of Transylvanian communities—the very aim of the program later known as "ethnic cleansing." Hungarian graduates were routinely assigned to jobs in remote regions, far from their ethnic communities in Transylvania, while in their place, ethnic Rumanians were resettled into Transylvania. A December 1984 Ministry of Education internal directive required enrolling students to sign a "contract" obliging them to accept what-

[20] "Memorandum" by Lajos Takács [former Rector of the Babes-Bolyai University], in *Witnesses to Cultural Genocide*, 153.

[21] "Authorities Restrict Hungarian-Language Higher Education to Two Percent Students," in *Hungarian Press of Transylvania*, Release No. 30/1986, dated May 28, 1986.

[22] Sándor Tonk, "Private Hungarian University in Rumania," in *Magyar Kisebbség* [Hungarian Minority], Kolozsvár/Cluj, vol. VI, no. 19, dated 1/2000.

ever position the government "guaranteed" them after gradua-
tion. Despite repeated protest by students and professors alike, the
practice was rigidly enforced.[23]

Repression of Culture

> *"The suppression of our culture is being*
> *carried out with unprecedented vehemence . . .*
> *Our language, in truth, has been*
> *forced out of public life entirely. . . ."*

In seeking to create a "unitary" (ethnically homogenous, "pure")
nation-state, the Rumanian government stifled the Hungarian minority's every forum for cultural expression. The Ceauşescu regime dismantled long-standing cultural institutions of Rumania's
Hungarian community, harassed prominent cultural figures, and
severed connections to Hungarian-language arts and letters originating outside Rumania.

In the 1970's four of the six Hungarian-language theaters became mere sections of Rumanian ones, where the management
and service personnel were exclusively Rumanian.[24] At one of the
two remaining Hungarian theaters, the one in Sepsiszent-
györgy/Sfintul Gheorghe (a 75 percent Hungarian-inhabited
town), the method employed in 1986 was to add a Rumanian section.[25]

One after another, Hungarian-language periodicals saw their
content absolutely censored and frequently reduced to verbatim
translations from Rumanian periodicals, their circulation cut back,
their editors summarily fired, or in some cases their existence
simply terminated. Beginning in 1973, six Hungarian-language
dailies were gradually reduced to weeklies, with their length also
cut, under the pretext of a "paper shortage." The Rumanian-
language papers also curtailed were subsequently restored to their
original state, while the Hungarian ones were not.[26] In 1983, edi-

---

[23] "Official Policy Forces Ethnic Minority University Graduates to Take Jobs Outside of Native Region," in *Hungarian Press of Transylvania*, Release No. 71/1984,
dated December 28, 1984.

[24] *Rumania's Violations of Helsinki Final Act Provisions*, 41.

[25] "Authorities Open Rumanian Section of Hungarian Theater In Sepsiszentgyörgy;
Brutal Intimidation of Hungarian Actors Continues," in *Hungarian Press of Transylvania*, Release No. 41/1986, dated June 21, 1986.

[26] *Rumania's Violations of Helsinki Final Act Provisions*, 39.

tors of *A Hét* [The Week], a Hungarian-language cultural weekly, were dismissed for unwillingness "to serve the Party leadership fully and without reservation."[27] In 1985, the government abolished the Hungarian cultural periodical *Művelődés*.[28]

At Hungarian cultural institutions, firings were commonplace. In 1984, forty Hungarian musicians at the Marosvásárhely [Tirgu Mureş] Philharmonic were summarily terminated and replaced with ethnic Rumanians.[29] In 1985, all performances of operettas or musicals in the Hungarian language were banned.[30]

The following chart[31] illustrates the limitations on use of the Hungarian language in naming the streets of Hungarian-inhabited cities.

Hungarian Street Names in Selected Larger Cities of Transylvania, 1992

City	Hungarian % of Population	Hungarian/All Street Names	Hungarian % of All Street Names
Arad/Arad	15.7%	11/680	1.6%
Nagyvárad/Oradea	33.2%	8/746	1.2%
Kolozsvár/Cluj	22.4%	21/750	2.8%
Marosvásárhely/Tirgu Mureş	51.1%	9/370	2.4%
Szatmárnémeti/Satu Mare	40.8%	9/350	2.6%

No trade magazines, scientific journals, drama or art journals were permitted in Hungarian.[32] At the same time, it was prohibited to subscribe to or bring into the country books, or other written material from Hungary. All such products were routinely confiscated as contraband at border crossings.

---

[27] "Authorities Fire Top Editors of Bucharest Hungarian-Language Cultural Weekly," in *Hungarian Press of Transylvania,* Release No. 11/1983, dated November 5, 1983.

[28] "Rumanian Authorities Plan to Eliminate the Periodical *Művelődés*," in *Hungarian Press of Transylvania* Release No. 81/1985, dated October 13, 1985.

[29] "Hungarian Musicians Fired in Marosvásárhely, Replaced by Ethnic Rumanians," in *Hungarian Press of Transylvania,* Release No. 46/1984, dated September 5, 1984.

[30] "New Measures Restrict the Performance of Hungarian-Composed Musical Works throughout Rumania," in *Hungarian Press of Transylvania,* Release No. 2/1985, dated January 5, 1985.

[31] Source: Bishop László Tőkés, *What's Behind a Statement: Ethnic Cleansing* (Nagyvárad/Oradea, Rumania: April 21, 1993), Appendix 20.

[32] *Rumania's Violations of Helsinki Final Act Provisions,* 38.

Undermining the use of the Hungarian language was an integral part of destroying the fabric of cultural life. The aim of official Rumanian measures was to force minority languages out of the public—indeed private—domain and promote Rumanian at their expense. While Article 22 of the Rumanian Constitution in effect at the time declared that the "coinhabiting nationalities shall be assured the free use of their mother tongue," the use of minority languages was, in fact, completely eliminated from all areas of official activity.[33] This included the removal of bilingual signs "identifying institutions, localities and so on in the native tongue of the local inhabitants," which had almost completely disappeared, wrote Communist Party dissident Károly Király in 1977.[34] The designation of place names in Hungarian was forbidden; for example, in the case of tenth grade Hungarian language and literature textbooks in 1986,[35] as was the granting of first names not translatable into Rumanian for newborns beginning January 1988.[36]

The other critical component of this objective was the elimination of minority-language radio and television broadcasting beginning January 1985 that had the added feature of rendering 600 minority individuals unemployed.[37] In 1981 Hungarian-language radio aired 6.5 hours a day, but in 1985, all local "foreign"–language programming (Hungarian, German and Serbian) was terminated. Similarly, Hungarian-language television broadcasting went from 2.7 hours a day in 1981 to zero in 1985 when German-language television was also eliminated.[38] Concurrently, the citizen who tried to gain access to information in their native language by listening to radio or television from Hungary faced

---

[33] *Rumania's Violations of Helsinki Final Act Provisions*, 35.

[34] *Witnesses to Cultural Genocide*, 175.

[35] "Use of Hungarian Place Names Banned Even in Hungarian Textbooks Used in Transylvania," in *Hungarian Press of Transylvania*, Release No. 67/1986, dated October 2, 1986.

[36] "Administrative Restrictions Expected on Hungarian Personal and Place Names," in *Hungarian Press of Transylvania*, Release No. 21/1988, dated March 31, 1988.

[37] "More on the Elimination of Hungarian and German Language Radio Broadcasts: 600 Employees Fired, Tape Archives Destroyed," in *Hungarian Press of Transylvania*, Release No. 17/1985, dated March 4, 1985.

[38] "All Minority-Language Radio Programming Eliminated in Rumania," in *Hungarian Press of Transylvania*, Release No. 6/1985, dated February 2, 1985.

severe punishment. For example, in August 1985 several people were sentenced to six months of forced labor to be served in their work place for retyping or rewriting schedules of Hungarian television programs while hundreds of individuals, including ethnic Rumanians, were interrogated by the police in Oradea [Nagyvárad] and Satu Mare [Szatmárnémeti] for their "viewing habits."[39]

The Rumanian authorities frequently harassed and persecuted prominent Hungarian intellectuals. In 1985, the actor Árpád Visky died under suspicious circumstances [40]and the prominent ethnographer Zoltán Kallós was arrested on the pretext of homosexuality while doing field work.[41] But simply banning a lecture, prohibiting the publication of a book, or slandering an artist's character in anonymous pamphlets, as with György Beke in 1984, served as a threatening example to the minority as a whole.[42]

Perhaps the most lasting damage in the regime's drive to purge Rumania of the culture, history and traditions of the Hungarians within its borders was achieved by removing folklore and other artifacts from local Transylvanian museums[43] and confiscating archival material from minority churches. This was authorized by Act No. 63 on the protection of the national cultural treasures, and Decree Law 207 amending decree law 472 of 1971 on the national archives.[44] The destruction fit exactly the definition of "cultural genocide" accepted in 1948 by the United Nations Ad Hoc Committee on Genocide (United Nations Document E/447),

---

[39] "Authorities Condemn Citizens to Six Months' Forced Labor for Distributing Hungarian Radio and TV Schedules," in *Hungarian Press of Transylvania* Release No. 64/1985, dated August 17, 1985.

[40] "Árpád Visky, Popular Transylvanian-Hungarian Actor, Dies Under Suspicious Circumstances," in *Hungarian Press of Transylvania*, Release No. 2/1986, dated January 13, 1986.

[41] *Rumania's Violations of Helsinki Final Act Provisions*, 45.

[42] "Authorities Fire Popular Hungarian Editor and Author György Beke; State Publishes Pamphlet Denigrating Beke and Threatening Hungarians," in *Hungarian Press of Transylvania*, Release No. 7/1985, dated February 10, 1985.

[43] *Hungarian Press of Transylvania*, Release No. 67/1984, dated December 11, 1984, reported, for example that in mid-November 1984 the directors of Hungarian community folk museums were summoned to a meeting and ordered to declare the percentage of "fakeries" in their collections and then to deliver the number of objects equal to the percentage they had declared, to the appropriate central museum, for destruction.

[44] *Rumania's Violations of Helsinki Final Act Provisions*, 47.

which speaks about the "systematic destruction of historical or religious monuments, or their diversion to alien uses, destruction or dispersion of documents and objects of historical, artistic, or religious value and of objects used in religious worship."

Restrictions on Human Contacts

"We demand to be considered an inseparable part of the entire Hungarian people."

Physical, intellectual and cultural isolation from their ethnic kin in Hungary and the West was a mechanism for weakening the Hungarian minority's ethnic identity. Restrictions on human contacts, of course, also served to keep information about the true situation of the Hungarian community from reaching beyond the country's borders.

In 1974, Decree/Law 224 made it a criminal offense to accommodate friends from abroad in private homes, except for immediate family members.[45] Of Rumania's population, the country's minorities were by far the most likely to be affected by this law and arguably its prime target. Of these, the minority with the most relatives and friends in bordering Hungary was the largest, Hungarian. Decree/Law 408, issued in January 1986, prohibited even the holding of conversations with foreign visitors in private homes.[46]

By the mid-*1980s*, Rumanian authorities had stepped up delays, restricted the numbers of tourists, and were systematically harassing Hungarians seeking to cross the border into Rumania from Hungary. During the first half of 1985, for example, an estimated 3,000 Hungarian citizens were refused permission to visit relatives in Transylvania.[47] Beginning January 1, 1986, an internal directive prohibited all individual tourist travel between Hungary and Rumania and permitted organized bus travel strictly on a reciprocal basis.[48]

---

[45] *Rumania's Violations of Helsinki Final Act Provisions*, page 53.

[46] "Rumanian Decree Imposes New Restrictions on Contacts with Foreigners," in *Hungarian Press of Transylvania*, Release No. 10/1986, dated February 10, 1986.

[47] "Visiting Hungarians Refused Entry," in *Agence France Presse*, Vienna, August 1, 1985.

[48] "Rumanian Government to Prohibit Individual Travel To and From Hungary," in *Hungarian Press of Transylvania*, Release No. 90/1985, dated November 2, 1985.

Limiting human contacts was closely linked to denying free access to information. On February 27, 1987, Ceauşescu publicly declared a ban, which had already long been in effect, on ordering or importing publications from Hungary.[49] No Hungarian-language printed material could enter or leave Rumania. All such materials were routinely and without exception confiscated at the border where visitors entering Rumania from Hungary were subjected to long waits and humiliating car searches.

Breakup of Compact Ethnic Communities and Forced Resettlement

"We demand an immediate end to measures aimed at artificially altering the ethnic composition of Transylvania."

With its near total control over the housing and labor market, for decades, Rumanian policy was to move ethnic Rumanians into minority-inhabited regions, forcibly resettle ethnic Hungarians to Rumanian-inhabited areas, and place ethnic minority professional workers far from their own people.[50] The aim was to break up the demographic unity of these areas, to weaken the minorities' sense of identity and to increase their sense of isolation from one another.

Citizens could not resettle into another city without official approval, therefore the policy was not to allow Hungarians to move between Transylvania but to move ethnic Rumanians from outside of the province into the region as codified under Article 201 of Law 5/1971.[51]

The reverse process moved ethnic Hungarians to Rumanian-inhabited areas. In September 1987, for example, hundreds of native Transylvanian workers were fired from their jobs and simultaneously hired to work at the Cernavoda nuclear power plant, located outside Transylvania, in a Rumanian-inhabited area. Workers who did not accept the job within five days, they were told, would be prosecuted for "parasitism." Since their work permits were already at the Cernavoda plant, they could not accept employment elsewhere. The men were further advised to take

---

[49] "Publications from Hungary are Officially Banned," in *Hungarian Press of Transylvania* Release No. 37/1987, dated March 2, 1987.

[50] *Rumania's Violations of Helsinki Final Act Provisions*, 31.

[51] "Memorandum" by György Lázár in *Witnesses to Cultural Genocide*, 133.

their families with them to Cernavoda, which would become their permanent assignment.[52]

By the mid-1980s the process of assigning ethnic Hungarian university graduates from Transylvania to jobs outside of the province while ethnic Rumanians graduates were hired to assume the vacant posts in Transylvania had become routine. For example, in 1988 all thirty–six graduates from the Tirgu Mureş [Mrosvásárhely] School of Medicine and Pharmacology were from Transylvania. All but ten of them were assigned jobs outside of Transylvania, while graduates from Iasi University were brought in to fill the vacancies[53]. In 1986, the county of Cluj [Kolozs], for example, had a Hungarian population of 350,000 but was allotted only twenty–two ethnic Hungarian doctors[54] so this measure had the added effect of depriving the community of native-speaking professionals as well. Another example, of the seventeen graduates from the Department of Hungarian Language and Literature of the Babeş-Bolyai University in 1988, only five were assigned jobs in Transylvania, and of those, only one to teach Hungarian.[55] Since the twelve posted to outside of Transylvania had no opportunity to practice their profession as there are no Hungarian communitie there, this measure also had the added effect of discouraging Hungarian youth from pursuing Hungarian studies.

In 1988 Ceauşescu announced a massive government urbanization plan, which would involve the bulldozing of fully 8,000 of the country's 13,000 rural communities.[56] The wholesale physical destruction of communities was now slated, including many 1,000 year-old, significant Hungarian settlements. Not only would have people been displaced, but thousands of historic churches, cemeteries, monuments, buildings, homes and landmarks would have been destroyed. For the Hungarian nation, a cultural heritage

---

[52] "Forced Labor Battalions in Rumania," in *Hungarian Press of Transylvania*, Release No. 101/1987, dated September 2, 1987.

[53] "Medical Students from Marosvásárhely and Temesvár Assigned to Internships," in *Hungarian Press of Transylvania*, Release No. 78/1988, dated December 3, 1988.

[54] Statement by the Committee for Human Rights in Rumania before the Subcommittee on International Trade of the Committee on Finance of the United States Senate, August 1, 1986, 13. (US GPO Doc. No. Doc. No. 65-139-0; S. HRG. 99-1008).

[55] "More than Two-Thirds of Recent Transylvanian-Hungarian Graduates Forced to Work Outside Transylvania," in *Hungarian Press of Transylvania*, Release No. 42/1988, dated July 15, 1988.

[56] "Details of So-Called 'Community Development Plan' Emerge in Rumania," in *Hungarian Press of Transylvania*, Release No. 10/1988, dated February 12, 1988.

carefully nurtured and defended for 1,000 years would have been irrevocably lost. By the time of the December 1989 overthrow of Ceauşescu, the plan's first stages were already being implemented.

The following charts quantify the legacy of Ceauşescu's policy of breaking-up compact ethnic communities. The first two show[57] the underrepresentation of ethnic Hungarians in local administration and the legal system, and the underemployment of ethnic Hungarians as managers of state-owned companies. The graphs[58] show the artificial alteration of the demographic composition in twelve cities and the sixteen counties forming Transylvania, 1910-1992. The long-term trends clearly (most dramatically in the case of the cities) point to the large-scale influx of ethnic Rumanians at the expense of the ethnic Hungarian population. The cities show the most radical change resulting in a shift in absolute majorities. These changes cannot be attributed to natural demographic changes resulting from a difference in the rural-to-urban migration patterns of ethnic Rumanians and Hungarians. In fact, at the end of March 1989, a jarring secret document surfaced in the Tirgu Mureş City Hall, which unequivocally substantiates the Rumanian authorities' goal to artificially alter the ethnic ratios of Transylvania in general, and its cities in particular. Dated November 1, 1985, the Notice by the First Secretary of the Mureş [Maros] County Committee of the Rumanian Communist Party details the plans for increasing the proportion of the ethnic Rumanian population in the city to over 50 percent. Containing exact statistics, the authorities calculated that bringing in 7,600 ethnic Rumanian workers, averaging three persons a family, would result in a 58-60 percent majority by the end of 1990.[59]

---

[57] Source: Tökés, *What's Behind a Statement: Ethnic Cleansing*, Appendices 18 and 19.

[58] Sources are the official census data published in *Anuarul statistic al Romaniei* (Bucharest, 1920-1939/40); *Anuarul statistic al Republicii Socialiste Romania* (Bucharest, 1966-1986, Directia Centrala de Statistica); and *Anuarul statistic al Romaniei* (Bucharest: Comisia Nationala pentru Statistica, 1990-1996).

[59] A facsimile copy of the original document was published in Tökés, *What's Behind a Statement: Ethnic Cleansing*, Appendix 14.

Underemployment of Ethnic Hungarians as Managers of State-Owned Companies, 1992

Locality	Hungarian % of Population	Hungarians/ All Managers	Hungarian % of Total
Bihar-Bihor County	12.5%	21/173	12.1%
Nagyvárad/Oradea City	33.2%	5/96	5.2%
Szatmár/Satu Mare County	35.0%	11/114	9.6%
Syilágy/Silaj County	23.7%		6.2%
Maros/Mureş County	41.3%		11.0%
Kovásyna/Covasna County	75.2%	50/88	56.8%

Underemployment of Ethnic Hungarians in
Local Administration and the Legal System in Rumania, 1992

COUNTY / County Seat	Hungarian % of Population	Prefect's Office Hung/Total	%	County Council Hung/Total	%	Mayor's Office Hung/Total	%	Notaries Public Hung/Total	%	Lawyers Hung/Total	%	Councillors Hung/Total	%	Judges Hung/Total	%
ARAD/ARAD	12.5%			5/110	4.5%										
Arad/Arad	15.7%					1/66	1.5%			4/66	6.1%			1/38	2.6%
BIHAR/BIHOR	28.5%	4/117	3.4%	3/183	1.6%			0	0.0%						
Nagyvárad/Oradea	33.2%					3/89	3.4%	0	0.0%	9/122	7.4%	1/12	8.3%	1/12	8.3%
SZATMÁR/SATU MARE	35.0%	18/129	14.0%									1/17	5.9%	2/21	9.5%
Szatmárnémeti/Satu Mare	40.8%							0	0.0%	8/36	22.2%	1/7	14.3%	1/7	14.3%
SZILÁGY/SALAJ	23.7%		9.0%					0	0.0%			1/10	10.0%	0/14	0.0%
Zilah/Zalau	19.8%						7.5%				17.4%				
KOLOZS/CLUJ	19.8%							0	0.0%						
Kolozsvár/Cluj	22.7%					8/168	4.8%								
HARGITA/HARGHITA	84.7%	14/41	34.1%	75/85	88.2%							2/5	40.0%	2/4	50.0%
Csíkszereda/Miercurea-Ciuc	83.0%					47/56	83.9%			26/65	38.4%	1/4	25.0%	2/7	28.6%
KOVÁSZNA/COVASNA	76.2%	13/41	39.4%									3/6	50.0%	1/6	16.7%
Sepsiszentgyörgy/Sf.Gh	74.8%									10/19	52.6%	2/5	40.0%	2/6	33.3%

Alterations in the Ethnic Composition of the Sixteen
Counties Forming Transylvania, 1910-1992

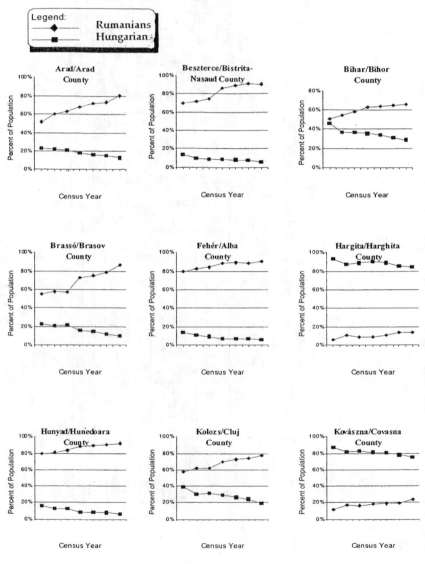

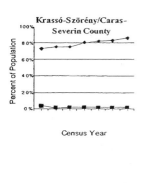

Krassó-Szörény/Caras-Severin County

Census Year

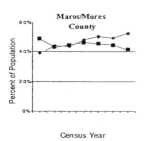

Maros/Mures County

Census Year

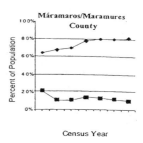

Máramaros/Maramures County

Census Year

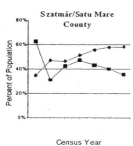

Szatmár/Satu Mare County

Census Year

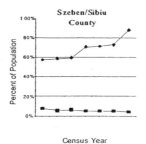

Szeben/Sibiu County

Census Year

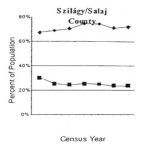

Szilágy/Salaj County

Census Year

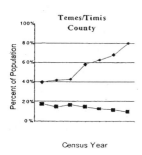

Temes/Timis County

Census Year

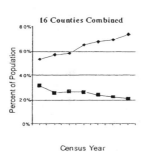

16 Counties Combined

Census Year

Alterations in the Ethnic Composition of 12 Cities
of Transylvania, 1910-1992

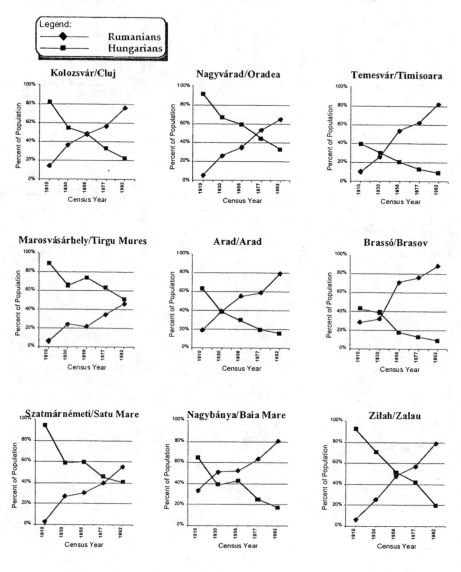

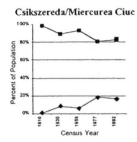

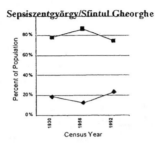

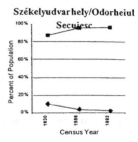

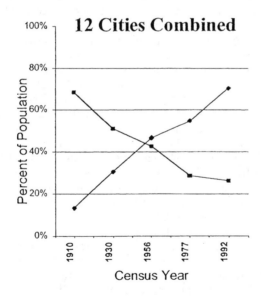

Ethnic Cleansing?

Finally, the question should be posed: what was the real impact of Ceauşescu's campaign to eradicate the Hungarian minority? Three results can be identified.

First, the campaign failed, at least as far as the Hungarians are concerned. Sadly, the same cannot be said of two other significant communities—Rumania's German and Jewish population, which once enriched the country's ethnic landscape, but during the Ceauşescu era left the scene forever. (Still numbering about 350,000 and 140,000 respectively after World War II, both groups fled in droves mostly under Ceauşescu, who extracted significant monetary ransom from the German and Israeli governments for each emigrant.)

Ceauşescu's impact then was to alter, probably irreversibly, the country's ethnic balance. His reign did succeed in reducing the diversity of the ethnic mix, increasing the dominance of the Rumanian majority and leaving only one significant historic minority where once there were several.

As for that minority, population statistics since the 1910 census show that the Ceauşescu regime continued and accelerated the artificial alteration, dramatic in urban areas, of replacing the ethnic Hungarian population with Rumanians. The evidence is overwhelming, including the illustrations in this paper, that the transformation was not due to spontaneous or voluntary internal migration patterns. During eighty–two years (1910-1992), the Rumanian population in the sixteen counties forming Transylvania doubled (from 2,830,000 in 1910 to 5,671,000 in 1992), while the Hungarian population declined by 65,000 (from 1,664,000 to 1,599,000). As a result, the Hungarian percentage of the total population of the province fell from nearly one-third (31.6 percent) to barely a fifth (20.7 percent). At the same time, although subject to continuing erosion, Ceauşescu did not succeed in breaking up the compact core of the Hungarian community at the center of Transylvania, the *Székelyland*.

Second, the campaign left a lasting legacy to Rumanian society that it is still acceptable, a decade after Ceauşescu's overthrow, to manipulate ethnic hostilities for political gain. The majority of Rumanian society today still subscribes to the misguided and obsolete falsehood that minorities serve to impoverish, not enrich the whole. The public discourse reveals a difficulty in understanding the basic concept that the granting of rights to a minority is not an undue favor or privilege which in some way deprives the "giftgiver," but that it is in the majority's own self-interest to strengthen internal cohesion and stability.

That Rumania's majority population has yet to come to terms with the nationalist Ceaușescu legacy is evident from the fact that two overtly xenophobic, anti-Hungarian and anti-Semitic parties participated in Ion Iliescu's governing coalition (1992-1996). The ultra-nationalist Mayor of Kolozsvár/Cluj Gheorghe Funar was reelected by popular vote to a third four-year term in May 2000. Even under the more democratic-minded forces governing the country since 1996, the state perpetuates the gross, discriminatory injustice of failing to provide restitution for 1,593 properties (schools, hospitals, charitable institutions) illegally confiscated from the Hungarian Catholic, Reformed and Unitarian churches under Communism. Despite a petition signed by 500,000 citizens, the state consistently refuses to reverse the forced merger of the Hungarian Bolyai university into its Rumanian counterpart which Ceaușescu personally executed in 1959. Instead of recognizing and seeking amends for past injustices, Rumanian society remains passive in the face of continuing discriminatory legislation and practice.

Third, instead of producing the Hungarian minority's passive submission or flight, Ceaușescu's campaign had the opposite effect of evoking the minority's greater internal cohesion and resistance to pressure.

Even under their worst years of victimization, the Hungarians of Rumania never allowed themselves to be defined purely as victims. In fact, while it was a widely propagated myth that there was no organized opposition in Ceaușescu's Rumania, the Hungarians of Rumania developed a well-organized, sophisticated underground movement to disseminate current information about their plight to the outside world and domestically as well. It was no accident that the overthrow of the Ceaușescu dictatorship in 1989 originated from the defiance of a Hungarian Protestant minister, Rev. László Tőkés, and his ethnic Hungarian congregation in the western Transylvanian city of Temesvár/Timișoara.

Its conduct during and since the 1989 overthrow of Communism clearly identifies the Hungarians of Rumania no longer as part of the problem, but as part of the solution.

Throughout the past decade, with the violence of ethnic-based conflict raging in neighboring states, this population, through its representative organization, the Democratic Alliance of Hungarians in Rumania, chose a different path. It pioneered a little-noticed, but equally dramatic experiment not to respond to aggression in kind, but to become an internally cohesive political entity and actively utilize peaceful, political means to promote not only its own particular interests, but the welfare of the broader community. Since 1990, the DAHR has worked within the system—between 1990 and 1992 as the largest party of the democratic opposition in the Rumanian Parliament, and since 1996 as partner in the country's democratic-minded governing coalition—to promote Western democratic values, moderation and regional security through lawful means.

As in the case of the large Hungarian minorities of Slovakia, Voivodina and Carpatho-Ruthenia or Subcarpathia, Rumania's Hungarian minority never reacted to the efforts aimed at its eradication with aggression, hostility or violence. When faced with threat, the record from the past century reveals a unique brand of nationalism, which was non-exclusivist, limited to seeking the protection and preservation of its culture, heritage, language, traditions and values, and proactive in the search for moderation and peaceful coexistence with the majority population. Even in response to greatest provocation—the decades-long, sustained persecution under the Ceauşescu regime—the pattern of ethnic heritage preservation which characterized Rumania's Hungarian minority must be distinguished from the Basque, Kurd or Palestinian avenues of violent resistance.

What was once a horror story is now one of reconstruction and hope. Perhaps enlightened Transylvania—the place where religious toleration was declared by law for the first time in history in 1568, at a time when wars of religious intolerance were raging throughout Europe—can embrace its true heritage and once again become the land of innovation and progress.

Ethnic Cleansing in the Former Yugoslavia in the 1990s: A Euphemism for Genocide?

KLEJDA MULAJ

Yugoslavia's disintegration produced new borders, renewed old nations, and established new polities. The wars of the 1990s that accompanied this process confirmed the revival of the old contests for territory and affected primarily that part of the civilian populations which the architects of these wars deemed unfit for their own ethno-national fabric. Yet, this time, there has not been much talk of "deportations" or "forced population transfers,"[1] but instead a new term has gained currency: "ethnic cleansing." The new term curiously incorporates both acts of deportation (ethnic cleansing of Albanians in Kosova and ethnic cleansing of Serbs in the Krajina region of Croatia) and acts of forced population transfer (ethnic cleansing of Bosnian Muslims in Bosnia Herzegovina).[2] Despite the wide use of the term, ethnic cleansing is neither recognized as a separate legal category nor prohibited as such by international criminal law. In addition, the meaning of the term remains blurred and is frequently used loosely and interchangeably with "genocide."

The interchangeable use of ethnic cleansing and genocide has been common not only amongst lay people but amongst many political practitioners, scholars of politics and international rela-

---

[1] While both deportation and forcible population transfers relate to the involuntary and unlawful evacuation of individuals from the territory in which they reside, customary international law differentiates between the two: deportation presumes transfer beyond state borders, whereas forcible transfer relates to displacement within a state. Prosecutor versus Radislav Krstić, Part III, Legal Findings, par. 521, http://www.un.org/icty/krstic/TrialC1/judgment/krs-tj010802e-3.htm.

[2] Throughout the multiple conflicts that accompanied the dissolution of former Yugoslavia however, instances of deportation and internal forced displacement were abundant and intermingled. Hence, the high number of both refugees and internally displaced peoples (IDPs). Although, expulsions aimed at ethno-national homogenization, their scale varied. It has been estimated that more than 150,000 Serbs were forced to leave Krajina and as many as 1 million Albanians were forced to leave Kosova in 1998-1999 although the latter case has been largely reversed following the ending of NATO's bombing of Serbia. In Bosnia Herzegovina the scale was even larger; up to 3 million people belonging to the three ethno-national communities—Bosnian Muslims, Croats and Serbs—were rendered homeless and up to 200,000 persons were killed.

tions, and journalists. For example, UN General Assembly Resolution 47/121 of 18 December 1992 states in paragraph 9 of the Preamble that "the abhorrent policy of 'ethnic cleansing' is a form of genocide...."[3] Genocide was recognized in Article 4 of the Statute of the International Tribunal for the Prosecution of Persons Responsible for Serious Violations of International Humanitarian Law Committed in the Territory of the Former Yugoslavia since 1991 (referred to as the International Tribunal for the former Yugoslavia (ICTY), established by the Security Council under Chapter VII of the United Nations Charter in May 1993.[4] In April 1995, a hearing before the Commission on Security and Cooperation in Europe of the U.S. Congress was entitled "Genocide in Bosnia Herzegovina."[5] From the very first week of the NATO bombing campaign against Serbia in March 1999, the U.S. State Department explicitly raised the possibility of genocide being under way in Kosova.[6] Helsinki Watch was the first non-governmental organisation to pronounce that its findings, at the very least *prima facie*, provided evidence that genocide was taking place in the region.[7] Amongst academics, Professor Rudolph J. Rummel of the University of Hawaii has no doubt that the Bosnian Serb massacre of Bosnian Moslems is genocide.[8] The same view was disseminated by Professor Helen Fein in her public lecture, "Denying Genocide: From Armenia to Bosnia" at the

---

[3] http://www.un.org/gopher-data/ga/recs/47/121.

[4] Security Council Resolution 827 of 1993 adopted on 25 May 1993. This Tribunal is empowered to prosecute persons committing genocide and other acts such as conspiracy to commit genocide, direct and public incitement to commit genocide, attempt to commit genocide or complicity in genocide. See, Adam Roberts and Richard Guelff, eds., *Documents on the Laws of War*, 3rd edition (Oxford, 2000), 569-570.

[5] U.S. Congress, *Genocide in Bosnia Herzegovina: Hearing before the Commission on Security and Cooperation in Europe*, 104th Congress, 4 April 1995, http://www.csce.gov/pdf/040495.pdf.

[6] Ivo H. Daalder and Michael E. O'Hanlon, *Winning Ugly—NATO's War to Save Kosovo* (Washington D.C., 2000), 111.

[7] *War Crimes in Bosnia and Herzegovina, A Helsinki Watch Report, Human Rights Watch* (August 1992), referred to as *Helsinki Watch 1st Report*. In addition, *Helsinki Watch 2nd Report, War Crimes in Bosnia and Herzegovina*, Helsinki Watch, A Division of Human Rights Watch (April 1993), maintains that "what is taking place in Bosnia Herzegovina is attempted genocide—the extermination of a people in whole or in part because of their race, religion or ethnicity."

[8] R. J. Rummel, *Death by Government* (New Brunswick and London, 1994), 31-32.

London School of Economics on 22 January 2001. Amongst journalists, Norman Cigar entitled a 1995 book *Genocide in Bosnia: The Policy of "Ethnic Cleansing"*, whereas Christopher Bennett, surveying the latest events in the region, has argued that ethnic cleansing, is, in fact, a euphemism for genocide.[9]

The following study is an attempt to explore the question of whether ethnic cleansing is a mild expression used to cover the horrors of genocide, or whether it is instead a notion in its own right used to express situations different from those of genocide. I will question here the legitimacy of the interchangeable use of these two terms in the context of the Yugoslav conflict of the last decade, and argue for the recognition of ethnic cleansing as a separate criminal offence.

What is Ethnic Cleansing?

The term "ethnic cleansing," unlike the practice it represents, is of relatively recent origin. Nevertheless, as yet, no coherent interpretation on the derivation of the term exists. Mary Kaldor, for instance, has asserted that the term "ethnic cleansing" was first used to describe the expulsion of Greeks and Armenians from Turkey in the early 1920s.[10] John McGarry and Brendan O'Leary argue rather that "ethnic cleansing" is a chilling expression coined by Serbs to describe forced mass population transfers, i.e. the physical transplantation of one (or more) ethnic community which is consequently compelled to live somewhere else.[11] Veljko Vujačić, on his part, traces the first use of the term in the late 1980s when Serbs complained of the alleged Albanian ethnic cleansing of Serbs from Kosova.[12] Whereas, Dražen Petrović has suggested that the expression "ethnic cleansing" has its origin in the Yugoslav military vocabulary of the 1990s—"ethnic cleansing"

---

[9] Christopher Bennett, "Ethnic cleansing in former Yugoslavia," in *The Ethnicity Reader: Nationalism, Multiculturalism and Migration*, ed. Monserrat Guibernau and John Rex (Cambridge, 1997), 122.

[10] Mary Kaldor, *New and Old Wars: Organised Violence in a Global Era* (Cambridge, 1999), 33.

[11] John McGarry and Brendan O'Leary, "Introduction: The Macro-Political Regulation of Ethnic Conflict," in *The Politics of Ethnic Conflict Regulation: Case Studies of Protracted Ethnic Conflicts*, ed. John McGarry and Brendan O'Leary (London and New York, 1993), 9.

[12] Quoted in Norman M. Naimark, "Ethnic Cleansing in Twentieth Century Europe," *The Donald W. Treadgold Papers, No. 19* (October 1998), 8.

696 Ethnic Cleansing in Yugoslavia in the 1990s

being a literal translation of the expression "*etničko čišćenje*" in Serbo-Croatian.[13]

The word "cleansing," however, was in use in the Balkan region much earlier. Philip J. Cohen states that during the Balkan Wars of 1912-1913 the term *cleansing* was used explicitly to describe Serbia's method of acquiring territories,[14] whilst Norman Cigar indicates that the term was used during the Second World War to connote the homogenous nature of the people inhabiting a given territory, particularly so in the framework of the fighting between Serbian nationalists of the Chetnik movement and Croatian nationalists of the Ustasha movement of the time and their efforts to forge their respective homogenous nation-states.[15] The addition of the age-old adjective "ethnic" indicates that people designated to be cleansed belong to other ethnic communities than that of the perpetrators.[16]

Ethnic cleansing is not what lawyers call "a term of art," lacking a legal definition and also a body of case law. It is instead a term used by soldiers, journalists, sociologists, social scientists, and others to describe a phenomenon which is not defined by law.[17] That said, a general agreement on a precise meaning of "ethnic cleansing" (despite the fact that several definitions of the term have been offered) is so far lacking. The Commission of Experts charged by the United Nations Security Council with inves-

---

[13] Dražen Petrović, "Ethnic Cleansing—An Attempt at Methodology," *European Journal of International Law* 5 (no. 4, 1994): 343.

[14] Philip J. Cohen, *Serbia's Secret War: Propaganda and the Deceit of History* (College Station, 1996), 7. Italics in original.

[15] Norman Cigar, *Genocide in Bosnia: The Policy of "Ethnic Cleansing"* (College Station, 1995), 18-19.

[16] Petrović, "Ethnic Cleansing—An Attempt at Methodology," 343. The adjective "ethnic" has been increasingly used since 1950s as a substitute for "racial" in the same way as "ethnicity" emerged in order to discard the concept of "race" and its sense of biological determinism. See, Paul Spoonley, *Racism and Ethnicity* (Oxford, 1993), 36. Generally speaking, "ethnicity" implies the quality of belonging to an ethnic community and it can mean the essence of an ethnic group or "what it is you have if you are in an 'ethnic group' ..."; this usually being so in "a context of relativities." "Ethnicity" has increasingly come to signify a marker of strangeness and unfamiliarity. See, "Introduction" in *History and Ethnicity*, ed. Elizabeth Tonkin, Maryon McDonald, and Malcolm Chapman (London and New York, 1989), 15-17.

[17] I am grateful to Professor Christopher Greenwood of the Law Department of the London School of Economics and Political Science for pointing this out to me during our conversation of 20 June 2000.

tigating war crimes in former Yugoslavia defined ethnic cleansing as "rendering an area ethnically homogeneous by using force or intimidation to remove from a given area persons of another ethnic or religious group."[18] This definition can nonetheless be criticized on the grounds that it does not qualify the perpetrators, ignores their intent, and also fails to emphasize the systematic character of ethnic cleansing.

UN Special Rapporteur Tadeusz Mazowiecki defined ethnic cleansing as "the elimination by the ethnic group exerting control over a given territory of members of other ethnic groups."[19] Mazowiecki added that "ethnic cleansing" may be equated with a systematic purge of the civilian population based on ethnic criteria, with a view to forcing the purged group to abandon the territories where it lives.[20] These two qualifications of ethnic cleansing suffer from a degree of ambiguity in so far as the meaning of "elimination" and "purge" is concerned. Indeed, it is not clear whether the author implies killing or removal of the targeted group(s). In addition, none of these two qualifications specifies ethnic cleansing as a policy of the perpetrators.

Andrew Bell-Fialkoff has argued that "ethnic cleansing can be understood as the expulsion of an 'undesirable' population from a given territory due to religious or ethnic discrimination, political, strategic or ideological considerations, or a combination of these."[21] This definition, however, seems to miss the deliberateness of the use of force or intimidation which is a standard feature of ethnic cleansing. Moreover, it fails to qualify the perpetrators and extends the range of the targeted group(s) beyond the ethnic, religious, and national criteria.

Finally, Dražen Petrović describes ethnic cleansing as "a well defined policy of a particular group of persons to systematically

---

[18] *Final Report of the United Nations Commission of Experts Established Pursuant to Security Council Resolution 780, 1990: Annex Summaries and Conclusions*, UN Doc. S/1994/674/Add.2 Vol. 1 (28 December 1994): 17.

[19] *Human Rights Questions: Human Rights Situations and Reports of the Special Rapporteurs and Representatives: Situation of Human Rights in the Territory of the Former Yugoslavia*, UN Doc. A/47/666-S/24809 (17 November 1992).

[20] *Situation of Human Rights in the Territory of the Former Yugoslavia, Sixth Periodic Report on the Situation of Human Rights in the Territory of the Former Yugoslavia submitted by Mr. Tadeusz Mazowiecki, Special Rapporteur of the Commission on Human Rights, pursuant to paragraph 32 of Commission resolution 1993/7 of 23 February 1993*, UN Doc. E/CN.4/1994/110 (February 1994) 44, par. 283.

[21] Andrew Bell-Fialkoff, "A Brief History of Ethnic Cleansing," *Foreign Affairs* 72 (1993), 110.

eliminate another group from a given territory on the basis of religious, ethnic or national origin."[22] The main flaw of this definition is that the characterization of the perpetrators as "a particular group of persons" is so vague as to render the definition inadequate.

If all of these definitions seem to fall short, one can still point out that they consistently recognize that in ethnic cleansing campaigns the bone of contention is *territory*, which is defined by the perpetrators in ethnic terms: the quest for territory inhabited only by the perpetrators' own people being the *modus operandi* of ethnic cleansing operations.[23] Consequently, ethnic minorities that inhabit contested territories are negatively affected as they are deemed by the perpetrators to impair the homogeneity of the dominant nation or ethnic community[24]--hence the recourse to force or intimidation to compel the targeted people to flee their homelands.

One must also say that while these definitions do a useful service in pointing out some important characteristics of ethnic cleansing, none of them on its own is sufficient in defining ethnic cleansing in all its complexity. Based on the above analysis, I propose that ethnic cleansing be defined as follows: Ethnic cleansing is a deliberate policy designed by and pursued under the leadership of a nation or ethnic community or with its consent, with the view to removing an "undesirable population" from a given territory on the basis of its ethnic, national, or religious origin, or a combination of these, by using systematically force or intimidation.

It should be pointed out that although the perpetrators of ethnic cleansing form a collectivity, this does not imply that each and every member of the perpetrators' community participates

---

[22] Petrović, "Ethnic Cleansing—An Attempt at Methodology," 351.

[23] Jennifer Jackson Preece, "Ethnic Cleansing as an Instrument of Nation-State Creation: Changing State Practices and Evolving Legal Norms," *Human Rights Quarterly*, 20 (no. 4, 1998): 821.

[24] An ethnic community, as defined by Anthony D. Smith, is "a named human population with myths of common ancestry, shared historical memories, one or more elements of common culture, a link with a homeland and a sense of solidarity among at least some of its members." See, John Hutchinson and A. D. Smith, eds., *Ethnicity* (Oxford and New York, 1996), 6. In general, a nation is more than an ethnic community; it has all the attributes of an ethnic community but, in addition, it is usually bigger in size, it is definitely attached to a clearly demarcated territory, has not only a common past but also a common vision for the future, and it has either acquired statehood or aspires to achieve statehood through self-determination.

directly in the acts of ethnic cleansing or even agrees with them. Instead, what makes ethnic cleansing a collective endeavour is that its perpetrators act under the directives of the leaders of their nation or ethnic community, or at least with the leadership's consent, and that the perpetrators have also the backing of a considerable section of their nation or ethnic community. Yet, there will be members of the perpetrators' community who do not approve the policy of ethnic cleansing, although they might be in minority or even if not so, they might be coerced into merging with the mainstream.[25]

As a practice, ethnic cleansing can consist of a range of different measures employed to create an atmosphere of fear and insecurity that compels the flight of the targeted ethnic population. This may include indirect, milder measures such as harassment, intimidation, and discrimination; denial of the means of employment, education, health care, public services, and public administration; prohibition of ethnic association; discriminatory and repressive legislation, etc.[26] Such indirect and milder measures were employed deliberately in Kosova, for instance, after 1989 when the autonomous status of the province was stripped off by Milošević until 1999 when Serb forces were forced out of the province in the aftermath of NATO's bombing of Serbia. Similar measures were adopted by Tudjman's regime towards the Serb community in Krajina following the independence of Croatia in 1992.

But more severe and violent measures may accompany ethnic cleansing, particularly when its policy and campaign are associated with times of war. Such violence may include outright expulsion of targeted ethnic communities from their places of residence (sometimes at gun point); shooting at their homes and properties or blowing them up with explosives; looting and

---

[25] Although statistics for all areas that underwent ethnic cleansing in former Yugoslavia are missing, a few researchers have provided some data about the size of the collectivity of perpetrators and their supporters. Peter Maass, for example, informs us that in Banja Luka, about 30 per cent of the Serb community opposed ethnic cleansing, while 60 per cent of them agreed or were confused and supported it overtly or tacitly. The latter went along with the 10 per cent who actually had the guns and controlled the television tower. Peter Maass, *Love thy Neighbour: A Story of War* (New York, 1995), 107. These statistics are supported also by Anthony Oberschall's research. See his article "The Manipulation of Ethnicity: from Ethnic Co-operation to Violence and War in Yugoslavia," *Ethnic and Racial Studies*, 23 (Nov. 2000), 986.

[26] Jackson Preece, "Ethnic Cleansing as an Instrument of Nation-State Creation," 822.

burning of residential areas including schools, medical facilities, churches, mosques to ensure that the targeted people do not return;[27] arbitrary mass arrests and arbitrary detention of civilians to provide a pool of prisoners for exchange, forced labour purposes, and/or use of detainees as human shields;[28] taking of hostages; forceful transfers of ethnic groups; use of landmines in locations of strategic significance such as border areas, roads, towns, villages, water-ways, industrial plants, and the like, situated in the contested territories;[29] deliberate torturing and other forms of cruel, inhuman and degrading maltreatment; killing of targeted persons or ethnic groups including their summary execution.[30] The siege of population centers, including shelling and sniper attacks and cutting off supplies of food and other essential goods, including the blocking of humanitarian aid is another tactic that has been used frequently to force the targeted ethnic

---

[27] In Kosova, for example, from March till May 1999, some 500 residential areas, including more than 300 villages, were burned in part or whole by Serb forces. http://www.state.gov/www/regions/eur/rpt_9905_ethnic_ksvo_exec.html.

[28] From July until the end of December 1992 alone the International Committee of the Red Cross (ICRC) registered approximately 10,800 detainees in more than 50 places of detention in Bosnia and Herzegovina. See E/CN.4/1993/50, *Report on the situation of human rights in the territory of the former Yugoslavia submitted by Mr. Tadeusz Mazowiecki, Special Rapporteur of the Commission on Human Rights, pursuant to Commission resolution 1992/S-1/1 of 12 August 1992* (February 1993), 11, par. 44.

[29] According to the United Nations Protection Force (UNPROFOR), whose mandate includes collection of information on minefields and coordination of mine clearance, an estimated four to six million landmines (mostly anti-personnel ones) may have been deployed by Serb, Croat and Bosnian Muslims forces throughout Bosnia-Herzegovina and the contested areas of Croatia. For the use of landmines in Croatia and Bosnia-Herzegovina, see Shawn Roberts and Jody Williams, *After the Guns Fall Silent: The Enduring Legacy of Landmines* (Washington, D.C.: Vietnam Veterans of America Foundation, 1995),181-205.

[30] A valuable contribution on the documentation of the practice and means of ethnic cleansing in former Yugoslavia in the 1990s has been made by Mr. Tadeusz Mazowiecki, who has written a number of reports on this matter. See, for instance, *Reports on the Situation of Human Rights in the Territory of the Former Yugoslavia submitted by Mr. Tadeusz Mazowiecki, Special Rapporteur of the Commission on Human Rights, pursuant to paragraphs 14 and 15 of Commission resolution 1992/S-1/1 of 14 August 1992,* respectively UN Doc. E/CN.4/1992/S-1/9 (28 August 1992) and UN Doc. E/CN.4/1992/S-1/10 (27 October 1992) and correspondingly *First,* UN Doc. E/CN.4/1994/3 (5 May 1993), *Second,* UN Doc. E/CN.4/1994/4 (19 May 1993), *Third,* UN Doc. E/CN.4/1994/6 (26 August 1993), *Fourth,* UN Doc. E/CN.4/1994/8 (6 September 1993) and *Fifth.* UN Doc. E/CN.4/1994/47 (17 November 1993) *Periodic Report(s) on the situation of human rights in the territory of the former Yugoslavia submitted by Mr. Tadeusz Mazowiecki, Special Rapporteur of the Commission on Human Rights, pursuant to paragraph 32 of Commission resolution 1993/7 of 23 February 1993.*

groups to flee. In particular, significant cities (Sarajevo, Mostar, and Vukovar included) were shelled on a regular basis in a deliberate attempt to spread terror among the civilian population.[31]

One of the most heinous features of ethnic cleansing campaigns, particularly in Bosnia Herzegovina and Kosova, has been the frequent occurrence of crimes against females. Sexual abuse of the targeted ethnic groups and in particular mass rape of women, including minors, have featured as a principle weapon of the perpetrators of ethnic cleansing.[32] While in principle ethnic cleansing can be a consequence of war, in the case of the former Yugoslavia in the 1990s, most regional analysts agree that war was initiated and carried out as a *means* of ethnic cleansing with the primary objective of the fighting being the establishment of ethnically homogenous regions. In this case, rather than being a by-product of war, ethnic cleansing constituted the war's most significant purpose.[33]

Genocide Defined

The term "genocide" is a synthesis of the Greek word *genos*, which means "race," and the Latin word *cide*, which means "to kill." It was coined by Raphael Lemkin in his 1944 book, *Axis Rule in Occupied Europe*, to describe the Holocaust. Four years later the legal definition of genocide was codified in Article 2 of the 1948 United Nations Convention on the Prevention and Punishment of Genocide (which entered into force in 1951), as follows:

> ...genocide means any of the following acts committed with intent to destroy, in whole or in part, a national, ethnical, racial or religious group, such as: (a) killing members of the group; (b)

---

[31] By early January 1994, for instance, there were on average 1,000 shell or rocket impacts per day in the city of Sarajevo launched by the Serb army. UN Doc. E/CN.4/1994/110 (n. 20 above), 11, par. 59. Similarly, Mostar has been subjected to constant shelling and sniping from Bosnian Croat forces. From May till September 1993, for instance, there were up to 400 shells impacting on the city every day. See, UN Doc. E/CN.4/1994/8, ibid., 4, par. 26.

[32] UN Doc. E/CN.4/1993/50, 19, paragraphs 82-85. Human Rights Watch Report: "Rape as a Weapon of "Ethnic Cleansing," http://www.hrw.org/reports/2000/fry. See also, UN Doc. E/CN.4/1994/5, *Rape and abuse of women in the territory of the former Yugoslavia: Report of the Secretary-General* (June 1993), 6, par. 17.

[33] For an endorsement of this view see, for instance, Kaldor (n. 10 above), 8; Noel Malcolm, *Bosnia A Short History*, New, Updated Ed. (London, 1996), 246; Bennett (n. 9 above), 122; Special Rapporteur Tadeusz Mazowiecki, E/CN.4/1992/S-1/10, 3, par. 6 and E/CN.4/1993/50 (n. 28 above), 7, par. 16.

causing serious bodily or mental harm to members of the group;
(c) deliberately inflicting on the group conditions of life calculated to bring about its physical destruction in whole or in part;
(d) imposing measures intended to prevent births within the
group; (e) forcibly transferring children of the group to another
group.[34]

Some scholars, however, have attempted to define genocide
in different terms than those of the United Nations Genocide
Convention. For example, Irving Horowitz defines genocide as "a
special form of murder: state-sanctioned liquidation against a
collective group, without regard to whether an individual has
committed any specific and punishable transgression."[35] Helen
Fein, for her part, has offered the following definition of genocide:
"Genocide is a series of purposeful actions by a perpetrator(s) to
destroy a collectivity through mass or selective murders of group
members and suppressing the biological and social reproduction
of the collectivity."[36] Frank Chalk and Kurt Jonassohn, on the
other hand, have proposed the following definition: "Genocide is
a form of one-sided mass killing in which a state or other authority intends to destroy a group, as that group and membership to it
are defined by the perpetrator."[37] Although these contributions
have the merit of pinpointing at least two important features of
genocide, namely that genocide is "state-sanctioned" and "a form
of one-sided mass killing." The most widely accepted definition of
genocide, nonetheless, remains that of the UN Genocide Convention.

According to the UN definition, genocide is characterized by
two legal ingredients, namely, the material element constituted
by one or several acts enumerated above, known in legal parlance
as *actus reus*, and the mental factor known in legal terminology as
mens rea, which consists of the special intent (or *dolus specialis*) to
destroy in whole or in part, a national, ethnic, or religious group,
as such. It should be emphasized that the special intent applies to
all genocidal acts mentioned in Article 2(a) to (e) above, that is, all
the enumerated acts must be part of a wider plan to destroy the

---

[34] Roberts and Guelff, eds., *Documents on the Laws of War*, 181.

[35] Irving Louis Horowitz, *Taking Lives: Genocide and State Power* (New Brunswick
and New York, 1980), 1-2.

[36] Helen Fein, *Genocide Watch* (New Haven and London, 1992), 3.

[37] Frank Chalk and Kurt Jonassohn, *The History and Sociology of Genocide: Analyses
and Case Studies* (New Haven, 1990), 23.

group as such. It is this special intent (*dolus specialis*) of the per-petrator/*genocidaire* to destroy a group in whole or in part that sets genocide aside from any other crime. As observed by the representative of Brazil during the *travaux preparatoires* of the Genocide Convention .

> genocide [is] characterized by the factor of particular intent to destroy a group. In the absence of that factor, whatever the degree of atrocity of an act and however similar it might be to the acts described in the convention, that act could still not be called genocide.[38]

While addressing the theoretical interpretation of genocide, both the United Nations' International Criminal Tribunal for Rwanda (ICTR) and the United Nations International Criminal Tribunal for the former Yugoslavia (ICTY) have held that the crime of genocide does not necessarily imply the actual extermination of the targeted group in its entirety.[39] Nor, according to the International Law Commission,[40] does it require the complete annihilation of a group from every corner of the globe.[41] Both Tribunals maintained that the geographical zone in which an attempt to eliminate a group is made may be limited in size (for example, it can be a region or even a municipality). We must ask the question: beyond what threshold could the crime be qualified as genocide?

The phrase "in whole or in part" was understood by both Tribunals to mean the destruction of a significant portion of the group from either a quantitative or a qualitative standpoint. Subsequently, genocidal intent may be manifested in two ways: it may consist of desiring the extermination of *a large majority* of the targeted group, in which case it would constitute an intention to destroy the group *en masse;* or it may consist of the destruction of a more limited number of persons, such as the leadership of the

---

[38] The Prosecutor versus Jean-Paul Akayesu, Case No. ICTR-96-4-T, decided on 2 September 1998, par. 519, http://www.ictr.org.

[39] The Prosecutor versus Clément Kayishema and Obed Ruzindana, Case No. ICTR-95-I-T decided on 21 May 1999, par. 95, http://www.ictr.org, and The Prosecutor versus Jean-Paul Akayesu, ibid., par. 497. The Prosecutor versus Goran Jelisic, Case No. IT-95-10 decided on 14 December 1999, paragraphs 79 to 83, http://www.un.org/icty/brcko/trialc1/judgment/index.htm.

[40] Quoted with approval by the ICTR in the Prosecutor versus Clément Kayishema and Obed Ruzindana, ibid.

[41] International Law Commission Draft Code of Crimes, 42, par. 8. Cf. http://www.un.org./law/ilc/reports/1996/chap02.htm.

group, chosen by the perpetrators for the impact that their disappearance would have upon the survival of the group as such.[42]

The definition of genocide, nevertheless, remains ambiguous for two reasons. The first has to do with the difficulty of determining the intent of the perpetrators and secondly, the targeted "group" is not quantified. Indeed, the genocide definition does not say what constitutes a "group"; in other words, how large must the targeted "group" be in order for the crime to be classified as genocide? Can it be the population of a district, a city, a town or even a village? In this context, the confusion on the meaning of "ethnic cleansing" is a corollary of the vagueness of the meaning of "genocide." Of course, it is only large-scale ethnic cleansing that strikes notes of similarities with genocide. To some extent, whether one decides to qualify such instances of ethnic cleansing as genocide will depend on whether one chooses to atomize the "group" to the level of the town or the village, in which case it will be easier to prove the intent of the perpetrators.

Indeed, while it might be difficult to prove that the Serbs had the intent to commit genocide against the whole community of Bosnian Muslims or Kosovar Albanians, it might be less difficult to do so should one look at particular villages or towns. In Prijedor in Bosnia, for instance, out of roughly 120,000 Bosnian Muslim residents, 56,000 were missing by 1995, and a large number were believed to have been killed. The Prijedor community's elite and leadership were particularly targeted. Perhaps even more striking as an example here, in Srebrenica between 13 and 19 July 1995 as many as 7,000-8,000 men of military age were systematically massacred while the remainder of the Bosnian Muslim population present at Srebrenica, some 25,000 people, were forcibly transferred outside the town.[43] The Trial Chamber in the Krstić case stated that it was "convinced beyond any reasonable doubt that a crime of genocide was committed in Srebrenica."[44] However, the Trial Chamber acknowledged that the central objective of the 1992-95 conflict between the Bosnian Muslims, Serbs, and Croats was ethnic cleansing, the use of military means to terrorize civilian population with the goal of forcing their flight.[45] In addition, the Trial Chamber expressed that except for

---

[42] The Prosecutor versus Goran Jelisić, par. 82. First italics added.

[43] Prosecutor versus Radislav Kristić, Part III, paragraphs 487, 519, 594.

[44] ICTY Press Release, The Hague, 2 August 2001, OF/P.I.S./609e. http://www.un.org/icty/pressreal/p609-e.htm.

[45] Prosecutor versus Radislav Krstić, Part III, Legal Findings, par. 562.

the high numbers of people that were executed, methodologically speaking, Srebrenica was no different from some other parts of Bosnia-Herzegovina.[46] The Krstić judgment also indicates that the strategic location of the enclave, situated between two Serb territories, may explain why the Bosnian Serb forces did not limit themselves to expelling the Bosnian Muslim population.[47] The judgment goes on to state that:

> The Bosnian Serbs' war objective was clearly spelt out, notably in a decision issued on 12 May 1992 by Momcilo Krajisnik, the President of the National Assembly of the Bosnian Serb People. The decision indicates that one of the strategic objectives of the Serbian people of Bosnia-Hercegovina was to reunite all Serbian people in a single State, in particular by erasing the border along the Drina which separated Serbia from Eastern Bosnia, whose population was mostly Serbian.[48]

The Krstić judgment indicates that the Trial Chamber has opted for the atomization of the "group" showing once more that the UN definition qualifies a situation as genocide or not depending upon one's interpretation of the word "group." Nevertheless, should one decide to atomize the "group" to the level of the town or village, one might run the risk of opting for countless genocides, which would look too pale when compared to the Holocaust, and the Armenian and the Tutsi genocides. In addition, should one atomize the "group" in this way, one will be faced with a number of violent conflicts which in their totality do not qualify as genocide and yet contain in themselves one or more cases of genocide, the former Yugoslavia in the 1990s being a recent case in point. In the opinion of this writer, when considering whether a situation qualifies as genocide or not, one should look at the big picture of the conflict and therefore avoid atomizing the (targeted) "group" to the level of a town or a village. This will allow for a conceptual consistency and also for an acknowledgement of a hierarchy of crimes in which genocide stands at the top.

---

[46] Prosecutor versus Radislav Krstić, Part II, Findings of Facts, par. 94, http://www.un.org/icty/Krstic/TrialC1/judgment/krs-tj010802e-1.htm.

[47] See Krstić Judgment: Part III, Prosecutor versus Radislav Krstić, Part III, Legal Findings, par. 597.

[48] Ibid., par. 562.

Comparing Ethnic Cleansing with Genocide

At a first glance, the UN definition of genocide may suggest that ethnic cleansing and genocide are coterminous. This can be attributed to the similarities of the *actus reus* ingredient of both phenomena. Indeed, the genocidal acts enumerated in the genocide definition can be present in ethnic cleansing campaigns although not in the same measure as in those of genocide. By the same token, the discriminatory character of both genocide and ethnic cleansing is conspicuous. In point of fact, both genocide and ethnic cleansing reject the egalitarian proposition that all men are equal and therefore should be so treated. On the contrary, their mission is to eliminate part of the population based on discriminatory grounds defined by the membership of individuals in a particular ethno-national group irrespective of their deeds. Consequently, both genocide and ethnic cleansing have served as political strategies which seek to eliminate ethnic differences within a given state and reinforce the hegemony of the perpetrators and their community.[49]

Although by no means only modern occurrences, both genocide and ethnic cleansing have assumed modernist traits: they are almost always sanctioned by the perpetrators' state. In turn, both these phenomena have served as instruments of nation-state creation or nation-state consolidation. It should be pointed out, nevertheless, that although by design, the beneficiary of the Holocaust was meant to be the German nation, Hitler did not undertake his grand endeavour in the name of the German nation but instead in the name of the Aryan race. He had come to realize that the concept of nation "could have only transient validity." In his own words, his aim was "to get rid of this false conception and set in its place the conception of race...." Indeed, his new order was to be conceived in terms of race and nations were to be fused into this higher order.[50] Ethnic cleansing in the Balkans, on the other hand, has not been motivated by grand fantasies of racial purification. Indeed, with so much intermingling and so many mixed marriages in the region the notion of "racial purification" sounds but a contradiction in terms. Instead, the campaigns of ethnic

---

[49] For a discussion of the methods employed to eliminate ethnic differences (as well as the methods used to accommodate such differences) see McGarry and O'Leary, eds., *The Politics of Ethnic Conflict Regulation,* especially "Introduction: The Macro-Political Regulation of Ethnic Conflict," 1-40.

[50] See, Alain Finkielkraut, *In the Name of Humanity: Reflections on the Twentieth Century,* trans. Judith Friedlander (New York, 2000), 51.

cleansing were carried out in the name of this or that nation, and for the benefit of this or that nation-state. Yet a note of similarity can be pointed out here. Like Hitler, who aspired for a "pure" German state that was to "embrace all Germans,"[51] so Milošević sought a state exclusively of the Serbian nation and Tudjman a state of the Croatian nation.[52] In employing ethnic cleansing and genocide to pursue the idea of the homogenous "nation-state," these countries followed a path beset with both danger and immorality. At the same time, they sought an end that could not be fully achieved. Indeed, despite the drastic efforts made to get rid of the targeted ethnic minorities, such minorities continue to exist in Serbia and Croatia, just like racial purification of Germany in the aftermath of the Holocaust had failed to be reached.

Andrew Bell-Fialkoff has pointed out that both ethnic cleansing and genocide are forms of what he calls "population cleansing," where "cleansing" is a euphemism that camouflages the ugly truth, namely, the human suffering of the targeted groups.[53] While both genocide and ethnic cleansing are concerned with getting rid of a targeted population, genocide constitutes an extreme case while ethnic cleansing a milder one, in so far as ethnic cleansing is more concerned with the removal of people while genocide is concerned with their physical extermination. One should note here that ethnic cleansing does not necessarily require the killing of people, although this does not imply that killing is not bound to happen, especially in case where the targeted people resist such removal with all their might. Ethnic cleansing that followed in the aftermath of the Balkan Wars of 1912-1913, for instance, did not occasion large scale killing.[54] This was particularly the case of the so-called population transfers that were conducted between Bulgaria and Turkey, Bulgaria and Greece,

---

[51] Adolf Hitler, *Mein Kampf*, trans. Ralph Manheim with an introduction by D. Cameron Watt (London, 1992), 362.

[52] This was explicitly expressed in Serbian and Croatian constitutions. See extracts from respective constitutions in Robert M. Hayden, "Imagined communities and real victims: self-determination and ethnic cleansing in Yugoslavia," *American Ethnologists*, 23(no. 4, 1996): 791. By pointing out this similarity, however, I am in no way drawing a sign of equality between these three figures.

[53] Andrew Bell-Fialkoff, *Ethnic Cleansing* (New York, 1996), 1-4.

[54] For a vivid description of the campaigns of ethnic cleansing in the course of the Balkan Wars of 1912-1913 see *The Other Balkan Wars: A 1913 Carnegie Endowment Inquiry in Retrospect with a New Introduction and Reflections on the Present Conflict*, Introduction by George F. Kennan (Washington, DC, 1993).

and Greece and Turkey.[55] Similarly, the case of ethnic cleansing of the Turkish minority in Bulgaria between 1984 to 1989 is another recent case in point. More than 300,000 ethnic Turks were expelled from the country that consequently fled to Turkey, but large-scale killing did not emerge.[56]

Nevertheless, in the case of the former Yugoslavia in the 1990s, one might argue that ethnic cleansing acquired a genocidal element given that killing, torture, rape, etc., were employed, at times in large scale, in order to spread such fear and insecurity as to compel the community of victims to take flight. Yet, the death tolls that resulted from ethnic cleansing of the 1990s in the former Yugoslavia do not compare with those that have resulted from the three genuine cases of genocide of the twentieth century namely: the Holocaust, the Armenian genocide of 1915 and the Rwandan genocide of 1994.[57] Killing can be viewed as a differentiating factor of the two phenomena: in genocide killing is an end in itself, while in ethnic cleansing killing is rather a means to an end. Subsequently, from a moral standpoint, the scale of extermination can be viewed as a measure of difference between ethnic cleansing and genocide.

In the three fully fledged cases of genocide referred to above, a large majority of the targeted groups was physically exterminated. It has been estimated that about six million Jews were killed by the Nazis, around one million Armenians were killed by the Young Turks, and at least eight hundred thousand Tutsis

---

[55] Although these population transfers were legalized in treaties, for all intent and purpose they qualify as *de facto* ethnic cleansing given that people were intimidated and compelled to leave, force was employed towards this end and hence human suffering was immense. For an analysis of these events see Stephen Ladas, *The Exchange of Minorities: Bulgaria, Greece and Turkey* (New York, 1932).

[56] See, Hugh Poulton, *Balkans: Minorities and States in Conflict* (London, 1991), 153-161.

[57] Using the UN definition of genocide and placing it in the context of the larger category of crimes against humanity, Alain Destexhe, has shown that in the course of the twentieth century there have been in fact only three genuine cases of genocide, namely: that of the Armenians by the Young Turks in 1915, that of Jews by the Nazis in 1939-1945, and that of the Tutsis by the Hutu between April-June 1994. See Alain Destexhe, *Rwanda and Genocide in the Twentieth Century*, trans. Alison Marschner with Foreward by William Shawcross (New York, 1995), 1-20. For Armenian genocide see, Vahakn N. Dadrian, *The History of the Armenian Genocide: Ethnic Conflict from the Balkans to Anatolian to the Caucasus* (Providence, 1997). For the Holocaust see Alvin H. Rosenfeld, ed., *Thinking about the Holocaust: After Half a Century* (Bloomington and Indianapolis, 1997). For the genocide of Tutsis see Philip Gourevitch, *We wish to inform you that tomorrow we will be killed with our families: Stories from Rwanda* (New York, 1998).

were killed in Rwanda, the latter at a rate three times faster than that of the Jews during the Holocaust. In former Yugoslavia, on the other hand, the scale of human destruction has been of smaller proportions with up to 200,000 victims in Bosnia Herzegovina (of all nationalities) and about 10,000 Albanians in Kosova. As pointed out in the ICTR hearings in the Kayishema and Ruzindana judgment, the scale of extermination measured with the number of dead victims from the targeted group can be an important factor when considering the intent of the perpetrators of the crime.[58] This implies that the scale of killing can serve as an indicator of the intent of the perpetrators to remove a targeted group from a given territory or physically destroy such group.

By extension, the targeted groups of ethnic cleansing, particularly in the case of the former Yugoslavia in the 1990s, have not proven to be completely defenceless. While it is impossible to argue for equivalency of action between the perpetrators and the victims, it is however the case that, in self-defense but not exclusively so, the targeted groups have also retaliated in kind. Indeed, a number of multiple conflicts took place following the disintegration of the federation of the former Yugoslavia. Serbs fought against Croats, Bosnian Muslims and Kosovar Albanians, and the latter three groups engaged in both defensive and offensive war against Serb forces. Moreover, in Bosnia-Herzegovina armed confrontations between Bosnian Muslims and Croats emerged in 1993. All parties are accused of war crimes and other human rights violations. This, again, stands in stark contrast with the case of the Holocaust, the Turkish genocide of Armenians, and also with the Rwandan genocide, which were one-sided mass killings sanctioned by the respective states with the goal of physically destroying the targeted group.

Although genocide acquires a territorial component in the sense that in its aftermath the perpetrators are likely to have greater control over the territory and its resources, the bone of contention in genocide remains the targeted people who, usually do not exercise sovereign claims over the territory. On the other hand, in ethnic cleansing what is primary at stake is *territory*, particularly when attempts are being made to redefine frontiers and contending parties claim disputed rights over given territories. In the case of former Yugoslavia in the 1990s, any one of the conflicting parties was a party in a territorial dispute. So long as a substantial number of people from an ethno-national community

---

[58] The Prosecutor versus Clément Kayishema and Obed Ruzindana, Case No. ICTR-95-I-T decided on 21 May 1999, par. 93.

lived in the disputed areas, such group could have some legitimate basis for its territorial claims. Hence, from the point of view of the perpetrators of ethnic cleansing, an efficient way to undermine the legitimacy of competing territorial claims was the removal of as many members of the rival group as possible. Indeed, given that in the case under consideration the aggressors were Serbs and that Serbs in contested territories were not a majority, they resorted to ethnic cleansing rather than genocide. Their tactics were to engage in the type of violence that would cause the targeted people to leave.[59]

Some analysts, nonetheless, have claimed that naming such events in the former Yugoslavia as ethnic cleansing and not genocide has to do with the reluctance to comply with the obligations put forward in the Genocide Convention.[60] Despite the fact that the initial unwillingness of the West to address the situation cannot be denied, one need not relate this only to whether the situation qualified as genocide. Although the Western reaction to the conflict was gradual, given the long-term commitment of international troops in Bosnia and the extent of NATO states' involvement in Kosova in 1999, it is imprudent to say that the West was not prepared to back up its responsibilities in the region. On the other hand, the nature of genocide in Rwanda was not contested, and yet the intervention there was so slow to come.

Conclusion

In this study I have sought to shed light on the puzzle of terminology that surrounds the notion of ethnic cleansing, with particular reference to the case of the former Yugoslavia in the 1990s. Despite the apparent similarities and the fact that ethnic cleansing is often used interchangeably with genocide, a comparison and contrast of the two terms suggests that ethnic cleansing is not necessarily a euphemism for genocide. The *prima facie* similarities between ethnic cleansing and genocide deriving from the *actus reus* ingredient of both phenomena may well tend to blind the observer to their dissimilarities. In particular, three differences between ethnic cleansing and genocide demand attention: whereas genocide is primarily concerned with the extermination of the targeted people, ethnic cleansing is concerned with the removal of the targeted groups from *disputed territories*; although more of-

---

[59] See the testimony of Professor Bassiouni, *Hearing before the Commission on the Security and Cooperation in Europe, 104ᵗʰ Congress, April 1995 (note 5 above)*.

[60] See, for instance, Cigar, *Genocide in Bosnia*, 115-116, 118.

ten than not killing accompanies ethnic cleansing, the scale of victimization during its campaigns has been smaller than in the case of genocide; and, while twentieth century's genocides have been one-sided mass killings, in ethnic cleansing of former Yugoslavia in the 1990s the targeted groups have retaliated in kind, though not in the same measure.

In our day, genocide has become a politically overloaded term; it is invoked to justify intervention on the part of the international community. Yet, ethnic cleansing as experienced in former Yugoslavia is already a very serious condition that demands outright action given the large scale violations of international human rights standards and humanitarian law. Verbal inflation risks deflating the meaning rather than justifying action.

The interchangeable use of the terms genocide and ethnic cleansing does not render justice to either term: genocide would be devalued and cheapened while the nature of ethnic cleansing would be obscured rather than explained. Hence the need to bring language into line with facts and develop an accurate concept that captures the essence of ethnic cleansing adequately. The time is therefore ripe to give ethnic cleansing the place it belongs in international criminal law by drawing up a convention on ethnic cleansing in an attempt to criminalize such a heinous crime and deter it from happening in the future.

Critique of the Concept of "Ethnic Cleansing": The Case of Yugoslavia

ROBERT H. WHEALEY

Yugoslavs began fighting among themselves in 1991. The collapse of the multinational Soviet Union in 1991 encouraged the Serbs, Croats, Muslim Slavs, and Albanians to renew their historic national-religious quarrels. The war, which began in Slovenia in the northwest, quickly spread to Croatia and then to Bosnia, and soon brought confrontation with the Serbian-dominated army. Bosnia declared its independence from the Socialist Federal Republic of Yugoslavia, and the North Atlantic Treaty Organization (NATO) powers legally recognized the independence of Bosnia in April 1992.

In February 1992, the United Nations (UN) sent a peacekeeping force to try to keep peace in disintegrating Yugoslavia; by November 1995, at the conclusion of the Dayton Truce in Ohio, the UN turned most of the responsibility for resolving the many problems in ruling Bosnia over to NATO.

Six great powers—the United States, Great Britain, France, Germany, Russia, and Italy—decided to intervene in the Yugoslavian civil wars for reasons that remain unclear to this day. Bosnia and Kosovo have been detached from Yugoslavia and Serbia, but as of early 1992 they were not in fact independent sovereign states which had the ability to collect taxes and balance their own national budgets.

Paris and London were more successful in helping the Croatians to defeat the Serbs than in helping the Bosnians. The French and the British promoted sending UN peacekeepers as a bluff, in 1991-1992, first to protect newly independent Croatia, and then, assuming that the Belgrade government would back down, to encourage recognition of a multinational Bosnian state where Muslims were the largest single group.

In 1991 Bosnia was nominally 44 percent Muslim, 31 percent (Orthodox Christian) Serbian, 17 percent (Roman Catholic) Croatian, and 8 percent mixed or other groups.[1] Serbs dominated the hills and valleys, while Muslims dominated most major towns.

---

[1] United States, Central Intelligence Agency, *The World Factbook 1991* (Washington, DC, 1991). See website: http://www.odci.gov/cia/publications/ 91fact/bk.html.

Prior to Sarajevo's declaration of independence, no Bosnian nation-state had ever existed.

Economically and militarily the U.S. clearly asserted its predominant status among the intervening six great powers, although politically and ideologically it appeared the most confused of those powers in defining its own interests in the Balkan region.

The Bush Administration and the State Department first blundered into the Bosnian affair because the major countries of the European Union—Germany, France, and Britain—acting through the mechanism of the UN, backed economic and political intervention in the Yugoslav civil wars. The Germans and Austrians had previously recognized breakaway Catholic Slovenia and Croatia in June 1991.

Most Americans, with little understanding of Balkan history, assumed that Catholic Croatians and Muslim Bosnians were victims of Orthodox Serbian aggression because the better-armed and militarily more successful Serbian elements in the Yugoslavia National Army (JNA) were presumed more brutal. Later, many in the American mass media assumed that an evil President Slobodan Milošević was mostly responsible for the ethnic cleansing of both Bosnian Muslims and Albanians from Kosovo. American journalistic opinion and secularized elites, especially in America, dismissed religion as politically insignificant to the conflict. Muslim civilians were certainly victims of a tragic war in Bosnia and in Kosovo, but like all wars, this war had multiple causes. To make Milošević the sole war criminal in a complex civil war was simplistic thinking, not the result of deep political analysis.

In reality, Milošević was a narrow-minded patriot born in rural Serbia and raised by an Orthodox priest. Milošević was educated in a Titoist, so-called socialist system to be a banker and state capitalist. From 1987 to 1989, he repudiated the heritage of the multiethnic, multinational Yugoslavia created by Josef Tito.

Inheriting a totalitarian bureaucracy, after 1989, Milošević was up to a point allowing a very disorganized and diverse opposition to run in elections and print hostile ideas. But in the end, his "Socialist Party," i.e. the bureaucracy, counted the votes. He dominated Belgrade through the secret police and mass media.[2]

U.S. senators and members of the House of Representatives at the time did not talk specifically about religion, but instead debated the Yugoslavian conflict in emotional terms, which seemed to take their prime cue from TV. In light of their advocacy of an expanded war in a distant nationalist-religious conflict, their hid-

---

[2] Robert Thomas, *The Politics of Serbia in the 1990s* (New York, 1998), 44-47, 93.

den values require clarification. Why should ordinary American citizens be asked to risk their money and lives for a Yugoslavian problem?

Although racial differences were important to the Nazis, who occupied Yugoslavia in World War II, members of various Yugoslav ethnic groups during that war generally did not consider themselves to be biologically different from one another. Both the Croatian-Serbian and the Serbian-Albanian quarrels of the 1990s were religious and linguistic in origin. During World War II, the big-three Allies (United States, Great Britain and the Soviet Union) generally viewed Serbs as friendly, pro-British, and pro-Soviet, while considering Croats as pro-German. The Albanians, who were pro-Italian, were largely ignored by the Big Three. Muslim Slavs, concentrated in Bosnia, remained a relatively powerless minority and were identified in Belgrade and in Zagreb with the despised remnants of the former Turkish Ottoman rulers.

Josip Tito's communist dictatorship, with its centralized bureaucracy in Belgrade, ruled Yugoslavia from the closing phases of World War II (1943-1945) until his death in 1980. A former guerrilla fighter, Tito rebuilt multiculturalism while running a police state. However, the anti-communists of the NATO bloc gave him little credit for his multinational policy until after his death. Tito's regime was tolerant of a variety of nationalities compared to the politicians who consolidated power under wartime conditions in 1991 and carried on military activities until at least June 1999.

From 1991 to 1995, a bellicose faction in America aimed to supply arms to both the Bosnian Muslims and Croat Catholics in order to defeat the traditional claims of a Serbian-led Belgrade government. Milošević was painted as a dictator when in fact he faced competition from four or five major Serbian nationalistic parties, some members of which wanted to intervene in Bosnia even more than Milošević himself. Actually, he exercised symbolic authority in Bosnia.[3] By 1998-1999, in the most recent round of the Yugoslav civil war, American interventionists wanted arms for ethnic Albanians, primarily in Kosovo, and loosely associated to the state of Albania and the Islamic faith.

From the outset in 1991, a number of Americans with ethnic ties to Europe, military pundits, and members of Congress had assumed that the Yugoslav quarrels could be quickly solved. Scenting potential headlines and sickened by reports of Serb atrocities, many journalists had become hawks and intervention-

---

[3] Ibid., 94-95, 97, 99-100, 178-180, 199-200, 208.

ists, who pressured the President to send American military assistance to the largely Muslim peoples in both Bosnia and Albania.

The Pentagon was reluctant to intervene in the Bosnian Civil War and also in the Albanian-Serbian War for fear that U.S. military involvement could become another Vietnam-style conflict with no clear goal or end in sight. And the generals realized that it would be hard to convince most Americans to fight in the Balkans for the idea of " multiculturalism." It would probably be even harder to drum up enthusiasm for two new Islamic states. Neither goal was worth risking war.

From 1993 to the summer of 1995, the Clinton Administration tried to enforce a strict arms embargo against the Milošević regime in Belgrade while allowing smugglers to import arms into Croatia. Paramilitary volunteers from Belgrade assisted the Pale government more than Milošević or the regular army, the JNA, to sustain the war against the Muslims and Croatians in Bosnia.[4]

Suddenly, in August of 1995, a well-supplied Croatian army invaded Bosnia and defeated the overextended Bosnian Serb military. NATO bombing for two weeks, in September, led by the U.S. Air Force, helped to force Milošević to sue for peace at Dayton in November of 1995. He thereby abandoned his allies among the Serbs living in Bosnia and the paramilitary volunteers from Belgrade fighting on the other side of the Drina River.

The second NATO bombing campaign, 24 March to 10 June 1999, destroyed much Yugoslavian industry and infrastructure, but had unclear consequences for the Kosovar refugees. As Ohio Representative Dennis Kucinich (a Democrat of mixed Yugoslav ancestry) put it on the floor of the U.S. House of Representatives on 28 April 1999: "Humanitarians do not bomb passenger trains. Humanitarians do not bomb refugees fleeing the battle. Humanitarians do not bomb residential areas. Humanitarians do not blow up water systems, electrical systems, sewage systems, and create an ecological catastrophe in the name of peace...No more bombing the villages to save the village...."

The Clinton Administration proclaimed in both wars, 1993-1995 and again in 1999, that it was trying to stop "ethnic cleansing" and was defending the Bosnian Muslim and the Albanian causes in the name of multiculturalism. In the hands of the American interventionist journalists, and hawkish Senators, "ethnic cleansing" became a slogan which justified air and some naval action against the Belgrade government. Belgrade's authoritarian ruler, Slobodan Milošević certainly shares some responsibility for

---

[4] Ibid., 148, 200, 208, 238.

ruthless "ethnic cleansing." Bosnian Serb leaders Radovan Karadžić and General Ratko Mladić, Croatia's Franjo Tudjman, and the rebellious Kosovo Liberation Army (KLA), also share responsibility for ethnic cleansing. Washington focused on Serb activities and suspected that President Milošević was behind them, while mostly overlooking killings and expulsions engaged in or encouraged by other Serbian party leaders like Vojislav Seslj, "Arkan" or Zeljko Ražnatović, Vuk Drašković, and Vojislav Kostunica. Yet, this recently elected politician who defeated Milošević in 2000, was, back in 1992, appealing to the Serbian monarchists and the Orthodox Church to save Bosnian Serbs. Kostunica was more democratic than most Serbs, but during the war in Bosnia, he was living in Serbia and allied to Chetnik militarists, who supported the Bosnian Serbs.[5] Washington was overlooking cruelties by Croats, Muslim Bosnians, and Albanian-speaking Kosovars, none of whom respected Tito's old borders or Tito's multi-ethnic ideals.

What is "ethnic cleansing?" The term is too broad, lumping together murder, arson, rape, expulsion, eviction, extortion, burglary and arbitrary arrest in hundreds of villages. This slogan had less of an echo in grassroots America than among the power-elite in Washington. Despite his brutal attacks, Milošević had his parallel more in Franjo Tudjman, President of Croatia, Hashim Thaqi (Thaci) guerrilla leader of the Kosovars, and Alija Izetbegović of Bosnia, rather than Adlof Hitler; the Serbs never engaged in systematic cremation or gassing in their concentration camps.

Paramilitary units, like *Frei Korps* and *fascio* gangs in Germany and Italy in 1919-1923, shot their neighbors. In addition, the Yugoslavs burned each other's houses, and in this sense, they were more barbaric than the early German and Italian fascists during the 1919-1922 period. Even the mass killings of military-age men in Srebrenica in July 1995 has been overdrawn in the American press. General Mladić was probably more responsible than anybody else for the dirty deed. In this notorious incident, perhaps 7,000 Muslim men of military age were shot in a mass execution.

The chain of command from 1941 to 1945, leading from Adolf Hitler to Adolf Eichmann, and on to six death camps in Poland, including Auschwitz, is clear. Just what responsibility Milošević had for the Srebrenica executions remains undocumented. Thus, Serbia had no Heinrich Himmler, and Milošević was no Hitler.

Ethnicity means a distinctive feeling of identity sometimes based on common geography, or language, race, religion, culture or class. Sometimes ethnicity is a unique combination of any two

---

[5] Ibid., 223.

of the six characteristics. Ethnicity is an especially slippery concept in Bosnia and among the disunited Albanian peoples, and also in the old Yugoslavia of 1991, where Serbs, Croats, and Bosniaks (which means Serbo-Croat-speaking Bosnian Muslims) were linguistically and biologically, if not identical, similar. Their key differences were religious. Religion is part of the ethnicity question, but a more flexible part than either language, heritage, or "race." In the words of Israeli historian David Vital, ethnicity is "modern and up-market and above all [an] elastic term."[6]

Rather than invoking a nineteenth-century Social Darwinist concept of race, it would be better to talk of biological or geographical inheritance. In the Yugoslav region, religion is a more accurate determinant of allegiances than "race," because Albanians, Croats, Serbs, and Bosnians are basically all of the same white or Caucasian race. When an "ethnic group" declares sovereignty, it proclaims itself a nation, not an "ethnicity." This is the reality which American journalists, who preferred references to the vague idea of "ethnicity," have tried to hide with fashionable language.

Albanians have a distinctive language but are diverse religiously. In the case of Albania, language is key. It may antedate the Greek and Roman civilizations, going back to before there were any Slavs in the area. The country is so wild that the Greek, Roman, Ottoman, and Stalinist Empires more or less left them alone. Enver Hoxha, Albania's repressive Communist dictator from 1943 until his death in 1985, provided no attraction for the dissatisfied Kosovars livings in Communist Yugoslavia. How much of his officially atheistic faith remains among present-day Albanians remains a mystery.

The United Nations Charter of 1945 was signed by fifty-one sovereign nations. That Charter was basically a restatement of the League of Nations Covenant. In the Paris Treaties signed in 1919, which redrew the map of Europe at the end of World War I, the concept of the nation implied sovereignty. Treaties which dealt with minorities recognized three kinds of minorities: religious, linguistic, and cultural. Ethnicity was not mentioned.[7] Nor was ethnicity mentioned in the classic study *Nationalism: Its Meaning and History* published in 1955 by Hans Kohn. Having lived through two world wars, he knew those wars had been more

---

[6] David Vital, "Irreversible Loss," *Times Literary Supplement* (5 May 1995), 10.

[7] I was reminded of this by Carole Fink of Ohio State University who is researching the treaties of 1919-1920s dealing with the concept of national self-determination in light of 2000.

about nationalism and sovereignty than ethnicity. Kohn identified six elements which contribute to the feeling of nationality—something for which people will risk their lives: language, religion, territory, a political entity (sovereignty), descent (a kind way of characterizing race) and customs (the vaguest concept).[8]

Only since 1992, when the NATO bloc put extraordinary pressure on Belgrade, could a Kosovo Liberation Army come into being to demand in April 1996 independence from Belgrade. Secretary of State Madeleine Albright eventually took a hand in stirring the Albanians up, which led to the inclusive February 1999 Rambouillet-Paris conference. Milošević refused to surrender the Kosovo province central to Serbian historical patriotism conquered by a Royal Serbian government in 1912. This led to the second round of American bombing. That decision masked poorly conceived war aims, and President Clinton could never have obtained a declaration of war from Congress for any lengthy conflict.

How has "ethnic *cleansing*" been defined? In 1922, Adolf Hitler spoke publicly about "cleansing" Germany of its last Jew. Clearly a racist, the Nazi party leader had even spoken of "cleansing" Austria of its "aliens" as early as February 1915.[9] The word "cleansing" became prominent in American mass media coverage of Serbia in the 1991-1995 period. Serbs were indeed using brutal tactics, trying to scare Muslims into leaving sections of Bosnia militarily controlled by Serbians. As far as is known, the *Chicago Tribune* coined the phrase "ethnic cleansing," using it on 21 May 1992 to associate Serbian military action with Hitler.[10] In a letter to the London *Guardian* on 30 July 1992, Bosnian Serb Radovan Karadžić, president of a so-called independent government in Pale, denied that he had set up concentration camps or was engaged in "ethnic cleansing." For the next three or four years, references to the term "ethnic cleansing" appeared in hundreds of articles written about the war. The Muslim cause was explained to American readers as a simple question of whether to support multicultural government. A later, generally judicious account published in 1999 does not give any figure for the number of peo-

---

[8] Hans Kohn, *Nationalism: Its Meaning and History* (Princeton, 1955), 9.

[9] John Lukacs, *The Hitler of History* (New York, 1997), 63. The 1922 date comes from John Toland's better known biography of Hitler.

[10] Ann McFeatters, Scrips Howard News Service, says that the term dated from 1992 but came from a State Department bureaucrat. Washington, "Clinton's Responsibilities in War," *Athens Messenger*, 18 April 1999. Former State Department official thinks the Department used the term in 1991.

ple killed in the Bosnian War, but Sarajevo estimates that 700,000 to 1.2 million Bosnians of all faiths were mostly *expelled,* based on guesses that pass for statistics.[11]

As the Bosnian War wound down militarily between May 1993 and May 1995, the outcries of U.S. media hawks gradually quieted. Although the many real war crimes in Bosnia were indeed deplorable, ethnic terror and expulsion in Bosnia were, in scale and thoroughness, nowhere near the intensity of the Nazis' "Final Solution," although various commentators have tried to make the comparison.[12]

---

[11] Not all Jews promote a Holocaust analogy for Yugoslavia. For a minority Jewish point of view, of one who did not support Sarajevo in its propaganda war, see Edward Serotta, "In Sarajevo: It is Jews Reaching out to save Christians and Muslims, providing needed Food, Shelter, and Medicine," *Los Angeles Times,* 27 April 1995. This account tells of the 1,200 members of Sarajevo's Jewish community, the majority of whom left for the U.S. in May 1992. Unlike many other American journalists, the author maintains strict neutrality between Catholic Croat, Muslim, and Orthodox Serb. Later substantiated by Paul Hockenos, "Peace without Pluralism?" *In These Times* (30 February 1996) 22, who adds that by 1996 only 400 Jews were left as devout rural Muslims fled to the city.

"Daily Report" (11 June 1997), Religion News Service, Washington, D.C. reprinted from Bosnet-Digest (vol. 5, No. 603) 18 June 1997, e-mail bosnet-digest@application.com, 18 June 1997, claims that after the war there were 1,000 Jews in all of Bosnia.

During the war another Jewish dove was also the pacifist Edwin Knoll, editor of *Progressive* before his untimely death. Henry Kissinger as a believer in the classic balance of power system also has no anti-Serb bias. A. M. Rosenthal sometime in 1995 began to write a series of columns that dissented from the major anti-Serb line of his newspaper, *The New York Times.* There is also a Jewish-Serbian Friendship Society of America in Chicago, e-mail, sii@moumee.calstatela.edu, 7 August 1995, but they have also been a small minority of Jewish opinion.

Carl Jacobsen, "Media Manipulation," *Mediterranean Quarterly* (1994), came to the same conclusion of a biased press independently.

An early important general survey is Florence Hamlish Levinsohn, *Belgrade—Among the Serbs* (Chicago, 1994). Levinsohn, a veteran Jewish journalist, defends the Serbs, detailing the effects of the long draconian UN embargo, through interviews with people in the city of Belgrade. From Chicago, she writes a travelogue which questions with different facts the New York and Washington elite and anti-Serb journalist. She is sensitive to the religious dimension of the Yugoslav problem.

Ramsey Clark, "International Action Center," 39 W 14th Street #206, New York, New York, sometimes in March and April 1999 was able to mobilize many Jews in a new anti-war movement. In contrast, Clinton and Albright were able to rekindle the New Left of the 1960s and belatedly drew the parallel to Indochina. The *Catholic Worker* supported this effort.

Robert Thomas, *Politics of Serbia,* in his survey of 35 Serbian newspapers and journals, does not use the term "ethnic cleansing."

[12] Steven L. Burg and Paul S. Shoup, *The War in Bosnia-Herzegovina: Ethnic Conflict and International Intervention* (New York, 1999), 171.

As a European mediator, Britain's Lord David Owen, understood from the beginning that the Yugoslavia problem was basically territorial, and that any solution would require getting all of the parties to agree to a redrawn map. The State Department was slow to endorse any map. Canadian General Lewis MacKenzie, a UN volunteer "peacekeeper" born in Nova Scotia, knew about historic tensions between Ontario and Quebec. He had a better understanding of the Bosnian problem than perceptions exhibited by New York-Washington journalists. When MacKenzie resigned from the UN peacekeepers in July 1992, "ethnic cleansing" to him meant that groups of Serbian soldiers chased Muslims out of their homes at gunpoint and then burned those houses down.[13] They or their compatriots also burned many mosques, shot males of military age, and raped many Muslim women, violently "encouraging" their neighbors to flee to Austria, Germany, Canada, the United States, or Switzerland. (Newly independent Slovenia and Croatia were not taking any of the Muslims.)

On 22 August 1995, senior Serb and Croat church leaders held their first formal meeting since war had broken out between their peoples in June 1991. These ecclesiastics proclaimed that their shared Christian faith could serve as a bridge between their nations. Noticeably absent from the conference were any Muslims, the third party in the wars of the former Yugoslavia. Meeting organizer Károly Tóth, a retired Hungarian Lutheran bishop, said that the conference aimed at breaking the ice between Croats and Serbs. Inviting the Muslims, according to Tóth, would have increased the likelihood that the meeting would have ended in fingerpointing instead of producing meaningful discussion.[14] News of the conference was welcomed by UN Secretary General Boutros-Ghali, who had worked for just such a dialogue for more than three years. In the aftermath of this event, the Orthodox Patriarch in Belgrade made a rare political intervention in August, symbolically abandoning the Pale Serb "president" Karadžić by backing Milošević's decision to negotiate with U.S. Ambassador Richard Holbrooke on a peace plan.[15]

The American mass media generally overlooked this type of news, much as they had ignored religion as an important political

---

[13] General Lewis W. MacKenzie, *Peacekeeper: The Road to Sarajevo* (Toronto, 1993), 153.

[14] Kecskemét, Hungary (Reuters), reprinted e-mail, sii@mounmee,calstatela.edu, 23 August 1995.

[15] Owen, *Balkan Odyssey*, 321.

factor in world affairs since the collapse of the USSR. In the words of historian Arthur Schlesinger, Jr., too many American journalists were in the business of "ethnic cheerleading," and the Serbs had few cheerleaders in America. As George Orwell put it, "some animals are more equal than others." Historians know that religion molds philosophy and ideology, and these shape politics. Religious partisanship has increased, a sign that something more than mere anti-communist sentiment and loosely-defined "morality" will be necessary to bring lasting peace to the Balkans.

In November 1995, Amnesty International released a report which held Serbs responsible for 80 percent of the atrocities in Bosnia, while Croats were responsible for only 15 percent and Muslims for 5 percent. But this may reflect a double standard. During the Croatian offensive of August 1995, columnists generally did not apply the term "ethnic cleansing" to Croat acts during their advance, but instead referred to "movement of troops." Some continued condemning ethnic cleansing, but they were speaking of Serb atrocities in Srebrenica the previous month. The use of the phrase "ethnic cleansing" helped mobilize first and second-generation Americans to become emotionally involved in the plight of the Bosnians.

Serb minorities also had fled from, or been expelled from, other parts of newly-independent Croatia during the 1991 Serbo-Croatian War. Serbs pointed out that by July 1992, 300,000 refugee Serbs had moved *into* areas ruled by Belgrade and Pale escaping out of places controlled by Croatia and Muslim-held Bosnia. Later, after official registration in Belgrade, this figure rose to 430,000.[16] Prewar Sarajevo had a 33 percent Serb population, but according to the Muslims, the city was only 7 percent Serbian as of the summer of 1995.[17] On the other hand, the number of Serbs driven out was smaller than the probable one million Muslims who fled, including refugees who had left Bosnia. These figures are frequently

---

[16] Press release from Belgrade by Stan Markotich, Prague, OMRI Special Report, I (9 January 1996) Reprinted e-mail, bosnet@grad.applicom.com, 11 January 1996.

[17] Letter Radovan Karadžić to Conservative MP John Kennedy, 15 July 1992, reprinted e-mail, Serbian Information Initiative, sii@mounee.calstatela.edu, 31 January 1995. He also lists seventeen concentration camps with 22,000 arrested Serbs. By March 1994 the Serbs told an American sociologist that the Serb refugee figure was approximately 450,000. Eric Markusen, "Report on Visit to Yugoslavia and Bosnia-Hercgovina, 2-11 March 1994," Marshall, Minn: Southwest State University, 31 March 1994, 6. Markusen, who made an independent investigation in Belgrade, claims 122,000 Serbian refugees there by June 1994, in "Report on Visits to Former Yugoslavia 19-29 April 1994 and 26 May 26-4 June 1994," Marshall, Minn: Southwest State University, 2 July 1994.

cited in U.S. press. Some 200,000 alleged deaths and casualties are included in this estimate.

Before the war, 50 percent of Sarajevo's population of 500,000 were Muslim.[18] During the war, Serbs loyal to Pale often seized Muslim-inhabited valleys for the strategic purpose of linking up the Serbian hills with one another. As time passed, Sarajevo became crowded with Muslim refugees chased out by the Bosnian Serb terrorist campaign. Together with the increase in their numbers, Sarajevo's Muslims became more religiously conscious. By April 1995 the Bosnian Army was more than 90 percent Muslim.[19] When Pope John Paul II visited the city two years later, Sarajevo itself was also 90 percent Muslim.

Croatia's August 1995 blitzkrieg in the Krajina was launched not from Croatian soil, but rather from Croat-inhabited Bosnian territory west of the city of Livno, with German and American-made arms. Nearly 200,000 Serbs lived in Krajina before the invasion and a total of 581,000 Serbs resided in all of Croatia, according to the 1991 census.[20] By the fall of 1995, fewer than 10,000 Serbs remained in the Krajina was geographically and politically part of Croatia, the conflict did not technically involve international aggression, but it certainly was an example of ethnicity-based expulsion and killing, if universal standards are applied.

The United States was directly, if clandestinely, involved with Croatia. The Department of Defense (DOD), CIA, and Defense Intelligence Agency (DIA) shared a secret operations staff, which occupied nineteen offices in the U.S. Embassy in Zagreb.[21] American Ambassador to Zagreb Peter Galbraith became another Croatian apologist on 9 August 1995 when he stated on the BBC that: "Ethnic cleansing is a practice supported by Belgrade and carried out by the Bosnian and the Krajinan Serbs, forcefully expelling local inhabitants and using terror tactics." In State Department "double think," he then denied that Croat actions in Knin were

---

[18] Paul Hockenos, "Sarajevo, Peace without Pluralism?" *In These Times* 20 (5 February 1996): 22-23.

[19] Paul Hockenos, "Sarajevo, Birth of a Nation," *In These Times* 19 (April 1995): 21.

[20] Daria Sito Sučić, Zagreb, Report of Helsinki Committee Round Table Committee proceedings. E-mail, Digest Vol. 5, No. 393, bosnet-digest@applicom.com, 29 October 1996.

[21] Ed Vulliamy, *The Guardian* (London) 29 January 1996, reprinted, e-mail, sii@moumee,calstatela.edu, 31 January 1996.

"ethnic cleansing."[22] However, privately Galbraith warned President Franjo Tudjman to enforce civilized standards within the Croatian Army and police forces, because they were reportedly beating up Serbians.[23]

In December 1998 the International Red Cross began more thorough research on the kinds of war crimes committed during the Bosnian war. Fortunately, they left out consideration of "ethnic cleansing," but looked instead for rape, expulsion, assaults, and other specific crimes defined as crime by civilized states. To celebrates the fiftieth anniversary of the modern Geneva Conventions, signed on 12 August 1949, the International Committee of the Red Cross/Red Crescent (ICRC) launched this project with the aim of building greater understanding of war crimes and respect for fundamental humanitarian principles. The idea of the project was to show that civilians and combatants alike would be able to share experiences and express opinions about what basic rules should apply, even in war, and discuss why those rules break down.

The questions lumped together Serbs, Croats, and Bosnian Muslims into one group of "ordinary citizens," as if Bosnia were already united. The ratio of respondents mirrored the proportions in the 1991 census. Although reports of deaths during the Bosnian war are still not statistically sound, the picture drawn is clearer than the sensationalism played up in the American mass media during the Yugoslav wars.

Highlights of the results of the Red Cross study are as follows: 80 percent of "ordinary citizens" of Bosnia-Herzegovina were familiar with the Geneva Conventions. More than 60 percent of the population lived in areas where the war took place, but less than 50 percent of the people were forced to move during the war. Fifty-three percent of people lost contact with family members, 44 percent of the population were forced to leave their homes and live elsewhere, 31 percent of Serbs said they had a close family member who had been killed, 30 percent of Bosniaks said a close family member had been killed, and 5 percent knew a woman who had been raped. (Appendix 55).

The New York Times editorials from 1994 to November 1995 began to tone down the previously anti-Serbian line that the paper had maintained during 1991-1993. Charges of ethnic terror and expulsion gradually diminished. *The New York Times* even pub-

---

[22] Patrick Moore, Open Media Research Institute, Inc., OMRI, Prague, reprinted e-mail, <sii.moumee.calstatela.edu>, 10 August 1995.

[23] Jane Perlez, *New York Times* Service, Banja Luka, Bosnia, 10 August 1995, reprinted e-mail, <sii.moumee.calstatela.edu>, 11 August 1995.

lished an August 1995 editorial voicing skepticism about Zabreb's retaking of Krajina.[24]

In December 1999, about the same time that the Red Cross study was published, Charles Krauthammer, a *New Republic* contributor and self-identified Zionist, exposed the hypocrisy of mainstream Jewish "moralists" in the American newsrooms, and in Congress, who fell mostly silent or even supported Croatian "ethnic cleansing" in Krajina.[25]

Serbian terror was not new or unique as war crime, apart from Irish and Basque political terror, but was unusual in Europe after World War II. Other East European peoples as well—Greeks, Turks, Bulgarians, Romanians, Poles, Czechs, Slovaks, and Hungarians—also expelled ethnic populations over the centuries. The last big European ethnic cleansing was the Germans in a large region from the Baltic to the Balkans in 1945.

In the Bosnian War, "ethnic cleansing" was often in the eye of the beholder. For instance, in May 1992, 1,200 to 1,500 Jews living in Sarajevo were evacuated to Israel.[26] The Izetbegović regime then confiscated their property while, ironically, Jewish leaders in the U.S. demonized the Serbs as racists and soft-pedaled Islamic Judeophobe measures.

Many American Jews identified with the Bosnian Muslims, perceiving them as a similarly-persecuted religious minority. This attitude seems to have been based upon insufficient study of Yugoslav conditions. In 1948, 38.8 percent of Yugoslavs were Serbs, 23.8 percent Croats, 8.5 percent Slovenes and 6 percent Bosnian Muslim. Thirteen "nationalities" were identified in Yugoslavia, including 64,159 Jews living in the cities. Like the USSR's leadership, Tito classified Muslims and Jews as nations, assuming with Marx, that religion would disappear, therefore, Jews ranked twelfth as a nationality, just ahead of Italians.[27] Apparently the Jewish population had declined to 6,000 by 1991, and Jews main-

---

[24] *The New York Times*, 8 August 1995.

[25] Charles Krauthammer, *Washington Post*, 11 August 1995, reprinted e-mail, sii@moumee,calstatela.edu, 8 September 1995.

[26] The figure 1,200 is from the note 11 above. *Liberation* in Paris has the figure 1,500 and the story of the Izetbegović confiscation. (1 September 1997).

[27] Ivo Banac, *The National Question in Yugoslavia: Origins, History, Politics* (Ithaca, 1984), 58.

tained a low assimilations profile in Serbia, Croatia, Slovenia, and Bosnia.[28]

The problem of the Jews of Yugoslavia was that those who survived World War II were content to be assimilated from 1945 to the 1980s. Those who were unhappy left for a more prosperous Western Europe, the U.S., or Israel. When the Yugoslav wars began to boil, Jews were split four ways—among Serbia, Croatia, Slovenia and Bosnia. Each of those four Yugoslav peoples accused "their Jews" of disloyalty but wanted to use them in mass media reports beamed to the West. Apparently Izetbegović won this aspect of the propaganda war in the American press. In 1991 Sarajevo was, in fact, more multicultural than other cities in Yugoslavia, particularly in comparison to Belgrade and Zagreb. Today, Sarajevo is more exclusively Muslim. In the 1990s, ethnic American Jews and Serbian nationalists have not been talking the same language. They defined morality quite differently.

Perhaps American journalists with sympathies toward Israel regret having pushed the "ethnic cleansing" line in the 1991-1995 period. By 1998, at the fiftieth anniversary of Israel's independence, one would have to acknowledge that Zionists may have "ethnically cleansed" Palestine of as many as 720,000 of its former residents.

Croats and Bosniaks as late as December 1996 were still fighting over control of Mostar. However, among Americans, similar events would be characterized as "evictions," rather than, "ethnic cleansing." In the 1990s, only Serbs received that accusation, on the basis of "rules" established in U.S. pressrooms.

American Bishops of the Roman Catholic Church called for US military intervention on the basis of the "just war theory."[29] Who decides what is "justice" or "injustice?" Would Catholics be judged as strictly as Orthodox Serbs?

Orthodox peoples generally identify Muslim Slavs and predominantly-Muslim Albanians culturally with the Turks, under whom the Serbs suffered from the fourteenth to the nineteenth

---

[28] Marko Živković, "The Wish to Be a Jew, or the Struggle Over Appropriating the Symbolic Power of 'Being a Jew' in the Yugoslav Conflict," 2. Paper read at Ninth International Conference of Europeanists, Chicago, 31 March-2 April 1994, Department of Anthropology, University of Chicago. The author is a Serbian Jew who handles the complex evidence with care on the "Jewish Question" in Slovenia, Kosovo, Serbia, Bosnia, and Croatia.

[29] Archbishop John Roach of the National Conference of Catholic Bishops, cited in James Turner Johnson, *Morality and Contemporary Warfare* (New Haven Conn: New Haven, Yale University Press, 1999), 95.

centuries. In the 1991-1995 crisis, Turkey unwisely reinforced this perception by supplying Albania with weapons.[30]

Belgrade's repression of Kosovars from about January to February 1998 on into 1999 could rightfully be called "ethnic cleansing." The main problem Americans had at that point was that the term so emotional in 1991-1992, had lost much of its punch. By 1999, historians were pointing out that many nations (including the U.S. during Andrew Jackson's administration) have long engaged in "ethnic cleansing."

American journalists writing for the mainstream newspapers in early 1998 had already built up biases against the Milošević regime. When in 1997-98, Albanian-speaking guerrillas began to fight more openly against Yugoslav officials in Kosovo, the American press and TV jumped in again with support for the victimized Kosovars against the repressive Milošević. They failed, however, to document the sources of the Kosovo Liberation Army's weapons. The anti-Serbian editorials of 1992-1995 were rerun in 1998-1999 to cheer on Albright's State Department to expand NATO jurisdiction over Pristina. Milošević was permanently charged again with "ethnic cleansing."

Most Americans were neither intellectually nor emotionally involved in the Albanian crisis, and the State Department began to run into passive resistance from Congress. Stories about the plight of the Albanian Yugoslav minority (a majority in Kosovo) made fewer headlines up to March 1999 than the Bosnian war had created in 1992-1995. Few cared except Clinton, Gore, and Holbrooke, Albright, and their sympathizers among the elite journalists on the *New York Times* and *Washington Post*.

Pro-Albanian Jews were fewer than Jews who had supported Sarajevo in 1992-1995. The Roman Catholic Church, once Croatian independence was achieved, distanced itself from the KLA. It was less willing to push the former stance of Milošević as the only "ethnic cleanser" in Yugoslavia. Only about one-third, at most, of the poorly-counted Albanians were presumed Catholic, with the rest largely Muslim. American Catholics and recent Albanian immigrants to the US put out misleading propaganda. Scholar Robert Thomas says there were only 55,000 Catholics in Kosovo out of 1.7 million.[31] The exception to the new media season of 1998-1999 was Senator Lieberman, who campaigned on TV in March and

---

[30] Misha Glenny, "Heading Off War in Southern Balkans," *Foreign Affairs* 74 (May/June 1995): 98-108, 105.

[31] Thomas, *Politics of Serbia*, 407-08.

April 1999 for a wider war against Milošević for the sake of Kosovo. His former colleague Senator Dole was out of the picture.

Democratic America has always been split in its views on creating and running an empire; many Americans have disapproved. Great Britain also had its internal opponents to empire, the "Little Englanders." What was unique about the Bosnian and the Kosovo occupations is that neo-colonialists had now begun to call themselves the "international community." Those administering Bosnia and Kosovo have been recruited from formerly imperialist nations, i.e., the six contact powers. Paradoxically, NATO protected Bosnia appears less like a traditional nation-state (as these have evolved since 1789), and more like a province or millet of the old Habsburg and Ottoman empires. Additionally, the Albanian speakers are still divided between the state of Albania, Kosovo, Macedonia, and Montenegro.

After all the fighting, the new rump Federal Republic of Yugoslavia was still more ethnically diverse than Croatia, Kosovo, and Slovenia. The real issue of the war appears to have been that no one wanted to wind up with minority status.

Henry Wadsworth Longfellow, one of American's classic poets, wrote a famous poem on the subject of ethnic cleansing. In *Evangeline*, the British rounded up French-speaking Acadians and shipped them to the French territory of Louisiana. Ironically, although the Clinton administration ostensibly sought to stop "ethnic cleansing" by bombing Yugoslavia, it served initially to increase the rate of expulsion of Kosovars.

Recent Developments in the Law of Genocide and Implications for Kosovo

JOHN CERONE[1]

In 1951, the International Court of Justice (ICJ) referred to genocide as:

> "a crime under international law" involving a denial of the right of existence of entire human groups, a denial which shocks the conscience of mankind and results in great losses to humanity, and which is contrary to moral law and to the spirit and aims of the United Nations.[2]

Nineteen fifty-one was also the year that saw the entry into force of the Genocide Convention—the international treaty that set forth the definition of genocide and required states to prevent and punish its occurrence. This definition has since been incorporated verbatim into all major international legal instruments dealing with genocide, including the statutes of the International Criminal Tribunal for the former Yugoslavia (ICTY), the International Criminal Tribunal for Rwanda (ICTR), and the International Criminal Court (ICC).

I. Introduction

A. The Definition—The 1948 Genocide Convention

Genocide, just like piracy and the slave-trade, is a crime under international law.[3] Its definition is set forth in Article 2 of the Genocide Convention, which states:

---

[1] Legal Advisor, Human Rights Policy Bureau, United Nations Interim Administration Mission in Kosovo. The views expressed in this paper are solely those of the author.

[2] *Reservations to the Convention on Genocide* (Advisory Opinion), ICJ Reports, 1951.

[3] Genocide is not, strictly speaking, a war-crime, as it may be committed in times of peace. It could be considered a crime against humanity, depending on the definition applied for "crime against humanity." At Nuremberg, however, crimes against humanity could only be prosecuted if they were committed along with other crimes within the subject matter jurisdiction of the tribunal (i.e. war crimes or crimes against peace (i.e. aggression)). One should note that by holding that the Nazis waged an aggressive war (crime against peace), all crimes against humanity committed subsequent to the start of the war were deemed to be committed in connection with the aggressive war, and were thus considered to be prosecutable.

In the present Convention, genocide means any of the following acts committed with intent to destroy, in whole or in part, a national, ethnic, racial, or religious group, as such:

(a) Killing members of the group;
(b) Causing serious bodily or mental harm to members of the group;
(c) Deliberately inflicting on the group conditions of life calculated to bring about its physical destruction in whole or in part;
(d) Imposing measures intended to prevent births within the group;
(e) Forcibly transferring children of the group to another group.

While there are slight variations under the laws of some states,[4] this definition is used verbatim in all relevant international legal instruments.

Broadly speaking, the definition could be divided into a mental requirement (the necessary intent to destroy the group as such) and a physical requirement (the commission of at least one of the enumerated acts).

1. Things to note about the definition

A few preliminary observations about the definition should be made. First, it is important to note that genocide is a specific intent crime. This special intent requirement (*or dolus specialis*) is an element of the crime. The perpetrator must have the intent to destroy, in whole or in part, the group as such. The intent "to de-

---

While the ICTY statute maintains the armed conflict requirement for crimes against humanity, it also includes the crime of genocide as a separately punishable act without imposing a requirement that it occur in the context of an armed conflict. The armed conflict requirement for crimes against humanity was dropped from the ICC and ICTR statutes. It is now widely accepted that crimes against humanity can be committed in times of peace.

[4] For example, under the domestic law of the Federal Republic of Yugoslavia, the enumerated acts include "forced displacement of the population." The US version of the definition states that the necessary intention must be to destroy the group in whole or "in substantial part." In the Canadian legislation, the target group may be "any identifiable group." This phrase is then defined as including groups identifiable by "colour."

stroy" is deemed to demonstrate the drafters' emphasis on the intended physical destruction of the group as a necessary element.

Second, it should also be noted that killing is not expressly required. The perpetrator need only commit one of the enumerated acts with the required intent, bearing in mind that the intent must generally be inferred from the acts committed and their context.

2. *Other things to note from the text of the Convention*

The rest of the Convention elaborates upon other legal aspects of genocide. First, the Convention makes clear that: (a) state parties are responsible for preventing and punishing genocide;[5] (b) genocide may be committed by private, or non-state actors (i.e. there is no requirement of state action);[6] (c) genocide may be committed in times of peace as well as in times of armed conflict;[7] and (d) there can be no immunity from prosecution and punishment of genocide, not even for a head of state.[8]

3. *Things to note from the background of the Convention*

The *travaux* make clear that the definition of genocide set forth in the Convention was not intended to encompass "cultural genocide"; nor was it to provide protection for political groups.[9]

B. Room for Growth

The ambiguity of the definition contained in the Convention left plenty of room for development, and created a need for jurisprudence to elaborate further upon the type of intentional conduct that would constitute genocide.

The ambiguities were many. First, the Convention does not expressly provide for individual criminal responsibility under in-

---

[5] Convention on the Prevention and Punishment of the Crime of Genocide ("Genocide Convention"), art. 1 ("The Contracting Parties confirm that genocide, whether committed in time of peace or in time of war, is a crime under international law which they undertake to prevent and to punish.").

[6] Genocide Convention, art. 4 ("Persons committing genocide or any of the other acts enumerated in article III shall be punished, whether they are constitutionally responsible rulers, public officials or private individuals.").

[7] Genocide Convention, art. 1.

[8] Genocide Convention, art. 4.

[9] See Josef L. Kunz, "The UN Convention on Genocide," 43 *American Journal of International Law* 4 (1949).

ternational law; it was therefore unclear whether individuals could be prosecuted for genocide under international law as such.[10] Second, the intent requirement is similarly vague (e.g., what constitutes the intent to destroy in part a group as such?). Third, although the list of protected groups is finite, did this imply that it was exhaustive? And this same question applied to the list of acts. Finally, it was also unclear whether a state could itself be held responsible for committing genocide.

C. The Character of the Norm Prohibiting Genocide

A few words should also be said about the character of the norm prohibiting genocide. First, the prohibition of genocide has entered the corpus of customary international law. Thus, the obligation to prevent and punish genocide exists independently of a state's treaty obligations (i.e. even states not parties to the Convention are bound by this obligation).[11]

Second, this norm has acquired the status of *jus cogens*, meaning that it is a higher-order norm overriding conflicting obligations and voiding conflicting treaties.[12]

Third, this norm gives rise to obligations *erga omnes*. This means that the obligation to prevent and punish genocide is owed to all members of the international community. Thus, all states have standing to protest irrespective of whether any material interest has been adversely affected by the failure to prevent or punish. In essence, states have a legal interest in adherence by all other states to these obligations.[13]

II. Elaboration of the Definition Through International Jurisprudence–The ICJ

Recent cases before the International Court of Justice have contributed to the elaboration of the law of genocide.

---

[10]   *Id.* See also Arthur K. Kuhn, "The Genocide Convention and State Rights," 43 AJIL 3 (1949).

[11] See *Reservations* case, *supra* note 2.

[12] Human Rights Committee, General Comment 6, 30 July 1982. See also *Case Concerning Application of the Convention on the Prevention and Punishment of the Crime of Genocide (Bosnia and Herzegovina v. Yugoslavia (Serbia and Montenegro))* (separate opinion of Judge Lauterpacht), 1993 I.C.J. 325.

[13] See *Barcelona Traction case*, ICJ Reports 1970.

A. *Bosnia & Herzegovina v. Yugoslavia* (Serbia & Montenegro) (1993-96)

In the *Case Concerning Application of the Convention on the Prevention and Punishment of the Crime of Genocide (Bosnia & Herzegovina v. Yugoslavia [Serbia & Montenegro])*[14], the ICJ made several important contributions in this regard. First, it confirmed that there is no territorial limitation to the obligation to prevent and punish acts of genocide (i.e. the obligation extends to acts committed beyond a state's own territory).[15]

Second, the ICJ also confirmed that a state can be responsible for itself committing genocide, rather than simply for violating the articles requiring it to prevent and punish acts of genocide.[16]

B. Legality of the Threat or Use of Nuclear Weapons (1996)

The issue of intent was broadly addressed in the *Nuclear Weapons Advisory Opinion* of 1996[17] where it had been argued that the use of nuclear weapons was illegal because they were inherently genocidal. The Court held that the specific intent to destroy a group was determinative, and that the question of genocide had to be determined on a case-by-case basis. Therefore, it could not rule out the possible use of nuclear weapons for a non-genocidal purpose.[18]

---

[14] Judgment of 11 July 1996, ICJ Reports 1996.

[15] *Id* at para. 31 ("It follows that the rights and obligations enshrined by the Convention are rights and obligations *erga omnes*. The Court notes that the obligation each State thus has to prevent and to punish the crime of genocide is not territorially limited by the Convention."). It should be noted, however, that the Court stated this in the context of a state party having control over a territory outside of its own.

[16] *Id* at para. 32 ("The Court would observe that the reference in Article IX to 'the responsibility of a State for genocide or for any of the other acts enumerated in Article III', does not exclude any form of State responsibility.").

[17] Advisory Opinion of 8 July 1996, ICJ Reports 1996.

[18] The dissent responded that given the well-known effects of using nuclear weapons, any party using them must intend the destruction, at least in part, of the national group of the country at which the weapons are directed. This question of whether certain acts are inherently genocidal was also addressed in the *Bosnia & Herzegovina v. Yugoslavia* case, *supra* note 14, where the court held that the annexation or incorporation of a state by another state was not necessarily a genocidal act.

III.   Elaboration of the Definition Through International Jurisprudence–The ICTs

The establishment of the International Criminal Tribunals (ICTs) for the former Yugoslavia and Rwanda enabled the further development of the law as applied to individual perpetrators.

A. Individual Criminal Responsibility

Early on the ICTY confirmed that there is individual criminal responsibility under international law for the commission of genocide.[19] Recall that at the time the Genocide Convention was adopted, it was not clear whether individuals could be prosecuted under international law as such.[20]

Now the dual nature of the prohibition of genocide becomes clear. Its violation can give rise to state responsibility as well as individual criminal responsibility.

B. Elaboration of the Definition

The ICTR has adopted a fairly expansive interpretation of the definition of genocide; the ICTY less so. However, there is a lot of cross-fertilization between the two tribunals, as each frequently cites cases of the other, leading to harmonization of their decisions.

1. Protected Groups

The ICTR has examined the nature of the groups listed in the definition and extracted what it deemed a common criterion -- "that membership in such groups would seem to be normally not challengeable by its members, who belong to it automatically, by birth, in a continuous and often irremediable manner."[21] It determined that any permanent, stable group should be protected.

---

[19] *The Prosecutor v. Dusko Tadić*, Decision on the Defence Motion for Interlocutory Appeal on Jurisdiction, 2 Oct. 1995, IT-94-1-AR72 (RP D6413-D6491).

[20] Kuhn, *supra* note 10; Kunz *supra* note 9.

[21] *Prosecutor v. Jean-Paul Akayesu*, Judgement, 2 September 1998, ICTR-96-4-T, para. 511.

The ICTY built upon this by holding that a group may be defined with reference to the perspective of the perpetrator. In the *Jelišić* case, the ICTY held:

> to attempt to define a national, ethnical or racial group today using objective and scientifically irreproachable criteria would be a perilous exercise whose result would not necessarily correspond to the perception of the persons concerned by such categorisation. Therefore, it is more appropriate to evaluate the status of a national, ethnical or racial group from the point of view of those persons who wish to single that group out from the rest of the community.[22]

The Tribunal stated further that a positive or negative approach could be used in making this determination.[23] A positive approach, as defined by the Tribunal, would involve distinguishing a group by characteristics which perpetrators deem particular to that group. A negative approach would be the case where perpetrators distinguish themselves as an ethnic, racial, religious, or national group distinct from the other group or groups.

2. Enumerated Acts

In the *Akayesu* case, the first genocide case by one of the ICTs, the ICTR elaborated upon the possible acts that constitute genocide when committed with the requisite intent.

(a) Killing members of the group

With regard to the first enumerated act—killing members of the group—the ICTR has employed a somewhat narrow interpretation by requiring that the killing amount to murder (a specific intent crime). However, there is nothing particularly new in this holding as killing with the intent to destroy the group will generally mean that the perpetrator intended to kill the victim in any case.[24]

(b) Causing serious bodily or mental harm to members of the group

---

[22] *Prosecutor v. Goran Jelišić*, Judgement, 14 December 1999, IT-95-10-T, para. 70.

[23] *Id* at para. 71.

[24] *Akayesu, supra* note 21, at para. 501.

Regarding the second enumerated act, the ICTR stated, "Causing serious bodily or mental harm to members of the group does not necessarily mean that the harm is permanent and irremediable."[25] In doing so, it cited the *Eichmann* case for the proposition that "serious bodily or mental harm of members of the group can be caused 'by the enslavement, starvation, deportation and persecution... and by their detention in ghettos, transit camps and concentration camps in conditions which were designed to cause their degradation, deprivation of their rights as human beings, and to suppress them and cause them inhumane suffering and torture'."[26] Ultimately, the Tribunal took serious bodily or mental harm, "without limiting itself thereto, to mean acts of torture, be they bodily or mental, inhumane or degrading treatment, persecution."[27] The *Akayesu* Tribunal expressly found that sexual violence fell into this category, and ultimately pointed to acts of rape in this case as genocidal acts.

(c) Deliberately inflicting on the group conditions of life calculated to bring about its physical destruction in whole or in part

The ICTR held that the means of deliberately inflicting on the group conditions of life calculated to bring about its physical destruction, in whole or part, "include, *inter alia*, subjecting a group of people to a subsistence diet, systematic expulsion from homes and the reduction of essential medical services below minimum requirement."[28]

(d) Imposing measures intended to prevent births within the group

Within this category of measures, the ICTR included sexual mutilation, the practice of sterilization, forced birth control, separation of the sexes, and prohibition of marriages.[29] In addition, it held that in a culture where membership in the group is deter-

---

[25] *Id* at para. 502. Compare the US legislation which requires "permanent impairment of the mental faculties through drugs, torture, or similar techniques." 18 USC 1091.

[26] *Id* at para. 503.

[27] *Id* at para. 504.

[28] *Id* at para. 506.

[29] *Id* at para. 507.

mined by the identity of the father, deliberate impregnation during rape by a man not of the group[30] could also constitute such a measure. The Tribunal further determined that such measures could be mental in nature. It stated: "For instance, rape can be a measure intended to prevent births when the person raped refuses subsequently to procreate, in the same way that members of a group can be led, through threats or trauma, not to procreate."[31]

(e)   Forcibly transferring children of the group to another group

In line with its expansive interpretation of the first four enumerated acts, the Tribunal opined that the objective of the fifth enumerated act "is not only to sanction a direct act of forcible physical transfer, but also to sanction acts of threats or trauma which would lead to the forcible transfer of children from one group to another."[32]

3. The Intent Requirement

It is important to recall that the bar was intentionally set very high in imposing a special intent requirement in the definition of the crime. For that reason, genocide is not easy to prove.

That is also why "in the absence of a confession from the accused, his intent can be inferred from a certain number of presumptions of fact." In the *Akayesu* case, the ICTR considered that it was possible to deduce the genocidal intent of a particular act from: (1) the general context of the perpetration of other culpable acts systematically directed against that same group, whether these acts were committed by the same offender or by others; (2) the scale of atrocities committed in a region or a country; and (3) the fact of deliberately and systematically targeting victims on account of their membership of a particular group, while excluding the members of other groups.[33]

---

[30] *Id* at para. 507 ("In patriarchal societies, where membership of a group is determined by the identity of the father, an example of a measure intended to prevent births within a group is the case where, during rape, a woman of the said group is deliberately impregnated by a man of another group, with the intent to have her give birth to a child who will consequently not belong to its mother's group.")

[31] *Id* at para. 508.

[32] *Id* at para. 509.

[33] *Id* at para. 523.

The ICTY has held that the requisite intent may be inferred from "the perpetration of acts which violate, or which the perpetrators themselves consider to violate the very foundation of the group—acts which are not in themselves covered by the list in Article 4(2) but which are committed as part of the same pattern of conduct." In that case, the ICTY found that "this intent derives from the combined effect of speeches or projects laying the groundwork for and justifying the acts, from the massive scale of their destructive effect and from their specific nature, which aims at undermining what is considered to be the foundation of the group."[34]

Going back to the ICTR, we've seen further examples of factors that can be used to infer genocidal intent. In the *Ruzindana* case, the Tribunal referred to a "pattern of purposeful action," which might include: (1) the physical targeting of the group or their property; (2) the use of derogatory language toward members of the targeted group; (3) the weapons employed and the extent of bodily injury; and (4) the methodical way of planning, the systematic manner of killing.

With respect to the "in whole or in part" aspect of the intended destruction, the ICTY affirmed the position of the International Law Commission by stating that complete annihilation from every corner of globe is not required.[35] In particular, the *Jelišić* Tribunal held that the intent may extend only to a limited geographical area. As with US domestic law, the ICTY held that "in part" means in *substantial* part. It further stated that a substantial part might include a large number or a representative faction. If the latter, that representative faction must be destroyed in such a way so as to threaten the survival of the group as a whole.[36]

The Tribunal in *Jelišić* also dealt with the issue of the role of the individual perpetrator in the commission of genocide. Generally, the tribunals have first determined whether genocide occurred in an area, and then proceeded to determine whether an individual has shared the genocidal intent. In *Jelišić*, the ICTY in-

---

[34] *Id* at para. 524.

[35] See *Jelišić, supra* note 22.

[36] *Jelišić, supra* note 22. See also 18 USC 1093 ("'Substantial part' means a part of a group of such numerical significance that the destruction or loss of that part would cause the destruction of the group as a viable entity within the nation of which such group is a part.").

dicated that an individual could be deemed responsible for genocide where he was one of many executing an over-all, higher level
planned genocide, or where he individually committed genocide.
Thus the Tribunal indicated that an individual alone could be
guilty of committing genocide; however, of course, the crime is
still very difficult to prove if the acts were not widespread and if
they were not backed by an organization or system.

It cannot be emphasized enough that the genocidal intent is
determinative of whether the crime occurred. In this context, it is
important to recall that although the tribunals have held that
forced expulsion may be one of the acts constituting genocide, this
expulsion must be carried out with the intent to destroy the group
if that act is to constitute genocide. Thus in the *Kupreškić* case, the
ICTY found that:

> The primary purpose of the massacre was to expel the Muslims
> from the village, by killing many of them, by burning their
> houses, by slaughtering their livestock, and by illegally detaining
> and deporting the survivors to another area. The ultimate goal of
> these acts was to spread terror among the population so as to
> deter the members of that particular ethnic group from ever re-
> turning to their homes.

The Tribunal thus held that this was a case of the crime
against humanity called Persecution, and that it was not a case of
genocide.[37]

IV. Ethnic Cleansing v. Genocide

Perhaps the most significant distinction between the terms
"genocide" and "ethnic cleansing" is that while the former is, in its
strict sense, a defined legal term, the latter is a political term without a precise definition. This must be borne in mind when attempting to compare the "elements" of each.

This ambiguity leads to serious confusion. In the recent application of Croatia to the World Court in a case against the Federal
Republic of Yugoslavia alleging genocide, the two terms are used
almost interchangeably. Further, this confusion is not limited to
unilateral declarations, as the United Nations Security Council
and General Assembly have used the phrase "ethnic cleansing" to
refer to a range of different acts. For example, while it is clear that
ethnic cleansing may take the form of genocide, the General As-

---

[37] But one should note that the Prosecutor agreed with the Tribunal in that case.
Genocide was not charged.

sembly passed a resolution stating that ethnic cleansing is a "form of genocide."[38]

In the jurisprudence of the ICTY, ethnic cleansing most closely corresponds to the crime against humanity of Persecution. In the *Kupreškić* case, the ICTY stated that ethnic cleansing was a form of the crime of Persecution. However, this is not to say that ethnic cleansing could not also amount to genocide. The Tribunal acknowledged as much when it stated clearly that ethnic cleansing is not a "term of art."[39]

From usage, there appear to be distinctions with respect to the mental and physical elements (*mens rea* and *actus reus*) of each. With respect to genocide the mental element is the intent to destroy. The physical element is commission of at least one of the enumerated acts. With respect to ethnic cleansing, the mental element appears to be the intent to remove. For a physical element, there is no finite range of acts. Acts constituting ethnic cleansing could possibly include systematic human rights abuses of a non-grave nature that encourage people to leave.[40]

With the broadening of the definition of genocide, however, it could be argued that it would apply to a greater number of cases of ethnic cleansing; perhaps indicating a degree of conceptual convergence. But it is important to remember that the most critical distinction between ethnic cleansing and genocide is in the nature of each term. Genocide is a legal term. Ethnic cleansing is not.[41]

V. Kosovo

In Kosovo, it was pretty clear that ethnic cleansing occurred in the forced expulsion of the Kosovo Albanian population; however, strong arguments can also be made that subsequent to the UN and

---

[38] GA Res. 47/121 (1992). It is important to note that the General Assembly is not an adjudicative body, and that its resolution are not legally binding instruments.

[39] *Prosecutor v. Zoran Kupreškić*, Judgement, 14 January 2000, IT-95-16-T.

[40] Open questions remain. What about offering incentives to relocate people elsewhere? Would that constitute ethnic cleansing? Would it be criminally wrongful? Presumably the underlying acts would have to be unlawful. If the definition of ethnic cleansing is an internationally wrongful act directed against a group with the intent to remove that group, then a positive financial incentive to leave would not constitute ethnic cleansing. Should it?

[41] One possible advantage of the term "ethnic cleansing" is that it may allow description of acts that occurred prior to the mid-twentieth century and that would have constituted genocide if they occurred after the crime of genocide was established in international law.

NATO taking control in Kosovo, ethnic cleansing has been occurring against the Serbs.

The genocide question is more complicated. With respect to the forced expulsion of the Kosovo Albanian population, it is pretty clear that the *actus reus* component is met, given the breadth of the definition as elaborated by the ICTs and as set forth in FRY domestic law. But again, the commission of the physical act is not sufficient to constitute genocide. Was the requisite *mens rea* present? To assess individual criminal responsibility, one would have to assess it on a case-by-case basis. In order to explore whether the campaign as a whole was genocidal, we would have to consider some of the possible mental states. If the intent was simply to destroy the Kosovo Liberation Army, then the accomplishment of that goal would probably not amount to genocide. If the intent was to remove all Kosovo Albanians from Kosovo, that also would not necessarily constitute genocide, particularly in light of the *Kupreškić* case. However, if the intent was to physically destroy the Kosovo Albanian population, then the campaign would have amounted to genocide.

Right now the only body likely to determine authoritatively whether genocide was committed in Kosovo is the ICTY. If it determines that genocide was committed, the decision would constitute a broadening of the definition, or at least confirm the breadth of the definition as found in the jurisprudence of the Rwanda Tribunal, and would indicate a departure from the *Kupreškić* decision.

The Shifting Interpretation of the Term "Ethnic Cleansing" in Central and Eastern Europe

JÁNOS MAZSU

The collapse of Yugoslavia and the tragic and shocking series of events accompanying its partitioning are well-known and well-documented. This is the result of the revolutionary changes in mass communication, which created a virtual "reality-channel" to cover these events for communities near and far in the world.

My objective in this brief presentation is not to offer a systematic outline or an analysis of the events, but to make some marginal comments and tentative reflections on the wide-spread practice of "ethnic cleansing" (Hungarian: *etnikai* or *nemzetiségi tisztogatás*). The main source for these marginal notes is this newly coined expression and its use in the region best kown to me: East Central Europe. Thus, the scope of my message is limited to that part of the world.

Ethnic Cleansing—A Philological Approach

The Hungarian expression *tisztogatás* was not unknown to those Hungarians who have served in the military service. It was the translation of the Russian word "chistka/chistenye," which in Soviet-Russian military vocabulary meant simply the demilitarization of a given area, i.e., the "clearing" or "cleaning" of a certain territory of enemy troops. Most probably it was a similarly uninteresting routine expression in all countries of the Soviet Bloc, and it had equivalents in the vocabularies of all other Warsaw Pact armed forces. Most Hungarian historians (Professor István Deák of Columbia University, among the most highly regarded experts) hold the view, that the expression "ethnic cleansing" has existed for a long period of time. But this compound word is in fact shockingly new, even if the phenomenon it covers is steeped in the distant past. I agree with Dražen Petrović (Sarajevo University, Law School) that the expression *"chistka"* in this special compound "ethnic cleansing" was not born in the Russian vocabulary, in spite of its Soviet-Russian origin. It was surely born in the related Slavic Serbo-Croatian or Croato-Serbian language as *"Etničko Čišćenje."*

Its emergence is likely to have preceded the decade of horrors in the Balkans by at least ten years.[1]

The mass media first discussed the creation of *ethnically clean territories* in Kosovo after 1981. At that time, the term related to administrative and nonviolent matters and referred mostly to the behavior of Kosovo Albanians towards the Serbian minority in that province. The term got its current meaning during the war in Bosnia and Herzegovina in the early 1990s, and was also used to describe certain events in Croatia. It is impossible to determine who was the first to use it, and in what context. The military officers of the former Yugoslav People's Army, who were given a role in the events, are the most likely to have coined it, and then put the expression "ethnic cleansing" into their military vocabulary. The simple expression "to clear the territory" originally was directed against enemies, and it was used mostly in the final phase of combat in order to take total control of a given territory. The word "ethnic" has been added to the military term primarily because the "enemies" are always considered to be members of the "other" ethnic communities.

Despite the widespread use of the term by the mass media during the new Balkan War, its exact meaning was never clearly defined. This was the reason why the expression "ethnic cleansing" was often preceded with the prefix "so-called."

In the course of the past decade or so, governments, international organizations, and experts in the media have employed diverse terminology to refer to forced population removals. The term "ethnic cleansing" has been described as a systematic process, a campaign, a conscious policy, or a sanctioned practice. All this may at first glance seem insignificant, but these descriptions may indicate substantial difference in attitudes toward ethnic cleansing.

Based on the attempt by Dražen Petrović and others[2] to summarize approaches and elements of meaning of ethnic cleansing, it is probably useful to differentiate among the various elements of this term—as a practice, or as a general policy.

---

[1] Dražen Petrović, "Ethnic Cleansing—An Attempt at Methodology," *European Journal of International Law,* 5 (1994): 342.

[2] Dražen Petrović, "Ethnic Cleansing," 342-360; Christine Chinkin, "Rape and Sexual Abuse of Women in International Law," *European Journal of International Law* 5 (1994): 326-342; Andrew Bell-Fialkoff, "A Brief History of Ethnic Cleansing," *Foreign Affairs,* 72 (1993): 110; Martin Terry, "The Origins of Soviet Ethnic Cleansing," *The Journal of Modern History* 70 (December 1998): 813–861.

As a practice, ethnic cleansing could mean a set of different actions, directly or indirectly related to military operations, committed by one group against members of another ethnic group living in the same territory. Mass media and international organizations used this description on several occasions, and they contained the following components:

a) *Administrative Measures:*
- forced removal of lawfully elected authorities,
- dismissal from work (especially from important public service positions),
- restrictions on the distribution of humanitarian aid,
- constant identity checking of members of minority ethnic groups,
- official notices to the effect that security of the members of other nations cannot be guaranteed,
- settlement of "appropriate" population (affiliated to the dominant nation, very often refugees) in the region,
- discriminatory and repressive legislation,
- refusal of treatment in hospital,
- making the departure of one member conditional upon the departure of the entire family,
- disconnection of telephones,
- forced labor, very often including work on the front-lines of armed conflict,
- prohibiting women of particular ethnic groups from giving birth in hospital,
- voluntary transfer of property by forcing people to sign documents stating that the property was permanently abandoned by the owner.

b) *Other Non-violent Measures:*
- local media inflaming fear and hatred,
- harassing phone-calls including death threats,
- publishing lists of citizens indicating their ethnic origin.

c) *Terrorizing Measures committed by soldiers or armed civilians:*
- robbery, terrorization and intimidation in the street,
- massive deportation, detention and ill-treatment of the civil population and their transfer to prisons and camps,
- shooting of selected civilian targets or blowing-up and setting fire to homes, shops and places of business,
- destruction of cultural and religious monuments and sites,
- mass displacement of communities,
- discrimination of refugees on the basis of ethnic differences.

Among the very specific elements of ethnic cleansing in this category are rape and other forms of sexual abuse, including castration. (Rape has been used most frequently and systematically against women of all ages, including very young women, with the intent of making them pregnant.)

d) Military Measures:
- executions, killing and torturing of leading citizens, religious and political leaders, intellectuals, policemen and members of the business community,
- holding towns and villages under siege,
- deliberate attacks and blocking of humanitarian aid,
- shelling of civilian targets,
- taking hostages and detention of civilians for exchange,
- use of civilians as human shields,
- attacks on refugees camps.

Ethnic Cleansing—A More Analytical Approach

The practice of ethnic cleansing did not always include all of the above-listed elements; nor does it now. But the wars following the dissolution of Yugoslavia contained all of them in different and strange combinations.

In order to identify various combinations of ethnic cleansing, the analysis of the aim of these actions must be made part of the defining process. Thus, the other necessary approach to identifying ethnic cleansing is to examine its goals and its motivations.

Reports by the mass media and documents by international organizations shared the view that the principal objective of the military conflict in Bosnia and Herzegovina was the establishment of ethnically homogeneous regions. Therefore, ethnic cleansing does not appear to be an unintended consequence of war, but rather the attainment of its planned goal.

The aim of ethnic cleansing could be defined both on the local and on the global level. On the local level, the aim of the policy of ethnic cleansing could be the creation of fear, humiliation, and terror toward the "other" community. The ultimate goal is to gain effective control over a given area, which may be achieved by provoking the members of the "other" community to flee because of possible reprisals. On the global level, the aim could be defined as an irreversible change of the demographic structure, the creation of ethnically homogeneous regions, and the achieving of more favorable positions for a particular ethnic group in the ensuing political negotiations, based on the logic of division along ethnic lines. The final aim could also be the extermination of certain groups of people from a particular territory, including the elimination of all physical traces of their presence.

Ethnic cleansing was also a deliberate effort to confront the international community with a *fait accompli*, in which the obtained advantages were expected to be politically and legally recognized. The fatal misinterpretation of the example of German reunification and its ramification tied down the attention of the West, and so did the disintegration of the Soviet Union. In addition the Serbs also assumed that the rebirth of a unified Germany could justify their own creation of a " Greater Serbia."

The fluidity of the international situation had a role in formulating objectives and in developing faith in their feasibility. Those favoring ethnic cleansing also assumed that in the well-to-do democracies there was a degree of reluctance to jeopardize the lives of military personnel, which could be endangered in spite of their vast technological superiority (but very little "readiness to die"). All these considerations led the participants to grab at the bounty with no regard to commitments of the international community to human rights norms and international legal conventions. They expected to achieve an ethnically clean territory without running too much risk. The various nations of the former Yugoslavia thus pursued the policy of area-maximization in creating old/new nation-states. They hoped to accumulate "bargaining chips" for later peace conferences, just as after World War I. Where borders, ethnic groups, and cultural and religions entities were hopelessly intermixed, without the hope of ever creating ethnically pure areas, the chief method was to carve up those territories and then to occupy or to annex parts of them in a more traditional fashion. Therefore, "ethnic cleansing" became a secondary method. (In a typical statement, the son of the Macedonian president, Vladimir Gligorov, declared: "Why should I be minority in your state, when you can be the same in mine?")

In summary, therefore, the method of ethnic cleansing cannot be interpreted merely as a tactical instrument. The formulated strategic goals and motivations behind them offer us a chance to understand and to define this horrible late twentieth-century phenomenon.

Do Historical and Politico-philosophical Analyses Help Us Understand the Concept of "Ethnic Cleansing"?

The popular concept has had a peculiar career even in scholarly circles owing to its visualization, and its favorable impact on TV viewership ratings. Within a decade or so extensive literature has been produced in which the term "ethnic cleansing" was applied to every historical phenomenon from tribal conflicts to future orbital conflicts in space. The authors of these books may have used this term with the hope that the media would popularize their books with the reading public. Many of them were on the level of sensationalist journalism. At the same time, many others represented serious historical scholarship. Thus, while we ignore cheap sensationalism, we have to acknowledge that the use of this term has also crept into "serious" historical writing.

The first noteworthy professional approach rests on the argument that such population displacements and massacres have also occurred in the past. These have been purposefully committed as part of an official state policy. Its goal was the total or partial extermination (genocide) of an ethnic or religious community. Such measures had also been used ostensibly for driving an ethnic group or some other community beyond the state borders. The terms formerly applied to these actions were expulsion, forced emigration, etc. The revived interest in doing research on these events and measures is the real positive byproduct of this horrible expression and of the shocking phenomenon that it covers. Revealing the similarities and continuities between historical past and the present helps our understanding, but the necessity of pointing out the differences also needs to be emphasized. Crucial in this effort—besides comparing techniques of genocide and expulsion—is the understanding of the context determined by the goals and motivations.

Prior to the spread of modern nationalism, Europe was the home of several supranational empires, which often resorted to forced migrations, and even some genocide. However, the legitimacy of these states differed basically from the later ones. In the age of monarchic rule (divine right kingship) the acquisition of legitimacy rarely required that "legitimacy deficit" should be the motivation for the expulsion or extermination of an ethnic group. The most important problem of supranational empires was the development and operation of a system of control and administration over a society characterized by diversity. The common feature of the diverse causes for genocide or expulsion was the action against those who ultimately jeopardized the empire's interests. In brief, we could say that in centuries prior to modern nationalism, the state punished the intractable troublemakers by using the above methods.

Industrial modernization did not only bring about technological development and the increased capability of state power to use force, but also replaced the legitimacy of empires by the principle of the sovereignty of the people and the sovereignty of the nation-state. This new legitimacy brought about different kinds of conflicts in the more fortunate West, where the transition was relatively smooth. As an example, the Anglo-Irish conflict produced characteristics easily comparable to ethnic cleansing. The new legitimation usually offered citizenship and equal rights before the law to those who were willing to assimilate into the dominant historical "state-creating" nation. The flourishing Western "nation-states" with their advanced civilizational developments became models for East-Central Europe. Thus, after the collapse of the Habsburg Monarchy and the Ottoman Turkish Empire, newly established "successor states" created their own nation-making mythologies by emulating the Western models. Although clearly multinational, these new states claimed to be ethnically homogeneous. This claim, however, was far from the truth. To create ethnically homogeneous regions out of the mingled ethnic and religious groups in the territorial cavalcade represented by ruins of medieval empires would have been an impossible task even for a rational landscape engineer. The political elites of these new states were not motivated by rationality, but by emotional commitments to their largely fictionalized past. During the transitional postwar period, prior to the signing of the peace treaties, they intended to retain or obtain sovereignty over as large a territory as possible. In my view, this produced, for the first time, an ethnic phenomenon similar in motivations, goals, and measures to those of late twentieth-century Yugoslavia.

The potential for genocide or expulsion was present, in spite of the remarkable backwardness of extermination technology. It was motivated by myths of ethno-genetic proofs concerning the individual nations' origins and their consequent rights to a given territory. This idea also secured the backing of the victorious great powers. With the implementation of the peace treaties, the winners recognized mythical historical legitimations, instead of the previously highly touted principle of national self-determination. They yielded to the principle of "one state one nation." On the basis of the aforesaid, and in the light of post-World War I developments, extending the use of this term to the period before the war resulted in the blurring—rather than in the clarification—of its meaning.

New works falling into the second group of historiography resort to the use of the concept of ethnic cleansing in order to understand better the anti-ethnic, anti-national, and genocidal activities of modern fascist and Communist dictatorships.[3] The So-

---

[3] Arch J. Getty, V. N. Zemskov, and Gábor Rittersporn, "Victims of the Soviet Penal System in the Pre-war years: A First Approach on the Basis of Archival Evi-

viet-Russian Communist state, which inherited extraordinary ethnic and religious diversity from Czarist Russia, applied violence
against various groups of its citizens throughout its existence,
from the moment of its birth, through the "Great Stalinist Terror,"
all the way to its demise. Apart from the transnational imperial
motives, the guiding logic of Communist dictatorship was to be
something else. The legitimacy of those possessing power was
made up of ideology of the proletarian revolution and the monumental vision of a social order free from oppression. For this reason, it was necessary to use terror against those who supposedly
hindered the fulfillment of Marxist-Leninist goals. These were
primarily the representatives of the old order who were considered as the barriers to communism, as well as those who opposed
the party-line as determined by the leaders of the Communist
Party. It is no secondary matter that this ideology-centered logic
promised "social justice" to the oppressed, which included the
"just" punishment of those who had "sinned against them." This
logic moved along the notion of social classes; therefore the
groups to be punished were primarily in certain social categories.
Owing to its legitimation and its motivating character, we could
actually speak about "social class cleansing." Naturally, the imperial scale and structure, never permitted a pure and complete application of "social-class cleansing." The collective identification of
the imperial troublemakers as unreliable class-traitors pointed—in
the eyes of Communist party leaders—at entire ethnic groups or
religious communities. That is why the term "ethnic cleansing"
can be applied only with methodological restrictions to the history
of the Communist system. (This feature is even more recognizable
in the countries of the Soviet Bloc after World War II.)

Unlike the Soviet-Russian state, the German fascist state inherited a nationalism based on exclusive racism. With the construction of fascist dictatorship, however, historical justice and
punishment became a dominant theme. Thus, the use of punitive
measures against political or racial "aliens," who allegedly stood
in the path of the fulfillment of the ideology of the totalitarian

---

dence," *American Historical Review* 98 (1993); E. Bacon, *The Gulag at War: Stalin's
Forced Labour System in the Light of the Archives* (London, 1994); Mihályné Fodor, *A
szovjet etnikai tisztogatások története* [The History of Soviet Ethnic Cleansings],
(Magyar Elektronikus Könyvtár [Hungarian Etlectronic Library]), www.mek.hu,
1998); Terry, "The Origins of Soviet Ethnic Cleansing," 813–861; Gerhard Simon,
*Nationalism and Policy toward the Nationalities in the Soviet Union: From Totalitarian
Dictatorship to Post-Stalinist Society*, trans. Karen Forster and Oswald Forster
(Boulder, Col., 1991).

power, was fully acceptable.[4] Beyond the racist foundations of German fascism, the actions against Jews and Gypsies actually intended to instigate and encourage alignment behind the totalitarian power against the "diabolic internal enemy." The use of the idea of "being constantly under siege from outside," while at the same time "being threatened by internal treason" was very similar to the practices of Communist dictatorship. The two differed only in their respective ideologically-based watch words, which in the case of the Communists always sounded like this: "The international situation and class struggle are intensifying!"

A horrible mixture of exclusivist racism and totalitarian logic was the primary cause of the Holocaust. Even if not in a clean-cut manner, "class cleansing" can be recognized as a dominant feature of German fascism. It appeared in a technologically perfected manner, which carried within itself the extremely cynical final solution of the Jewish question. This concept was shared by the facist leaders of Italy, Croatia, Rumania, Slovakia, and even France.

Social class distinction about this question was also to be observed in Hungary. Thus, while the country's political leadership turned a blind eye to the deportation of half a million rural Jews in the spring and early summer of 1944, the deportation of Jews from the capital was halted in July of that year. Regent Nicholas Horthy justified this by saying that the Jews of the capital city were better educated, more assimilated, and better-off than those in the countryside.

All this appears to suggest that—when it comes to the conceptualizing ethnic cleansing—there are some overlappings between the Communist and the fascist models. Thus the Holocaust is a strange combination of authoritarian "class cleansing," nationalist "ethnic cleansing," and the old fashioned imperial "troublemaker cleansing."

Perplexing were the steps taken by allegedly democratic Czechoslovakia against its German and Hungarian citizens. Although the Czechoslovak Republic was viewed even by Western standards as a model democracy, its leadership acted very undemocratically against the German and Hungarian minorities. President Edward Beneš was fully aware that there was a willingness among the victorious great powers to apply the principle of collective punishment to the defeated. He decided to take advantage of it, and thereby make up for some historical omissions. The

---

[4] István Deák, "Holokaust und ethnische Säuberung," *Europäische Rundschau,* no. 4 (1999): 81–94; in Hungarian: *Kisebbségkutatás* [Minority Research], 9 (2000).

two related Slavic nationalities—the Czechs and the Slovaks—that lacked independence before World War I (the Slovaks had no history of independent statehood at all) were fully aware of the tenuous and fragile nature of their legitimacy. Consequently, they opted to embrace not the principle of national self-determination (which was the rationale behind the creation of Czechoslovakia), but to move against the minority nationalities, which, in their view, jeopardized the existence of their alleged "nation-state." They were able to follow this path with the tacit approval of the victorious great powers that wished to punish fascist atrocities and war crimes. President Beneš openly proclaimed that "the Czechs and Slovaks do not wish to live in the same country with the Germans and the Hungarians!"

It should also be noted here, however, that the Czechoslovak leadership did not endeavor to exterminate the members of the targeted groups physically, as has been done by the fascist and Communist dictatorships. At the same time, Czechoslovakia was no less efficient in repressing and expelling its minorities than had been the fascist and Communist dictatorships earlier. The Czechoslovak political elite accomplished after World War II what it had failed to do after World War I: the territorial maximization of their nation-state. With the expulsion of minorities they created an ethnically nearly pure nation-state. In my view, the execution, logic, and goals of the Beneš decrees constitute a typical ethnic cleansing, even if a somewhat less violent and somewhat more pacified version of the same. Some scholars have called this action "ethnic cleansing in kid gloves."

Conclusions: A Methodological Proposal

The use of the term "ethnic cleansing" can be explained by drawing up a set of coordinates. At the top of the vertical axis of this coordinate one finds the supranational empire, while at the lower end one finds the homogenous (ethnically pure) nation-state. The left end of the horizontal coordinate marks totalitarian (Communist or fascist) dictatorship, while the opposite end signifies democracy. Within this context, the supranational empire corresponds to "troublemaker cleansing," totalitarian dictatorship to "social-class cleansing," the homogenous nation-state to "ethnic cleansing," and democracy to "social-class and ethnic tolerance." Naturally, none of these types can be found in pure form. The recognition of typology and dominance, however, can help one's orientation in the mixed and transitional cases, when one is trying to assess certain given historical processes.

BIBLIOGRAPHY

Printed Sources

Bacon, E. *The Gulag at War: Stalin's Forced Labour System in the Light of the Archives*. London, 1994.

Bell-Fialkoff, Andrew. "A Brief History of Ethnic Cleansing." *Foreign Affairs*, 72 (1993).

Chinkin, Christine. "Rape and Sexual Abuse of Women in International Law." *European Journal of International Law*, 5 (1994): 326-342.

Coakley, John, ed. *The Social Origins of Nationalist Movements: The Contemporary West European Experience*. London, 1992.

Coakley, John, ed. *The Territorial Management of Ethnic Conflict*. London, 1993.

Deák, István. "Holokaust und ethnische Säuberung," *Europäische Rundschau* , 1999/4, pp. 81-94. Its Hungarian version is in: *Kisebbségkutatás* [Minority Research], 9/3 (2000).

De Varennes, Ferdinand. *Language, Minorities and Human Rights*. The Hague, 1996.

De Zayas, Alfred M. *The German Expellees: Victims in War and Peace*, trans. John A. Koehler. New York, 1993.

Fejős, Zoltán. *Hungary Facing Internal and External Minority Problems*. Occasional Papers, No. 5 (Budapest: Teleki László Foundation Institute for Central European Studies, 1996). In association with International Centre for Ethnic Studies, Kandy, Sri Lanka. The PEW Charitable Trust, Philadelphia, PA, USA. Tulane Institute for International Development, Arlington, Va., USA.

Fodor, Mihályné. *A szovjet etnikai tisztogatások története* [The History of Soviet Ethnic Cleansings] (Magyar Elektronikus Könyvtár = Hungarian Electronic Library – WWW.mek.hu, 1998).

Getty, J. Arch., Zemskov, V.N., and Rittersporn, Gábor, "Victims of the Soviet Penal System in the Pre-war Years: A First Approach on the Basis of Archival Evidence," *American Historical Review*, 98 (1993).

Grizold, Anton. "Nemzetközi biztonság és etnikai konfliktusok" [International Security and Minority Conflicts], *Kisebbségkutatás* [Minority Research], 9 (1999).

Lapidoth, Ruth. *Autonomy: Flexible Solutions to Ethnic Conflicts.* Washington D. C., 1996.

Lewin, Moshe. *Russia/USSR/Russia: The Drive and Drift of a Superstate.* New York, 1995.

Terry, Martin. "The Origins of Soviet Ethnic Cleansing." *The Journal of Modern History,* 70 (December 1998): 813–861.

McGarry, John and O'Leary, Brendan, eds. *The Politics of Ethnic Conflict Regulation.* London-New York, 1993.

Mézes, Zsolt: *Az európai kisebbségvédelmi gyakorlat adaptációjának lehetőségei Közép-Kelet-Európában* [The Possibilities of the Adaptation of European Minority Protection Practices in East Central Europe]. Publication of the European Comparative Minority Research Program Office [Európai Összehasonlító Kisebbségkutatások Programiroda]. Budapest, 2000.

Ong Hing, Bill. "Beyond the Rhetoric of Assimilation and Cultural Pluralism: Addressing the Tension of Separatism and Conflict in an Immigration-Driven Multiracial Society," *California Law Review* 81 (1993): 863-925, 864.

Petrović, Dražen. "Ethnic Cleansing—An Attempt at Methodology." *European Journal of International Law,* 5 (1994): 342-360.

Pohl , J. Otto. *Ethnic Cleansing in the USSR, 1937-1949.* (Westport, Conn., 1999).

Popély, Gyula. *Az elnyomott kisebbség.* Magyarok Csehszlovákiában / Szlovákiában [An Oppressed Minority. Hungarians in Czechoslovakia / Slovakia]. An unpublished political memorandum, 1993.

Poulton, Hugh. *Minorities in Southeast Europe: Inclusion and Exclusion.* Minority Rights Group International 1998. Reprinted in *Kisebbségkutatás* [Minority Research], 9 (1999).

Simon, Gerhard. *Nationalism and Policy toward the Nationalities in the Soviet Union: From Totalitarian Dictatorship to Post-Stalinist Society.* Trans. Karen Forster and Oswald Forster. Boulder, Col., 1991.

Smith, Anthony D. *National Identity.* London, 1991.

Szarka, László. "A közép-európai kisebbségek tipológiai besorolhatósága" [The Typological Characterizaton of Central European Minorities]. Paper presented a the Conference of National Minority Research, Budapest, April 8-9, 1999.

Volkogonov, Dmitri. *Stalin: Triumph and Tragedy.* Ed. and trans. Harold Shukman Rocklin, Cal., 1992. A political biography of the Soviet leader responsible for the mass deportation of whole nations.

Internet Sources

The Careless Usage of "Genocide." Florian Bieber, PhD student, University of Vienna H-Net Discussion Logs—The Careless Usage of Genocide.htm

http://www.slovakia.org/history-magyarization.htm. Hungarian Ethnic Cleansing (Magyarization).

"Ten Untaught Lessons about Central Europe: An Historical Perspective by Charles Ingrao," HABSBURG Occasional Papers, No. 1. 1996.

H-Net Discussion Logs—Comment: "Ten Untaught Lessons about Central Europe. How to Learn from Charles Ingrao's 'Untaught Lessons'" by István Deák, Columbia University.

The Evolving Definitions of IDPs and Links to Ethnic Cleansing in Europe

GABRIEL S. PELLÁTHY

The problem of dealing with refugees and displaced persons has been a major international issue in Europe through two World Wars and into this new century. One major cause of migration, including ethnic cleansing, has been a continual feature of the political upheavals that have caused mass movements of peoples on that continent in the twentieth century. From the efforts of Hungary to "Hungarianize" the ethnic minorities within its borders during the Austro-Hungarian empire before World War I to the more recent, much publicized "cleansings" in Bosnia, Herzegovina, and Kosovo and the lesser known persecution and displacement of Hungarian minorities in Romanian Transylvania and elsewhere, the history of the region is rife with examples of dominant ethnic groups or nations forcibly displacing, expelling, or otherwise persecuting or decimating minority ethnic populations. Thus by definition victims of ethnic cleansing form a significant subgroup of those populations who are on the move. These migrations have stimulated various responses from the international community, but much more needs to be done in defining and focusing on the needs of particularly vulnerable groups, those who are displaced but have not crossed international borders, the Internally Displaced Persons (hereinafter IDPs). The most vulnerable of this group may be those groups or individuals that have been or continue to be subject to ethnic cleansing.

Recognition of IDPs Linked to Ethnic Cleansing

The history of the recognition of IDPs by the international community can be linked to the perceptions that have been formed by the special exigencies of groups and people threatened, persecuted, or decimated by "ethnic cleansing." The Holocaust under Nazi Germany, which has been defined as "genocide," also falls into the broad category of "ethnic cleansing." As a result of these "ethnic cleanings" many groups left their homes for another part of their country, others crossed borders, and became refugees. The causes for their flight were not specifically addressed, but the international community under the Geneva Conventions and other protocols based on humanitarian law recognized their plight since the time of World War I. Special problems of victims of ethnic cleansing arose during repatriation efforts, when these groups refused to return to their homeland, showing the difficulties of re-integration. Again the same situation arose during and after World War II when the category of "refugees" in United Nations protocol was expanded to include those who "fled from persecution," a category of persons not fully covered under previous conventions. The idea of "persecution" and not just "war or armed conflict" as one criterion for refugee status certainly recognized those groups threatened by ethnic cleansing, as well as those subject to political persecutions.[1] Not only were these groups or people "displaced persons" (DPs), subject to the turmoil and dangers that caused the across-border displacement of all DPs, but they were subject to special dangers within their borders which kept them from being able to return to their previous areas of residence. Thus, Hungarians fleeing certain areas of Yugoslavia and Romania, because of World War II, for example, faced the threat of being "ethnically cleansed" upon their return to their former residences. Eventually the problem of ethnic cleansing had to be faced by the international community, which recognized that "the refugee problem was not temporary..." since many [refugees who had crossed international borders] refused repatriation on the basis of fear of persecution...."[2]

IDPs and Ethnic Cleansing Ignored

After World War II, the failure of many groups to repatriate was ascribed to economic and political hardships that they would face if they returned to their home countries, and new international conventions were drawn on the basis of humanitarian law to ease their plight. With the hardening of political lines during the Cold War, political threats to refugees were more readily recognized by the "West" (giving rise in fact to several new categories in the definitions of refugees). The international community largely downplayed the fear of ethnic cleansing, even though such policies were carried on by dominant ethnic groups against other groups within their own borders, both during and after World War II[3]

---

[1] Pirkko Kourula, *Broadening the Edges: Refugee Definition and International Protection Revisited* (The Hague, 1997), 171-201.

[2] Ibid., 172.

[3] Andreas Wesserle, "Ethnic Cleansing of the Germans in Slovakia, 1944-1950" (Conference Paper, Duquesne University, November 2000), and Alexander Prusin, "Ethnic Cleansing of Poles by Ukrainian Nationalists in Western Ukraine during World War II" (Conference Paper, Duquesne University, November 2000); see these elsewhere in this volume.

(e.g., the ethnic cleansing of Germans in Slovakia or of the Poles by Ukranian Nationals during World War II). Because of the lack of information emanating from the eastern bloc nations, the news of ethnic cleansing was not disseminated to the West. Particularly, what went on within the so-called "Iron Curtain countries" in terms of IDPs was also unknown to the "West."[4] Thus the plight of the internally displaced, often comprising groups or people who face the dangers of ethnic cleansing, has not been addressed again until fairly recently, since the fall of Soviet control, because Communist countries resisted "intervention" in their domestic affairs and appealed to principles of state sovereignty.

The International Scope and Problems of IDPs

Since the late 1980s, the challenge posed by the displacement of populations and groups who—despite their dire needs and mistreatment—remain within the boundaries of their own countries has reached crisis proportions. In 1982 there were a little more than 1.2 million of such "internally displaced persons" in only eleven countries. By 1998, however, there were over 25 million such persons in thirty-five to forty countries. Many of them are displaced by internal conflicts—civil wars, guerrilla wars, insurrections, irredentism, political persecution and the like; a significant portion are displaced by ethnic or tribal cleansing. No region of the world is free from this phenomenon. Indeed the plight of these IDPs has been called by many the "great human tragedy of our time."[5] Estimates of these numbers have grown dramatically; the problem is widespread and the plight of the victims is almost insurmountable.[6] In fact, the refugee population after World War II was one-tenth the size of those that are internally displaced today. The real problem is to identify and deliver aid to the many problematic situations that IDPs find themselves mired in.

There seem to be three convenient ways to view this critical problem, one definitionally and theoretically, the second, administratively, and the third, practically. The theoretical and administrative views, while quite complicated, have their basis in international law, United Nations practice and definitions. The practical application of the theory, however, is so fraught with uncertainties that an orderly process addressing the issues around IDPs, often

---

[4] "Report on the World Today: Hungary," *Atlantic Monthly* (1952): 4, 10.

[5] Roberta Cohen and Francis Deng, *Masses in Flight* (Washington, D.C., 1998), xix.

[6] Ibid., cf. charts pp. 30, 32 and 33 and map on frontispiece.

predicated on issues surrounding the practice of ethnic cleansing, may not be achieved in the foreseeable future. Many calls for an international convention on IDPs have gone unheeded, but because of the magnitude and seriousness of the problem, solutions must be attempted in both theory and practice.

Historical Overview

Developing Definitions of Refugees

The definition of IDPs has still not been fully achieved by the international community, though the precedent for helping people or groups displaced by wars, disasters (both natural and man-made), famines, or other catastrophes has a firm grounding in international humanitarian law, beginning with the Geneva conventions, then the League of Nations, and now continuing through the United Nations. Until recently, IDPs have not been recognized as a group apart, but have been considered only in times of major disasters, or within the definitions of "refugees" when they were on the verge of crossing borders.

Historically, refugees were defined as "those persons who were forced to flee their homeland because of armed conflict and had *crossed international borders in so doing*"[7] (emphasis supplied). They were regarded as a special group protected by international law as early as 1920 through the League of Nations, when Fridtjof Nansen was appointed the first High Commissioner for Refugees.[8] Under the League's supervision, 1.5 million people were designated as refugees and given international oversight, because they fell into certain designated groups or definitions. Half a million prisoners of war from twenty-six countries were also repatriated under the High Commission for Refugees (HCR). [9]

In a very interesting development, mass humanitarian relief through the HCR was given to some 30 million people within the borders of the USSR during the winter of 1921. Humanitarian law was used as justification when relief was offered to so-called "refugees" within the borders of the Soviet Union, even though these individuals or groups had not crossed any political or na-

---

[7] UN, *Convention Relating to the Status of Refugees* (189/UNTS/150,1951).

[8] His tenure lasted until 1930.

[9] Gil Loescher, *Beyond Charity: International Cooperation and the Global Refugee Crisis* (London, 1996) 37-40.

tional borders.[10] These were in fact internally displaced persons, though not yet so defined. This close association of "humanitarian aid" and the definition of "refugee" status still holds, and forms the basis of analogies in law for the present definitions of IDPs. Without the recognition and definition of refugees, the definition of IDPs would have been impossible.

Expanding the Definitions of Refugees

The UN continued and expanded the definitions of refugees. In its first session in 1946, the UN pledged "collective responsibility for those fleeing persecution"[11] adding persecution to war relief as another basis for assistance. This was in recognition of the dangers of political persecution and ethnic cleansing that caused certain groups of people to flee. The International Refugee Organization (IRO) was formed, offering special protection for those groups formerly designated by the League of Nations, as well as setting up new categories of refugees from the war torn nations of World War II, within the existing definition of "Displaced Persons." At first these refugees were defined as those "people or groups who had been displaced from their habitual place of residence by the exigencies of armed conflict and other dangers such as war, violence and violation of human rights,"[12] who had crossed international borders, but were not fully covered by earlier conventions.

In 1950 the Office of UN Commissioner for Refugees (UNHCR) was created and tried to fulfill its mandate in covering the following ways:

1. The administration of the 1951 Convention Relating to the Status of Refugees on refugees which mandated "providing international protection ... and seeking permanent solutions for the problem of refugees."

2. The definition, monitoring and reduction of "statelessness" (i.e. policies and mandates for repatriation or other immigration options).

3. The formation of an Executive Committee of the High Commissioner's Program (EXCOM) which still functions today as policy and budget consultant to the UNHCR.[13]

---

[10] Ibid., 38.

[11] Kourula, *Broadening the Edges,* fn. 1, 173.

[12] Ibid., fn. 1, 181.

[13] UNHCR, *General Assembly Resolution 319 (IV) and General Assembly Resolution 428 (V), with Annex* (Statute of the High Commission for Refugees, 1948).

Early Definitions of IDPs

At the outset, IDPs were seen as an extension of the refugee problem, not as a totally different group with unique requirements of their own. Since the end of World War II, and particularly since the end of the Cold War, large populations of people have been forced to leave their homes for a variety of reasons, such as disasters (both natural and manmade), internal strife, armed conflicts, revolutions, systematic violations of human rights, including genocide, and ethnic cleansing or a variety of combinations of these reasons. Some of the reasons are also reflected in the UN definition of DPs (see above). Rather than crossing internationally recognized boundaries, these people faced displacement within the recognized boundaries of their own nations. Had they left their homelands, the status and definition of these people as refugees would have given them access to an established system of international protection and assistance. By contrast, these internally displaced persons (IDPs) remain within their own countries, without legal or institutional standing for receiving protection from the international community. However, since there were no comprehensive protocols or international conventions on those who were displaced within their own borders, these persons continued to be viewed as "potential refugees" who would have become refugees if only the borders had been just a little closer, or more conveniently placed so they could have crossed them.

The dangers faced by those groups who had to flee to avoid ethnic cleansing, yet stayed within their own borders. (e.g. Hungarians in Yugoslavia, Germans in Poland, Turks in Bulgaria) became particularly evident to the international community. Since these groups remained within their borders, wholesale violations of human rights, often ethnic cleansing and genocide, continued to occur, and no international protocols applied. The international community, as such, could not afford them aid within their own borders, if not invited to do so officially by the various countries within which they were being displaced, which of course would be most unlikely.

Thus by the 1950s, the international community began to recognize IDPs as separate from refugees. IDPs were defined as "persons who have been forced to flee their homes suddenly or unexpectedly in large numbers, as a result of armed conflict, internal strife, systematic violation of human rights, or natural or manmade disasters, and who are within the territory of their own country."[14] Although this definition no longer linked IDPs to refugees, the ways in which IDPs could be aided were still very restricted.

Application of Humanitarian Law

New Definitions of IDPs

On March 5, 1992, the UN Commission on Human Rights adopted resolution 1992/73, which asked the secretary-general to appoint a representative to explore the applicability of human rights law, humanitarian law, laws on refugees, and other international laws and conventions to the growing and pressing problem of IDPs.[15] By now, the fact and extent of ethnic cleansing had also made an impact on the international community. Deng, the UN representative, and others criticized the previous UN definition as too narrow, pointing out that many IDPs do not necessarily leave their home "suddenly or unexpectedly." The term "forced to flee" is also too narrow, and would exclude, for example persons expelled from their homes on ethnic or religious grounds, witness the movement of Bosnian Muslims subject to ethnic cleansing.

Various International Definitions of IDPs

In 1998 a broader and more comprehensive definition of IDPs known as the Guiding Principles on Internal Displacement redefines IDPs in this way:

1. IDPs are "persons or groups of persons who have been forced or obliged to flee or leave their homes or places of habitual residence, in particular as a result of or in order to avoid the effects of armed conflict, situations of generalized violence, violations of human rights, or natural or human-made disasters, and who have not crossed an internationally recognized state border."[16]

2. IDPs are persons within the borders of their own countries who suffer coerced displacement "...by armed conflicts, internal strife, and systematic violation of human rights... [and are] dis-

---

[14] Cf. "Guiding Principles on Internal Displacement" in Roberta Cohen and Francis Deng, *Masses in Flight* (Washington, D.C., 1998), 305. See also Ibid., 2.

[15] Francis Deng, *Protection the Dispossessed* (Washington, D.C., 1993), 2, 149-153.

[16] Cohen and Deng, *Masses in Flight*, 305-306.

possessed by their own governments and other controlling authorities."[17]

3. An IDP is any person who is in need of emergency humanitarian aid because of any type of upheaval, whether political or natural as determined on a case by case basis. This definition focuses on upheavals or significant changes of environment.[18]

International Humanitarian Law

Moving toward a more comprehensive definition gives IDPs more standing in international law and practice. These new definitions recognize that human rights violations across the globe are the responsibility of the international community, whether occurring inside or outside a nation-state. For example existing international standards already consider that displacement of all kinds generally entails multiple human rights violations in and of itself.[19] Legal scholars have pointed out that both Article 13 of the Universal Declaration of Human Rights, and Article 12 of the covenant on Civil and Political Rights[20] militate against forced displacement and promote the individual's right to freedom of residence as part of the right to life. Thus by definition IDPs and its subgroup of victims of ethnic cleansing are immediately covered by these international conventions. When humanitarian law forms the basis of defining IDPs there are stronger and more immediate actions that can be taken, even without waiting for invitation by the governing bodies of the territories where these violations may be occurring.

Another way to strengthen protections for IDPs is to examine the currently existing international laws and protocols on human rights violations, humanitarian law, and refugee law and to test their applicability to IDPs. For example, in the Draft Code of Crimes against Peace and Security of Mankind adopted in 1996,

---

[17] Roberta Cohen and Francis Deng, eds., *The Forsaken People* (Washington, D.C., 1998), 1. See also: UNHCR, *General Assembly Official Records*, 47th Session, Supplement No. 12 (A/47/12), 1991, where the seven categories of persons seeking refugee and asylum status include internally displaced persons, thus broadening the competence of the UNHCR, fn. 17 continued. See also: *UN General Assembly Resolution* (48/116, 1993), encouraging the UNHCR to undertake the protection of IDPs.

[18] Paraphrase of the definition used by the United Nations High Commissioner for Refugees (UNHCR).

[19] Displacements may violate the right to a stable residence, the right to voluntary movement, the right to economic opportunities, etc.

[20] Adopted in 1966, entered into force in 1976.

Article 18 prohibits the systematic and "forcible transfer of population." This is one of the chief means that ethnic cleansing occurs. Populations are forcibly expelled so that the ethnic "purity" of the region may be protected (E.g. the expulsion of the Sudeten Germans). Article 17 of the Covenant on the Civil and Political Rights provides that "no one shall be subjected to arbitrary and unlawful interference with his ...home." This refers to all kinds of residential property. This deprivation will be "arbitrary" if it is done with injustice, unpredictability and unreasonableness. The 1966 Covenant on Economic, Social and Cultural Rights speaks directly to the right of housing, addressing forced displacements. The deprivation of housing may render persons subject to intense police supervision, or may prohibit their movement from location to location. Police and political harassment or even violence, unpunished assaults (again read ethnic cleansing), forced labor and the like may be part of the matrix of deprivation of housing. Often subsistence needs such as water, food, medicine and shelter are also denied, leading to loss of health or loss of life. Internal displacement may result in the loss of personal papers and documentation, leaving IDPs almost in a state of official "non-personhood." Family ties, family unity, religious needs may be severely restricted or subverted. International humanitarian law recognizes all these violations under various categories, definitions and protocols. The body of law is very complex and far from systematic, although it addresses these and a variety of other issues, and affords the basis of some forms of international intervention.

Conclusions

Thus by a combination of definitions and a gathering up of already existing laws, the plight of IDPs can be more favorably addressed. This is all theoretical and the difficulty is the very complex nature of international laws, protocols and conventions. Nevertheless, there has been considerable progress in the theoretical recognition of IDPs and the attempt to apply the three type of existing categories of law, humanitarian, human rights and refugee law that are the sources of protection of IDPs. [21] The application of human rights law is by far the more fruitful approach to the

---

[21] Various international conventions may be cited, including the four 1949 Geneva Conventions, the 1951 Refugee Convention, the 1966 Convention on Civil and Political Rights, and the 1966 Convention on the Elimination of All Forms of Racial Discrimination, and the 1984 Convention against Torture. In general see Cohen and Deng, *Masses in Flight*, chap. 3, and Deng, *Protecting the Dispossessed*, 4-10.

problems of IDPs, since much greater latitude is given to the international agencies for enforcement. In fact, the linking of IDPs with refugees often is detrimental to addressing the special needs that arise when peoples do not cross international borders.

Administrative Remedies

Change in Viewing IDPs

Administratively, there has been an evolution: the appointment of a High Commissioner on Refugees, the 1951 Convention on Refugees, the long-awaited 1966 Convention on Civil and Political Rights, the constitution of the High Commissioner on Human Rights in 1984, and the appointment of Francis Deng as special representative for IDPs in 1992. This shows the internationalization of human rights and the humanization of international law. Today, human rights are no longer exclusively governed by national law. Reference should also be made to the Helsinki Final Act, which contains human rights pledges. Also the monitoring of asylum applications has provided a mechanism for gathering evidence on human rights violations within a country's borders.

Administrative Obstacles

However, the single biggest obstacle to the resolution of the issue of aid to IDPs is the concept of national sovereignty and the inviolability of national borders. International interference is not well tolerated and national borders are difficult to cross and are vigorously guarded, typically by regimes where violations are occurring. There is a need today to pierce borders and to extend the reach of the international community. Because of issues of national sovereignty, there has been considerable discrepancy between the ability to afford international aid to refugees and to IDPs. Aid and resources often do not reach the IDPs and the states where they are located often resent outside interference in a world where the concept of nation state is still strong, and state sovereignty is jealously guarded.

Forward steps have already been taken in several European situations, like the intervention in Kosovo on humanitarian grounds, since under Articles 55 and 56 of the United Nations Charter, member states are obliged "to promote universal respect for and observance of human rights and fundamental freedoms for all." [22] This "respect" is required to avoid the slippery slope of

---

[22] Cohen and Deng, *Masses in Flight*, 305.

persecution and ethnic cleansing that could lead to torture and genocide. The obligation of these articles has been reaffirmed by the World Court.

Special Protections Against Ethnic Cleansing

The Guiding Principles speak to their foundations in international human rights law and humanitarian law. They spell out rights protecting against arbitrary displacement from one's home, particularly citing unacceptable reasons for such displacement, such as "ethnic cleansing" and "collective punishment."[23]

Using these Guiding Principles of 1992 four categories of IDPs may be identified:

a) *Deslocados* — "dislocated persons"—internally displaced persons compelled to abandon their home for reasons beyond their control;

b) *Afectados*—"affected persons"—persons who live in an area impacted by war or by natural disaster;

c) *Recuperados* — "liberated persons"—persons retaken from insurgent control and moved to government controlled areas;

d) *Regresados* — "returning persons"—persons returning either to their homeland or to their former domiciles.[24]

These new definitions put IDPs squarely under the protection of international agencies where humanitarian aid can be given and international conventions applied.[25] The special nature of IDPs legal standing can be addressed by enforcing human rights conventions and by applying international standards under the Universal Declaration of Human Rights and the covenant on Civil and Political Rights, since loss of domicile is implied by each of the above designations. Freedom of residence is a basic tenant of an individual's "right to life" under humanitarian law. The World Court has upheld this right, and international agencies are thereby given supranational powers to enforce protections against such

---

[23] Ibid., 307. See especially Principle 6 of "Guiding Principles" (1992).

[24] See UNHCR, *Operational Experience with Internally Displaced Persons* (September 1994), 64. Also see UNHCR, *The Protection Aspects of UNHCR Activities on Behalf of Internally Displaced Persons* (EC/1994/SCP/CRP 2), 4 May 1994. The Spanish designations were forged in the case of Mozambique.

[25] The operation of NGO's such as the International Red Cross for the delivery of aid, and the Center for Migration and Policy Development (ICMPD) for monitoring, research and advising, are also important. Supranational agencies, such as UNICEF have extended their activities in meaningful ways.

violations, particularly in cases of ethnic cleansing, where forcible transfer from one's domicile is a basic feature.

Practical Considerations

Political Complexities

Despite the theoretical underpinnings, the delivery of aid can be very difficult on a practical level, even in response to natural or man-made disasters, which are the most politically neutral situations that effect IDPs. While most governments are willing to co-operate in receiving emergency aid, some totalitarian states will not release information about disasters or population displacements that may result, and will not seek international aid. A good example of this was the Chernobyl disaster in the USSR where no emergency aid was sought and the extent of the dislocation of populations is still not known.

Paramount in these disaster situations is the delivery of aid, rather than affirmation of human rights doctrines, even though all humanitarian aid is predicated on concepts of basic human rights such as freedom from want, i.e. the right to food, water and basic shelter. Thus international relief organizations have often participated in floods, earthquakes and the like throughout the world, because these various non-governmental organizations (NGOs), such as the International Red Cross, or supragovernmental bodies, such as the United Nations Children Fund (UNICEF), or the World Health Organization (WHO), try to maintain a neutral stance toward the host government's political policies.

However, there may be other political considerations of closed borders or totalitarian control, where outside observers into the country are unwelcome, even though humanitarian aid would be helpful. The international community will therefore not respond, since information is not forthcoming. At various points, the former USSR and China have both followed the model of closing their borders and controlling the flow of all information out of, and aid into their borders. Thus even the most direct cause of population displacement can also have political ramifications.

Aid delivery to non-disaster affected IDPs

The real difficulty remains the delivery of aid to those persons displaced because of armed conflict, internal strife, revolution, a systematic violation of human rights, ethnic cleansing, and a variety of other more political causes within a matrix of other considerations. For example:

1. IDPs are often made up of rootless, marginal and "invisible" populations in their own countries. They move frequently and quietly within the borders of their countries to avoid conflicts, strife or persecution and ethnic cleansing. They try not to bring attention to themselves to avoid persecution. They move frequently either to avoid detection or to find more suitable locations to survive. They are often from the countryside where keeping accurate counts of populations has never been accomplished. They meld into the stable population of the nation for their own protection. Unlike refugees who leave their country, and thus are more readily identifiable, the population comprising these IDPs would probably not have been counted in their own nation with any accuracy even if they had never been uprooted. These marginal peoples are often very vulnerable to the threat of ethnic cleansing, (as for example the gypsy populations in Romania).

2. Often IDPs, particularly in the populations who fear ethnic cleansing, wish to remain unidentified. The only protection they have is their anonymity. Often large numbers of people are forced to be on the move to avoid genocide by other groups within their country, so they try to remain hidden, moving about secretly, not calling attention to themselves in hopes of eluding their persecutors. If the international community suddenly focuses attention on them in attempting to deliver aid, their very lives are at stake. This became very apparent during the conflicts between the Serbs and the Bosnians in the former Yugoslavia.

3. Often governments will deliberately close their borders so that fleeing populations will not be identified as refugees. This closing of the borders bodes ill for these governments' cooperation with international aid delivery within their borders. Here these governments are often motivated by a desire to avoid bad publicity, or to maintain their propaganda image that would be called into question by a mass exodus. Thus they keep their citizens confined within their borders, but refuse any international relief, because that would also necessitate international scrutiny. Instead of being allowed aid, these IDPs are instead targeted for various political or ethnic reprisals, because of their desire to flee.

4. Often IDPs are faced with chaotic border disputes, where boundaries are shifting. Then IDPs are often left with no aid delivery as the international community struggles with questions of definition. As boundaries change, fleeing populations may at any given time have crossed international boundaries, making them refugees by international standards, and subject to one set of rules, or they may still be within the boundaries of their own country, which would then give them IDP status. As the international

community tries to reach a definition, the displaced population is barred from aid.

5. Governments may favor certain groups within their borders and allow international aid to those people as IDPs, while ignoring the oftentimes more pressing needs of displaced groups who are not in favor with the given regime. Often those populations out of favor may even be targeted for ethnic cleansing. Some hostile governments may use the aid meant for their citizens to enrich themselves and refuse to allow the delivery of aid to any IDPs within their borders.

6. Resettlement of IDPs is a stated long-term goal of international relief organizations. Often this resettlement is attempted before proper steps have been taken to insure the viability of the resettled populations. The threats that these populations face can be many: unstable conditions, ongoing hostilities, and lack of necessary requirements to carry on daily life caused by destruction, devastation and conflicts. Victims of ethnic cleansing become particularly vulnerable when resettlement is hasty or lacking precautions.

Discussion and Recommendations

These problems do not lend themselves to easy solutions. The population of IDPs continues to grow, and their needs continue to mount. The international community has become more aware of their needs and problems, but has not yet found adequate solutions for them. Nations do not readily give up their powers to supragovernmental or international bodies for a variety of reasons, among them nationalism, political or policy considerations, severe unrest or strife within their borders. In fact, often nations that have the greatest population displacements, and who would benefit most from international interventions, are those most unwilling or unable to expedite such aid for their IDPs. Theoretical considerations have progressed and definitions have been refined. Still the practical problem of delivering aid and insuring human rights remains almost insurmountable in the foreseeable future. With the notable exception of disaster relief, other types of relief rarely find the target populations. Human rights violations often have to be countered by "peacekeeping" military efforts on the part of the international community, or by embargoes and other sanctions. Neither force nor deprivation as means of enforcement of humanitarian measures or human rights issues seems to make much sense logically. Bombing a nation in order to force it to accept humanitarian aid or to prevent human rights violations seems very punitive on the part of the international community, although this has happened in Serbia and Kosovo. To cut off all trade or aid in order to force a nation to comply with humanitarian goals, (as in Iraq and Libya), also seems somewhat punitive. Those who suffer most are always those who have the least, and are, as often as not, the very persons that the international community is purporting to help. Nations must slowly be educated to the notion that their citizens are their most valuable commodity and that violating humanitarian or human rights violates the most fundamental right of all persons. This cannot be accomplished by force, but by repeated visible action on the part of the international community in dealing with its nation states, and by delivering the most basic relief to any population in need with no political polarization and power considerations.

This type of neutral apolitical stance by the international community is the only way in which there is hope of delivering humanitarian aid to the victims of ethnic cleansing, as well. Given their already vulnerable position, victims of ethnic cleansing should not be identified as a specific subgroup of IDPs, but rather just as any IDP in need of humanitarian aid. The victims of ethnic cleansing can only be aided if political judgments are withheld and the focus is on the humanitarian needs of the victims. Nations will not readily agree to international interventions that might polarize the factions within the country, or that call into question the values and policies of the governing parties.

It is also arguable that the "disaster victims" category, although allowing international relief measures most readily, beclouds the real focus of the problem with IDPs, and should be handled by special protocols with the appointment of a Commission for Emergency Management. However there is a serious *caveat* to this proposal. Exactly because disasters are very recognizable and allow the entrance of aid on humanitarian grounds most readily, relief agencies once *in locus quo* may also see and respond to other more hidden and less obvious human rights violations and humanitarian needs. In fact, the broader the definition, and the more specific and limited the reasons for intervention, the more easily international relief agencies may find it to enter. In other words, victims of ethnic cleansing and other IDPs might be given aid by agencies that first responded to a natural or man made disaster. It is much more difficult to insist on entry to remedy human rights and other violations initially, but once channels for delivering aid have been established for a specific disaster, for example, then these and other violations may be more readily addressed. If disaster relief is the avenue by which international aid is first given, then it may be used as a first step to the delivery of aid to other categories of victims.

The question remains, how can aid be readily delivered to the most number of people in the shortest possible time, and with the least loss of life? This is an all important question when addressing the needs of victims of ethnic cleansing who are among the most vulnerable IDPs, since either by their own governments' policies or by other unofficial groups violence is being perpetrated on them. These groups are the targeted victims who draw attention to themselves if they flee, and who remain ready victims if they do not. As always, solutions are hard to achieve, but the international community must attempt to aid the most vulnerable in the name of humanity.

The Long-Term Consequences of Forced Population Transfers: Institutionalized Ethnic Cleansing as the Road to New (In-) Stability? A European Perspective

STEFAN WOLFF

Striving for internal stability and external security, many states have sought to minimize the impact of minorities with an ethnic affiliation to other, often neighboring, states by expelling them or exchanging them against ethnic kins. Such forced population transfers in Europe are primarily linked with two phenomena which in themselves are interrelated: the collapse of (multinational) states and the redrawing of state boundaries. From the First and Second Balkan Wars, to the two World Wars, and finally to the violent breakup of Yugoslavia, Europe has seen numerous expulsions and exchanges of populations that did not fit into the concept of "relatively homogeneous wholes" sought to be created by states which had gone to war with each other "at the high noon of ethnic nationalism."[1] As a consequence of the ethnic mobilization of polities and because of the hostilities that existed between them before and during wars, and that continue to do so after an often unstable peace had been made, "the ethnic mosaics which were the pride of empires, became liabilities."[2]

Forced Population Transfers: Expulsions and Population Exchanges as Forms of Ethnic Cleansing

Following Bell-Fialkoff, I understand ethnic cleansing as a "planned, deliberate removal from a certain territory of an undesirable population"[3] on the basis of ethnic criteria. In the specific context of this paper, the term can be further specified as state policies resulting in the forced transfer of members of an ethnic group across borders to the territory of an ethnic kin-state, or as

---

[1] Rogers Brubaker, "Aftermaths of Empire and the Unmixing of Peoples: Historical and Comparative Perspectives," *Ethnic and Racial Studies* 18 (1995): 192.

[2] Karen Barkey, "Thinking about Consequences of Empire," in *After Empire: Multi-Ethnic Societies and Nation-Building*, ed. Karen Barkey and Mark von Hagen (Boulder, Co., 1997), 102.

[3] Andrew Bell-Fialkoff, *Ethnic Cleansing* (London, 1996), 3.

the refusal to allow refugees to return. As such, these policies violate norms of international law as they "are collective in nature, ...carried out by force or threat of force, ...are involuntary, ...deliberate on the part of the ...party conducting [them], ...systematic, ...discriminatory, and ...take place without due process."[4]

The goal of such ethnic cleansings is to rid states of troublesome minorities that are considered (mostly on the basis of selectively interpreting historical and recent evidence) as threats to the internal stability and/or external security of their pre-cleansing host-state on whose territory they may have lived for some time or by whom they may have been "acquired" as part of war and/or postwar territorial gains. The two predominant forms ethnic cleansing takes in this context are population exchanges and expulsions. Population exchanges imply an agreement between the two states involved; an expulsion is usually a unilateral act of one state. In many cases, both forms of ethnic cleansing have the explicit consent, or at least tacit approval, of relevant regional and world powers.

Thus, the forced population transfers I am considering in this paper are institutionalized at the level of states: they are forms of ethnic cleansing that are legally and/or "contractually" regulated; they are openly declared and pursued policies; and their objectives are apparently legitimate from the perspective of internal stability and external security of the states involved.

Forced Population Transfers before 1945

The first relevant cases are the minority exchanges in the Balkans after the 1912-13 Balkan wars and the territorial reorganization of the region after the First World War between 1919 and 1923. The first instance in this context was the Bulgarian-Turkish exchange of populations agreed upon by the two states in an Annex to the Peace Treaty of Constantinople on 15 November 1913. Together with the Treaty of Bucharest, this instrument meant the second partition of Bulgaria after its brief Greater Bulgarian existence following the treaty of San Stefano in 1878. A third partition followed in November 1919 with the Treaty of Neuilly between the Allied Powers (including Greece) and Bulgaria, and again it included a convention that affected a massive cross-border migra-

---

[4] United Nations Commission on Human Rights, *Human Rights and Population Transfer: Final Report of the Special Rapporteur* (E/CN.4/Sub.2/1997/23), paragraph 10.

tion of Bulgarians to their kin-state. The relevant convention annexed to this treaty was concluded with Greece and effectively stipulated the reciprocal migration of Greeks and Bulgarians. It was implemented over a period of thirteen years and involved the migration of 92,000 Bulgarians and 46,000 Greeks.[5] Bulgaria was ill-equipped to integrate the refugees, many of whom were housed in camps for years and burdened public spending. The loss, within forty years, of significant territories considered part of a Greater Bulgaria and the fact that, as a consequence of this, by 1919 approximately one million Bulgarians (or 16 percent of all Bulgarians at the time) lived outside the country[6] has been a source of Bulgarian irredentism ever since.[7] Despite the fact that the forced population transfers before and after the First World War only marginally rectified this part of the problem, Bulgarian irredentism has never posed a serious threat to regional stability, although relations between Bulgaria and Macedonia were uneasy for a number of years before the two countries finally signed a series of bilateral treaties and agreements in 1999. However, the incompleteness of the population transfers up to 1919 kept the issue of emigration on the agenda: Bulgaria and Turkey signed various bilateral agreements after 1945 resulting in the emigration of almost 270,000 Turks from Bulgaria between 1950 and 1978,[8] and in 1989 around 370,000 Turks left the country when the opportunity to do so arose.[9]

Because of its magnitude and impact, the Convention Concerning the Exchange of Greek and Turkish Populations of 30 January 1923 warrants separate and more extensive treatment.

---

[5] Kalliopi K. Koufa and Constantinos Svolopoulos, "The Compulsory Exchange of Populations between Greece and Turkey: the Settlement of Minority Questions at the Conference of Lausanne and Its Impact on Greek-Turkish Relations", in *Ethnic Groups in International Relations*, ed. Paul Smith (Dartmouth, 1991), 281. However, it should be noted that this Greek-Bulgarian Convention on Reciprocal Emigration recognized the *right* of emigration and emphasized the *voluntary* character of any emigration in its context; it was drafted by the Commission on New States and for the Protection of the Rights of Minorities and subsequently accepted by Greece and Bulgaria (Stephen P. Ladas, *The Exchange of Minorities. Bulgaria, Greece, and Turkey* [New York, 1932], 40ff.).

[6] R. J. Crampton, *A Concise History of Bulgaria* (Cambridge, 1997), 149.

[7] Cf. Duncan M. Perry, "Bulgarian Nationalism: Permutations on the Past," in *Contemporary Nationalism in East Central Europe*, ed. Paul Latawski (London, 1995), 38-50.

[8] Ali Eminov, *Turkish and Other Muslim Minorities in Bulgaria* (London, 1997), 78.

[9] Brubaker, 194.

According to the convention, approximately 1.1 million Greeks from Asia Minor and Eastern Thrace were expelled to Greece, and between 350,000 and 500,000 Muslims, primarily from the Greek provinces of Macedonia and Epirus, were expelled to Turkey. The implementation of the convention showed that at least some effort was made to handle the whole exchange with efficiency and a minimum of dignity. The transferred individuals automatically lost the citizenship of the country they left, but gained that of their country of destination; neither state charged duty on possessions carried by those part of the exchange; property left behind (houses, land) was either exchanged or became part of a lengthy and not wholly satisfactory compensation process; and both governments provided transport facilities and sought to administer the exchange as quickly as possible.[10] Furthermore, both governments agreed to the setting-up of a Mixed Commission, including representatives of the League of Nations, to supervise and facilitate the population exchange and to liquidate the property left behind by the transferred people.

International efforts to ease the transfer and its domestic consequences extended particularly to Greece, which had to deal with the far larger number of people transferred. An international loan worth £12.3 million was granted for the resettlement of those transferred in 1923/24, resulting, within two years, of half of those part of the population exchange being resettled and economically self-sufficient.[11] Additionally, Greece received a stabilization loan of £7.5 million in 1928, and private organizations, such as the Red Cross, funded aid projects of various kinds throughout Greece.[12] Nevertheless, the integration process was very prolonged. In Greece, for example, the resettlement program continued throughout the 1960s, and the housing situation of many of the children and grandchildren of those transferred from Turkey remains unsatisfactory to date.[13]

As a consequence of the population exchange, the Greek domestic political landscape was fundamentally changed, since approximately 300,000 voters were added to the existing 800,000.

---

[10] Koufa and Svolopoulos, 290f.

[11] Dimitri Pentzopoulos, *The Balkan Exchange of Minorities and Its Impact upon Greece* (Paris, The Hague, 1962), 75ff.

[12] Ibid.

[13] Cf. Renée Hirschon, *Heirs of the Greek Catastrophe: The Social Life of Asia Minor Refugees in Piraeus* (Oxford, 1989).

Those transferred from Turkey were generally more republican and cosmopolitan in their attitudes. As they never formed their own party, they were gradually absorbed into the existing party-political system, but particularly into the ranks of supporters of the Liberal Party, which returned to power in 1928 primarily because of the backing of those transferred from Turkey. Equally important was the economic impact: under the influence of the newly arrived population segments, both agriculture and industry were modernized, new production methods introduced, and diversification of the output achieved.[14]

However, the transfer impacted Greece in negative ways. Although here, as in Turkey, the "exchange of populations was a drastic, but largely effective way of eliminating frictions that had been caused by the multiethnic character of certain regions,"[15] new tensions arose from the republican political orientations of the so-called New Provinces, largely inhabited by those transferred, which contrasted sharply with the pro-monarchy views of the Old Provinces.

The gradual rapprochement between Greece and Turkey, exemplified by the Ankara Convention of 1930, was not very well received among those transferred to Greece seven years earlier. They particularly resented the fact that there would be no possibility of return to their ancestral homelands, that compensation for lost properties was inadequate, and that they had no say in the negotiation process.[16] Ironically, their electoral support had made possible the victory of the Liberal Party and with it a certain normalization of Greek-Turkish relations[17] in the interwar period, which in turn necessitated concessions at the expense of that part of the core electorate of the Liberal Party.

In Turkey, most of those transferred from Greece were resettled either in the big cities or on the Aegean and Mediterranean coasts.[18] In contrast to the situation in Greece, the particular facts

---

[14] Douglas Dakin, *The Unification of Greece, 1770-1923* (London, 1972), 269-270.

[15] Mark Mazower, *Greece and the Inter-War Economic Crisis* (Oxford, 1991), 43.

[16] Pentzopoulos, 183ff.; Mazower, 129.

[17] The Ankara Convention of 1930 was followed in the same year by an Agreement on Friendship, Neutrality, and Arbitration, a Trade Agreement, and a Convention on Naval Armaments; and in 1933, both countries signed a security agreement (Koufa and Svolopoulos, 300ff.).

[18] Nicole Pope and Hugh Pope, *Turkey Unveiled: Atatürk and after* (London, 1997), 114.

of this forced migration are relatively unknown in Turkey, even among the descendants of those originally uprooted.[19] The reasons for this are primarily three. The number of Muslims expelled from Greece to Turkey was relatively small (only about one-third of those sent to Greece, and only between three and five per cent of the number of ethnic Germans expelled after the Second World War), and thus benefited from the reciprocity of the exchange.[20] As Turkey had a long tradition of categorizing population groups by religion, the Muslims from Greece blended in relatively well with mainstream society, even though there were linguistic problems because many of them had Greek as their mother tongue.[21] Finally, the territorial and demographic changes that took place as a consequence of the Treaty of Lausanne were and are seen in Turkey "as only part of the complicated story of the collapse of the Ottoman empire and the growth of the Turks' own sense of national identity."[22] Thus, it is not surprising that, from a Turkish point of view, irredentism has played no role in the country's relationship with either the expellees it received or their former homelands.[23]

Tensions between Greece and Turkey in the post-1945 period, although they were often interpreted in both countries in the context of the Treaty of Lausanne and the compulsory population exchange, had their sources elsewhere. The reasons for the tense relationship between the two countries were, and to some extent are, mainly three: border disputes and the economic ramifications of border demarcations in the Aegean Sea, the unresolved Cyprus conflict, and the treatment of reciprocal minorities by both states. This last issue, interestingly, is related to the Treaty of Lausanne, in particular to the allowances that were made for specific sectors of the "qualifying" population to be exempted from the forced exchange. Yet, it would be wrong to argue that a more complete population exchange would have prevented tensions between the two countries. The situation of the reciprocal minorities is not deplorable because they were excluded from the population trans-

---

[19] Ibid.

[20] For example, more village communities from Greece were able simply to swap land and properties with their "counterparts" expelled from Turkey.

[21] Bernard Lewis, *The Emergence of Modern Turkey* (Oxford, 1968), 355.

[22] Pope, 114. In contrast, the loss of its territories in Asia Minor was nothing less than a national catastrophe for many Greeks.

[23] Landau, 79ff.; Pope, 115ff.

fer, but because both Greece and Turkey have persistently denied basic minority rights and even the recognition of minority status to various minority communities. Likewise, the constant exploitation and manipulation of this issue in bilateral relations has done little to change the plight of the two minority groups. Despite the difficult relationship between Greece and Turkey, international pressure and international integration have thus far successfully prevented the outbreak of military hostilities.

In other instances in post-1919 Eastern Europe, similar situations did not lead to a massive forced population transfer – the territorial losses incurred by Hungary following the Treaty of Trianon in 1920, particularly in relation to Transylvania, the Vojvodina, and southern Slovakia, and the incorporation of the predominantly German-speaking *Sudetenland* into the newly established Czechoslovak state. The cases of Transylvania and southern Slovakia would have allowed population exchanges, the former probably in connection with further border revisions. In Vojvodina and the *Sudetenland*, a lack of "sufficient" numbers of Serbs in Hungary and even more so of Czechs in Germany or Austria would have only left the measure of expulsions to create ethnically more homogenous polities. For various reasons, neither the Great Powers nor the states involved had a particular interest in pursuing a policy of ethnic cleansing either against Hungarians or against Germans at the time. The consequences are well-known: The Munich Agreement of 1938 dismembered Czechoslovakia, annexing the *Sudetenland* to Germany that had already "united" with Austria earlier the same year. The two Vienna Awards of 1939 and 1940 granted Hungary large parts of southern Slovakia and led to a north-south partition of Transylvania between Hungary and Romania. A forced population transfer of ethnic Germans initially did also not occur in the emerging Soviet Empire. After the "clearing" of ethnic Germans from front zones during the German campaign against Russia in World War I, Soviet nationality policy created an Autonomous Soviet Socialist Republic of the Germans in the Volga area. Yet, after Germany's attack on the Soviet Union in 1941, the Republic was dissolved and all ethnic Germans were deported to camps in Siberia and Central Asia. Only in 1955 were they reinstated in their citizenship rights, but even after 1989 a restoration of the Volga Republic was ruled out.

Forced Population Transfers after 1945

Here I am particularly interested in two cases, namely the expulsions of ethnic Germans from Poland and Czechoslovakia between 1945 and 1950.[24] Initially, I will treat these two cases of expulsion jointly, primarily in terms of their origins, administration, and impact on Germany. Thereafter, I will distinguish the development of bilateral relations between Germany and Poland from that between Germany and Czechoslovakia/the Czech Republic.

Origins, Administration, and Impact of the Expulsions

The expulsions can be seen to have their origins in three distinct but related policy considerations: to prevent the instrumentalization of external minorities for an irredentist foreign policy; to accommodate the westward shift of Poland; and to punish ethnic German minorities for the role they had played in German occupation policy in Central and Eastern Europe. The Allies gave their consent in the Potsdam Agreement to the "humane and orderly transfer" of almost 3.5 million ethnic Germans from the *Sudetenland* and of over 7 million Germans from the "formerly German territories" of Poland, plus another more than 1.5 million people from other parts of Poland and the Free City of Danzig.[25] These numbers include a significant percentage of refugees who had escaped with the retreating German armies. As a consequence of flight and expulsion, approximately 2 million ethnic Germans died from exhaustion, starvation, and attacks by local mobs and regular and irregular military units.

Despite the magnitude of the expulsions and further emigration after 1950, there are still ethnic Germans living in both countries: some 50,000 in the Czech Republic, who, far from being united, are too dispersed to preserve a distinct ethno-cultural identity; in Poland, in contrast, the majority of the approximately half million ethnic Germans live relatively compactly in Upper Silesia, being a local majority in a number of communities and districts.

---

[24] Two other phenomena of forced population transfers in the post-1945 period need to be noted as well. One is the Slovak-Hungarian population exchange, affecting about 75,000 on each side. The other is the deportation of ethnic Germans from countries in Central and Eastern Europe to forced labor camps in the Soviet Union as part of a policy of collective victimization and punishment.

[25] The introductory paragraph to section "XII. Orderly Transfer of German Populations" of the Potsdam Agreement reads as follows: "The Three Governments, having considered the question in all its aspects, recognize that the transfer to Germany of German populations, or elements thereof, remaining in Poland, Czechoslovakia and Hungary, will have to be undertaken. They agree that any transfers that take place should be effected in an orderly and humane manner."

About two-thirds of the refugees and expellees were resettled in the American and British occupation zones; of the remaining third sent to the Soviet zone, approximately forty per cent left for West Germany before 1961. The integration problem initially involved shortages of housing, food, and employment, a situation which could lead to hostile receptions locally for the expellees and refugees. Subsequently, however, the positive contribution to the economic and social modernization, especially of relatively backward and underdeveloped areas in the later West Germany, such as Bavaria and Schleswig-Holstein, was both appreciated and publicly acknowledged, and contributed to the overall successful economic and social integration of the expellees. Apart from a brief period in the 1950s, expellees never had their own political party. Their integration in the political process of West Germany helped its stabilization and consolidation into a three party system by the early 1960s.

The experience of many injustices during the expulsions and of the loss of home and property, however, also meant that particularly the older generation held deep feelings of resentfulness against Poland and Czechoslovakia and resisted, until the early 1990s, any form of reconciliation. Younger generations—who had no memories, or only vague ones, of flight and expulsion—proved to be more flexible and open to the idea of constructive reconciliation.

German-Polish Bilateral Relations

The major problem of German-Polish relations was the uncertain status of territories that had belonged to Germany before 1937 and that were placed under Polish administration in the Potsdam Agreement. The potential claim that Germany could make to these territories strained the relations between West Germany and Poland until finally, in 1990, the united Germany and Poland signed the so-called border treaty, finally and formally guaranteeing Poland's western borders. Prior to this, the danger of German irredentism had been confined by two factors. One was the Cold War division of Europe. The other was the fact that West Germany, following the partition of Germany in 1949, no longer had a common border with Poland, while East Germany had recognized the border as early as 1950 in the Treaty of Görlitz, despite the fact that in the Potsdam Agreement the Allies had postponed a settlement of the border question subject to a peace treaty with Germany. However, the deep cultural and personal significance attached to their ancestral homelands by many expellees, and their rhetoric to this effect, kept the danger clear and present in the view of successive Polish governments, but also in the eyes of many ordinary Poles, affecting negatively the policies towards the remaining ethnic Germans in Poland.

Another problem in the relationship between the two countries was the interpretation and teaching of their common history. This problem involved aspects as complex as the origin and existence of German minorities in Poland, officially disputed by successive Polish governments until 1989, German occupation of Poland during the Second World War, the reasons and extent of the expulsions of ethnic Germans, etc. In many ways, all these issues directly affected the ethnic German expellees and their identity, cultural heritage, and future. Thus, the more conciliatory an approach German governments took towards Poland from the late 1960s onwards, the more resentful expellees from Poland grew—towards their own government and towards Poland. This culminated in open defiance of German government policy and led increasingly to the marginalization of political representatives of the expellees in the political and social spheres of German public life. Only after the expellee community began to embrace the new opportunities for constructive reconciliation with Poland from about 1993 onwards was this process reversed. This development led to a rapprochement not only domestically, but surprisingly also between German expellees and Poland. This process was facilitated both by the democratization of Poland and its desire for integration into NATO and the European Union, as well as by a generational and change of mind within the expellee community. However, in the context of Poland's NATO and EU integration, two still unresolved issues have regained some prominence: the question of compensation for expellees and their right to resettle in their former homelands.

German-Czech Bilateral Relations

German-Czech relations were never burdened by unresolved territorial issues, as the territory in question, the *Sudetenland*, had never been part of a German state before 1937, and the invalidity of the Munich Agreement of 1938 was not disputed by any German government after 1949.[26]

---

[26] There have been arguments, however, about whether the Munich Agreement had been invalid from the beginning, or whether it became invalid in the course of World War II. While the Czech point of view has always been that, as a treaty signed under the threat of use of force, it had never been valid, successive German governments up to 1973 (and mostly for practical reasons, such as citizenship issues and the legality of marriage certificates) insisted on a later date of invalidation in accordance with the declarations by France and the United

The issue in German-Czech relations was and still is primarily one of the legitimacy of the expulsions and the way in which they were conducted. The collective victimization of ethnic Germans (and of ethnic Hungarians) in the so-called Beneš Decrees and the subsequent amnesty granted for all crimes committed during the expulsions are deeply resented within the expellee community as well as within significant parts of the German political establishment.

After years of negotiations and crises, the 1997 German-Czech Declaration was the smallest common denominator the two governments could find on the two most critical issues – the role of the Sudeten Germans in the breakup of Czechoslovakia in 1938 and their collective victimization and expulsion after the end of the Second World War. Yet, in the vagueness of its wording, it satisfied neither side completely. The German government accepted the responsibility of Germany in the developments leading up to the Munich Agreement and the destruction of Czechoslovakia, expressed its deep sorrow over the suffering of Czechs during the Nazi occupation of their country, and acknowledged that it was these two issues that prepared the ground for the post-war treatment and expulsion of members of the German minority in the country. The Czech government, on the other side, regretted the post-war policy vis-à-vis ethnic Germans, resulting in the expulsion and expropriation of a large section of the German minority, including many innocent people. In some sense, the German-Czech Declaration can be seen as an attempt to deal with "the fundamental question of the innate antagonism between peace and justice."[27] The greater compromise here was made by the German government,[28] to the dismay of many expellees. This

---

Kingdom in 1942 and of Italy in 1943 about the invalidity of the Agreement (Radko Břach, "Die Bedeutung des Prager Vertrages von 1973 für die deutsche Ostpolitik," in *Im geteilten Europa Tschechen, Slowaken und Deutsche und ihre Staaten 1948-1989*, ed. Hans Lemberg, Jan Křen, and Dušan Kováč [Essen, 1998], 172).

[27] United Nations Commission, paragraph 63.

[28] Two further developments since 1997 also testify to this rather uneven compromise: German Chancellor Schröder and Czech Minister-President Zeman issued a joint declaration on 8 March 1999, stating that "neither government will re-introduce property issues [into their bilateral relationship] either today or in the future" (Deutscher Bundestag, *Antrag der Fraktionen SPD und Bündnis 90/Die Grünen: Weiterentwicklung der deutsch-tschechischen Beziehungen*, Drucksache 14/1873, 26 October 1999). In the context of the conclusion of the negotiations on the compensation of forced laborers in 2000, the Czech government welcomed the US government declaration that the German-American agreement on the compensation of forced laborers did not affect any potential reparation claims by the Czech Republic against Germany. This is seen as a Czech insurance against private claims of Sudeten German expellees for compensation and/or restitution.

is particularly the case because the Beneš Decrees remain part of the legal order of the Czech Republic, and there is strong public resistance against rescinding them, despite a call from the European parliament to do so.[29]

Similar to Poland, the country's integration in NATO and EU has brought to the fore two still unresolved issues: compensation and the right for expellees to return to their former homeland.

Assessing the Consequences of Forced Population Transfers

On the one hand, the previous case studies have illustrated that forced population transfers can contribute, although in differing degrees, to the internal stability and external security of the states involved, achieving, for the most part, two essential objectives: to avoid internal ethnic strife and to prevent external minorities from being used as instruments of irredentist foreign policies. At the same time, however, bilateral relations between expelling and receiving states have often been poisoned for decades as a consequence of an expulsion, especially when unresolved issues of compensation and restitution remain. On the other hand, avoiding population transfers has had even more disastrous longterm consequences in at least one case—the German dismemberment of Czechoslovakia in 1938/39, and has resulted in severe discrimination and ethnic tensions, occasionally escalating in violence in other cases—such as in Transylvania and southern Slovakia. Incomplete population transfers often lead to further emigration under pressure, such as in Bulgaria in the late 1980s, or when opportunities to do so arise, such as in post-1970 and post-1989 Poland and Romania, and the organization representing ethnic Germans expelled from Czechoslovakia, the *Sudetendeutsche Landsmannschaft*, which has recently made several attempts to achieve for its own members the same international recognition (including from countries like Poland and the Czech Republic) as victims of injustice as has been accorded to Kosovar Albanians and forced laborers with all its legal consequences.[30]

---

[29] In April 1999, a resolution was passed by the European Parliament in which its the members called "on the Czech Government, in the same spirit of reconciliatory statements made by President Havel, to repeal the surviving laws and decrees from 1945 and 1946, insofar as they concern the expulsion of individual ethnic groups in the former Czechoslovakia" (European Parliament, *Resolution on the Regular Report from the Commission on the Czech Republic's Progress towards Accession*, COM(98)0708—C4-0111/99).

[30] To give just three examples of how the Sudeten German Regional Association has tried to capitalize on recent debates on human rights: On 7 April 1999, all Sudeten Germans in Germany were urged to donate money for Kosovo refugees,

Assessing the long-term consequences of forced population transfers is a complex endeavor, and generalizations are dangerous to make. What can be said with relative certainty is that there are a number of common factors and that these factors must be sought at five different levels – in the nature of the transfer itself, in the domestic processes in both the receiving and expelling states, in the development of their bilateral relationship, and in the wider international context. (see Table appended, following page.)

Finally, it should be noted that forced population transfers are, in one way or another, the culmination of complex inter-ethnic relations within and across borders. However, they are only one element in the continued development of such relations, and not every renewed escalation of inter-ethnic/inter-state conflict can necessarily be attributed to them and their consequences.

---

with the reasoning that solidarity with the people in Kosovo would sensitize the German and international public to the fate of the Sudeten Germans as well (Sudetendeutsche Landsmannschaft, "Neubauer ruft Sudetendeutsche zu Spenden für Kosovo auf", Press Release, 7 April 1999). The motto chosen for the annual Sudeten German Day in 2000 was: "For a Worldwide Ban on Expulsions" (*Vertreibung weltweit ächten*) (Sudetendeutsche Landsmannschaft, "Aufruf des Sprechers Franz Neubauer zum 51. Sudetendeutschen Tag 2000", Press Release, no date). On 24 March 2000, Neubauer welcomed the fact that Czech victims of Nazi forced labor camps were to receive compensation and noted that this implied a recognition of the fact that crimes "of a certain dimension do not fall under the statute of limitations" and that their victims have to be compensated sooner or later. This was seen as "good news for the German expellees" (Sudetendeutsche Landsmannschaft, "Neubauer: Entschädigung für Zwangsarbeiter ist zu begrüßen und bestätigt Offenheit der sudetendeutschen Frage", Press Release, 24 March 2000).

Appendix—Table: The Factors Determining the Long-Term Consequences of Forced Population Transfers

Population Transfer	Receiving State[1]	Expelling State[1]	Bilateral Relationship	International Context
• Completeness • Degree of violence • Compensation for lost property/assets • Resolution of the citizenship issue	• Political stability • Degree to which the population transfer and its consequences can be instrumentalized in domestic politics • Availability and commitment of resources to pursue/resist irredentist policies and international backing for this • Success of political/economic/social integration o expellees • Cultural and historical significance of/rootedness in the territory from which expellees come for them/their identity, and for the national identity of the receiving state	• Size, territorial concentration and location, and political and economic significance of remaining members of the expelled ethnic group • Perceived internal and external vulnerability as a result of territorial disputes and remaining ethnic minorities and responses to it (minority and security policy)	• Long-term policy agenda vis-à-vis each other (reconciliation; revisionism) • Substance of territorial claims, and vigor with which they are pursued • Significance of ethnic issues • Shared interests and/or conflicts in other spheres	• Approval/facilitation of population transfer • Post-transfer involvement to regional stability • Likelihood of spill-over effects (i.e., one "agreed" transfer triggers further expulsions)

[1] In the case of population exchanges, both states are receiving and expelling at the same time.

Ethnic Cleansing 1945 and Today: Observations on Its Illegality and Implications

ALFRED DE ZAYAS[1]

The phenomenon of ethnic cleansing in Kosovo and in the former Yugoslavia is not the first manifestation of forced population transfers in this century of refugees. "Ethnic cleansing" is but a new term to describe the old State practice of expelling minorities on racial or religious grounds.

In analyzing the issue of the legality of mass population transfers, it is useful to look at it from various angles: a) from the perspective of public international law and the rules on State responsibility, in particular the obligation to make reparations, and b) from the perspective of international criminal law and the penal liability of individuals who order or carry out mass expulsions.

The issue should also be considered from the perspective of the authority of the sources of law[2]: "hard law" or treaties in force (*leges latae*) and "soft law" or not yet legally binding political principles (*de lege ferenda*), including resolutions and declarations of international organizations such as the League of Nations and the United Nations. Of course, most of the crimes that necessarily accompany ethnic cleansing are common crimes in the civil codes of all civilized nations: murder, manslaughter, battery, enslavement, theft, destruction of private property, violation of religious sites, desecration of cemeteries, etc. However, in times of ethnic or religious conflict, States rarely prosecute their own citizens for abuses committed against the targeted groups, especially when the abuses are part of official policy.

The following words of the first United Nations High Commissioner for Human Rights, José Ayala Lasso (Ecuador) were spoken at the *Paulskirche* in Frankfurt-am-Main on 28 May 1995 on the occasion of the solemn ceremony to remember fifty years since the expulsion of fifteen million Germans from Eastern and Central Europe, the largest forced population transfer in history.

---

[1] The opinions expressed in this article are the author's in his personal capacity and do not necessarily represent those of his employers or of the organizations with which he is associated.

[2] For the sources of law see article 38 of the Statute of the International Court of Justice: treaties and conventions, customary international law, general principles of law, and the jurisprudence of the highest tribunals.

The right not to be expelled from one's homeland is a fundamental right.... I submit that if in the years following the Second World War the States had reflected more on the implications of the enforced flight and the expulsion of the Germans, today's demographic catastrophes, particularly those referred to as "ethnic cleansing," would, perhaps, not have occurred to the same extent.... There is no doubt that during the Nazi occupation the peoples of Central and Eastern Europe suffered enormous injustices that cannot be forgotten. Accordingly they had a legitimate claim for reparation. However, legitimate claims ought not to be enforced through collective punishment on the basis of general discrimination and without a determination of personal guilt.[3]

Among legal experts there is no question today that the practice of "ethnic cleansing" or mass population transfers is doubly illegal—as an internationally wrongful act, giving rise to State responsibility and as a grave violation of international criminal law, giving rise to personal criminal liability.[4]

The expulsions by Germany's National Socialist government of one million Poles from the Warthegau in 1939/40 and of 105,000 Frenchmen from Alsace in 1940 were listed in the Nürnberg indictment as "war crimes" and "crimes against humanity." The transcripts of the Nürnberg trials contain much discussion on the Nazi practice of deporting enemy aliens for purposes of demographic manipulation and also for purposes of forced enlistment into the labor force of the Nazi war machine. The Nürnberg judgment held several Nazi leaders guilty of having committed these crimes and sentenced them to death by hanging. It is an anomaly that in spite of this clear condemnation of mass deportations, the Allies themselve carried out even greater expulsions in the last few months of the Second World War and the years that followed. Article XIII of the Potsdam Protocol attempts to throw a mantle of legality over the expulsions carried out by Czechoslovakia, Hungary, and Poland. Nothing is said about the expulsions from other countries like Yugoslavia and Romania.

---

[3] The complete text in German was published in Bonn, 1995, in Dieter Blumenwitz, ed., *Dokumentation der Gedenkstunde in der Paulskirche zu Frankfurt-am-Main am 28. Mai 1995: 50 Jahre Flucht, Deportation, Vertreibung*, 4. Excerpts from the English original are quoted in de Zayas "The Right to One's Homeland, Ethnic Cleansing, and the International Criminal Tribunal for the Former Yugoslavia," *Criminal Law Forum* 6 (1995): 257-314 at 291-92; see the documentary appendix to this book.

[4] De Zayas, "Population, Expulsion and Transfer," in R. Bernhardt, ed., *Encyclopaedia of Public International Law*, vol. III, 1997, 1062-1067.

However, the victorious Allies at Potsdam were not above international law and thus could not legalize criminal acts by common agreement. There is no doubt that the mass expulsion of Germans from their homelands in East Prussia, Pomerania, Silesia, East Brandenburg, Sudetenland, Hungary, Romania, and Yugoslavia constituted "war crimes," to the extent that they occurred during wartime, and "crimes against humanity," whether committed during war or in peacetime.

Moreover, the slave labor imposed on nearly one million ethnic Germans as "reparations in kind," which was agreed on 11 February 1945 by Churchill, Roosevelt, and Stalin at the Yalta Conference,[5] also constituted a particularly heinous crime, which led to hundreds of thousands of deaths during the deportation to slave labor, during the years of hard work with little food, and as sequel of this inhuman and degrading treatment.[6]

International Law Standards Applicable on the Eve of the Second World War

As far as public international law is concerned, a "right to one's homeland"[7] had not been formulated *expressis verbis* by the outbreak of the Second World War, but many of its components were already part of "hard law." The violation of the right to remain in one's homeland would necessarily have violated many binding principles of international law, notably the Fourth Hague Convention of 1907, and articles 42-56 of the Hague Regulations on Land Warfare, which limit the rights of a belligerent occupant and forbid collective punishment. These principles formed the basis for the indictment and conviction of Nazi leaders at Nürnberg for the mass deportation of civilians from occupied territories.

Minorities have been frequent targets of expulsion and spoliation. Their international protection has given rise to important

---

[5] *Foreign Relations of the United States—The Conferences at Malta and Yalta*, 979.

[6] Kurt Böhme, *Gesucht Wird* (Munich, 1965), 275; de Zayas, *A Terrible Revenge: The Ethnic Cleansing of the East European Germans 1944-1950* (New York, 1994), 81.

[7] Christian Tomuschat, "Das Recht auf die Heimat: Neue rechtliche Aspekte," in *Des Menschen Recht zwischen Freiheit und Verantwortung—Festschrift für Karl Josef Partsch*, ed. J. Jekewitz (Berlin, 1989), 183-212; Christa Meindersma, "Legal Issues Surrounding Population Transfers in Conflict Situations," *Netherlands International Law Review* 41 (1994): 31-93; J. M. Henckaerts, *Mass Expulsion in Modern International Law and Practice* (Dordrecht, 1985).

precedents[8] which are relevant in the context of "ethnic cleansing." The treaties of Versailles and St. Germain of 1919 had imposed certain obligations on those States that following the redrawing of frontiers at the Paris Peace Conference had assumed responsibility for large numbers of persons belonging to other ethnic, religious, or linguistic groups. Unfortunately, most of these States ignored the rights of their minorities, as thousands of petitions addressed to the League of Nations amply document. There were two areas of frequent conflict between the German minorities and Polish authorities, one being the question of obtaining Polish citizenship and the other the widespread confiscation of German farms and eviction of the German owners through discriminatory Polish legislation. Some cases even reached the Permanent Court of International Justice at The Hague. A typical case was decided on 10 September 1923, when the Court delivered an advisory opinion on the merits and held unanimously

> that the measures complained of were a virtual annulment of legal rights possessed by the farmers under their contracts, and, being directed in fact against a minority and subjecting it to discriminating and injurious treatment to which other citizens holding contracts of sale or lease were not subject, were a breach of Poland's obligations under the Minorities Treaty.[9]

Many such cases came before the Court and the League of Nations until 1934 when the Polish government unilaterally repudiated the League's minority system.

The violation of international norms by governments does not, however, mean that the norms do not exist. It only means that there is no proper mechanism of implementation and no sanctions for enforcing compliance. In any event, the existence of minority treaties in the interwar period does show that the international community had intended to give minorities special protection. It is obvious that ethnic cleansing, mass expulsion, and mass spoliation grossly violated the letter and spirit of the entire treaty regime for the protection of minority rights.

---

[8] De Zayas, "The International Judicial Protection of Peoples and Minorities," in *Peoples and Minorities in International Law*, ed. C. Brölmann, R. Lefeber, and M. Zieck (Amsterdam, 1993), 253-287.

[9] *British Yearbook of International Law*, 5 (1924): 207-8. See also Lauterpacht, *Annual Digest of Public International Law Cases*, vol. 2, cases 167, 168. Compare the Chorzow (Königshütte) Factory Case, Judgement No. 9, delivered 26 July 1927, PCIJ, *British Yearbook*, 1928, vol. 9, 135ff.

Moreover, since President Woodrow Wilson's many pronouncements on the right to self-determination, and since his famous Fourteen Points, the concept of self-determination had become a much discussed—if rarely implemented—principle of international law. Again, the existence of a norm is not negated by its violation. Unfortunately, however, self-determination has seldom been obtained peacefully as a right, but rather it has been won or lost by force of arms.

So it was, for instance, that in 1919 the Turkish State refused to sign the Treaty of Sèvres, to grant autonomy or independence to the Kurds, the Armenians, or the Greeks, or even to accept an obligation to ensure the rights of its minorities. The leader of the nationalist forces, Mustafa Kemal Atatürk, established a provisional rebel government and in 1921-22 the Turkish armies under his leadership expelled the Greek communities from Anatolia, where they had lived for more than two thousand years. This grotesque ethnic cleansing was largely condemned by public opinion of the time, but in the face of brute force, the Western Allies, who had other priorities at the time, condoned the *fait accompli*. In the Treaty of Lausanne of 1923 the Allies accepted the expulsion, which established a population transfer commission to oversee future population exchanges between Greece and Turkey and to ensure a measure of compensation for private property. This horrendous precedent went against the principles of self-determination and the right to one's homeland, but no one in Europe was prepared to use force against Turkey in order to reverse the expulsions. Prominent critics of the Treaty of Lausanne were Lord Curzon, British Foreign Minister from 1919 to 1924 and participant at the Lausanne Conference, who at that time warned that "the world will pay a heavy penalty for a hundred years to come" for such a "thoroughly bad and vicious solution."[10] Later commentators agreed that the population exchange was not a model of either humanity or wisdom, and its repercussions, economic and political, were considerable. Sir John Hope-Simpson, who also had been intimately involved with the Lausanne Treaty process, observed in 1946 that the exchange of Greeks and Turks had meant an appalling amount of misery and hardship to everyone concerned.[11]

---

[10] Quoted in Great Britain, 1944, *Parliamentary Debates*, Lords, 5s, 130:1120 (speech of Lord Noel-Buxton); see also Georgios S. Streit, *Der Lausanner Vertrag und der griechisch-türkische Bevölkerungsaustausch...* (Berlin, 1929), 24.

[11] G.B., 1946, *Parl. Deb.*, 5s, 139:68.

International treaties on human rights as such did not exist in 1939 or 1945. There were, however, treaties regulating the conduct of belligerents in time of war and providing, *inter alia,* for the protection of civilians in occupied territory and for the protection of prisoners of war pursuant to the Geneva Convention of 1929. To the extent that Nazi Germany subjected foreign prisoners of war to forced labor, the Convention was violated. Similarly, the Allies violated the Geneva Convention when they did not release German prisoners of war until several years after the conclusion of the war and in the meantime exploited them mercilessly as forced laborers.

Standard-Setting After the Second World War—International Norms

While the Convention on the Prevention and Punishment of the Crime of Genocide of 9 December 1948 (one day before the UN General Assembly adopted the Universal Declaration on Human Rights) does not by its terms prohibit population transfers and the implantation of settlers in occupied territory, this practice may well constitute genocide not only under the terms of the convention but also as a matter of customary international law.[12]

Article 2 defines genocide as encompassing any of the following acts committed with intent to destroy, in whole or in part, a national, ethnic, racial, or religious group:

(a) killing members of the group;
(b) causing serious bodily or mental harm to members of the group;
(c) deliberately inflicting on the group conditions of life calculated to bring about its physical destruction in whole or in part;
(d) imposing measures intended to prevent births within the group;
(e) forcibly transferring children of the group to another group.

It is not difficult to prove that population transfers have frequently led to enormous loss of life, in direct violation of article 2(a) or 2(c). As noted earlier, the expulsion and enforced flight of some fifteen million ethnic Germans caused the deaths of over two million of them, and there is ample evidence that numerous lead-

---

[12] 78 U.N.T.S. 277 (entered into force 12 January 1951). The Interntional Court of Justice held in its advisory opinion on Reservations to the Convention on the Prevention and Punishment of the Crime of Genocide that the principles underlying the Convention are declaratory of customary international law; 1951 I.C.J. 15.

ers of the Soviet Union, Poland, and Czechoslovakia intended that loss of life.[13]

Moreover, the traumatic experience of losing their homes and every link to the land where they were born and where their parents and grandparents were buried certainly also caused serious bodily and mental harm to the surviving members of the group, in violation of article 2(b). It is hardly tenable that those who order or carry out such expulsions do not intend their foreseeable consequences.

The first attempt after the war expressly to criminalize population transfers was taken in the context of the protection of civilians in armed conflict. Indeed, most population transfers have occurred in the context of war. Article 49 of the Fourth Geneva Convention of 1949 stipulates:

> Individual or mass forcible transfers, as well as deportations of protected persons from occupied territory to the territory of the Occupying Power or to that of any other country, occupied or not, are prohibited, regardless of their motive..... Nevertheless, the Occupying Power may undertake total or partial evacuation of a given area if the security of the population or imperative military reasons so demand. Such evacuations may not involve the displacement of protected persons outside the bounds of the occupied territory except when for material reasons it is impossible to avoid such displacement. Persons thus evacuated shall be transferred back to their homes as soon as hostilities in the area in question have ceased.... The Occupying Power shall not deport or transfer parts of its own civilian population into the territory it occupies.

In order to put some teeth into the convention, the drafters stipulated in article 146 that the high contracting parties must enact legislation providing effective penal sanctions for persons committing, or ordering to be committed, the grave breaches listed in article 147 of the convention. The category of "grave breaches" includes the unlawful deportation or transfer of persons.

---

[13] According to former U.S. statesman and professor George Kennan: "The disaster that befell this area [Eastern Germany] with the entry of the Soviet forces has no parallel in modern European experience. There were considerable sections of it where ... scarcely a man, woman or child of the indigenous population was left alive after the initial passage of the Soviet forces; and one cannot believe that they all succeeded in fleeing to the West." George F. Kennan, *Memoirs, 1925-1950* (Boston, 1967), 265.

The prohibitions spelled out in the Fourth Geneva Convention apply, in principle, only in situations of international warfare. In situations of armed conflict not of an international character, article 3 of the convention stipulates that the high contracting parties must respect, and suppress the violation of certain minimum rules; notably, to treat humanely all persons taking no active part in the hostilities, to spare them violence to life and person, to refrain from hostage taking, and to refrain from committing outrages upon personal dignity. Deportation surely falls within article 3.

It took nearly three more decades to codify the prohibition of forced removal of civilians in internal armed conflicts. Under Additional Protocol II (1977) to the 1949 Geneva Conventions:

> The displacement of the civilian population shall not be ordered for reasons related to the conflict unless the security of the civilians involved or imperative military reasons so demand. Should such displacements have to be carried out, all possible measures shall be taken in order that the civilian population may be received under satisfactory conditions of shelter, hygiene, health, safety and nutrition.[14]

Ethnic Cleansing and the United Nations

The Universal Declaration of Human Rights is a supreme achievement in standard setting. It has been followed by at least fifty conventions, covenants, and protocols.

As the first High Commissioner for Human Rights repeatedly stated, mass population transfers violate the gamut of human rights. Among the hard law provisions that are frequently invoked against such population transfers is article 12 of the Covenant on Civil and Political Rights, which stipulates a right to freedom of movement, to reside in and to return to one's country.

Par. 4 of article 12 provides: No one shall be arbitrarily deprived of the right to enter his own country." In its commentary to this provision, the Human Rights Committee has stated:

> The right of a person to enter his or her own country recognizes the special relationship of a person to that country. The right has

---

[14] Protocol Additional to the Geneva Convention of 12 August 1949, and Relating to the Protection of Victims of Non-International Armed Conflicts (Protocol II), adopted 8 June 1977, art. 17, 1125 U.N.T.S. 609 (entered into force 7 December 1978). For the history of this article, see Howard Levie, The Law of Non-international Armed Conflict: Protocol II to the 1949 Geneva Conventions 529/43 (1987).

various facets. It implies the right to remain in one's own country. It includes not only the right to return after having left one's own country; it may also entitle a person to come to the country for the first time if he or she was born outside the country (for example, if that country is the person's State of nationality). The right to return is of the utmost importance for refugees seeking voluntary repatriation. It also implies prohibition of enforced population transfers or mass expulsions to other countries.[15]

Thus, according to the most prestigious United Nations expert Committee on human rights, enforced population transfers are illegal, and all refugees and expellees are entitled to voluntary repatriation.

In the context of the expulsion of 175,000 ethnic Greeks from northern Cyprus by the Turkish Army in 1974 and more recently during the war in the former Yugoslavia, the Security Council, the General Assembly and the United Nations Commission on Human Rights have repeatedly condemned the practice of terrorizing the civilian population to force them to flee and the actual mass expulsions. Such practices have been condemned not just as violations of general international law, but also as international crimes.

The most recent "hard law" development in the area of criminalizing population transfers is manifested in the Statute of the International Criminal Court, which was adopted in Rome on 17 July 1998 by the United Nations Diplomatic Conference of Plenipotentiaries on the Establishment of an International Criminal Court." It was stipulated that "deportation or forcible transfer of population" constitutes crimes against humanity under article 7, and that "unlawful deportation or transfer" constitutes war crimes under article 8 of the Statute.[16]

The most important "soft law" development is contained in the final report of Special Rapporteur Awn Shawkat Al-Khasawneh (a distinguished jurist from Jordan, member of the International Law Commission, and since 1999 a Judge at the International Court of Justice at the Hague) to the United Nations

---

[15] Human Rights Committee, General Comment No. 27, adopted on 18 October 1999, Official Records of the General Assembly, Report of the Human Rights Committee, Supplement No. 40 (A/55/40), vol. I.

[16] Rome Statute for the International Criminal Court (A/Conf.183/9, 1998). See also M. Cherif Bassiouni, ed., *The Statute of the International Criminal Court. A Documentary History* (Ardsley, N.Y., 1998), 41f. The Rome Statute entered into force on 1 July 2002.

Sub-Commission on the Promotion and Protection of Human Rights.

This Sub-Commission has traditionally played a seminal role by undertaking studies that have shaped the thinking of policy makers and contributed not only to progressive standard-setting but also to the development of monitoring mechanisms and preventive strategies. For instance, following the 1977 Sub-Commission Study by Francesco Capotorti on The Rights of Persons belonging to Ethnic, Religious and Linguistic Minorities,[17] a working group was set up to draft the Declaration on the Rights of Persons Belonging to National or Ethnic, Religious, and Linguistic Minorities,[18] which was adopted by the General Assembly in 1992 and led to the establishment of the Commission's Working Group on Minorities.

True to its mandate, the Sub-Commission designated in 1992 two of its members, Mr. Awn Shawkat Al-Khasawneh and Mr. Ribot Hatano (Japan), as Special Rapporteurs, entrusting them with a study on the human rights dimensions of population transfers, including the implantation of settlers.[19] In their preliminary report (E/CN.4/Sub.2/1993/17 and Corr.1) Al-Khasawneh and Hatano found that forced population transfers were *prima facie* unlawful and violated important provisions of humanitarian law and human rights law vis-à-vis both the transferred and the receiving populations. The study was carried on by Mr. Al-Khasawneh, who submitted a progress report in 1994 (E/CN.4/Sub.2/1994/18 and Corr.1).

From 17 to 21 February 1997 an expert seminar was held in Geneva to assist the Rapporteur in preparing his final report. At the opening of the expert meeting on 17 February 1997 the then High Commissioner for Human Rights, José Ayala Lasso, said:

> The right to live in one's native land is a very precious and fundamental right. Compulsory population transfers, including the implantation of settlers and settlements, are a serious matter, not only because they affect many people, but also because they violate the whole gamut of civil and political rights, economic, social and cultural rights. Let us remember, human rights are not exercised in a vacuum, but quite concretely where one lives. Expulsion by its very nature deprives victims of the exercise of many

---

[17] UNP Sales No. E. 91.XIV.2.

[18] GA Res. 47/135, U.N. GAOR, 47th Session, Supp. No. 49, at 210, U.N. Doc. A/47/49 (1992).

[19] Resolution 1992/28 of 27 August 1992.

rights and is frequently accompanied by physical abuses and even by the ultimate violation of the right to life.[20]

In his final report, to the forty-ninth session of the Sub-Commission in July 1997 (E/CN.4/Sub.2/1997/23 and Corr.1), Special Rapporteur Al-Khasawneh affirms the fundamental human right to live and remain in one's homeland as a prerequisite to the enjoyment of other rights. The report observes that

> collective expulsions or population transfers usually target national, ethnic, religious or linguistic minorities and thus, *prima facie*, violate individual as well as collective rights contained in several important international human rights instruments, in particular the International Covenant on Civil and Political Rights, the International Covenant on Economic, Social and Cultural Rights, the Convention on the Elimination of All Forms of Racial Discrimination and the Convention on the Rights of the Child .... Specific rights which population transfers violate include the right to self-determination, the right to privacy, family life and home, the prohibition on forced labor, the right to work, the prohibition of arbitrary detention, including internment prior to expulsion, the right to nationality as well as the right of a child to a nationality, the right to property or peaceful enjoyment of possession, the right to social security, and protection from incitement to racial hatred or religious intolerance.[21]

In its conclusions Al-Khasawneh's report underlines the fact that forced population transfers not only violate international law but also engage State responsibility and the criminal responsibility of individuals.[22] The report annexes a "Draft Declaration on Population Transfer and the Implantation of Settlers," which, like the Declaration on the Rights of Persons Belonging to National or Ethnic, Religious and Linguistic Minorities, should one day be adopted by the General Assembly. Al-Khasawneh's draft declaration has enormous relevance not only to ethnic cleansing but also to the phenomenon of internal displacement. It is worth quoting several of its provisions:

---

[20] Unpublished conference papers available from the Office of the High Commissioner for Human Rights, Geneva. A German translation of excerpts of the statement was published in A. de Zayas, *Heimatrecht ist Menschenrecht* (Munich, 2001).

[21] Al-Khasawneh Report, C/CN.4/Sub.2/1997/23, paras. 14 and 15.

[22] Ibid., para. 65.

[Art. 4 stipulates] 1. Every person has the right to remain in peace, security and dignity in one's home, or on one's land and in one's country. 2. No person shall be compelled to leave his place of residence. 3. The displacement of the population or part thereof shall not be ordered, induced or carried out unless their safety or imperative military reasons so demand. All persons thus displaced shall be allowed to return to their homes, lands, or places of origin immediately upon cessation of the conditions which made their displacement imperative.

[Article 7 stipulates] Population transfers or exchanges of population cannot be legalized by international agreement when they violate fundamental human rights norms or peremptory norms of international law.

[Article 8 sets forth the remedies available to victims] Every person has the right to return voluntarily, and in safety and dignity, to the country of origin and, within it, to the place of origin or choice. The excercise of the right to return does not preclude the victim's right to adequate remedies, including restoration of properties of which they were deprived in connection with or as a result of population transfers, compensation for any property that cannot be restored to them, and any other reparations provided for in international law.

[Article 10 places specific obligations on all States] Where acts or omissions prohibited in the present Declaration are committed, the international community as a whole and individual States, are under an obligation: (a) not to recognize as legal the situation created by such acts; (b) in ongoing situations, to ensure the immediate cessation of the act and the reversal of the harmful consequences; (c) not to render aid, assistance or support, financial or otherwise, to the State which has committed or is committing such act in the maintaining or strengthening of the situation created by such act.

Not surprisingly, this draft declaration has been frequently quoted by political leaders of many countries as well as in academic circles. Whereas it undoubtedly constitutes an important step forward in standard-setting, it is, however, "soft law". Conceivably "hard law" in the form of a Protocol on the Right to One's Homeland could be added to the International Covenant on Civil and Political Rights, or a Convention on the Prevention and Punishment of the Crime of Mass Expulsion could be negotiated and adopted by the General Assembly. But no declaration, protocol or convention can effectively ban the occurrence of ethnic cleansing, unless preventive strategies are developed and effective implementation machinery is in place.

In his final report Al-Khasawneh makes an interesting recommendation that "the Sub-Commission should consider establishing a working group to monitor compliance with the declaration, in particular by developing early-warning and preventive
mechanisms and coordinating advisory services and technical assistance, as required."[23] Al-Khasawneh's report was subsequently
endorsed by the Commission on Human Rights and by the Economic and Social Council, but it has not yet been adopted by the
General Assembly.

It is interesting to note that the International Law Commission
in its Draft Code on Crimes Against the Peace and Security of
Mankind lists deportation or forcible transfer of population as a
"crime against humanity" under article 18 and as a "war crime"
under article 20. The ILC commentary observes that

> a crime of this nature could be committed not only in time of
> armed conflict but also in time of peace ...[Deportation] implies
> expulsion from the national territory, whereas the forcible
> transfer of population could occur wholly within the frontiers
> of one and the same State.... Transfers of population under the
> draft article meant transfers intended, for instance, to alter a
> territory's demographic composition for political, racial, relig-
> ious or other reasons, or transfers made in an attempt to up-
> root a people from their ancestral lands. One member of the
> Commission was of the view that this crime could also come
> under the heading of genocide."[24] The ILC commentary further
> observes that "establishing settlers in an occupied territory
> constitutes a particularly serious misuse of power, especially
> since such an act could involve the disguised intent to annex
> the occupied territory. Changes to the demographic composi-
> tion of an occupied territory seemed to the Commission to be
> such a serious act that it could echo the seriousness of geno-
> cide.[25]

The International Criminal Tribunal for the Former Yugoslavia has indicted the former Serb leader Radovan Karadžić and the
Bosnian Serb military commander Ratko Mladić on counts of
genocide and crimes against humanity. Paragraph 19 of the in-

---

[23] Ibid. para. 72.

[24] Draft Code of Crimes against the Peace and Security of Mankind, Report of the
International Law Commmission on Its Forty-third Session, U.N. GAOR, 46th
session Supp. No. 10, at 268.

[25] Ibid. para. 271.

dictment charges them with the "unlawful deportation and transfer of civilians." Paragraph 25 specifically charges:

> Thousands of Bosnian Muslims and Bosnian Croats from the areas of Vlasenica, Prijedor, Bosanski Samac, Brcko and Foco, among others, were systematically arrested and interned in detention facilities established and maintained by the Bosnian Serb military, police and their agents and thereafter unlawfully deported or transferred to locations inside and outside of the Republic of Bosnia and Herzegovina. In addition, Bosnian Muslim and Bosnian Croat civilians, including women, children and elderly persons, were taken directly from their homes and eventually used in prisoner exchanges by Bosnian Serb military and police and their agents under the control and direction of Radovan Karadžić and Ratko Mladić. These deportations and others were not conducted as evacuations for safety, military necessity or for any other lawful purpose and have, in conjunction with other actions directed against Bosnian Muslim and Bosnian Croat civilians, resulted in a significant reduction or elimination of Bosnian Muslims and Bosnian Croats in certain occupied regions.[26]

Admittedly, neither Karadžić nor Mladić have as yet been arrested and brought before the Tribunal. But it is clear that the offence of forcible population transfers has been recognized by the international community to be an international crime for which political and military leaders are liable and for which there should be no impunity. The indictment on 27 May 1999 of the then President of Yugoslavia, Slobodan Milošević, for crimes committed in Kosovo, underlines this point.

As far as remedies for the victims, the Dayton Accords of December 1995 have provided in its annex VII for the refugees' right to return to their places of origin. Unfortunately, there has been only partial implementation of this section of the Dayton Accords. On the other hand, the return of ethnic Albanians to their homes in Kosovo gives hope to refugees and expellees the world over.

Regional International Law

On the regional level, collective expulsions violate several provisions of the European Convention for the Protection of Human Rights and Fundamental Freedoms. Article 1 binds the states parties to "secure to everyone within their jurisdiction the rights and freedoms" defined and guaranteed in the convention, which

---

[26] Prosecutor v. Karadzic Case No. IT-95-5-I (ICTY 25 July 1995).

largely tracks the Universal Declaration of Human Rights. Protocol 4 to the convention specifically provides:

1. No one shall be expelled, by means either of an individual or of a collective measure, from the territory of the State of which he is a national.
2. No one shall be deprived of the right to enter the territory of the State of which he is a national.[27]
The protocol also expressly prohibits the "collective expulsion of aliens."[28]

Expulsions would similarly violate many of the civil and political rights protected by the American Convention on Human Rights.[29] Most important in terms of the right to one's homeland are article 22(5), which provides that "no one can be expelled from the territory of the state of which he is a national or be deprived of the right to enter it," and article 22(9) which prohibits "the collective expulsion of aliens."

Likewise, the Banjul (African) Charter on Human and Peoples' Rights expressly prohibits the "mass expulsion of non-nationals," which is defined as deportation "aimed at national, racial, ethnic or religious groups."[30]

Regional Case-Law

Regional international jurisprudence also provides some hope for the victims. With regard to the expulsion by Turkish Cypriot forces of some 175,000 Cypriots of Greek ethnic origin from Northern Cyprus to Southern Cyprus, the European Commission on Human Rights found that "the transportation of Greek Cypriots to other places, in particular the excursions within the territory controlled by the Turkish army, and the deportation of Greek Cypriots to the demarcation line ... constitute an interference with

---

[27] Protocol No. 4 to the Convention for the Protection of Human Rights and Fundamental Freedoms, 26 September 1963, art. 3, Eur. T.S. 46 (entered into force 2 May 1968).

[28] Id., art. 4.

[29] American Convention on Human Rights, 1144 U.N.T.S. 123 (in force 18 July 1978).

[30] Banjul (African) Charter on Human and Peoples' Rights, adopted 27 June 1981, art. 12(5), O.A.U. Doc. CAB/LEG/67/3/Rev.5, in force 21 October 1986.

their private life, guaranteed in article 8(1) which cannot be justified on any ground under paragraph 8(2)." The Commission furthermore considered that the prevention of the physical possibility of the return of Greek Cypriots, transferred to the south of Cyprus under various intercommunal agreements, to the homes in the north of Cyprus amounts to an infringement of their right to respect of their homes as guaranteed in article 8(1). The Commission further noted that the acts violating the Convention were directed exclusively against members of the Greek Cypriot community and concluded that Turkey had failed to secure the rights and freedoms set forth in the Convention without discrimination on the grounds of ethnic origin, race, and religion as required by article 14 of the Convention.[31]

Still more significant is the Judgment of the European Court of Human Rights of 18 December 1996 in the *Loizidou v. Turkey* case, in which the Court held that the right to property of Mrs. Loizidou, a displaced person from Northern Cyprus, had been violated and in a further Judgment of 28 July 1998 ordered compensation in the in the amount of the equivalent of one million US dollars.

In Case 7964, the Inter-American Commission on Human Rights found that the refusal by Nicaragua to allow the Miskito Indians to return to their ancestral lands would amount to an impermissible restriction of movement and choice of residence in violation of Art. 22 of the American Convention on Human Rights.[32] Moreover, in the context of internal displacement of indigenous populations, the Inter-American Commission found that the spoliation of Brazilian Amerindios and their compulsory transfer into the "Yanomani Indian Park" violated their rights under the Convention.[33]

These regional decisions and reports further illustrate that forced population transfers are universally deemed to be illegal and that the violation of the right to one's homeland is justiciable.

---

[31] In 1983 the European Commission again dealt with the issue and found that the displacement of persons, separation of families and discrimination violated the European Convention. See Christa Meindersma, "Population Transfers in Conflict Situations," *Netherlands International Law Review*, 1994, 31-83 at 71-72.

[32] Report on the Situation of Human Rights of a Segment of the Nicaraguan Population of Miskito Origin, OEA/Ser.L/V/III.62 (1984).

[33] Report of 5 March 1985, Annual Report of the Inter-American Commission on Human Rights, 1984-85, OEA/Ser.L/V/II.66, doc. 10 rev. 1, Res. 12/85, Case 7615 (Brazil).

Conclusion

As already noted in the study of Special Rapporteur Al-Khasawneh, the norms and the jurisprudence on the illegality of forced population transfers are consistent. It has now become an imperative to develop an international machinery for ensuring compliance with these norms, to prevent future outbreaks of "ethnic cleansing" and for making effective remedies available to the victims. A clearly demonstrated political will on the part of the international community is a prerequisite for translating law into action and for eradicating the phenomenon of forced population transfers.

DOCUMENTARY APPENDIX

Selected Documents on Ethnic Cleansing in Twentieth-Century Europe

1.
The Lausanne Treaty and the Greek-Turkish Exchange of Populations

Treaty of Peace with Turkey Signed at Lausanne, July 24, 1923[1] — Convention Concerning the Exchange of Greek and Turkish Populations, Signed at Lausanne, January 30, 1923

The Government of the Grand National Assembly of Turkey and the Greek Government have agreed upon the following provisions:

Article 1
As from the 1st May, 1923, there shall take place a compulsory exchange of Turkish nationals of the Greek Orthodox religion established in Turkish territory, and of Greek nationals of the Moslem religion established in Greek territory.

These persons shall not return to live in Turkey or Greece respectively without the authorisation of the Turkish Government or of the Greek Government respectively.

Article 2
The following persons shall not be included in the exchange provided for in Article 1: a) The Greek inhabitants of Constantinople. b) The Moslem inhabitants of Western Thrace.

All Greeks who were already established before the 30th October, 1918, within the areas under the Prefecture of the City of Constantinople, as defined by the law of 1912, shall be considered as Greek inhabitants of Constantinople.

Moslems established in the region to the east of the frontier line laid down in 1918 by the Treaty of Bucharest shall be considered as Moslem inhabitants of Western Thrace.

Article 3
Those Greeks and Moslems who have already, and since the 18th October, 1912, left the territories the Greek and Turkish inhabitants of which are to be respectively exchanged, shall be considered as included in the exchange provided for in Article 1.

The expression "emigrant" in the present Convention includes all physical and juridical persons who have been obliged to emigrate or have emigrated since the 18th October, 1912.

---

[1] *The Treaties of Peace 1919-1923*, 2 (New York, 1924).

Article 4
All able-bodied men belonging to the Greek population, whose families have already left Turkish territory, and who are now detained in Turkey, shall constitute the first instalment of Greeks sent to Greece in accordance with the present Convention.

Article 5
Subject to the provisions of Articles 9 and 10 of the present Convention, the rights of property and monetary assets of Greeks in Turkey or Moslems in Greece shall not be prejudiced in consequence of the exchange to be carried out under the present Convention.

Article 6
No obstacle may be placed for any reason whatever in the way of the departure of a person belonging to the populations which are to be exchanged. In the event of an emigrant having received a definite sentence of imprisonment, or a sentence which is not yet definitive, or of his being the object of criminal proceedings, he shall be handed over by the authorities of the prosecuting country to the authorities of the country whither he is going, in order that he may serve his sentence or be brought to trial.

Article 7
The emigrants will lose the nationality of the country which they are leaving, and will acquire the nationality of the country of their destination, upon their arrival in the territory of the latter country.

Such emigrants as have already left one or other of the two countries and have not yet acquired their new nationality, shall acquire that nationality on the date of the signature of the present Convention.

Article 8
Emigrants shall be free to take away with them or to arrange for the transport of their movable property of every kind, without being liable on this account to the payment of any export duty or any other tax.

Similarly, the members of each community (including the personnel of mosques, tekkes, meddresses, churches, convents, schools, hospitals, societies, associations and juridical persons, or other foundations of any nature whatever) which is to leave the territory of one of the Contracting States under the present Convention, shall have the right to take away freely or to arrange for the transport of the movable property belonging to their communities.

The fullest facilities for transport shall be provided by the authorities of the two countries, upon the recommendation of the Mixed Commission provided for in Article 11.

Emigrants who may not be able to take away all or part of their movable property can leave it behind. In that event, the local authorities shall be required to draw up, the emigrant in question being given an opportunity to be heard, an inventory and valuation of the property left by him. Procès-verbaux containing the inventory and the valuation of the movable property left by the emigrant shall be drawn up in four copies, one of which shall be kept by the local authorities, the second transmitted to the Mixed Commission provided for in Article 11 to serve as the basis for the liquidation provided for by Article 9, the third shall be handed to the Government of the country to which the emigrant is going, and the fourth to the emigrant himself.

Article 9
Immovable property, whether rural or urban, belonging to emigrants, or to the communities mentioned in Article 8, and the movable property left by these emigrants or communities, shall be liquidated in accordance with the following provisions by the Mixed Commission provided for in Article 11.

Property situated in the districts to which the compulsory exchange applies and belonging to religious or benevolent institutions of the communities established in a district to which the exchange does not apply, shall likewise be liquidated under the same conditions.

Article 10
The movable and immovable property belonging to persons who have already left the territory of the High Contracting Parties and are considered, in accordance with Article 3 of the present Convention, as being included in the exchange of populations, shall be liquidated in accordance with Article 9. This liquidation shall take place independently of all measures of any kind whatever, which, under the laws passed and the regulations of any kind made in Greece and in Turkey since the 18th October, 1912, or in any other way, have resulted in any restriction on rights of ownership over the property in question, such as confiscation, forced sale, &c. In the event of the property mentioned in this Article or in Article 9 having been submitted to a measure of this kind, its value shall be fixed by the Commission

provided for in Article 11, as if the measures in question had not been applied.

As regards expropriated property, the Mixed Commission shall undertake a fresh valuation of such property, if it has been expropriated since the 18th October, 1912, having previously belonged to persons liable to the exchange of populations in the two countries, and is situated in territories to which the exchange applies. The Commission shall fix for the benefit of the owners such compensation as will repair the injury which the Commission has ascertained. The total amount of this compensation shall be carried to the credit of these owners and to the debit of the Government on whose territory the expropriated property is situated.

In the event of any persons mentioned in Articles 8 and 9 not having received the income from property, the enjoyment of which they have lost in one way or another, the restoration of the amount of this income shall be guaranteed to them on the basis of the average yield of the property before the war, and in accordance with the methods to be laid down by the Mixed Commission. The Mixed Commission provided for in Article 11. when proceeding to the liquidation of Wakf property in Greece and of the rights and interests connected therewith, and to the liquidation of similar foundations belonging to Greeks in Turkey, shall follow the principles laid down in previous Treaties with a view to fully safeguarding the rights and interests of these foundations and of the individuals interested in them.

The Mixed Commission provided for in Article 11 shall be entrusted with the duty of executing these provisions.

Article 11
Within one month from the coming into force of the present Convention a Mixed Commission shall be set up in Turkey or in Greece consisting of four members representing each of the High Contracting Parties, and of Nations from among nationals of Powers which did not take part in the war of 1914-1918. The Presidency of the Commission shall be exercised in turn by each of these three neutral members.

The Mixed Commission shall have the right to set up, in such places as it may appear to them necessary, Sub-Commissions working under its order. Each such Sub-Commission shall consist of a Turkish member, a Greek member and a neutral President to be designated by the Mixed Commission. The Mixed Commission shall decide the powers to be delegated to the Sub-Commission.

Article 12

The duties of the Mixed Commission shall be to supervise and facilitate the emigration provided for in the present Convention, and to carry out the liquidation of the movable and immovable property for which provision is made in Articles 9 and 10.

The Commission shall settle the methods to be followed as regards the emigration and liquidation mentioned above.

In a general way the Mixed Commission shall have full power to take the measures necessitated by the execution of the present Convention and to decide all questions to which this Convention may give rise.

The decisions of the Mixed Commission shall be taken by a majority.

All disputes relating to property, rights and interests which are to be liquidated shall be settled definitely by the Commission.

Article 13

The Mixed Commission shall have full power to cause the valuation to be made of the movable and immovable property which is to be liquidated under the present Convention, the interested parties being given a hearing or being duly summoned so that they may be heard.

The basis for the valuation of the property to be liquidated shall be the value of the property in gold currency.

Article 14

The Commission shall transmit to the owner concerned a declaration stating the sum due to him in respect of the property of which he has been dispossessed, and such property shall remain at the disposal of the Government on whose territory it is situated.

The total sums due on the basis of these declarations shall constitute a Government debt from the country where the liquidation takes place to the Government of the country to which the emigrant belongs. The emigrant shall in principle be entitled to receive in the country to which he emigrates, as representing the sums due to him, property of a value equal to and of the same nature as that which he has left behind.

Once every six months an account shall be drawn up of the sums due by the respective Governments on the basis of the declarations as above.

When the liquidation is completed, if the sums of money due to both sides correspond, the accounts relating thereto shall be balanced. If a sum remains due from one of the Governments to the other Government after a balance has been struck, the debit balance shall be paid in cash. If the debtor Governments requests a postponement in making this payment, the Commission may grant such postponement, provided that the sum due be paid in three annuities at most. The Commission shall fix the interest to be paid during the period of postponement.

If the sum to be paid is fairly large and requires longer postponement, the debtor Government shall pay in cash a sum to be fixed by the Mixed Commission, up to a maximum of 20 per cent of the total due, and shall issue in respect of the balance loan certificates bearing such interest as the Mixed Commission may fix, to be said off within 20 years at most. The debtor Government shall assign to the service of these loans pledges approved by the Commission, which shall be administered and of which the revenues shall be encashed by the International Commission in Greece and by the Council of the Public Debt at Constantinople. In the absence of agreement in regard to these pledges, they shall be selected by the Council of the League of Nations.

Article 15
With a view to facilitating emigration, funds shall be advanced to the Mixed Commission by the States concerned, under conditions laid down by the said Commission.

Article 16
The Turkish and Greek Governments shall come to an agreement with the Mixed Commission provided for in Article 11 in regard to all questions concerning the notification to be made to persons who are to leave the territory of Turkey and Greece under the present Convention, and concerning the ports to which these persons are to go for the purpose of being transported to the country of their destination.

High Contracting Parties undertake mutually that no pressure direct or indirect shall be exercised on the populations which are to be exchanged with a view to making than leave their homes or abandon their property before the date fixed for their departure. They likewise undertake to impose on the emigrants who have left or who are to leave the country no special taxes or dues. No obstacle shall be placed in the way of the inhabitants of the districts excepted from the exchange under Article 2 exercising freely their right to remain in or return to those districts and to enjoy to the full their liberties and rights of property in Turkey and in Greece. This provision shall not be invoked as a motive for preventing the free alienation of property belonging to inhabitants of the said regions which are excepted from the exchange, or the voluntary departure of those among these inhabitants who wish to leave Turkey or Greece.

Article 17
The expenses entailed by the maintenance and working of the Mixed Commission and of the organisations dependent on it shall be borne by the Governments concerned in proportions to be fixed by the Commission.

Article 18
The High Contracting Parties undertakes to introduce in their respective laws such modifications as may be necessary with a view to ensuring the execution of the present Convention.

Article 19
The present Convention shall have the same force and effect as between the High Contracting Parties as if it formed part of the Treaty of Peace to be concluded with Turkey. It shall come into force immediately after the ratification of the said Treaty by the two High Contracting Parties.

In faith whereof, the undersigned Plenipotentiaries, whose respective full Powers have been found in good and due form, have signed the present Convention.

Done at Lausanne, the 30th January, 1923, in three copies, one of which shall be transmitted to the Greek Government, one to the Government of the Grand National Assembly of Turkey, and the third shall be deposited in the archives of the Government of the French Republic, which shall deliver certified copies to the other Powers signatory of the Treaty of Peace with Turkey.

(L.S.) E.K.Veniselos, (L.S.) D.Caclamanos, (L.S.) Ismet, (L.S.) Dr.Ryza Nour , (L.S.) Hassan

Protocol
The undersigned Turkish Plenipotentiaries, duly authorised to that effect, declare that, without waiting for the coming into force of the Convention with Greece of even date, relating to the exchange of the Greek and Turkish populations, and by way to exception to Article 1 of that Convention the Turkish Government, on the signature of the Treaty of Peace, will release the able-bodied men referred to in Article 4 of the said Convention, and will provide for their departure.

Done at Lausanne, the 30th January, 1923.
Ismet, Dr. Ryza Nour, Hassan

2.

Atlantic Charter: Joint Statement by President Roosevelt and Prime Minister Churchill, August 14, 1941[2]

The President of the United States of America and the Prime Minister, Mr. Churchill, representing His Majesty's Government in the United Kingdom, being met together, deem it right to make known certain common principles in the national policies of their respective countries on which they base their hopes for a better future for the world.

First, their countries seek no aggrandizement, territorial or other;

Second, they desire to see no territorial changes that do not accord with the freely expressed wishes of the peoples concerned;

Third, they respect the right of all peoples to choose the form of government under which they will live; and they wish to see sovereign rights and self government restored to those who have been forcibly deprived of them;

Fourth, they will endeavor, with due respect for their existing obligations, to further the enjoyment by all States, great or small, victor or vanquished, of access, on equal terms, to the trade and to the raw materials of the world which are needed for their economic prosperity;

Fifth, they desire to bring about the fullest collaboration between all nations in the economic field with the object of securing, for all, improved labor standards, economic advancement and social security;

Sixth, after the final destruction of the Nazi tyranny, they hope to see established a peace which will afford to all nations the means of dwelling in safety within their own boundaries, and which will afford assurance that all the men in all lands may live out their lives in freedom from fear and want;

---

[2] Published in *Department of State Executive Agreement Series*, No. 236.

Seventh, such a peace should enable all men to traverse the high seas and oceans without hindrance;

Eighth, they believe that all of the nations of the world, for realistic as well as spiritual reasons must come to the abandonment of the use of force. Since no future peace can be maintained if land, sea or air armaments continue to be employed by nations which threaten, or may threaten, aggression outside of their frontiers, they believe, pending the establishment of a wider and permanent system of general security, that the disarmament of such nations is essential. They will likewise aid and encourage all other practicable measures which will lighten for peace-loving peoples the crushing burden of armaments.

Franklin D. Roosevelt

Winston S. Churchill

3.
United Nations Declaration, January 1, 1942[3]

A Joint Declaration by the United States, the United Kingdom, the Union of Soviet Socialist Republics, China, Australia, Belgium, Canada, Costa Rica, Cuba, Czechoslovakia, Dominican Republic, El Salvador, Greece, Guatemala, Haiti, Honduras, India, Luxembourg, Netherlands, New Zealand, Nicaragua, Norway, Panama, Poland, South Africa, Yugoslavia[4]

The Governments signatory hereto,

Having subscribed to a common program of purposes and principles embodied in the Joint Declaration of the President of the United States of America and the Prime Minister of the United Kingdom of Great Britain and Northern Ireland dated August 14, 1941, known as the Atlantic Charter.

Being convinced that complete victory over their enemies is essential to defend life, liberty, independence and religious freedom, and to preserve human rights and justice in their own lands as well as in other lands, and that they are now engaged in a common struggle against savage and brutal forces seeking to subjugate the world,
  DECLARE:
 (1) Each Government pledges itself to employ its full resources, military or economic, against those members of the Tripartite Pact and its adherents with which such government is at war.

(2) Each Government pledges itself to cooperate with the Governments signatory hereto and not to make a separate armistice or peace with the enemies.

The foregoing declaration may be adhered to by other nations which are, or which may be, rendering material assistance and contributions in the struggle for victory over Hitlerism.

Done at Washington, January First, 1942

---

[3] From U. S. State Department, *A Decade of American Foreign Policy: Basic Documents, 1941-49.*

[4] Over the next three years, the original signatories to the United Nations Declaration were joined by Mexico, Ecuador, the Philippines, Peru, Ethiopia, Chile, Iraq, Paraguay, Brazil, Venezuela, Bolivia, Uruguay, Iran, Turkey, Colombia, Egypt, Liberia, Saudi Arabia, and France.

4.
Facsimile—Ilya Ehrenburg leaflet, "Ubei"—"Kill"

УБЕЙ!

Вот отрывки из трех писем, найденных на убитых немцах:

Управляющий Рейнгардт пишет лейтенанту Отто фон Шираху:

"Французов от нас забрали на завод. Я выбрал шесть русских из Минского округа. Они гораздо выносливей французов. Только один из них умер, остальные продолжают работать в поле и на ферме. Содержание их ничего не стоит и мы не должны страдать от того, что эти звери, дети которых может быть убивают наших солдат, едят немецкий хлеб. Вчера я подверг легкой экзекуции двух русских бестий, которые тайком пожрали снятое молоко, предназначавшееся для свиных маток..."

Матиас Цимлих пишет своему брату ефрейтору Генриху Цимлиху:

"В Лейдене имеется лагерь для русских, там можно их видеть. Оружия они не боятся, но мы с ними разговариваем хорошей плетью..."

Некто Отто Эссман пишет лейтенанту Гельмуту Вейганду:

"У нас здесь есть пленные русские. Эти типы пожирают дождевых червей на площадке аэродрома, они кидаются на помойное ведро. Я видел, как они ели сорную траву. И подумать, что это люди..."

Рабовладельцы они хотят превратить наш народ в рабов. Они вывозят русских к себе, издеваются, водят их голодом до безумия, до того, что, умирая, люди едят траву и червей, а поганый немец с тухлой сигарой в зубах философствует: "Разве это люди?.."

Мы знаем все. Мы помним все. Мы поняли: немцы не люди. Отныне слово "немец" для нас самое страшное проклятье. Отныне слово "немец" разряжает ружье. Не будем говорить. Не будем возмущаться. Будем убивать. ЕСЛИ ТЫ НЕ УБИЛ ЗА ДЕНЬ ХОТЯ БЫ ОДНОГО НЕМЦА, ТВОЙ ДЕНЬ ПРОПАЛ. Если ты думаешь, что за тебя немца убьет твой сосед, ты не понял угрозы. Если ты не убьешь немца, немец убьет тебя. Он возьмет твоих и будет мучить в своей окаянной Германии. Если ты не можешь убить немца пулей, убей штыком. Если на твоем участке затишье, если ты ждешь боя, убей немца до боя. Если ты оставишь немца жить, немец повесит русского человека и опозорит русскую женщину. ЕСЛИ ТЫ УБИЛ ОДНОГО НЕМЦА, УБЕЙ ДРУГОГО — НЕТ ДЛЯ НАС НИЧЕГО ВЕСЕЛЕЕ НЕМЕЦКИХ ТРУПОВ. Не считай дней. Не считай верст. Считай одно: убитых тобою немцев. Убей немца! — это просит тебя старуха-мать. Убей немца! — это кричит родная земля. НЕ ПРОМАХНИСЬ. НЕ ПРОПУСТИ. УБЕЙ!

Илья ЭРЕНБУРГ.

Facsimile of the leaflet authored by Ilya Ehrenburg, entitled: "Ubei"—"Kill." The leaflet was distributed by military authorities to Soviet troops as they approached Germany in late 1944, early 1945. Photo of the original courtesy of Alfred de Zayas and with his permission.

Documentary Appendix

5.

The AVNOJ Regulations Concerning the German Minority in Yugoslavia, 1944-1948[5]

The AVNOJ (*Antifašističko veče narodnog oslobodjenja Jugoslavije*; Anti-Fascist Council for the National Liberation of Yugoslavia) was founded by Partisan forces (and though including other parties, dominated by Yugoslav Communists) in Bihać, in Bosnia, at the end of 1942 . The body expanded its claim to be sole authority for the creation of a new Yugoslav government in 1943, naming Josip Broz Tito as commander of the Yugoslav armed forces and marginalizing the Yugoslav government-in-exile in London. In 1944 and 1945, the AVNOJ assumed the role of authoritative body which created a permanent government and bureaucracy.

Provisions of the AVNOJ, 1943
On 21 Nov. 1943, the AVNOJ issued the following provisions, "On the Deprivation of Civil Rights," which formed the legal basis for the treatment of the Germans in Yugoslavia in the following years:

1. All persons of German nationality living in Yugoslavia automatically lose their Yugoslavian citizenship as well as all civil rights.

2. The entire movable and immovable possessions of all persons of German nationality are confiscated by the state and henceforth its property.

3. Persons of German nationality are neither allowed to claim or exercise any rights, nor to use courts or other institutions for their personal or legal protection.

Provisions of the AVNOJ, 1944

The 1943 provisions formed the basis for the stipulations of the third meeting of the AVNOJ in Belgrade on 21 November 1944, which dealt with the "Transfer of Enemy Property into State

---

[5] Provisions selected and compiled by Peter Wassertheurer; translated by Simon Coles.

Property" and the deprivation of civil rights of persons of German nationality. These provisions were as follows:

Article 1
Upon the coming into force of this resolution, the following will become state property:

1. the entire property of the German Reich and its citizens situated in Yugoslavian territory;

2. the entire property of persons belonging to the German people, with the exeption of those Germans who have fought in the National Liberation Army, and the Yugoslavian Partisan units or those who are citizens of neutral states, who have not shown any hostility during the occupation.

3. the entire property of war criminals and their accomplicies, without any consideration to their citizenship and the property of each person, who have been condemned to give up their property in favour of the state by civil or military law courts.

....

Article 3
Property, in the sense of this act, is seen as: immovable goods, movable goods and rights such as the possession of land, house, furniture, forests, mining rights, enterprises with all fixtures and fittings and stock, bonds, jewelery, shares, companies, societies of all types, funds, beneficiary rights, modes of payment of all types, claims, shares of businesses and enterprises, copyright laws, industrial property rights, and all rights from the items mentioned above.

Article 1, clause 2 was interpreted by AVNOJ according to the law from the 8 June 1945, as follows:

1. The decision of the Anti-fascist Council of the National Liberation of Yugoslavia of 21 Nov. 1944 (article 1, point 1) concerns those Yugoslavian citizens of German nationality who, during the occupation, declared themselves as Germans, or were known as such, disregarding if they had acted as such before the war, or had been considered assimilated Croats, Slovenians, or Serbs.
2. Not deprived of their civil rights or their property are Yugoslavian citizens of German nationality or German descent or with German surnames who, as partisans or soldiers, took part in the national fight for liberation or were active in the national liberation movement;

a. who, before the war, had been assimilated as Croats, Slovenians, or Serbs, and had neither joined the Cultural Union (*Kulturbund*) nor acted as members of the German ethnic group during the war;

b. who, during the occupation, refused to declare themselves members of the German ethnic group, even when demanded by the occupation or collaborator authorities;

c. who (be it man or woman), despite their German nationality, contracted a mixed marriage with a person of one of the Yugoslavian nationalities or a person of Jewish, Slovak, Ukrainian, Magyar, Romanian, or any other recognized nationality.

3. Persons who, during the occupation, offended against the fight for liberation of the Yugoslavian peoples through their behaviour and who were helpers of the occupation forces, are not entitled to the protection provided by the previous article, points....

Deprivation of Citizenship

Based on the provisions of the AVNOJ, the Provisional Plenary Assembly of the Democratic Federal Yugoslavia passed a citizenship law on 23 Aug. 1945 which stipulated in Article 16 that members of those nationalities "whose states had taken part in the war against the peoples of the Democratic Federal Yugoslavia, and who, during or before the war, by disloyal behaviour, had violated the national and public interests of the peoples of the Democratic Federal Yugoslavia, and hence their duty as citizens thereof," could be deprived of their Yugoslavian citizenship.

Thus, as a first step, all Germans who, since the autumn of 1943, had fled or had been driven away were deprived of their Yugoslavian citizenship. The collective deprivation of the Yugoslavian citizenship was then based on the Amendment of 1 Dec. 1948, which, in article 35, stipulated as follows:

All persons who, according to the valid regulations, were citizens of the Federal People's Republic of Yugoslavia on 28 Aug. 1945 are considered citizens of the Federal People's Republic of Yugoslavia. Not considered citizens of the Federal People's Republic of Yugoslavia are (...) persons of German nationality who live abroad and who, during or before the war, violated their civic duties through disloyal behaviour towards the national or public interests of the peoples of the Federal People's Republic of Yugoslavia.

Those Germans, however, who were kept in Yugoslavian camps until 1948 could be deprived of their Yugoslavian citizenship through decisions made by the Belgrade Ministry of the Interior or according to the Amendment to the Yugoslavian Citizenship Act of 1 July 1946.

6.

The Potsdam Declaration—Excerpt on Transfer of Populations

Article XIII from "Tripartite Agreement by the United States, the United Kingdom, and Soviet Russia concerning Conquered Countries, 2 August 1945"

Orderly Transfers of German Populations

The conference reached the following agreement on the removal of Germans from Poland, Czechoslovakia and Hungary:

The three Governments having considered the question in all its aspects, recognize that the transfer to Germany of German populations, or elements thereof, remaining in Poland, Czechoslovakia and Hungary, will have to be undertaken. They agree that any transfers that take place should be effected in an orderly and humane manner.

Since the influx of a large number of Germans into Germany would increase the burden already resting on the occupying authorities, they consider that the Allied Control Council in Germany should in the first instance examine the problem with special regard to the question of the equitable distribution of these Germans among the several zones of occupation. They are accordingly instructing their respective representatives on the control council to report to their Governments as soon as possible the extent to which such persons have already entered Germany from Poland, Czechoslovakia and Hungary, and to submit an estimate of the time and rate at which further transfers could be carried out, having regard to the present situation in Germany.

The Czechoslovak Government, the Polish Provisional Government and the control council in Hungary are at the same time being informed of the above and are being requested meanwhile to suspend further expulsions pending the examination by the Governments concerned of the report from their representatives on the control council.

7.
Anti-German and Anti-Hungarian Discriminatory Edicts, Decrees, and Statutes: The "Beneš Decrees"[6]

Presidential and Constitutional Edicts
Laws and Statutes
Government Decrees
Decrees of the Slovak National Council (Bratislava)
Ministerial Decrees
Decrees of the Slovak Commissioners (Bratislava)

Presidential and Constitutional Edicts

005/1945
Edict of the President of the Republic concerning the invalidity of transactions involving property rights from the time of the occupation and concerning the National Administration of property assets of Germans, Magyars, traitors and collaborators and of certain organizations and associations. (May 19, 1945)

012/1945
Edict of the President of the Republic concerning the confiscation and early re-allotment of agricultural property of Germans, Magyars, as well as traitors and enemies of the Czech and Slovak people. (June 21, 1945)

016/1945
Presidential edict concerning the establishment of special People's Courts for traitors and collaborators. (June 19, 1945)

017/1945
Presidential edict concerning People's Courts for unfaithful citizens. (June 19, 1945)

021/1945
Presidential edicts concerning legislative power during the time of transition. The president had temporary power to exercise legislative function. Reprint from the *Uredni Vestnik* (Official Gazette) in exile in London, England. (February 27, 1945)

---

[6] Compiled by Professor Charles Udvardy.

027/1945
Presidential edict concerning domestic colonization. (Colonization of the Slavic population in German and Hungarian districts). (June 27, 1945)

028/1945
Presidential edict concerning the settlement of Czech, Slovak or other Slavic farmers on the confiscated properties of Germans, Hungarians and other enemies of the state. (May 20, 1945)

033/1945
Presidential edict concerning the right of Czechoslovak citizenship. German and Hungarian nationals lost their citizenship. (August 2, 1945)

050/1945
Presidential edict concerning films. (August 11, 1945)

059/1945
Presidential edict concerning the repeal of civil servant appointments during the occupation. (August 20, 1945)

071/1945
Presidential edict concerning forced labor services of persons who had lost Czechoslovak citizenship. (September 19, 1945)

081/1945
Presidential edict concerning the dissolution of all German and Hungarian clubs and cultural, social and sports associations in Czechoslovakia. Their confiscated properties were transferred to the state and, in most cases, their libraries were destroyed. (September 25, 1945)

088/1945
Presidential edict concerning public labor. This edict ordered the deportation of the Hungarian nationals to the evacuated German districts in Bohemia. (October 1, 1945)

091/1945
Presidential edict freezing bank deposits belonging to Germans and Hungarians and prohibition of withdrawals even for personal expenses. Total losses suffered by the Hungarians in Czechoslovakia were estimated to be 1.102 billion Czech crowns as of July 16, 1948. (October 19, 1945)

100/1945
Presidential edict concerning the nationalization of mines and some other industrial plants. (October 24, 1945)

101/1945
Presidential edict concerning the nationalization of the feed industry. (October 24, 1945)

102/1945
Presidential edict concerning the nationalization of banks of stock corporations. (October 24, 1945)

103/1945
Presidential edict concerning the nationalization of private insurance companies. (October 24, 1945)

105/1945
Presidential edict concerning the purging committees reviewing civil servant activities. (October 24, 1945)

108/1945
Presidential edict concerning the confiscation of enemy property and the funds for national regeneration. Hungarian property was confiscated with the exception of their personal belongings. (October 25, 1945)

Presidential edicts concerning nationalization excluded all Hungarians from any compensation.

143/1945
Presidential edict concerning civil action limitations in criminal proceedings. (October 27, 1945)

Laws and Statutes

026/1946
Concerning voter lists. (February 21, 1946)

065/1946
Constitutional law concerning the National Constituent Assembly. It effectively abolished the franchise of Hungarians in Czechoslovakia. (April 11, 1946)

083/1946
Concerning the employment of Germans, Hungarians, traitors and collaborators. This law went so far as to terminate employment of Hungarians. (April 11, 1946)

128/1946
Concerning the nullification of all property transactions through which a Hungarian acquired property after September 29, 1938, the date of the Munich Four-Power Agreement. Subsequently such property, although legally transacted and fully paid by a Hungarian, was either returned to its previous non-Hungarian owner or transferred to the state. (May 16, 1946)

It is noteworthy that on February 12, 1942, four years after the first Vienna arbitral award, the Hungarian government concluded a bilateral treaty which compensated and thoroughly satisfied the individuals involved.

130/1946
Concerning the addenda and changes to Presidential edict 105/1945 dealing with Purging Committees. (May 16, 1946)

163/1946
Concerning extraordinary provisions which permitted the termination of a transaction between a Hungarian and a real estate owner. (July 18, 1946)

164/1946
Concerning relief to victims of war and fascist persecution. Hungarians became ineligible for relief due to the loss of their Czechoslovak citizenship, as a result of Presidential edict 033/1945. (July 18, 1946)

232/1946
Concerning the disenfranchisement of Czechoslovak citizens of ethnic Hungarian origin. Government decree 216/1946 also prohibited the election of a Hungarian to factory committees even in situations where almost all the workers in certain agricultural or industrial workplaces were Hungarian. Hungarians were excluded from trade unions in post World War II Czechoslovakia. (December 10, 1946)

247/1946
Concerning the modification of Presidential edict 105/1945 dealing with Purging Committees. (December 19, 1945)

252/1946
Concerning employee compensation in the event of employment loss as a result of confiscation or land reform. Hungarian workers held no claim to compensation. (December 20, 1946)

090/1947
Concerning legal procedures in the land registry office for the distribution of confiscated property. (May 8, 1947)

107/1947
Concerning provisions against unauthorized border crossings. (May 29, 1947)

114/1948
Concerning additional nationalization of industrial plants. (April 28, 1948)

115/1948
Concerning additional nationalization of feed industry plants. (April 28, 1948)

118/1948
Concerning nationalization of wholesale commerce. (April 28, 1948)

119/1948
Concerning nationalization of foreign trade and international shipping. (April 28, 1948)

120/1948
Concerning nationalization of enterprises of over fifty employees. (April 28, 1948)

121/1948
Concerning nationalization of the construction industry. (April 28, 1948)

122/1948
Concerning nationalization of travel agencies. (April 28, 1948)

123/1948
Concerning nationalization of printing shops. (April 28, 1948)

124/1948
Concerning nationalization of restaurants and hotels. (April 28, 1948)

125/1948
Concerning nationalization of spas. (April 28, 1948)

126/1948
Concerning nationalization of certain seed improvement enterprises. (April 28, 1948)

138/1948
Concerning landlord/tenant proceedings. This allowed for the cancellation of agreements with tenants regarded as disloyal from a state security standpoint. By May 1948, the implementation of this law in Pressburg (Bratislava) alone resulted in over four hundred Hungarian families receiving notices to vacate their premises with two to five hours' notice. Similar expulsions also occurred in the countryside. (April 28, 1948)

Government Decrees (Prague)

048/1945
Concerning Provisional National Assembly elections. This decree disenfranchised Czechoslovak citizens of Hungarian descent until 1949. (August 25, 1945)

216/1946
Concerning the enforcement of the provisions of decree 104/1945, enacted on August 23, 1945 by the Slovak National Council, regarding factory councils, excluding ethnic Hungarians from those councils. (November 5, 1946)

030/1948
Concerning the administration and distribution of property, belonging to Hungarians who were transferred to Hungary, among patriotic Czechoslovak citizenry. (March 19, 1948)

Decrees of the Slovak National Council (Bratislava)

006/1944
Concerning Hungarian school closings as well as the banning, in many places, of Catholic and Protestant religious services conducted in Hungarian. This decree was issued during the first

Slovak Republic (1939-1945) by the then illegitimate Slovak National Council in exile. (September 6, 1944)

004/1945
Concerning the confiscation and accelerated distribution of immovable landed property belonging to Germans, Hungarians, traitors and enemies of the Slovak nation. (February 27, 1945)

008/1945
Concerning the restriction on service in the armed forces to Czech, Slovak or Ukrainian nationals. (March 6, 1945)

016/1945
Concerning freezing bank deposits of Hungarian nationals. (March 23, 1945)

020/1945
Concerning granting authority to local industrial boards to review and cancel trade licenses to individuals considered to hold questionable political loyalty. (March 29, 1945)

026/1945
Concerning the prohibition of organizing administrative councils, called People's Councils (*Narodny Vybor*), in Hungarian populated villages, towns and districts. In these places, local government was executed by centrally appointed non-Hungarians organized as Administrative Commissions (*Spravna Komisia*) whose members were reliable Slovak communists who received their instructions directly from the Communist Party of Slovakia. (April 7, 1945)

033/1945
Criminalizing any political, economic and cultural activity having any connection with Hungarian government administration of former southern Slovakia subsequent to the September 1938 Munich Agreement. This decree also regulated procedures of the People's Courts in Slovakia. (May 15, 1945)

043/1945
Concerning rules for membership renewal for attorneys to the Bar of Slovakia. The Bar Association of Pressburg (Bratislava), the only one in Slovakia, refused membership applications from Hungarian lawyers, referring to the Yalta Conference resolutions. (May 25, 1945)

044/1945
Concerning civil servant employment and the dismissal of all Hungarian civil servants, with immediate effect or no later than July 31, 1945, without any claims or compensation, including the loss of retirement benefits. (May 25, 1945)

050/1945
Concerning the National (State) Administration to be established on properties owned by Hungarians, regarded collectively as politically unreliable from the point of view of the Czechoslovak state and the people's democracy. The resultant damage caused by the government-appointed Slovak or Czech administrators was enormous: at least 6120 administrators were imposed to oversee Hungarian properties, resulting in an estimated financial loss between 1945-1948 of 600 million Czech crowns. (June 5, 1945)

051/1945
Concerning the dissolution of Hungarian clubs and cultural, social and sports associations in Slovakia as well as the confiscation and transfer of Hungarian-owned property to the state and the destruction of Hungarian libraries. (May 25, 1945)

This decree was identical in content with Presidential edict 081/1945 of September 25, 1945.

052/1945
Concerning the nullification of all property transactions through which a Hungarian acquired property after September 28, 1938. (June 6, 1945).

This decree was identical to Law 128/1946.

062/1945
Concerning the freezing of bank deposits of Hungarians and the prohibition against withdrawals, even for personal expenses. (July 3, 1945)

Identical to Presidential edict 091/1945 of October 19, 1945.

067/1945
Concerning reporting of war damages. (July 3, 1945)

069/1945
Concerning the dismissal of all employees of Hungarian origin with immediate effect, without notice and without claim to compensation. (July 3, 1945)

082/1945
Concerning restricting legal and notarial professional practice to Slovaks. (July 25, 1945)

097/1945
Concerning the prohibition against compensation to Hungarians for war damages. (August 23, 1945)

099/1945
Concerning the dismissal of Hungarian civil servants. Only a very small percentage of discharged Hungarians received social relief of 1,000 Czech crowns, roughly twenty dollars. (August 23, 1945)

104/1945
Concerning the confiscation and accelerated distribution of immovable Hungarian-owned property without compensation. The objective was to insure that the confiscated property, including cultivated land, forests, livestock, farms and farm implements, would devolve to those considered to be politically reliable. These confiscation commissions, were involved in 4538 such cases between 1945 and 1948. (August 23, 1945)

105/1945
Concerning the establishment of labor camps for those considered to be unreliable. Enforcement responsibility was delegated to national committees at the local and county levels. (August 23, 1945)

107/1945
Concerning the provision of benefits to elderly, disabled and poor Czechoslovak citizens. Hungarians and stateless individuals were ineligible for consideration to receive social benefits. (August 23, 1945)

130/1945
Concerning compensation for war damages. See also decrees 67/1945 and 97/1945. Hungarians were ineligible to receive compensation, even though the destruction due to military action in southern Slovakia during 1944-1945 occurred in districts which were populated mainly by Hungarians. (November 15, 1945)

054/1946
Concerning the termination of agreements between Hungarians and landlords. See also laws 163/1946 and 138/1948. (April 23, 1946)

062/1946
Concerning the removal from office of all notaries public of Hungarian origin. (May 10, 1946)

064/1946
Concerning the modification of the confiscation and accelerated distribution of agricultural properties of Germans, Hungarians, traitors and enemies of the Slovak nation. (May 14, 1946)

065/1946
Concerning mortgaging of immovable property. (May 14, 1946)

069/1946
Addenda to decrees concerning the confiscation and accelerated distribution of Hungarian-owned property. (December 19, 1946)

005/1948
Concerning the recognition of bar examinations for judges and attorneys completed in Hungary for individuals not of Hungarian descent. (March 15, 1948)

Ministerial Decrees (Prague)

043/1945
Concerning the force of Presidential edict 004/1944 (in exile in London) on the National Councils and Provisional National Assembly. (August 3, 1945)

045/1945
Concerning the official powers and elections of the National Councils. Minister of the Interior. (August 24, 1945)

2139/1946
Concerning the partial release of frozen bank deposits. Minister of Finance. (December 6, 1946)

077/1948
Concerning the deadline for changes regarding eligibility to Czechoslovak citizenship. Minister of the Interior. (April 16, 1948)

Decrees of the Slovak Commissioners (Bratislava) and the Presidium of the Board of Commissioners (Provincial Government)

082/1948
Concerning compensation to employees who were terminated as a result of decrees of the Slovak National Council 104/1945 and 64/1946. (May 31, 1946)

109/1946
Concerning the discontinuation of compensation to retired miners who had their citizenship revoked on grounds of disloyalty to the state. (September 10, 1946)

Commissioner of the Interior

253/1945
Concerning the regulation of the status of the Lutheran Church in Slovakia. (September 10, 1945)

287/1945
Concerning the regulation of Czechoslovak citizenship in accordance with Presidential edict 033/1945 dated August 2, 1945. (October 22, 1945)

297/1945
Concerning the issuance to any Hungarian of the certificate of political reliability. This certificate was required to seek employment in post-World War II Czechoslovakia. (November 12, 1945)

20000/1946
Concerning the forced Slovakization of Hungarians in Slovakia, referred to as Reslovakization. In addition to dispersion, expulsion and transfer, a segment of the Hungarian population was forced to solemnly declare itself as Slovak. This was the reason for the establishment of so-called Reslovakization Commissions throughout southern Slovakia by the Commissioner of the Interior. (June 17, 1946)

126/1948
Concerning a nationality requirement for inclusion in the permanent voters list. (January 23, 1948)

A—311/18-II   /3-1948
Contains a long list of places whose names had been "Slavified."
(June 11, 1948)

Commissioner of Industry and Commerce

1104/1946
Concerning the establishment of a national governmental agency
overseeing patent and intellectual property rights and protections
for Hungarians, considered by the regime to be people of
questionable reliability. See also Presidential edict 005/1945 and
Slovak National Council decree 050/1945. (May 8, 1946)

Commissioner of Social Welfare

751/1946
Concerning the ineligibility to receive social benefits of disabled
war veterans, war widows and orphans of Hungarian descent due
to the collective revocation of their Czechoslovak citizenship (see
Presidential edict 033/1945). (March 13, 1946)

8.

Facsimile—Charter of the German Expellees

CHARTA DER DEUTSCHEN HEIMATVERTRIEBENEN

Im Bewußtsein ihrer Verantwortung vor Gott und den Menschen,
im Bewußtsein ihrer Zugehörigkeit zum christlich-abendländischen Kulturkreis,
im Bewußtsein ihres deutschen Volkstums und in der Erkenntnis der gemeinsamen Aufgabe aller europäischen Völker
haben die erwählten Vertreter von Millionen Heimatvertriebener, nach reiflicher Überlegung und nach Prüfung ihres Gewissens
beschlossen, dem Deutschen Volk und der Weltöffentlichkeit gegenüber eine

feierliche Erklärung

abzugeben, die die Pflichten und Rechte festlegt, welche die deutschen Heimatvertriebenen als ihr Grundgesetz und als unumgängliche Voraussetzung für die Herbeiführung eines freien und geeinten Europa ansehen.

1. Wir Heimatvertriebenen verzichten auf Rache und Vergeltung. Dieser Entschluß ist uns ernst und heilig im Gedenken an das unendliche Leid, welches im besonderen das letzte Jahrzehnt über die Menschheit gebracht hat.

2. Wir werden jedes Beginnen mit allen Kräften unterstützen, das auf die Schaffung eines geeinten Europas gerichtet ist, in dem die Völker ohne Furcht und Zwang leben können.

3. Wir werden durch harte, unermüdliche Arbeit teilnehmen am Wiederaufbau Deutschlands und Europas.

Wir haben unsere Heimat verloren. Heimatlose sind Fremdlinge auf dieser Erde. Gott hat die Menschen in ihre Heimat hineingestellt. Den Menschen mit Zwang von seiner Heimat trennen, bedeutet ihn im Geiste töten.

Wir haben dieses Schicksal erlitten und erlebt. Daher fühlen wir uns berufen zu verlangen, daß das

Recht auf die Heimat

als eines der von Gott geschenkten Grundrechte der Menschheit anerkannt und verwirklicht wird.

Solange dieses Recht für uns nicht verwirklicht ist, wollen wir aber nicht zur Untätigkeit verurteilt beiseite stehen, sondern in neuen geläuterten Formen verständnisvollen und brüderlichen Zusammenlebens mit allen Gliedern unseres Volkes schaffen und wirken. Darum fordern und verlangen wir heute wie gestern:

1. Gleiches Recht als Staatsbürger, nicht nur vor dem Gesetz, sondern auch in der Wirklichkeit des Alltags.

2. Gerechte und sinnvolle Verteilung der Lasten des letzten Krieges auf das ganze deutsche Volk und eine ehrliche Durchführung dieses Grundsatzes.

3. Sinnvollen Einbau aller Berufsgruppen der Heimatvertriebenen in das Leben des Deutschen Volkes.

4. Tätige Einschaltung der deutschen Heimatvertriebenen in den Wiederaufbau Europas.

Die Völker der Welt sollen ihre Mitverantwortung am Schicksal der Heimatvertriebenen als der vom Leid dieser Zeit am schwersten Betroffenen empfinden.

Die Völker sollen handeln, wie es ihren christlichen Pflichten und ihrem Gewissen entspricht.

Die Völker müssen erkennen, daß das Schicksal der deutschen Heimatvertriebenen, wie aller Flüchtlinge, ein Weltproblem ist, dessen Lösung höchste sittliche Verantwortung und Verpflichtung zu gewaltiger Leistung fordert.

Wir rufen Völker und Menschen auf, die guten Willens sind, Hand anzulegen an das Werk, damit aus Schuld, Unglück, Leid, Armut und Elend für uns alle der Weg in eine bessere Zukunft gefunden wird.

Stuttgart, den 5. August 1950

Charta der Heimatvertriebenen, *Charter of the German Expellees, August 5, 1950. Facsimile of an original leaflet in possession of Alfred de Zayas and printed here with his permission.*

9.

United Nations Draft Declaration on Population Transfer and the Implantation of Settlers[7]

Article 1
This Declaration sets standards which are applicable in all situations, including peacetime, disturbances and tensions, internal violence, internal armed conflict, mixed internal-international armed conflict, international armed conflict and public emergency situations. The norms contained in this Declaration must be respected under all circumstances.

Article 2
These norms shall be respected by, and are applicable to all persons, groups and authorities, irrespective of their legal status.

Article 3
Unlawful population transfers entail a practice or policy having the purpose or effect of moving persons into or out of an area, either within or across an international border, or within, into or out of an occupied territory, without the free and informed consent of the transferred population and any receiving population.

Article 4
1. Every person has the right to remain in peace, security and dignity in one's home, or on one's land and in one's country. 2. No person shall be compelled to leave his place of residence. 3. The displacement of the population or parts thereof shall not be ordered, induced or carried out unless their safety or imperative military reasons so demand. All persons thus displaced shall be allowed to return to their homes, lands, or places of origin immediately upon cessation of the conditions which made their displacement imperative.

Article 5
The settlement, by transfer or inducement, by the Occupying Power of parts of its own civilian population into the territory it

---

[7] Source: Appendix to Sub-Commission on Prevention of Discrimination and Protection of Minorities, 49th session, Item 10 of the Provisional Agenda, *Freedom of Movement, Human rights and Population Transfer, Final Report of the Special Rapporteur, Mr. Al-Khasawneh* (E/CN.4/Sub.2/ 1997/ 23, 27 June 1997).

occupies or by the Power exercising de facto control over a disputed territory is unlawful.

Article 6
Practices and polices having the purpose or effect of changing the demographic composition of the region in which a national, ethnic, linguistic, or other minority or an indigenous population is residing, whether by deportation, displacement, and/or the implantation of settlers, or a combination thereof, are unlawful.

Article 7
Population transfers or exchanges of population cannot be legalized by international agreement when they violate fundamental human rights norms or peremptory norms of international law.

Article 8
Every person has the right to return voluntarily, and in safety and dignity, to the country of origin and, within it, to the place of origin or choice. The exercise of the right to return does not preclude the victim's right to adequate remedies, including restoration of properties of which they were deprived in connection with or as a result of population transfers, compensation for any property that cannot be restored to them, and any other reparations provided for in international law.

Article 9
The above practices of population transfer constitute internationally wrongful acts giving rise to State responsibility and to individual criminal liability.

Article 10
Where acts or omissions prohibited in the present Declaration are committed, the international community as a whole and individual States, are under an obligation: (a) not to recognize as legal the situation created by such acts; (b) in ongoing situations, to ensure the immediate cessation of the act and the reversal of the harmful consequences; (c) not to render aid, assistance or support, financial or otherwise, to the State which has committed or is committing such act in the maintaining or strengthening of the situation created by such act.

Article 11
States shall adopt measures aimed at preventing the occurrence of population transfers and the implantation of settlers, including the prohibition of incitement to racial, religious or linguistic hatred.

Article 12
Nothing in these articles shall be construed as affecting the legal status of any authorities, groups or persons involved in situations of internal violence, disturbances, tensions or public emergency.

Article 13
1. Nothing in these articles shall be construed to restrict or impair the provisions of any international humanitarian or human rights instruments. 2. In case of different norms applicable to the same situation, the standard offering maximum protection to persons and groups subjected to population transfers, shall prevail.

10.

Statement of the United Nations High Commissioner for Human Rights to the German Expellees at the Ceremony held at the Paulskirche, Frankfurt am Main, on 28 May 1995, on the Occasion of the Fiftieth Anniversary of the Expulsion of Ethnic Germans from Eastern and Central Europe, 1945-1948

At this historic Church of St. Paul many have already spoken about human rights and democracy. This is good, because our commitment to the *dignitas humana* needs reaffirmation everywhere and on every occasion.

Fifty years after the end of the Second World War, we see that new wars and grave human rights violations continue to take their toll in lives, cause major refugee movements, deprive men and women of their rights and render them homeless.

Also fifty years ago the United Nations Organization was founded with the noble aims of maintaining international peace and security and promoting and protecting human rights throughout the world. The Organization has worked hard, achieved many successes, but also experienced serious disappointments. The United Nations and I myself as High Commissioner for Human Rights will devote all of our energies to make these goals reality.

Over the past fifty years the General Assembly has adopted *inter alia* the Universal Declaration of Human Rights, the Covenant on Civil and Political Rights, the Covenant on Economic, Social and Cultural Rights, the Convention on the Elimination of all Forms of Racial Discrimination, and the Convention against Torture. In this perspective, it is clear that ethnic cleansing, expulsion and involuntary transfers of population violate many of the fundamental human rights enshrined in these Conventions.

The right not to be expelled from one's homeland is a fundamental right. The Sub-Commission on Prevention of Discrimination and Protection of Minorities is currently seised of the question of the human rights dimensions of population transfers. The newest report of Special Rapporteur Awn Shawkat Al-Khasawneh concludes that population transfers violate the human rights of both transferred and receiving populations (E/CN.4/Sub.2/1994/18).

The United Nations International Law Commission is also currently examining this important question. In Article 21 of the Draft Code of Crimes against the Peace and Security of Mankind the expulsion of persons from their homeland is referred to as a gross and systematic violation of human rights and as an international crime. In Article 22 of the Code population expulsions and collective punishments against the civilian population are listed among the gravest war crimes.

The most recent statement of the United Nations on the Right to the homeland was given on 26 August 1994 by the Sub-Commission, which in its Resolution 1994/24 affirmed the right of persons to remain in peace in their own homes, on their own lands and in their own countries. Moreover, the Resolution affirms the right of refugees and displaced persons to return in safety and dignity, to their country of origin.

I submit that if in the years following the Second World War the States had reflected more on the implications of the enforced flight and the expulsion of the Germans, today's demographic catastrophes, particularly those referred to as "ethnic cleansing," would, perhaps, not have occurred to the same extent.

In this context I should like to refer to the Charter of the German Expellees. It is good that men and women who have suffered injustice are prepared to break the vicious circle of revenge and reprisals and devote themselves in peaceful ways to seek the recognition of the right to the homeland and work toward reconstruction and integration in Europe. One day this peaceful approach will receive the recognition it deserves.

There is no doubt that during the Nazi occupation the peoples of Central and Eastern Europe suffered enormous injustices that cannot be forgotten. Accordingly they had a legitimate claim for reparation. However, legitimate claims ought not to be enforced through collective punishment on the basis of general discrimination and without a determination of personal guilt. In the Nuremberg and Tokyo trials the crucial principle of personal responsibility for crimes was wisely applied. It is worth while to reread the Nuremberg protocols and judgment.

Our goal remains the universal recognition of human rights, which are based on the principle of the equality of all human beings. Indeed, all victims of war and injustice deserve our respect and compassion, since every individual human life is precious. It is our duty to continue our endeavors in the name of the *dignitas humana.*

Jose Ayala Lasso
High Commissioner for Human Rights

Paulskirche, 28 May 1995

11.
Dayton Peace Accords Excerpts[8]

General Framework Agreement
Annex 6, Human Rights, Article 1
Annex 7, Refugees and Displaced Persons, Article 1

General Framework Agreement for Peace in Bosnia and Herzegovina

The Republic of Bosnia and Herzegovina, the Republic of Croatia and the Federal Republic of Yugoslavia (the "Parties"),

Recognizing the need for a comprehensive settlement to bring an end to the tragic conflict in the region,

Desiring to contribute toward that end and to promote an enduring peace and stability,

Affirming their commitment to the Agreed Basic Principles issued on September 8, 1995, the Further Agreed Basic Principles issued on September 26, 1995, and the cease-fire agreements of September 14 and October 5, 1995,

Noting the agreement of August 29, 1995, which authorized the delegation of the Federal Republic of Yugoslavia to sign, on behalf of the Republika Srpska, the parts of the peace plan concerning it, with the obligation to implement the agreement that is reached strictly and consequently,

Have agreed as follows:

Article I
The Parties shall conduct their relations in accordance with the principles set forth in the **United Nations Charter**, as well as the Helsinki Final Act and other documents of the Organization for Security and Cooperation in Europe. In particular, the Parties shall fully respect the sovereign equality of one another, shall settle disputes by peaceful means, and shall refrain from any action, by

---

[8] Source: *U. S. Department of State Dispatch, The Dayton Peace Accords, March 1996*, vol. 7, Supplement No. 1 (Washington, DC, 1996).

threat or use of force or otherwise, against the territorial integrity or political independence of Bosnia and Herzegovina or any other State.

Article II
The Parties welcome and endorse the arrangements that have been made concerning the military aspects of the peace settlement and aspects of regional stabilization, as set forth in the Agreements at **Annex 1-A** and **Annex 1-B**. The Parties shall fully respect and promote fulfillment of the commitments made in **Annex 1-A**, and shall comply fully with their commitments as set forth in **Annex 1-B**.

Article III
The Parties welcome and endorse the arrangements that have been made concerning the boundary demarcation between the two Entities, the Federation of Bosnia and Herzegovina and Republika Srpska, as set forth in the Agreement at **Annex 2**. The Parties shall fully respect and promote fulfillment of the commitments made therein.

Article IV
The Parties welcome and endorse the elections program for Bosnia and Herzegovina as set forth in **Annex 3**. The Parties shall fully respect and promote fulfillment of that program.

Article V
The Parties welcome and endorse the arrangements that have been made concerning the Constitution of Bosnia and Herzegovina, as set forth in **Annex 4**. The Parties shall fully respect and promote fulfillment of the commitments made therein.

Article VI
The Parties welcome and endorse the arrangements that have been made concerning the establishment of an **arbitration tribunal**, a **Commission on Human Rights**, a **Commission on Refugees and Displaced Persons**, a **Commission to Preserve National Monuments**, and **Bosnia and Herzegovina Public Corporations**, as set forth in the Agreements at Annexes 5-9. The Parties shall fully respect and promote fulfillment of the commitments made therein.

Article VII
Recognizing that the observance of human rights and the protection of refugees and displaced persons are of vital

importance in achieving a lasting peace, the Parties agree to and shall comply fully with the provisions concerning human rights set forth in Chapter One of the Agreement at **Annex 6**, as well as the provisions concerning refugees and displaced persons set forth in **Chapter One of the Agreement at Annex 7.**

Article VIII
The Parties welcome and endorse the arrangements that have been made concerning the implementation of this peace settlement, including in particular those pertaining to the civilian (non-military) implementation, as set forth in the Agreement at **Annex 10**, and the international police task force, as set forth in the Agreement at **Annex 11**. The Parties shall fully respect and promote fulfillment of the commitments made therein.

Article IX
The Parties shall cooperate fully with all entities involved in implementation of this peace settlement, as described in the Annexes to this Agreement, or which are otherwise authorized by the United Nations Security Council, pursuant to the obligation of all Parties to cooperate in the investigation and prosecution of war crimes and other violations of international humanitarian law.

Article X
The Federal Republic of Yugoslavia and the Republic of Bosnia and Herzegovina recognize each other as sovereign independent States within their international borders. Further aspects of their mutual recognition will be subject to subsequent discussions.

Article XI
This Agreement shall enter into force upon signature.

DONE at Paris, this [21st] day of [November] , 1995, in the Bosnian, Croatian, English and Serbian languages, each text being equally authentic.

For the Republic of Bosnia and Herzegovina
For the Republic of Croatia
For the Federal Republic of Yugoslavia
Witnessed by:
European Union Special Negotiator
For the French Republic
For the Federal Republic of Germany
For the Russian Federation
For the United Kingdom of Great Britain and Northern Ireland
For the United States of America

Annex 6–Human Rights

Article I—Fundamental Rights and Freedoms

The Parties shall secure to all persons within their jurisdiction the highest level of internationally recognized human rights and fundamental freedoms, including the rights and freedoms provided in the European Convention for the Protection of Human Rights and Fundamental Freedoms and its Protocols and the other international agreements listed in the **Appendix to this Annex**. These include:

(1) The right to life.

(2) The right not to be subjected to torture or to inhuman or degrading treatment or punishment.

(3) The right not to be held in slavery or servitude or to perform forced or compulsory labor.

(4) The right to liberty and security of person.

(5) The right to a fair hearing in civil and criminal matters, and other rights relating to criminal proceedings.

(6) The right to private and family life, home, and correspondence.

(7) Freedom of thought, conscience and religion.

(8) Freedom of expression.

(9) Freedom of peaceful assembly and freedom of association with others.

(10) The right to marry and to found a family.

(11) The right to property.

(12) The right to education.

(13) The right to liberty of movement and residence.

(14) The enjoyment of the rights and freedoms provided for in this Article or in the international agreements listed in the Annex to

this Constitution secured without discrimination on any ground such as sex, race, color, language, religion, political or other opinion, national or social origin, association with a national minority, property, birth or other status.

Annex 7--Refugees and Displaced Persons

Article I—Rights of Refugees and Displaced Persons

1. All refugees and displaced persons have the right freely to return to their homes of origin. They shall have the right to have restored to them property of which they were deprived in the course of hostilities since 1991 and to be compensated for any property that cannot be restored to them. The early return of refugees and displaced persons is an important objective of the settlement of the conflict in Bosnia and Herzegovina. The Parties confirm that they will accept the return of such persons who have left their territory, including those who have been accorded temporary protection by third countries.

2. The Parties shall ensure that refugees and displaced persons are permitted to return in safety, without risk of harassment, intimidation, persecution, or discrimination, particularly on account of their ethnic origin, religious belief, or political opinion.

3. The Parties shall take all necessary steps to prevent activities within their territories which would hinder or impede the safe and voluntary return of refugees and displaced persons. To demonstrate their commitment to securing full respect for the human rights and fundamental freedoms of all persons within their jurisdiction and creating without delay conditions suitable for return of refugees and displaced persons, the Parties shall take immediately the following confidence building measures:

(a) the repeal of domestic legislation and administrative practices with discriminatory intent or effect;

(b) the prevention and prompt suppression of any written or verbal incitement, through media or otherwise, of ethnic or religious hostility or hatred;

(c) the dissemination, through the media, of warnings against, and the prompt suppression of, acts of retribution by military, paramilitary, and police services, and by other public officials or private individuals;

(d) the protection of ethnic and/or minority populations wherever they are found and the provision of immediate access to these populations by international humanitarian organizations and monitors;

(e) the prosecution, dismissal or transfer, as appropriate, of persons in military, paramilitary, and police forces, and other public servants, responsible for serious violations of the basic rights of persons belonging to ethnic or minority groups.

4. Choice of destination shall be up to the individual or family, and the principle of the unity of the family shall be preserved. The Parties shall not interfere with the returnees' choice of destination, nor shall they compel them to remain in or move to situations of serious danger or insecurity, or to areas lacking in the basic infrastructure necessary to resume a normal life. The Parties shall facilitate the flow of information necessary for refugees and displaced persons to make informed judgments about local conditions for return.

5. The Parties call upon the United Nations High Commissioner for Refugees ("UNHCR") to develop in close consultation with asylum countries and the Parties a repatriation plan that will allow for an early, peaceful, orderly and phased return of refugees and displaced persons, which may include priorities for certain areas and certain categories of returnees. The Parties agree to implement such a plan and to conform their international agreements and internal laws to it. They accordingly call upon States that have accepted refugees to promote the early return of refugees consistent with international law.

Contributors

ARCHDUKE OTTO VON HABSBURG is the oldest son of Emperor-King Charles (r. 1916-1918), the last ruler of the former Austro-Hungarian Empire, and as such he is the head of the House of Habsburg-Lothringen [Habsburg-Lorraine]. He celebrated his ninetieth birthday on November 20, 2002, yet he is still active in European politics. He is President of the Paneuropa-Union (since 1973), represents Bavaria in the European Parliament in Strassburg (since 1979), and writes weekly columns for several dozen European and North American newspapers and periodicals. He is a native speaker of three languages (German, French, and Hungarian) and is fluent in several others, including English, Spanish, Italian, and to a lesser degree Czech and Croatian.

Otto von Habsburg is also a scholar and the author of three dozen books in several languages. Among them are books on European and international politics, and historical biographies of a number of his own ancestors. The most important of these are his biographies of Rudolf I (r. 1273-1291), the first Holy Roman Emperor of the Habsburg dynasty; Charles V (r. 1519-1558), whose empire was so vast that he rightfully claimed that "the sun never sets in my empire;" and Emperor Francis Joseph (r. 1848-1916) of Austria-Hungary, whose reign lasted for sixty-eight years.

Archduke Otto von Habsburg is a member of the French Academy, and of the Royal Academy of Ethics and Political Sciences of Madrid. He is the recipient of honorary doctorates from Loránd Eötvös University of Budapest and the Medical University of Pécs (Hungary). In recognition for his political and social activities, he has been awarded the German Charlemagne Prize, the Robert Schuman Gold Medal, the Konrad Adenauer Prize, the Louis Weiss Prize, the Grand Cross of the Order of Pope Gregory, the Bavarian Medal of Honor, and the Hungarian Medal of Honor.

He resides in the town of Pöcking, near Munich, while his seven children are scattered throughout the world, including the United States. Of his two sons, the older, Karl (b. 1961), is Austria's representative in the European Parliament, and the younger, George (b. 1964), is Hungary's ambassador-at-large, while also heading Archduke Otto's office in Budapest.

ANGI, JÁNOS, Ph.D., is Assistant Professor of History at Debrecen University, Hungary (1992), director of Multiplex Media–Debrecen University Press(1995-1999), co-editor of the scholarly quarterly, *Debreceni Szemle* [Debrecen Review], and editor of a series of history monographs and textbooks for Kossuth University in Debrecen. He is the author of several scholarly articles, co-author of a number of history textbooks, and the author of a forthcoming major monograph: *Catherine II [the Great] and the Problem of Enlightened Absolutism in Russia*. In addition to his native Hungarian, he is fluent in Russian and English.

BARBER, CHARLES M., Ph.D., is Professor Emeritus of History at Northeastern Illinois University in Chicago, where he taught from 1967 to 2000. He received his B.A. from Princeton University and his Ph.D. from the University of Wisconsin with thesis on Kurt von Schleicher. He has published articles on German-Americans in the *Yearbook of German-American Studies* and the German-American weekly, *Eintracht*. His other writings include articles on William Langer and North Dakota History ("A Diamond in the Rough: William Langer reexamined"), and a paper at the 1998 Northern Great Plains History Conference in Bismarck, ND ("The Real Mr. Smith was from North Dakota: William Langer vs. Frank Capra.") Dr. Barber also has been active as an ensemble singer and soloist with two German-language singing groups and has participated in several European concert tours with the *Rheinischer Gesang Verein*, and the German American Singers of Chicago. He is active in Illinois Democratic politics.

BARTA, RÓBERT, Ph.D., is Lecturer in History at Debrecen University in Hungary and a specialist in modern Hungarian and modern European diplomatic history. He is the author of a number of scholarly articles on the British Conservative Party, British foreign policy in the Balkans, Winston Churchill, and the Hungarian Numerus Clausus Act of 1921. His forthcoming monographs include "The History of the Hungarian Unity Party" and "Hungarian-British Relations in the Twentieth Century."

BLANKE, RICHARD, Ph.D., is Adelaide and Alan Bird Professor of History at the University of Maine. A graduate of the University of California-Berkeley, Dr. Blanke has published several books on various aspects of German-Polish relations, including *Prussian Poland in the German Empire, 1871-1900*(1981), *Orphans of Versailles: Germans in Western Poland, 1918-1939* (1993), *Polish-Speaking Germans?: Language and National Identity Among the Masurians Since 1871* (2001). He is currently working on the ethnic cleansing of Germans from East-Central Europe after World War II.

BRUNSTETTER, SCOTT, is a Ph.D. Candidate in International Studies at Old Dominion University in Norfolk, Virginia. He was a Fulbright Scholar at the German Council on Foreign Relations (DGAP) in Berlin during 2001 and 2002, working on a dissertation examining German Bundeswehr out-of-area operations and the politics of the Green party. He has taught German at West Virginia University, History at Old Dominion University, and Peacekeeping at the Joint Forces Staff College. He made many scholarly presentations and a number of media appearances to discuss international relations issues on German and American TV and radio. He holds an M.A. in History (1997) from West Virginia University and a B.A. in History from the Pennsylvania State University (1993).

BUSCHER, FRANK, Ph.D., is Professor of History at Christian Brothers University in Memphis, Tennessee. He received his Ph.D. in 1988 from Marquette University, where he had worked with Professor Mike Phayer. Dr. Buscher's publications *include The U.S. War Crimes Trial Program in Germany 1946-55* and articles in *German Studies Review, Holocaust and Genocide Studies, German History, Central European History,* and *Duquesne University Studies in History.* He has presented papers at numerous conferences. From 1992 until 2000 Buscher also worked as a contract historian for the Crimes Against Humanities and War Crimes Section of the Canadian Department of Justice. At present, his research interests focus on the Catholic Church and German expellees and refugees after the Second World War.

CARMICHAEL, CATHIE, Ph.D., teaches European history at the University of East Anglia in Norwich. She was a student at the London School of Economics and the University of Ljubljana before completing a Ph.D. at the University of Bradford in 1993. She is the author of articles on Southeast European history, two bibliographical studies on Croatia and Slovenia, and most recently a short volume entitled *Ethnic Cleansing in the Balkans: Nationalism and the Destruction of Tradition* (2002). She has also co-authored *Slovenia and the Slovenes: A Small State in the New Europe* (2000) with James Gow, and co-edited *Language and Nationalism in Europe* (2000) with Stephen Barbour. She is currently co-editing a book on Montenegrin National Identity which will be published in 2003.

CERONE, JOHN, J.D., is Executive Director of the War Crimes Research Office at American University's Washington College of Law, as well as an adjunct faculty member of the University's School of International Service. He has worked as a human rights legal advisor with the United Nations Mission in Kosovo, and as a legal consultant for the International Secretariat of Amnesty International. He has also served as Legal Advisor to the Attorney General of Sierra Leone in negotiations with the United Nations on the establishment of the Special Court for Sierra Leone. He holds a J.D. from Notre Dame Law School and an advanced law degree in Public International Law from New York University. Recent publications include "Minding the Gap: Outlining KFOR Accountability in Post-Conflict Kosovo" (*European Journal of International Law,* June 2001), and "Legal Constraints on the International Community's Responses to Gross Violations of Human Rights and Humanitarian Law" (*Human Rights Review,* September 2001).

CHÁSZÁR, EDWARD, Ph.D., is Professor Emeritus of Political Science at Indiana University of Pennsylvania, where he taught International Law, International Organizations, World Politics, and Latin American Politics. An internationally known political scientist, during his tenure at IUP he was also director of IUP's Minority Rights Research Program. A native of Hungary, he studied law and political science at Pázmány Péter University of Budapest. Following his emigration he earned an M.A. at Case Western Reserve University and a Ph.D. at George Washington University. Professor Chászár has been engaged in research on the human rights of national and ethnic minorities throughout his career, and he has also been an observer at several UN sessions devoted to human rights, including the session devoted to Soviet actions against the Hungarian Revolution in 1956. His books include *Decision in Vienna: The Czechoslovak-Hungarian Border Dispute of 1988* (1978), and *Hungarians in Czechoslovakia Yesterday and Today* (1988). The revised and enlarged edition of his book, *The International Problem of National Minorities*, was recently published by the Matthias Corvinus Publishing.

CLARK, ELIZABETH MORROW, Ph.D., is Assistant Professor in Department of History and Political Science at West Texas A&M University, Canyon, TX. Dr. Clark's research interests include Polish-German relations at the local and international levels, interwar Polish politics, *Aussiedler* and *Vertriebenen* questions, and the history of Gdańsk/Danzig. Her dissertation, "Poland and the Free City of Danzig, 1926-1927: Foundations for Reconciliation," explores the identity of the Free City and the relationship between Danzig and Warsaw. She completed her doctorate at the University of Kansas under Anna M. Cienciala. Dr. Clark has been a Fulbright Scholar to Poland, a Title VI Foreign Language and Area Studies Fellow, and a Woodrow Wilson Center Junior Scholar. She has published in *Polish Review*, and the *Rocznik polsko-niemiecki* and has contributed an essay *to Danzig, sein Platz in Vergangenheit und Gegenwart* (Warsaw: Friedrich Ebert Stiftung, 1998).

DREISZIGER, NÁNDOR F., Ph.D., is Professor of History at the Royal Military College of Canada. He has earned his graduate degrees at the University of Toronto. Since 1970 he has been teaching courses in modern European and Canadian history at RMC. He has published extensively on North American and East Central European subjects. Among the journals that featured his articles are *the Journal of Modern History, Canadian Historical Association Historical Papers, Canadian Slavonic Papers, New York History, War and Society, Canadian Review of Studies in Nationalism*, and the *Journal of Canadian Studies*. Over the years he has received numerous research grants, including several from SSHRCC, as well as a Senior Fellowship in Canadian Ethnic Studies. In recent years his research has focused on the wartime interaction of the state and immigrant ethnic minorities in North America. He is the founding editor of *Hungarian Studies Review* (1974-) and a recipient of the Officer's Cross of the Republic of Hungary (1993).

ELEFTHERIOU, ELENI, is a Lecturer at the University of Michigan, where she teaches International Human Rights. She is also writing her dissertation on international human rights law and state practice. She received her M.A. in National Security Studies from Georgetown University and is a Ph.D. Candidate in Political Science at Vanderbilt University. She is a past recipient of the National Security Education Fellowship, which she used to conduct field research on Cyprus. Her research interests include International Negotiations and Conflict Resolution, Ethnic conflict and Minority Rights, and Human Rights Education.

HÁMOS, LÁSZLÓ, is founding president of the Hungarian Human Rights Foundation in New York, which since the mid-1970s has devoted considerable attention to the problems of the Hungarian minorities in Romania, Yugoslavia, Czechoslovakia/Slovakia, and Carpatho-Ruthenia (Soviet Union, then Ukraine). During the tenure of the Orbán Government in Hungary (1998-2002), Hámos also served as an officially appointed adviser to Prime Minister Viktor Orbán on matters relating to Hungarian foreign policy and the issue of the Hungarian minorities in the surrounding states. He is the author or co-author of several testimonies and documentary collections on minority oppression in the above countries, prepared for the U.S. Congress and the U.S. State Department.

HARSÁNYI, NICOLAE, Ph.D., is Research Associate of the Center for Slavic, Eurasian, and East European Studies at the University of North Carolina at Chapel Hill. A graduate from the University of Timişoara, Romania, he obtained his M.A. from the same university, where he taught English. In 1989 he got his Ph.D. from the University of Cluj, Romania. An active participant in the Timişoara uprising on December 17-22, 1989, which brought down Ceauşescu's dictatorial regime, he was among the founding members of the "Timişoara Society," an NGO that had as its goal the defense of human rights in post-Communist Romania. From 1990 to 1993 he taught at the University of Michigan. In the 1993-1994 academic year he was a fellow at the National Humanities Center, in North Carolina, where he did research on populism and nationalism in twentieth-century Romania. Since 1995, he has taught courses on nationalism at North Carolina.

HAUSNER, KARL AND HERMINE, were both born and raised in the Sudetenland. They were expelled from their homeland just after World War II. He immigrated to the United States in 1952. Hermine Hausner's family ended up in Bavaria, where she and her future husband met in 1954, while he was there on a business trip. They saw each other only five times during weekends, became engaged, and two years later, after Hermine had received immigration papers, were married in New York at the Cathedral of St. Paul the Apostle. He worked for International Harvester until he lost his eyesight at age forty, the result of an untreated eye infection while performing forced labor before his expulsion. Karl and Hermine then started their own company, which produces medical instruments, and which they still operate. Their company eventually came to represent Siemens Medical, which Karl Hausner joined in 1964. He was a division manager and later a member of the executive. In addition to these activities, the Hausners own and run a dairy farm and an animal feed analysis company. According to Karl Hausner: "Our principles are human rights, the right to life, the right to liberty, the right to homeland, the right to property within a free market economy and Western culture." The Hausners currently devote much time to issues of expellee rights, church restoration in the Czech Republic, and related matters.

HAYDEN, LT. GEN. MICHAEL V., M.A., is Director of the National Security Agency. He entered active duty in 1969 after earning a bachelor's degree in history in 1967 and a master's degree in modern American history in 1969, both from Duquesne University. He has served as commander of the Air Intelligence Agency and Director of the Joint Command and Control Warfare Center. He has served in senior staff positions in the Pentagon, Headquarters U.S. European Command (Stuttgart), National Security Council, Washington, D.C., and the U.S. Embassy in the People's Republic of Bulgaria. Prior to his current assignment, the general served as deputy chief of staff for United Nations Command and U.S. Forces Korea, Yongsan Army Garrison.

HELFERT, ERICH A., DBA, is chairman and CEO of Modernsoft, Inc., developers of an advanced financial analysis software product, Financial Genome. He is also president of Helfert Associates, management consultants in executive development in financial/economic decision-making and shareholder value creation. Previously he had a corporate management career of twenty years, and before that served on the faculty of the Harvard Business School for seven years. He has published several books on finance, including the best-selling *Techniques of Financial Analysis*, 11th edition, with over 500,000 copies in print in ten languages. His first literary work, *Valley of the Shadow–After the Turmoil My Heart Cries No More*, describes his and his family's experiences during the ethnic cleansing of the Sudetenland in 1945 and 1946. A sequel, to be entitled *Journey to Forgiveness*, is in the writing stage. Dr. Helfert came to the United States as a student in 1950 and earned his advanced degrees (M.B.A. and D.B.A.) at Harvard University. He is an active public speaker about the postwar events in his former homeland, and about current financial topics. Dr. Helfert's web site (www.heleassoc.net) contains background material about his publications and other activities.

HUPCHICK, DENNIS P., Ph.D., is Associate Professor of History at Wilkes University in Pennsylvania. He is the author (or co-author) of seven books on Bulgarian, Balkan, and East European topics, including: *The Balkans: From Constantinople to Communism* (2002); *The Palgrave Concise Historical Atlases of the Balkans* (2001) and *Eastern Europe* (2001); *Conflict and Chaos in Eastern Europe* (1995); *Culture and History in Eastern Europe* (1994); and *The Bulgarians in the Seventeenth Century* (1993). Hupchick has edited (or co-edited) three other books treating Bulgarian and Hungarian history and has authored a number of articles and encyclopedia entries dealing with Bulgarian and Balkan ethnic, religious, and political cultural history. He is a former Fulbright (1989) and IREX (1976-77) research scholar to Bulgaria, past president of the Bulgarian Studies Association (1993-95), and past director of the Wilkes University East European and Russian Studies Minor Degree Program (1991-2002).

JESZENSZKY, GÉZA, Ph.D., was born in Budapest in 1941. He studied history, English, and library science at Eötvös Loránd University, Budapest, receiving an M.A. in 1966 and a Ph.D. in 1970. He was a teacher, then librarian, before joining, in 1976, the Budapest (formerly Karl Marx) University of Economic Sciences, where he was appointed Reader in the History of International Relations in 1981. He is the author of numerous scholarly publications, including a book on Hungary's changing image in the Anglo-Saxon world at the end of the nineteenth century and beginning of the twentieth. Committed to the ideas of the 1956 Hungarian Revolution, Dr. Jeszenszky became a founding member of the Hungarian Democratic Forum (1988), which challenged the Communist system and won the free elections in April 1990. He was Hungary's Minister for Foreign Affairs in the first post-Communist Hungarian Government, headed by József Antall (1990-1994). Following the elections of 1994 he became a Member of the Opposition in the Hungarian Parliament. From 1998 to 2002, he was Ambassador to the United States representing the government headed by Viktor Orbán. Dr. Jeszenszky has received many awards, among them the C.I.E.S. Fulbright Grant (1984-1986 at the University of California, Santa Barbara) and a Guest Scholar Grant from the Woodrow Wilson International Center for Scholars (1985). In 1996 he was Helen DeRoy Visiting Professor at the University of Michigan. He can be reached at geza@jeszen.hu

KAMUSELLA, TOMASZ, Ph.D., graduated in English Philology, English and European Studies from the University of Silesia, Katowice, Poland; Potchefstroom University, Potchefstroom, South Africa; and the Central European University, Prague, the Czech Republic. Since 1995 he has taught International Relations (in English) at Opole University and during the year 2002-2003 held a postdoctoral Jean Monnet fellowship at the European University Institute (Department of History and Civilization), Florence, Italy. In addition he has acted as the Regional Governor's Plenipotentiary on European Integration (1996-1999) and, later, as the Regional President's Advisor on Foreign Affairs (1999-2002). In 2001, he defended his Ph.D. dissertation, "The Emergence of the National and Ethnic Groups in Silesia, 1848-1918," in the Institute of Western Affairs (*Instytut Zachodni*), Poznań, Poland. He has published widely on the problem of ethnic relations in the multicultural region of Silesia, European integration, and the theory of ethnicity and nationalism.

KENT, MARTHA, Ph.D., has been a clinical neuropsychologist in hospital-based patient care for over two decades. Her formative experience includes four years of captivity in Poland before expulsion. Her formal education began at age nine in Germany and continued three years later in Alberta, Canada. Later, she attended the University of Michigan, Indiana University, and Michigan State University, where she earned a Ph.D. in psychology. Post-doctoral training in clinical neuropsychology followed later. As a neuropsychologist, she has developed evaluation and treatment programs for patients with neurological illnesses with sequelae affecting thinking, memory, and behavior. Her professional life has encompassed work with posttraumatic stress disorder and with populations that have included Vietnam era veterans, and former prisoners of war as well as work on torture with the Human Rights Clinic in Phoenix. Her academic teaching and research have been published in Volume III of "Primary Prevention of Psychopathology" and curriculum volumes of "The Vermont Competency Program." Her present writing is focused on recovery from trauma, the role of positive emotions, and the neurobiology of affiliation as an anti-stress system. She has written about her own experiences in the autobiographical work, *Eine Porzellanscherbe im Graben: Eine deutsche Flüchtlingskindheit* (Bern: Scherz, 2003).

KOPPER, CHRISTOPHER, Ph.D., is DAAD Professor of German Studies at the University of Pittsburgh. Born in 1964 in Bergisch Gladbach, Germany, he received his M.A. in Modern History at Ruhr-Universität Bochum in 1988 and his Ph.D. in Modern History at the same institution in 1992. His dissertation on German banking policies during the Third Reich was published in Germany in 1995.

LIEBERMAN, BEN, Ph.D., is Associate Professor at Fitchburg State College in Massachusetts. With a Ph.D. from the University of Chicago, he is the author of *From Recovery to Catastrophe: Municipal Stabilization and Political Crisis* (Berghahn Books, 1998) as well as several articles on Weimar politics. After teaching a course on the Holocaust for years, Dr. Lieberman is now conducting research on the history of ethnic cleansing.

LOHNE, RAYMOND, M.A., was born in Frankfurt, Germany in 1956. He holds a Master's Degree in History from Northeastern Illinois University and is currently a Ph.D. student in the History Department of the University of Illinois at Chicago. He is the author of *The Great Chicago Refugee Rescue* (Rockport, Maine: Picton Press, 1997), *German Chicago: The Danube Swabians and the American Aid Societies* (Charleston, SC: Arcadia Publishers, 1999), and *German Chicago Revisited* (Charleston, S.C., Arcadia Publishers, 2001). He is currently teaching at Columbia College of Chicago.

LUDÁNYI, ANDREW, Ph.D., is Professor of Political Science at Ohio Northern University, where he is teaching Comparative Politics and International Relations. Dr. Ludányi received his B.A. from Elmhurst College (Illinois) in 1963 and his M.A.(1965) and Ph.D.(1971) from Louisiana State University. His research has focused on interethnic and internationality relations in East Central Europe, with particular attention to developments in the Transylvanian region of Romania and the Voivodina region of northern Yugoslavia. He has published numerous articles and reviews and edited three books about this subject, including studies on nationalities policies in Titoist Yugoslavia as well as in Ceauşescu's Romania. These include *Transylvania: The Roots of Ethnic Conflict* (1983), and *Hungary and Hungarian Minorities* (2000).

MAZSU, JÁNOS, Ph.D., A Professor of History at Debrecen University in Hungary, Dr. Mazsu has also been Visiting Professor at Gent University in Belgium (1993), as well as at Indiana University in Bloomington (Spring 1999, Fall 2000). In the past he had served as Vice Mayor of the City of Debrecen (1991-1994), and since 1995 he has been the Program Director of Csokonai Press in the same city. His publications include two dozen articles, two co-edited books, and a major English-language monograph entitled *The Social History of the Hungarian Intelligentsia, 1825-1914* (1997).

MENTZEL, PETER, Ph.D., is Associate Professor of History at Utah State University. He earned his doctorate at the University of Washington. His articles have appeared in *Nationalities Papers, Arkeoloji ve Sanat*, and several edited volumes, most recently *Evolutionary Theory and Ethnic Conflict* (Praeger, 2001). He was guest editor of a special issue of *Nationalities Papers* devoted to "Muslim Minorities in the Balkans" (March, 2000). He was a Fulbright Senior Research Fellow in Istanbul during the 1998/1999 academic year. His work focuses on the social and political history of the Ottoman Empire.

MULAJ, KLEJDA, is a Ph.D. Candidate in International Relations at the London School of Economics and Political Science. Her thesis deals with ethnic cleansing in the former Yugoslavia during the 1990s. Her academic interests include conflict and peace studies with particular reference to the Balkan region, and the ethics of war and interstate relations.

NAGENGAST, EMIL, Ph.D., is Associate Professor of Politics at Juniata College in Huntingdon, Pennsylvania. He received his Ph.D. from the University of Pittsburgh in 1996. He has written several articles on German *Ostpolitik*. The focus of his research has been on the role of the expellees in German foreign policymaking. He is currently preparing a book manuscript about the evolution of German *Ostpolitik* in the twentieth century.

PELLÁTHY, GABRIEL S. J.D., Ph.D., is Professor of Political Science in the McKenna School of Business, Economics and Government, Saint Vincent College, Latrobe, Pennsylvania. He earned a Ph.D. in Political Science, an LL.M. in International Law from New York University, and a J.D. degree from the Cornell University School of Law. He has published a number of articles in his areas of specialization: Political Theory, East European Politics, and American Government.

PRUSIN, ALEXANDER V., Ph.D., is Assistant Professor of History at New Mexico Institute of Mining and Technology, Socorro. Dr. Prusin defended his doctoral dissertation at the University of Toronto in 2000. His research interests are focussed on modern European history, Russia, Eastern Europe, and international relations.

ROUDOMETOF, VICTOR, Ph.D., is Visiting Assistant Professor of Sociology at Miami University. He received his B.A. in Economics at University of Macedonia, Greece, in 1987, his M.A. in Sociology at Bowling Green State University in 1990, both an M.A. in History and his Ph.D. in Sociology and Cultural Studies at the University of Pittsburgh (1994 and 1996). His articles on globalization, nationalism, and national identity in the Balkans have been published in the *Journal of Modern Greek Studies, Diaspora, East European Quarterly, Mediterranean Quarterly, the Journal of Political and Military Sociology*, and the *European Journal of Social Theory*. He is the author of *Nationalism, Globalization and Orthodoxy: The Social Origins of Ethnic Conflict in the Balkans* (Westport, CT: Greenwood, 2001) and *Collective Memory, National Identity and Ethnic Conflict: Greece, Bulgaria and the*

Macedonian Question (Westport, CT: Praeger, 2002). He is also the editor or co-editor of several volumes on transnationalism, ethnic conflict, Americanization, and post-1989 Balkan politics.

SCHINDLER, JOHN R., Ph.D., works for the U.S. Department of Defense in Washington, DC, where he is also a researcher at the American Enterprise Institute for Public Policy Research. Additionally, he teaches History at the University of Maryland, Baltimore County. He holds three degrees in History—B.A. (Hons. 1990), M.A. (1991), University of Massachusetts; and Ph.D. (1995), McMaster University—and is an expert in the Balkans and East Central Europe and military and intelligence history. He is the author of numerous articles, as well as two books: *Isonzo: The Forgotten Sacrifice of the Great War* (2001), and, forthcoming, *Tito's Ghost: Inside the World's Most Dangerous Intelligence Agency*. He is currently writing his third book, *Redl: Spy of the Century*.

STARK, TAMÁS, Ph.D., is a senior research fellow at the Institute of History of the Hungarian Academy of Sciences. His specialization is forced population movements in East Central Europe in the period of 1938-1956, with special regard to the history of the Holocaust, the fate of prisoners of war and civilian internees, and the postwar migrations. He teaches courses on the history of the Holocaust at the Eötvös Lóránd University, and at the Jewish University in Budapest. He has held visiting fellowships at the United States Holocaust Memorial Museum, at the American Jewish Archives at Hebrew Union College in Cincinnati, and at the Institute of Contemporary Jewry at the Hebrew University in Jerusalem. His major publications include *Hungarian Jews During the Holocaust and after the Second World War, 1939-1949: A Statistical Review* (New York, 2000) and *Hungary's Human Losses in World War II* (Uppsala, 1995).

THUM, GREGOR, Ph.D., was born in Munich and studied History and Slavic Studies at the Freie Universität, Berlin, and in Moscow. In 2002, he defended his doctoral dissertation, a study of Wrocław/Breslau after the Second World War ("Die Fremde Stadt: Breslau nach dem Bevölkerungsaustausch") at the Europa-Universität Viadrina in Frankfurt/Oder, where he has been a scholarly associate (*wissenschaftlicher Mitarbeiter*) since 1995. He has published a number of articles and co-authored *Chronik russischen Lebens in Deutschland 1918-1941* (Berlin, 1999).

TOOLEY, T. HUNT, Ph.D., is Professor of History at Austin College in Sherman, Texas. He received his Ph.D. in Modern European History from the University of Virginia in 1986, after having earned his B.A. and M.A. degrees from Texas A&M University. A former DAAD/Fulbright Fellow, he specializes in the history of Central Europe in the twentieth century, particularly the German-Polish borderlands, and in broader issues of war, revolution, and peacemaking. He is the author of *National Identity and Weimar Germany: Upper Silesia and the Eastern Border, 1918-1922* (1997) and *The Western Front: Battle Ground and Home Front in the First World War* (2003). He has published numerous articles and reviews on the history of the modern world. He was co-organizer of the Conference on Ethnic Cleansing at Duquesne University in November 2000 and is co-editor of this volume.

VÁRDY, ÁGNES HUSZÁR, Ph.D., is Professor of Communications and Comparative Literature at Robert Morris University, and the author or co-author of eight books and over eighty articles and essays, among them two books on Austro-German Romanticism, and a social-historical novel, *Mimi*, that is now being used in history classes at several American universities. A frequent traveler and lecturer in Europe, Dr. Várdy is an invited member of the International P.E.N. (1985), and of the Hungarian Writers' Federation (1997), as well as a Board Member of the International Association of Hungarian Language and Culture, and of the Institute for German-American Relations. She is the recipient of Hungary's Berzsenyi Prize (1992) and of the Árpád Academy's Gold Medal (1998), in recognition of her scholarly achievements and for her promotion of Hungarian culture abroad. She is the associate editor of this volume.

VÁRDY, STEVEN BÉLA, Ph.D., is McAnulty Distinguished Professor of European History at Duquesne University, long-time Director of the Duquesne University History Forum, and former Chairman of the Department of History. He is an invited member of the Hungarian Writers' Federation and of the International P.E.N., Board Member of the World Federation of Hungarian Historians, and of the International Association of Hungarian Language and Culture, as well as the President of the Pittsburgh-based Institute for German American Relations. Professor Várdy is the author or coauthor of seventeen books and nearly five-hundred articles, essays, and reviews. His books deal with Hungary, the Habsburg Empire, historiography, liberalism, national minorities, and other related topics concerning Central and Southeastern Europe, as well as with American immigration. His writings have been published in about a dozen countries in several languages. Professor Várdy is the recipient of over two dozen fellowships and major research grants, including Duquesne University's "Presidential Award for Excellence in Scholarship" (1984), Hungary's Berzsenyi Prize (1992), the Árpád Academy's "Gold Medal"(1997), and the Officer's Cross awarded by the President of the Republic of Hungary (2001). He is the co-editor of this volume.

WESSERLE, ANDREAS ROLAND, Ph.D., earned his M.A. at Marquette University and his Ph.D. at the University of Southern Illinois. He has published articles and op-ed pieces on philology and modern history. He is retired and lives in Milwaukee.

WHEALEY, ROBERT H., Ph.D., retired in 2001 from his post as Associate Professor of History at Ohio University in Athens, where he had taught since receiving his doctorate from the University of Michigan in 1964. Still actively writing, he is a specialist on twentieth-century European diplomatic history and the history of modern Spain. His books include: *Hitler and the Spanish Civil War* (1989), and the recent *American Intervention in Yugoslavia: A Diplomatic History Since 1991* (Prometheus, 2002). His articles include: "Goebbels and the Spanish Civil War" (1999), and short biographical essays on General Francisco Franco (1995) and Admiral Wilhelm Canaris (1992).

WIGGERS, RICHARD DOMINIC, Ph.D., holds a B.A. in History and Journalism from Carleton University, an M.A. in History and International Relations from the University of Ottawa, and a Ph.D. in History and International Law from Georgetown University. He has worked as a Senior Historian with the Crimes Against Humanity and War Crimes Section of the Canadian Department of Justice (1987-93) and as a research consultant on native lands claims (1993-94). Since 2001 he has been employed as a Senior Policy Analyst coordinating university policy for the New Brunswick Department of Education, and he continues to teach university courses on World War II and Popular Memory, Modern War Crimes, and U.S. Foreign Policy. Dr. Wiggers has won numerous scholarships, awards, and research grants and has published several dozen chapters, articles and book reviews. The article that he contributed to this publication is derived from his Ph.D. dissertation (2000), which is entitled "Creating International Humanitarian Law: World War II, the Allied Occupations, and the Treaties that Followed."

WILLIAMS, BRIAN GLYN, Ph.D., is Associate Professor of History at the University of Massachusetts, Dartmouth. Formerly a lecturer at the University of London SOAS, he specializes in the ethnic cleansing of Muslim ethnic groups in Eastern Europe. His articles on the deportation of the Chechens and Crimean Tatars appear in such journals as *Central Asian Survey*, *History and Memory*, and *Middle East Policy*. His field research on these topics extends from Kosovo to Uzbekistan and has provided the background for his recent book entitled *The Crimean Tatars: The Diaspora Experience and the Forging of a Nation*. He is currently writing a book on the ethnic cleansing of the Chechens during World War II.

WOLFF, STEFAN, Ph.D., is Lecturer in the Department of European Studies at the University of Bath, England. He holds an M.Phil. in Political Theory from Magdalene College, Cambridge, and a Ph.D. in Political Science from the London School of Economics. He has extensively written on forced migration, ethnic conflict, and conflict resolution. Dr. Wolff is author of *Disputed Territories: The Transnational Dynamics of Ethnic Conflict Settlement* (Berghahn 2002) and is co-editor, with Jörg Neuheiser, of *Peace at Last? The Impact of the Good Friday Agreement on Northern Ireland* (Berghahn, 2002), and with Ulrich Schneckener, *of Managing and Settling Ethnic Conflicts* (Hurst, 2002). His other publications include two edited collections on *German Minorities in Europe: Ethnic Identity and Cultural Belonging* (Berghahn, 2000) and *Coming Home to Germany? The Integration of Ethnic Germans in the Federal Republic* (Berghahn, 2002), with David Rock. Dr. Wolff is one of the editors of a book series on ethnopolitics for Berghahn Books, a member of the Programme Committee of the Association for the Study of Nationalities, co-chair of the Specialist Group on Ethnic Politics of the Political Studies Association of the United Kingdom, and owner-moderator of an electronic discussion list on ethnopolitics.

ZAYAS, ALFRED de, J.D., Dr.Phil., has served as Senior Human Rights Officer, United Nations High Commissioner for Human Rights, Geneva, as well as the Secretary of the UN Human Rights Committee and Chief of the UN Human Rights Petitions Unit. A graduate of Harvard Law School, he is a member of both the New York and the Florida Bars, and has been an attorney with the firm of Cyrus Vance in New York. He was a Fulbright Fellow in Germany and received the Dr.Phil. in modern European history at the University of Göttingen, where he also served as law lecturer, 1974-1979. He has also taught as Professor of Law at DePaul University and has been a Senior Fellow at the Max Planck Institut für Völkerrecht in Heidelberg and member of the Editorial Committee of the *Encyclopaedia of Public International Law* and a contributor to the encyclopedia. Dr. de Zayas is the author of *Nemesis at Potsdam* (first published 1977), *A Terrible Revenge* (1994), *The Wehrmacht War Crimes Bureau, Heimatrecht ist Menschenrecht* (2001), *Human Rights in the Administration of Criminal Justice* (with Professor Cherif Bassiouni, 1994), and many other edited volumes, scholarly articles, reviews, and op-ed pieces. He has published poetry in English, French, German, and Spanish and is currently Secretary-General of the Swiss-French P.E.N. Club.

860

* * *

"This collective volume contains excellent studies about one of the most dramatic events of the past century, the expulsion of fifteen million Germans from their centuries-old homelands in Eastern Germany beyond the Oder and Neisse rivers, and from East Central Europe. Knowledge about this great Central European catastrophe... has remained meager in the world at large. It is therefore important that the English-speaking world should learn about this episode in a measured tone and — as in the present volume — grounded in solid historical scholarship.

The initiators of this project — Professors S. B. Várdy and T. Hunt Tooley, Dr. Alfred M. de Zayas, and Dr. Marianne Bouvier of the Institute for German American Relations — deserve our thanks for their energy and determination to inform the North American public, within the necessary historical context, about this well-documented but little known mega-crime in history."

Erika Steinbach
Member of the German Parliament

DONORS TO THE CONFERENCE ON ETHNIC CLEANSING AND THE RESULTING VOLUME

Institutions

Duquesne University Grant for the Conference on Ethnic Cleansing $5,000
Austin College $3,350
Danube Swabian Aid and Youth Society of New York $2,000
Advanced Machinery and Engineering $1,000
Milwaukee Danube Swabians $600
American Aid Society of German Descendants $500
Carpathia German Club, Sterling Heights, Michigan $500
Danube Swabian German American Cultural Center $500
Danube Swabians, Santa Ana, California $500
Danube Swabians, Philadelphia, Pennsylvania $250
Verband der Rumänien-Deutschen in America $175
National Federation of Hungarian Americans $100
Danube Swabians, Olmstead, Ohio $50

Individuals

Mr. Arthur C. Schwotzer $1,000
Mr. Erwin Jausz $750
Dr. Helmut Awender $700
Dr. William Everett $200
Mr. & Mrs. N. Ippach $200
Mr. Lawrence Niemann $200
Dr. Konstantin Siegmund $200
Mr. & Mrs. Michael Walter $200
Mr. Karl Hausner $150
Dr. Tanya Augsburg $130
Mr. Dietrich Schaupp $125
Mr. Thomas Au $100
Mrs. Katarina Heide $100
Mr. Theo Junker $100
Mr. Frank Knisley $100
Dr. Nikolaus Marx $100
Dr. Robert Neeson $100
Mr. E. Reidel $100
Dr. Dan Wagner $100